MONOGRAPHS ON THE PHYSICS AND CHEMISTRY OF MATERIALS

MONOGRAPHS ON THE PHYSICS AND CHEMISTRY OF MATERIALS

Theory of dielectrics M. Frohlich
Strong solids (Third edition) A. Kelly and N.H. Macmillan
Optical spectroscopy of inorganic solids B. Henderson and G.F. Imbusch
Quantum theory of collective phenomena G.L. Sewell
Experimental high-resolution electron microscopy (Second edition) J.C.H. Spence
Experimental techniques in low-temperature physics (Third edition) Guy K. White
Principles of dielectrics B.K.P. Scaife
Surface analytical techniques J.C. Rivière
Basic theory of surface states Sydney G. Davison and Maria Stęślicka
Acoustic microscopy Andrew Briggs
Light scattering: principles and development W. Brown
Quasicrystals: a primer (Second edition) C. Janot
Interfaces in crystalline materials A.P. Sutton and R.W. Balluffi
Atom probe field ion microscopy M.K. Miller, A. Cerezo, M.G. Hetherington, and G.D.W. Smith
Rare-earth iron permanent magnets J.M.D. Coey
Statistical physics of fracture and breakdown in disordered systems B.K. Chakrabarthi and L.G. Benguigui
Electronic processes in organic crystals and polymers (Second edition) M. Pope and C.E. Swenberg
NMR imaging of materials B. Blümich

NMR Imaging of Materials

BERNHARD BLÜMICH
Institut für Technische Chemie und Makromolekulare Chemie
Rheinisch-Westfälische Technische Hochschule
Aachen

CLARENDON PRESS • OXFORD

2000

OXFORD
UNIVERSITY PRESS

Great Clarendon Street, Oxford OX2 6DP

Oxford University Press is a department of the University of Oxford
It furthers the University's objective of excellence in research, scholarship,
and education by publishing worldwide in

Oxford New York

Auckland Bangkok Buenos Aires Cape Town Chennai
Dar es Salaam Delhi Hong Kong Istanbul Karachi Kolkata
Kuala Lumpur Madrid Melbourne Mexico City Mumbai Nairobi
São Paulo Shanghai Taipei Tokyo Toronto

Published in the United States
by Oxford University Press Inc., New York

First published 2000

First published New as Paperback 2002

A catalogue record for this title is available from the British Library

Library of Congress Cataloging in Publication Data

Blümich, Bernhard. NMR imaging of materials / Bernhard Blümich.
p. cm.—(Monographs on the physics and chemistry of materials; 57)
Includes bibliographical references and index.
1. Nuclear magnetic resonance. 2. Materials science. I. Title. II. Series.
QD96.N8 B58 2000 538′.362—dc21 99-087343
1 3 4 5 7 9 10 8 6 4 2

ISBN 0 19 852676 8

Typeset by Newgen Imaging Systems (P) Ltd., Chennai, India
Printed in India
by Thomson (India) Ltd

Preface

More than 25 years after its discovery, the use of NMR (nuclear magnetic resonance) imaging is still expanding. In addition to clinical and biomedical applications, more and more materials-oriented topics are being investigated. This book deals with non-clinical NMR imaging, in particular, with imaging of materials. Nevertheless, some of the concepts pursued in this book may be of interest to medical imaging as well, for example, when the material properties of biological tissues are of interest. Materials pose different challenges to NMR than human patients. They come in a larger variety of shapes and features. Little may be known *a priori* about the heterogeneous structures to be investigated, and the aggregational states of materials range from liquid to rigid solid with all intermediate shades of molecular mobility in between.

The first question usually asked when confronting an unbiased audience with NMR imaging of materials is about the resolution: the spatial resolution of NMR imaging is bad. Structures of less than 50 μm in solids and less than 5 μm in liquids can hardly be resolved. But NMR imaging is not restricted to surfaces. It can look inside materials, even into pieces of carbon-black filled rubber like car tyres. But that is not enough to justify the use of an expensive method of image-forming analysis like NMR. The real advantage, where NMR outperforms all other methods is the large variety of image contrast which can be created based on the various different molecular parameters accessible by NMR spectroscopy and relaxometry in liquids and in solids. The image contrast is not restricted to just material density, but can be set to monitor molecular motion in different time windows ranging from essentially 10^{-12} to 1 s and more. Furthermore, chemical composition can be imaged by analysing signal intensities at different chemical shifts in each volume element of the object. Diffusion constants, particle velocity, and acceleration components can be displayed, and variations in molecular order and orientation can be converted to image contrast.

The tremendous arsenal of parameters opens up a wide variety of applications, but it also requires the development of methods suitable for a particular application, and the communication about the achievements of NMR imaging from the developer to the potential user. Particularly, interesting applications of NMR imaging outside the medical field are in elastomer analysis, localizing and identifying filler inhomogeneities, deterioration processes, internal stresses, and composite structures like belts and rubber layers in car tyres. Other important applications are in chemical engineering for analysis of flow patterns of rheologically complex fluids in pipes, mixers, extruders, and other processing devices.

This book is intended for graduate students and scientists interested in methods and applications of nonmedical NMR imaging. It poses a strong emphasis on the development of NMR methods to develop relevant applications. The book has been written primarily for the students in my research group, who are graduate students in chemistry, physics, and engineering. The fluctuation of students associated with their educational progress necessitates continuous updating of the know-how of the group. To facilitate this task, an extensive set of lecture notes was written in 1988. Over the years these notes have been converted into this book. The contents reflect a personal perception of the field, biased by my own interests. I apologize to all my colleagues whose work is not properly credited in this book.

Chapters 1–4 are introductory to imaging of solid materials in different aspects. They cover some fundamentals about magnetic fields, NMR theory and hardware, an introduction to NMR spectroscopy of solids, and a short chapter about mathematical prerequisites. The treatment of NMR spectroscopy is limited to the parts relevant to NMR imaging. Thus, many of the exciting modern solid-state techniques cannot be found in this book. These introductory parts lead up to Chapters 5–8 on concepts of spatial resolution, imaging methods for liquids and solids, and contrast. Following the conviction that the prime asset of NMR imaging is the abundant wealth of contrast criteria, particular attention is paid not to increasing the spatial resolution, but to increasing the parameter contrast. This is achieved by the use of magnetization filters prior to the actual imaging experiment and by acquisition of a spectroscopic dimension for each picture element. Several different magnetization filters are treated in Chapter 7. This chapter is considered to be the most important one for successful use of NMR imaging in materials science. By providing concepts for materials imaging in this way, the book may serve not only as an introduction to imaging methods, but also as a source for ideas to try. Chapter 9 treats spatially resolved NMR, that is, methods by which NMR information is acquired from a restricted volume section of a larger object. Such local analysis may be based either on the principles of NMR imaging or on completely new hardware approaches. Applications are in diverse fields, ranging from well logging to quality and process control. The book concludes with a collection of the author's top ten application areas of nonmedical imaging. It is hoped that the book is of interest to scientists and engineers who are curious about NMR imaging, the different methods, and its application potential for materials research.

Numerous cross-references are given in the text, so that related topics can readily be found. Key words are written in italic. The citation of references is according to the first three letters of the first author's surname followed by a number. In this way some information on the literature cited is provided already in the text, and work by one author is readily identified in the list of references for each chapter at the end of the book. But as a consequence, references are not cited in chronological order. Most figures have been adapted from the literature. The respective references are also listed at the end of the book for quick access to the original literature as well as for acknowledgement to the authors.

Writing the book proved to be a far more extensive effort than originally anticipated. I am grateful to H.W. Spiess (Mainz) for the possibility of establishing a materials-oriented research group in NMR imaging at the Max-Planck-Institut für Polymerforschung, to

W.J. MacKnight (Amherst) for his hospitality when spending a sabbatical at the Polymer Science and Engineering Department of the University of Massachusetts in 1992 to continue writing this book, to P.T. Callaghan (Palmerston North) for his encouragement to complete this book and his hospitality and support while staying at his laboratory for two months in 1998, to the Deutsche Forschungsgemeinschaft (DFG) for continuing support of research and providing creative freedom, to D.E. Demco (Aachen) for proof reading, many enlightening discussions, and his contributions in particular to Chapter 8, to S. Stapf for his critical proof reading, to I. Koptyug for his contribution to imaging of catalysts pellets, to my students and collaborators R. Savelsberg, S. Laukemper-Ostendorf, K. Rombach, V. Göbbels, M. Schneider, R. Haken, A. Guthausen, G. Zimmer, P. Blümler, K. Weingarten, L. Gasper, N. Paus, and U. Schmitz for their contributions to Chapter 10, to Frau G. Nanz (Mainz) for her dedicated artwork with many figures, and to C. Bucciferro and G. Blümich for their help in the final stages of finishing this book. Special thanks go to P. Blümler, my longtime companion in academia and NMR without whose dedicated work this book would have been much thinner and less interesting. Last but not least, I thank my wife Mary-Joan and my children Gwendolyn, Franziska, and Max for support and tolerance during the many years of writing this text.

Aachen Bernhard Blümich
1999

Contents

Symbols and abbreviations

0Q	zero quantum
2Q	double quantum
$\boldsymbol{a}$	acceleration vector, orientation vector of the average effective field
a	arbitrary constant, magnitude of the acceleration vector, length of a Kuhn segment
A	area
$a(\sigma)$, $a(\tau)$	auto-correlation function
$A(\omega)$	real part of the complex Lorentz line
A, B, C	expansion coefficients
ADC	analog-to-digital converter
ADRF	adiabatic demagnetization in the rotating frame
af	audio-frequency
a_{H}	amplitude of the Hahn echo
a_m	expansion coefficient
A_{nm}, B_{nm}	expansion coefficients
ARRF	adiabatic remagnetization in the rotating frame
a_{s}	amplitude of the stimulated echo
$\boldsymbol{B}$	vector of the magnetic induction, referred to as magnetic field vector
$\boldsymbol{B}_0$	static magnetic field vector
$\boldsymbol{B}_1$	rf magnetic field in the rotating frame
$B_{1\mathrm{SL}}$	amplitude of the spin-lock field
B_{1xy}	transverse components of the B_1 field
$\boldsymbol{B}_{\mathrm{eff}}$	effective magnetic field
$\boldsymbol{B}_{\mathrm{fic}}$	fictitious magnetic field in the rotating frame
$\boldsymbol{B}_{\mathrm{loc}}$	local magnetic field
BR	Burum, Rhim
$\boldsymbol{B}_r$	magnetic field in the rotating frame
$\boldsymbol{B}_{\mathrm{rf}}$	rf magnetic field in the laboratory frame
BURP	band-selective uniform-response pure-phase
B_z	z-component of the magnetic field
c	arbitrary constant
C	capacitance
CE-FAST	contrast-enhanced Fourier-acquired steady state
CHESS	chemical-shift selection
C_{lm}	probability density of molecules diffusing from pore l to pore m

c_n	correlation function of order n
C_n	Fourier transform of a correlation function of order n
COSY	correlation spectroscopy
CP MAS	cross-polarization and magic-angle spinning
CP	cross-polarization
CPMG	Carr, Purcell, Meiboom, Gill
CRAMPS	combined rotation and multi-pulse spectroscopy
CTE	coherence-transfer echo
CW	continuous wave
CYCLCROP	cyclic cross-polarization
CYLCPOT	cyclic polarization transfer
C^{λ}	coupling factor for interaction λ
D	coefficient of translational self-diffusion
d	diameter
$\mathbf{D}$	dipole–dipole coupling tensor
$D(\omega)$	imaginary part of the complex Lorentz line
DANTE	delays alternating with nutation for tailored excitation
DAS	dynamic angle spinning
dc	direct current
DD	high-power dipolar decoupling
DEC	decoupling
D_{eff}	effective diffusion coefficient
DEPT	distortionless enhancement by polarization transfer
DEPTH	depth-selective excitation for surface coils; not an acronym
DIGGER	discrete isolation from gradient-governed isolation of resonances
D^{λ}_{lm}	elements of the Wigner rotation matrix of order λ
DOR	double rotation
DOSY	diffusion ordered spectroscopy
E	energy, modulus of elasticity
$\boldsymbol{E}$	vector of ones
E'	storage modulus
E''	loss modulus
E_{c}	energy of a coil
E_m	quantum-mechanical energy of state m
EPDM	ethylene–propylene–diene monomer
EPI	echo-planar imaging
EPR	electron paramagnetic resonance
EPSM	echo-planar shift mapping
ESR	electron spin resonance
EVI	echo-volumnar imaging
f	arbitrary function
F	Fourier transform of f
$f(\gamma)$	orientation function in MAS
$\mathbf{F}$	Fourier-transformation operator
FAST	Fourier-acquired steady state

FFT	fast Fourier transformation
FID	free-induction decay
F_L	Laplace transform of f
FLASH	fast low-angle shot
f_N	Nyquist frequency
FONAR	field-focused nuclear magnetic resonance
FT	Fourier transformation
F_z	z transform of f
g	Fourier-conjugate variable of the gradient G
$\mathbf{G}$	gradient tensor
$\boldsymbol{G}$	gradient vector
G	magnitude of the gradient vector, shear modulus
$G(\omega)$	Fourier transform of $g(t)$
$g(t)$	arbitrary function of time
G_i	arbitrary components of the gradient vector
GRASS	gradient-recalled acquisition in the steady state
G_x, G_y, G_z	specific components of the gradient vector
G_{xx}, G_{yy}	field gradient components in the rotating frame
$\mathbf{H}$	Hamilton operator
$\boldsymbol{H}$	magnetic field vector
$\hbar$	Planck's constant divided by 2π
HDPE	high-density poly(ethylene)
HMB	hexamethylbenzene
HMQC	heteronuclear multi-quantum coherence
h_n	Wiener kernel of order n
HPLC	high-pressure liquid chromatography
HSI	Hadamard spectroscopic imaging
HYCAT	hydrocarbon tomography
$\mathbf{I}_-$	$\mathbf{I}_x - \mathrm{i}\mathbf{I}_y$
$\mathbf{I}$	spin operator
I	spin quantum number, magnitude of electric current
$\mathbf{I}_+$	$\mathbf{I}_x + \mathrm{i}\mathbf{I}_y$
INEPT	insensitive nuclei enhanced by polarization transfer
ISIS	image-selected *in vivo* spectroscopy
$\mathbf{I}_x, \mathbf{I}_y, \mathbf{I}_z$	components of the spin operator
$\boldsymbol{j}$	current density
J	indirect spin–spin coupling constant
$\mathbf{J}$	tensor of indirect spin–spin coupling
$J(\omega)$	spectral density of motion at frequency ω
J_n	Bessel function of order n
k	magnitude of the wave vector, counting index
K	ratio of radii
$\boldsymbol{k}$	wave vector corresponding to space
k_1	linear impulse response function, wave number at time t_1
K_1	linear transfer function or 1D susceptibility

k_2	second-order impulse response function, wave number at time t_2
K_2	second-order transfer function or 2D susceptibility
k_3	third-order impulse response function
K_3	third-order transfer function or 3D susceptibility
k_B	Boltzmann constant
k_x, k_y, k_z	wave numbers in reciprocal space
l	counting index
L	free-evolution operator in the toggling frame
L	inductance, length
$L(z)$	lattice-correlation function
$\mathsf{L}\{\}$	Laplace-transformation operator
L_0	initial length
LC	liquid chromatography
LCF	laboratory-coordinate frame
LDPE	low-density poly(ethylene)
LOSY	localized spectroscopy, lock-pulse selective spectroscopy
$[M_c]$	cross-link density
m	magnetic quantum number, counting index
$\boldsymbol{M}$	magnetization vector
M^+	complex transverse magnetization
M_0	thermodynamic equilibrium magnetization
M_2	second moment of the lineshape
MAGROFI	magnetization-grid rotating-frame imaging
MARF	magic-angle rotating frame
MAS	magic-angle spinning
MASSEY	modulus addition using spatially separated echo spectroscopy
MEPSI	magic-echo phase-encoding solid-state imaging
$\boldsymbol{m}_n$	nth moment of a time-dependent magnetic field contribution
M_n	nth order spectral moment of the lineshape
MOIST	mismatch optimized IS transfer
MOSY	mobility ordered spectroscopy
MOUSE	mobile universal surface explorer
$\boldsymbol{M}_p$	magnetic polarization
MQ	multi-quantum
MREV	Mansfield, Rhim, Elleman, Vaughan
MW4	Mansfield–Ware sequence with four time intervals
M_x	x-component of transverse magnetization
M_y	y-component of transverse magnetization
M_z	longitudinal magnetization
n	counting index
N	number
n_{data}	number of data points
N_e	effective number of Kuhn segments
n_m	population of energy level m
NMR	nuclear magnetic resonance

NOE	nuclear Overhauser effect
NOESY	nuclear Overhauser effect spectroscopy
NR	natural rubber
$\mathbf{O}$	arbitrary operator
OMAS	off magic-angle spinning
OW4	Ostroff–Waugh sequence with four time intervals
$\mathbf{P}$	arbitrary coupling tensor
p	coherence order, variable of the Laplace transform, Fourier transform of a projection
P	electric power, conditional probability density, projection
P	population
$\mathbf{P}$	pulse-rotation operator
$P(\cos\theta)$	orientational distribution function
$P(x)$	projection of the spin density onto the x-axis
$P(\alpha, \beta, \gamma)$	orientational distribution function
PC	poly(carbonate)
PDMS	poly(dimethylsiloxane)
PE	poly(ethylene)
PFG	pulsed-field gradient
phr	parts per hundred rubber
PI	poly(isoprene)
P_{ij}	components of irreducible spherical coupling tensors
P_l	Legendre moment of order l
$P_l(\cos\theta)$	Legendre polynomial of order l
P_{nm}	associated Legendre polynomial
PP	poly(propylene)
PPS	poly(phenylene sulphide)
PRAWN	pulsed rotating-frame transfer sequences with windows
PRESS	point-resolved spectroscopy
PS	poly(styrene)
PTFE	poly(tetrafluoro ethylene) or Teflon
PVC	poly(vinyl chloride)
P_{XY}	components of the coupling tensor in the principal axes system
q	magnitude of the wave vector corresponding to displacement, scaling fact
$\mathbf{Q}$	quadrupole coupling tensor
Q	quality factor
$\boldsymbol{q}$	wave vector corresponding to displacement
$\boldsymbol{R}$	displacement vector, end-to-end vector of a cross-link chain
R	magnitude of displacement, reduced radial variable
r	radius, magnitude of the space vector
$\mathbf{R}$	relaxation matrix
R	relaxation rate, radius of a coil, variable of spatial displacement, resistance
$\boldsymbol{r}$	space vector
R_λ	isotropic average of the spin coupling tensor for the interaction λ

RARE	rapid acquisition with relaxation enhancement
RCF	rotating coordinate frame
Re	Reynolds number
rf	radio-frequency
r_{IJ}	distance between spins I and J
r_{ij}	internuclear distance
RX	receiver
R_σ	chemical-shift scaling factor
S	Fourier transform of s, surface
s	space axis orthogonal to r
$s(\boldsymbol{k})$	signal in reciprocal space
$S(\boldsymbol{r})$	signal in physical space
$s(t)$	time-domain signal
$S(\omega)$	frequency-domain signal
S/N	signal-to-noise ratio
SBR	styrene-co-butadiene rubber
SEDOR	spin-echo double resonance
SFC	superfluid chromatography
SL	spin lock
SLISE	spin-lock induced slice excitation
$\mathrm{Sp}\{\cdots\}$	spur, trace
SPACE	a method of spatially resolved spectroscopy, not an acronym
SPARS	spatially resolved spectroscopy
SPINOE	spin-polarization induced nuclear Overhauser effect
SPLASH	spectroscopic low-angle shot
sPP	syndiotactic poly(propylene)
SPREAD	saturation pulses with reduced amplitude distribution
SPRITE	single-point ramped imaging with T_1 enhancement
SQUID	superconducting quantum-interference device
SSFP	steady-state free precession
STEAM	stimulated-echo acquisition mode
STRAFI	stray-field imaging
S_w	spectral width
T	absolute temperature, period of an oscillation, total acquisition time, tortuosity
T	Dyson time-ordering operator
t	time
t_1	evolution time
T_1	longitudinal relaxation time
$T_{1\rho}$	longitudinal relaxation time in the rotating frame
$T_{1\mathrm{D}}$	relaxation time of dipolar order
$T_{1\mathrm{Q}}$	relaxation time of quadrupolar order
t_2	detection time in a 2D experiment
T_2	transverse relaxation time
$T_{2\rho}$	transverse relaxation time in the rotating frame

T_2^*	transverse relaxation time with inhomogeneous broadening
T_{2e}	effective transverse relaxation time
t_c	duration of a current pulse, cycle time
TCF	toggling-coordinate frame
T_{CH}	dipolar relaxation time between ^{13}C and ^{1}H
t_{CP}	cross-polarization time
t_E	echo time
t_f, t_{f1}, t_{f2}	filter times
T_g	glass-transition temperature
$\mathbf{T}_{lm}$	spherical tensor components of spin operators
t_m	mixing time, spin-diffusion time
TOP	two-dimensional one-pulse spectroscopy
TOSS	total suppression of spinning sidebands
t_p	duration of an rf pulse
TPPI	time-proportional phase increments
t_R	repetition time, rotor period
TREV	time reversal
t_w	time delay: time to wait
TX	transmitter
$\mathbf{U}$	evolution operator
$\mathbf{U}_p$, $\mathbf{U}_m$	time evolution operators for preparation and mixing (reconversion) of multi-quantum coherences
UV	ultra violet
$\boldsymbol{v}$	velocity vector
v	magnitude of velocity vector
V	volume
v_φ	azimuthal velocity
VEST	selected volume excitation by stimulated echoes
VOSINER	volume of interest by selective inversion, excitation, and refocusing
VOSING	volume-selective editing
VOSY	volume-selective spectroscopy
VSE	volume-selective excitation
W	combined probability density, probability density
WAHUHA	Waugh, Huber, Haeberlen
W_c	energy density of a coil
WIM	windowless isotropic mixing
W_{nm}	quantum-mechanical transition probability
X	excitation spectrum
x	time-domain excitation signal
x, y, z	components of the space vector
X, Y, Z	coordinates in the principal axes frame of a tensor
Y	response spectrum
y	time-domain response signal
z	space coordinate, variable of the z transform
Δ	anisotropy of an interaction, time delay

$\Delta k_x,\ \Delta k_y$	$\boldsymbol{k}$-space increments in k_x- and k_y-directions
Δt	time increment, time delay, dwell time
Δt_s	signal delay
$\Delta x,\ \Delta y$	space increments in x- and y-directions
$\Delta \Omega$	frequency range in the rotating-coordinate frame
$\Delta \nu$	frequency increment, linewidth
$\Delta \omega_{1/2}$	linewidth at half height in angular units
$\mathbf{\Gamma}$	relaxation superoperator
Γ_i	orientation angles at time t_i
Λ	elongation ratio, stretching ratio
$\mathbf{\Pi}$	exchange matrix
Θ	orientation angle of a magnetization component in MAS
Ω	angular-dependent resonance frequency on resonance in the rotating frame
$\mathbf{\Omega}$	diagonal matrix of the angle-dependent resonance frequencies
Ω_0	offset frequency in the rotating frame without chemical shift
Ω_L	offset frequency in the rotating frame with chemical shift
$\Omega_L,\ \Omega_M,\ \Omega_N$	chemical-shift defined offset frequencies in the rotating frame
$\hat{\Omega}$	angular-dependent resonance frequency off-resonance in the rotating frame
Ψ	wave function
ψ_m	magic angle
α	flip angle of an rf pulse, wave function of a spin $\frac{1}{2}$
α, β, γ	Euler angles
α_E	Ernst angle
α_{eff}	effective tip angle of an rf pulse
β	wave function of a spin $\frac{1}{2}$, parameter of the hyperbolic secant pulse
χ_l	Legendre expansion coefficient
δ	largest principal value of a coupling tensor
ε	Fourier conjugate variable to acceleration
ϕ	receiver phase
γ	gyromagnetic ratio, shear strain
η	asymmetry parameter, viscosity
φ	polar angle, transmitter phase, phase angle
φ_0, φ_1	variables of the frequency-linear phase correction
λ	index denoting a spin interaction
$\boldsymbol{\mu}$	magnetic permeability tensor
μ	magnetic permeability, parameter of the hyperbolic secant pulse
μ_0	magnetic permeability of vacuum
$\boldsymbol{\mu}_I$	magnetic moment
ν	frequency
ν_0	nominal NMR frequency
ν_L	Larmor frequency
ν_R	sample-rotation frequency

ν_s	Larmor frequency of a reference standard
θ	polar angle
ρ	density matrix
ρ_0	density matrix in thermodynamic equilibrium
ρ_{mn}	density-matrix element
σ	chemical-shielding constant, electrical conductivity, time delay
$\boldsymbol{\sigma}$	chemical-shielding tensor
σ_{11}	tensile stress
σ_{12}	shear stress
τ, τ_i	time delays
τ_c	correlation time
τ_{SL}	spin-lock delay
ω	circular frequency in angular units
ω_0	nominal NMR frequency in angular units
ω_c	cut-off frequency
ω_{cs}	chemical-shift frequency
ω_{eff}	amplitude of $\boldsymbol{B}_{eff}$ in units of angular frequency
ω_G	gradient oscillation frequency in angular units
ω_L	Larmor frequency in angular units
ω_R	sample-rotation frequency in angular units
ω_{rf}	frequency of the rf carrier in angular units
$\xi(t)$	signal phase
ξ_1, ξ_2, ξ_3	evolution phases of transverse magnetization components
ψ	orientation angle of the spinning axis with respect to the magnetic field

1

Introduction

Nuclear magnetic resonance (NMR) [Abr1] is the most important analytical technique for structural analysis of molecules in solution [Ern1, Bov1]. For a long time, NMR has been associated with chemistry. With the invention of *NMR imaging* [Lau1, Man2], magnetic resonance has become a familiar name even to the nonscientific community, and more medical personnel and engineers are now working with NMR than chemists. NMR imaging of materials is much less known than medical NMR imaging, but the progress in methodical development along with the identification of significant applications, for example, to material flow and soft matter analysis, leads to increasing use of the method. This book intends to provide the methodical background for the development of imaging methods and applications in materials science.

1.1 NUCLEAR MAGNETIC RESONANCE

NMR is a method which probes molecular properties by interrogating atomic nuclei with magnetic fields and radio-frequency (rf) irradiation. More specifically, the phenomenon of *magnetic resonance* [Abr1] denotes the resonant interaction of magnetic moments in a time-invariant magnetic field with the magnetic component of an electromagnetic wave. In a way, NMR denotes the rf communication between laboratory transmitters and receivers on one side and the magnetic polarization of atomic nuclei exposed to a magnetic field on the other side. This principle is illustrated in Fig. 1.1.1. The rf signal of the nuclei is stimulated and received by the laboratory spectrometer. From this, information about the chemical structure of molecules, molecular order in solid matter, molecular dynamics, morphology of semicrystalline polymers, macroscopic sample heterogeneities, and molecular transport can be derived. The frequency ω_L with which the atomic nuclei respond is called the *Larmor frequency*. It is related to the strength of the magnetic field $|\boldsymbol{B}_{loc}|$ at the site of the nucleus by the fundamental equation

$$\omega_L = -\gamma |\boldsymbol{B}_{loc}| = -\gamma B_{loc}, \tag{1.1.1}$$

where γ is the gyromagnetic ratio. It is a constant for a particular nuclear isotope. For a magnetic field of 2.35 T, the Larmor frequency of the most sensitive stable nucleus,

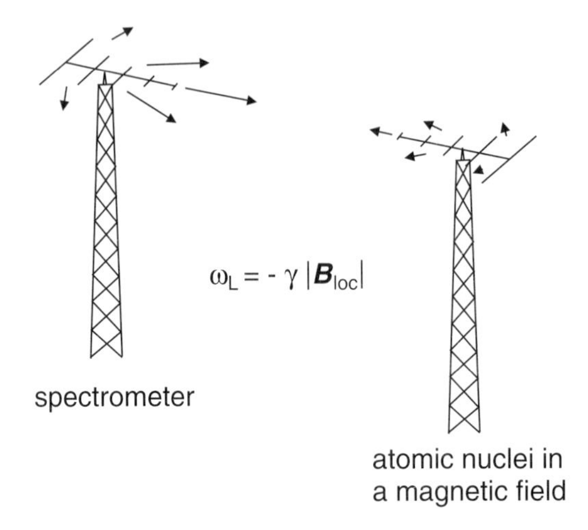

FIG. 1.1.1 Magnetic resonance denotes the rf communication between a laboratory spectrometer and atomic nuclei in a magnetic field.

the proton, is near 100 MHz. Thus, it is in the frequency modulation (FM) band of radio stations.

The magnetic resonance phenomenon is also observed for unpaired electrons. It is called *electron spin resonance* (ESR) or *electron paramagnetic resonance* (EPR) [Abr2, Ath1, Gor1, Ike2, Kev1, Pil1, Poo1, Sch2, Wei1], and the corresponding frequencies are in the GHz or microwave regime (Fig. 1.1.2) [Kre1]. Of both techniques, NMR is by far the more wide-spread method, because radio waves can technically more easily be handled than microwaves, and because nuclear spins are in abundance in nearly every material. Free electrons, on the other hand, often are chemically reactive, so that they need to be trapped or incorporated into the material in terms of spin labels. For this reason, EPR methods are excluded from this book, although exciting progress has been achieved in EPR imaging in recent years [Ber1, Eat1, Eat2, Ike1, Ike2, Zwe1].

1.1.1 NMR and the spinning top

The description of proton NMR in simple fluids without couplings among the protons is analogous to that of a *spinning top* with angular momentum $\boldsymbol{L}$ in a gravitational field $\boldsymbol{G}$ (Fig. 1.1.3) [Blo1, Kle1]. If the spinning top is not aligned with the gravitational field, then the spinning axis precesses around the field direction (Fig. 1.1.3, top). The nuclear magnetization $\boldsymbol{M}$ is proportional to the sum of all nuclear magnetic dipole moments of the sample. These moments are also referred to as *spins*, because the nuclear magnetic moment is proportional to a quantum-mechanical angular momentum, and a classical angular momentum is associated with rotation or with spinning. For the simple fluid with uncoupled protons, the nuclear magnetization precesses around the direction of the magnetic field $\boldsymbol{B}$ with the Larmor frequency ω_L, once it has been disturbed from alignment

	frequency		wavelength		
	$3\ 10^{14}$	cosmic rays < 10 nm^{-5}	10^{-6}		
	$3\ 10^{13}$		10^{-5}		
	$3\ 10^{12}$		10^{-4}		
	$3\ 10^{11}$		10^{-3}		
	$3\ 10^{10}$	x ray sand γ rays:	10^{-2}	nm	
	$3\ 10^{9}$	10^{-5} nm - 10 nm	10^{-1}		
	$3\ 10^{8}$		1		
GHz	$3\ 10^{7}$	ultraviolet: 10 nm - 400 nm	10		
	$3\ 10^{6}$	visible: 400 nm - 780 nm	100		
	$3\ 10^{5}$		1		
	$3\ 10^{4}$	infrared: 780 nm - 1 mm	10	µm	optical regime
	$3\ 10^{3}$		100		
	300		1		
	30	microwaves: 1 mm - 1 m	10	mm	EPR
	3	very high frequencies:	100		
	300	1 m - 10 m	1		NMR
MHz	30	television: 2 m	10	m	
	3	short waves : 10 m - 80 m	100		
	300	medium waves: 200 m - 600 m	1		
kHz	30	long waves: > 600 m	10		
	3		10^{2}		
	300		10^{3}	km	
Hz	30		10^{4}		
	3		10^{5}		

FIG. 1.1.2 Electromagnetic waves. Frequency (left) and wavelength (right) scales for optical, EPR, and NMR analysis. Adapted from [Kre1] with permission from Publicis MCD.

with the field direction. Thus, the motion of the nuclear magnetization is in complete analogy with the motion of a spinning top in a gravitational field. In a nearby coil, the precessing magnetization can induce a current which provides the measurement signal.

1.1.2 NMR spectroscopy

NMR spectroscopy [Abr1] thrives from the fact that the local field $\boldsymbol{B}_{\text{loc}}$ at the site of the nucleus in a molecule differs from the applied field $\boldsymbol{B}_0$. In NMR spectroscopy, highly homogeneous magnetic fields

$$\boldsymbol{B}_0 = \begin{pmatrix} 0 \\ 0 \\ B_0 \end{pmatrix} = (0,\ 0,\ B_0)^{\text{t}} \tag{1.1.2}$$

are applied. Here the superscript t denotes the transpose of a vector or a matrix. This symbol is not consistently carried along in the text. It is only used in situations where the notation is not quite obvious.

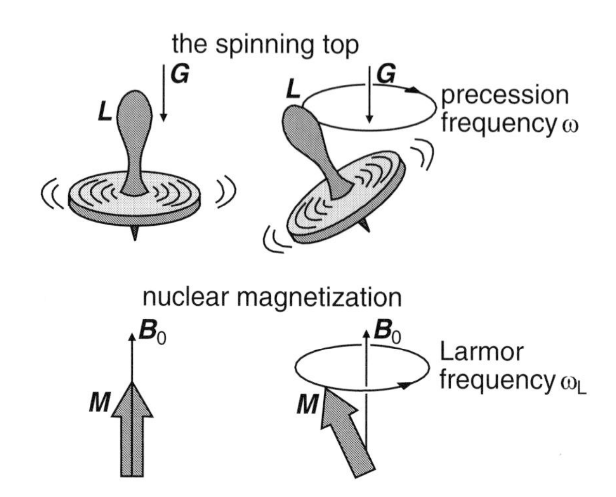

FIG. 1.1.3 [Blü5] NMR and the spinning top. A spinning top with angular momentum $\boldsymbol{L}$ precesses around the direction of the gravitational field $\boldsymbol{G}$ with frequency ω (top). The nuclear magnetization is the vector sum of nuclear spins. It precesses around the direction of a magnetic field $\boldsymbol{B}_0$ with the Larmor frequency ω_{L}.

In isotropic fluids, the applied field is shielded from the nucleus by the magnetic fields arising from the electrons moving around the nucleus,

$$B_{\mathrm{loc}} = (1 - \sigma) B_0, \tag{1.1.3}$$

so that the Larmor frequency is given by

$$\omega_{\mathrm{L}} = (1 - \sigma)\omega_0, \tag{1.1.4}$$

and $\omega_0 = -\gamma B_0$ is the NMR frequency without shielding.

The quantity σ measures the degree of *magnetic shielding*. It is determined by the binding electrons between atoms in a molecule and, thus, by the structure of the molecule. Because the absolute values of ω_0 cannot easily be obtained for nuclei bare of electrons, values of σ relative to a reference compound are tabulated as *chemical shifts* for various chemical groups and compounds [Bre1]. Often, magnetic shielding and chemical shift are used synonymously.

The distribution of Larmor frequencies of all atoms in a molecule gives rise to the *NMR spectrum*. For many nuclei like the ^{1}H and ^{13}C nuclei in organic molecules, the NMR spectrum is a fingerprint of the molecular structure. For this reason, NMR spectroscopy is the most significant analytical tool in chemical analysis [Bov1, Ern1]. The NMR spectrum can be compared to the spectrum of radio stations in a particular area, where the rf spectrum is a fingerprint of the geographic position (Fig. 1.1.4).

For NMR measurements, pulsed excitation is routinely used. It exploits the multiplex advantage by simultaneous acquisition of the response at all frequencies, and it enables convenient manipulation of the excitation by conducting the experiment in the time domain. The time axis can, for instance, be separated into different intervals, during which the nuclear spins of the sample can be manipulated by rf irradiation and mechanical

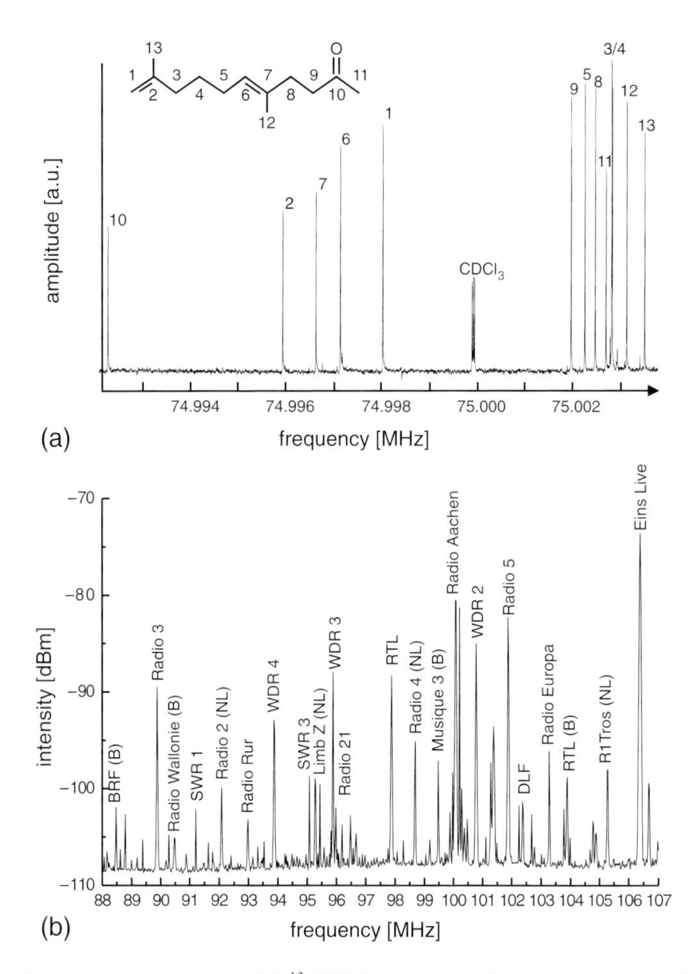

FIG. 1.1.4 Radio-frequency spectra. (a) ^{13}C NMR spectrum of the organic molecule geranyl acetone. It is a fingerprint of the chemical structure. Each carbon atom gives rise to a signal at a different frequency. (b) Spectrum of rf signals in the FM band in Aachen. The spectrum is a fingerprint of the geographic position of Aachen.

sample reorientation to provide selected information about the sample. This approach immediately leads to multi-dimensional NMR spectroscopy and to spectra, in which the signal intensity is a function of more than one frequency [Ern1, Sch1]. A wealth of detailed information about molecular structure, order and dynamics can be accessed in this way in liquids as well as in solids [Mar1, Meh1, Sch1, Sli1].

1.1.3 NMR imaging

NMR imaging is a noninvasive analytical technique, which is capable of producing images of arbitrarily oriented slices through optically nontransparent objects. Biological tissue, plants, foodstuffs and many synthetic materials can be penetrated by rf waves, and the signal is hardly attenuated by absorption and emission of rf energy at the resonance frequencies of the nuclear spins. In this sense, the objects appear transparent to radio waves, while they are nontransparent to electromagnetic waves at optical frequencies, where the human eye is a sensitive detector, or exhibit different absorption properties at higher frequency irradiation, for instance, as provided by X-rays and electron beams (Fig. 1.1.2) [Kre1].

NMR imaging can be perceived as a particular form of multi-dimensional spectroscopy, where the frequency axes have been converted to space axes by application of inhomogeneous magnetic fields. For convenience, space-invariant or constant *magnetic-field gradients* are generally used (Fig. 1.1.5). A field gradient is the spatial derivative of the field. A constant gradient denotes a linear variation of the field with space.

In a space-dependent magnetic field, the Larmor frequency depends on position. A sufficiently weak space dependence of the magnetic field can be expanded into a Taylor series. For example, a variation along the x coordinate is described by

$$B_z(x) = B|_{x=0} + \left.\frac{\partial B_z}{\partial x}\right|_{x=0} x + \frac{1}{2}\left.\frac{\partial^2 B_z}{\partial x^2}\right|_{x=0} x^2 + \cdots . \tag{1.1.5}$$

In most imaging experiments, the second and higher order derivatives in this expansion are small and negligible. However, this is not a necessity for obtaining spatial resolution

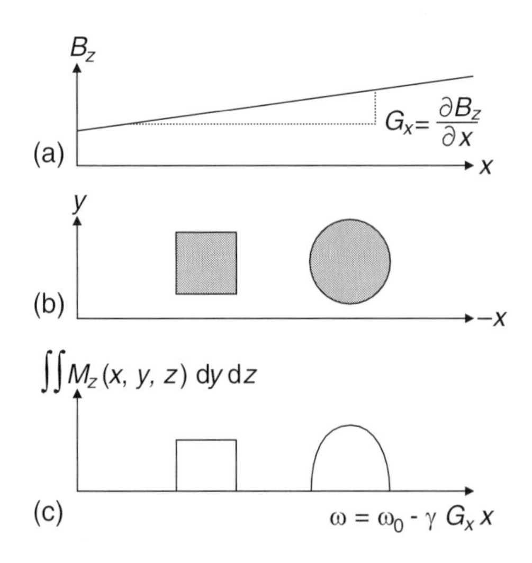

FIG. 1.1.5 The principle of NMR imaging. (a) The magnetic field varies linearly across the sample by application of a field gradient G_x in x-direction. (b) Sample shapes in the two-dimensional (2D) xy plane. (c) The NMR spectrum acquired in the presence of a magnetic-field gradient provides a projection of the sample. The signal amplitude is proportional to the number of nuclear spins at a given value of the magnetic field.

in NMR. It is only convenient, because the frequency and space coordinates scale in proportion as can be seen by combining eqns (1.1.1), (1.1.3), and (1.1.5),

$$\omega_{\mathrm{L}} = -\gamma(1-\sigma)\left[B|_{x=0} + \left.\frac{\partial B_z}{\partial x}\right|_{x=0} x + \frac{1}{2}\left.\frac{\partial^2 B_z}{\partial x^2}\right|_{x=0} x^2 + \cdots \right]. \tag{1.1.6}$$

In imaging experiments, the space variation of the external field must be made strong enough to override the spread in chemical shift or linewidth. In this case (1.1.6) can be approximated by

$$\omega_{\mathrm{L}} = (1-\sigma)\omega_0 - \gamma\left[G_x|_{x=0}\, x + \frac{1}{2}\left.\frac{\partial^2 B_z}{\partial x^2}\right|_{x=0} x^2 + \cdots \right]. \tag{1.1.7}$$

The linear relationship between NMR frequency and space coordinate is obtained by restricting the space dependence of the magnetic field to the field gradient $G_x = \partial B_z/\partial x$ in (1.1.7) and by neglecting chemical shift as illustrated in Fig. 1.1.5. In this case, each point along the x-axis of the sample is characterized by a different resonance frequency. The total signal intensity is proportional to the number of nuclei with a given NMR frequency. It is obtained by integration of the sample magnetization along the y- and z-coordinates and is thus given by the *projection* of the signal onto the x-axis. From a set of projections acquired with magnetic field gradients pointing in different directions, an image of the object can be reconstructed.

For a linear space dependence of the Larmor frequency, the *spatial resolution* $1/\Delta x$ is related to the width of the NMR absorption line or the spread $\Delta\nu = \Delta\omega_{\mathrm{L}}/2\pi$ in Larmor frequencies ω_{L} according to (1.1.7) by

$$\frac{1}{|\Delta x|} = \left|\frac{\gamma G_x}{2\pi\,\Delta\nu}\right|. \tag{1.1.8}$$

This expression applies to direct detection of the NMR signal in the presence of a magnetic field gradient G_x, also called *frequency encoding* of the space information. It states that the larger the linewidth $\Delta\nu$, the worse the spatial resolution $1/\Delta x$. Liquids exhibit narrow linewidth of the order of 1 Hz, so that the spatial resolution may be said to be good. Because of the dipole–dipole interaction between neighbouring nuclear spins and restrictions in molecular motion, linewidths of up to 100 kHz are observed in solids, so that the spatial resolution is bad. Soft matter, like tissue in medicine, and elastomers in materials science exhibit sufficiently narrow lines between 10 Hz and 3 kHz, and spatial resolution is acceptable, i.e. between 10 and 100 μm in one dimension. Rigid solids like glassy polymers are particularly difficult to image.

Increasing the gradient strength is one way to achieve higher spatial resolution, but this reduces the number of spins in a given frequency interval and thus the penalty is a reduction in *signal-to-noise ratio* (S/N) or, equivalently, in *sensitivity*. The other solution is to manipulate the *linewidth* by appropriate techniques like *multi-pulse excitation* or *magic-angle spinning* (MAS) which are well established in solid-state NMR spectroscopy [Mar1, Meh1, Sli1]. Here the penalty is increased experimental complexity and the restriction to samples with diameters of 10 mm and less due to limitations in present-day rf-amplifier and NMR-probe technology. Therefore, imaging of solid materials challenges the development of new methods suitable to handle the problems posed by linewidth, sensitivity, and sample size.

In summary, the spatial resolution of NMR imaging can be said to be just marginally better than that of the human eye. Thus, the name *NMR microscopy* appears misleading when comparing the spatial resolution achievable by NMR to that achievable by light microscopy, electron microscopy, or even scanning tunnel microscopy. Within the NMR community, however, it is well established and expresses the pride of resolving structures by NMR which are slightly below the spatial resolution limit of the human eye [Ecc1, Kuh1].

1.1.4 NMR imaging in biomedicine

The most significant application of NMR imaging is in biomedicine where the method has become an invaluable diagnostic tool complementing X-ray tomography [And1, Bud1, Haa1, Hau1, Hen1, Man4, Mor1, Mor2, Sta1, Vla1, Weh1], although the first reports of NMR imaging to medical [Lau1] and materials [Man2] research were almost coincidental. Though the achievable spatial resolution was originally believed to be inferior to X-ray tomography, it was known already at the start that NMR provides image *contrast* which is fundamentally different from that of X-rays [Dam1, Dam2]. In particular, superior contrast can be achieved in soft matter by NMR imaging. For X-rays, the signal attenuation at the given frequency is the only source of contrast available. For this reason, the use of contrast agents has become common practice in X-ray tomography.

1.1.5 Contrast in NMR imaging

In NMR, the image *contrast* is determined not only by the density of the observed nucleus, but also by the numerous other parameters which are measured in NMR spectroscopy to determine the molecular characteristics of condensed matter [Blü10, Blü11, Xia1]. These parameters include the *relaxation times* T_1 for energy dissipation and T_2 for dephasing of signal coherence, the *chemical shifts* and mutual *couplings of nuclear spins* which are characteristic of molecular configuration and conformation, and the size and orientation dependence of different spin interactions which are effective in the solid state whenever the molecular motion is restricted to rates slower than the size of the interaction. The *lineshapes* of the solid-state resonance signals provide information on the degree of *molecular order*, and the timescale and the mechanism of slow *molecular reorientation* [Bov1, Meh1, Sch1, Sli1].

Relaxation-time contrast is particularly useful. For example, it is exploited in biomedical imaging to differentiate between soft tissues. This is illustrated in Fig. 1.1.6 by an image through the head of a healthy volunteer [Blü4]. The image contrast is determined by the spin density weighted with an exponential function of the space dependent T_1 relaxation time. In such an image, the structure of the brain can be revealed in unsurpassed detail. An image in which the contrast is determined by the spin density weighted with a function of an NMR parameter is called a *parameter-weighted image*. Images of NMR

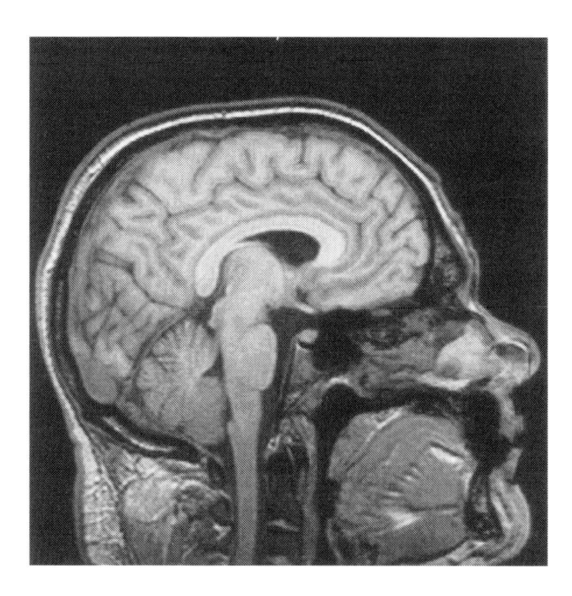

FIG. 1.1.6 Sagital image through a human head. The contrast is determined by the local values of the spin density and the T_1 relaxation time. Adapted from [Blü4] with permission from Springer-Verlag.

pure parameters, like relaxation times or diffusion constants, can also be obtained. They are called *parameter images*.

To increase the information contents of NMR images even further, it is advantageous to attach an NMR spectrum to each volume element and extract different contrast parameters for the image from the spectrum. Examples of such parameters are signal intensities at given chemical shifts, and linewidths. This form of imaging is called *spectroscopic imaging*. It is expensive in terms of measurement time, because essentially another dimension is attached to the space dimensions of the image. Nevertheless, the time can be afforded in particular in materials applications, because as opposed to clinical diagnostics, overnight experiments are feasible.

1.1.6 NMR imaging in materials science

The access to a large variety of *contrast parameters* in combination with the transparency of many materials to rf irradiation is the prime justification for the exploitation of NMR imaging in materials science. The achievable *spatial resolution* is secondary, because compared to many forms of microscopy it is rather bad. The nondestructiveness of the method is only partially significant for materials applications, because many material properties are preserved upon cutting. Exceptions are temperature profiles, stress distributions, and flow patterns of fluids. In fact, at this stage many objects still must be cut for examination to fit the limited sample volume available in present day imaging equipment. However, novel magnets with shapes suitable for investigation of surfaces, sheets and bulgy objects are being designed (cf. Section 9.3.4). High spatial resolution promises to be attainable by the NMR force microscope. This device is being developed based on the principles of force microscopy in order to exploit the chemical sensitivity of NMR for mapping of material surfaces [Rug1, Rug2, Sch3, Sid1, Sid2].

The use of NMR imaging in materials science is illustrated in Fig. 1.1.7 [Blü2] with an image of a *stress distribution*. It shows a section of a strained, filled poly(dimethylsiloxane) strip with a cut. The image has been derived from a T_2 parameter image obtained by spatially resolved measurement of the T_2 relaxation time. The T_2 time becomes shorter as the segmental motion in the elastomer network becomes anisotropic upon stretching it to 200%. Subsequent recalibration of the T_2 values in terms of uniaxial strain and stress produced the image (cf. Fig. 7.1.11). Different stress values are found for the lips of the cut (0 MPa) and for the inside of the cut (of the order of 2 MPa), while the average stress was 1.5 MPa. In addition to this, heterogeneities are observed under stress which are attributed to *filler inhomogeneities*. The sizes of the heterogeneities are larger than those observed by light microscopy of the surface. This is attributed mainly to the high sensitivity of the particular NMR imaging technique used towards small differences in molecular mobility and partially to the distortions of the magnetic field by variations in magnetic susceptibility of the sample.

The image illustrates two important features of materials imaging. First, novel contrast features can be exploited to locate previously unknown material heterogeneities.

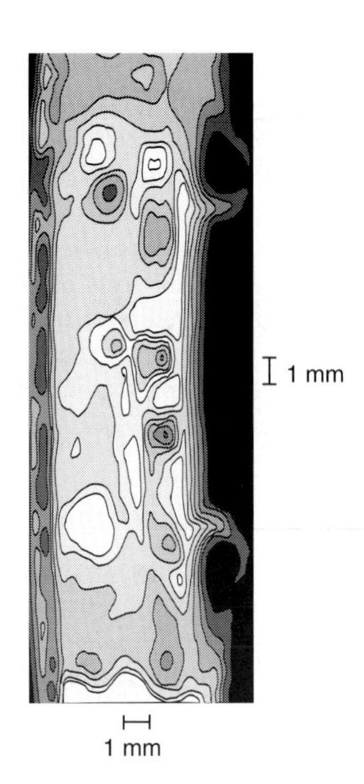

FIG. 1.1.7 [Blü2] Stress image of a stretched polydimethyl-siloxane band with a cut (right). A heterogeneous stress distribution is observed which results from the cut as well as from filler inhomogeneities. The grey scale indicates local stress in the range from 0 to 2.4 MPa.

Second, typical NMR parameters like the T_2 relaxation time can be found to scale with well-known macroscopic material properties such as strain. Like biological tissue and fluid matter, elastomers are materials particularly suited for NMR imaging, because the segmental mobility is high, leading to motionally narrowed resonances and thus to reasonable spatial resolution. Nevertheless, applications of NMR imaging to materials are less spectacular and more diverse than in medicine. Compared to medical imaging materials imaging poses higher experimental complexity and a wider range of potential applications. This prevents the development of a standard imager with just a limited set of different acquisition schemes, and requires operators well trained in NMR methods as well as in materials science.

1.1.7 Equipment

Equipment typical for use in material imaging is depicted in Fig. 1.1.8. It consists of a solid-state NMR spectrometer console, a high-field wide-bore magnet (4–14 T), high-power rf amplifiers ($P = 1$ kW) for ^{1}H and for other nuclei, also called X nuclei, three gradient amplifiers, one for each space direction, and an rf receiver. One computer

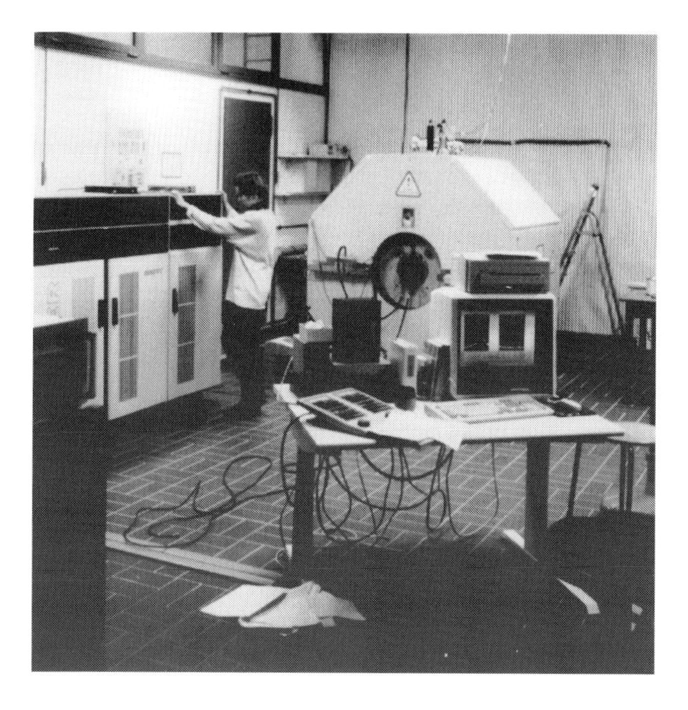

FIG. 1.1.8 A high-power solid-state NMR spectrometer with a 7 T horizontal-bore magnet for imaging of solid materials. A set of gradient coils is part of the imaging probe inside the magnet (right). The spectrometer console consists of high-power transmitters (TX, in cabinet) for ^{1}H and X nuclei, a receiver (RX, in cabinet), and of computers for controlling the experiment and for data evaluation.

controls the sequence of events for rf and gradient signals during the experiment, while another is used for data evaluation and display of the results.

1.2 ABOUT THIS BOOK

In this book, NMR is viewed from the perspective of imaging. NMR spectroscopy, relaxometry, and transport measurements are considered to be useful for defining image contrast. Clearly, such an approach is likely to be foreign to an NMR spectroscopist, who may consider NMR imaging a modification of multi-dimensional NMR. These different perspectives can be related to each other by considering the time dependence of the Larmor frequency (1.1.7).

1.2.1 The NMR signal in the presence of slow molecular motion

A phenomenological expression for the single-pulse response in NMR can be derived from (1.1.7) in terms of an attenuated precession of the transverse magnetization around the direction of the magnetic field in a given volume cell, or *voxel*, in the 3D case, or a picture cell, or *pixel*, in the 2D case,

$$M^{+}(t) = M_z(t=0)\exp\left\{-\frac{t}{T_2} + \mathrm{i}\int_0^t \omega_{\mathrm{L}}(t')\,\mathrm{d}t'\right\}. \qquad (1.2.1)$$

Here T_2 is the effective transverse relaxation time, and the Larmor frequency ω_{L} is taken to be time dependent in order to account for effects of slow molecular motion as well as for the effects of time-dependent magnetic field gradients. The initial magnetization before the 90° excitation pulse at $t = 0$ is taken to be M_z which can be different from the thermodynamic equilibrium value M_0 as a result of incomplete relaxation or the use of a filter for longitudinal magnetization.

By inspection of (1.1.7), it can be seen that the time dependence of the Larmor frequency can be introduced by three sources. The first term, the NMR frequency, may be modulated because of slow *molecular rotation* in solids, where the Larmor frequency depends on the orientation of the molecule with respect to the applied magnetic field (cf. Section 3.1.4). The origin of the time dependence of the following terms may derive from two sources. One is molecular *translational motion*. Then the space coordinate x and the respective powers of x are time dependent. This modulation is a sample property like rotational molecular motion in solids. The other source may be the time dependence of the field inhomogeneity expressed by the gradient G_x and higher-order spatial derivatives $\partial B_z^n/(\partial x)^n$ of the magnetic field. If generated separately from the homogeneous field, the inhomogeneous field may be pulsed. In fact, pulsing of magnetic field gradients is standard technology in NMR imaging. But even as an integral part of the polarization field, the time dependence of the inhomogeneous field effective for the spin precession can be manipulated through the application of rf pulses. This is done routinely, for example, in measurement of molecular diffusion in time-invariant field gradients (cf. Section 7.2.6). Thus, rf pulses and pulsed magnetic-field gradients are used to interrogate rotational and translational molecular motion.

To understand where in this picture *NMR imaging* comes in, the time dependence of the position coordinates is expanded into a Taylor series, where the expansion coefficients at $t = 0$ denote the starting position x_0, the starting velocity v_{x0} in x-direction, the starting acceleration a_{x0} in x-direction, etc.:

$$x(t) = x_0 + v_{x0}t + \tfrac{1}{2}a_{x0}t^2 + \cdots \tag{1.2.2}$$

and

$$x^2(t) = x_0^2 + 2x_0 v_{x0} t + (v_{x0}^2 + x_0 a_{x0})t^2 + \cdots . \tag{1.2.3}$$

Combination of eqns (1.2.1)–(1.2.3) with (1.1.7) provides the following expansion for the NMR signal:

$$M^+(t) = M_{z0} \prod_k \prod_l M_{kl}, \tag{1.2.4}$$

where the lowest-order phase terms M_{kl} are given by

$k\backslash l$	0	1	2
0	$\exp\{-t/T_2\}$ $\times \exp\left\{\mathrm{i}\,\omega_0 \int_0^t [1-\sigma(t')]\mathrm{d}t'\right\}$	$\times \exp\left\{-\mathrm{i}\,\gamma \int_0^t G_x(t')\mathrm{d}t' x_0\right\}$	$\times \exp\left\{-\mathrm{i}\,\gamma \int_0^t F_x(t')\mathrm{d}t' x_0^2\right\}$
1		$\times \exp\left\{-\mathrm{i}\,\gamma \int_0^t G_x(t')t'\mathrm{d}t' v_{x0}\right\}$	$\times \exp\left\{-\mathrm{i}\,2\gamma \int_0^t F_x(t')t'\mathrm{d}t' x_0 v_{x0}\right\}$
2		$\times \exp\left\{-\mathrm{i}\frac{\gamma}{2} \int_0^t G_x(t')t'^2\mathrm{d}t' a_{x0}\right\}$	$\times \exp\left\{-\mathrm{i}\,\gamma \int_0^t F_x(t')t'^2\mathrm{d}t' \times (v_{x0}^2 + x_0 a_{a0})\right\}$

Here the short-hand notation $F_x = \frac{1}{2}\partial^2 B_z/\partial x^2$ has been used.

The term M_{00} defines *NMR spectroscopy* including *relaxometry* [Kim1]. This is the type of NMR applied most often in chemistry and physics. It forms a subject in itself [Abr1, Ern1, Sli1] but is not the focus of this book. Here NMR spectroscopy and relaxometry are considered to be a highly important asset to imaging, because they provide most of the contrast features exploited in NMR imaging of materials. Although NMR spectroscopy and relaxometry are introduced in Chapters 2 and 3 of this book, because they constitute the most important term in the perturbation expansion (1.2.4), they are treated in more variety from the point of view of contrast in Chapter 7.

The term M_{01} describes the contribution of the voxel position x_0 to the signal phase. This term defines the point of view taken in this book. The term is most important in *NMR imaging*, because it reveals how to obtain spatial resolution: The time integral of the gradient has to be varied over a sufficient range of values, and for each value a single data point is acquired. Fourier transformation with respect to the gradient integrals directly produces a signal amplitude as a function of position x_0, that is, a projection of the object onto the x-axis. Following the term M_{01}, the Fourier-conjugate variable to the position vector $\boldsymbol{r}$ is defined as

$$\boldsymbol{k} = -\gamma \int_0^t \boldsymbol{G}(t')\mathrm{d}t', \tag{1.2.5}$$

where the gradient vector is given by

$$\boldsymbol{G} = \left(\frac{\partial B_z}{\partial x}, \frac{\partial B_z}{\partial y}, \frac{\partial B_z}{\partial z} \right)^{\mathrm{t}}. \tag{1.2.6}$$

The quantity $\boldsymbol{k}$ is called the *wave vector*, and one component of it is a wave number. It measures the oscillations of a wave in space (cf. Fig. 2.2.4).

As the phase of the precessing magnetization increases with time under the influence of the magnetic-field gradient, the magnitude of the $\boldsymbol{k}$ vector changes accordingly. The alignment of $\boldsymbol{k}$ is parallel to $\boldsymbol{G}$, and the sign of $\boldsymbol{k}$ depends on the time dependence of $\boldsymbol{G}$. Thus, measurements of the NMR signal as a function of time t in the presence of the gradient $\boldsymbol{G}$ or as a function of $\boldsymbol{G}$ for fixed time intervals t provide the image information in $\boldsymbol{k}$ *space*. Therefore, the NMR imaging methods are designed in such a way that information about all points in $\boldsymbol{k}$ space is acquired, so that the actual image is retrieved by simple Fourier transformation of the $\boldsymbol{k}$-space signal [Kum1, Man3]. The use of *reciprocal space* or $\boldsymbol{k}$ *space* is common practice not only in NMR imaging but also in the description of scattering experiments (cf. Sections 5.4.2 and 5.4.3) [Cal1, Fle1, Man3].

Returning to the matrix (1.2.4) of phase terms, the quantities M_{11} and M_{21} describe the phase contributions from translational motion of nuclear spins from constant flow velocity v_{x0} and constant acceleration a_{x0} in space-invariant field gradients. This type of NMR is usually not considered to be part of spectroscopy, but has important applications in chemical engineering [Cal1, Cap1] for imaging of flow profiles. This contribution to the signal phase, therefore, forms part of the *contrast* resources to be exploited in NMR imaging (cf. Section 7.2.6). The phase terms M_{11} and M_{21} are manipulated in the experiment by suitable time-modulation of the gradient wave form $\boldsymbol{G}(t)$. The important parameters for encoding of molecular transport properties in the signal phase are the gradient moments

$$\boldsymbol{m}_k = \int_0^t \boldsymbol{G}(t')t'^k \,\mathrm{d}t'. \tag{1.2.7}$$

Clearly, the wave vector $\boldsymbol{k}$ and $\boldsymbol{m}_0$ are related by $\boldsymbol{k} = -\gamma \boldsymbol{m}_0$.

The phase terms M_{k2} in (1.2.4) refer to quadratic field profiles. Such a profile is encountered in good approximation in single-sided NMR, for example with the *NMR-MOUSE* (MObile Universal Surface Explorer, cf. Section 9.3.4) [Blü6]. Use of nonlinear field profiles appears to be restricted to exceptional cases; nevertheless, it also provides access to molecular transport parameters by appropriate manipulation of the moments of $F_x(t) = \frac{1}{2}\partial^2 B_z(t)/\partial x^2$.

1.2.2 Literature

Nonclinical NMR imaging is a rapidly expanding field with steady progress in methodical developments and innovative applications. A number of reviews covers different aspects of the method. *NMR microscopy* denotes NMR imaging at high spatial resolution [Ecc1, Kuh1]. For reasons of the NMR linewidth, it aims primarily at imaging of liquids in different environments. Examples are cancer and drug discovery research, plant studies, food quality control, and flow and diffusion studies [Man1, Blü1, Blü7]. Excellent books on the principles of NMR microscopy as well as NMR imaging including diffusion and relaxation have been written by Callaghan [Cal1] and Kimmich [Kim1]. Technical and methodical aspects of NMR imaging are treated in the book by Vlaardingerbroek and den Boer [Vla1], and details on imaging hardware can be found in the books by Chen and Hoult [Che1] and Krestel [Kre1]. Imaging work on methods and applications to solid materials has been published in edited conference proceedings [Ack1, Blü1, Blü7, Bor1, Bor2, Man1] and in review articles [Blü3, Blü4, Blü8, Blü9, Bot1, Cha1, Cor1, Jez1, Jez2, Mil1].

1.2.3 The contents of this book

This book focusses on the methodical aspects of nonclinical solid- and liquid-state imaging with applications in materials science and chemical engineering. However, interesting developments in biomedical imaging are included, as they may become of use to materials applications in the future. Chapter 2 reviews some elementary physics of magnetic fields and NMR. Because most materials are solids, an introduction to solid-state NMR is given in Chapter 3. Chapter 4 summarizes selected mathematical concepts on transformations, convolution, and correlation, which are helpful for understanding many of the imaging techniques in later chapters. Chapter 5 covers the concepts of spatial resolution, such as selective excitation to reduce the process of mapping a 3D volume by one or two dimensions, and magnetization in time-varying gradients, which leads to the introduction of reciprocal space. Chapter 6 deals with the fundamental imaging methods. Phase- and frequency-encoding techniques are treated including backprojection, Fourier, and fast imaging methods. Although back-projection imaging does not attract much interest in present day clinical imaging, it exhibits features which are advantageous to exploit in materials applications. Ways to increase the information contents and the contrast in imaging are explored in Chapter 7. Here the concept of magnetization filters applied in preparation of the initial magnetization used for space encoding is elaborated. This is a key topic. The availability of magnetization filters for generation of image contrast is the most outstanding feature of NMR imaging compared to other imaging methods. Chapter 8 is devoted to imaging techniques suitable for solid objects. Different approaches to overcome the spatial resolution limit imposed by large linewidths in the solid state are treated. The analysis of selected regions localized within a heterogeneous object is covered in Chapter 9. For this purpose, gradient methods as well as surface coils and surface magnets can be used. Nonmedical applications of these techniques are in well logging, the analysis of large objects by mobile NMR devices, and in process and quality control. Chapter 10 features a selection of ten remarkable application areas of NMR imaging outside medicine. NMR imaging is expensive in terms of operator know-how and instrument cost. Its use needs to be well justified. Contrast and noninvasiveness are the most outstanding features of the method but not the quality of spatial resolution. Particularly useful applications are in soft-matter analysis: NMR is unique in its capability for analysis of mass transport phenomena by diffusion and flow, as well as in characterizing distributions of properties in elastomers and biological materials, for example, temperature distributions and macroscopic molecular order from processing and applied mechanical load.

2

Fundamentals

Nuclear magnetic resonance (NMR) exploits the interaction of nuclei with magnetic fields [Abr1, Ern1]. A strong static field is applied to polarize the nuclear magnetic moments, time-dependent magnetic rf fields are used to stimulate the spectroscopic response, and magnetic-field gradients are needed to obtain spatial resolution [Cal1]. Following the description of the different magnetic fields used in NMR spectroscopy and imaging, the behaviour of magnetic nuclei exposed to these fields is treated first in terms of the classic *vector model*, and then the *density-matrix* concept is introduced. The latter is required to describe the couplings among nuclei, for example, the dipole–dipole interaction which dominates the ^{1}H NMR spectrum of most solid materials [Meh1, Sch2, Sli1]. Also, knowledge of the density matrix is helpful to understand *multi-quantum coherences* as well as the imaging methods developed for investigations of solid materials.

2.1 MAGNETIC FIELDS

In NMR spectroscopy, the polarizing magnetic field is required to be highly homogeneous, that is, constant over the sample. In NMR imaging, the polarizing field needs to be inhomogeneous. In most cases, the inhomogeneous part of the field is linearly dependent on space, so that the field gradient is constant. This space-dependent part of the field is often referred to as the *gradient field*. It introduces spatial resolution to the NMR experiment. Typically, it is generated by a separate set of current-bearing coils. The strong, homogeneous static magnetic field is denoted by $\boldsymbol{B}_0$. By convention, it is oriented along the z-direction of the laboratory frame (Fig. 2.1.1).

The strength of $\boldsymbol{B}_0$ is of the order of 0.5–21 T. It defines the NMR frequency $\omega_0 = 2\pi\nu_0$ by

$$\omega_0 = -\gamma B_0, \tag{2.1.1}$$

where γ is the *gyromagnetic ratio* (cf. Section 2.2.1), and B_0 is the magnitude of the strong magnetic field $\boldsymbol{B}_0$ (cf. eqn (1.1.2)). For excitation of the spectroscopic response, a weak, time-dependent magnetic field $\boldsymbol{B}_{\mathrm{rf}}$ perpendicular to the static field is required.

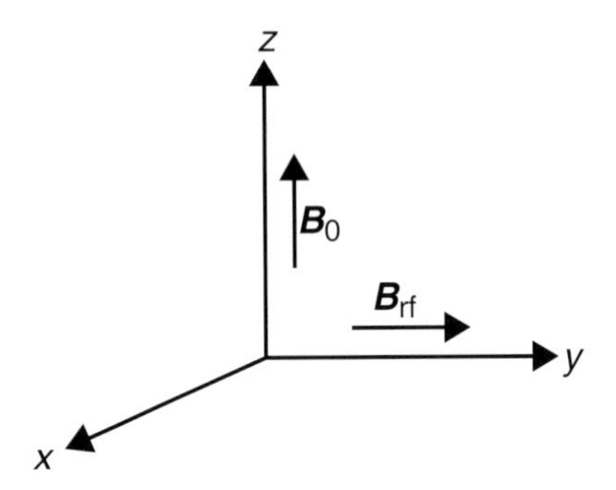

FIG. 2.1.1 Magnetic fields used in NMR spectroscopy and imaging. A strong magnetic field $\boldsymbol{B}_0$ is oriented along the z-direction of the laboratory frame, and a weak field $\boldsymbol{B}_{\text{rf}}$ is oscillating perpendicular to it in the rf regime.

When the weak field $\boldsymbol{B}_{\text{rf}}(t)$ oscillates with the nuclear resonance frequency, energy can be transferred from the oscillating field to the nuclei and *vice versa*. Typical NMR frequencies are in the rf regime between 10 and 900 MHz. The strength of the excitation field is of the order of 1 mT and less.

Field gradients can be used with either the strong, static field or the weak, time-dependent field to obtain spatial resolution. Therefore, both static and time-dependent homogeneous and inhomogeneous magnetic fields have to be considered.

2.1.1 Homogeneous magnetic fields

Magnetic induction versus magnetic field

For the description of magnetic fields, the *magnetic induction* $\boldsymbol{B}$ is used as well as the *magnetic field* $\boldsymbol{H}$. These names are historic. It turns out that the magnetic field acting in matter is described by the quantity $\boldsymbol{B}$, while $\boldsymbol{H}$ is the part of the magnetic field which is generated by macroscopic currents $\boldsymbol{I}$ in vacuum. Thus, the magnetic induction also takes account of the effects of electron motion and nuclear dipoles in matter. In vacuum, the magnetic induction $\boldsymbol{B}$ and the magnetic field $\boldsymbol{H}$ are equivalent. They differ only by a scalar factor, the permeability of vacuum, $\mu_0 = 4\pi\ 10^{-7}$ V s/A m,

$$\boldsymbol{B}_{\text{vacuum}} = \mu_0 \boldsymbol{H}. \tag{2.1.2}$$

In matter, the value of the magnetic induction is changed by the *magnetic polarization* $\boldsymbol{M}_{\text{p}}$,

$$\boldsymbol{B} = \mu_0(\boldsymbol{H} + \boldsymbol{M}_{\text{p}}). \tag{2.1.3}$$

Both $\boldsymbol{H}$ and $\boldsymbol{M}_{\text{p}}$ are vectors, which in general are not parallel. Therefore, the permeability $\boldsymbol{\mu}$ of matter must be introduced as a tensor,

$$\boldsymbol{B} = \mu_0\, \boldsymbol{\mu}\, \boldsymbol{H}. \tag{2.1.4}$$

Magnetization

The quantity relevant to NMR is the contribution of the nuclei to the *magnetic polarization* $\boldsymbol{M}_{\mathrm{p}}$. This contribution multiplied by the sample volume is referred to simply as *magnetization*. In thermodynamic equilibrium the magnetization $\boldsymbol{M}_0$ established in the polarizing magnetic field $\boldsymbol{B}_0$ is given by the *Curie law*,

$$\boldsymbol{M}_0 = N\frac{\gamma^2\hbar^2 I(I+1)}{3k_{\mathrm{B}}T}\boldsymbol{B}_0, \tag{2.1.5}$$

where I is the nuclear spin quantum number (Section 2.2.1), k_{B} the Boltzmann constant, T the temperature, and N is the number of nuclei with spin I in the sample. The magnetization $\boldsymbol{M}_0$ is manipulated by the weak, time-dependent rf magnetic field $\boldsymbol{B}_{\mathrm{rf}}$ to generate the response signal for NMR spectroscopy and imaging.

Magnetic field energy

The energy density $W_{\mathrm{c}} = E_{\mathrm{c}}/V$ stored in the magnetic field generated by a current in a coil of volume V is given by

$$W_{\mathrm{c}} = \frac{\boldsymbol{B}\boldsymbol{H}}{2}. \tag{2.1.6}$$

Using (2.1.4), the coil energy for isotropic substances simplifies to

$$E_{\mathrm{c}} = \frac{\mu\mu_0 H^2 V}{2}. \tag{2.1.7}$$

It is this energy which eventually limits the energy deposition in the material inside the coil. When H denotes the magnitude of a rf magnetic field $\boldsymbol{H}_{\mathrm{rf}}$ in particular, the *sample heating* depends on this equation, where the permeability μ is frequency dependent.

2.1.2 Magnetic-field gradients

Definition

The components G_{kl} of the *magnetic-field gradient tensor* $\mathbf{G}$ are defined as the spatial derivatives of the magnetic field,

$$G_{kl} = \frac{\partial B_k}{\partial x_l}. \tag{2.1.8}$$

If they are independent of space, they are constant and the magnetic field varies linearly with space. Because the magnetic field $\boldsymbol{B}$ is a vector with components B_x, B_y, and B_z, the magnetic-field gradient is a second-rank *tensor* with nine components. It can be written as the dyadic product of the gradient operator ∇ and the magnetic field,

$$\mathbf{G} = \nabla\boldsymbol{B} = \begin{bmatrix} G_{xx} & G_{xy} & G_{xz} \\ G_{yx} & G_{yy} & G_{yz} \\ G_{zx} & G_{zy} & G_{zz} \end{bmatrix}. \tag{2.1.9}$$

The linearly space-dependent part of the magnetic field is called the *gradient field*. It is generated by an extra set of coils and is added to the homogeneous magnetic fields given by the strong static field $\boldsymbol{B}_0$ in the z-direction and the weak time-dependent field $\boldsymbol{B}_{\mathrm{rf}}$ perpendicular to $\boldsymbol{B}_0$ (cf. Fig. 2.1.1),

$$\boldsymbol{B} = \boldsymbol{B}_0 + \boldsymbol{B}_{\mathrm{rf}} + \mathbf{G}\boldsymbol{r}, \tag{2.1.10}$$

where $\boldsymbol{r}$ is the space vector with components x, y, and z.

The gradient components are simply referred to as the *gradients* as such in the following. They can be applied either quasi-static to modify $\boldsymbol{B}_0$ or oscillating with rf to modify $\boldsymbol{B}_{\mathrm{rf}}$. The first case is standard in most imaging experiments. The gradients are said to be applied in the laboratory frame. Here the maximum value of the gradient field $\mathbf{G}\boldsymbol{r}$ typically is less than 1% of the strength of the homogeneous magnetic field $\boldsymbol{B}_0$. However, in some cases the gradients oscillate at the NMR frequency to provide a component which rotates about the static field $\boldsymbol{B}_0$; the gradients are then said to be applied in the rotating frame. The maximum values of the gradient field in this case are of the same order of the strength of $\boldsymbol{B}_{\mathrm{rf}}$.

Relationships between field-gradient components

The components of the *gradient tensor* are not independent. They are related by Maxwell's field equations [Sch1]

$$\nabla \cdot \boldsymbol{B} = 0 \tag{2.1.11}$$

and

$$\nabla \times \boldsymbol{B} = \mathbf{0}. \tag{2.1.12}$$

The first equation always applies, while the second only applies for samples without electric currents inside and through their surface. Thus, the second equation is valid for nonconducting materials, but not necessarily for conductors. It establishes a relation between the Cartesian gradient components which are obtained by permutation of indices,

$$G_{kl} - G_{lk} = 0, \tag{2.1.13}$$

while eqn (2.1.11) requires the trace (spur) of the gradient tensor to vanish,

$$\mathrm{Sp}\{\mathbf{G}\} = G_{xx} + G_{yy} + G_{zz} = 0. \tag{2.1.14}$$

Space dependence of the NMR frequency

In general, the maximum strength of the gradient field is small enough to be treated as a perturbation of the magnetic field $\boldsymbol{B}_0$, and the gradients are applied in a coordinate frame the symmetry axis of which is parallel to the z-axis of the magnetic field $\boldsymbol{B}_0$. In this case, only three components of the gradient tensor determine the NMR frequency to first order. Using (2.1.10) with $\boldsymbol{B}_{\mathrm{rf}} = \boldsymbol{0}$, the NMR frequency (2.1.1) becomes dependent on space,

$$
\begin{aligned}
\omega_0(\boldsymbol{r}) &= -\gamma|\boldsymbol{B}| = -\gamma|\boldsymbol{B}_0 + \mathbf{G}\boldsymbol{r}| \\
&= -\gamma[(B_{0x} + G_{xx}x + G_{xy}y + G_{xz}z)^2 \\
&\quad + (B_{0y} + G_{yx}x + G_{yy}y + G_{yz}z)^2 \\
&\quad + (B_{0z} + G_{zx}x + G_{zy}y + G_{zz}z)^2]^{1/2}.
\end{aligned} \tag{2.1.15}
$$

With $B_{0x} = 0$ and $B_{0y} = 0$, the terms dominating the resonance frequency are given in the last row. Thus only three elements of the *gradient tensor* determine the resonance frequency in first order. The others, though nonzero, by (2.1.13) and (2.1.14), can be neglected. The three relevant terms are often concatenated to form the *gradient vector* $\boldsymbol{G}$,

$$
\begin{pmatrix} G_{zx} \\ G_{zy} \\ G_{zz} \end{pmatrix} \equiv \boldsymbol{G} = \begin{pmatrix} G_x \\ G_y \\ G_z \end{pmatrix}, \tag{2.1.16}
$$

which is used in the majority of the magnetic resonance imaging literature. If the gradient-coordinate system, however, is tilted in the laboratory frame, then the other gradient elements in (2.1.15) may no longer be neglected. A situation where this is the case is *MAS imaging*, that is imaging in combination with *magic-angle spinning* (cf. Section 8.5). Here the z-axes of laboratory- and gradient-coordinate systems enclose the magic angle of 54.7° [Meh1]. For parallel gradient- and laboratory-system z-axes, (2.1.15) is well approximated by

$$
\omega_0(\boldsymbol{r}) \cong -\gamma(B_{0z} + G_{xz}x + G_{yz}y + G_{zz}z) = -\gamma(B_0 + \boldsymbol{G}\boldsymbol{r}), \tag{2.1.17}
$$

where $B_0 = B_{0z}$ has been used.

Projections

The relationship between the spatial domain and the frequency domain expressed by (2.1.17) forms the basis of magnetic resonance imaging. In the presence of a gradient, planes of constant field strength become planes of constant frequency $\omega_0(\boldsymbol{r})$. This can readily be seen when particular gradient fields are chosen in (2.1.17), for instance, $\boldsymbol{G}\boldsymbol{r} = G_x x$. The NMR spectrum obtained in the presence of a field gradient therefore shows a distribution of NMR frequencies. The spectral amplitude at each frequency is proportional to the number of contributing nuclei in the respective constant-frequency plane. As a result, the NMR spectrum of an object placed in a magnetic field gradient corresponds to the projection of the number of nuclear spins onto the gradient direction [Osh1] (Fig. 1.1.5). One way of obtaining images by NMR is by *reconstruction from projections* (Fig. 2.1.2, cf. Section 6.1) [Lau1]. The computational procedures for this form of NMR imaging are the same as for *computed X-ray tomography* (CT).

Higher-order gradients

So far, gradients that are constant in space have been assumed, which produce space variables directly proportional to frequency in first order. If the space dependence of the

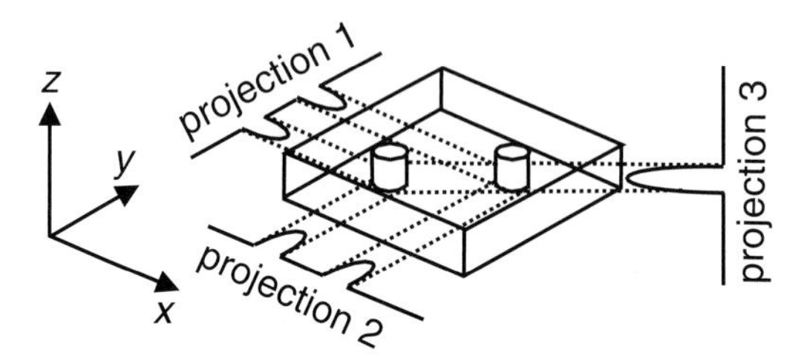

FIG. 2.1.2 The NMR spectrum acquired in the presence of a field gradient corresponds to a projection of the object. An image can be reconstructed from several such projections acquired under different angles.

magnetic field is weakly nonlinear, B_z can be expanded into a series of space-dependent terms. This has been formulated in (1.1.5) for Cartesian coordinates. The cylindrical symmetry of superconducting magnets favours an expansion in terms of the associated *Legendre polynomials* $P_{nm}(\cos\theta)$ [Kre1, Mor1],

$$B_z(r, \theta, \varphi) = \sum_{n=0}^{N} \sum_{m=0}^{M \leq N} \left(\frac{r}{R}\right)^n P_{nm}(\cos\theta)\, [A_{nm}\cos(m\varphi) + B_{nm}\sin(m\varphi)]\,, \quad (2.1.18)$$

where r, θ and φ are the spherical coordinates, and R is the radius of the volume of interest. The Legendre polynomial P_{00} is unity. Thus the coefficient A_{00} is the nominal strength B_0 of the constant magnetic field $\boldsymbol{B}_0$. The other terms correspond to the *gradient field*. In particular, the coefficient A_{10} defines the z gradient and the coefficients A_{11} and B_{11} define the x and y gradients. The other components define *higher-order gradients* which produce nonlinear field variations.

The orthogonality of terms in (2.1.18) makes the expansion particularly suitable for design of higher-order gradient coils which contribute components to the total magnetic field independent of those generated by the other coils. Such coils are needed when the homogeneity of the polarizing magnetic field $\boldsymbol{B}_0$ is to be optimized for spectroscopy. The associated procedure is called *shimming* [Chm1, Fuk2]. It consists of careful adjustment of the various correction fields corresponding to the individual terms of the expansion in order to cancel the residual inhomogeneities of the magnet. Shim coils are part of the magnet of every high-resolution NMR spectrometer. Clearly, those for the constant gradients can be used for imaging as well, provided they can withstand the currents necessary to obtain sufficient spatial resolution.

2.2 PRINCIPLES OF NMR

The first successful detection of NMR in condensed matter was achieved independently in 1945 by the groups of Felix Bloch [Blo1] and Edward Purcell [Pur1]. With the discovery of the *chemical shift* in 1951, that is the fine structure of the resonance line depending on the electronic environment of the nuclei, NMR was rapidly becoming a tool for chemical analysis of molecules dissolved in liquids [Arn1]. In solids this fine structure is not as easily observed. Here the molecular motion is severely restricted. As a consequence, spin interactions like the *dipole–dipole coupling*, the *quadrupole coupling*, and the *anisotropy of the chemical shift* are not averaged out. This leads to severe line broadening and obscures the effects of chemical shift differences unless special techniques are employed [Ger1, Meh1, Ste3].

Originally, the NMR spectrum was observed by sweeping the magnetic field under irradiation with a *continuous* rf *wave* (CW NMR). This time consuming technique was eventually replaced by the pulsed *Fourier transform* methods (FT NMR) following their introduction by Richard Ernst in 1966 [Ern2]. Here, all frequency components of the NMR spectrum are observed simultaneously by measuring the response to an excitation pulse. The NMR spectrum is the Fourier transform of the pulse response. The success of *Fourier NMR* was aided by the rediscovery of the fast Fourier transformation (FFT) algorithm, the development of affordable laboratory computers, and the availability of superconducting magnets.

The next milestone, in the history of NMR [Fre1], was the extension of the NMR spectrum to more than one frequency coordinate. It is called *multi-dimensional spectroscopy* and is a form of nonlinear spectroscopy. The technique was introduced by Jean Jeener in 1971 [Jee1] with two-dimensional (2D) NMR. It was subsequently explored systematically by the research group of Richard Ernst [Ern1] who also introduced Fourier imaging [Kum1]. Today such techniques are valuable tools, for instance, in the structure elucidation of biological macromolecules in solution in competition with X-ray analysis of crystallized molecules as well as in solid state NMR of polymers (cf. Fig. 3.2.7) [Sch2].

The use of NMR for imaging was demonstrated in 1973 by Paul Lauterbur for medical applications [Lau1] and by Peter Mansfield [Man1] for materials. Before that the potential of NMR for medical diagnostics had already been recognized by Raymond Damadian in 1971 [Dam1].

Most aspects of NMR can be described in terms of the classical vector model of magnetization precessing in a magnetic field similar to a spinning top precessing in a gravitational field (cf. Fig. 1.1.3). Many of the advanced methods, however, require the use of quantum mechanics. Here the density matrix is the appropriate tool for the semiclassical description of the motion of an ensemble of interacting magnetic moments. Therefore both the classical and the quantum mechanical description of NMR are introduced.

2.2.1 An NMR primer

This section summarizes primarily the *classical description of NMR* based on the *vector model* of the *Bloch equations*. Important concepts like the *rotating frame*, the effect of *rf pulses*, and the *free precession* of transverse magnetization are introduced. More detailed accounts, still on an elementary level, are provided in textbooks [Der1, Far1, Fuk1].

The origin of nuclear magnetization

The origin of nuclear magnetization is quantum mechanical. Many nuclei possess a property similar to *angular momentum* which is called *spin* [Zum1]. The spin $\mathbf{I}$ is a quantum

Table 2.2.1 Nuclei and their NMR properties [Bru1]

Isotope	Spin	Nat. abundance (%)	Quadrupole moment[a]	Rel. sensitivity[b]	Freq. (MHz) at 2.3488 T	Chemical shift range (ppm)	Chemical shift reference
^{1}H	1/2	99.98	—	1.0	100.000	12 to −1	$SiMe_4$
^{2}H	1	$1.5 \cdot 10^{-2}$	0.002875	$9.65 \cdot 10^{-3}$	15.351	12 to −1	$SiMe_4$
^{6}Li	1	7.42	−0.000644	$8.50 \cdot 10^{-3}$	14.716	5 to −10	1 M LiCl
^{7}Li	3/2	92.58	−0.040	0.29	38.863	5 to −10	1 M LiCl
^{11}B	3/2	80.42	0.040	0.17	32.084	100 to −120	BF_3OEt_2
^{13}C	1/2	1.108	—	$1.59 \cdot 10^{-2}$	25.144	240 to −10	$SiMe_4$
^{15}N	1/2	0.37	—	$1.04 \cdot 10^{-3}$	10.133	1200 to −500	$MeNo_2$
^{17}O	5/2	$3.7 \cdot 10^{-2}$	−0.026	$2.91 \cdot 10^{-2}$	13.557	1400 to −100	H_2O
^{19}F	1/2	100	—	0.83	94.077	100 to −300	$CFCl_3$
^{23}Na	3/2	100	0.108	$9.25 \cdot 10^{-2}$	26.451	10 to −60	1 M NaCl
^{27}Al	5/2	100	0.150	0.21	26.057	200 to −200	$[Al(H_2O)_6]^{3+}$
^{29}Si	1/2	4.7	—	$7.84 \cdot 10^{-3}$	19.865	100 to −400	$SiMe_4$
^{31}P	1/2	100	—	$6.63 \cdot 10^{-2}$	40.481	230 to −200	H_3PO_4
^{43}Ca	7/2	0.145	< 0.23	$6.40 \cdot 10^{-3}$	6.728	40 to −40	$CaCl_2$
^{51}V	7/2	99.76	−0.0515	0.38	26.289	0 to −2000	$VOCl_3$
^{67}Zn	5/2	4.11	0.150	$2.85 \cdot 10^{-3}$	6.254	100 to −2700	$ZnClO_4$
^{77}Se	1/2	7.58	—	$6.93 \cdot 10^{-3}$	19.067	1600 to −1000	$SeMe_2$
^{93}Nb	9/2	100	−0.28	0.48	24.442	0 to −2000	$NbCl_6^-$
^{99}Ru	3/2	12.72	0.076	$1.95 \cdot 10^{-4}$	3.389	3000 to −3000	RuO_4/CCl_4
^{199}Sn	1/2	8.58	—	$5.18 \cdot 10^{-2}$	37.272	5000 to −3000	$SnMe_4$
^{121}Sb	5/2	57.25	−0.33	0.16	23.930	1000 to −2700	Et_4NSbCl_6
^{129}Xe	1/2	26.44	—	$2.12 \cdot 10^{-2}$	27.660	2000 to −6000	$XeOF_4$
^{133}Cs	7/2	100	−0.003	$4.74 \cdot 10^{-2}$	13.117	300 to −300	CsBr
^{195}Pt	1/2	33.8	—	$9.94 \cdot 10^{-3}$	21.499	9000 to −6000	Na_2PtCl_6
^{199}Hg	1/2	16.84	—	$5.67 \cdot 10^{-3}$	17.827	500 to −3000	$HgMe_2$

[a] Electric quadrupole moment Q in multiples of $|e|\ 10^{-24}\ cm^2$.

[b] At constant field and equal number of nuclei.

mechanical operator. The eigenvalue of $\mathbf{I}^2$ is $I(I+1)$, where I is the *spin quantum number* which can assume integral and half-integral values. Often the spin quantum number itself is referred to as spin, for instance ^{1}H, ^{13}C, ^{29}Si, and ^{31}P are spin-$\frac{1}{2}$ nuclei with $I = \frac{1}{2}$, and ^{2}H, and ^{6}Li are spin-1 nuclei with $I = 1$. A list of nuclei with nonzero spin relevant to materials science is given in Table 2.2.1 together with other information pertinent to NMR spectroscopy.

The *magnetic moment* $\boldsymbol{\mu}_I$ is proportional to the spin $\mathbf{I}$ of the nucleus,

$$\boldsymbol{\mu}_I = \gamma \hbar \mathbf{I}, \tag{2.2.1}$$

where $\hbar$ is Planck's constant h divided by 2π. This equation defines the *gyromagnetic ratio* γ. It is a specific constant of the nucleus and can assume positive and negative values. By default γ is taken to be negative [Ern1].

The macroscopic thermodynamic equilibrium magnetization $\boldsymbol{M}_0$ is formed by the sum of projections of all nuclear magnetic moments onto the axis of the magnetic field $\boldsymbol{B}_0$. Its value is expressed by the *Curie law* in (2.1.5) as a function of magnetic field and temperature. The magnetization $\boldsymbol{M}$ is a macroscopic magnetic dipole moment. Its potential energy in a magnetic field $\boldsymbol{B}_0$ depends on the angle θ between the dipole moment and the field (Fig. 2.2.1(a)),

$$E = -\boldsymbol{M}\boldsymbol{B}_0 = -|\boldsymbol{M}||\boldsymbol{B}_0| \cos\theta = -M_z B_0. \tag{2.2.2}$$

Because in NMR the orientation of the field defines the z-axis of the laboratory coordinate frame of reference, $\boldsymbol{B}_0 = (0, 0, B_0)^{\mathrm{t}}$ and $M_z = |M| \cos\theta$ is the projection of the magnetization vector on to the direction of the magnetic field.

The quantum mechanical operator corresponding to the energy is the *Hamilton operator*. The potential energy of a single magnetic moment in a magnetic field is given

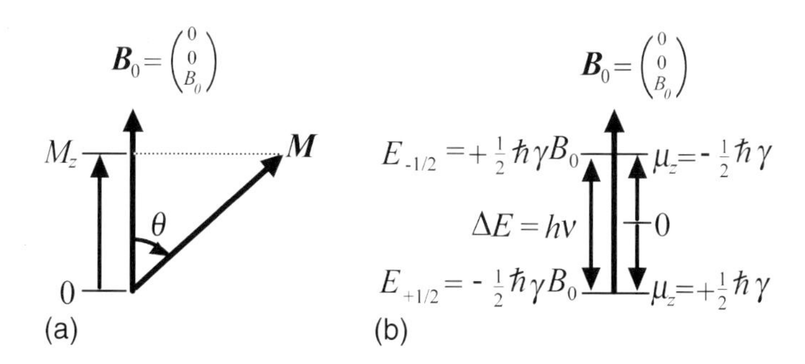

FIG. 2.2.1 Classical and quantum mechanical energies of magnetic dipoles in a magnetic field. (a)The potential energy of the macroscopic magnetization $\boldsymbol{M}$ in a magnetic field $\boldsymbol{B}_0$ is a product of the magnetic field with the projection of the magnetization onto the axis of the field. It depends on the angle θ between the magnetization and the field. (b)For a quantum mechanical magnetic moment with spin $I = 1/2$ there are two stationary states in a magnetic field. One has its projectiion parallel, the other antiparallel to the direction of the field. Both states differ in energy. The diagram applies for nuclei with positive values of γ, where the magnetization aligns antiparallel to $\boldsymbol{B}_0$ in thermodynamic equilibrium. For simplicity of notation the thermodynamic equilibrium magnetization is taken parallel to $\boldsymbol{B}_0$ in the following.

in analogy to the last part of (2.2.2) by

$$\mathbf{H}_Z = -\gamma \hbar \mathbf{I}_z B_0. \tag{2.2.3}$$

The interaction of magnetic moments with a magnetic field gives rise to a splitting of energy levels. It is called *Zeeman interaction*. For this reason the index Z is used in (2.2.3).

Clearly, the unit of the Hamilton operator is that of energy. In the other chapters of this book standard NMR nomenclature is followed by expressing energies and Hamilton operators in frequency units. Frequency units are obtained from energy units by division by $\hbar$. The energy levels E_m are defined as the *eigenvalues* of the Hamilton operator,

$$E_m = -\gamma \hbar m B_0. \tag{2.2.4}$$

Here m is the *magnetic quantum number*. It can assume the values

$$-I \leq m \leq +I. \tag{2.2.5}$$

Thus a nuclear spin with quantum number I can be in one of $2I + 1$ stationary states in a magnetic field. Nuclei like ^1H and ^{13}C with spins $I = 1/2$ have two *eigenstates*. These are referred to as spin-up and spin-down, depending on whether the z-component of the magnetic moment is parallel or antiparallel to the magnetic field (Fig. 2.2.1(b)). Nuclei like ^2H with spins $I = 1$, possess three eigenstates. The energy difference ΔE between neighbouring energy levels is absorbed or emitted by a nuclear spin when it reorients and moves from one energy level to the next. This energy difference determines the NMR frequency $\omega_0 = 2\pi \nu_0$,

$$\Delta E = E_m - E_{m-1} = -\hbar \gamma B_0 = \hbar \omega_0 = 2\pi \hbar \nu_0. \tag{2.2.6}$$

In Table 2.2.1 the NMR frequencies ν_0 of different nuclei are given in MHz for a magnetic field of 2.3488 T.

By (2.2.6) the NMR frequency is proportional to the strength B_0 of the magnetic field. So is the fine structure of the resonance which results from shielding of the magnetic field at the site of the nucleus by the surrounding electrons. It is called *chemical shift* (cf. Fig. 1.1.4(a)). Thus, higher field strengths provide better spectroscopic resolution. But, they also provide better sensitivity (Fig. 2.2.2) [Wol1]. Apart from instrumental parameters this is due to an increase in magnetic polarization (cf. eqn (2.1.5). As mentioned above, the polarization is the sum of all components of the nuclear magnetic moments parallel to the applied field. From the quantum mechanics above it is known that in thermodynamic equilibrium, all magnetic moments are found in one of the energy eigenstates E_m having one of the $2I + 1$ allowed projections along the z-axis. Thus the *nuclear magnetic polarization* is determined by the differences in population of the energy levels. The relative number n_{m-1}/n_m of spins in these states is given by the *Boltzmann distribution*

$$\frac{n_{m-1}}{n_m} = \exp\left\{-\frac{\hbar \omega_0}{k_B T}\right\}. \tag{2.2.7}$$

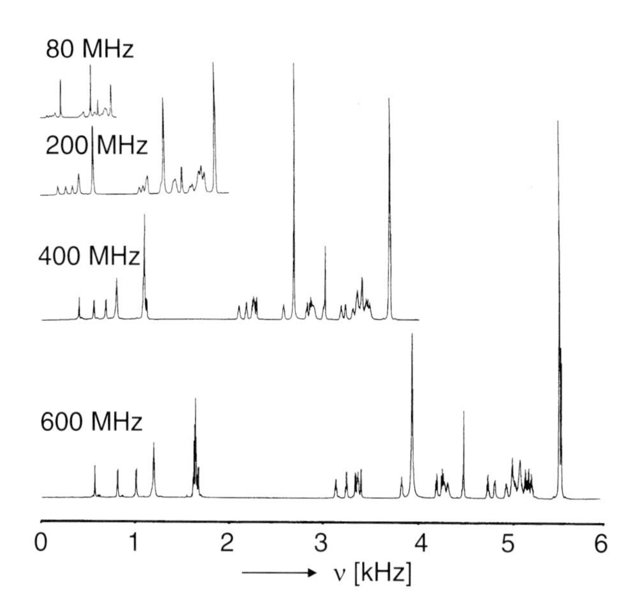

FIG. 2.2.2 [Wol1] ^{1}H NMR spectra of gramicidine at different strength of the magnetic field corresponding to the NMR frequencies given in the spectra. The signal intensity and the chemical shift are proportional to the strength of the magnetic field. The hyperfine structure due to the indirect coupling of spins is independent of the magnetic field strength.

From this the population difference $\Delta n = n_m - n_{m-1}$ is calculated. For ^{1}H at room temperature and a frequency of 100 MHz the exponent is given by $\hbar\omega_0/k_B T = 1.6 \cdot 10^{-5}$. At this temperature $k_B T \gg \hbar\omega_0$, so that the exponential in (2.2.7) can be expanded, and the expansion can be truncated after the second term. This so-called *high- temperature approximation* can be used down to rather low temperatures. In this limit the population difference corresponding to the magnitude of the magnetization is proportional to the strength of the magnetic field. This relationship is expressed by the *Curie law* in (2.1.5). Given 10^{20} spins in a sample only $1.6 \cdot 10^{15}$ of them make up for the nuclear magnetization. For this reason NMR spectroscopy is a method insensitive compared to infrared and optical spectroscopy with respect to the amount of sample needed. This lack of sensitivity translates directly into the limited spatial resolution achievable by NMR imaging. But this disadvantage is offset by the unsurpassed manifold of information accessible by NMR.

When an initially unmagnetized sample is exposed to the magnetic field, the formation of the thermodynamic equilibrium magnetization requires the transfer of energy from the spins to the surrounding lattice. This energy transfer takes place in a characteristic time T_1. The *energy dissipation time* is denoted by T_1. A typical value for T_1 of ^{1}H is 1 s at high magnetic fields.

Classical equation of motion: the Bloch equation

The equation of motion of the macroscopic magnetization vector has been derived by Felix Bloch [Blo1] by identifying $\boldsymbol{M}/\gamma$ as angular momentum, which experiences a torque $\boldsymbol{M} \times \boldsymbol{B}$ in the magnetic field $\boldsymbol{B}$. As a result any magnetization component not parallel to the magnetic field precesses around it. This situation is completely analogous to a top spinning in a gravitational field, which precesses around the direction of the field (cf. Fig. 1.1.3). Neglecting the shielding of the applied field by the electrons, the nuclear precession proceeds with the NMR frequency ω_0. By equating the torque to the rate of change of angular momentum, and by adding a relaxation term which allows the establishment of thermodynamic equilibrium with time, the *Bloch equation* is obtained [Blo1, Ern1]:

$$\mathrm{d}\boldsymbol{M}/\mathrm{d}t = \gamma \boldsymbol{M}(t) \times \boldsymbol{B}(t) - \mathbf{R}[\boldsymbol{M}(t) - \boldsymbol{M}_0]. \tag{2.2.8}$$

The time-dependent magnetization vector $\boldsymbol{M}(t)$ has the thermodynamic equilibrium value $\boldsymbol{M}_0 = (0, 0, M_0)$ which is determined by the Curie law (2.1.5), and $\mathbf{R}$ is the relaxation matrix,

$$\mathbf{R} = \begin{pmatrix} 1/T_2 & 0 & 0 \\ 0 & 1/T_2 & 0 \\ 0 & & 1/T_1 \end{pmatrix}, \tag{2.2.9}$$

with the longitudinal and transverse relaxation times T_1 and T_2.

The *longitudinal relaxation time* T_1 is the *energy dissipation time* characteristic for build up of the magnetization parallel to the magnetic field. It is also called *spin–lattice relaxation time*. The *transverse relaxation time* T_2 is the time constant for disappearance of magnetization components orthogonal to the magnetic field. T_2 is generally shorter than or equal to T_1. In liquids, it is close to T_1, while in solids it can be orders of magnitude shorter. Transverse magnetization components can be generated by application of resonant rf irradiation. This can be seen by solving the Bloch equations. In the following, the most important conclusions obtained from solutions of the Bloch equations are summarized.

The rotating coordinate frame

To solve the Bloch equations the magnetic field $\boldsymbol{B}(t)$ is written explicitly as the sum of the strong static magnetic field $\boldsymbol{B}_0$ and a weak, time-dependent rf field $\boldsymbol{B}_{\mathrm{rf}}(t)$ perpendicular to $\boldsymbol{B}_0$ (cf. Fig. 2.1.1),

$$\boldsymbol{B}(t) = \boldsymbol{B}_0 + \boldsymbol{B}_{\mathrm{rf}}(t). \tag{2.2.10}$$

The rf field is usually applied with linear polarization,

$$\boldsymbol{B}_{\mathrm{rf}}(t) = \begin{pmatrix} 2B_1 \cos(\omega_{\mathrm{rf}} t + \varphi) \\ 0 \\ 0 \end{pmatrix}, \tag{2.2.11}$$

where φ describes a phase offset which can be manipulated by the transmitter electronics. By using the relationship $2\cos(\omega t) = \exp\{i\omega t\} + \exp\{-i\omega t\}$, $\boldsymbol{B}_{\mathrm{rf}}(t)$ can be decomposed into two counter-rotating components. One component follows the precession of the magnetization. It is retained. The other can be discarded, because it is out of resonance by twice the NMR frequency. So instead of (2.2.11)

$$\boldsymbol{B}_{\mathrm{rf}}(t) = \begin{pmatrix} B_1 \cos(\omega_{\mathrm{rf}} t + \varphi) \\ B_1 \sin(\omega_{\mathrm{rf}} t + \varphi) \\ 0 \end{pmatrix} \tag{2.2.12}$$

can be used without penalty.

The calculation is simplified considerably by transforming the Bloch equations into a coordinate system which rotates with the rf magnetic field vector (2.2.12) around the z-axis of the laboratory frame. In this *rotating frame* the magnetic field including the rf field component appears static, but the magnitude of the B_0 field in z-direction is changed, and (2.2.10) turns into

$$\boldsymbol{B}_{\mathrm{r}} = \begin{pmatrix} B_1 \cos\varphi \\ B_1 \sin\varphi \\ B_0 + \omega_{\mathrm{rf}}/\gamma \end{pmatrix}. \tag{2.2.13}$$

The change of the magnetic field in the rotating frame can be rationalized when considering a magnetization vector which has been placed perpendicular to the magnetic field $\boldsymbol{B}_0$. Following the Bloch equations (2.2.8) the transverse magnetization precesses around the applied field in the laboratory frame. When rotating exactly with the same frequency as the magnetization, the magnetization appears static in the rotating frame and the cause for rotation of the magnetization seems to have vanished. So the magnetic field must be zero in this case. If the rf frequency ω_{rf} does not match the NMR frequency ω_0 the magnetization rotates at a frequency offset Ω_0 in the rotating frame,

$$\Omega_0 = -\gamma B_0 - \omega_{\mathrm{rf}} = \omega_0 - \omega_{\mathrm{rf}} \equiv -\gamma B_{\mathrm{fic}}, \tag{2.2.14}$$

where the definition (2.1.1) of the NMR frequency has been used. The offset frequency is ascribed to a *fictitious magnetic field* along the z-axis with magnitude B_{fic}. In the following, all computations will be carried out in the rotating frame, so that the index r used in (2.2.13) is omitted.

Radio-frequency pulses

If the rf frequency ω_{rf} matches the NMR frequency ω_0 then the fictitious magnetic field along the z-axis vanishes, and the nuclear magnetization appears static in the rotating frame. This situation changes, when an rf field is applied. The rf field appears static in the rotating frame as well. But if it is applied perpendicular to the thermodynamic equilibrium magnetization $\boldsymbol{M}_0$, the magnetization is exposed to a nonvanishing magnetic field in the rotating frame. Consequently, it experiences a torque and rotates around this field with frequency

$$\omega_1 = -\gamma B_1, \tag{2.2.15}$$

where B_1 is the magnitude of the rotating field component (Fig. 2.2.3(a)). The duration t_{p} for which the rf field is turned on is adjustable in NMR spectrometers, so that the angle

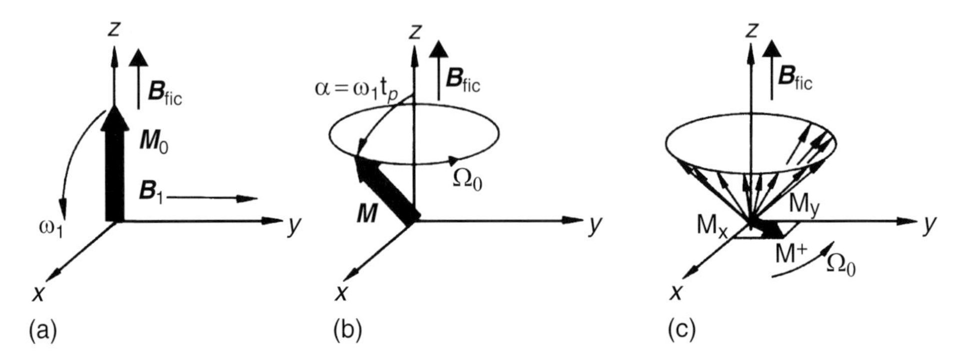

FIG. 2.2.3 Magnetization in the rotating coodinate frame. (a) The frame rotates with the rf frequency ω_{rf}. On resonance $\Omega_0 = 0$, and the rotating rf field component $\boldsymbol{B}_1$ appears static in this frame. The magnetization $\boldsymbol{M}_0$ rotates around the $\boldsymbol{B}_1$ field with frequency ω_1. (b) When the rf field is turned off the magnetization rotates around the z-axis of the rotating frame with frequency Ω_0 if the rf frequency is set off resonance. (c) The phase coherence among the magnetization components making up the xy part of the vector sum $\boldsymbol{M}$ is lost with time as a result of differences in local NMR frequencies which fluctuate with time. The characteristic time for coherence loss is the transverse relaxtion time T_2. The magnetization components are not drawn to scale.

α of precession around the axis of the rf field can be manipulated,

$$\alpha = \omega_1 t_{\mathrm{p}}. \tag{2.2.16}$$

In this way so-called 90° and 180° pulses as well as pulses with arbitrary *flip angles* can be applied. The nomenclature used is that for mathematically positive rotations [Ern1]. Here the right hand rule applies, where the thumb of the right hand points into the direction of the rotation axis and the fingers point into the direction of the rotation.

Depending on the phase φ of the rf in the laboratory frame (cf. eqn (2.2.12)), the direction of the $\boldsymbol{B}_1$ field can be set anywhere within the xy plane of the rotating frame. For example, when choosing $\varphi = 0°, 90°, 180°$, and 270° the $\boldsymbol{B}_1$ field is parallel to the $+x$, $+y$, $-x$, and $-y$ axes, respectively. A pulse of flip angle α along the y-axis is commonly called an α_y pulse.

Free induction decay

Immediately after an rf pulse with flip angle α has been applied at $t = 0$ along the y-direction of the rotating frame, the magnetization

$$\boldsymbol{M}(0_+) = M_0 \begin{pmatrix} \sin\alpha \\ 0 \\ \cos\alpha \end{pmatrix} \tag{2.2.17}$$

has been generated (Fig. 2.2.3(b)). If the coordinate system rotates exactly on resonance with $\Omega_0 = 0$, this magnetization appears static, except for a decay of its component in the transverse plane. If the rotation frequency of the coordinate frame is offset from the

NMR frequency, the magnetization precesses around the z-axis with the offset frequency Ω_0 . In this case the transverse magnetization components are described by

$$M_x(t) = M_0 \sin\alpha \cos(\Omega_0 t) \exp\left\{-\frac{t}{T_2}\right\}, \tag{2.2.18a}$$

$$M_y(t) = M_0 \sin\alpha \sin(\Omega_0 t) \exp\left\{-\frac{t}{T_2}\right\}. \tag{2.2.18b}$$

Both components are conveniently combined in complex notation,

$$M^+(t) = M_x(t) + \mathrm{i}M_y(t) = M_0 \sin\alpha \exp\left\{-\left(\frac{1}{T_2} - \mathrm{i}\Omega_0\right)t\right\}. \tag{2.2.19}$$

The decay of the pulse response is described by the *transverse relaxation time* T_2, which accounts for loss of the phase coherence of the precessing magnetization components as a result of time-dependent differences in local NMR frequencies. These components accumulate different precession phases as time goes on, so that the vector sum of all magnetization components eventually vanishes (Fig. 2.2.3(c)).

The signal $s(t)$ measured in response to an excitation pulse by simultaneous observation of both the x- and the y-components is directly proportional to the complex magnetization $M^+(t)$. Because the signal is induced in a receiver coil in the absence of an rf field, it is referred to as *free induction decay* (FID). It is given by the derivative of $M^+(t)$ and is thus proportional to $M^+(t)$ and phase shifted by 90°. For convenience, the proportionality constant is set to 1 and the phase shift is ignored in the following, so that $s(t) = M^+(t)$.

Space encoding

For heterogeneous objects the equilibrium magnetization M_0 depends on space $\boldsymbol{r}$, and $M_0(\boldsymbol{r})$ is called the spin density. For a 90° pulse, $\sin\alpha = 1$, and the magnetization from the volume element at position $\boldsymbol{r}$ is given by

$$M^+(t) = M_0(\boldsymbol{r}) \exp\{\mathrm{i}\omega_0(\boldsymbol{r})t\}, \tag{2.2.20}$$

where $\omega_0(\boldsymbol{r})$ is the space-dependent NMR frequency (2.1.17). This equation neglects relaxation and spectral distributions of resonance frequencies, but it is a good approximation for signals in strong gradients. In practice, the signal is detected in the rotating coordinate frame. On resonance the rf ω_{rf} coincides with the centre frequency $\omega_0 = -\gamma B_0$ of the gradient field, so that any signal modulation arises from frequency offset with respect to ω_0 induced by the gradients. Furthermore, the sum of signals originating from the nuclei at all positions $\boldsymbol{r}$ is observed, so that (2.2.20) needs to be integrated,

$$M^+(t) = \int_{-\infty}^{\infty} M_0(\boldsymbol{r}) \exp\{-\mathrm{i}\gamma \boldsymbol{G}\boldsymbol{r}t\}\, \mathrm{d}\boldsymbol{r}. \tag{2.2.21}$$

If in addition the gradient is time dependent, the phase of the exponential in (2.2.21) is determined by the time integral of $\boldsymbol{G}$,

$$M^+(t) = \int_{-\infty}^{\infty} M_0(\boldsymbol{r}) \exp\left\{-\mathrm{i}\gamma \boldsymbol{r} \int_0^t \boldsymbol{G}(t')\,\mathrm{d}t'\right\} \mathrm{d}\boldsymbol{r}$$

$$\equiv \int_{-\infty}^{\infty} M_0(\boldsymbol{r}) \exp\{-\mathrm{i}\boldsymbol{k}(t)\boldsymbol{r}\}\,\mathrm{d}\boldsymbol{r}. \qquad (2.2.22)$$

This equation formulates the basic Fourier relationship between the NMR signal acquired in the time domain in the absence of an rf field $\boldsymbol{B}_{\mathrm{rf}}$ and the spin density $M_0(\boldsymbol{r})$ by introducing the *wave vector* $\boldsymbol{k}$ as the Fourier conjugate variable to the space coordinate $\boldsymbol{r}$ (cf. eqn (1.2.5)) [Man4, Man5],

$$\boldsymbol{k}(t) = -\gamma \int_0^t \boldsymbol{G}(t')\,\mathrm{d}t'. \qquad (2.2.23)$$

The wave vector is inversely proportional to the wavelength λ by $|\boldsymbol{k}| = 2\pi/\lambda$. This wavelength is not that of the rf excitation, but it is defined by the time integral of the applied gradient (Fig. 2.2.4). Transverse magnetization, which is generated by a nonselective pulse, will be in phase at all space coordinates immediately after the pulse. But in the presence of a magnetic-field gradient the precession frequency varies in space, and a different precession phase is accumulated over time at each space coordinate. For a linear field dependence the phase modulation is a harmonic function of space, the period of which defines the wavelength λ. As time proceeds, the accumulated phase and the pitch of the magnetization winding around the space axis changes and so do the wavelength and the wave vector. Because the magnetization can be imagined to warp around the space axis, the name *spin-warp imaging* was chosen for imaging methods where only the pitch of the magnetization is changed in a constant evolution time (cf. Section 6.2.1) [Ede1].

To obtain an image, the gradients $\boldsymbol{G}(t)$ must be varied in such a way that all values of $\boldsymbol{k}$ are sampled which are relevant to the image. The image itself is then derived by *Fourier transformation* of the NMR signal (2.2.22). There are two ways to encode the space information to the NMR signal (Fig. 2.2.5) [Kum1]. First, the gradients can be turned on to a constant value G_x during data acquisition. In this case, k_x scales with the acquisition time t_2 , and the space information in x-direction is frequency encoded. Second, the gradient G_y can be turned on for a fixed time t_1 before data acquisition to yield a certain value of the gradient integral k_y at the start. In this case, the space information in y-direction is encoded in the phase of the signal acquired during t_2, and the experiment needs to be repeated for different initial phases obtained by varying $\boldsymbol{G}$ for fixed t_1. This approach to space encoding is called *spin-warp imaging*. It is less sensitive to the effects of magnetic field inhomogeneity, relaxation, and spin interactions than the alternative approach of varying t_1 at fixed $\boldsymbol{G}$ [Ede1]. A 2D image is typically obtained by

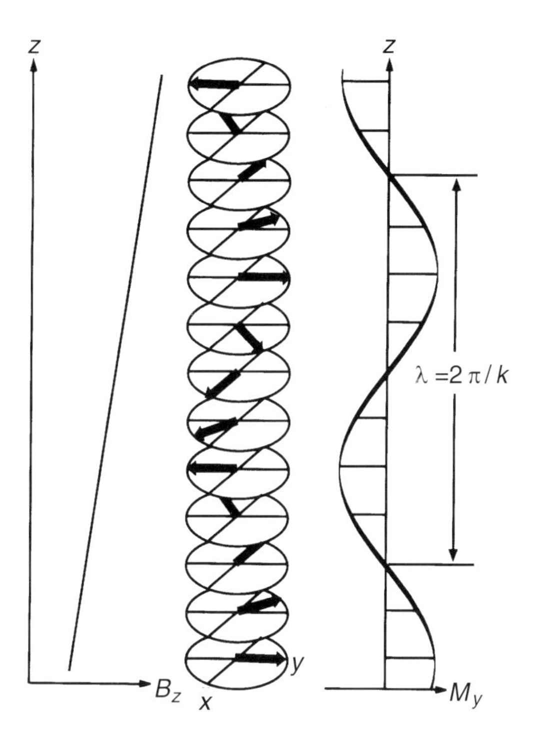

FIG. 2.2.4 Illustration of the wavelength λ in NMR imaging. It is defined by the pitch of the transverse magnetization which changes with time in the presence of a linearly space-dependent magnetic field.

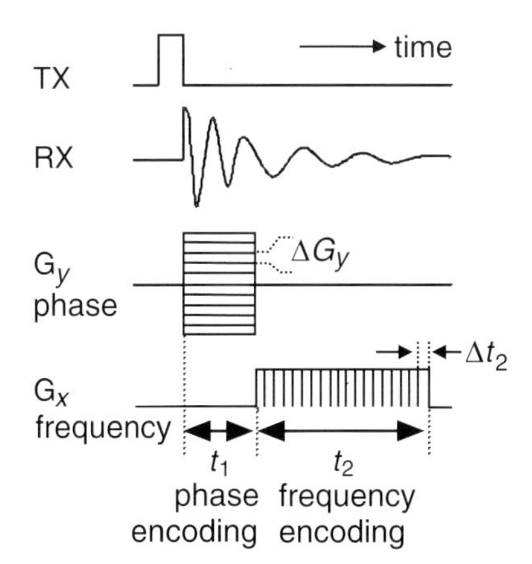

FIG. 2.2.5 Phase and frequency encoding of the NMR signal (RX : receiver) following an rf excitation pulse (TX : transmitter). The space information in the y-direction is encoded in the signal phase by a gradient pulse G_y of length t_1. The phase-modulated signal is aquired at discrete time intervals Δt_2 in the presence of a gradient G_x which encodes the space information in x-direction in the frequency of the signal.

a combination of both, *phase encoding* in one dimension (y in Fig. 2.2.5) and *frequency encoding* in the other (x in Fig. 2.2.5).

The NMR spectrum

The experimental pulse response $M^+(t)$ (cf. eqn (2.2.19)) acquired in the absence of a field gradient is Fourier transformed to obtain the NMR spectrum,

$$S(\omega) = \int_0^\infty s(t) \exp\{-\mathrm{i}\omega t\} \, \mathrm{d}t = U(\omega) + \mathrm{i}V(\omega) \,. \tag{2.2.24}$$

The spectrum consists of real and imaginary parts, $U(\omega)$ and $V(\omega)$, respectively:

$$U(\omega) = M_0 \sin\alpha \, A(\omega), \tag{2.2.25a}$$

$$V(\omega) = M_0 \sin\alpha \, D(\omega). \tag{2.2.25b}$$

Here ω denotes the frequency axis of the spectrum, while Ω_0 is the precession frequency of the magnetization in the rotating frame. Clearly, the maximum signal amplitude is obtained for a flip angle $\alpha = 90°$ of the rf pulse.

$S(\omega)$ is called a complex *Lorentz line*. Its real and imaginary parts, $A(\omega)$ and $D(\omega)$, denote the *absorption signal* and the *dispersion signal*, respectively (Fig. 2.2.6):

$$A(\omega) = \frac{1/T_2}{1/T_2^2 + (\Omega_0 - \omega)^2}, \qquad D(\omega) = \frac{\Omega_0 - \omega}{1/T_2^2 + (\Omega_0 - \omega)^2}. \tag{2.2.26}$$

In NMR spectroscopy, absorption parts $A(\omega)$ are usually displayed and referred to as the phase-sensitive NMR spectrum. The *linewidth* $\Delta\omega_{1/2}$ at half height of $A(\omega)$ is determined by T_2,

$$\Delta\omega_{1/2} = \frac{2}{T_2}. \tag{2.2.27}$$

Alternatively, the magnitude spectrum $|A(\omega) + \mathrm{i}D(\omega)|$ can be displayed. Here, however, the linewidth at half height is broader by a factor of $3^{1/2}$, so that the spectral resolution is

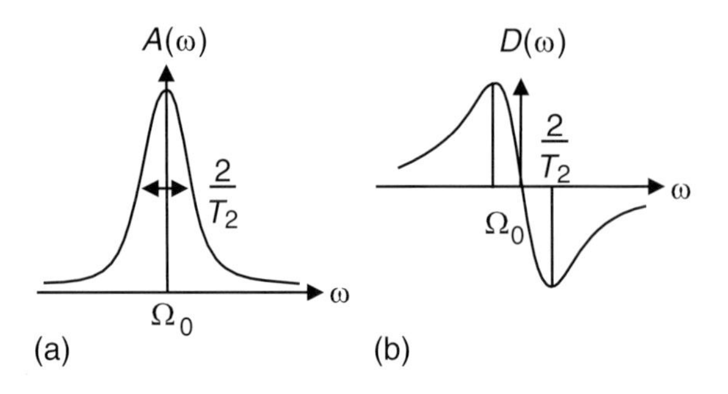

FIG. 2.2.6 The complex Lorentz line. (a) Absorptive real part $A(\omega)$. (b) Dispersive imaginary part $D(\omega)$.

reduced accordingly. These values apply to Lorentz lines, which are commonly observed in the liquid state. In solid-state NMR of abundant nuclei Gaussian lineshapes are encountered most frequently.

In practice, the transverse magnetization $M^+(t)$ is detected with an instrumental phase offset ϕ_0 with respect to the phase of the rotating coordinate frame. Another phase offset is introduced by signal delays Δt_s from propagation times through various electronic components including cables and amplifiers in the spectrometer as well as from delayed acquisition due to finite-length rf pulses and receiver deadtime. Thus, the receiver signal is proportional to

$$\begin{aligned} M^+(t) &= M_0 \sin\alpha \exp\left\{-\left(\frac{1}{T_2} - \mathrm{i}\Omega_0\right)(t + \Delta t_s) + \mathrm{i}\phi_0\right\} \\ &= M_0 \sin\alpha \exp\left\{-\left(\frac{1}{T_2} - \mathrm{i}\Omega_0\right)t\right\} \exp\left\{\frac{-\Delta t_s}{T_2}\right\} \exp\{\mathrm{i}(\phi_0 + \Omega_0 \Delta t_s)\} \\ &\cong M^+(t) \exp\{\mathrm{i}(\phi_0 + \Omega_0 \Delta t_s)\}. \end{aligned} \tag{2.2.28}$$

In the last step, the exponential $\exp\{-\Delta t_s/T_2\}$ has been approximated by 1 because $\Delta t_s \ll T_2$ applies. It is seen from (2.2.28) that the phase shift of the recorded signal is linear in frequency, and two variables ϕ_0 and Δt_s are required for its determination. To obtain pure absorption mode real parts of experimental spectra, their phase is adjusted by multiplication with the $\exp\{-\mathrm{i}(\phi_0 + \Omega_0 \Delta t_s)\}$ to cancel the phase shift. This process of mixing real and imaginary parts of the spectrum is called *phase correction*. It is a routine operation in obtaining phase-sensitive NMR spectra. In liquid-state NMR the time delay Δt_s is sometimes called ϕ_1 to indicate its function as a parameter for frequency–linear phase correction.

Off-resonance effects and the effective field

In the previous section the effect of the rf pulses has been described for the case that the rf frequency matches the NMR frequency, that is for $\Omega_0 = 0$. Typical NMR spectra, however, comprise a number of resonance lines within an often narrow spectral window (cf. Fig. 2.2.2). Thus when chemical shift is included, the resonance condition $\Omega_0 = 0$ cannot be fulfilled for all signals. But if the maximum frequency offset Ω_0 is small compared to the amplitude ω_1 of the rf field,

$$|-\gamma B_1| = |\omega_1| \gg |\Omega_0| = |\omega_0 - \omega_{\mathrm{rf}}|, \tag{2.2.29}$$

essentially all magnetization components are rotated through the same angle α by the rf pulse, and the pulse is called a *nonselective pulse*. In reality, the magnetization is rotated around an *effective field* $\boldsymbol{B}_{\mathrm{eff}}$ which is the vector sum of the *fictitious field* $\boldsymbol{B}_{\mathrm{fic}}$ along the z-axis (cf. eqn (2.2.14)) and the rf field $\boldsymbol{B}_1$ along the y-axis of the *rotating coordinate frame* (Fig. 2.2.6),

$$\boldsymbol{B}_{\mathrm{eff}} = \boldsymbol{B}_{\mathrm{fic}} + \boldsymbol{B}_1 = \begin{pmatrix} 0 \\ B_1 \\ B_0 + \omega_{\mathrm{rf}}/\gamma \end{pmatrix} \tag{2.2.30}$$

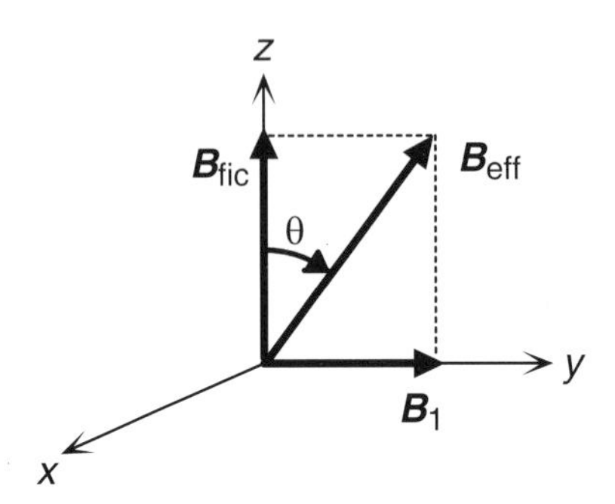

FIG. 2.2.7 Magnetic fields in the rotating frame. Depending on the offset Ω_0 of the NMR frequency from the rotation frequency ω_{rf} of the rotating frame, a fictitious magnetic field of magnitude $B_{fic} = -\Omega_0/\gamma$ acts along the z-direction. Along the y-direction the rf magnetic field of amplitude $B_1 = -\omega_1/\gamma$ is applied. The vector sum of both forms the effective field $\boldsymbol{B}_{eff}$, around which the magnetization is rotating with frequency $\omega_{eff} = -\gamma B_{eff}$.

Taking the effective field into consideration is, particularly, important for pulses of weak amplitude ω_1. Such pulses are *selective pulses*, because the rotation angle depends on the offset Ω_0 of the NMR frequency from the rf frequency, which serves as a reference for the rotating frame. The angle θ by which the effective field is tilted from the z-direction (Fig. 2.2.7) is given by

$$\tan\theta = \frac{B_1}{B_0 + \omega_{rf}/\gamma}, \tag{2.2.31}$$

and the amplitude of the effective field derives from (2.2.30) as

$$B_{eff} = \sqrt{B_1^2 + (B_0 + \omega_{rf}/\gamma)^2} = -\gamma\omega_{eff}. \tag{2.2.32}$$

Thus the effective flip angle α_{eff} of the pulse is obtained as

$$\alpha_{eff} = \omega_{eff} t_p. \tag{2.2.33}$$

This angle increases with offset Ω_0, while the rotation of a magnetization component originally along the z-axis describes a narrower cone.

Signal averaging

To improve the *signal-to-noise ratio* in NMR, the data acquired under similar conditions are added. The signal gains in proportion with the number N_s of scans, while the variance of the noise scales with $N_s^{1/2}$. Thus the S/N scales with $N_s/N_s^{1/2} = N_s^{1/2}$. To achieve similar initial conditions, the magnetization is often allowed to recover along the z-axis between successive scans. This requires a repetition time t_R of the order of a few T_1.

If the repetition time is made shorter, more scans can be measured in a given time, but the initial magnetization will be smaller than the thermodynamic equilibrium value M_0. This effect is called *partial saturation*. Depending on the excitation flip angle α the initial transverse magnetization after the excitation pulse is obtained as [Ern1]

$$M_x = M_0 \frac{1 - \exp\{-t_R/T_1\}}{1 - \cos\alpha \exp\{-t_R/T_1\}} \sin\alpha. \tag{2.2.34}$$

This equation is valid as long as there are no interference effects of the transverse magnetization before and after the pulse. Such a situation is encountered in fast NMR imaging, so that the residual transverse magnetization is destroyed by application of a homogeneity-spoil pulse before application of the next pulse. From (2.2.34) it is clear that the maximum amplitude of the transverse magnetization is obtained not for a 90° pulse. Instead it is obtained for a pulse the flip angle of which is given by the *Ernst angle*

$$\cos\alpha_E = \exp\left\{-\frac{t_R}{T_1}\right\}. \tag{2.2.35}$$

For $t_R/T_1 = 3$ the Ernst angle is close to 90°. In a repetitive pulse experiment in which t_R is chosen much shorter than this and α is set to α_E an S/N advantage of about $2^{1/2}$ is obtained.

Phase cycling

When adding the response of successive scans for signal averaging, excitation and signal acquisition are often manipulated in such a way that unwanted signal contributions arising from spectrometer imperfections are cancelled. Such imperfections are, for instance, misadjustments in the pulse flip angle, the pulse phase, and in amplification and phase of the receiver quadrature channels (cf. Section 2.3.4). Among these the latter are of primary concern. They can be eliminated by appropriate phase cycles such as the *CYCLOPS* sequence (cyclically ordered phase sequence) [Hou1] listed in Table 2.2.2. The transmitter phase φ assumes all four nominal quadrature values of 0°, 90°, 180°, and 270° corresponding to $+x$, $+y$, $-x$, and $-y$, respectively.

If the signals in two orthogonal receiver channels are denoted with $\mathrm{Re}\{s(t)\}$ and $\mathrm{Im}\{s(t)\}$, the real and imaginary parts of the acquired signal provide the x- and y-components of the transverse magnetization $M^+(t)$ with changing signs in alternating channels as indicated in the second and third column of the table. For coherent addition of the experimental data the signal $s(t) = \mathrm{Re}\{s(t)\} + \mathrm{i}\,\mathrm{Im}\{s(t)\}$ needs to be phase shifted by the receiver reference phase ϕ given in the last column. This is equivalent

Table 2.2.2 The CYCLOPS sequence for transmitter and receiver phase cycling

Transmitter phase φ	$\mathrm{Re}\{s(t)\}$	$\mathrm{Im}\{s(t)\}$	Effective rec. phase ϕ	
$+y$	$+M_x$	$+M_y$	$+x$	0°
$-x$	$-M_y$	$+M_x$	$+y$	90°
$-y$	$-M_x$	$-M_y$	$-x$	180°
$+x$	$+M_y$	$-M_x$	$-y$	270°

to multiplication by $\exp\{i\phi\}$ but is done in practice by changing the sign of the signal components and by swapping real and imaginary parts.

Saturation and inversion recovery

The inversion and saturation recovery pulse sequences are used for measurement of the T_1 relaxation time and for partial suppression of signals in samples with distributions of T_1 relaxation times. These pulse sequences can be employed for contrast enhancement in imaging (cf. Section 7.2.1).

For *saturation recovery*, any initial magnetization is destroyed in the beginning by application of an aperiodic series of 90° pulses (Fig. 2.2.8(a)) [Mar1]. Longitudinal magnetization M_z is then allowed to build up for a time t_1. It is subsequently converted into detectable transverse magnetization by a 90°_y pulse. The amplitude of the transverse magnetization acquired during t_2 displays the recovery of the longitudinal magnetization towards its thermodynamic equilibrium value M_0 as a function of t_1,

$$M^+(t_1, t_2) = M_0\left(1 - \exp\left\{-\frac{t_1}{T_1}\right\}\right)\exp\left\{-\left(\frac{1}{T_2} - i\Omega_0\right)t_2\right\}. \tag{2.2.36}$$

The advantage of using the saturation recovery technique for measurement of T_1 is, that no recycle delays t_R have to be included between successive scans. This is not the case for the *inversion recovery* technique [Vol1]. Here the initial magnetization at the start of the experiment is the thermodynamic equilibrium magnetization M_0 (Fig. 2.2.8(b)). It is inverted by a 180° pulse, so that the range of recovery during the subsequent build-up of M_0 is extended to $2M_0$. Again, the z magnetization is interrogated at different times t_1 by a 90°_y pulse, which converts the longitudinal into measurable transverse magnetization,

$$M^+(t_1, t_2) = M_0\left(1 - 2\exp\left\{-\frac{t_1}{T_1}\right\}\right)\exp\left\{-\left(\frac{1}{T_2} - i\Omega_0\right)t_2\right\}. \tag{2.2.37}$$

In contrast to the saturation recovery technique, negative z magnetization can be generated, but care must be taken to start from thermodynamic equilibrium. If these techniques are used for generation of T_1 contrast in NMR images, different contrast is obtained because signal is suppressed by the saturation recovery technique while it is preserved but inverted by the inversion recovery technique.

Echoes

A signal which has first vanished with time and then reappears some time later is called an *echo*. In spectroscopy, the echo is formally associated with a reversal of time, so that the reappearing signal can be understood in terms of time running backwards for a sufficiently isolated ensemble of molecules or spins (Fig. 2.2.9) [Blü1]. For uncoupled spins in simple liquids an NMR echo of the FID is generated by a 180° flip of the phase of all magnetization components. Since the discovery of the original two-pulse echo [Hah1], many other echoes (cf. Section 3.4) have been discovered in spectroscopy based

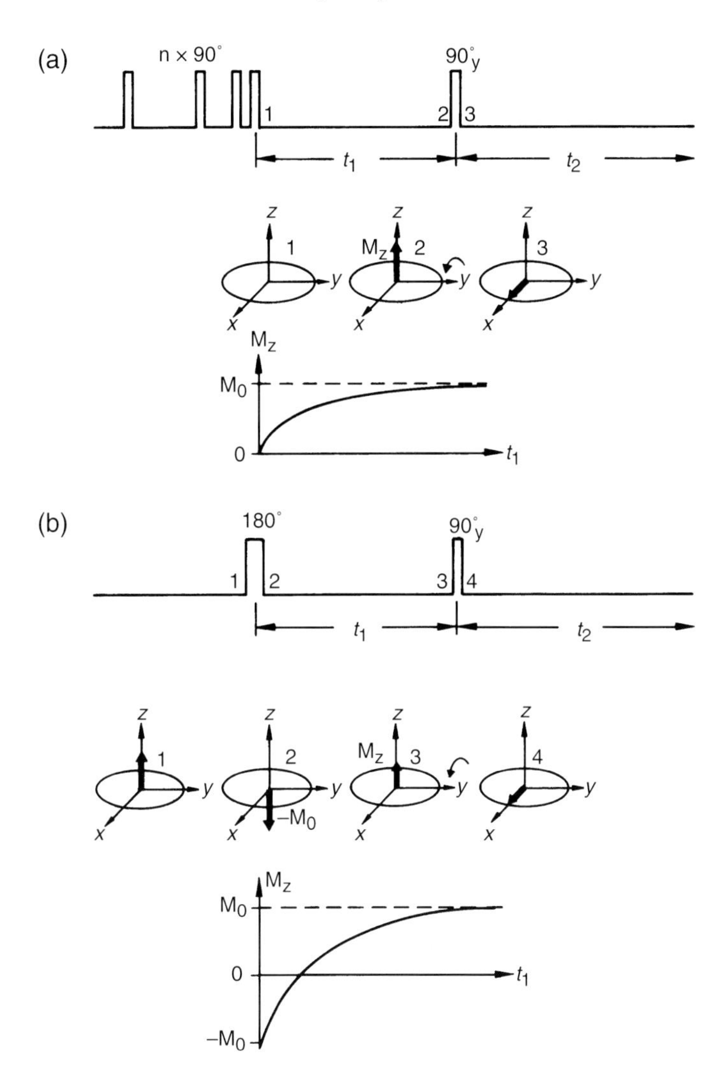

FIG. 2.2.8 Pulse sequences for measurement of T_1 relaxation times by (a) saturation recovery and (b) inversion recovery. The effect of the pulse sequences is illustrated in terms of the vector model of the nuclear magnetization and by graphs showing the evolution of the longitudinal magnetization M_z as a function of t_1.

on the principle of time reversal which can be realized by manipulating the effective spin interactions in different ways. For example, in solids, the solid echo (cf. Fig. 3.2.6(a)), the alignment echo (cf. Fig. 3.2.6(b)), the magic echo (cf. Fig. 3.4.3), and rotational echoes (cf. Fig. 3.3.7(a)) can be generated. They are exploited for line narrowing in solid-state NMR spectroscopy (cf. Section 3.3), for imaging (cf. Chapter 8), and for generation of parameter contrast (cf. Chapter 7). In this section only the basic *two-pulse echo* and the *three-pulse echo* observed in liquids as well as the *gradient echo* are treated.

FIG. 2.2.9 [Blü1] Contrary to the acoustic echo the spectroscopic echo is associated with time reversal. Copyright Wiley-VCH, Weinheim.

The first two are also referred to as *spin echo* or *Hahn echo*, and as *stimulated echo*, respectively [Hah1].

The Hahn echo

The *Hahn echo* is generated by two pulses applied a time $t_E/2$ apart (Fig. 2.2.10(a)). The first pulse is a 90°_y pulse generating an FID signal. Right after the pulse all components of the transverse magnetization are precessing with approximately the same phase. The magnetization undergoes a coherent rotation. As a result of differences in local magnetic fields, the coherence of the rotation is lost, and different magnetization components accumulate different precession phases with time. Thus the vector sum of all magnetization components eventually vanishes. But as long as the precession frequency of each magnetization component remains unchanged, a simple permutation of fast against slow components will generate a reoccurrence of the signal after another time $t_E/2$. This permutation is achieved on a circular path by a 180° phase jump, which is the result of a 180° pulse applied along an axis orthogonal to that of the first pulse. Thus, counting from the first pulse, an echo of the FID signal will appear after a time corresponding to twice the pulse separation. This time is called the *echo time* t_E in imaging. An echo with negative amplitude is obtained if both pulses are applied with the same phase.

In the centre of the Hahn echo, all magnetization components are refocused, although the precession frequencies differ because of different shielding of the magnetic field $\mathbf{B}_0$ and inhomogeneities in the static magnetic field. Thus, the Hahn echo can also be observed in the presence of a magnetic field gradient. But, even under perfect excitation conditions, its amplitude is somewhat smaller than that of the FID following the first pulse, because part of the transverse magnetization has disappeared by T_2 relaxation. In fact, the amplitude a_H of the Hahn echo as a function of the echo time t_E is used to measure T_2,

$$a_H(t_E) = M_0 \exp\left\{-\frac{t_E}{T_2}\right\}. \tag{2.2.38}$$

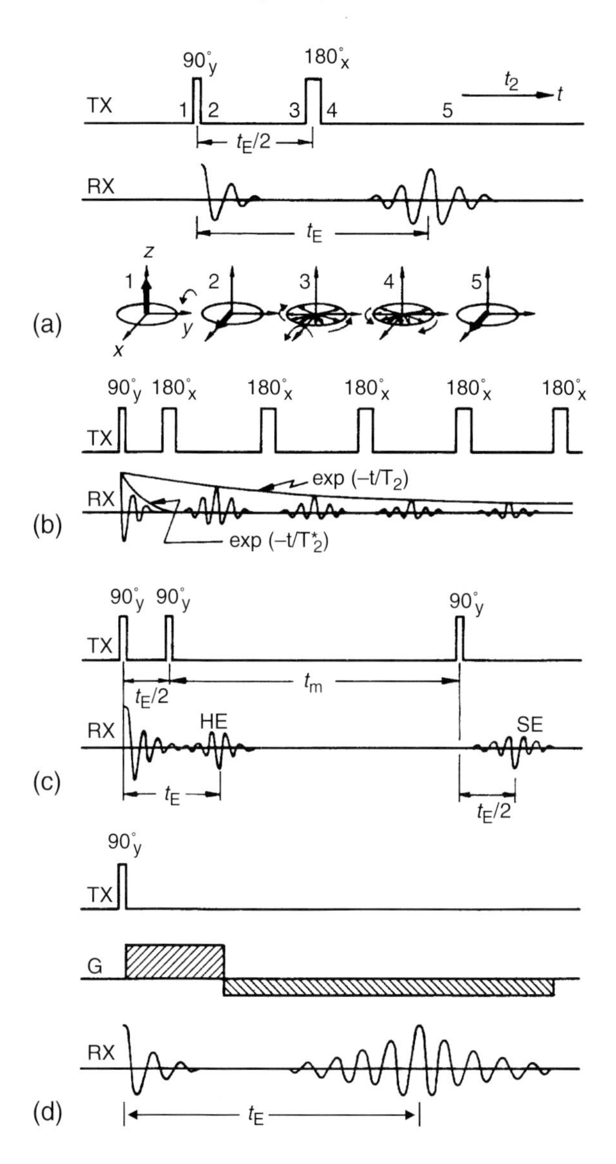

FIG. 2.2.10 Echoes in NMR. (a) Two-pulse Hahn echo. (b) CPMG sequence with multiple refocusing pulses. (c) Stimulated echo sequence showing both, the Hahn echo (HE) or primary echo and the stimulated echo (SE). (d) Gradient echo.

This measurement can be accomplished in a single shot, if the magnetization is repeatedly refocused [Car1] (Fig. 2.2.10(b)). In liquids, rapid refocusing attenuates the effect of molecular diffusion into different regions of an inhomogeneous magnetic field. If the 180° pulses are shifted in phase by 90° with respect to the first pulse, the effects of imperfections in the flip angle of the refocusing pulses on the echo formation are reduced

[Mei1]. As a tribute to the inventors (Carr, Purcell, Meiboom, and Gill) of this technique, the pulse train is named the *CPMG sequence*.

The Fourier transform of the decay of the echo maxima is a narrow line, the width of which is determined by irreversible T_2 relaxation (cf. Section 3.4). This line is said to be *homogeneously broadened*. The linewidths observed in the Fourier transforms of signals which can be refocused by formation of Hahn echoes are broader. Such lines are said to be *inhomogeneously broadened*. The generation of narrow, homogeneously broadened lines by repeated echo formation is important for high spatial resolution in imaging with frequency encoding. To distinguish the decay time of signals from inhomogeneously broadened lines from that of homogeneously broadened lines, the symbol T_2^* is used for the former and T_2 for the latter.

The stimulated echo

The *stimulated echo* can be understood to derive from the Hahn echo, if the 180° refocusing pulse is split into two 90° pulses (Fig. 2.2.10(c)). As a consequence, two echoes are observed, each with half the amplitude of the initial magnetization. The first echo is the *Hahn echo* or *primary echo*, the second one is the stimulated echo. The succession of two 90° pulses for refocusing acts like a 180° pulse on half of the magnetization, because the other half is lost by dephasing during the time in between the pulses. Starting from longitudinal magnetization before the first pulse, longitudinal magnetization is obtained again after the second pulse during the time t_m. This magnetization relaxes with T_1, while the transverse magnetization of the Hahn echo disappears with T_2^*. Because T_1 is often longer than T_2, the stimulated echo can be used to store one magnetization component while instrumental parameters like the gradient strength are being adjusted. Systematic measurement of stimulated echo amplitudes a_s for different times t_m yields access to the T_1 relaxation time,

$$a_s(t_m) = \frac{1}{2} M_0 \exp\left\{-\frac{t_m}{T_1}\right\} \exp\left\{-\frac{t_E}{T_2}\right\}. \tag{2.2.39}$$

Both the Hahn and the stimulated echo are basic elements of many imaging methods. Because the stimulated echo consists of three pulses and three time periods, a greater variety of imaging methods exists for the stimulated echo (cf. Section 6.2.5) [Bur1].

Molecular self-diffusion

The Hahn and the stimulated echo are used to study molecular self-diffusion in fluids with field gradients which are active during the pulse sequence [Cal2, Kär1, Kim1, Sti1]. These gradients can be static [Hah1] or pulsed [Ste1]. For a static gradient of known magnitude G the *self-diffusion constant* D can be determined from the amplitude a_H of the Hahn echo as a function of the echo time t_E [Hah1, Car1],

$$a_H(t_E) = M^+(0) \exp\left\{-\frac{t_E}{T_2}\right\} \exp\left\{-\frac{-\gamma^2 D G^2 t_E^3}{12}\right\}, \tag{2.2.40}$$

whereas the amplitude of the stimulated echo is given by [Hah1]

$$a_s(t_E, t_m) = \frac{M^+(0)}{2} \exp\left\{-\frac{t_m}{T_1}\right\} \exp\left\{-\frac{t_E}{T_2}\right\} \exp\left\{-\gamma^2 D G^2 \frac{(t_E^3/3) + t_m t_E^2}{4}\right\}. \tag{2.2.41}$$

Pulsed gradients are particularly suitable for combination of diffusion studies with imaging [Cal1, Xia1]. In contrast to imaging of structures in real space, diffusion studies provide information about structures in *displacement space* in terms of average dimensions of morphological features, which typically are on the micrometer scale (cf. Section 5.4.3). This form of mapping average spatial dimensions is also referred to as $\boldsymbol{q}$-space imaging [Cal1].

The gradient echo

By the use of gradients, echoes can be generated without rf refocusing. If the field gradient is applied during the signal decay, an echo is generated by reversing the sign of the gradient. In the *gradient-echo* centre, any dephasing as a result of the applied gradients must be zero for all space coordinates. According to (2.2.22) the initial signal is regained if the exponential becomes unity, that is, $\boldsymbol{k}(t)$ is required to vanish. Therefore, in the gradient-echo maximum at $t = t_E$,

$$\boldsymbol{k}(t_E) = -\gamma \int_0^{t_E} \boldsymbol{G}(t)\, dt = 0, \tag{2.2.42}$$

and the gradient echo arises when the areas under the gradient envelopes are matched under sign reversal (Fig. 2.2.10(d)).

The echo amplitude is attenuated compared to that of the initial FID by T_2 relaxation, as well as by dephasing from magnetic field inhomogeneities and chemical shift dispersion. The latter can be refocused in the presence of a field gradient, if the gradient echo is combined with a Hahn echo. This is a situation typical for NMR imaging. However, if both types of echo are combined, the sign of the gradient must be the same, and the gradient integrals must be identical during the defocusing and the refocusing times before and after the 180° refocusing pulse of the Hahn echo.

Steady-state free precession

In the CPMG pulse train the transverse magnetization eventually vanishes by transverse relaxation. Alternative multi-echo techniques can be designed, where the transverse magnetization is retained all the time in a steady state driven by continuous application of rf pulses with flip angles α. The magnetization is said to be in the *steady-state free precession* (SSFP) mode [Car2]. In principle, all pulses can be identical, but it is common to alternate their phases between 0° and 180° [Hin1], so that the cycle time is twice the pulse separation τ (Fig. 2.2.11). In this way magnetization is retained for components the resonance frequencies Ω_L of which are an integral multiple of $2\pi/\tau$. Its amplitude

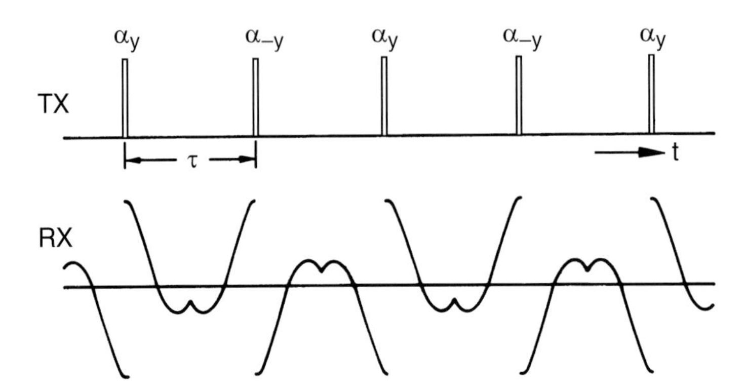

FIG. 2.2.11 The SSFP technique. A steady state of transverse magnetization is maintained by rapid application of rf pulses. A typical flip angle α is 90°. The equilibrium value of the SSFP magnetization depends on the resonance offset Ω_0 and the pulse spacing τ. Adapted from [Cal1] with permission from Oxford University Press.

is given by [Man2]

$$M_x = M_0 \frac{\left[1 - \exp\{-\tau/T_1\}\right] \sin\alpha}{1 + \exp\{-\tau/T_1\}\exp\{-\tau/T_2\} - \cos\alpha\left[\exp\{-\tau/T_1\} + \exp\{-\tau/T_2\}\right]}. \tag{2.2.43}$$

In the limit of long pulse separations τ this equation reduces to (2.2.34) derived for partial saturation and a pulse spacing t_R long enough for the signal to disappear by T_2 relaxation. The flip angle α is often set to a value in the vicinity of 90°. Under optimum conditions, half of the thermodynamic equilibrium magnetization can be maintained in the steady state [Ern1]. The use of SSFP methods is of particular importance to fast imaging methods, where recycle delays are to be avoided and signal intensity is to be maximized [Gyn1].

2.2.2 Spin-system response from quantum mechanics

For a more thorough understanding of NMR the use of quantum mechanics becomes inevitable [Mun1, Far2]. In fact, much of the jargon of NMR derives from the quantum mechanical description of the NMR phenomenon. The most important quantity is the *Hamilton operator*. It describes the interactions of the nuclear spins with the static and the rf magnetic fields as well as the various other types of interactions, such as the *chemical shift*, the *indirect spin–spin coupling* or *J coupling*, the *dipole–dipole coupling*, and the *quadrupole coupling*. The particular significance of these interactions for different aspects of NMR is the topic of Chapter 3. At this point, it is sufficient to know that the *eigenvalues* of the Hamilton operator determine the nuclear energy levels, and thus the transition frequencies between them. A sizable part of NMR spectroscopy deals with the probing of transition frequencies and the subsequent analysis of the arrangement of energy levels.

The Hamilton operator is part of the *Schrödinger equation*, which determines the evolution of the *wave function* of a nuclear spin as time goes on. But, in an NMR experiment some 10^{20} spins are manipulated, and the observables are ensemble averages over the properties of all these spins. Only if a sufficient number of spins undergoes a coherent motion, the ensemble average is different from zero, unless populations of energy levels are considered. In this context, the term *coherence* is used. It is a generalization of the concept of magnetization. *Magnetization* is a term reserved for transverse and longitudinal magnetization. It addresses those elements of the density matrix which are directly observable and are included in the vector model of the Bloch equations. In coupled spin systems, coherences other than transverse magnetization exist which precess with multiples of the resonance frequency. These terms are called *multi-quantum coherences* as opposed to the *single-quantum coherences*, which form the transverse magnetization. The appropriate way of describing coherences is by means of the *density matrix*, which is nothing else but a collection in matrix form of the ensemble averages of all twofold products of the expansion coefficients of the wave function. The density matrix is the central tool for calculating the effects of pulse sequences on nuclear spins. Similar to the Schrödinger equation for wave functions, an equation of motion exists for the density matrix. This is the *von Neumann equation*. For an ensemble of noninteracting spins $\frac{1}{2}$ it is equivalent to the Bloch equations (2.2.8).

The wave function

A quantum mechanical particle in equilibrium can be found in one of a number of discrete states. For instance, the z-component of the magnetic moment of a spin-1/2 nucleus in a magnetic field is found in either the parallel or the antiparallel orientation with respect to the magnetic field (cf. Fig. 2.2.1(b)). The energies of these states are the *eigenvalues* of the *energy operator*, which is the *Hamilton operator*. Each state is associated with an *eigenfunction* $|m\rangle$. Here the magnetic quantum number m is used to characterize the eigenfunctions, because the Hamilton operator is proportional to the z-component $\mathbf{I}_z$ of the angular momentum operator if spin–spin couplings are neglected (cf. eqn (2.2.3)). For spins $\frac{1}{2}$, the eigenfunctions are also denoted by $|\alpha\rangle$ and $|\beta\rangle$ for the spin orientations parallel and antiparallel to the magnetic field. Any wave function $|\Psi\rangle$ can be expressed as a superposition of the eigenfunctions $|m\rangle$ of an operator,

$$|\Psi\rangle = \sum_m a_m |m\rangle. \tag{2.2.44}$$

The expansion coefficients a_m are complex numbers consisting of amplitude and phase.

The Schrödinger equation

The eigenfunctions are often chosen to depend on time, while the operators are chosen time independent. This choice is referred to as the *Schrödinger representation*. The change of time of a quantity is called *evolution*. The evolution of the wave function towards equilibrium is determined by the *Schrödinger equation*,

$$\mathrm{i}\frac{\partial}{\partial t}|\Psi(t)\rangle = \mathbf{H}(t)\,|\Psi(t)\rangle, \tag{2.2.45}$$

where the Hamilton operator $\mathbf{H}(t)$ is written in units of $\hbar$ so that frequency units are obtained instead of energy units. If $\mathbf{H}$ is time independent, (2.2.45) can readily be solved for $|\Psi(t)\rangle$,

$$|\Psi(t)\rangle = \mathbf{U}(t,0)|\Psi(0)\rangle, \tag{2.2.46}$$

where

$$\mathbf{U}(t,0) = \exp\{-\mathrm{i}\mathbf{H}t\} \tag{2.2.47}$$

is the *evolution operator* of the wave function. Thus, given the initial wave function $|\Psi(0)\rangle$, and $\mathbf{U}(t)$, $|\Psi(t)\rangle$ can be predicted for all times t.

If the Hamilton operator depends on time in a harmonic fashion, the time dependence can be eliminated by transformation into a rotating reference frame in analogy to the transformation of the Bloch equations. A representation in the rotating frame is also called *interaction representation* in quantum mechanics. If the time dependence is more general, the Schrödinger equation is solved for small enough time increments, during which $\mathbf{H}$ is approximately constant. For each of the n time increments Δt a solution of the form (2.2.46) applies. The complete evolution operator is the time ordered product of the incremental evolution operators. This operator is written in short hand as

$$\mathbf{U}(t,0) = \mathbf{U}(t, t-\Delta t)\cdots\mathbf{U}(\Delta t, 0) = \mathbf{T}\exp\left\{-\mathrm{i}\int_0^t \mathbf{H}(t')\,\mathrm{d}t'\right\}. \tag{2.2.48}$$

Because the time order of the incremental evolution operators is destroyed by the integral in the exponent, the so-called *Dyson time ordering operator* $\mathbf{T}$ is introduced to reestablish the lost time order.

The density matrix

Measurable quantities are expressed by expectation values of operators. The *expectation value* of the operator $\mathbf{O}$ is given by

$$\langle\mathbf{O}\rangle = \langle\Psi|\mathbf{O}|\Psi\rangle. \tag{2.2.49}$$

To work with this equation the wave function $|\Psi\rangle$ is expressed in terms of the *eigenfunctions* $|m\rangle$ of some operator, usually the dominant part of the Hamilton operator. By use of (2.2.44) the expectation value is expressed in terms of a double sum,

$$\langle\mathbf{O}\rangle = \sum_m\sum_n a_m a_n^*\langle n|\mathbf{O}|m\rangle \tag{2.2.50}$$

This expectation value is of quantum mechanical nature. It does not account for the average over all spins, which is purely statistical. Expressing this average by an overbar, the expectation value over the 10^{20} spins in the sample is obtained,

$$\langle\bar{\mathbf{O}}\rangle = \sum_m\sum_n \overline{a_m a_n^*}\langle n|\mathbf{O}|m\rangle = \sum_m\sum_n \rho_{mn}O_{mn} = \mathrm{Sp}\left\{\boldsymbol{\rho}\mathbf{O}^{\mathrm{t}^*}\right\}, \tag{2.2.51}$$

where * denotes the complex conjugate, and t denotes transposition. This equation defines the *density matrix* ρ in terms of its elements

$$\rho_{mn} = \overline{a_m a_n^*}. \tag{2.2.52}$$

It can be read as the trace Sp$\{\cdots\}$ over the corresponding product $\boldsymbol{\rho}\mathbf{O}^{t^*}$, where the $\langle n|\mathbf{O}|m\rangle$ are the elements of the operator matrix $\mathbf{O}$.

The density matrix is Hermitean, that is,

$$\rho_{mn} = \rho_{nm}^*. \tag{2.2.53}$$

The diagonal elements

$$\rho_{nn} = \overline{|a_n|^2} \tag{2.2.54}$$

of the density matrix are the populations of the energy levels if $\boldsymbol{\rho}$ is expressed in the eigenbasis of the Hamilton operator. This can immediately be seen if the expectation value of the $\mathbf{I}_z$ operator is formed, which determines the Zeeman splitting of the nuclear energy levels.

Example: longitudinal and transverse magnetization

For spins $\frac{1}{2}$ the vector components of the spin operator $\mathbf{I}$ are given by the *Pauli spin matrices*,

$$\mathbf{I}_x = \frac{1}{2}\begin{bmatrix} 0 & 1 \\ 1 & 0 \end{bmatrix}, \quad \mathbf{I}_y = \frac{\mathrm{i}}{2}\begin{bmatrix} 0 & -1 \\ 1 & 0 \end{bmatrix}, \quad \mathbf{I}_z = \frac{1}{2}\begin{bmatrix} 1 & 0 \\ 0 & -1 \end{bmatrix}. \tag{2.2.55}$$

The longitudinal magnetization is proportional to the expectation value of $\mathbf{I}_z$,

$$M_z = \gamma\hbar \mathrm{Sp}\left\{\boldsymbol{\rho}\mathbf{I}_z^{t^*}\right\} = \frac{\gamma\hbar}{2}(\rho_{11} - \rho_{22}) = \frac{\gamma\hbar}{2}\left(\overline{|a_{1/2}|^2} - \overline{|a_{-1/2}|^2}\right). \tag{2.2.56}$$

It is determined by the population difference expressed by the difference in diagonal elements of the density matrix. The sum of diagonal density-matrix elements is always unity, as it counts the relative number of nuclei in all states,

$$\mathrm{Sp}\{\boldsymbol{\rho}\} = 1. \tag{2.2.57}$$

In *thermodynamic equilibrium* the off-diagonal density-matrix elements are zero. The diagonal elements are determined by the Boltzmann distribution in the eigenbasis of the Hamilton operator $\mathbf{H}_Z$,

$$\boldsymbol{\rho}_0 = \frac{\exp\{-\mathbf{H}_Z/k_B T\}}{\mathrm{Sp}\left\{\exp\left[-\mathbf{H}_Z/k_B T\right]\right\}} \tag{2.2.58}$$

Off-diagonal elements are created by rf pulses. They describe coherent transitions between energy levels, which a large number of the spins undergo, with the same initial phase. Therefore, these matrix elements are called *coherences*. The transition frequencies correspond to precessions at multiples p of the resonance frequency. The factor p is the *coherence order*,

$$p = m_\mathrm{f} - m_\mathrm{i}, \tag{2.2.59}$$

where m_f and m_i are the magnetic quantum numbers of the final and initial energy states, respectively. In general, the magnetization measured directly during the detection time in NMR experiments is associated with a change $|p| = 1$ in magnetic quantum number. Such magnetization is termed *single quantum coherence*. If the coherence order differs from 1, the off-diagonal matrix elements are called *multi-quantum coherences*.

An ensemble of noninteracting spins $\frac{1}{2}$ exhibits only single quantum coherences. This case can be treated classically by the Bloch equations. The transverse magnetization is given by

$$M^+ = \gamma\hbar \mathrm{Sp}\{\boldsymbol{\rho}(\mathbf{I}_x + \mathrm{i}\mathbf{I}_y)^{\mathrm{t}^*}\} = \gamma\hbar\rho_{12}. \tag{2.2.60}$$

Thus the transverse magnetization is proportional to the off-diagonal density-matrix element ρ_{12} for spins $\frac{1}{2}$.

Multi-quantum coherences

Multi-quantum coherences are off-diagonal elements of the density matrix, which cannot be observed directly. They arise in systems with spin $I > \frac{1}{2}$ and in systems of coupled spins. Consider, for example, an ensemble of two spins $\frac{1}{2}$ which are coupled magnetically by the polarization of the electrons surrounding the spins. This coupling is the *indirect coupling* or *J coupling*. It is the most important form of spin coupling observed in liquid-state NMR. It is expressed by a splitting of the energy levels and, in the weak coupling limit of two spins $\frac{1}{2}$, by a doubling of lines in the spectrum (Fig. 2.2.12).

For such a coupled spin system the density matrix is conveniently expressed in the *product space* of the *eigenfunctions* of the $\mathbf{I}_z$ operators of the uncoupled spins. Denoting the eigenfunctions for the spin-up and the spin-down states as $|\alpha\rangle$ and $|\beta\rangle$, respectively, the following eigenvalue equations apply for each of the coupled spins,

$$\mathbf{I}_z|\alpha\rangle = \tfrac{1}{2}|\alpha\rangle, \qquad \mathbf{I}_z|\beta\rangle = -\tfrac{1}{2}|\beta\rangle, \tag{2.2.61}$$

where $\frac{1}{2}$ and $-\frac{1}{2}$ are the values of the magnetic quantum number m for these states. The product functions are written as $|\alpha\alpha\rangle$, $|\alpha\beta\rangle$, $|\beta\alpha\rangle$, and $|\beta\beta\rangle$, where the first function refers to spin A and the second to spin X. They are the eigenfunction of the $\mathbf{I}_z$ operator of the coupled AX system. In the limit of weak coupling they approximate the eigenfunctions of the Hamilton operator. In the following, they will be used to number rows and columns of the density matrix in the sequence given above. Thus, the density matrix has 4×4 elements in this case.

The magnetic quantum number m of the coupled system is given by the sum of magnetic quantum numbers of the individual spins, $m = m_\mathrm{A} + m_\mathrm{X}$. Thus the magnetic quantum number of a coupled spin pair can assume the values 1, 0, and -1, where zero

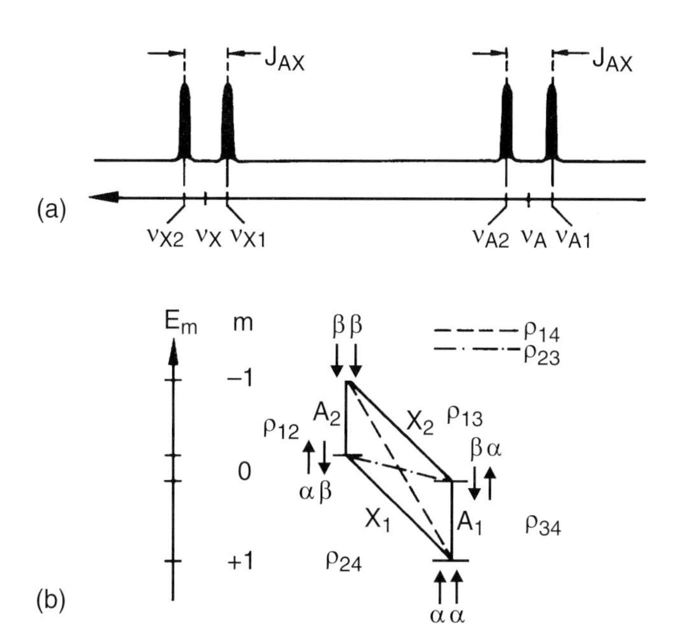

FIG. 2.2.12 Spectrum (a) and energy level diagram (b) of two coupled spins $\frac{1}{2}$, A and X. The arrows indicate the orientation of the magnetic moments in the magnetic field. The continuous lines connecting the energy levels correspond to observable single-quantum transitions. The broken lines indicate forbidden multi-quantum transitions. The splitting of the resonance lines gives the strength J_{AX} of the interaction.

appears twice, once for the state $|\alpha\beta\rangle$ and once for the state $|\beta\alpha\rangle$. Both states differ only very little in energy, so that the same magnetic quantum number applies in zeroth order. The possible coherence orders $p = m_{\mathrm{f}} - m_{\mathrm{i}}$ are then ± 2, ± 1, and 0. A transition, where only one of the two coupled spins changes its orientation, is a *single-quantum transition*, because m changes by ± 1 (Fig. 2.2.12(b)). They are directly observable and can be assigned to the resonance lines in the NMR spectrum. Transitions, where more than one spin flip simultaneously are *multi-quantum transitions*. For the *double-quantum transition* ($p = \pm 2$) both spins flip in the same direction, for the *zero-quantum transition* ($p = 0$) they flip in opposite directions.

Multi-quantum transitions can only be observed indirectly by a modulation of the detected signal with the phase of the multi-quantum coherence. This modulation is achieved in an experiment by variation of an evolution time prior to detection. Repetitive detection of the signal for different evolution times provides the information about the evolution of the multi-quantum coherence. The *indirect detection* of spectroscopic information based on phase or amplitude modulation of the detected signal is the principle of multi-dimensional NMR spectroscopy [Ern1]. Thus multi-quantum NMR is a special form of 2D NMR. Also, NMR imaging can be viewed as a special form of multi-dimensional NMR spectroscopy, where the frequency axes have been coded by the use of magnetic field gradients to provide spatial information.

The organization of the density matrix of the AX system with its different types of elements is recognized by writing down the matrix explicitly and by using the coherence order in place of the matrix elements and the letter P for populations,

$$\rho = [\rho_{mn}] = \begin{array}{c} \\ \alpha\alpha \\ \alpha\beta \\ \beta\alpha \\ \beta\beta \end{array} \begin{array}{c} \begin{array}{cccc} \alpha\alpha & \alpha\beta & \beta\alpha & \beta\beta \end{array} \\ \begin{bmatrix} \mathrm{P} & +1 & +1 & +2 \\ -1 & \mathrm{P} & 0 & +1 \\ -1 & 0 & \mathrm{P} & +1 \\ -2 & -1 & -1 & \mathrm{P} \end{bmatrix} \end{array} \tag{2.2.62}$$

Evolution

In the *Schrödinger representation* the expansion coefficients a_m of the wave function are time dependent. So are the density matrix elements. From the *Schrödinger equation* (2.2.45) the *equation of motion of the density-matrix* derives. It is known as the *von Neumann equation*,

$$\mathrm{i}\frac{\partial}{\partial t}\boldsymbol{\rho} = [\mathbf{H}, \boldsymbol{\rho}]. \tag{2.2.63}$$

From it, the density matrix at time t is obtained in terms of the *time-evolution operator* (2.2.48), given the density matrix at time t_0

$$\boldsymbol{\rho}(t) = \mathbf{U}(t, t_0)\, \boldsymbol{\rho}(t_0)\, \mathbf{U}^{-1}(t, t_0). \tag{2.2.64}$$

For the *Zeeman interaction* of the nuclear spins with the static magnetic field, the Hamilton operator is time-independent, $\mathbf{H}_Z = -\gamma \mathbf{I}_z \boldsymbol{B}_0 = \omega_0 \mathbf{I}_z$. Then the evolution operator (2.2.47) applies and (2.2.64) reduces to the free precession signal, which eventually determines the FID *via* the expectation value of the transverse angular momentum operators $\mathbf{I}_x$ and $\mathbf{I}_y$,

$$\boldsymbol{\rho}(t) = \exp\{-\mathrm{i}\omega_0 \mathbf{I}_z\,(t - t_0)\}\boldsymbol{\rho}(t_0)\exp\{\mathrm{i}\omega_0 \mathbf{I}_z\,(t - t_0)\}. \tag{2.2.65}$$

The Bloch equations describe the motion of the transverse magnetization in the static magnetic field in terms of a precession around the axis of the field. Similarly (2.2.65) describes a rotation of the density matrix around the z-axis by an angle $\omega_0(t - t_0)$. The effects of rf pulses are consequently described by rotations of the density matrix around axes in the transverse plane. For instance, a rotation around the y-axis by an angle α is expressed by

$$\boldsymbol{\rho}(t_{0+}) = \exp\{-\mathrm{i}\alpha \mathbf{I}_y\}\boldsymbol{\rho}(t_{0-})\exp\{\mathrm{i}\alpha \mathbf{I}_y\}, \tag{2.2.66}$$

where t_{0-} and t_{0+} are the times immediately before and after application of the pulse.

So far, the spins have been considered as isolated entities, which are completely decoupled from the surrounding environment. This is, in fact a good approximation for short evolution times. For longer times, the coupling to the lattice must be considered. This can be done either by including the lattice in the density matrix or by introducing a relaxation operator $\boldsymbol{\Gamma}$ which summarily accounts for the effects of the lattice on each density-matrix element [Ern1]. It is sufficient to use the latter approach in most NMR applications. With the *relaxation operator* $\mathbf{\Gamma}$ the von Neumann equation is rewritten in close analogy to the Bloch equations (2.2.8),

$$\frac{\mathrm{d}\boldsymbol{\rho}(t)}{\mathrm{d}t} = -\mathrm{i}\left[\mathbf{H}(t), \boldsymbol{\rho}(t)\right] - \mathbf{\Gamma}\left[\boldsymbol{\rho}(t) - \boldsymbol{\rho}_0\right]. \tag{2.2.67}$$

Because $\mathbf{\Gamma}$ operates on each element of a matrix it is called a *superoperator*. In fact, the *Hilbert-space* formulation of quantum mechanics leading to the von Neumann equation of motion of the density matrix can be simplified considerably by introduction of a superoperator notation in the so-called *Liouville space*. Furthermore, for the analysis of NMR experiments with complicated pulse sequences it is of great help to expand the density matrix into products of operators, where each *product operator* exhibits characteristic transformation properties under rotation [Ern1].

2.3 HARDWARE

The *hardware* applicable to NMR imaging is largely similar to that needed for pulsed Fourier NMR spectroscopy [Che1, Ell1, Fuk1, Hou1, Red1, Red3]. In addition, however, strong gradient coils are needed as well as the hardware to drive them in a pulsed fashion [Kre1, Man2, Mor1]. A detailed description of imaging hardware is given by Callaghan [Cal1] and by Chen and Hoult [Che1]. The basic components of an NMR imaging spectrometer are a magnet with field-gradient coils, a high-power rf transmitter, a gradient amplifier, a phase-sensitive rf detector, a computer for controlling the experiment as well as for data processing and display, and the rf antenna for excitation and detection. In nonmedical NMR imaging, the rf antenna and gradient coils are often part of the probe which carries the object and rests in the centre of the magnet [Dot1, Dot2, Sch3].

The actual rf signal received from the NMR probe is converted to an audio-frequency signal in the receiver. This process is equivalent to the transformation into the *rotating frame*. Therefore, NMR signals are measured in this frame and not in the laboratory frame. The response data are sampled and stored in digital form. The sampling rate has to be larger than the width of the spectrum to get a faithful representation of the analogue receiver signal. For noise reduction and elimination of unwanted signals the response data are filtered usually twice, once in an electronic analogue filter before digitization and once after digitization by numerical data manipulation.

2.3.1 Overview of basic components

The basic components of an NMR imaging spectrometer are marked in the block diagram of Fig. 2.3.1 [Kre1, Mor1]. The *magnet* is equipped with *shim coils* for adjustment of the homogeneity of the basic field $\boldsymbol{B}_0$, with a set of gradient coils to generate constant gradients G_x, G_y, and G_z, and with an rf resonator or coil for excitation and detection of the response. A switch connects the coil either to the *transmitter* or the *receiver* side of the spectrometer. Phase φ and amplitude (gain) of the transmitter signal can be

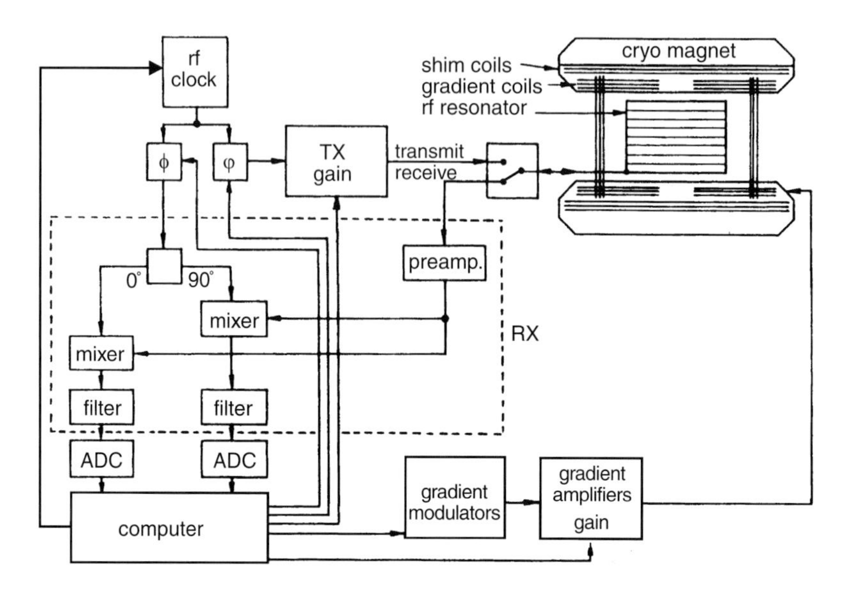

FIG. 2.3.1 Block diagram of an NMR imaging spectrometer. The main components are the magnet with gradient coils, transmitter TX, receiver RX, gradient amplifier, and a computer for control of the experiment and for data processing and display. Adapted from [Kre1] with permission from Publicis MCD.

adjusted, so that the excitation pulses can be shaped in both components. The phase control is needed to apply rf pulses in all directions of the transverse plane in the rotating coordinate system. Amplitude control is necessary, for instance, when defining a slice by selective excitation (cf. Chapter 5). Also, under computer control are the shape and the amplification of the gradient signals, because they must be pulsed in concert with the rf excitation.

The receiver side is somewhat more complex than the transmitter side. After preamplification, the response signal is split into two channels for phase-sensitive detection of two orthogonal components. To this end the signal is mixed with a sine and with a cosine reference wave, respectively, which are derived from the same clock as the rf excitation. The common phase ϕ of these reference signals can be adjusted during spectrometer tune-up. The mixing of rf response and reference waves generates signals at sum and difference frequencies in both channels. The signals at the sum frequency are blocked by low pass filters. The passing signals are the two quadrature components of the transverse magnetization in the rotating frame. They are digitized, sampled and stored in a computer for further processing and display.

2.3.2 The magnet

The *magnet* generates the magnetic field which polarizes the nuclear magnetic moments. Its strength determines the degree of polarization and thus enters the sensitivity of the measurement. Furthermore, it determines the measurement frequency as well as the spread in resonance frequencies (cf. Fig. 2.2.2). The volume of the magnetic field limits the size of the samples which can be investigated. For imaging superconducting magnets with bore diameters ranging from 5 to 120 cm and field strengths of 14–0.5 T are being used.

The magnetic fields employed in most NMR-imaging experiments are composed of a strong homogeneous field $\boldsymbol{B}_0$ and a much smaller *gradient field* which provides the spatial resolution. The maximum strength of the gradient field is less than 1% of the homogeneous field. The gradient fields often need to be switched on and off in times short compared to the effective signal decay time T_2^*, that is, in times of typically less than 1 ms.

The starting point for the calculation of magnetic field distributions for static fields, field gradients, and rf fields is the *Biot–Savart law*. It determines the magnetic field $\boldsymbol{B}$ at a distance $\boldsymbol{r}$ from the wire element $\mathrm{d}\boldsymbol{l}$ created by a current of strength I

$$\boldsymbol{B} = \frac{\mu_0 I}{4\pi} \int \frac{\mathrm{d}\boldsymbol{l} \times \boldsymbol{r}}{r^3}. \tag{2.3.1}$$

By integration, the field at the centre of a current loop of radius a can be obtained. Two such current loops spaced a apart in parallel and concentric configuration form the *Helmholtz coil*, which exhibits a uniform field about the centre.

Generation and homogeneity of the B_0 field

In most spectroscopy and imaging NMR equipment, today, *superconducting magnets* are used for generation of the strong, homogeneous magnetic field $\boldsymbol{B}_0$. The operating costs are considerably lower than for most electromagnets. Instead of electric energy and water cooling, only the costs for liquid helium and liquid nitrogen are significant in sustaining the magnetic field. A 2 T whole body magnet is produced from about 60 km of superconducting multi-filament niobium–titanium wire embedded in a copper matrix. The wire is wound onto 6 coils, and the number of windings is calculated, so that the homogeneity is optimized inside a sphere with a radius of about 50 cm for medical applications. The *homogeneity* achieved, in practice, over this volume is approximately 100 ppm, so that the use of additional shimming is required for improvement.

Shimming of the magnetic field can be done by insertion of iron plates on the walls of the magnet bore and by use of shim coils. In practice, the basic magnet homogeneity is first improved by iron plates. Also *superconducting shim* coils are used, which are part of the superconducting magnet assembly. Further refinement is obtained by adjusting the currents in a set of *room temperature shim coils* [Kon1], which is inserted inside the magnet bore [Chm1].

The shim coils are constructed based on an expansion of the magnetic field inhomogeneity in terms of spherical harmonics. The fields produced by the coils are orthogonal and can be adjusted independent of each other. Following (2.1.18) the shim fields B_{snm} are characterized by the associated Legendre polynomials $P_{nm}(\cos\theta)$ and the expansion coefficients A_{nm} and B_{nm},

$$B_{snm} = \left(\frac{r}{R}\right)^n P_{nm}(\cos\theta)\,[A_{nm}\cos(m\varphi) + B_{nm}\sin(m\varphi)] \tag{2.3.2}$$

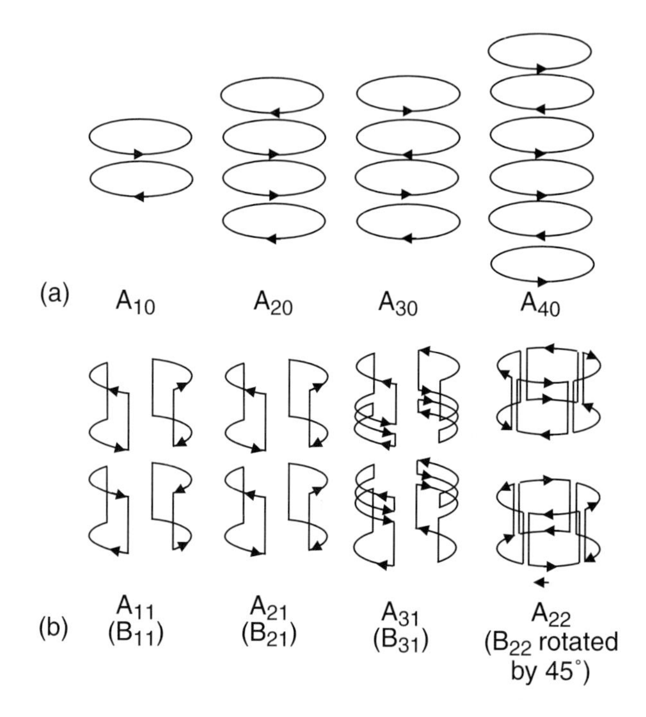

FIG. 2.3.2 Schematic drawings of shim coils. (a) Zonal windings. (b) Tesseral windings. Adapted from [Kre1] with permission from Publicis MCD.

The fields with $m = 0$ are independent of the angle φ. They are called *zonal fields*. The others are called *tesseral fields*. Some shim coils for production of tesseral and zonal fields are sketched in Fig. 2.3.2 [Kre1, Mor1]. Room temperature shims may provide correction fields up to $n = 5$, while cryoshims frequently provide corrections only up to $n = 1$. In this way homogeneities of ± 2 ppm can be reached over 30 cm and ± 5 ppm over 50 cm diameter volumes.

Depending on the field strength and the diameter of the magnet bore, the *fringe field* needs to be shielded, because it can pose a hazard for people with cardiac pacemakers or magnetic implants, for magnetic data storage devices, computer screens, and for credit cards with magnetic coding, wherever the magnetic field strengths supercedes that of the earth magnetic field by a factor of 10, that is, near the 0.5 mT line. To keep this region small, strong fringe fields can be shielded. This is done either with iron plates in the walls of the magnet room, with iron plates surrounding the magnet directly (self-shielding), or by active self-*shielding*. The latter is based on external, current bearing coils around the magnet, the fields of which are adjusted to compensate the fringe field of the magnet.

The gradient system

The *gradient coils* are inserted inside the room temperature shim coils in the magnet (Fig. 2.3.1). Depending on the type of spectrometer, they can be part of the magnet assembly or part of the removable probe assembly. The former are typical for medical imagers suitable for large objects, the latter are used preferentially for small objects including solid samples. To keep the access to the sample along the axis of the magnet parallel to the direction of the $\boldsymbol{B}_0$ field, a pair of *Maxwell coils* is used for generation of the G_{zz} gradient [Tan1]. It is similar to a Helmholtz configuration consisting of two current loops, but the currents are flowing in opposite directions as illustrated in Fig. 2.3.2 by the A_{10} configuration (zonal gradient). The tesseral gradients, G_{xz} and G_{yz}, are generated by *saddle coils* or *Golay coils* [Hou1]. This is depicted in Fig. 2.3.2 by the $A_{11}(B_{11})$ configuration. Another way of generating tesseral gradients is by using straight wires parallel to the field direction [Man2].

Because the gradient fields are identical in symmetry to fields of the shim coils, they may couple to the shim coils and degrade the shim parameters, in particular, during pulsed operation of the gradients. Furthermore the gradients and the connecting wires may act as antennas for exterior rf sources, which are then coupled from the gradient coils to the rf receiver circuit *via* the rf coil in the probe. This problem is particularly acute for imaging of rare or insensitive nuclei. It can be alleviated in part by the use of filters in the gradient-connecting wires [Gün1].

Unless simple backprojection techniques are used for imaging, the gradients need to be switched during the pulse sequence. The changing magnetic fields lead to the formation of *eddy currents* in the cryostat dewar of the magnet and in the rf shields of the probe, which counteract the field changes according to Lenz's law in addition to the inductance of the gradient coil. As a result, dynamic fields build up and decay during gradient switching, and the switching of the gradient field is slowed down [Maj1]. This is illustrated in Fig. 2.3.3(a) for rectangular current pulses [Kre1, Mor1]. The resultant time-dependence of the gradient field is approximated by a multi-exponential function with different time constants for fast, intermediate, and slow changes. A more rectangular gradient shape is obtained if the current pulse is appropriately distorted (Fig. 2.3.3(b)). In practice, the gradient currents are modified in a *preemphasis* unit, which, for example, provides five variable time constants for each of the three gradient channels. These time

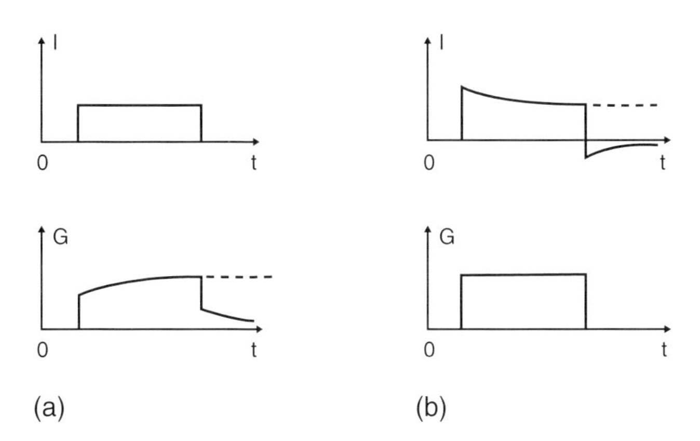

FIG. 2.3.3 Current (top) and gradient (bottom) pulse shapes. (a) Without preemphasis. (b) With preemphasis. Adapted from [Kre1] with permission from Publicis MCD.

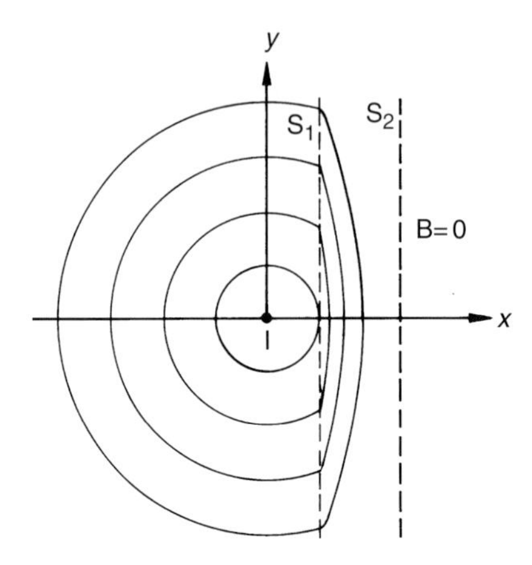

FIG. 2.3.4 [Man3] Magnetic field lines around an infinite, straight wire carrying a current I, which is shielded by a double-active screen. The screens are labelled S_1 and S_2. The field outside the second screen is zero.

constants need to be adjusted for optimum gradient shape for each gradient and probe geometry prior to the imaging experiment.

The transient fields generated by induced eddy currents can be reduced if gradient coils are employed with a diameter significantly smaller than that of the magnet bore. However, this makes inefficient use of the available magnet volume. Alternatively, the gradient coils can be shielded [Cal1, Bow1]. This is done by screens made of discrete wires, the positions of which are calculated to attenuate the gradient fields outside the gradient coils. By using two sets of screens, the exterior field can even be cancelled completely, and the interior field can be tailored to some predetermined profile. The compression of field lines achievable with *shielded gradients* is illustrated in Fig. 2.3.4 by example of the field of an infinite, straight wire along the z-direction [Man3]. *Gradient screens* can be operated passively by induced currents or actively by externally driven currents. The concept of active screening has also been exploited to reduce acoustic noise from gradient switching [Cha1].

The minimization of transient gradient fields and associated eddy currents is particularly important for solid-state imaging with multi-pulse line-narrowing (cf. Sections 8.7 and 8.8). Here gradient pulses just a few microseconds long are applied repetitively within the time windows of the rf pulse sequence, and transient effects will be accumulative. The generation of such short gradient pulses requires dedicated hardware, which is not available on ordinary imaging spectrometers [Con1].

2.3.3 The transmitter side

On the transmitter side, the rf excitation is modulated to form pulses of given shape and phase. These pulses are amplified and fed into the resonator or rf oscillator of the probe, which is usually shared with the receiver. For material applications, most of these components are required in duplicate if rare nuclei such as ^{13}C and ^{29}Si are imaged.

Double irradiation (cf. Section 7.2.12) of rare and abundant (^{1}H) nuclei is necessary for sensitivity enhancement as well as for decoupling of the heteronuclear dipole–dipole interaction.

The rf modulator

The rf pulses required for excitation are generated in a *modulator*, which multiplies the continuous rf wave by the pulse envelope and shifts the rf phase φ as determined by the pulse-shape memory. Here the phase and amplitude values are stored in digital form and read out during generation of the pulse. The number of stored values and the read-out speed determine the resolution of the pulse shape. For a slice-selective pulse 256 amplitude values are typical, one being read out every 10 μs. For the generation of simple rectangular pulses, the pulse-shape memory is bypassed. Here the rf phase φ determines the axis in the transverse plane of the rotating coordinate frame around which the density matrix is rotated by the pulse.

The transmitter

The *transmitter power* needs to be adjusted to provide pulses short enough to obtain nonselective 180° pulses. In applications requiring large coils, resonators can be constructed which are fed with the sine and the cosine component of the rf signal [Hou2]. By using such circularly polarized excitation a factor of two is saved in the rf power in comparison to linearly polarized excitation [Glo1].

The rf power P is proportional to the square of the NMR frequency ω_0 and the fifth power of the sample radius. For a cylindrical sample of length l parallel to $\boldsymbol{B}_0$, diameter d, and conductivity σ the rf power is approximated by [Kre1, Mor1]

$$P = \frac{\pi}{512}\omega_0^2 \sigma B_1^2 \left[d^4 l + \frac{4}{3} d^2 l^3 \right]. \tag{2.3.3}$$

Typical values for medical imaging are $d = l = 0.4$ m, $\sigma = 0.5/\Omega$ m (physiological saline solution), a selective 180° pulse of duration 1 ms, and an NMR frequency of $\omega_0 = 2\pi 42.6$ MHz, resulting in a calculated power of 2.8 kW. This is by a factor of about 2 too large compared to experimental values on patients [Kre1]. But to admit shorter rf pulses larger rf power is used in practice.

For *solid-state imaging* the rf pulse amplitude needs to be stronger than the spin interactions, most importantly the homonuclear dipole–dipole coupling for ^{1}H. Given a dipole–dipole coupling strength of 50 kHz, the 90° pulse width needs to be shorter than 1 μs. Conventional amplifier and probe technology can handle such short pulses only with powers of a few kW. The reason is, that the residual rf signal must be smaller than the response signal (1 μV) to be detected within less than 5 μs after the end of the pulse (1 kV). Therefore, the diameters of real solid objects for NMR imaging usually are restricted to less than 10 mm. However, often solids with high molecular mobility are to be imaged or liquids confined in solids or flowing in pipes and other devices. In this case, the rf power demands are similar to those encountered in medical imaging, because the dipole–dipole interaction is partially averaged out by the molecular motion, so that pulses with similar amplitudes can be applied.

Radio-frequency antennas

Radio-frequency antennas are coils and resonators employed to emit and receive the rf radiation to and from the sample. They are used in a time-shared fashion by the transmitter and the receiver side of the spectrometer. Both sides are decoupled by an electronic switch, the *transmit–receive switch* (Fig. 2.3.5(a)) [Kre1, Mor1]. *Crossed diodes* are inserted at both ends of the switch. For voltage levels lower than the break-through voltage of the diodes (0.7 V), the crossed diodes act as a resistor with infinitely high resistivity. For larger voltage levels they are highly conducting. When the transmitter is in the off-state, the pair at the transmitter side blocks the transmitter noise, and the diodes at the receiver side exert no influence other than a residual capacitance. Therefore, the receiver sees only the low signal from the coil in the probe. When the transmitter is on, both pairs of crossed diodes become conducting. Now the pair at the transmitter side exerts no influence, but the pair at the receiver side is placed at a voltage node and short circuits any residual voltage from the pulse to protect the receiver.

The *antenna* consists of a resonant rf circuit with parallel and series capacitors, C_p and C_s , for tuning of the resonance frequency and for matching of the impedance, respectively (Fig. 2.3.5(a)). The *impedance* is a complex quantity which needs to be adjusted to 50 Ω magnitude and 0° phase for optimum transfer of rf power. Depending on the equivalent resistance R, inductance L, and capacitance C of the components of the antenna (Fig. 2.3.5(b)), the *quality factor*

$$Q = \frac{\omega L}{R} \tag{2.3.4}$$

is defined, where

$$\omega = \sqrt{\frac{1}{LC} - \frac{R^2}{(2L)^2}} \tag{2.3.5}$$

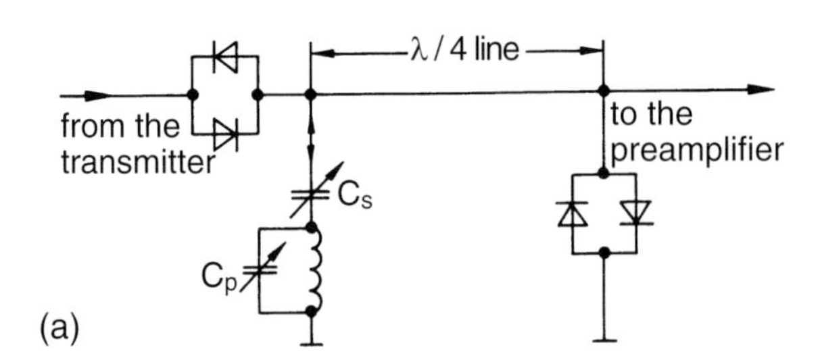

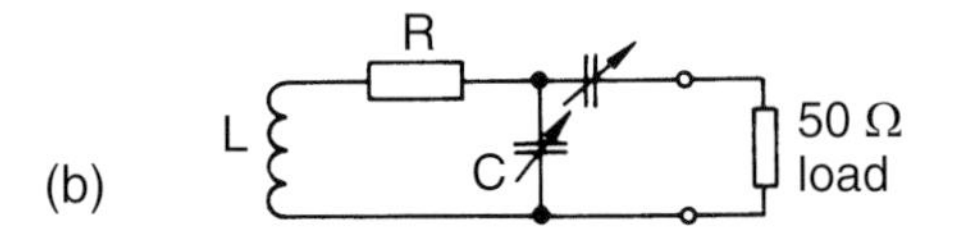

FIG. 2.3.5 (a) Transmit–receive switch for time-sharing of the antenna between transmitter and receiver. (b) Equivalent circuit diagram of an NMR antenna with loss. Adapted from [Kre1] with permission from Publicis MCD.

is the *resonance frequency of the antenna circuit.* A large Q characterizes a sensitive antenna with narrow *bandwidth*

$$\Delta\omega_{1/2} = \frac{R}{L} = \frac{\omega}{Q}, \tag{2.3.6}$$

while a low Q characterizes a less sensitive antenna with a larger bandwidth. The bandwidth is the width at half height of the resonance curve of the antenna, equivalent to the absorption part $A(\omega)$ of a complex Lorentz curve (cf. Fig. 2.2.6). For imaging with broadband detection, the bandwidth of the antenna must be broader than the bandwidth of the magnetization response. Antenna bandwidths of 100 kHz to 1 MHz are typical for solid-state NMR and for NMR imaging. Thus, the design of antennas for NMR requires a compromise between bandwidth and *sensitivity*. Also, antennas with large Q have a longer *ring-down time* after being exposed to an rf pulse or to rf pulse break through. The ring-down time causes a *deadtime* between the end of the pulse and the beginning of the data acquisition. During this period valuable parts of the response signal may get lost irreversibly. Therefore, tuned circuits with high Q values need to be avoided where possible in broadband spectrometers.

The magnetic fields effective in excitation and detection of the NMR response are perpendicular to the $\boldsymbol{B}_0$ field of the magnet. In superconducting magnets the $\boldsymbol{B}_0$ field is parallel to the long axis of the cylindrical bore. Thus the rf field $\boldsymbol{B}_1$ must be created perpendicular to this axis. At the same time a high filling factor of the coil is desired for high sensitivity. The best solution for the choice of antenna is a long *solenoid*. In this case, however, the access is restricted, and the cross-section must be small for tuning to typical NMR frequencies. This geometry, is used for small objects only, that is for NMR microscopy and spectroscopy.

A cylindrical structure fitting the geometry of the magnet bore is the *saddle coil*. It can provide homogeneous $\boldsymbol{B}_1$ fields at moderate NMR frequencies (25 MHz) for larger volumes (up to 30 cm diameter). However, its sensitivity is lower by a factor of $\sqrt{3}$ compared to that of a solenoid. But a factor of $\sqrt{2}$ can be recovered, if two saddle coils are arranged in quadrature for independent detection of the x- and y-components of the induced signal.

The saddle coil is an approximation of a long cylinder which carries a current distribution $I(\varphi)$ (Fig. 2.3.6(a)) [Kre1],

$$I(\varphi) = I_0 \sin\varphi. \tag{2.3.7}$$

Such a current distribution can be generated by a large number of parallel wires along the surface of the cylinder. In the lowest order approximation, six wires spread evenly about the circumference of the cylinder would be needed (Fig. 2.3.6(b)), two of which do not carry any current. The current in the others has the same strength but flows in changing directions. The saddle coil (Fig. 2.3.6(c)) is obtained from this by keeping only the current bearing wires and connecting them at the ends to obtain the desired current flow.

For large objects or at high frequencies, the dimension of the antenna is no longer short compared to the wavelength. Then *resonators* have to be employed. Their dimensions

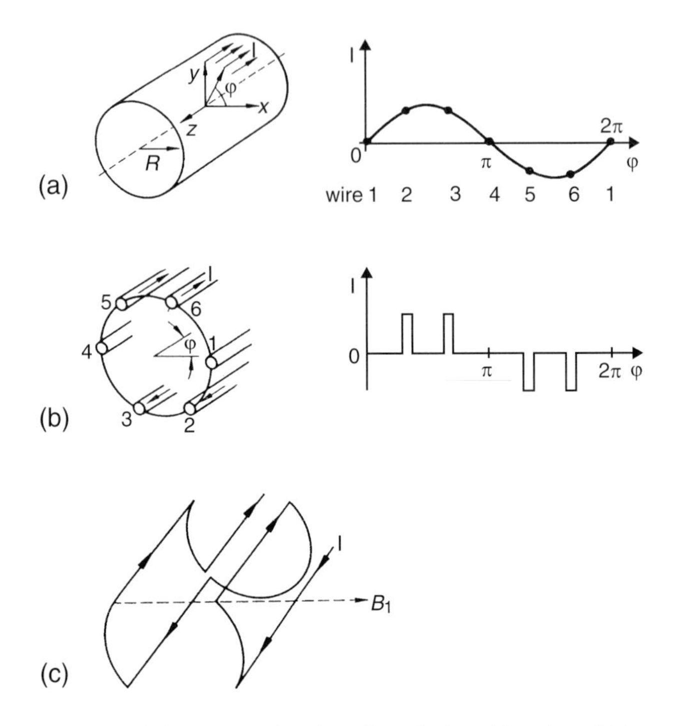

FIG. 2.3.6 The saddle coil is an approximation of a cylinder with a sinusoidal current distribution. (a) Sinusoidal current distribution on a cylinder. (b) Discrete approximation of the cylinder by six parallel wires carrying current I. (c) Saddle coil obtained from (b) by connecting the current bearing wires. Adapted from [Kre1] with permission from publicis MCD.

are matched to the effective wavelength in order to trap a standing wave inside [Ald1, Che1, Kre1, Mor1]. To optimize the homogeneity of the $\boldsymbol{B}_1$ field, they often consist of a network of distributed capacitances and inductances. This network must be strong enough to withstand high rf power because for conducting samples the rf power scales with the fifth power of the diameter and the second power of the NMR frequency (cf. eqn (2.3.3)). Current resonators used in practice are based on either the design by *Alderman and Grant resonator* [Ald1] or the *birdcage resonator* (Fig. 2.3.7) [Har1, Hay1, Wat1].

The power problem becomes less relevant if *surface coils* are applied to large objects (cf. Section 9.2) [Ack1]. They can be used for detection only or for excitation and detection together. The coil must be oriented so that its $\boldsymbol{B}_1$ field is orthogonal to the $\boldsymbol{B}_0$ field. In the simplest case a surface coil is formed by a single wire loop of radius a, where the tuning elements are attached right at its end (Fig. 2.3.8) [Kre1, Mor1]. Along the y-axis, perpendicular to the face of the coil and through its centre, the magnetic field of the current loop is given by

$$B_1(y) = \frac{\mu_0 a^2}{2\left(a^2 + y^2\right)^{3/2}} I. \tag{2.3.8}$$

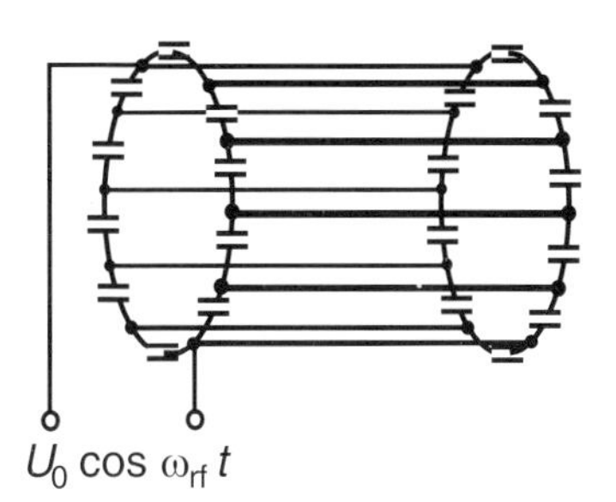

FIG. 2.3.7 Birdcage resonator. The capacitances are distributed for cylindrical surface currents (cf. Fig. 2.4.6(a)) to generate an rf field transverse to the axial direction.

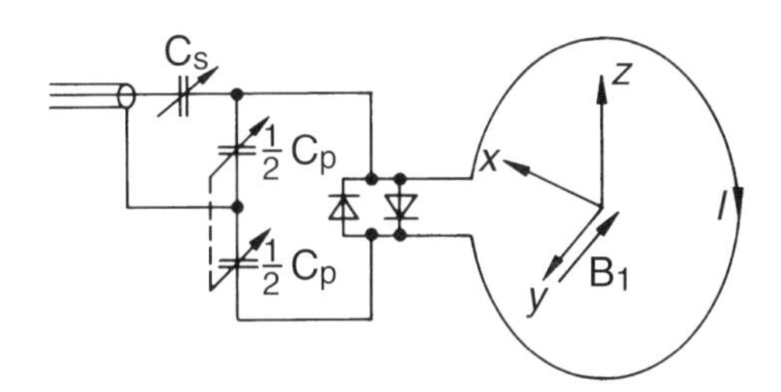

FIG. 2.3.8 Surface coil with symmetrical matching network and detuning diodes. Adapted from [Kre1] with permission from Publicis MCD.

The sensitivity along the coil axis and the field about the wire loop decrease rapidly with distance y from the coil. As a rule of thumb, the effective penetration depth for single-pulse excitation is about the radius of the coil (cf. Fig. 9.2.2). The $\boldsymbol{B}_1$ field gradient associated with the surface coil can be exploited for generation of spatial resolution by employing so-called rotating frame techniques (cf. Sections 6.3 and 9.2.2). Also, the geometry of surface coils can be tailored to fit the geometry of the object. Despite the fact that their sensitivity decreases with distance, their use provides interesting solutions to imaging of large objects and surface layers. Improvements in sensitivity and signal-to-noise ratio can be achieved by optimization of the inductive coupling between sample and receiver by using *microcoils* [Mcf1] for microscopy (cf. Fig. 9.3.2(a)), cooled receiver and preamplifier circuits [Zho1, Mcf1, Web1], and by employing super-conducting resonators [Bla1, Bla2].

The axis of the receiver surface coil must be orthogonal to both the static magnetic field and the excitation field. Because the latter is difficult to achieve, inductive coupling of the receiver to the transmitter can be reduced by *detuning diodes* (Fig. 2.3.8), which are switched into the conducting state by the induced voltage [Ben1, Ede2].

2.3.4 The receiver side

The current induced in the coil during signal detection is amplified immediately after passing the transmit–receive switch (Fig. 2.3.1). It is then split into two channels by a power divider. The resultant two signals are fed to mixers, which multiply the received signal with a cosine reference wave in one channel and with a sine reference wave in the other. The reference waves oscillate with the rf frequency ω_{rf} , and the received signal oscillates with the NMR frequency ω_0 . The multiplication of reference and received signals together with the subsequent elimination of the signal at the sum frequency by filtering corresponds to the transformation into the rotating frame.

Transformation into the rotating coordinate frame

If the received signal

$$s''(t) = 2s_0 \cos(\omega_0 t). \tag{2.3.9}$$

is multiplied by the reference signals

$$s'_x(t) = \cos(\omega_{\mathrm{rf}} t - \phi) \quad \text{and} \quad s'_y(t) = \sin(\omega_{\mathrm{rf}} t - \phi), \tag{2.3.10}$$

the following product signals are obtained:

$$s'_x(t)s''(t) = s_0\{\cos[(\omega_0 + \omega_{\mathrm{rf}})t - \phi] + \cos[(\omega_0 - \omega_{\mathrm{rf}})t + \phi]\}, \tag{2.3.11a}$$

$$s'_y(t)s''(t) = s_0\{\sin[(\omega_0 + \omega_{\mathrm{rf}})t - \phi] + \sin[(\omega_0 - \omega_{\mathrm{rf}})t + \phi]\}. \tag{2.3.11b}$$

The low-pass filter following the signal multipliers eliminates the contribution at the sum frequency. Hereby half of the signal is lost. The other half, which oscillates at the difference frequency $\Omega_0 = \omega_0 - \omega_{\mathrm{rf}}$, is retained. The complex sum of the respective x- and y-components is the quadrature receiver signal,

$$s^+(t) = s_x(t) + \mathrm{i}s_y(t) = s_0 \exp\{\mathrm{i}(\Omega_0 t + \phi)\}. \tag{2.3.12}$$

Both quadrature components can be obtained in the rotating frame, although only one component is received in the laboratory frame. If both quadrature laboratory components were derived from the resonator, no signals at the sum frequency would be observed after multiplication with the reference wave. Consequently, no signal intensity would be lost, and the full signal would appear at the difference frequency Ω_0.

The phase ϕ appearing in eqns (2.3.10)–(2.3.12) is the receiver phase, which is adjustable under computer control. Ordinarily it is tuned to provide the pure cosine component of the signal in the real part of $s^+(t)$ and the sine component in the imaginary part. Although this phase could be manipulated in phase cycles such as the CYCLOPS sequence, this is not done if only the four quadrature values 0°, 90°, 180°, and 270° are needed. Then, it is technically simpler and less prone to shortcomings in tuning to permute real and imaginary parts of the signal in combination with sign changes as indicated in Table 2.2.2.

Data sampling

The signal (2.3.12) is in the audio-frequency regime, where it can be sampled via *analogue-to-digital converters* (ADCs) and stored in a computer for further processing and display. Generally, the response is acquired at equidistant sampling intervals Δt. The sampling rate $1/\Delta t$ has to be fast enough, so that the signal at the maximum response frequency is correctly digitized. If two ADCs are used, one for each of the two signal channels, real and imaginary parts of the response can be sampled simultaneously. Then positive and negative frequencies can readily be discriminated. Therefore, the centre frequency ω_{rf} can be positioned in the middle of the NMR spectrum, so that a receiver filter, with its bandwidth adjusted to the half of the spectral width, can be used for blocking of high frequency receiver noise.

The available spectral width S_{W} is determined by the sampling rate

$$\frac{1}{\Delta t} = S_{\text{W}}. \tag{2.3.13}$$

Equation (2.3.13) states the *Nyquist sampling theorem*. If Δt is set too high, the spectral width is too low, and signals at higher frequencies will appear at false positions (Fig. 2.3.9). This phenomenon is called *signal aliasing* [Der1].

In some older spectrometers only one ADC is available for sampling of both channels of the complex response. There, real and imaginary parts must be sampled alternately, and the sampling rate is doubled for a given spectral width. Discrimination of positive and negative frequencies is still possible, if the sign of every second complex data pair is inverted before Fourier transformation [Red2]. The net effect amounts to a shift of the receiver phase by 90° after each sampling interval Δt. This technique is known as the *Redfield method* or the method of *time proportional phase increments* (TPPI) [Ern1].

Each acquired data string has finite length. Cyclic Fourier transformation of a record of length T produces a finite set of Fourier coefficients on a discrete frequency grid. The grid spacing determines the *digital resolution*

$$\frac{2\pi}{\Delta\omega} = T = n_{\text{data}}\Delta t, \tag{2.3.14}$$

where n_{data} is the number of acquired complex data points. The digital resolution $2\pi/\Delta\omega$ is readily increased, if zeros are appended to the time signal before Fourier transformation so as to increase n_{data}.

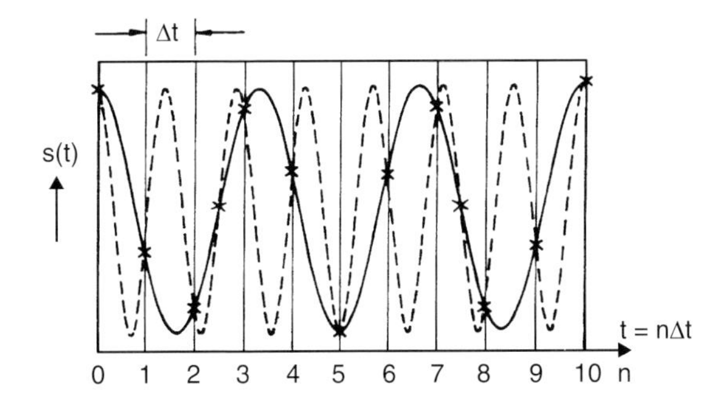

FIG. 2.3.9 [Der1] Digital sampling of analog signals. The continuous signal is digitized at an adequate sampling rate, the broken signal is not. After sampling it appears at the frequency of the continuously drawn signal.

This process is called *zero filling*. It should be performed only if the signal has already decayed to zero before, otherwise, a stepfunction is introduced and $(\sin x/x)$ oscillations appear in the resultant spectrum. These oscillations can be avoided by one of two ways. The standard method is the application of a filter to the spectrum. This is equivalent to multiplication of the time-domain signal by the Fourier transform of the filter function, a procedure called *apodization*. The most common procedure here is *exponential multiplication*, which denotes the multiplication of the FID by a decaying exponential. It effectively eliminates high-frequency components of the spectrum, but it also increases the linewidth, because it shortens the FID (cf. eqn (2.2.27)). Other filters have a less unfavourable influence on the lineshape [Ern1]. The second way is to avoid zero filling and Fourier transformation all together. The spectrum can be derived by numerical methods of time series analysis [Kan1, Pri1], like the *linear prediction*, the *maximum entropy* [Ste2] and the *wavelet transform* [Coi1] (cf. Section 4.4.6) methods.

On modern spectrometers the time-domain data are sampled much faster than required by the Nyquist theorem for subsequent digital filtering [Bes1]. *Digital filtering* provides two advantages: First, symmetric time-domain filter functions can be used to avoid phase distortions in the filtered response, and second, the filter transfer function can be designed with less ripple and steeper edges for better signal-to-noise ratios.

Spatial resolution

A simplified illustration of the static magnetic fields used for NMR spectroscopy and NMR imaging is presented in Fig. 2.3.10. In spectroscopy, the field is required to be homogeneous over the sample (a), while in imaging gradients are applied (b). In the simple imaging scheme of Fig. 2.2.5 the space information in the y-direction is encoded indirectly by phase modulation of the signal detected during t_2. The space information in x-direction is detected directly by a change of the NMR frequency as a result of the applied gradient. The latter case is referred to as *frequency encoding*. Here, the gradient strength determines the achievable *spatial resolution* $1/\Delta x$, which is limited by the width of the NMR resonance line or by the width $\Delta\omega_L = 2\pi\,\Delta\nu$ of the NMR spectrum. Assuming a gradient in x-direction the result (1.1.8) applies:

$$\frac{1}{|\Delta x|} = \left|\frac{\gamma G_x}{2\pi\,\Delta\nu}\right|. \tag{2.3.15}$$

The larger the linewidth, the stronger the gradient must be to resolve a given structure.

This limitation of the spatial resolution by the NMR linewidth does not apply for indirect detection of the space information by *phase encoding* (in y-direction in Fig. 2.2.5). For phase encoding the spatial resolution is determined by the sensitivity. It only depends on the maximum value of $k_{y\max} = n_y \Delta k_y$ (cf. eqn (2.3.16)) and on the signal-to-noise ratio, because the signal strength decreases with increasing $\boldsymbol{k}$. In principle, infinitely high resolution could be obtained by phase encoding [Emi1]. Imaging by phase encoding is also referred to as *constant-time imaging* (cf. Section 8.3).

Digital image resolution

The *digital resolution* or the *pixel resolution* of an NMR image can be much higher than the actual spatial resolution. The former is essentially defined like the digital resolution

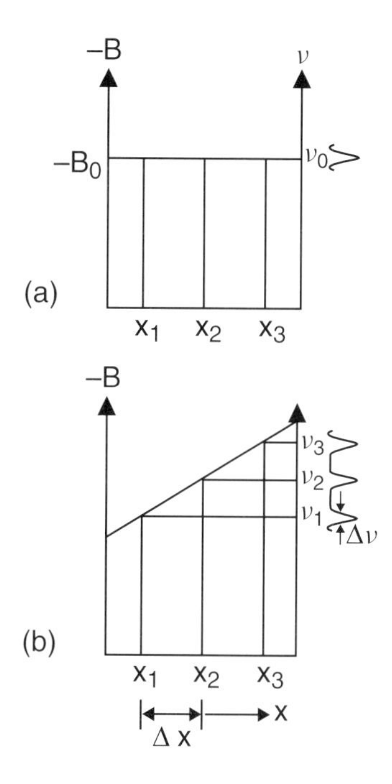

FIG. 2.3.10 [Blü1] Magnetic fields used in NMR spectroscopy and imaging. (a) In spectroscopy the field is required to be homogenous over the sample. (b) In imaging gradients are applied. Their strength determines the change of magnetic field with space and thus the spatial resolution.

(2.3.14) in spectroscopy. In imaging the $\mathbf{k}$-space values are sampled on a discrete mesh with a finite number of points in each dimension. The digital resolution $1/\Delta y$ in the indirectly detected dimension y of the image is determined by the maximum value of k_y or equivalently by the number n_y of complex signal values and by the $\mathbf{k}$-space sampling interval $\Delta k_y = \Delta G_y t_1$ (cf. Fig. 2.2.5),

$$\frac{1}{\Delta y} = \frac{n_y \Delta k_y}{2\pi}. \tag{2.3.16}$$

To avoid signal aliasing, k_y has to be stepped in small enough intervals so that the signal-phase increment $\Delta k_y y_{\text{max}}$ for the maximum object coordinate y_{max} never exceeds 2π. Similar considerations apply for the other space dimensions in x- and z-directions.

The total *imaging time* $T \cong n_y t_{\text{R}}$ for a 2D image is determined primarily by the number n_y of steps in the indirectly detected dimension, because each acquisition of a frequency-encoded signal is followed by a comparatively long repetition time t_{R} needed for signal recovery. Therefore, higher digital resolution along the indirectly detected dimension requests a proportionate increase in imaging time.

Field of view

The *field of view* (FOV) is the maximum spatial distance which can be displayed in the image. It corresponds to the spectral width (2.3.13) in spectroscopy. It is determined by the inverse of the $\boldsymbol{k}$ space sampling interval $\Delta \boldsymbol{k}$. Different expressions apply for *phase encoding* and for *frequency encoding* of the signal. Assuming constant gradients during phase and frequency-encoding times t_1 and t_2, respectively (cf. Fig. 2.2.5), the following equations are valid:

Frequency encoding:

$$x_{\max} = \left| \frac{2\pi}{\gamma G_x \Delta t_2} \right|, \quad G_x = \text{const}, \quad t_{2\max} = n_x \Delta t_2. \tag{2.3.17a}$$

Phase encoding:

$$y_{\max} = \left| \frac{2\pi}{\gamma \Delta G_y t_1} \right|, \quad G_{y\max} = n_y \Delta G_y, \quad t_1 = \text{const}. \tag{2.3.17b}$$

3

NMR spectroscopy

The appearance of NMR spectra is determined by various interactions of the nuclear spins with each other, as well as with quantities like the local and applied magnetic fields, the electric field gradient, and the coupling to the surroundings or the lattice. These interactions not only determine the particular *resonance frequencies*, but also *lineshapes* and *relaxation times* [Abr1, Ern1, Sch9, Sli1]. In solids, the description of the spin interactions is far more complicated than in liquids, because molecular motion is slow on the NMR timescale [Blü7, Eck1, Eck2, Ger1, Gri2, Hae1, Mcb2, Meh1, Sli1, Spi1, Ste2]. As a consequence, many interactions have an effect on frequency and intensity of the resonance, which are ineffective under fast motion as in liquids. For NMR imaging in materials science, these interactions are important, because most materials are solids and their characterization and image contrast can be explained. On the other hand, it is these interactions which complicate NMR imaging of solids and often require experimental techniques completely different from those used for liquids, plastic solids, and many biomedical objects. This chapter reviews the basic nuclear spin interactions and some elementary techniques of solid-state NMR spectroscopy relevant to imaging.

The most important interaction in NMR is the coupling of the nuclear spins to the applied magnetic field. This is the *Zeeman interaction* (cf. Section 2.2.1). In a strong magnetic field, it determines the value of the resonance frequency in zeroth order. For a more accurate description of the NMR frequency, the *chemical shift* and the *indirect coupling* of spins have to be considered in liquids. Both depend on the details of the electron states in the neighbourhood of the nuclei. Therefore, they are used as fingerprints of the chemistry of the material.

However, spins, that is, magnetic moments, and their coupling partners are vectors. Thus, they are quantities with magnitude and orientation. The interaction of two vectors is described by a *tensor* which is represented by a matrix in Cartesian coordinates. Examples are the *direct* or *dipole–dipole coupling*, the indirect coupling between two spins, and the couplings of the spin to the local magnetic field (chemical shift) and the applied rf magnetic fields (Zeeman interaction, stimulating rf field). A similar description holds for the *quadrupole coupling*, which is quadratic in the coupling spin and denotes the interaction of the spin with the tensor of the electric field gradient at the site of the nucleus. In liquids, the tensorial notation can be discarded, because the anisotropic contributions are averaged to zero by fast isotropic molecular motion, so that only the orientation-independent parts corresponding to the traces of the coupling tensors are effective. From all the interactions mentioned above, only the coupling tensors of the chemical shift and the indirect coupling have nonvanishing traces. In solids, however, the slow molecular motion is ineffective in averaging out the anisotropic parts of the coupling tensors, so that the tensorial character of the spin interactions leads to resonance frequencies which, in the end, depend on the orientation of the molecules in the applied magnetic field. In this case, a description of the interaction *Hamilton operator* in terms of *irreducible spherical tensors* is helpful [Meh1, Sli1, Spi1].

If the *orientation dependence of the resonance frequency* of a spin $\frac{1}{2}$ is determined by just one interaction, it can be exploited for use as a protractor to measure angles of molecular orientation. In powders and materials with partial molecular orientation, the orientation angles and, therefore, the resonance frequencies are distributed over a range of values. This leads to the so-called *wideline spectra*. From the lineshape, the *orientational distribution function* of the molecules can be obtained. These lineshapes need to be discriminated from temperature-dependent changes of the lineshape which result from slow *molecular reorientation* on the timescale of the inverse width of the wideline spectrum. The lineshapes of wideline spectra, therefore, provide information about *molecular order* as well as about the type and the timescale of slow *molecular motion* in solids [Sch9, Spi1].

Usually, the effects of several interactions are observed simultaneously in an NMR spectrum. Without *isotope enrichment* or the application of special techniques, the effect of a single interaction often cannot be observed independent of others with similar or even higher strength. Then a quantitative evaluation of the wideline spectrum can be cumbersome or even impossible. However, there are a number of techniques for manipulation of nuclear spin interactions, for example, by mechanical spinning of the sample at the magic angle (*MAS: magic angle spinning*) [Meh1], and by cyclic irradiation with well-defined rf-pulse sequences during data acquisition (selective averaging by *multi-pulse sequences*) [Hae1, Ger1]. In this way, spin interactions can be isolated, and the resonance lines in wideline spectra can sufficiently be narrowed to produce liquid-like NMR spectra. For example, in high-resolution ^{13}C solid-state NMR spectroscopy high-power decoupling of ^{1}H and MAS are applied on a routine basis. The narrowing of wideline resonances provides a practical approach to solid-state NMR imaging with good spatial resolution (cf. Chapter 8). The longer the FID signal, the better the achievable spatial resolution.

The decay of the free induction signal is governed not only by destructive interference of harmonic oscillations with a common initial phase and a distribution of frequencies, but also by the *spin–spin relaxation time* or *phase relaxation time* T_2. All *relaxation times*, including the *spin–lattice* or *energy relaxation time* T_1, the energy relaxation time $T_{1\rho}$ *in the rotating frame*, and the phase relaxation time T_2, are determined by randomly driven molecular motions which modulate the spin interactions [Abr1, Ern1]. Therefore, in addition to the lineshapes of wideline spectra, the relaxation times are parameters which provide information about molecular mobility on a wide range of timescales. In particular, the relaxation times T_1 and T_2 can be exploited for contrast generation in

NMR images by conventional saturation recovery, inversion recovery, and spin-echo techniques (cf. Chapter 7).

3.1 ANISOTROPIC NUCLEAR SPIN INTERACTIONS

The interaction energies of the spins determine the resonance frequencies and, thus, the separation of the energy levels of the nuclear spin states. The energy levels are the eigenvalues of the *Hamilton operator* $\mathbf{H}$ of the spin system. This operator is the sum of operators $\mathbf{H}_\lambda$ for each individual interaction λ,

$$\mathbf{H} = \underset{100\,\text{MHz}}{\mathbf{H}_Z} + \underset{250\,\text{kHz}}{\mathbf{H}_Q} + \underset{100\,\text{kHz}}{\mathbf{H}_{rf}} + \underset{50\,\text{kHz}}{\mathbf{H}_D} + \underset{1\,\text{kHz}}{\mathbf{H}_\sigma} + \underset{10\,\text{Hz}}{\mathbf{H}_J}. \tag{3.1.1}$$

The frequencies under each operator indicate the size of the interaction. The largest interaction next to the Zeeman interaction $\mathbf{H}_Z$ is the quadrupole interaction $\mathbf{H}_Q$, followed by the coupling $\mathbf{H}_{rf}$ of the spins to the exciting rf field, the dipole–dipole coupling $\mathbf{H}_D$, the chemical shift $\mathbf{H}_\sigma$, and the indirect coupling $\mathbf{H}_J$.

3.1.1 Interaction Hamiltonians

In the following, the individual spin interactions are summarized together with some of their properties.

The Zeeman interaction

The largest interaction is the *Zeeman interaction*. It essentially defines the nuclear polarization (cf. eqn (2.2.3)),

$$\mathbf{H}_Z = C^Z \mathbf{I} \mathbf{1} \boldsymbol{B}_0. \tag{3.1.2}$$

Here the coupling tensor between the spin vector operator $\mathbf{I}$ and the applied magnetic field $\boldsymbol{B}_0$ is given by the unit matrix $\mathbf{1}$. The prefactor C^Z is listed in Table 3.1.1. The energy level splitting resulting from $\mathbf{H}_Z$ defines the NMR frequency ω_0 by (2.2.6). Typical values

Table 3.1.1 [Spi1] Factors C^λ and symbols of the coupling parameters

Interaction	λ	C^λ	R^λ	δ_λ	η_λ
Zeeman	Z	$-\gamma$	1	–	–
Quadrupole	Q	$eQ/[2I(2I-1)\hbar]$	0	eq	η_Q
rf	rf	$-\gamma$	1	–	–
dipole–dipole	D	$-\mu_0\gamma_i\gamma_j/(4\pi)$	0	r_{ij}^{-3}	0
Magnetic shielding	σ	γ	σ	$\frac{2}{3}\Delta_\sigma$	η_σ
Indirect coupling	J	1	J	$\frac{2}{3}\Delta_J$	η_J

of $\omega_0/2\pi$ are summarized in Table 2.2.1 for a number of nuclei. The other interactions can often be viewed as mere perturbations of the energy splitting defined by the Zeeman interaction, and perturbation theory is applied to calculate the shifts and splittings of the energy levels. In high magnetic fields, first-order perturbation theory is sufficient in most cases, except for the quadrupole interaction of nuclei other than ^{2}H. The observed isotropic resonance frequency including all shifts is referred to as the *Larmor frequency* ω_L in this book.

The quadrupole interaction

The second largest interaction is the quadrupole interaction. It is expressed by the operator

$$\mathbf{H}_Q = C^Q \mathbf{IQI}. \tag{3.1.3}$$

Nuclei with spin quantum number $I > \frac{1}{2}$ exhibit an electric *quadrupole moment*, which couples to the *electric field gradient* established by the electrons surrounding the nucleus. The quadrupole interaction, therefore, is a valuable sensor of the electronic structure. Following ^{6}Li, ^{2}H possesses the smallest quadrupole moment (cf. Table 2.2.1). In rigid aromatic and aliphatic compounds, the quadrupole splitting of ^{2}H (Fig. 3.1.1(a)) is of the order of 130 kHz. The interaction is quadratic with respect to the spin vector $\mathbf{I}$. It is described by the *quadrupole coupling tensor* $\mathbf{Q}$, which is proportional to the tensor of the electric field gradient. The average of $\mathbf{Q}$ is determined by its trace *Spur* (in German). Following the Laplace equation, the trace is zero,

$$\tfrac{1}{3}\mathrm{Sp}\{\mathbf{Q}\} = 0. \tag{3.1.4}$$

Therefore, a quadrupole splitting of the energy levels cannot be observed under fast isotropic motion as in liquids. The quantity C^Q in (3.1.3) is a proportionality constant defined in Table 3.1.1.

The interaction with the applied radio-frequency field

The interaction of a nuclear spin with the applied *rf field* enables excitation of a detectable signal in NMR experiments. It has the same general form as the Zeeman interaction (3.1.2),

$$\mathbf{H}_{rf} = C^{rf} \mathbf{I1} \boldsymbol{B}_1. \tag{3.1.5}$$

The difference is that $\boldsymbol{B}_1$ is applied in the transverse plane, for instance along the x-axis of the laboratory coordinate frame. Consequently, only the x-component of the spin angular momentum operator $\mathbf{I}$ defines the interaction energy together with the magnitude and time dependence of the x-component of $\boldsymbol{B}_1$ (cf. Section 2.2.1).

It is desirable to apply $\boldsymbol{B}_1$ fields of strong enough amplitude so that $\mathbf{H}_{rf}$ dominates all other interaction Hamiltonians except for the Zeeman interaction. The rf pulses can then be treated as infinitely short delta pulses, and the analysis of the experimental spectra becomes comparatively simple. However, arcing in the probe limits useful amplitudes to the order of 200 kHz, so that in solid-state NMR the delta-pulse approximation must be treated with care for the dipole–dipole interaction among protons, and it breaks down for the quadrupole interaction.

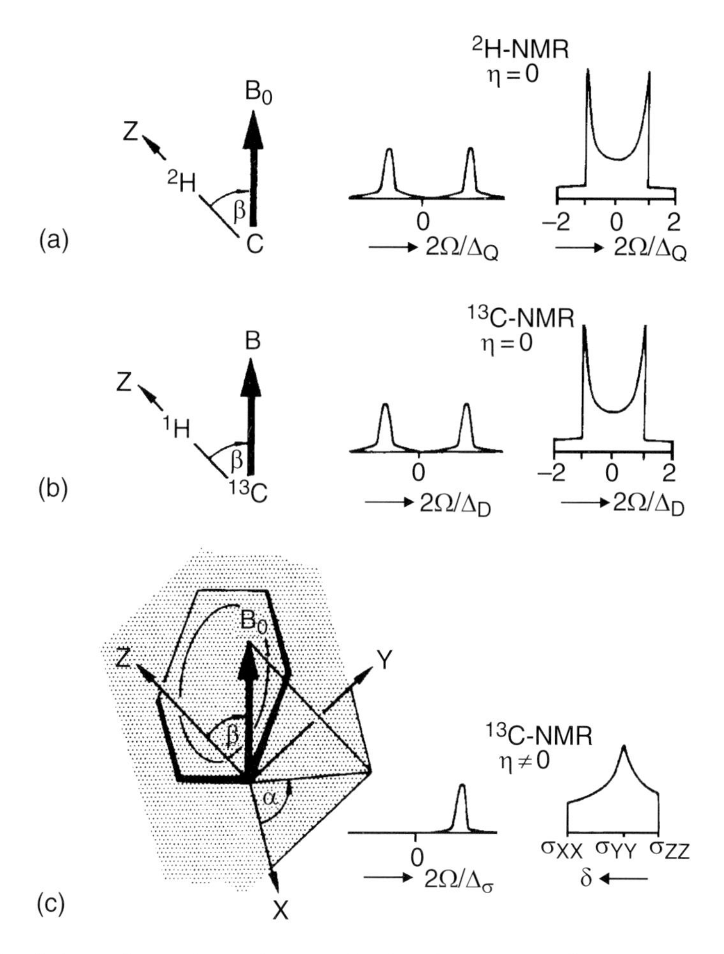

FIG. 3.1.1 Anisotropic spin interactions. (a) Quadrupole coupling of ^{2}H nucleus to the electric field gradient of a C–^{2}H bond. (b) Dipole–dipole coupling between ^{13}C and ^{1}H. (c) Anisotropic magnetic shielding of ^{13}C nuclei. Left: Geometry of the interaction and principal axes of the coupling tensor. Middle: NMR spectrum for a single molecular orientation. Right: The average over all orientations is the powder spectrum. The parameters Δ_λ denote the anisotropy of the interaction λ. δ is the chemical shift. Adapted from [Blü3] with permission from Wiley-VCH.

The dipole–dipole interaction

The *dipole–dipole interaction* describes the through-space coupling of two magnetic moments (Fig. 3.1.1(b)). The Hamilton operator for the homonuclear dipole–dipole interaction of two spin-$\frac{1}{2}$ nuclei formally looks like that for the quadrupole coupling of a spin with $I = 1$, because in this case the total spin quantum number is also 1. But the perturbation of the energy levels is often smaller by more than one order of magnitude. Typical values are 50–100 kHz for the coupling of abundant like spins as ^{1}H and 10 kHz for the heteronuclear coupling between different nuclei like ^{1}H and ^{13}C.

The Hamilton operator for the coupling between two spins is written as

$$\mathbf{H}_{\mathrm{D}} = C^{\mathrm{D}}\mathbf{I}^{i}\mathbf{D}\mathbf{I}^{j}, \tag{3.1.6}$$

where $\mathbf{I}^i$ and $\mathbf{I}^j$ are the spin vector operators of the coupling nuclei i and j. The entries in Table 3.1.1 reveal that the coupling energy scales with the cube of the inverse separation r_{ij} of the coupling partners. This dependence on distance is a highly valuable source of information about the structural geometry of molecules. In fact, the dipole–dipole coupling is exploited in various ways to obtain internuclear distances in solids as well as in liquids.

In an ensemble of spins eqn (3.1.6) has to be summed over all coupling pairs. The coupling tensor is denoted by $\mathbf{D}$. Its trace is zero,

$$\tfrac{1}{3}\mathrm{Sp}\{\mathbf{D}\} = 0, \tag{3.1.7}$$

so that the dipole–dipole coupling has no effect on the resonance frequencies in liquids. Nevertheless, in many solid and liquid samples it is the dominating mechanism for relaxation.

Magnetic shielding

The externally applied magnetic field is shielded at the site of the nucleus by the surrounding electrons (Fig. 3.1.1(c)). The resultant *local field* is given by (cf. eqn (1.1.3))

$$\boldsymbol{B} = (\mathbf{1} - \boldsymbol{\sigma})\boldsymbol{B}_0, \tag{3.1.8}$$

where $\boldsymbol{\sigma}$ is the shielding tensor [Ane1]. This shielding is specific of the particular electronic environment and thus of the chemistry. Contrary to the quadrupole and the dipole–dipole interactions, the shielding is dependent on the strength of the magnetic field $\boldsymbol{B}_0$, and the corresponding Hamilton operator is

$$\mathbf{H}_{\sigma} = C^{\sigma}\mathbf{I}\boldsymbol{\sigma}\boldsymbol{B}_0. \tag{3.1.9}$$

Values for the principal axes components of the shielding tensor are tabulated in the literature for various compounds [Dun1, Meh1, Vee1].

The average value of σ,

$$\tfrac{1}{3}\mathrm{Sp}\{\boldsymbol{\sigma}\} = \sigma, \tag{3.1.10}$$

is observed naturally in liquids and by use of line-narrowing techniques also in solids (cf. Section 3.3). This is the isotropic *magnetic shielding*, which determines the *chemical shift* in high-resolution NMR. Following (3.1.9), the resonance frequency ω_{L} differs from the NMR frequency ω_0 of the nucleus by (cf. eqn (1.1.4))

$$\omega_{\mathrm{L}} = (1 - \sigma)\omega_0 = 2\pi\nu_{\mathrm{L}}. \tag{3.1.11}$$

The index L indicates that the resonance frequency ω_{L} is usually referred to as the Larmor frequency. It is the significance of σ and, thus, of the chemical shift for structure elucidation, which makes NMR spectroscopy the standard analytical tool in preparative chemistry (cf. Fig. 2.2.2) [Bec1, Bov1, Gün1, Kal1].

Because the NMR frequency of free nuclei cannot be readily measured, the chemical shift is tabulated with reference to a standard. The field dependence is eliminated by using relative values, which give the frequency difference to the standard compound normalized by the NMR frequency of the nucleus,

$$\delta = \frac{\nu_{\mathrm{L}} - \nu_{\mathrm{S}}}{\nu_{\mathrm{S}}} \qquad (3.1.12)$$

By convention, the symbol δ is used for the relative chemical shift, and literature values are given in ppm (parts per million). The chemical shift ranges of a variety of nuclei and reference standards are listed in Table 2.2.1. The chemical shift range of ^{1}H encompasses only 13 ppm [Gün1], while that of ^{13}C covers 250 ppm [Kal1]. For this reason, ^{13}C spectra are far more sensitive to chemical information than ^{1}H spectra. However, for reasons of sensitivity, ^{1}H NMR is most popular in organic chemistry. With increasing field strength, the spread of resonance frequencies and, thus, the spectral resolution increases (cf. Fig. 2.2.2).

The indirect coupling

In addition to the direct dipole–dipole coupling of magnetic moments, there is also an *indirect coupling*, which is mediated by a polarization of the orbital angular momentum of the electrons. This coupling is rather weak in the homonuclear case among ^{1}H (1–10 Hz) but can reach values of several hundred hertz or even kilo-hertz in heteronuclear cases for heavy atoms [Mas1]. As a consequence of the larger linewidth in solids, the indirect coupling is difficult to observe in solid-state NMR of organic compounds.

Similar to the dipole–dipole coupling, the indirect coupling is written as

$$\mathbf{H}_{\mathrm{J}} = C^{\mathrm{J}}\mathbf{I}^{i}\mathbf{J}\mathbf{I}^{j}, \qquad (3.1.13)$$

where $\mathbf{I}^{i}$ and $\mathbf{I}^{j}$ are the spin vector operators of the coupling nuclei, but contrary to the dipole–dipole coupling the trace of the coupling tensor is different from zero,

$$\tfrac{1}{3}\mathrm{Sp}\{\mathbf{J}\} = J, \qquad (3.1.14)$$

so that the indirect coupling is observed also under fast isotropic motion such as in liquids. Because the symbol J is used to denote the coupling constant, the indirect coupling is also called the *J coupling*.

In liquid-state NMR the J coupling is responsible for the hyperfine structure of the high-resolution spectra (cf. Fig. 2.2.2) and the underlying, often highly intricate *energy-level diagrams* (cf. Fig. 2.2.11). Similar to the chemical shift, the J couplings are fingerprints of the chemical structure. To analyse even complicated molecular conformations, the respective energy level diagrams bearing the information about the chemical shift, J coupling, and resonance connectivities need to be unravelled. For this purpose, an abundant number of NMR methods has been developed primarily in liquid-state NMR, which often lead to *multi-dimensional spectra* [Ern1].

3.1.2 General formalism

The coupling tensors $\mathbf{Q}$, $\mathbf{D}$, σ, and $\mathbf{J}$ are tensors $\mathbf{P}$ of rank 2 [Hae1, Meh1, Spi1]. For each of the corresponding interactions, λ, the tensor, can be separated into an isotropic part $\mathbf{P}_\lambda^{(0)}$, an antisymmetric part $\mathbf{P}_\lambda^{(1)}$, and a traceless symmetric part $\mathbf{P}_\lambda^{(2)}$. For simplicity of notation, the index λ is not carried along in the next six equations. In the principal axes system *XYZ* of the symmetric part of the coupling tensor, the generic coupling tensor $\mathbf{P} = \{P_{ij}\}$, where $i, j = X, Y, Z$, is written as

$$\mathbf{P} = R\begin{bmatrix} 1 & 0 & 0 \\ 0 & 1 & 0 \\ 0 & 0 & 1 \end{bmatrix} + \begin{bmatrix} 0 & P_{XY} & P_{XZ} \\ -P_{XY} & 0 & P_{YZ} \\ -P_{XZ} & -P_{YZ} & 0 \end{bmatrix} + \delta\begin{bmatrix} -\frac{1}{2}(1+\eta) & 0 & 0 \\ 0 & -\frac{1}{2}(1-\eta) & 0 \\ 0 & 0 & 1 \end{bmatrix}$$
$$= \qquad \mathbf{P}^{(0)} \qquad + \qquad \mathbf{P}^{(1)} \qquad + \qquad \mathbf{P}^{(2)} \tag{3.1.15}$$

Equation (3.1.15) is written in Cartesian coordinates. Because the tensorial properties of the interactions lead to an angular dependence of the resonance frequency, a representation in spherical coordinates is preferred [Spi1],

$$\begin{aligned} &P_{00} = R, \\ &P_{10} = -\mathrm{i}\sqrt{2}P_{XY}; \qquad P_{1\pm1} = P_{XY} \pm P_{YZ}; \\ &P_{20} = \sqrt{\frac{3}{2}}\delta; \qquad P_{2\pm1} = 0; \qquad P_{2\pm2} = -\frac{1}{2}\delta\eta. \end{aligned} \tag{3.1.16}$$

The isotropic part is given by

$$R = \tfrac{1}{3}(P_{XX} + P_{YY} + P_{ZZ}) = \tfrac{1}{3}\mathrm{Sp}\{\mathbf{P}\}. \tag{3.1.17}$$

It is always observed, even under fast molecular reorientation. For the different interactions λ the expressions for R are listed in Table 3.1.1.

To first order, the resonance frequency is determined by the isotropic and the symmetric parts only and not by the antisymmetric part of $\mathbf{P}$ [Ane1]. Because the trace of the symmetric part $\mathbf{P}^{(2)}$ is zero, two parameters are sufficient for its determination. They are the largest principal value δ or the anisotropy Δ,

$$\delta = P_{ZZ}^{(2)} = \frac{2\Delta}{3} = \frac{2}{3}\left[P_{ZZ} - \frac{P_{XX} + P_{YY}}{2}\right] \tag{3.1.18}$$

and the asymmetry parameter

$$\eta = \frac{P_{YY}^{(2)} - P_{XX}^{(2)}}{\delta} = \frac{P_{YY} - P_{XX}}{P_{ZZ} - R}, \tag{3.1.19}$$

where the principal values are ordered following the convention of Haeberlen [Hae1],

$$|P_{ZZ}^{(2)}| \geq |P_{XX}^{(2)}| \geq |P_{YY}^{(2)}|. \tag{3.1.20}$$

Using the generic coupling tensor representing the different interactions λ, the generic Hamilton operator $\mathbf{H}_\lambda$ is expressed in the notation of irreducible spherical tensors

$$\mathbf{H}_\lambda = C^\lambda \sum_{l=0}^{2} \sum_{m=-l}^{l} (-1)^m \mathbf{T}_{lm}^\lambda \sum_{m'=-l}^{l} P_{lm'}^\lambda D_{m'-m}^{(l)} \left(\alpha^\lambda, \beta^\lambda, \gamma^\lambda\right). \tag{3.1.21}$$

The space- and spin-dependent parts of the Hamilton operator are described by the tensor components P_{lm} (3.1.16) in physical space and by the operator components $\mathbf{T}_{lm}$ in spin space. The $\mathbf{T}_{lm}$ are tabulated in Table 3.1.2 for the different interactions λ [Spi1]. Instead of Cartesian vectors, first rank tensors are used, for example, $\mathbf{I}_{\pm 1} = \mp(\mathbf{I}_x \pm \mathrm{i}\mathbf{I}_y)/2^{1/2}$, and $\mathbf{I}_0 = \mathbf{I}_z$. The prefactors C^λ are summarized in Table 3.1.1 together with the parameters R^λ, δ^λ, and η^λ which are needed in (3.1.16) for calculation of the P_{lm}^λ [Spi1].

Equation (3.1.21) is valid in the laboratory coordinate frame, where the z-axis is defined by the direction of the magnetic field $\boldsymbol{B}_0$. This, however, is rotated against the principal axes frame of a coupling tensor. In order to use the principal axes values of the coupling tensor, the tensor $\mathbf{P}$ in (3.1.21) is rotated by the use of *Wigner rotation matrices* $\mathbf{D}_{m'm}^{(l)}(\alpha, \beta, \gamma)$ with *Euler angles* α, β, and γ from the principal axes frame into the laboratory frame. For practical applications of (3.1.21) the elements of the Wigner rotation matrix are listed in Table 3.1.3 for $l = 2$. The contributions to $\mathbf{H}_\lambda$ for $l = 1$ can be neglected within the validity of first-order perturbation theory.

3.1.3 Strong magnetic fields

In *strong magnetic fields*, the resonance frequencies are determined largely by the *Zeeman interaction* ($\lambda = \mathrm{Z}$ in Tables 3.1.1 and 3.1.2). The other interactions can be treated as perturbations (cf. eqn (3.1.1)). The coupling to the rf field, the dipole–dipole interaction, the chemical shift, and the J coupling can be readily treated by *first-order perturbation theory*. For the quadrupole interaction, this approximation holds true only for small quadrupole moments like those of ^{6}Li and ^{2}H.

When truncating the perturbation expansion after the first-order term, only the parts of the interaction Hamiltonians are kept which are diagonal in the eigenbasis of the Zeeman Hamiltonian [Hae1, Meh1, Spi1]. These are the tensor components $\mathbf{T}_{00}$ and $\mathbf{T}_{20}$ (Table 3.1.2). Then the angular-dependent part of (3.1.21) simplifies to

$$\mathbf{H}_\lambda = C^\lambda \mathbf{T}_{20}^\lambda \frac{1}{2}\sqrt{\frac{3}{2}}\delta_\lambda \left[3\cos^2\beta^\lambda - 1 - \eta_\lambda \sin^2\beta^\lambda \cos(2\alpha^\lambda)\right]. \tag{3.1.22}$$

The two Euler angles α^λ and β^λ are the polar angles which specify the orientation of the magnetic field $\boldsymbol{B}_0$ in the principal axes system of the coupling tensor $\mathbf{P}^{(2),\lambda}$ (Fig. 3.1.2).

Within first-order perturbation theory, spectral lines are observed only for changes of the magnetic quantum number m by $|\Delta m| = 1$. These are $2I + 1$ lines for the interaction of a spin-I nucleus, for example, two lines for the quadrupole interaction of the deuteron (Fig. 3.1.1(a)). The same is true for the dipole–dipole coupling between two spins with $I_i = I_j = \frac{1}{2}$, because the total spin quantum number $I = I_i + I_j$ is equal to 1 in this

Table 3.1.2 [Spi1] Spin operators $\mathbf{T}_{lm}^{\lambda}$ in the laboratory system

λ	$\mathbf{T}_{00}$	$\mathbf{T}_{10}$	$\mathbf{T}_{1\pm1}$	$\mathbf{T}_{20}$	$\mathbf{T}_{2\pm1}$	$\mathbf{T}_{2\pm2}$
Z	$\mathbf{I}_0 B_0$	–	–	–	–	–
Q	–	–	–	$6^{-1/2}[3\mathbf{I}_0^2 - I(I+1)]$	$2^{-1/2}(\mathbf{I}_{\pm1}\mathbf{I}_0 + \mathbf{I}_0\mathbf{I}_{\pm1})$	$\mathbf{I}_{\pm1}^2$
rf	$(\mathbf{I}_+ + \mathbf{I}_-)/2 \times (B_+ + B_-)$	–	–	–	–	–
D	–	–	–	$6^{-1/2}(3\mathbf{I}_0^i\mathbf{I}_0^j - \mathbf{I}^i\mathbf{I}^j)$	$2^{-1/2}(\mathbf{I}_{\pm1}^i\mathbf{I}_0^j + \mathbf{I}_0^i\mathbf{I}_{\pm1}^j)$	$\mathbf{I}_{\pm1}^i\mathbf{I}_{\pm1}^j$
σ	$\mathbf{I}_0 B_0$	–	$2^{-1/2}(\pm\mathbf{I}_{\pm1} B_0)$	$(2/3)^{1/2}\mathbf{I}_0 B_0$	$2^{-1/2}\mathbf{I}_{\pm1} B_0$	–
J	$\mathbf{I}^i\mathbf{I}^j$	$2^{-1/2}(\mathbf{I}_{\pm1}^i\mathbf{I}_{-1}^j - \mathbf{I}_{-1}^i\mathbf{I}_{+1}^j)$	$2^{-1/2}(\pm\mathbf{I}_{\pm1}^i\mathbf{I}_0^j \mp \mathbf{I}_0^i\mathbf{I}_{\pm1}^j)$	$6^{-1/2}(3\mathbf{I}_0^i\mathbf{I}_0^j - \mathbf{I}^i\mathbf{I}^j)$	$2^{-1/2}(\mathbf{I}_{\pm1}^i\mathbf{I}_0^j + \mathbf{I}_0^i\mathbf{I}_{\pm1}^j)$	$\mathbf{I}_{\pm1}^i\mathbf{I}_{\pm1}^j$

Table 3.1.3 Wigner rotation matrix $\mathbf{D}^{(2)}(\alpha, \beta, \gamma)$

m'	m				
	2	1	0	-1	-2
2	$\frac{1}{4}(1+\cos\beta)^2$ $\times e^{-2i(\alpha+\gamma)}$	$-\frac{1}{2}(1+\cos\beta)\sin\beta$ $\times e^{-i(2\alpha+\gamma)}$	$\left(\frac{3}{8}\right)^{1/2}\sin^2\beta$ $\times e^{-i2\alpha}$	$\frac{1}{2}(1-\cos\beta)\sin\beta$ $\times e^{i(-2\alpha+\gamma)}$	$\frac{1}{4}(1-\cos\beta)^2$ $\times e^{2i(-\alpha+\gamma)}$
1	$\frac{1}{2}(1+\cos\beta)\sin\beta$ $\times e^{-i(\alpha+2\gamma)}$	$\left[\cos^2\beta-\frac{1}{2}(1-\cos\beta)\right]$ $\times e^{-i(\alpha+\gamma)}$	$-\left(\frac{3}{8}\right)^{1/2}\sin 2\beta$ $\times e^{-i\alpha}$	$\left[\frac{1}{2}(1+\cos\beta)-\cos^2\beta\right]$ $\times e^{i(-\alpha+\gamma)}$	$-\frac{1}{2}(1-\cos\beta)\sin\beta$ $\times e^{i(-\alpha+2\gamma)}$
0	$\left(\frac{3}{8}\right)^{1/2}\sin^2\beta$ $\times e^{-i2\gamma}$	$\left(\frac{3}{8}\right)^{1/2}\sin 2\beta$ $\times e^{-i\gamma}$	$\frac{1}{2}(3\cos^2\beta-1)$	$-\left(\frac{3}{8}\right)^{1/2}\sin 2\beta$ $\times e^{i\gamma}$	$\left(\frac{3}{8}\right)^{1/2}\sin^2\beta$ $\times e^{i2\gamma}$
-1	$\frac{1}{2}(1-\cos\beta)\sin\beta$ $\times e^{i(\alpha-2\gamma)}$	$\left[\frac{1}{2}(1+\cos\beta)-\cos^2\beta\right]$ $\times e^{i(\alpha-\gamma)}$	$\left(\frac{3}{8}\right)^{1/2}\sin 2\beta$ $\times e^{i\alpha}$	$\left[\cos^2\beta-\frac{1}{2}(1-\cos\beta)\right]$ $e^{i(\alpha+\gamma)}$	$-\frac{1}{2}(1+\cos\beta)\sin\beta$ $\times e^{i(\alpha+2\gamma)}$
-2	$\frac{1}{4}(1-\cos\beta)^2$ $\times e^{2i(\alpha-\gamma)}$	$\frac{1}{2}(1-\cos\beta)\sin\beta$ $\times e^{i(2\alpha-\gamma)}$	$\left(\frac{3}{8}\right)^{1/2}\sin^2\beta$ $\times e^{i2\alpha}$	$\frac{1}{2}(1+\cos\beta)$ $\times e^{i(2\alpha+\gamma)}$	$\frac{1}{4}(1+\cos\beta)^2\sin\beta$ $\times e^{2i(\alpha+\gamma)}$

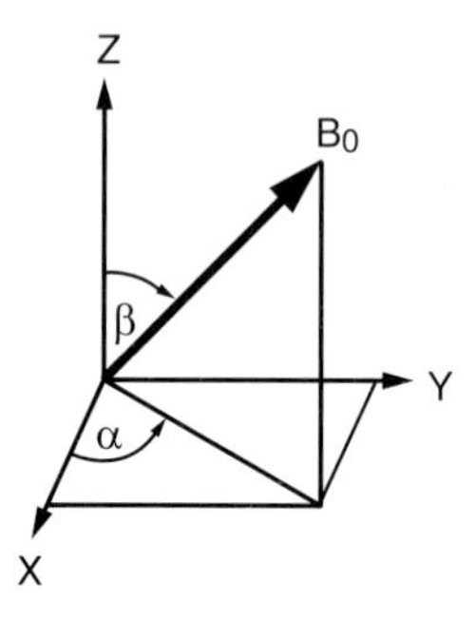

FIG. 3.1.2 Definition of the Euler angles α and β in terms of the polar angles specifying the orientation of the magnetic field in the principal axes system XYZ of the coupling tensor.

case also (Fig. 3.1.1(b)). For the magnetic shielding of a nucleus with $I = \frac{1}{2}$, however, only one line is observed (Fig. 3.1.1(c)).

For aliphatic deuterons, the quadrupole coupling tensor typically possesses *axial symmetry* ($\eta \cong 0$). As a consequence its Z principal axis usually coincides with the axis of the C–^{2}H bond, and the angle between this bond axis and the magnetic field determines the separation of the resonance frequencies (Fig. 3.1.1(a)). The coupling tensor of the dipole–dipole interaction is always axially symmetric ($\eta = 0$), and its Z principal axis is aligned with the internuclear vector (Fig. 3.1.1(b)). The orientation of the principal axis of the magnetic shielding tensor, however, depends on the anisotropy of the electron density surrounding the nucleus and is determined by the molecular geometry (Fig. 3.1.1.(c)). Here, generally, η differs from zero [Kut1, Meh1, Mor1, Vee1].

3.1.4 Orientation dependence of the resonance frequency

From the Hamilton operator (3.1.22), the general form of dependence of the resonance frequency on the orientation of the magnetic field in the principal axes system of the coupling tensor is calculated to be

$$\Omega = \omega - \omega_{\mathrm{L}} = \frac{\delta}{2}\left(3\cos^2\beta - 1 - \eta\sin^2\beta\cos 2\alpha\right). \tag{3.1.23}$$

Here Ω denotes the angular-dependent resonance frequency centred at the isotropic mean value or the Larmor frequency ω_{L}.

In *powders* and *polycrystalline materials* all orientations arise with equal probability. For $\eta = 0$ the powder average of the resonance frequency (3.1.23) leads to a spectrum (Fig. 3.1.3(a)) with the lineshape function

$$S(\Omega) = \left[\sqrt{3}\Delta\sqrt{\frac{2\Omega}{\Delta} + 1}\right]^{-1} \tag{3.1.24}$$

Following (3.1.23) Ω can assume the values $-\Delta/2 \leq \Omega \leq \Delta$ for $\eta = 0$. Therefore, the powder spectrum exhibits a singularity at $\Omega_{\perp} = -\Delta/2$, which, by (3.1.23), corresponds

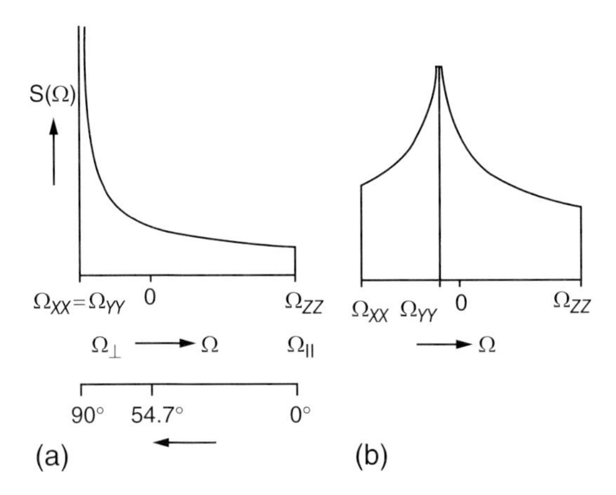

FIG. 3.1.3 Solid-state wideline spectra. (a) Powder spectrum as isotropic average for an axially symmetric coupling tensor ($\eta = 0$). The resonance frequency is related to the orientation of the coupling tensor by (3.1.23) and can serve as a protractor for molecular orientations relative to the magnetic field. (b) Powder spectrum for $\eta = 2/3$.

to the angle $\beta = 90°$ between the magnetic field $\boldsymbol{B}_0$ and the Z-axis of the coupling tensor. For, this angle $\boldsymbol{B}_0$ is in the XY plane of the principal axes system of the coupling tensor. The cut-off at $\Omega_\parallel = \Delta$ corresponds to an orientation of $\boldsymbol{B}_0$ parallel to the Z principal axis. Thus from the resonance frequency of a magnetization component of a wideline spectrum the orientation of the magnetic field $\boldsymbol{B}_0$ relative to the principal axis Z of the coupling tensor can be read off. When the orientation of the coupling tensor in the molecule-fixed coordinate frame is known the *wideline resonance* can be understood as a protractor for molecular orientations (Fig. 3.1.3).

For $\eta > 0$, a lineshape like the one depicted in Fig. 3.1.3(b) results. From the high- and low-frequency cut-offs and from the singularity, the principal values Ω_{XX}, Ω_{YY}, and Ω_{ZZ} of the coupling tensor can be obtained. At the same time, the largest principal value δ and the asymmetry parameter η can be determined by using (3.1.18) and (3.1.19).

In general, the Hamilton operator $\mathbf{H}$ applicable to an experimentally observed resonance is the sum (3.1.1) of operators $\mathbf{H}_\lambda$ of different interactions. The lineshape, therefore, is the result of all spin interactions. For observation of just one dominant interaction, special techniques need to be applied such as *isotope enrichment*, homonuclear multi-pulse and heteronuclear high-power decoupling. Nevertheless, in ^{13}C NMR, for instance, despite high-power ^{1}H decoupling an overlap of chemical shielding powder spectra centred at different chemical shifts is observed in most cases without site-specific isotope enrichment.

3.2 WIDELINE NMR

Until MAS, multi-pulse, and Fourier techniques became available on a commercial basis, *wideline NMR* was the most frequently used method of solid-state NMR spectroscopy.

Originally, this meant the measurement and evaluation of structureless, wideline solid-state NMR spectra in terms of frequency averages or *moments* [Abr1, Sli1]. Today, however, the lineshapes of the anisotropic chemical shielding, the dipole–dipole, and the quadrupole interaction can be isolated by isotope enrichment and by homo- and heteronuclear decoupling, so that a single interaction dominates the shape of the wide-line spectrum. Then a *lineshape analysis* reveals far more detailed information about molecular order and mobility than a simple moment analysis [Bec2, Hoa1, Jel1, Mül1, Mül2, Sch9, Spi1, Spi2].

For ^{1}H in organic solids, a wideline-NMR lineshape analysis is an exception. Even if the strong dipole–dipole interaction among the protons is sufficiently reduced by multi-pulse sequences, the chemical shift range of ^{1}H is so small that the powder spectra of the anisotropic chemical shielding, which are centred at different chemical shifts, strongly overlap. This overlap also applies to ^{13}C spectra. But here, the homonuclear dipole–dipole coupling is often negligible because of the low natural abundance (1%) of ^{13}C, and the heteronuclear dipole–dipole coupling can be effectively eliminated by high-power ^{1}H decoupling (cf. Section 3.3.2 below). The overlap of wideline resonances can be reduced by the use of 2D techniques [Ern1, Sch9] or fast sample rotation at an angle different from the magic angle [Ste1] (cf. Section 3.3.3 below). The most straight forward approach to wideline spectra of a single interaction is also the most labour intensive one. It requires site-selective isotope enrichment. The best known representative of this approach is ^{2}H wideline NMR spectroscopy [Jel1, Mül1, Mül2, Spi1, Spi2]. Here the quadrupole coupling dominates all other interactions, and wideline spectra can be measured by relatively simple techniques on modern solid-state NMR spectrometers.

Because ^{2}H possesses a spin with $I = 1$, the deuteron resonance for a single molecular orientation exhibits two lines, which correspond to the two transitions with $\Delta m = \pm 1$ in the energy level diagram (Fig. 3.2.1). The lines appear symmetric with respect to the Larmor frequency ω_L,

$$\omega_{\pm} = \omega_L \pm (\delta/2)[3\cos^2\beta - 1 - \eta\,\sin^2\beta\cos(2\alpha)]. \tag{3.2.1}$$

The ^{2}H wideline spectrum, consequently, consists of two overlapping powder spectra, which possess mirror symmetry with respect to ω_L . This is the *Pake doublet*. In addition to the two observable *single-quantum transitions* ($|\Delta m| = 1$), a *double-quantum transition* ($|\Delta m| = 2$) exists which can be detected indirectly. Its transition frequency is the sum of both single quantum transitions,

$$\omega_{2Q} = \omega_+ + \omega_- = 2\omega_L. \tag{3.2.2}$$

With the help of (3.2.1), it is readily seen that ω_{2Q} does not depend on the orientation angles α and β of the principal axes system. Therefore, the transition corresponds to a narrow resonance which can be favourably exploited for space encoding in imaging.

3.2.1 Molecular order

Molecular order is determined conventionally by X-ray diffraction and in some other cases by neutron scattering. Highly ordered structures lead to sharp reflexes, and weakly

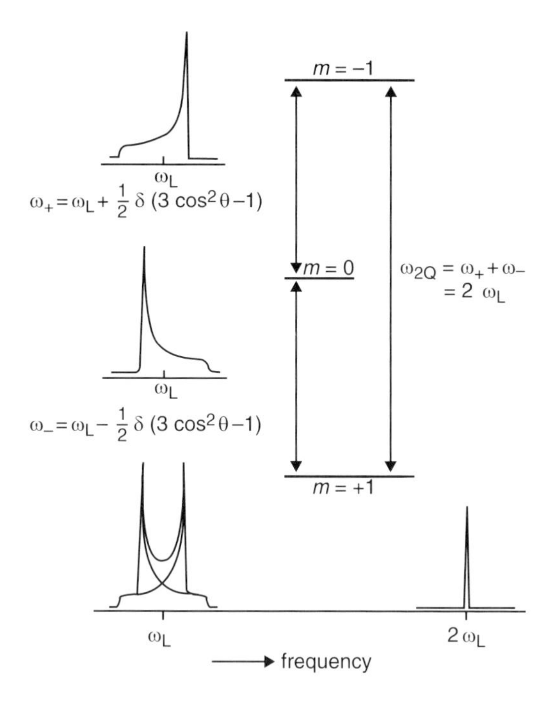

FIG. 3.2.1 [Gün2] Energy level diagram and powder spectra of the deutron. The wideline spectrum consists of two superimposed powder spectra. The double-quantum resonance at frequency ω_{2Q} appears independent of molecular orientation.

ordered ones to diffuse reflexes. Diffraction methods are, therefore, primarily used for analysis of high order. For the characterization of molecular order by solid-state NMR lineshape analysis [Hen1, Spi3], such a restriction does not apply. But compared to scattering techniques, the strength of NMR methods is the analysis of weak order. It can arise in many materials, for instance, during processing by straining, plastic deformation, and moulding.

The orientational distribution function

Molecular order is described by the *orientational distribution function* $P(\theta)$ [Mcb1]. This is the probability density of finding a preferential direction $\boldsymbol{n}$ in the sample under an angle θ in a molecule-fixed coordinate frame (Fig. 3.2.2(a)). For simplicity, macroscopically uniaxial samples with cylindrically symmetric molecules are considered. Then, one angle is sufficient to characterize the orientational distribution function. In practice, not the angle θ itself but its cosine is used as the variable and for weak order the distribution function is expanded into *Legendre polynomials* $P_l(\cos\theta)$,

$$P(\cos\theta) = \sum_{l=0}^{\infty} \chi_l P_l(\cos\theta). \tag{3.2.3}$$

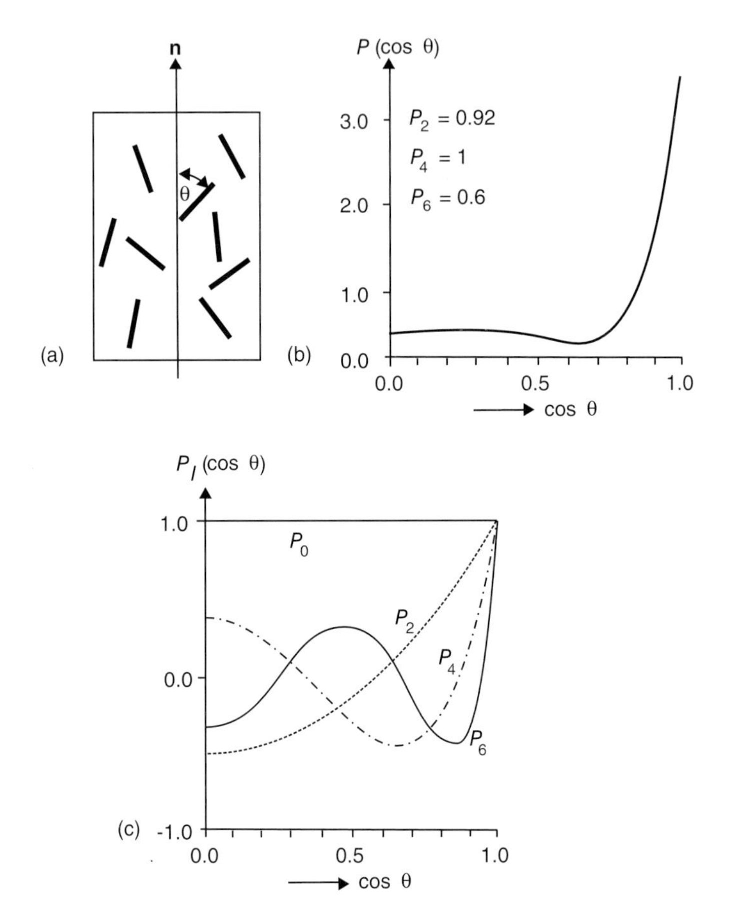

FIG. 3.2.2 The orientational distribution function. (a) Definition of the orientation angle θ. (b) Angular distribution of $\cos\theta$. (c) Legendre polynomials.

The expansion coefficients χ_l are related to the moments $P_l = \langle P_l(\cos\theta)\rangle$ of the distribution function according to

$$\chi_l = \frac{(2l+1)\langle P_l(\cos\theta)\rangle}{8\pi^2}, \tag{3.2.4}$$

where $\langle\cdots\rangle$ denotes the average weighted by the orientational distribution function. For an isotropic distribution $\langle P_0\rangle = 1$ and $\langle P_l\rangle = 0$ for all $l > 0$. For a completely ordered sample, $\langle P_l\rangle = 1$. $\langle P_0\rangle$ quantifies the isotropic part of the orientational distribution function. $\langle P_2\rangle$ is the *second moment* or the *degree of order*. With increasing order, higher values of $\langle P_l\rangle$ become significant. Because the expansion treats molecular order as a perturbation of the isotropic state, it converges for weak order. To achieve convergence at high order, the orientational distribution function may be expanded, for instance, into planar or conical distributions [Spi3].

Many methods for determination of molecular order can only measure $\langle P_2 \rangle$ [Mcb1, War1]. Among them are measurements of the infrared dichroism and of the refractive index. $\langle P_2 \rangle$ and $\langle P_4 \rangle$ can be determined by Raman fluorescence depolarization and from ^{1}H wideline NMR spectra. However, X-ray diffraction and lineshape analysis of resonances from isolated NMR interactions can also provide higher moments.

Determination of the orientational distribution function

The orientational distribution function $P(\cos\theta)$ enters the shape of the wideline spectrum $S(\Omega)$ in a slightly hidden way. The angular dependence of the resonance frequency is given by (3.1.23) via the orientation of the magnetic field in the principal axes system XYZ of the coupling tensor (cf. Fig. 3.1.2), while the orientational distribution function specifies the distribution of the preferential direction $\boldsymbol{n}$ in a molecule-fixed coordinate frame (Fig. 3.2.2(a)). Figure 3.2.3 shows the relationship between the different coordinate frames and the definition of the relative orientation angles.

The *orientational distribution function* $P'(\cos\beta)$ of axially symmetric coupling tensors ($\eta = 0$) in the laboratory frame can be read directly from the NMR spectrum $S(\Omega)$. Its expansion coefficients χ'_l can be transformed to those, χ_l, of the orientational distribution function $P(\cos\theta)$ of the molecules in the sample fixed frame by using the known orientation angles Θ and Φ of the *principal-axes frame* in the molecule-fixed coordinate frame, and the orientation angle β_0 of the sample frame in the laboratory frame. By convention, the preferential axis $\boldsymbol{n}$ of the sample is parallel to the z_s-axis of the sample frame.

The wideline NMR spectrum $S(\Omega)$ can be written as a convolution of the resonance frequency with the orientational distribution $P'(\cos\beta)$ [Wef1],

$$S(\Omega') = \int_0^\pi \delta(\Omega' - \Omega(\cos\beta))\, P(\cos\beta)\sin\beta\, d\beta, \tag{3.2.5}$$

where $\Omega(\beta) = (\delta/2)(3\cos^2\beta - 1)$ is the angular-dependent resonance frequency (3.1.23) for an axially symmetric coupling tensor ($\eta = 0$). Because of the dependence on

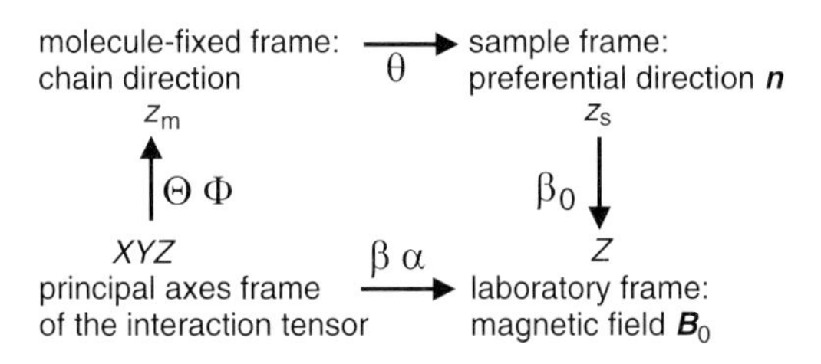

FIG. 3.2.3 Relationships between coordinate systems for the description of molecular order. The orientation of the laboratory coordinate system in the principal axes system of the coupling tensor determines the angular dependence of the resonance frequency. The orientations of the preferential sample direction in a molecule-fixed coordinate frame determines the orientational distribution function.

$\cos^2 \beta$ only angles β can be discriminated which are found in the first quadrant between 0 and $\pi/2$,

$$S(\Omega') = -\frac{1}{2}\int_0^1 \delta\left(\Omega' - \frac{\delta}{2}(3\cos^2\beta - 1)\right)[P(\cos\beta) + P(-\cos\beta)]\,\mathrm{d}\cos\beta. \quad (3.2.6)$$

Thus, only the even moments $\langle P_l \rangle$, $l = 0, 2, 4, \ldots$, of the orientational distribution function $P(\cos\beta)$ can be determined by NMR. Using an expansion similar to (3.2.3), one arrives at

$$S(\Omega') = \sum_{l=0,2,4,\ldots}^{\infty} \chi_l' S_l(\Omega'). \quad (3.2.7)$$

This equation defines the Legendre subspectra $S_l(\Omega')$,

$$S_l(\Omega') = \int_0^1 \delta\left(\Omega' - \frac{\delta}{2}(3\cos^2\beta - 1)\right) P_l(\cos\beta)\,\mathrm{d}\cos\beta. \quad (3.2.8)$$

The expansion coefficients χ_l and, therefore, also the moments $\langle P_l \rangle$ (cf. eqn (3.2.4)) of the desired orientational distribution function $P(\cos\theta)$ are related to the expansion coefficients χ_l' of (3.2.7) by the orientation angles Θ and β_0 (Fig. 3.2.3) [Hen1],

$$\chi_l' = \frac{2l+1}{8\pi^2}\langle P_l(\cos\theta)\rangle P_l(\cos\Theta) P_l(\cos\beta_0). \quad (3.2.9)$$

Thus, from the expansion coefficients χ_l' of a Legendre *subspectral analysis* according to (3.2.8) the moments $\langle P_l(\cos\theta)\rangle$ of the orientational distribution function can be determined.

The validity of (3.2.9) is restricted to the symmetries mentioned above, that is to cylindrical molecules, macroscopically uniaxial samples, and $\eta = 0$. For many samples, these conditions are fulfilled when using ^{2}H NMR, because the quadrupole coupling tensor of aliphatic deuterons is often found to be axially symmetric. In ^{13}C wideline NMR, the anisotropy of the magnetic shielding is used. Here the angular resolution is lower, and the calculation has to be extended to include $\eta > 0$ [Hen1]. In combination with MAS (cf. Section 3.3), the Legendre *subspectral analysis* has been used successfully for the determination of molecular order in partially ordered polymers [Har1].

As an example of a Legendre subspectral analysis, orientation-dependent NMR spectra of a ^{13}C-labelled liquid crystalline polymer film are shown in Fig. 3.2.4 [Wie1]. The mesogenic sidechain of this polymer contains azobenzene as a dye, which changes its conformation from *trans* to *cis* upon irradiation with light. This transition is associated with a change in order of the liquid crystalline matrix. The resonance of the ^{13}C-labelled cyano group in the phenyl benzoate of the mesogen has been analysed in terms of Legendre subspectra. Switching of the dye by irradiation with light reduces the Legendre moments P_2, P_4, and P_6. This indicates a reduction in molecular order after switching.

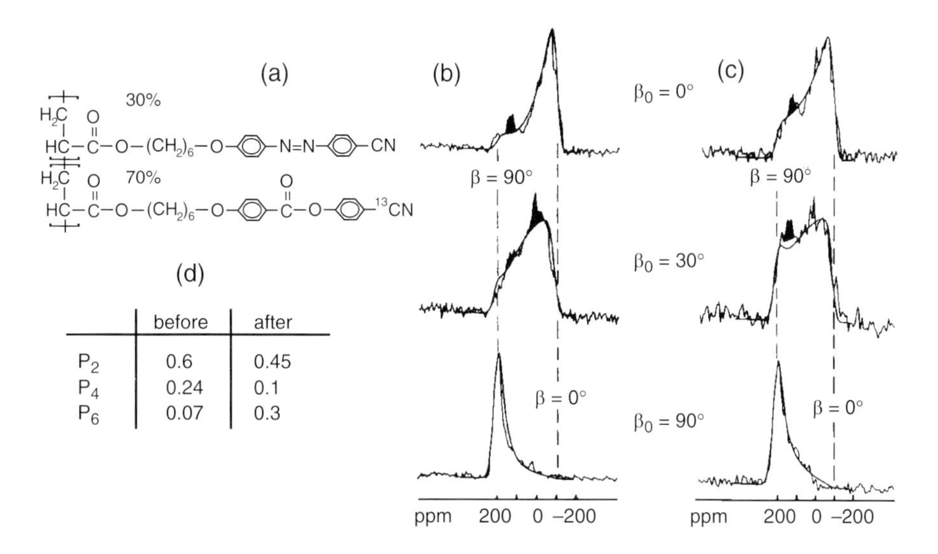

FIG. 3.2.4 [Wie1] Orientation dependent ^{13}C NMR spectra of the liquid crystalline sidechain polymer (a) before (b) and after (c) irradition with light. The shaded regions are ^{13}C resonances from nuclei in natural abundance. The angle β_0 measures the orientation of optical axis of the film relative to the magnetic field $\boldsymbol{B}_0$. (d) Moments of the orientational distribution function before and after irradition.

3.2.2 Molecular reorientation

NMR can provide detailed information about type and timescale of *slow molecular motion*. Slow molecular reorientational processes can be probed by making use of the angular dependence (3.1.23) of the resonance frequency. Slow molecular translation can be investigated with NMR by measuring the particle diffusion and flow in magnetic field gradients (cf. Section 7.2.6) [Cal1, Cal2, Kär1, Kim1, Sti1].

The correlation time

Molecular motions generally appear incoherent and are described by a normalized stochastic process $f(t)$. One important quantity to characterize stationary stochastic processes is the *auto-correlation function* $a(\sigma)$ (cf. Section 4.3),

$$a(\sigma) = \lim_{T\to\infty} \frac{1}{T} \int_0^T f(t) f(t-\sigma)\,\mathrm{d}t = \exp\left\{-\frac{\sigma}{\tau_\mathrm{c}}\right\}. \tag{3.2.10}$$

In many cases the auto-correlation function is an exponential function with a time constant τ_c, which is called the *correlation time* of the process. Following the definition of the correlation time for an exponential correlation function (3.2.10) the correlation time for a nonexponential correlation function is defined as

$$\tau_\mathrm{c} = \int_0^\infty a(\sigma)\,\mathrm{d}\sigma. \tag{3.2.11}$$

In polymeric materials distributions of correlation times are observed for molecular motions. Such a distribution can be interpreted in two ways. Either different molecules exhibit different correlation times during the time of observation (heterogeneous distribution) or a single molecule exhibits different correlation times in different observation intervals (homogeneous distribution).

Solid-echo spectra

For reorientations with correlation times in the range of the inverse spectral width of the powder spectrum, temperature-dependent changes of the lineshape are observed which are characteristic of the motional process [Jel1, Mül1, Spi1, Spi2]. As an example, Fig. 3.2.5 shows ^{2}H NMR spectra for different motional mechanisms and different correlation times [Mül1]. However, such wideline spectra cannot be readily measured with single-pulse excitation, because the beginning of the FID will decay within the

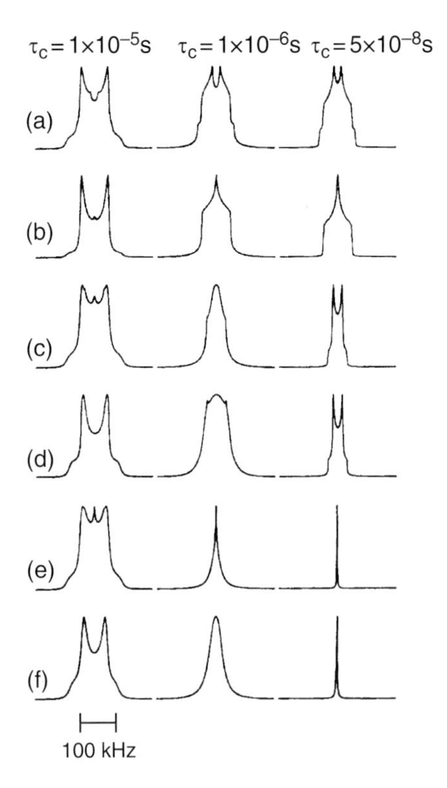

FIG. 3.2.5 [Mül1] Deuteron wideline spectra for different motional mechanisms and times τ_c. The angle between the axis of rotation and C$-^2$H bond (principal axis Z of the quadrupole coupling tensor) is denoted by θ. (a) Twofold jump with $\theta = 60°$ (b) Twofold jump with $\theta = 180°$ (flips of p phenylene). (c) Three-fold jump with $\theta = 109°$ (rotation of a methyl group). (d) Rotational diffusion on a cone $\theta = 109°$. (e) Tetrahedral jump. (f) Isotropic rotational diffusion.

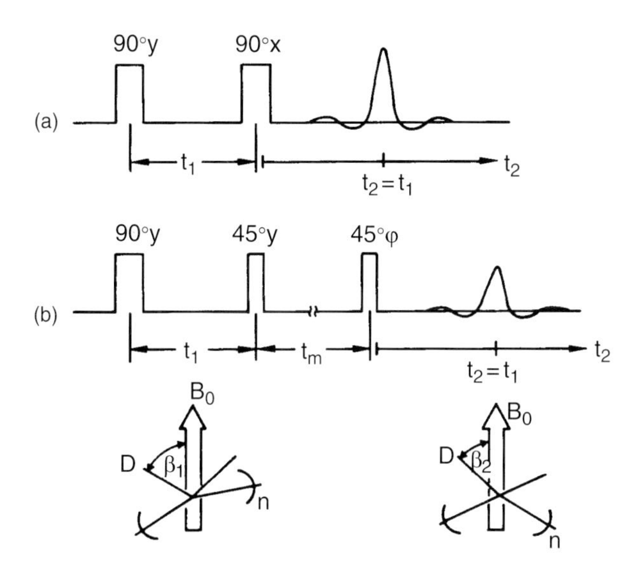

FIG. 3.2.6 Pulse sequences for magnetization echoes of spin-1 nuclei in the solid state. (a) Quadrupole or solid echo. (b) Jeener–Broekaert or alignment echo.

receiver deadtime, and only the end can be acquired. Then, after Fourier transformation, only the narrow horns of the singularities ($\beta = 90^\circ$ in Fig. 3.1.3(a)) will fully be observed but not the broad shoulders ($\beta = 0^\circ$ in Fig. 3.1.3(a)).

To bypass receiver deadtime effects, wideline spectra are derived by Fourier transformation of the decay of an echo. By use of the Hahn echo and the stimulated echo (Section 2.2.1), wideline spectra of ^{13}C and other spin-$\frac{1}{2}$ nuclei can be measured, for example, but not the spectra of dipolar coupled spins and of quadrupolar nuclei like ^{2}H. The magnetization of nuclei with spin $I = 1$ can be refocused by the *quadrupole echo* or the *solid echo*, and by the *Jeener–Broekaert echo* or the *alignment echo* [Sli1] (Fig. 3.2.6).

The quadrupole and the Jeener–Broekaert echo

A *quadrupole echo* is generated similar to a Hahn echo by two pulses which are separated by half the echo time $t_1 = t_E/2$ (Fig. 3.2.6(a)). However, the second pulse is a 90° and not a 180° pulse, and it is shifted in phase by 90° with respect to the first. The quadrupole echo appears at a time $t_2 = t_E/2$ after the second pulse. The spectra shown in Fig. 3.2.5 have been simulated for the quadrupole echo technique with acquisition of the echo decay during $t_2 - t_1$. Clearly, the lineshape strongly depends on type and time scale of the motion. In this way, molecular reorientation with correlation times in the range of 10^{-8} s $< \tau_c < 10^{-4}$ s can be characterized by ^{2}H NMR [Lau1, Weh1]. Faster motion with correlation times in the range of 10^{-12} s $< \tau_c < 10^{-8}$ s can be investigated by measurements of the spin–lattice relaxation time T_1. For investigations of slower processes with correlation times in the range of 10^{-4} s $< \tau_c < 10$ s the *Jeener–Broekaert echo* [Jee1] can be employed [Mül1, Spi1, Spi4].

The Jeener–Broekaert echo is stimulated with three pulses (Fig. 3.2.6(b)). The second and the third pulse are each a 45° pulse, whereby the phase of the second pulse is shifted against that of the first pulse by 90°. In addition to double-quantum coherences the second pulse generates a long-lived magnetization state along the z-axis which is similar to z magnetization, but exhibits no net magnetization. This state is best explained by considering two spins with $I = \frac{1}{2}$, which are coupled by the dipole–dipole interaction. In this state the z projection of one spin is parallel to the z-axis, that of the other is antiparallel. Thus the spins produce no common polarization but an antiparallel order, that is, they exhibit *alignment*. The alignment state decays with relaxation times T_{1D} and T_{1Q}, for *dipolar order* and *quadrupolar order*, respectively. The magnitude of these relaxation times is comparable to that of T_1.

At sufficiently long time after the second pulse, the double-quantum coherences will have decayed, and only spin alignment remains. Then, the memory of the magnetization to the phase of the first two pulses has been lost. Therefore, the phase ϕ of the third pulse is arbitrary. It transforms the spin alignment into single-quantum coherences, which refocus to form the Jeener–Broekaert echo after a time $t_2 = t_1$ following the third pulse. The timescale of the molecular motion which is interrogated by the alignment echo is determined by the separation t_m between the second and the third pulse. It is limited at long times by the alignment relaxation times T_{1D} or T_{1Q} and at short times by the separation $t_1 = t_E/2$ of the first two pulses.

2D exchange spectroscopy

Contrary to the description of echoes in Fig. 2.2.7, two time variables are introduced in Fig. 3.2.6. This has been done because the echo is a fundamental phenomenon upon which many *two-dimensional NMR* techniques are based, and the nomenclature used here is that of 2D spectroscopy [Ern1]. The time t_1 is the *evolution time*, t_m the *mixing time*, and t_2 the *detection time*. In fact, the Jeener–Broekaert echo can be executed as a 2D experiment by repeating the data acquisition with a systematic variation of the evolution time t_1. The data acquired as a function of $t_2 - t_1$ are written into a matrix, the rows of which are labelled by t_1. The acquired alignment signal is modulated by the sine component of the magnetization, which exists during the evolution time t_1 between the first and the second pulse. The cosine part can be measured with the stimulated echo $90^\circ_x - t_1 - 90^\circ_x - t_m - 90^\circ_x - t_2$, which creates true longitudinal magnetization during the mixing time t_m . For generation of purely absorptive 2D spectra, both components have to be acquired, and the respective data matrices need to be combined [Sch1]. 2D Fourier transformation over the evolution time t_1 and the detection time t_2 produces a 2D spectrum with frequency axes ω_1 and ω_2.

As an example, Fig. 3.2.7 depicts ^{2}H–2D spectra of polypropylene obtained in this way [Sch2]. Signals are identified which spread over the 2D frequency plane. These signals characterize the reorientation of the molecules during the mixing time t_m. Because detailed balance requires that frequency components are exchanged by reorientation, this type of 2D spectrum is referred to as a *2D exchange spectrum*.

Isotactic polypropylene is semicrystalline. In the crystalline regions the molecules form a 3_1 helix, that is three monomer units form one turn of a helix. This helix performs rotational jumps around its own axis. After one jump, each monomer occupies the place of its next neighbour before the jump. Molecular reorientation by discrete jumps

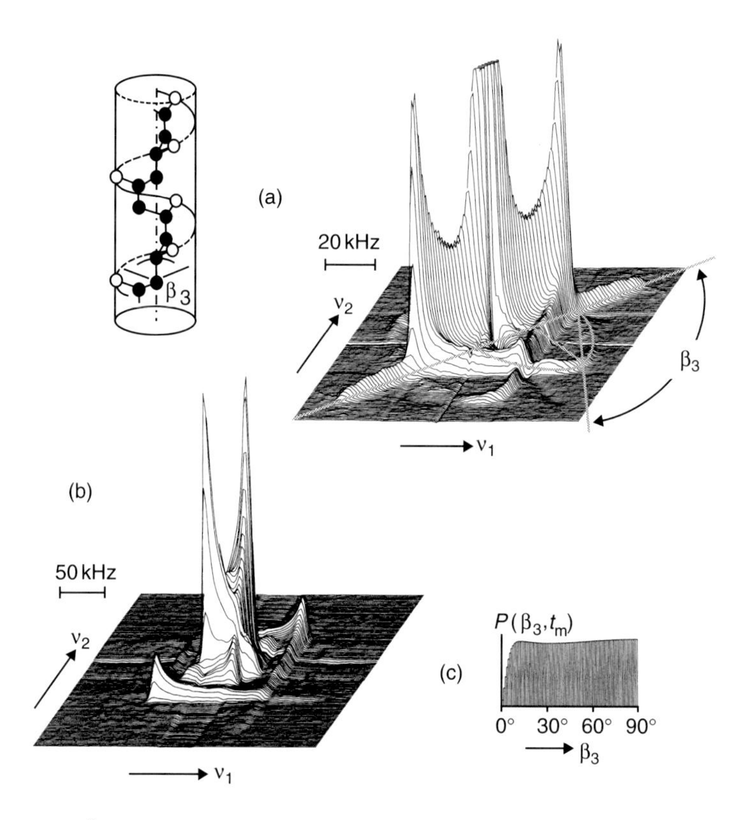

FIG. 3.2.7 2D – ^{2}H exchange spectra of polypropylene (PP). (a) Methyl deuterated isotactic PP; $t_m = 150$ ms, $T = 380$ K. The helix molecules in crystalline regions perform discrete reorientational jumps. The principal axis Z of the axially symmetric quadrupole coupling tensor of the methylene groups is parallel to the axis of the C–C^{2}H$_3$ bond. The reorientation angle β_3 of this axis can be read directly from the 2D spectrum. (b) Perdeuterated atactic PP; $t_m = 25$ ms, $T = 275$ K. The powder spectra of the deuterons at different chemical shifts overlap. The material is completely amorphous. The molecular segments undergo reorientation by isotropic rotational diffusion. (c) Distribution of reorientation angles for atactic PP established within the mixing time t_m. Adapted from [Blü4] with permission from Wiley-VCH.

leads to characteristic ridges in the 2D plane, which assume the shape of an ellipse for axially symmetric coupling tensors. From the excentricity of the ellipse, the reorientation angle β_3 can be read off directly, by which the Z-axis of the coupling tensor has been reorienting during the mixing time t_m (Fig. 3.2.7(a)). Diffusive small angle reorientation, on the other hand, leads to broad exchange signals in the 2D plane. Such motion is characteristic for amorphous polymers above the glass transition. This is illustrated for atactic polypropylene in Fig. 3.2.7(b). By numerical simulation of the 2D spectrum the distribution of reorientation angles (Fig. 3.2.7(c)) can be extracted from the measured data [Wef1]. The thermally driven reorientational motion can be replaced by mechanical sample reorientation leading to intriguing 2D experiments in wideline solid-state NMR [Bec2, Huj1].

Statistical theory

For simplicity, it is assumed that the evolution time t_1 and the detection time t_2 are short compared to the mixing time t_m . Then it is a good approximation to say, that the molecules reorient primarily during the mixing time. When relaxation during the mixing time is neglected, the combined 2D time-domain signal $s(t_1, t_m, t_2)$ measured with the stimulated and the Jeener–Broekaert echoes can be written as ensemble average $\langle \cdots \rangle$ over all molecules in the sample,

$$\begin{aligned} s(t_1, t_m, t_2) &= \langle \exp\{i\omega_1 t_1\} \exp\{i\omega_2 t_2\} \rangle \\ &= \iint \exp\{i\omega_1(\Gamma_1)t_1\} \exp\{i\omega_2(\Gamma_2)t_2\} W(\Gamma_1, \Gamma_2; t_m)\, d\Gamma_1\, d\Gamma_2. \end{aligned} \qquad (3.2.12)$$

Here Γ_1 and Γ_2 denote the orientation angles (α, β) of the magnetic field $\boldsymbol{B}_0$ in the principal axes frame of the interaction tensor (Fig. 3.1.2) during the evolution time t_1 and the detection time t_2, respectively. The quantity which characterizes the reorientation process is the *combined probability density*

$$W(\Gamma_1, \Gamma_2; t_m) = W(\Gamma_1) P(\Gamma_1 \mid \Gamma_2, t_m), \qquad (3.2.13)$$

to find a frequency ω_1 in the interval $[\omega_1, \omega_1 + d\omega_1]$ during the evolution time t_1 and a frequency ω_2 in the interval $[\omega_2, \omega_2 + d\omega_2]$ during the detection time t_2. It can be written as the product of the *probability density* $W(\Gamma_1)$ to find a molecule with orientation Γ_1 and the *conditional probability density* $P(\Gamma_1 \mid \Gamma_2, t_m)$, that a molecule is found with orientation Γ_1 at the beginning of the mixing time t_m and with orientation Γ_2 at the end of the mixing time [Abr1]. From the conditional probability density $P(\Gamma_1 \mid \Gamma_2, t_m)$ with initial and final orientations Γ_1 and Γ_2 , the distribution $P(\Gamma_3, t_m)$ of reorientation angles Γ_3 can be calculated [Wef1]. For axially symmetric coupling tensors this angle distribution is a function of the angle β_3 only, which specifies the reorientation of the angle between the principal axis Z of the coupling tensor and the magnetic field. The angle distribution $P(\beta_3, t_m)$ is depicted in Fig. 3.2.7(c) for the rotational diffusion in atactic polypropylene corresponding to a mixing time of $t_m = 25$ ms and a temperature of $T = 275$ K. For a unique characterization of the motion the angle distribution must be determined for different mixing times and at different temperatures. The method is not restricted to ^{2}H NMR. It can also be applied to other nuclei like ^{31}P and ^{13}C [Blü1, Hag1].

Computer simulations

A detailed quantitative analysis of timescale and angle distribution requires numerical simulation of the spectra acquired with the solid and the alignment echoes. To this end the time-domain signal is written as a vector. The components of which correspond to the different molecular orientations [Abr1, And1],

$$s(t_1, t_m, t_2) = \boldsymbol{E}^t \exp\{(\boldsymbol{\Pi} + i\boldsymbol{\Omega})t_2\} \exp\{\boldsymbol{\Pi} t_m\} \exp\{(\boldsymbol{\Pi} \pm i\boldsymbol{\Omega})t_1\} \boldsymbol{W}. \qquad (3.2.14)$$

Here $\boldsymbol{\Pi}$ denotes the *exchange matrix* associated with the reorientation process, $\boldsymbol{\Omega}$ is a diagonal matrix of the angle dependent resonance frequencies, $\boldsymbol{W}$ is a vector which determines the probability density to find a molecule at orientation $\boldsymbol{\Gamma}$ within the interval $[\boldsymbol{\Gamma}, \boldsymbol{\Gamma} + \mathrm{d}\boldsymbol{\Gamma}]$, and $\mathbf{E}^{\mathrm{t}}$ is a row vector, the elements of which are all equal to one.

Equation (3.2.14) is the starting point for numerical simulation of dynamic wideline NMR spectra. With the Jeener–Broekaert echo the imaginary part of the exponential depending on t_1 is measured, and with the stimulated echo the real part is measured. The lineshapes of solid-echo spectra follow from (3.2.14) with $t_{\mathrm{m}} = 0$.

3.3 HIGH-RESOLUTION SOLID-STATE NMR

Although in solids the orientation dependence of the resonance frequency, as well as the simultaneous presence of several interactions, gives rise to wide, overlapping resonances, by use of special techniques high-resolution spectra can be measured, which exhibit narrow lines at the positions of the isotropic chemical shifts similar to liquid-state spectra. This is true, in particular, for spectra of rare nuclei with spin $\frac{1}{2}$ like ^{13}C and ^{29}Si. As an example, Fig. 3.3.1 shows ^{13}C spectra of a benzil derivative [Yan1]. A signal in response to conventionally pulsed excitation cannot be detected for the solid even after 12 h acquisition time in the presence of ^{1}H *decoupling* (a), while in the same time an excellent signal can be acquired from a dilute solution (e) under similar conditions. The situation improves if magnetization is transferred from ^{1}H to ^{13}C by *cross-polarization* (CP). Then a wideline spectrum is obtained already after only 1.2 h (b). When decreasing the ^{1}H–^{13}C dipole–dipole coupling by stronger ^{1}H decoupling, the spectrum gains structure (c). This structure results from overlap of the wideline resonances of the *chemical shielding anisotropies*, which are centred at different chemical shifts. For spins $\frac{1}{2}$, the wideline resonances can be collapsed into narrow lines at the isotropic means by fast *magic angle spinning* (*MAS*) (d) [And3, Low1, Mar1]. This denotes mechanical rotation of the sample around an axis which is inclined by the *magic angle* of 54.7° against the magnetic field $\mathbf{B}_0$. It is associated with a substantial increase in sensitivity. The linewidths in such CP MAS spectra can be just a few hertz. Then, the resolution on the chemical shift scale is comparable to that of liquid-state spectra (e).

Heteronuclear decoupling and CP MAS are techniques which are primarily applied to rare nuclei with spin $\frac{1}{2}$. For abundant spin-$\frac{1}{2}$ nuclei such as ^{1}H the homonuclear dipole–dipole coupling dominates all other interactions, so that broad, unstructured resonances with linewidths of up to 100 kHz are observed. The strong homonuclear dipole–dipole interaction can, however, be reduced by *multi-pulse irradiation* [Hae1, Hae2, Meh1, Wau1]. This technique is a type of averaging by rotation in spin space instead of in real space by MAS. Multi-pulse excitation can be tailored for selective averaging of different interactions. Like MAS it is used for imaging in solids. But the so-called high-resolution ^{1}H spectra require ultra-fast MAS or the combination of MAS with homonuclear multi-pulse decoupling. Combined rotation and multi-pulse spectroscopy is called *CRAMPS* [Bur2, Ger2, Sch6]. Nevertheless, in comparison to ^{1}H spectra of liquids the resolution of solid-state ^{1}H CRAMPS spectra of rigid solids is still low. In rigid amorphous polymers only up to three resonances can be resolved: signals from aliphatic, carbonyl, and aromatic segments. The most promising route to probe dipolar and J connectivities in solid-state NMR and to high-resolution NMR of quadrupolar

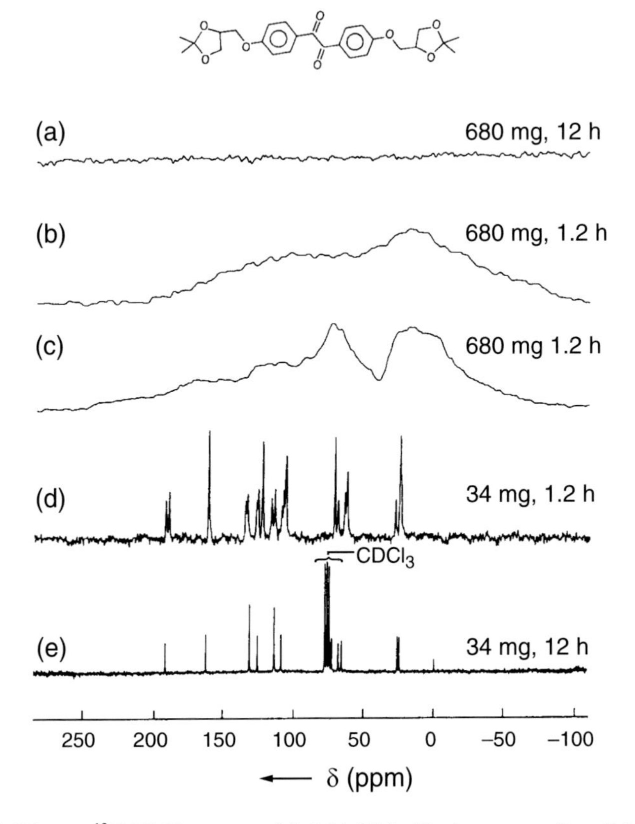

FIG. 3.3.1 Solid-state ^{13}C NMR spectra of 4,4′-bis[(2,3-dihydroxypropyl)oxyl]-benzil: (a) 680 mg sample, 60° flip angle excitation every 10 s at 6 kHz ^{1}H decoupling; (b) same as (a) but with cross-polarization and a repetition time of 1 s between excitation pulses; (c) same as (b) but 43 kHz ^{1}H decoupling; (d) 34 mg sample, same parameters as in (c) but with magic angle spinning; (e) 34 mg sample dissolved in $CDCl_3$, same acquisition parameters as in (a). Acquisition times: 12 h for (a) and (e), 1.2 h for (b)–(d). Adapted with permission from [Yan1]. Copyright 1982 American Physical Society.

nuclei is *multi-quantum MAS spectroscopy* [Fer1, Fri1, Fry1]. In this section, only those concepts of high-resolution solid-state NMR are discussed, which have been explored in NMR imaging.

3.3.1 Cross-polarization

When observing rare nuclei like ^{13}C and ^{29}Si, a significant gain in signal-to-noise ratio can be achieved if magnetization is transferred to the rare spins S from abundant spins I like ^{1}H [Pin1]. The gain can be twofold: a higher polarization of the S spins is achieved, and the often longer relaxation times of the S spins are circumvented in magnetization build-up. *Cross-polarization* is routinely used in magic-angle spinning NMR of rare nuclei [Mic1, Sli1].

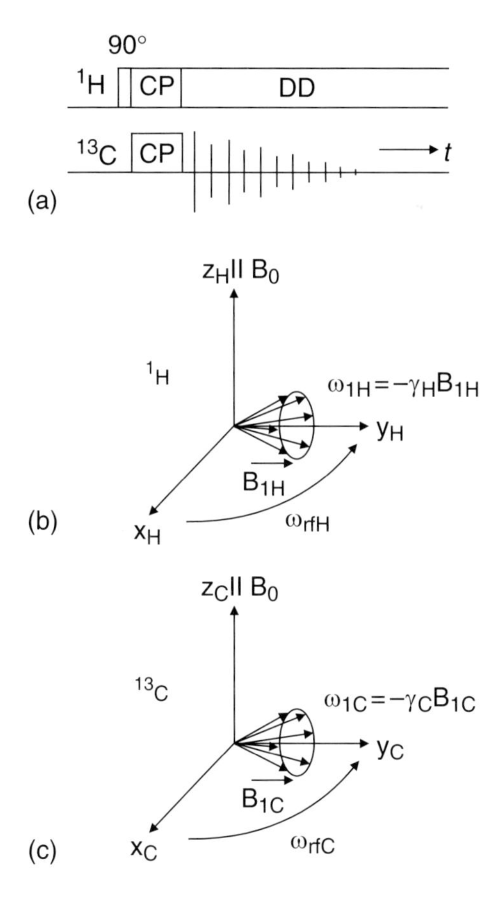

FIG. 3.3.2 Illustration of cross-polarization (CP): (a) Timing diagram of rf excitation for cross-polarization of ^{1}H magnetization to ^{13}C. DD denotes heteronuclear dipolar decoupling (b) Precession of ^{1}H magnetization components during cross-polarization with a frequency $\omega_{1\mathrm{H}}$ around the magnetic field $\boldsymbol{B}_{1\mathrm{H}}$ in the coordinate system, which rotates with frequency ω_{rfH}. (c) Precession of ^{13}C magnetization components during cross-polarization with a frequency $\omega_{1\mathrm{C}}$ around the magnetic field $\boldsymbol{B}_{1\mathrm{C}}$ in the coordinate system, which rotates with frequency ω_{rfC}. Adapted from [Blü3] with permission from Wiley-VCH.

The Hartmann–Hahn condition

Cross-polarization is achieved in a *double-resonance* experiment (Fig. 3.3.2(a)). Transverse magnetization of the I spins is generated by a 90° pulse at frequency ω_{rfI}. However, the transmitter is not turned off afterwards, only the rf phase is shifted by 90°. Thus, the $\boldsymbol{B}_{1\mathrm{I}}$ field is now applied parallel to the I magnetization. In the frame rotating with frequency ω_{rfI} around the z-axis, the $\boldsymbol{B}_{1\mathrm{I}}$ field is the dominant magnetic field which the I spins experience if its amplitude $\omega_{1\mathrm{I}}$ is larger than the frequency offset $\omega_{\mathrm{LI}} - \omega_{\mathrm{rfI}}$ and the other interactions of the I spin. Then the I magnetization is locked along this field in the rotating frame. This is why the technique is called *spin-locking*. While the I spins are locked in the transverse plane, another rf field $\boldsymbol{B}_{1\mathrm{S}}$ is applied at frequency ω_{rfS} but to the S

spins. If the magnitudes B_{1S} and B_{1I} of both rf fields are matched by the *Hartmann–Hahn condition* [Har2],

$$\gamma_S B_{1S} = \gamma_I B_{1I}, \tag{3.3.1}$$

each spin species precesses with the same frequency $\omega_1 = -\gamma B_1$ around the axis of its rf field in its own rotating frame. But because both rotating frames share the same z-axis, there is an oscillation of local I and S magnetization components along the z-axis with the same frequency ω_1 (Fig 3.3.2(b), (c)). In a classical way, one can argue that by this frequency match magnetization can be exchanged between both spin species. But because only the I spins were polarized at the beginning, magnetization is transferred from the I spins to the S spins.

Selectivity in cross-polarization

The efficiency of cross-polarization is determined by the size of the dipole–dipole interaction between I and S spins, and by the relaxation times $T_{1\rho I}$ and $T_{1\rho S}$ of the spin-locked I and S magnetizations. $T_{1\rho}$ is the *longitudinal relaxation time in the rotating frame*. It is an energy relaxation time similar to T_1 (cf. Section 3.5.1). By variation of the contact time t_{CP}, that is the length of the spin-lock time, local differences in the dipole–dipole couplings and in the $T_{1\rho}$ relaxation times can be exploited to selectively polarize different chemical and morphological structures. For short *contact times*, only the strongly coupled nuclei are polarized, for long contact times the weakly coupled ones are also polarized.

During data acquisition, the dipole–dipole interaction between I and S spins is made ineffective by high-power *dipolar decoupling* (DD) (cf. Fig. 3.3.2(a) and Section 3.3.2). But if the decoupler is turned off for a short time t_D , the magnetization of the S spins strongly coupled to the I spins dephases and only the magnetization of the weakly coupled S spins survives. In this way, for example, magnetization can be selectively transferred from ^1H to ^{13}C of rigid crystalline and of mobile amorphous regions, or of protonated and unprotonated nuclei. For rigid segments and for short internuclear distances the dipole–dipole coupling is strong, and for mobile segments and long internuclear distances the coupling is weak. The use of *selective magnetization transfer* by cross-polarization is illustrated in Fig. 3.3.3 for the polarization of protonated and unprotonated ^{13}C in polyethylene terephthalate. The sum of spectra (b) of the protonated carbons and (c) of the unprotonated carbons agrees with spectrum (a), which was acquired with polarization transfer to all carbons.

Cross-polarization efficiency

The S magnetization $M(t_{CP})$ is built up in the beginning with a time constant T_{CH} which is characteristic for the strength of the dipole–dipole coupling between I and S spins. With increasing cross-polarization time t_{CP} the magnetization passes through a maximum and is then attenuated by the influence of the relaxation times $T_{1\rho H}$ and $T_{1\rho C}$ in the rotating frames of ^1H and ^{13}C [Meh1, Sli1],

$$M(t_{CP}) = \frac{M_0}{\lambda}\left[1 - \exp\left\{-\frac{\lambda t_{CP}}{T_{CH}}\right\}\right]\exp\left\{-\frac{t_{CP}}{T_{1\rho H}}\right\}, \tag{3.3.2}$$

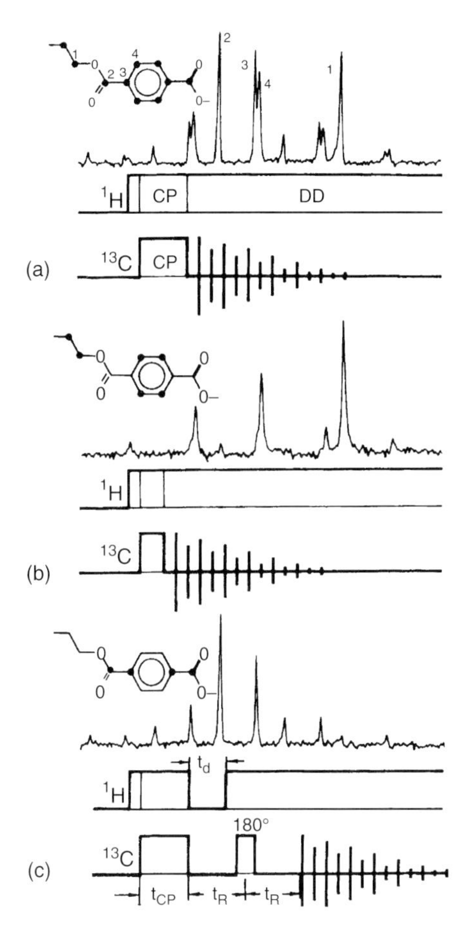

FIG. 3.3.3 Selectivity of cross-polarization demonstrated for polyethylene terephthalate at room temperature and a spinning frequency of $\omega_R/2\pi = 4090$ Hz. (a) Polarization of protonated and unprotonated ^{13}C nuclei with $t_{CP} = 5$ ms. (b) Polarization of protonated ^{13}C nuclei for $t_{CP} = 50\,\mu$s. (c) Polarization of unprotonated ^{13}C nuclei for $t_{CP} = 5$ ms and subsequent dephasing for $t_d = 100\,\mu$s under the influence of the ^{1}H–^{13}C dipole–dipole coupling. Adapted from [Blü3] with permission from Wiley-VCH.

where

$$\lambda = 1 + \frac{T_{CH}}{T_{1\rho C}} - \frac{T_{CH}}{T_{1\rho H}}. \tag{3.3.3}$$

While $T_{1\rho H}$ is about the same for all protons throughout the sample as a result of spin diffusion from multiple homonuclear dipole–dipole couplings, $T_{1\rho C}$ and T_{CH} are different for different parts of the molecule. They need to be known for a quantitative analysis of CP signals [Voe1].

The cross-polarization technique described above is the basic one most often used in practice [Mic1]. Several variations of it exist [Lev1, Teg1, Tek1], and other techniques can be used as well [Pin1]: in particular, the spin-lock fields can be ramped off and on adiabatically, so that magnetization of abundant nuclei can be transferred from a spin-lock state along $\boldsymbol{B}_1$ in the rotating frame into local heteronuclear dipolar fields and from there back into another spin-lock state along the $\boldsymbol{B}_1$ field of rare nuclei. This procedure is called *adiabatic cross-polarization* by demagnetization (ADRF) and remagnetization (ARRF) in the rotating frame [Meh1, Mei1, Sli1]. It is more efficient than Hartmann–Hahn cross-polarization, but also experimentally more demanding. Another example is the use of multi-pulse sequences (cf. Section 3.3.4). They can be applied simultaneously to both I and S nuclei. Then, different interactions can selectively be averaged out during polarization transfer. This approach is taken in cross-polarization by *isotropic mixing*. Here the homonuclear dipole–dipole interaction and the chemical shift are scaled to zero under the influence of a multi-pulse sequence, so that the frequency dependence of the cross-polarization efficiency is reduced [Car1, Car2].

3.3.2 High-power decoupling

To measure spectra of rare spins S in the presence of abundant spins I, the heteronuclear dipole–dipole coupling between I and S spins has to be suppressed. The most effective tool for this is *high-power dipolar decoupling*. Rare spins are ^{13}C and ^{29}Si, for example. Abundant spins are most often ^{1}H. For I decoupling during acquisition of S magnetization a strong continuous or pulsed rf field of amplitude ω_{1I} is applied at the I spin frequency ω_{rfI} (Figs 3.3.2 and 3.3.3).

The decoupling efficiency depends on two factors [Eng1, Meh1]: (1) The amplitude $\omega_{1I} = -\gamma B_{1I}$ of the decoupling field in comparison with the strength of the heteronuclear dipole–dipole interaction. (2) The modulation of the heteronuclear dipole–dipole coupling by flip-flop transitions in the system of the abundant I spins, which communicate by the homonuclear dipole–dipole interaction. In addition to this, the influence of thermal motion has to be considered [Meh1].

A measure for the decoupling efficiency is the linewidth in the spectrum of the S spins. It increases with the square of the rf frequency offset from the resonance of the I spins. The amplitude of the decoupler field has to be set large compared to the strength of the heteronuclear dipole–dipole interaction. For decoupling of ^{13}C and ^{1}H, this means that $|\gamma B_{1H}| > 2\pi\, 100$ kHz should be chosen. In practice, this requires transmitter power in excess of 1 kW for coil diameters of 5–10 mm, and problems with sample heating can be encountered. Continuous high-power decoupling not only averages the heteronuclear dipole–dipole coupling to zero, but also the heteronuclear indirect coupling (J coupling).

3.3.3 Sample spinning

The effect of high-power decoupling can be visualized as a forced rotation of the magnetization of the abundant I spins around the decoupler field. Then, in zeroth order, the rare spins can couple only to the time average of the rotating I magnetization. Thus, decoupling is obtained by time-dependent modulation of the interaction. Such a modulation can be achieved not only in spin space by high-power decoupling (Section 3.3.2) and by multi-pulse excitation (Section 3.3.4), but also in real space by mechanical rotation of the sample. The last point is the fundamental idea underlying

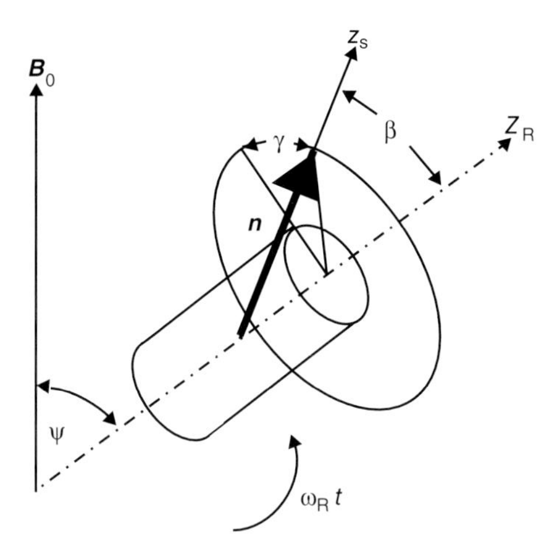

FIG. 3.3.4 Geometry and angle definition for mechanical sample spinning. ***n*** denotes the orientation of a crystallite in the spinner or the preferential axis of the sample. Adapted from [Har1] with permission from the American Institute of Physics.

the *magic angle spinning* technique [And3, Gri1, Low1, Meh1, Sam1], which is applied routinely in high-resolution NMR of rare spin-$\frac{1}{2}$ nuclei, often in combination with cross-polarization [Eng2] (cf. Fig. 3.3.1(d)). It is also used for imaging of ^{1}H, ^{2}H, and ^{13}C in solids (cf. Section 8.5), because chemical-shift resolution can be achieved in this way.

The sample is placed into a gas driven spinner, the rotation axis of which is inclined by an angle ψ against the magnetic field $\boldsymbol{B}_0$ (Fig. 3.3.4). For MAS this angle is adjusted to the magic angle of 54.7°. Then the anisotropic parts of all interactions which are described by second-rank tensors can be averaged. These are the anisotropy of the chemical shift, the dipole–dipole interaction, and the first-order quadrupole interaction.

For averaging of higher rank coupling tensors, the sample must be spun around more than one axis [Sam1, Chm1]. Examples are techniques such as DOR (double rotation) [Ywu1] and DAS (dynamic angle spinning) [Mue1]. By the DOR technique the sample is spun around two axes simultaneously. This is sufficient to average the first and the second-order quadrupole interaction, so that high-resolution spectra of many quadrupolar nuclei with half integral spin like ^{11}B, ^{17}O, ^{23}Na, and ^{27}Al can be obtained. For these nuclei the linewidth of the central transition ($m = 1/2 \leftrightarrow m = -1/2$) is determined by the quadrupole interaction in second order, which prevents the measurement of high-resolution spectra by conventional MAS. The DAS and DOR techniques can be replaced by methods of 2D spectroscopy with MAS and multi-quantum excitation [Fer1, Fry1]. These approaches to high-resolution solid-state spectroscopy are important in particular for investigations of many ceramic materials.

MAS and heteronuclear high-power decoupling

Solid-state NMR spectra of dilute nuclei in organic molecules are broadened by chemical-shift anisotropy and dipole–dipole couplings between ^{1}H and ^{13}C [Van1]. The effect of these interactions can be eliminated from the NMR spectrum by MAS in combination with *high-power dipolar decoupling* by CW rf irradiation of the abundant nucleus. For example, the ^{13}C spectrum (e) in Fig. 3.3.1 has been obtained by MAS and CW irradiation of ^{1}H.

Because the optimum decoupler frequency depends on the proton resonance frequencies, it is difficult to decouple protons over a large range of frequencies [Vau2]. This problem is aggravated at high fields, because the frequency spread due to chemical shift increases, and the resultant line broadening increases nonlinearly with the off-set of the resonance frequency from the decoupler frequency. The *decoupling efficiency* can be improved by phase and frequency modulation of the rf decoupler field [Ben2, Ern2, Gan1, Tek2]. By a two-pulse phase-modulation sequence the spectroscopic resolution can significantly be improved, and a gain of the signal amplitudes in the spectrum is observed [Ben2]. This increase of signal amplitude has been attributed to a reduction of decoupling sidebands [Ern2]. Use of the phase-modulation technique is useful, when the available decoupler power is insufficient to overcome the homonuclear dipole–dipole couplings among ^{1}H.

MAS and cross-polarization

The combination of cross-polarization and MAS (CP MAS) together with high-power decoupling of the abundant nucleus [Ste3] is common practice in analytical solid-state NMR studies of ^{13}C in organic molecules (cf. Fig. 3.3.1). Cross-polarization enhances the signal of the rare spins and MAS averages the anisotropy of the chemical shielding so that signals are observed in solids at the isotropic chemical shifts similar to liquid-state NMR spectra. The polarization enhancement of low-γ nuclei by CP under MAS from high-γ nuclei becomes difficult when the sample spinning speed is sufficiently large to effectively modulate the homonuclear and heteronuclear dipole-dipole couplings [Mei1 and references therein]. At high spinning speeds (typically 15 kHz and above) the efficiency of the original Hartmann–Hahn sequence of Fig. 3.3.2 is reduced from a broad frequency region to narrow frequency bands spaced at the spinning frequency, and the CP efficiency becomes sensitive to chemical shift, rf inhomogeneities, and fluctuations in the overall rf power and the spinning speed. The CP process in solids then becomes similar to CP in liquids, where the stability of the *Hartmann–Hahn match* (3.3.1) is critical for the efficiency of the polarization transfer [Kim3 and references therein]. The CP process can be made less sensitive to these effects by the use of time-dependent modulations of phase and amplitude of the spin-lock fields applied for polarization transfer [Bar1, Pee1, Sch10, Sun1, Wux1, Zha2].

Magic angle spinning and off-magic angle spinning

In MAS spectra of ^{13}C in natural abundance, the lineshape is dominated by the anisotropies of the chemical shielding at different chemical shifts (Fig. 3.3.5(a)). By use of MAS with spinning speeds which are larger than the anisotropies Δ (cf. eqn (3.1.23)) of the interaction these wideline signals are compacted into narrow lines at the isotropic chemical shifts resulting in high resolution and an associated gain in sensitivity (b). The price paid is a loss of information about the anisotropic character of the interaction. Therefore, the resonances can no longer be used as a protractor, and such spectra are bare of information about molecular order and slow molecular reorientation (Section 3.2).

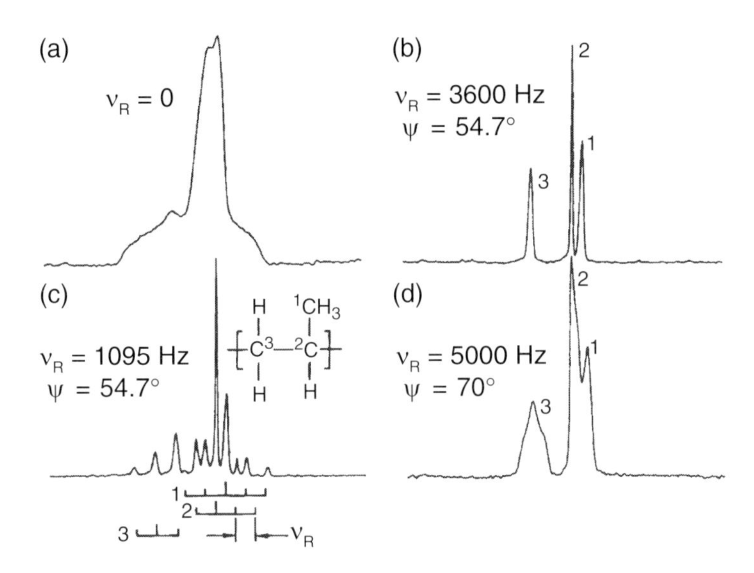

FIG. 3.3.5 ^{1}H decoupled ^{13}C spectra of isotactic polypropylene for different spinning frequencies $\omega_R = 2\pi\nu_R$ and orientation angles ψ of the rotation axis. (a) Static sample. The wideline resonances of the different carbons overlap. (b) MAS spectrum with fast sample spinning. Narrow signals are observed at the isotropic chemical shifts only. (c) MAS spectrum with slow sample spinning. In addition to the centre line, sideband signals are observed at seperations $n\omega_R$ from centre lines. (d) OMAS spectrum with fast sample spinning. The orientation of the axis deviates from the magic angle. Each resonance forms a powder spectrum with reduced width, which can serve as a protractor (cf. Fig. 3.1.3). Adapted from [Blü4] with permission from Wiley-VCH.

This information, however, can be partly regained in two ways by compromising the spectral resolution. If the sample is spun with a frequency ω_R which is small compared to the anisotropy Δ, *rotational sidebands* or *spinning sidebands* arise at separations ω_R from each centre frequency ω_L (isotropic chemical shift), the envelope of which approaches the shape of the wideline spectrum for vanishing rotation speeds ω_R (c). Information about the principal axes of the coupling tensor, and about molecular order and mobility can then be derived by an analysis of rotational sidebands using tables published in the literature [Her1]. The other option is fast sample spinning at an angle ψ different from the magic angle (*OMAS*: *off-magic angle spinning*) (d). In this case the anisotropies Δ of the coupling tensors are scaled by the factor $(3\cos^2\psi - 1)/2$. By partial narrowing of the wideline resonances the overlap can be reduced while the protractor property can be retained although with reduced angular resolution.

Time dependence of the resonance frequency

The formal description of MAS involves different coordinate frames when samples with a preferential axis of *molecular order* are considered (Fig. 3.3.6). On the molecular level a *sample frame* is introduced, the z_S-axis of which is parallel to the preferential direction of the sample. In some cases, this frame can be attached to the molecule or the crystallite

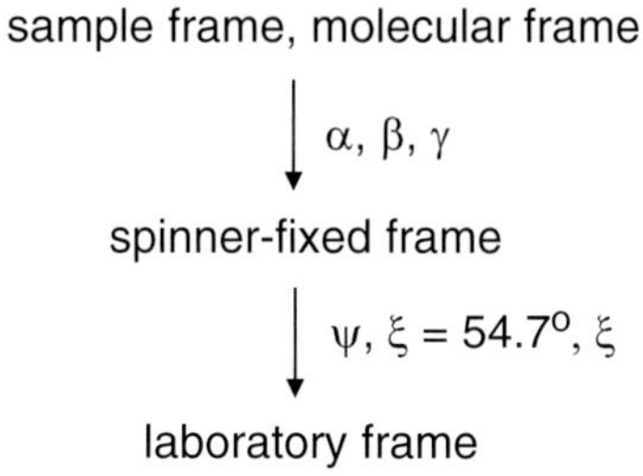

sample frame, molecular frame

↓ α, β, γ

spinner-fixed frame

↓ ψ, ξ = 54.7°, ξ

laboratory frame

FIG. 3.3.6 Relationship between coordinate frames in sample spinning.

instead and assumes the role of a *molecular frame*. Thus the *coupling tensor* of the interaction appears nondiagonal within this frame. The orientation of the sample frame in the *spinner-fixed frame* is specified by the Euler angles α ,β, and γ. The orientation of the spinner-fixed frame in the *laboratory frame* is determined by the polar coordinates ψ and ξ. In the spinner-fixed frame the magnetic field $\boldsymbol{B}_0$ precesses around the rotor axis at an angle ψ, which is adjusted to 54.7° for MAS [Hag2].

The phase, which the spinner accumulates during sample rotation with constant rotation speed ω_{R} is given by

$$\xi(t) = \xi_0 + \omega_{\mathrm{R}} t. \tag{3.3.4}$$

The vector $\boldsymbol{n}$ in Fig. 3.3.4, which defines the preferential axis of the sample in the magnetic field $\boldsymbol{B}_0$, then becomes time dependent via the angle $\xi(t)$. In this way, the resonance frequency

$$\hat{\Omega}(t) = \Omega(t) - \sigma\omega_0 \tag{3.3.5}$$

also becomes dependent on time. Here $\hat{\Omega}(t)$ denotes the angular-dependent resonance frequency centred off-resonance with respect to ω_0 (cf. eqns (3.1.11) and (3.1.23)). It can be split into a time-independent and a time-dependent part [Meh1]

$$\hat{\Omega}(t) = \hat{\Omega}' + \hat{\Omega}''(t) \tag{3.3.6}$$

The time-independent part is given by

$$\begin{aligned} \hat{\Omega}' =& -\omega_0 \left\{ \sigma + \frac{3\cos^2\psi - 1}{2} \left[\frac{3\cos^2\beta - 1}{2} (\sigma_{33} - \sigma) \right. \right. \\ & \left. \left. + \frac{\sigma_{11} - \sigma_{22}}{2} \sin^2\beta \cos(2\alpha) \right] \right\}, \end{aligned} \tag{3.3.7}$$

and the time-dependent part is

$$\begin{aligned} \hat{\Omega}''(t) =& A_2 \cos(2\gamma + 2\omega_{\mathrm{R}} t) + B_2 \sin(2\gamma + 2\omega_{\mathrm{R}} t) \\ & + A_1 \cos(\gamma + \omega_{\mathrm{R}} t) + B_1 \sin(\gamma + \omega_{\mathrm{R}} t). \end{aligned} \tag{3.3.8}$$

The quantity σ is the isotropic chemical shift (3.1.10). The coefficients A_1, B_1, A_2, and B_2 are given at the magic angle by [Meh1, Sch9]

$$A_1 = -\frac{2\sqrt{2}\,\omega_0}{3}\left\{\sin(2\beta)\left[\frac{\sigma_{11}+\sigma_{22}-2\sigma_{33}}{4} - \frac{\sigma_{22}-\sigma_{11}}{4}\cos(2\alpha) + \frac{\sigma_{12}\sin(2\alpha)}{2}\right] + \cos(2\beta)[\sigma_{13}\cos\alpha + \sigma_{23}\sin\alpha]\right\}, \quad (3.3.9a)$$

$$B_1 = -\frac{2\sqrt{2}\,\omega_0}{3}\left\{\sin(\beta)\left[(\sigma_{22}-\sigma_{11})\frac{\sin(2\alpha)}{2} + \sigma_{12}\cos(2\alpha)\right] - \cos\beta[\sigma_{13}\sin\alpha - \sigma_{23}\cos\alpha]\right\}, \quad (3.3.9b)$$

$$A_2 = \frac{\omega_0}{3}\left\{(-\sigma_{11}-\sigma_{22}+2\sigma_{33})\frac{\sin^2\beta}{2} - (\sigma_{22}-\sigma_{11})\cos(2\alpha)\frac{1+\cos^2\beta}{2} + \sigma_{12}\sin(2\alpha)(1+\cos^2\beta) - 2\sin(2\beta)[\sigma_{13}\cos\alpha + \sigma_{23}\sin\alpha]\right\}, \quad (3.3.9c)$$

$$B_2 = -\frac{\omega_0}{3}\{\cos\beta(\sigma_{22}-\sigma_{11})\sin(2\alpha) + 2\sigma_{12}\cos(2\alpha) + 2\sin\beta[\sigma_{13}\sin\alpha - \sigma_{23}\cos\alpha]\}. \quad (3.3.9d)$$

Because the principal axes of the coupling tensor are not aligned with the axes of the sample frame, all elements of the coupling tensor appear in (3.3.9). Though the calculation explicitly deals with the chemical shielding, similar expressions are valid for other anisotropic second-rank interactions like the dipole–dipole coupling between two spins and the first-order quadrupole interaction.

At the magic angle $\psi = \arccos(3^{-1/2}) = 54.7^\circ$ the time-independent part (3.3.7) of the resonance frequency becomes independent of the orientation angles α and β. Deviations from the magic angle lead to powder spectra, the widths of which are scaled by the factor $(3\cos^2\psi - 1)/2$ compared to the spectra of nonspinning samples. This applies to rotation frequencies ω_R which are larger than the anisotropy Δ of the interaction. Otherwise rotational sidebands appear which arise from the time-dependent part $\Omega''(t)$ given by (3.3.8). Fast OMAS is used when spectral overlap is to be reduced, yet the anisotropy of the resonance frequency is to be maintained (cf. Fig. 3.3.6(d)) [Ste1]. Examples from ^{13}C NMR are the determination of the principal values of chemical shielding tensors (cf. Fig. 3.1.3), investigations of molecular dynamics on the timescale of the inverse linewidth [Sch3], and 2D exchange NMR [Blü1].

The free induction decay under magic angle spinning

For calculation of the *FID under MAS*, the precession phase ζ of an arbitrary magnetization component corresponding to a fixed orientation of the coupling tensor in the rotor frame is written as the time integral over the frequency $\hat{\Omega}(t)$,

$$\xi(t_1, t_2) = \int_{t_1}^{t_2} \hat{\Omega}(t)\,\mathrm{d}t = \Theta(t_2) - \Theta(t_1). \tag{3.3.10}$$

The integral is determined by the difference in angles $\Theta(t_2)$ and $\Theta(t_1)$ which the magnetization vector forms with the x-axis of the coordinate frame at times t_1 and t_2, respectively. Then the signal $s'(t)$ of the free induction decay of this component is given by

$$\begin{aligned} s'(t) &= \exp\{-\mathrm{i}\omega_0\sigma t\}\exp\left\{-\mathrm{i}\int_0^t \hat{\Omega}''(t')\mathrm{d}t'\right\} \\ &= \exp\{-\mathrm{i}\omega_0\sigma t\}\exp\{\mathrm{i}\Theta(t)\}\exp\{-\mathrm{i}\Theta(0)\}. \end{aligned} \tag{3.3.11}$$

Because $\hat{\Omega}''(t)$ depends on the orientation of the principal axis frame of the coupling tensor relative to the magnetic field, each magnetization vector in the sample follows an individual phase trajectory in the transverse plane, which is described by (3.3.10) [Gri1]. Therefore, the total magnetization dephases rapidly. But $\hat{\Omega}''(t)$ also exclusively depends on terms which oscillate periodically with ω_R and $2\omega_R$. For this reason, the integral in (3.3.10) vanishes for each magnetization vector after an integral multiple of the rotor period T_R, and *rotational echoes* are observed at these times in the FID of a powder sample.

The f functions

The computation of MAS time-domain signals is facilitated by use of the so-called *f functions* [Meh1, Hag1]. For spinning at the magic angle $\psi = 54.7°$ the FID (3.3.11) of a given magnetization component is rewritten with the help of (3.3.4) as

$$s'(t) = \exp\{-\mathrm{i}\omega_0\sigma t\} f(\gamma + \xi(t)) f^*(\gamma + \xi_0), \tag{3.3.12}$$

where

$$f(\gamma) = \exp\{\mathrm{i}[A_2' \sin(2\gamma) - B_2' \cos(2\gamma) + A_1' \sin\gamma - B_1' \cos\gamma]\}, \tag{3.3.13}$$

and

$$A_1' = \frac{A_1}{\omega_R} \qquad B_1' = \frac{B_1}{\omega_R}, \tag{3.3.14a}$$

$$A_2' = \frac{A_2}{2\omega_R} \qquad B_2' = \frac{B_2}{2\omega_R}. \tag{3.3.14b}$$

For simplicity of notation, the arbitrary but fixed spinner phase ξ_0 and the rotation angle $\omega_R t$ are not always carried along explicitly in the *f* functions.

Spinning sidebands

For calculation of the FID decay signal during MAS, (3.3.11) is rewritten by use of the delta function

$$\delta(\vartheta - \gamma - \omega_R t) = \frac{1}{2\pi} \sum_{N=-\infty}^{+\infty} \exp\{\pm iN(\vartheta - \gamma - \omega_R t)\} \tag{3.3.15}$$

to yield

$$\begin{aligned} s'(t) &= \exp\{-i\omega_0 \sigma t\} f^*(\gamma) \int_0^{2\pi} \delta(\vartheta - \gamma - \omega_R t) f(\vartheta)\, d\vartheta \\ &= \frac{1}{2\pi} \exp\{-i\omega_0 \sigma t\} \sum_{N=-\infty}^{+\infty} \exp\{iN\omega_R t\} \exp\{iN\gamma\} f^*(\gamma) \\ &\quad \times \int_0^{2\pi} \exp(-iN\vartheta) f(\vartheta)\, d\vartheta \end{aligned} \tag{3.3.16}$$

The signal of the total magnetization derives from (3.3.16) by summation of all magnetization components. This can be achieved by integration over all orientations α, β, γ of the sample frame in the rotor frame weighted by the *orientational distribution function* $P(\alpha, \beta, \gamma)$ of the sample frame,

$$\begin{aligned} s(t) &= \int_0^{2\pi} \int_0^{\pi} \int_0^{2\pi} P(\alpha, \beta, \gamma) \exp\left\{-\left(\frac{1}{T_2} - i\omega_0 \sigma\right) t\right\} s'(t)\, d\gamma \sin\beta\, d\beta\, d\alpha \\ &= \frac{1}{2} \exp\left\{-\left(\frac{1}{T_2} + i\omega_0 \sigma\right) t\right\} \\ &\quad \times \sum_N \int_0^{2\pi} \int_0^{\pi} \int_0^{2\pi} P(\alpha, \beta, \gamma) \exp\{iN\gamma\} f^*(\gamma)\, d\gamma \sin\beta\, d\beta\, d\alpha \\ &\quad \times \int_0^{2\pi} \exp\{-iN\vartheta\} f(\vartheta)\, d\theta. \end{aligned} \tag{3.3.17}$$

With the procedures outlined in Section 3.2.1, $P(\alpha, \beta, \gamma)$ can be expressed in terms of the molecular orientation in the sample frame and the orientation of the coupling tensor in the molecular frame. By use of (3.3.17) the FID signal can be calculated for oriented samples under MAS conditions, and MAS spectra are obtained by subsequent Fourier transformation. The finite width of centre lines and sideband signals has been accounted for in (3.3.17) by multiplication of the FID with the exponential line function $\exp\{-(t/T_2)\}$.

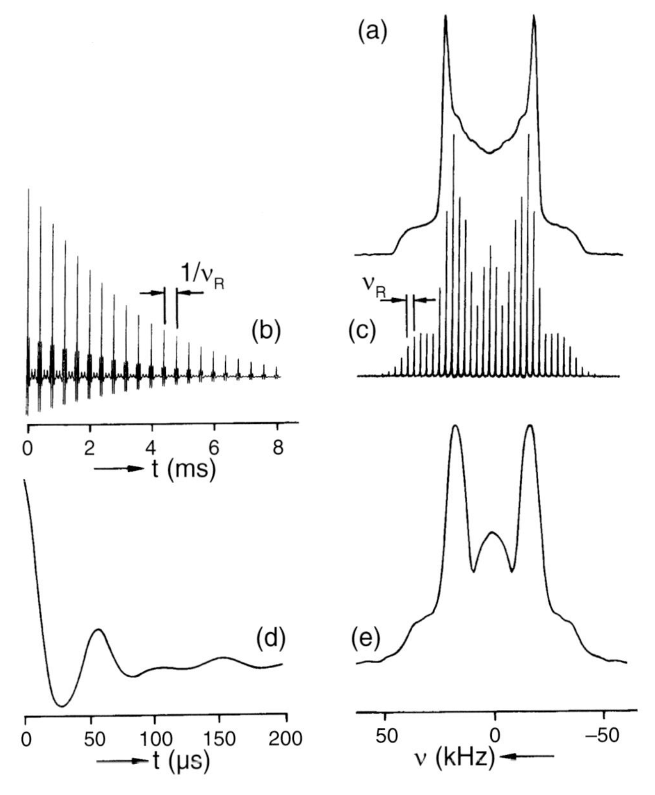

FIG. 3.3.7 ^{2}H MAS NMR of dimethyysulfone-d$_0$ at a rotor frequency of $\omega_R/2\pi = 2500$ Hz. (a) Solid-echo spectrum of the nonspinning sample. (b) On-resonance FID with rotary echoes. (c) MAS sideband spectrum. (d) Decay of a rotational echo signal. (e) Fourier transform of a rotary echo. Adapted from [Gün3] with permission from Elsevier Science.

For an isotropic orientational distribution $P(\alpha, \beta, \gamma) = (8\pi^2)^{-1}$, so that

$$s(t) = \frac{1}{(4\pi)^2} \exp\left\{-\left(\frac{1}{T^2} + \mathrm{i}\omega_0\sigma\right)t\right\} \sum_N \exp\{\mathrm{i}N\omega_R t\} F_N^* F_N, \tag{3.3.18}$$

where

$$F_N = \int_0^{2\pi} \exp\{-\mathrm{i}N\vartheta\} f(\vartheta)\, \mathrm{d}\vartheta. \tag{3.3.19}$$

Thus by Fourier transformation of (4.3.17) a sideband spectrum is obtained with signals at frequencies $\omega_0\sigma + N\omega_R$ with intensities $F_N^* F_N$, where N assumes positive and negative integer values.

The MAS signals are illustrated in Fig. 3.3.7 with experimental ^{2}H data of dimethylsulfone [Gün3]. The FID (b) consists of series of *rotational echoes* (d) which are formed every rotor period $T_R = (2\pi)/\omega_R$. The Fourier transform (c) of the FID is the sideband

MAS spectrum, consisting of narrow lines separated by the rotor frequency ω_R. The maxima of the rotational echoes in (b) follow the evolution of the isotropic mean. If the MAS FID is sampled at the echo maxima only, no sidebands will appear in the MAS spectrum. A similar situation is encountered for fast spinning, when the rotational echoes are no longer resolved. The *envelope of the sideband spectrum* is the Fourier transform of the rotational echo decay (c) [Mar2]. For slow spinning speeds, it approaches the shape of the wideline spectrum (a) of the nonspinning sample.

Rotor synchronization

If a macroscopically *ordered sample* is rotated, the preferential axis $\boldsymbol{n}$ of which is not aligned with the spinning axis, the phase of the sideband spectrum is modulated by the phase ξ_0 of the spinner (cf. eqn (3.3.4)). This modulation can be interrogated experimentally by synchronizing the data acquisition with the spinner phase in a 2D experiment (Fig. 3.3.8(a)) [Har1, Tan1]. To this end, the spinner phase is incremented in typically 16 steps through one rotor period in the evolution time t_1,

$$\xi_0 = \omega_R t_1. \tag{3.3.20}$$

A 2D spectrum is obtained by 2D Fourier transformation over the spinner phase and the acquisition time t_2. It exhibits spinning sideband signals in both dimensions. Those in the additional dimension ω_1 are characteristic of the molecular order. No sidebands appear in this direction if the sample is isotropic. The 2D sideband signals can be analysed to obtain the *orientational distribution function*.

An example is depicted in Fig. 3.3.8 with ^{13}C data of the high modulus fiber *Vectra® B900*. The sideband signals of the quarternary carbon marked in the formula (c) and in the cross-sections (d) of the spectrum are well resolved and can be used for quantitative analysis. Given the orientation of the chemical shielding tensor in the molecule fixed frame and the orientation of the fiber axis in the rotor frame, the orientational distribution function can be calculated from the experimental 2D spectrum. In analogy to the procedure outlined in Section 3.2.1, an analysis in terms of 2D *Legendre subspectra* can be followed for weak molecular order. The resulting distribution function is depicted in Fig. 3.2.2(b). It is well approximated by the Legendre moments P_2, P_4, and P_6.

Sideband suppression

Rotational sideband signals contain important information about anisotropies of spin interactions. Nevertheless, they can be a nuisance in crowded high-resolution spectra. Sideband signals do not arise if the spinner frequency ω_R is larger than the anisotropy Δ of the interaction. This can be achieved with the currently available spinning frequencies of $\omega_R/(2\pi) = 5$–35 kHz for the anisotropic shielding of most aliphatic and many aromatic carbons even at high field. In other cases, the magnetization can be prepared before data acquisition by use of certain pulse sequences, in such a way that the sideband signals are modulated or even suppressed [Ant1, Dix1, Gri1].

The pulse sequence used most often for this purpose is the *TOSS sequence* (total suppression of spinning sidebands, Fig. 3.3.9) [Dix1, Gri1]. Before the start of data acquisition at $t = t_5$ the spin system is prepared by four 180° pulses at well-defined times

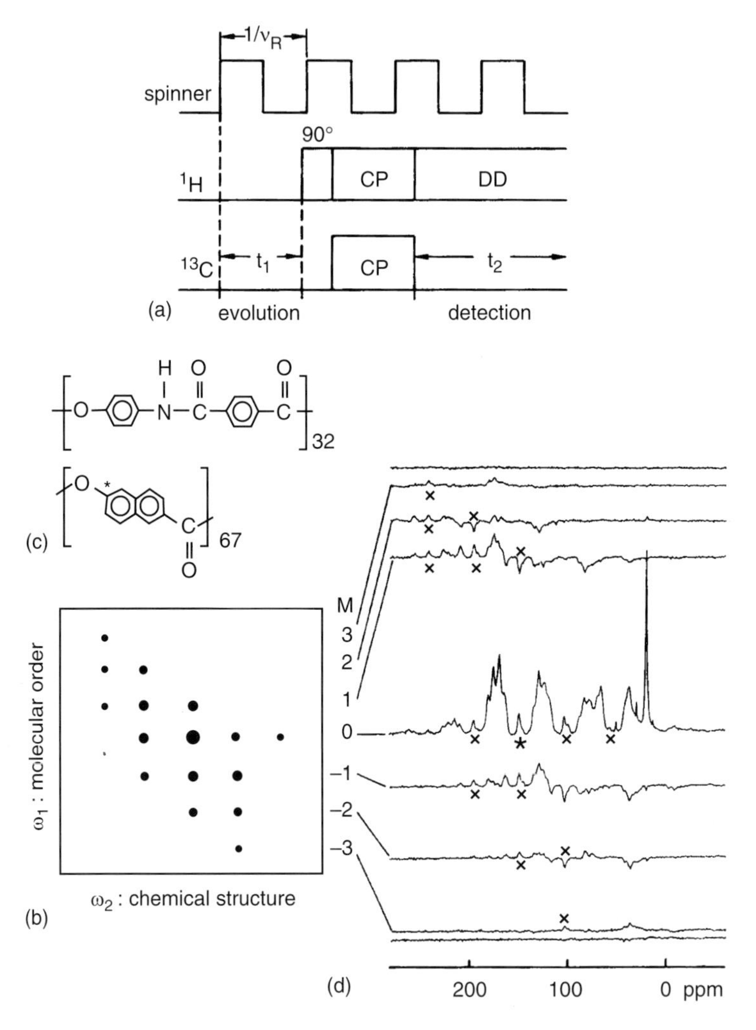

FIG. 3.3.8 2D MAS NMR with rotor synchronization for the detection of molecular order. (a) Spinner signal and rf excitation. $T_R = 2\pi/\omega_R$ denotes the rotor period. (b) Arrangement of signals in the 2D sideband spectrum for ^{13}C nucleus marked in the chemical formula (c) of the high performance fiber VECTRA® B900. (d) 1D cross-sections through the 2D sideband spectrum. Rotor frequency and angle between fibre and spinner axes were set to $\omega_R/2\pi = 3500$ Hz and 45°, respectively. The corresponding orientational distribution function is depicted in Fig. 3.2.2(b). Adapted from [Blü4] with permission from Wiley-VCH.

t_1, t_2, t_3, and t_4. The timing of the pulses is chosen in a way so that the magnetization components giving rise to the sidebands in the MAS spectrum interfere to zero during data acquisition, and only the signals at the isotropic chemical shifts remain. The TOSS sequence works in this way only for isotropic samples. For macroscopically ordered samples, sidebands are still observed which are modulated by the spinner phase. This can be exploited, for instance, for separation of signals from the isotropic and the anisotropic parts of the interaction [Ant2, Tit1].

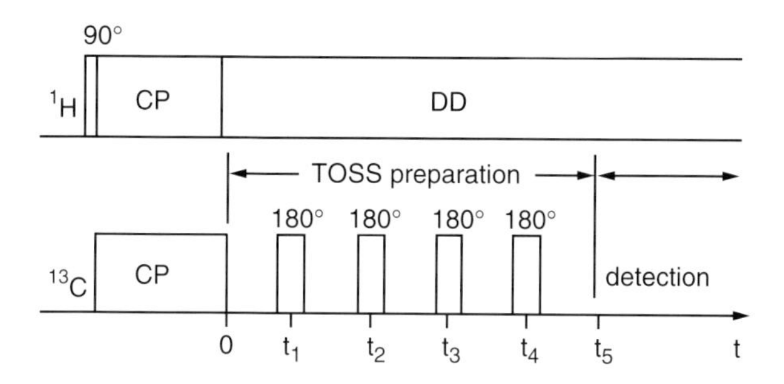

FIG. 3.3.9 TOSS preparation in MAS NMR for suppression of spinning sidebands. The 180° TOSS pulses are not equally spaced in practice.

3.3.4 Multi-pulse methods

Under MAS all those interactions are nonselectively averaged to their isotropic means, the anisotropy Δ of which (cf. eqn (3.1.18)) is smaller than the spinning speed ω_R. With fast spinning speeds of 20 kHz and more the heteronuclear dipole–dipole interaction can be averaged. This is not so for the homonuclear dipole–dipole interaction, which can be up to 100 kHz among ^{1}H. Here, however, *multi-pulse techniques* can be employed, which, depending on the pulse sequence, selectively average different interactions [Abr2, Ger1, Hae1, Meh1]. A *multi-pulse sequence* consists of a cycle of rf pulses, which is repetitively applied to the spin system. Such sequences are more demanding but also more versatile than the first approach taken by *Lee and Goldburg*, who supported locking of the magnetization at the magic angle in the rotating frame for elimination of the homonuclear dipole–dipole interaction [Lee1].

An important application of multi-pulse techniques is the elimination of the homonuclear dipole–dipole interaction for determination of the orientation dependence of the chemical shift in single crystals. Other applications are isolation of the heteronuclear dipole–dipole interaction between abundant and rare spins, isotropic magnetization transfer independent of chemical shielding effects [Car1, Car2] instead of cross-polarization, filtering of magnetization according to chemical [Sch4] and morphological criteria [Sch5] for studies of magnetization spreading by spin diffusion (cf. Sections 3.5.3 and 7.2.9), and multi-pulse imaging of rigid solids [Chi1, Cor1, Man1, Man2] (cf. Section 8.7). Multi-pulse excitation is experimentally demanding. It requires high power pulses with extraordinary phase and amplitude stability.

Multi-pulse sequences can be viewed as an extension of multi-echo excitation with solid echoes (cf. Fig. 3.2.6(a)) or with magic echoes (cf. Fig. 3.4.3), by which echoes are formed in the presence of the dipole–dipole interaction between two or more spins and in the presence of the quadrupole interaction. Most multi-pulse sequences are sequences of solid-echo pulse pairs with well-defined relative phases. When applied to a system of spins, the time average of the dipole–dipole coupling vanishes, while the chemical shielding is scaled or reduced to zero as well. In the first case, the amplitude of the solid echoes is modulated by the scaled chemical shielding and by relaxation only. The echo maxima are measured stroboscopically during the pulse spacings. They form the time-domain signal acquired under multi-pulse excitation (Fig. 3.3.10). The most

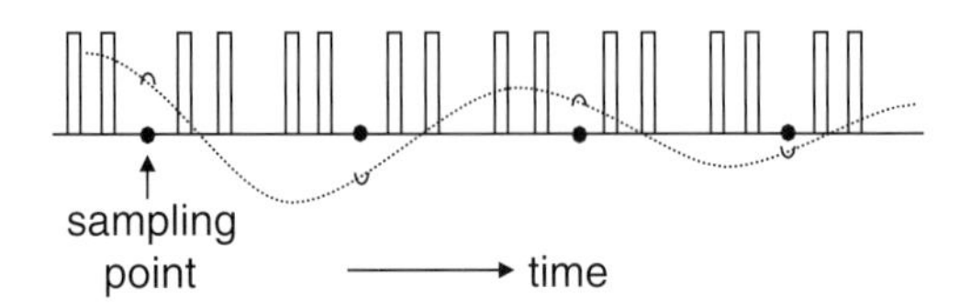

FIG. 3.3.10 Stroboscopic data acquisition in multi-pulse NMR. Adapted from [Sch9] with permission from Academic Press.

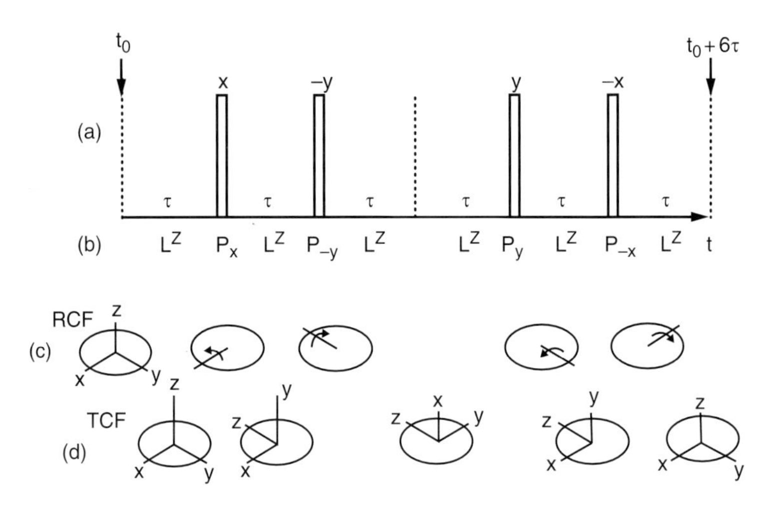

FIG. 3.3.11 The WAHUHA pulse cycle. (a) Timing diagram. (b) Time-evolution operators in the rotating coordinate frame. (c) Rotations in the rotating coordinate frame (RCF). (d) Orientation of the toggling coordinate frame (TCF). Adapted from [Sch9] with permission from Academic Press.

simple multi-pulse sequence is the WAHUHA sequence named after Waugh, Huber, and Haeberlen (Fig. 3.3.11) [Wau1].

Elementary multi-pulse NMR

To gain some insight into the theory of multi-pulse NMR, the WAHUHA sequence $(\tau, +x, \tau, -y, \tau, \tau, +y, \tau, -x, \tau)$ is applied in the rotating coordinate frame to a pair of spins coupled by the dipole–dipole interaction [Sch9]. The pulses are considered to be infinitely short. Their flip angle is 90°, and they are applied in $+x$, $-y$, $+y$, and $-x$ directions of the rotating coordinate frame. The density matrix $\boldsymbol{\rho}(t_0+6\tau)$ after completion of a WAHUHA cycle is obtained from the density matrix $\boldsymbol{\rho}(t_0)$ before application of the cycle by transformation with the evolution operator $\mathbf{U}(t)$ of the cycle,

$$\boldsymbol{\rho}(t_0 + 6\tau) = \mathbf{U}(6\tau)\boldsymbol{\rho}(t_0)\mathbf{U}^{-1}(6\tau). \tag{3.3.21}$$

It is composed of a sequence of operators $\mathbf{P}_\alpha = \exp\{-i\mathbf{I}_\alpha\theta\}$ and $\mathbf{L}^\beta(\tau) = \exp\{-i\mathbf{H}^\beta\tau\}$ for rf pulses with flip angles θ around the α-axis, and for respectively. $\mathbf{H}^\beta$ denotes the Hamilton operator of the interaction λ (cf. eqn (3.1.1)), and the index β indicates the quantization axis of the reference coordinate system. For $\beta = z$ this is the rotating coordinate frame. Following the convention that time increases from right to left in quantum mechanical notations, one obtains for one cycle of the WAHUHA sequence (Fig. 3.3.11)

$$\mathbf{U}(6\tau) = \mathbf{L}^z\mathbf{P}_{-x}\mathbf{L}^z\mathbf{P}_{+y}\mathbf{L}^z\mathbf{L}^z\mathbf{P}_{-y}\mathbf{L}^z\mathbf{P}_{+x}\mathbf{L}^z. \tag{3.3.22}$$

By insertion of the unit operator $\mathbf{1} = \mathbf{P}_\alpha^{-1}\mathbf{P}_\alpha$, pulse operators $\mathbf{P}_\alpha$ and free precession operators $\mathbf{L}^\beta$ can be separated,

$$\begin{aligned}
\mathbf{U}(6\tau) &= \mathbf{L}^z\mathbf{P}_{-x}\mathbf{L}^z(\mathbf{P}_{-x}^{-1}\mathbf{P}_{-x})\mathbf{P}_{+y}\mathbf{L}^z\mathbf{L}^z(\mathbf{P}_{+y}^{-1}\mathbf{P}_{+y})\mathbf{P}_{-y}\mathbf{L}^z(\mathbf{P}_{-y}^{-1}\mathbf{P}_{-y})\mathbf{P}_{+x}\mathbf{L}^z(\mathbf{P}_{+x}^{-1}\mathbf{P}_{+x})\\
&= \mathbf{L}^z(\mathbf{P}_{-x}\mathbf{L}^z\mathbf{P}_{-x}^{-1})\mathbf{P}_{-x}(\mathbf{P}_{+y}\mathbf{L}^z\mathbf{L}^z\mathbf{P}_{+y}^{-1})\mathbf{P}_{+y}(\mathbf{P}_{-y}\mathbf{L}^z\mathbf{P}_{-y}^{-1})\mathbf{P}_{-y}(\mathbf{P}_{+x}\mathbf{L}^z\mathbf{P}_{+x}^{-1})\mathbf{P}_{+x}\\
&= \mathbf{L}^z\mathbf{L}^y\mathbf{P}_{-x}\mathbf{L}^x\mathbf{L}^x\mathbf{P}_{+y}\mathbf{L}^{-x}\mathbf{P}_{-y}\mathbf{L}^{-y}\mathbf{P}_{+x}\\
&= \cdots = \mathbf{L}^z\mathbf{L}^y\mathbf{L}^x\mathbf{L}^x\mathbf{L}^y\mathbf{L}^z\mathbf{P}_{-x}\mathbf{P}_{+y}\mathbf{P}_{-y}\mathbf{P}_{+x}.
\end{aligned} \tag{3.3.23}$$

Application of the pulse operator $\mathbf{P}_\alpha$ to the free precession operator $\mathbf{L}^\beta$ transforms the free precession operator into a coordinate frame, which toggles with the rf pulse phase. This is the *toggling coordinate frame*. Using positive rotations (right-hand rule), one obtains for example $\mathbf{P}_{-x}\mathbf{I}_z\mathbf{P}_{-x}^{-1} = \mathbf{I}_y$, so that $\mathbf{P}_{-x}\mathbf{L}^z\mathbf{P}_{-x}^{-1}$ can be replaced by $\mathbf{L}_y$ in (3.3.23), etc.

Multi-pulse sequences are designed in such a way, that the product of pulse operators of a cycle is the unit operator $\mathbf{1}$. Then, for sufficiently small pulse separations τ, the evolution operator in (3.3.33) can be approximated by

$$\begin{aligned}
\mathbf{U}(6\tau) &= \mathbf{L}^z\mathbf{L}^y\mathbf{L}^x\mathbf{L}^x\mathbf{L}^y\mathbf{L}^z\\
&= \exp\{-\mathrm{i}\mathbf{H}^z\tau\}\exp\{-\mathrm{i}\mathbf{H}^y\tau\}\exp\{-\mathrm{i}2\mathbf{H}^x\tau\}\exp\{-\mathrm{i}\mathbf{H}^y\tau\}\exp\{-\mathrm{i}\mathbf{H}^z\tau\}\\
&= \mathbf{1} - \mathrm{i}\mathbf{H}^z\tau - \mathrm{i}\mathbf{H}^y\tau - \mathrm{i}2\mathbf{H}^x\tau - \mathrm{i}\mathbf{H}^y\tau - \mathrm{i}\mathbf{H}^z\tau - \cdots\\
&= \mathbf{1} - \mathrm{i}2\tau(\mathbf{H}^x + \mathbf{H}^y + \mathbf{H}^z) - \cdots \approx \exp\{-\mathrm{i}\mathbf{H}^{\mathrm{A}}6\tau\}.
\end{aligned} \tag{3.3.24}$$

The last equation defines the *average Hamilton operator* $\mathbf{H}^{\mathrm{A}}$ in the toggling frame, which is given in zeroth order by the average $(\mathbf{H}^x + \mathbf{H}^y + \mathbf{H}^z)/3$. For $\mathbf{H}^\beta$ the Hamilton operators of the *chemical shielding* ($\lambda = \sigma$ in Section 3.1.1) and of the *dipole–dipole interaction* ($\lambda = \mathrm{D}$) are inserted. They are given in the rotating frame by

$$\mathbf{H}_\sigma = \omega_0\sigma_{zz}\mathbf{I}_z, \tag{3.3.25}$$

and by

$$\mathbf{H}_D = \frac{\mu_0\gamma_i\gamma_j}{8\pi r_{ij}^3}\left(3\cos^2\beta_{ij} - 1\right)\left(3\mathbf{I}_z^i\mathbf{I}_z^j - \mathbf{I}^i\mathbf{I}^j\right). \tag{3.3.26}$$

In the average Hamilton operator $\mathbf{H}^{\mathrm{A}}$, therefore, the dipolar terms cancel in lowest order,

$$\begin{aligned}
&\frac{\mathbf{H}_{\mathrm{D}}^x + \mathbf{H}_{\mathrm{D}}^y + \mathbf{H}_{\mathrm{D}}^z}{3}\\
&= \frac{\mu_0\gamma_i\gamma_j}{8\pi r_{ij}^3}\left(3\cos^2\beta_{ij} - 1\right)\frac{3\mathbf{I}_x^i\,\mathbf{I}_x^j - \mathbf{I}^i\mathbf{I}^j + 3\mathbf{I}_y^i\,\mathbf{I}_y^j - \mathbf{I}^i\mathbf{I}^j - 3\mathbf{I}_z^i\,\mathbf{I}_z^j - \mathbf{I}^i\mathbf{I}^j}{3} = 0,
\end{aligned} \tag{3.3.27}$$

while the magnetic shielding is reduced to

$$\frac{\mathbf{H}_\sigma^x + \mathbf{H}_\sigma^y + \mathbf{H}_\sigma^z}{3} = \frac{\omega_0 \sigma_{zz} (\mathbf{I}_x + \mathbf{I}_y + \mathbf{I}_z)}{3}. \tag{3.3.28}$$

For the average Hamilton operator of the chemical shift, the *effective field* is in the direction of the space diagonal along $\boldsymbol{a} = (1, 1, 1)$. This means that the magnetization precesses around an axis which is oriented at the magic angle Ψ_m relative to the magnetic field $\boldsymbol{B}_0$. Thus the magnetic field, which determines the resonance frequency is reduced by the factor $R_\sigma = \cos \Psi_\mathrm{m} = 3^{-1/2}$. The chemical shift is then scaled by this factor and the resonance frequencies appear to be compressed.

Further multi-pulse sequences

Elimination of the homonuclear dipole-dipole coupling, while preserving the chemical shielding, is the most frequent requirement for multi-pulse sequences. Different sequences vary in the number of pulses per cycle and in the pulse phases. In this way the interactions can be made to cancel in the average Hamilton operator not only in zeroth order ($\mathbf{H}_\mathrm{D}^{\mathrm{A}(0)} = \mathbf{0}$), but also in first and higher orders, even up to fifth order [Hoh1]. Also the chemical shielding can be scaled by different factors R_σ, and effects of finite pulse lengths t_p, of B_1 inhomogeneities, as well as effects of cross terms between different interactions can be compensated for to different degrees of accuracy [Hoh1]. Further improvement of the averaging can be achieved by *second averaging*. This denotes the introduction of a second axis of averaging, which is orthogonal to the first [Abr2, Cor1, Ger1, Hae1, Meh1].

Some properties of important multi-pulse sequences are collected in Table 3.3.1. The vector $\boldsymbol{a}$ indicates the direction of the average effective field. For homonuclear decoupling the MREV8 [Rhi1] sequence is used most often in practice. The WIM24 sequence [Car1, Car2] is applied for polarization transfer by isotropic mixing. It not only averages the homonuclear dipole–dipole coupling to zero, but also averages the chemical shielding ($R_\sigma = 0$) [Man1]. Then, apart from relaxation, there is no evolution of the transverse magnetization. For this reason, multi-pulse sequences which average the chemical shift and the homonuclear dipole–dipole interaction to zero are referred to as *time suspension sequences* [Cor2].

Illustration of homonuclear decoupling

The use of multi-pulse excitation for high-resolution solid-state NMR of ^{1}H is illustrated in Fig. 3.3.12 by an MREV8 spectrum of a calcium formiate crystal acquired with a minimum pulse separation of $\tau = 4\,\mu$s and a width of the 90° pulses of $t_\mathrm{p} = 0.75\,\mu$s [Pri1, Pri2]. The line splittings are revealed only under decoupling of the homonuclear dipole–dipole interaction. For acquisition of spectrum *a*, a crystal sphere was glued to the tip of a glass rod. The glass rod served to fix the sample and to rotate it around well-defined angles for determination of the chemical shift tensor elements from measurement of the angular dependence of the resonance frequency. For acquisition of spectrum *b* the crystal was polished to the shape of a rotational ellipsoid, which was placed inside a glass tube. The resulting improvement in resolution demonstrates that the observed

Table 3.3.1 Multi-pulse sequences

Cycle	Ref.	Length	R_σ	$\mathbf{H}_D^{A(0)}$	$\mathbf{H}_D^{A(1)}$	$\mathbf{H}_D^{A(2)}$	Compensation of t_p	Compensation of B_1 inhom.	$\boldsymbol{a}$
WAHUHA	Wau1	6τ	$1/\sqrt{3}$	0	0	$\neq 0$	No	No	(1, 1, 1)
MREV8	Man3 Rhi1	12τ	$\sqrt{2}/3$	0	0	$\neq 0$	Yes	Yes	(1, 0, 1)
BR24	Bur1	36τ	$2/(3\sqrt{3})$	0	0	0	Yes	Yes	(1, 1, 1)
WIM24	Car2	24τ	0	0	0	–	–	–	(0, 0, 0)

Cycle	Sequence
WAHUHA	$\tau, +x, \tau, -y, \tau, \tau, +y, \tau, -x, \tau$
MREV8	WAHUHA, $\tau, -x, \tau, -y, \tau, \tau, +y, \tau, +x, \tau$
BR24	MREV8, $\tau, -y, \tau, +x, \tau, \tau, -x, \tau, +y, \tau, \tau, +y, \tau, +x, \tau, \tau, -y, \tau, +x, \tau, \tau, -x, \tau, +y, \tau, \tau, +y, \tau, +x, \tau, \tau, -x, \tau, -y, \tau, \tau, -x, \tau, -y, \tau$
WIM24	$-y, +x, -y, -y, +x, -y, +y, +x, +y, +y, +x, +y, -y, -x, -y, -y, -x, -y, +y, -x, +y, +y, -x, +y$

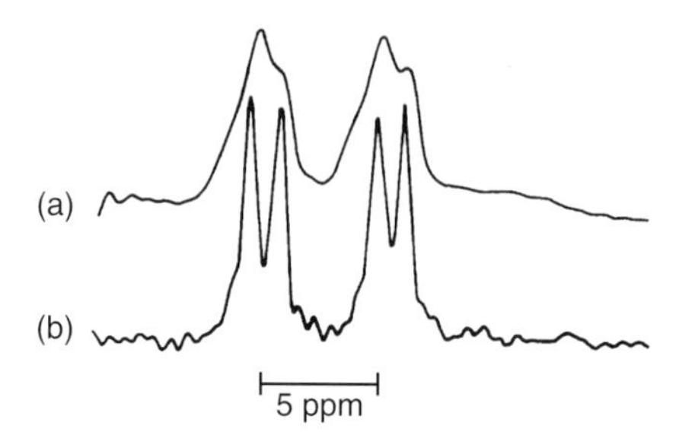

FIG. 3.3.12 [Pri2] ^{1}H-MREV8 spectra of a calcium formiate crystal for a pulse seperation of $\tau = 4\,\mu$s and a pulse width of $t_p = 0.75\,\mu$s. (a) Crystal sphere glued to the tip of a glass rod. (b) Rotational ellipsoid inside a long glass tube without a glass rod. Adapted from [Pri1] with permission.

linewidths are the result of magnetic field distortions by susceptibility differences rather than of shortcomings in the efficiency of the homonuclear decoupling caused by finite pulse widths t_p and pulse separations τ. Thus, sample shape is an important factor in multi-pulse experiments, in addition to reliable and well-tuned hardware.

Combined rotation and multi-pulse spectroscopy: CRAMPS

To measure high-resolution spectra of disordered samples, the anisotropy of the chemical shift must be eliminated in addition to the homonuclear dipole–dipole interaction. This can be achieved for abundant nuclei by *combined rotation and multi-pulse spectroscopy (CRAMPS)*, which is the combination of homonuclear multi-pulse spectroscopy and MAS [Bur2, Ger2, Sch6]. In Fig. 3.3.13, ^{1}H spectra of monoethyl fumarate are compared for different NMR techniques [Bro1]. Only the combination of MAS and multi-pulse excitation produces a high-resolution spectrum of the polycrystalline powder.

Amorphous and semicrystalline polymers are structurally more heterogeneous than polycrystalline materials. The associated spread in chemical shifts reduces the achievable spectral resolution in such samples. This is illustrated in Fig. 3.3.14 for a number of different synthetic polymers [Bro1]. However, wide lines in CRAMPS spectra can also be observed if dynamic processes are present on a timescale which matches the cycle time, or if the spinning frequency ω_R or a multiple thereof matches the off-set frequency [Bro1]. At fast rotation speeds, interference effects between MAS and multi-pulse excitation become increasingly important [Dem2, Haf1]. By these and other processes [Oas1, Ral1] including suitable multi-pulse sequences, the dipole–dipole interaction can be recoupled into the signal [Ben1, Lee2, Tyc1, Nie1].

3.4 ECHOES

Many NMR techniques rely on the formation of magnetization *echoes* [Bag1, Blü5]. Examples are (1) the measurement of T_2 relaxation times by the Hahn echo (Section 2.2), (2) the analysis of slow motion by stimulated echoes for spins with $I = 1/2$, and for nuclei with spin $I = 1$ or pairs of coupled spins with $I = 1/2$ by the solid echo and the alignment echo (Section 3.2), (3) MAS, where echoes are generated by mechanical

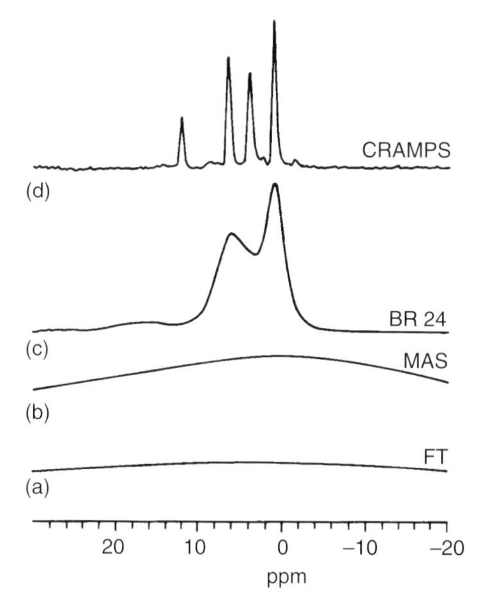

FIG. 3.3.13 ^{1}H NMR spectra of $CH_3CH_2O_2CCH{=}CHCO_2H$ at 187 MHz for different excitation techniques. (a) Single-pulse excitation of the nonspinning sample. (b) Single-pulse excitation and MAS. (c) Multi-pulse excitation of the nonspinning sample with BR24 sequence. (d) Combination of BR24 and MAS. Adapted from [Bro1]. Copyright 1988 American Chemical Society.

sample rotation for line narrowing (Section 3.3.3), (4) multi-pulse NMR, where intricate cycles of solid echoes and magic echoes are employed for instance for homonuclear decoupling in high-resolution NMR of abundant spins in solids (Section 3.3.4), and (5) gradient echoes, where magnetization dephased in a magnetic field gradient is rephased by inversion of the gradient (Section 2.2.1).

Homogeneously and inhomogeneously broadened lines

Echoes are associated with an apparent *time reversal* for the evolution of most of the spins, that is, whatever spin evolution has taken place is retraced in reverse order. A simple explanation rests on the treatment of uncoupled magnetization components. The FID signal decays by accumulation of phase differences between individual magnetization components in the transverse plane, starting from a common initial phase at the beginning of the signal. If these differences can be made to cancel, an echo forms. The fact that this can be done has been demonstrated by Erwin Hahn in 1949 (cf. Section 2.2.1) [Hah1]. Nevertheless, the echo amplitude is usually somewhat smaller than the amplitude of the initial FID, so the spin precession cannot completely be reversed. Therefore, the echo amplitude decays even under continuous refocussing (Fig. 3.4.1(a)). This decay is due to a time-dependent modulation of the precession phases of essentially all transverse magnetization components. The slow signal decay transforms into the so-called *homogeneously broadened line* (b). The fast signal decay of the FID or of the echo itself is predominantly

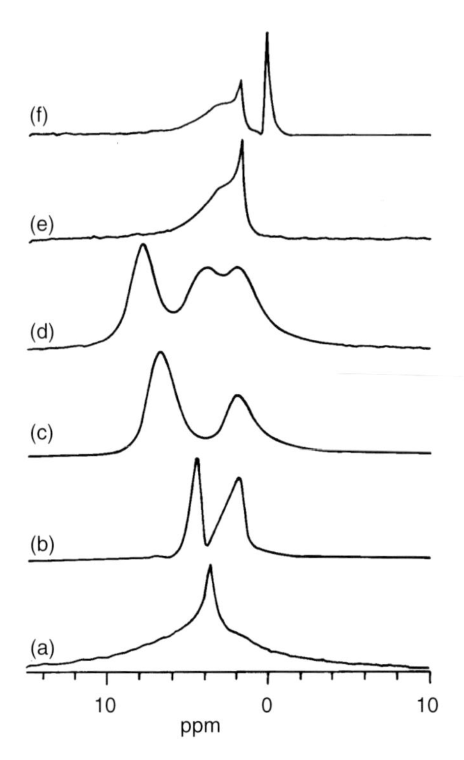

FIG. 3.3.14 ^{1}H CRAMPS spectra at 187 MHz for different amorphous samples. (a) Polyethylene oxide. (b) Polymethylmethacrylate. (c) Polystyrene. (d) Polybutanediolterephthalate. (e) Silicagel. (f) A silicagel $(CH_3)_3$ derivative. Adapted from [Bro1]. Copyright 1988 American Chemical Society.

caused by a time-independent distribution of resonance frequencies. Its Fourier transform is the so-called *inhomogeneously broadened line* (c). The inhomogeneously broadened line is the sum of all homogeneously broadened lines, each of which results from a magnetization component with a different frequency. Thus a solid-state powder spectrum (cf. Fig. 3.1.3) is an example of an inhomogeneously broadened line, and so is the spectrum of magnetization measured in an inhomogeneous magnetic field, for instance in a field gradient. In addition to homogeneously and inhomogeneously broadened lines, *heterogeneously broadened lines* can be considered as an intermediate case [Meh1]. Heterogeneously broadenend lines contain coupled homogeneously broadened lines. Such lines can be observed for rare spins coupled to strongly interacting abundant spins.

Echo modulation

The apparent change in sign of the time axis required for formation of the echo is achieved in practice by sign inversion of the complete Hamilton operator, which can be manipulated by rf pulses, sample reorientation, and magnetic field manipulations. But these manipulations can be carried out selectively for just some parts of the complete Hamilton operator. Then the evolution of the magnetization under the influence of the

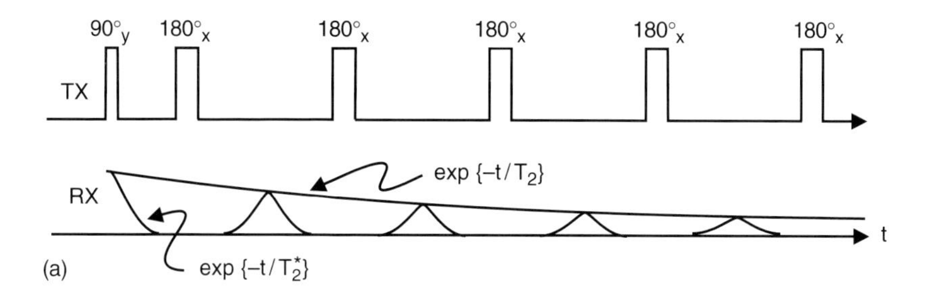

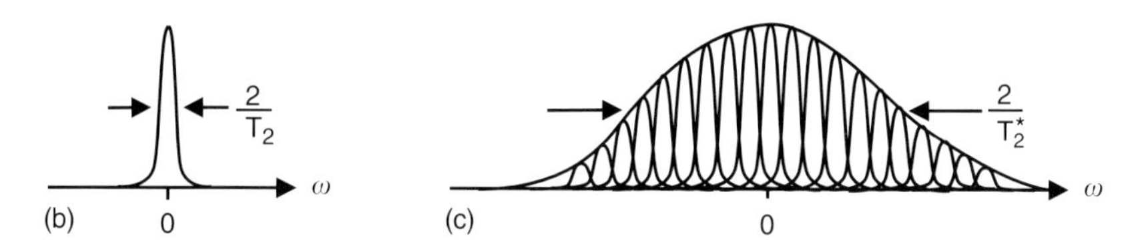

FIG. 3.4.1 Homogeneously and inhomogeneously broadened lines. (a) Echo train generated by repeated refocussing of the FID (CPMG method, cf. Fig 2.2.10(b)). (b) The Fourier transform of the slowly decaying echo envelope is the homogeneously broadened line. (c) The Fourier transform of the fast decaying echo is the inhomogeneously broadened line.

other parts will continue and modulate the echo amplitude. For example, the evolution of magnetization under the influence of the homonuclear coupling is unaffected by the rf pulses generating the Hahn echo. Only the phase evolution resulting from different chemical shifts and magnetic field inhomogeneities is reversed. Thus, the echo amplitude is modulated by the homonuclear coupling. This fact is exploited in *2D J spectroscopy* [Aue1]. Here a 2D spectrum is measured [Ern1], where one frequency axis represents the NMR spectrum with the effects of chemical shift and J coupling included, the other frequency axis represents the effects of only the J coupling. One says that the echo refocusses the chemical shift, and not the J coupling in this case. Similarly, the gradient echo refocusses the effects of frequency dispersion by the applied gradient only, and not the phase evolution resulting from the Zeeman and other spin interactions including the inhomogeneity of the static magnetic field. Another example of selective refocussing is the rotational echo observed in MAS NMR. Here only the anisotropic parts of the interactions are refocussed, but not the isotropic parts. Therefore, narrow lines are produced, for instance at the isotropic chemical shifts in ^{13}C MAS NMR, which are flanked by spinning sidebands. The remaining information about the anisotropies contained in the spinning sidebands can be separated from the information about the isotropic parts in a 2D experiment, by stacking the rotational echoes in a 2D matrix and applying a 2D Fourier transformation (Fig. 3.4.2) [Blü2]. In such a 2D spectrum, measured in response to a single pulse (*TOP spectroscopy*: 2D one-pulse spectroscopy), the modulation of the echo maxima appears on one axis of the spectrum, and the echo decay on the other. Thus isotropic chemical shifts and anisotropic wideline information are separated.

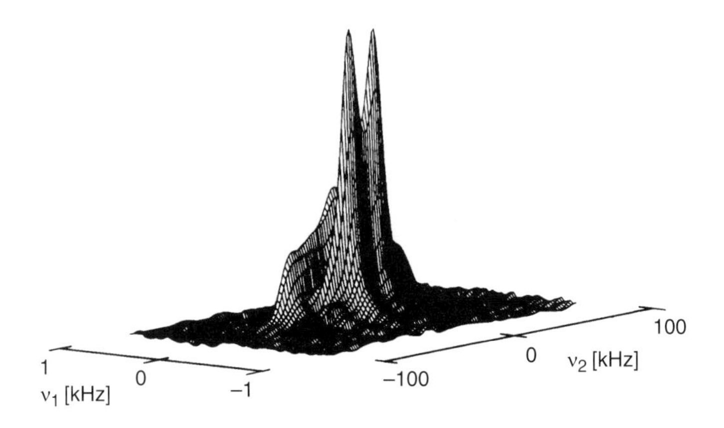

FIG. 3.4.2 2D one-pulse ^{2}H MAS spectrum of bisphenyl-A polycarbonate, deuterated at the ortho positions of the phenylene rings and at 5% of the methyl groups. Along ν_1-axis the isotropic chemical shift is resolved with the homogeneous linewidth. Phenyl and methyl resonances are clearly discriminated. Along ν_2-axis the inhomogeneous linewidth appears, which is partially averaged by MAS. The wider line derives from the phenyl deuterons, the narrower one from the methyl deuterons. Adapted from [Blü2] with permission from Elsevier Science.

Echoes and NMR imaging

The importance of echoes in imaging is based on the fact that narrow signals can be created, the linewidths of which are often as narrow as the homogeneous widths. If a field gradient can be applied in such a way that it effectively acts only on this narrow line (the chemical shift dimension $\omega_1 = 2\pi\nu_1$ in Fig. 3.4.2, for instance), then high spatial resolution can be achieved, even in solids. When extending the spatial axis by a spectroscopic axis (the wideline axis $\omega_2 = 2\pi\nu_2$ in Fig. 3.4.2), detailed information about molecular configuration, conformation, order, and mobility can be obtained for each point in space. Many liquid- and solid-state imaging techniques are indeed based in one way or another on the exploitation of echoes for generation of narrow lines.

There is a second reason for the use of echoes in imaging. In most cases magnetic-field gradients need to be switched, and the signal response in the presence of magnetic field gradients decays fast. After switching a gradient, the gradient needs to settle to obtain a magnetic field free of transient phase and amplitude distortions. Also after applying a strong rf pulse, the receiver needs to recover from the impact it received by the inevitable break-through from the transmitter. For both reasons, the signals need to be delayed in time for acquisition to overcome instrumental *deadtime*.

The third reason for the use of echoes is that signals need to be measured for negative and positive $\boldsymbol{k}$ values (cf. eqn (2.2.23)) in *frequency encoding*, which correspond to a phase evolution of the magnetization components from negative values to zero and to positive values. At zero phase, $\boldsymbol{k} = 0$ and the gradient echo is formed. Thus, the goal is to acquire the build-up and decay of gradient echoes. For optimum signal strength, these are often made to coincide with echoes, which refocus other parts of the Hamilton operator, for instance with the Hahn echo for refocussing of the chemical shift. Another NMR technique where two types of echoes are made to coincide is the TOSS experiment. Here, the rotational echo generated by sample rotation and the Hahn echo are coincident at the beginning of data acquisition (cf. Section 3.3.3).

The magic echo

In addition to the echoes already mentioned, there is one other echo in particular, which is used successfully in imaging. This is the *magic echo* [Rhi2, Sch7, Sch8, Sli1]. With it the dephasing under the homonuclear dipole–dipole interaction can be refocussed completely also in homonuclear spin systems, even after a time as long as T_2. The solid and alignment echoes are capable of doing this only for isolated spin pairs. If more than two spins couple, refocussing by solid and alignment echoes is incomplete. For a long time it was believed that under multiple homonuclear dipole–dipole couplings the magnetization is lost after a time of the order of T_2 and no echo can be formed because the intuitive picture of individual magnetization components no longer applies. This was proven wrong with the discovery of the magic echo.

The two-centre dipole–dipole coupling scales with the second Legendre polynomial $P_2(\cos\theta)$, where θ is the angle between the static field $\boldsymbol{B}_0$ and the effective field $\boldsymbol{B}_{\text{eff}}$. For $\theta = 0$, $P_2 = 1$, and for $\theta = 90^\circ$, $P_2 = -1/2$. Therefore, sign and magnitude of the dipole–dipole interaction change as the effective field is rotated from the z-axis into the xy plane. In this way, the evolution under the dipole–dipole interaction can be reversed, so that an echo is formed.

The pulse sequence is depicted in Fig. 3.4.3. Time reversal is achieved by the *magic sandwich*, which essentially consists of a lock field with duration 4τ. Before and after the lock field the magnetization precesses freely for a time 2τ in total. Given the different strengths of the dipole–dipole interaction during free precession and in the presence of the $\boldsymbol{B}_1$ field, refocussing occurs at time 6τ. After a time 3τ from the beginning of the pulse sequence, another echo occurs, which is also known as *rotary echo*. It arises from rotation of the magnetization around the effective field [Gol1]. This rotary echo can be observed if the continuous spin lock field is broken down into pulses separated by free precession windows for data acquisition. In fact, several possibilities exist to generate magic echoes by different pulse sequences [Rhi2, Sch7, Sch8]. For example, a tetrahedral magic echo can be generated in two-spin systems [Dem3].

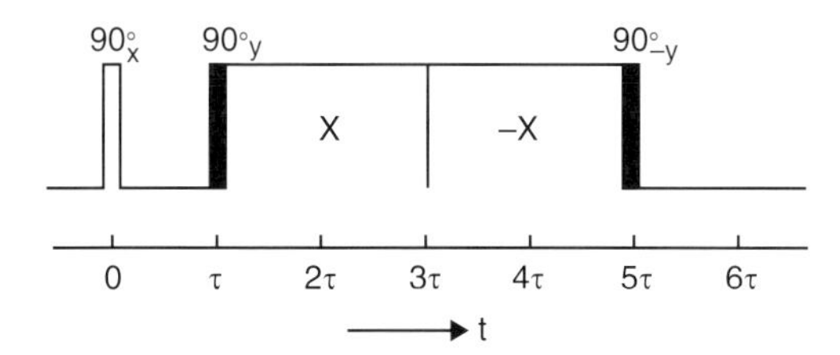

FIG. 3.4.3 Pulse sequence for generation of the magic echo. The magnetization dephases for a time 2τ by free precession before and after refocussing by the magic sandwich of length 4τ.

3.5 RELAXATION

Relaxation is the process by which the density matrix elements approach their thermodynamic equilibrium values. Following the simple picture of the Bloch equations, one can distinguish two types of relaxation times. First, there are the energy or T_1-type relaxation times, which determine the rate at which the populations, that is the diagonal elements of the density matrix adjust to thermodynamic equilibrium. If the density matrix has dimension n, then there are $n-1$ energy relaxation times, because the sum of all relative populations is always equal to one. Second, there are the phase or T_2-type relaxation times of the transverse magnetization, or, more generally, of the off-diagonal density matrix elements. They determine the rate at which these matrix elements relax to zero.

Relaxation theory is concerned with a formal description of the processes which achieve relaxation. In general these are statistical, rotational and translational molecular motions, which lead to fluctuating local fields at the sites of the nuclei. These fields can modulate the resonance frequencies and induce transitions. Therefore, the relaxation times can provide important information about molecular motion on different timescales. They are highly effective for use in magnetization weights and as parameters to generate contrast in NMR images (cf. Chapter 7). The origin of the local fields can be manifold. The dominating mechanism is the *intra*- and *inter*-molecular dipole–dipole interaction. Others are the quadrupole interaction, the chemical shift anisotropy, and the spin rotation interaction. A rigorous analysis is beyond the scope of this book [Abr1, Sli1, Spi1, Sud1]. Instead the prominent features are summarized focussing on the dipolar relaxation [Ern1] and a simple description of longitudinal and transverse relaxation times T_1 and T_2, respectively [Cal1, Far1].

3.5.1 Liquids

Molecules in liquids are in rapid motion with respect to the inverse Larmor frequency. Relaxation in the rapid motion limit was originally treated by Bloembergen, Purcell, and Pound [Blo1]. Their theory of relaxation is called the *BPP theory* [Red1].

Spectral densities and longitudinal relaxation

Following the treatment of Slichter [Cal2, Sli1] the *spin–lattice relaxation* is written as

$$\frac{1}{T_1} = \frac{\sum_{nm} W_{nm}\left(E_n^2 - E_m^2\right)}{2\sum_n E_n^2}, \tag{3.5.1}$$

where the transition rate between eigenstates n and m with energies E_n and E_m is calculated by second-order perturbation theory,

$$W_{nm} = C\sum_{qq'}\left[\int_0^\infty \exp\left\{\mathrm{i}(E_n - E_m)\frac{\tau}{\hbar}\right\}\overline{P_{2q}(0)P_{2-q'}(\tau)\mathrm{d}\tau}\right]\langle n|\mathbf{T}_{2q}|m\rangle\langle m|\mathbf{T}_{2-q'}|n\rangle. \tag{3.5.2}$$

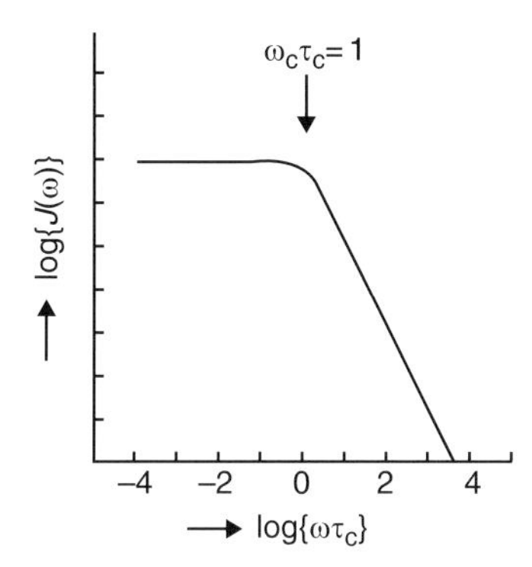

FIG. 3.5.1 Double logarithmic representation of the spectral density J as a function of frequency ω for a fixed correlation time τ_c. Adapted from [Cal2] with permission from Oxford University Press.

Here the Hamiltonian responsible for relaxation ($\lambda = \mathrm{D}$ for the dipole–dipole interaction) has been written in the notation of *irreducible spherical tensors* introduced in Section 3.1.2. For isotropic motion, the ensemble average reduces to

$$\overline{P_{2q}(0) P_{2-q'}(\tau)} = \delta_{qq'} a^q(\tau) \tag{3.5.3}$$

where $a^q(\tau)$ is the *auto-correlation function* of the spatial component q. Because the $P_{2q}(\tau)$ are dipolar fields, $a^q(\tau)$ is proportional to r^{-6}. With the definition (3.5.3) of $a^q(\tau)$ the integral in (3.5.2) is the Fourier transform of the auto-correlation function. It is known as the *spectral density function* $J^{(q)}(\omega)$. It represents the intensity of fluctuation in P_{2q} at frequency ω. As the auto-correlation function is characterized by the correlation time τ_c , where $a^q(\tau)$ approaches zero as $\tau \gg \tau_c$, the spectral density function is characterized by the frequency $\omega_c = 1/\tau_c$, where $J^{(q)}(\omega)$ approaches zero as $\omega \gg \omega_c$ (Fig. 3.5.1). Clearly the fastest motion possible is the free reorientation of a molecule, so that, depending on the type of motion, the *cut-off frequency* ω_c is anywhere between zero and 10^{13} Hz.

The selection rules defined by the tensor-operator matrix elements $\langle n|\mathbf{T}_{2q}|m\rangle$ restrict the relevant transitions to the Larmor frequency and twice the Larmor frequency. The result of the calculation yields for the T_1 relaxation of an isolated spin pair by random dipolar fields

$$\frac{1}{T_1} = \left(\frac{\mu_0}{4\pi}\right)^2 \gamma^4 \hbar^2 I(I+1) \frac{3}{2} \left[J^{(1)}(\omega_\mathrm{L}) + J^{(2)}(2\omega_\mathrm{L})\right]. \tag{3.5.4}$$

Spin–spin relaxation

Spin–spin relaxation is handled by second order perturbation theory [Abr1] of the density matrix equation of motion (2.2.62) in the rotating frame (RCF) [Abr1],

$$\frac{\mathrm{d}\boldsymbol{\rho}_\mathrm{RCF}(\mathrm{t})}{\mathrm{dt}} = -\int_0^\infty \overline{\left[\mathbf{H}^\mathrm{D}(t), \left[\mathbf{H}^\mathrm{D}(t-\tau), \boldsymbol{\rho}_\mathrm{RCF}(0)\right]\right]}\, d\tau. \tag{3.5.5}$$

The resulting transverse relaxation time also depends on the spectral density at frequency zero,

$$\frac{1}{T_2} = \left(\frac{\mu_0}{4\pi}\right)^2 \gamma^4 \hbar^2 I(I+1) \frac{3}{2} \left[\frac{J^{(0)}(0)}{4} + \frac{5J^{(1)}(\omega_L)}{2} + \frac{J^{(2)}(2\omega_L)}{4} \right]. \tag{3.5.6}$$

The key assumption in the derivation of (3.5.6) is that the strength δ_D of the dipole–dipole field is small compared to the cut-off frequency ω_c. This is the *fast motion limit*, where the anisotropy of the lineshape is *motionally narrowed*.

Relaxation in the rotating frame

A similar expression can be derived for the relaxation time $T_{1\rho}$ in the rotating frame. The spins are locked by the $\boldsymbol{B}_1$ field applied parallel to the transverse magnetization, if the amplitude ω_1 of the lock field is larger than the strength δ_D of the dipole–dipole coupling. The magnetization parallel to the lock field decays at a rate

$$\frac{1}{T_{1\rho}} = \left(\frac{\mu_0}{4\pi}\right)^2 \gamma^4 \hbar^2 I(I+1) \frac{3}{2} \left[\frac{J^{(0)}(\omega_1)}{4} + \frac{5J^{(1)}(\omega_L)}{2} + \frac{J^{(2)}(2\omega_L)}{4} \right]. \tag{3.5.7}$$

The expression for $T_{1\rho}$ equals that for T_2 in the limit as ω_1 approaches zero, although the spin-lock condition breaks down with decreasing ω_1. Nevertheless, from the experimental point of view $T_{1\rho}$ is a very handy parameter for the characterization of slow molecular motion on the time scale of ω_1^{-1}, because ω_1 can be adjusted via the spin-lock power so that $T_{1\rho}$ can be determined as a function of frequency.

Isotropic rotational diffusion

For many cases of molecular reorientation in liquids, the isotropic rotational diffusion model is applicable and the dipole–dipole interaction provides the dominant relaxation mechanism. Then, the spectral densities are given by [FAR1]

$$J^{(0)}(\omega) = \frac{24}{15 r_{ij}^6} \frac{\tau_c}{1 + \omega^2 \tau_c^2}, \tag{3.5.8a}$$

$$J^{(1)}(\omega) = \frac{4}{15 r_{ij}^6} \frac{\tau_c}{1 + \omega^2 \tau_c^2}, \tag{3.5.8b}$$

$$J^{(2)}(\omega) = \frac{16}{15 r_{ij}^6} \frac{\tau_c}{1 + \omega^2 \tau_c^2}, \tag{3.5.8c}$$

Upon substitution of these functions into the expressions (3.5.4) and (3.5.6) for T_1 and T_2, respectively, the dependence of the relaxation times on the correlation time τ_c depicted in Fig. 3.5.2 is obtained for a given Larmor frequency ω_L. Regimes of slow and fast motion are discriminated by the T_1 *minimum*, where the correlation time is of the order of the Larmor period. In the regime where $\tau_c^{-1} \gg \omega_L$ molecular motion is fast on the NMR timescale, and the homogeneous linewidth is highly reduced. This situation is typical for small molecules at room temperature. It is called the *extreme narrowing limit*, where both T_1 and T_2 coincide.

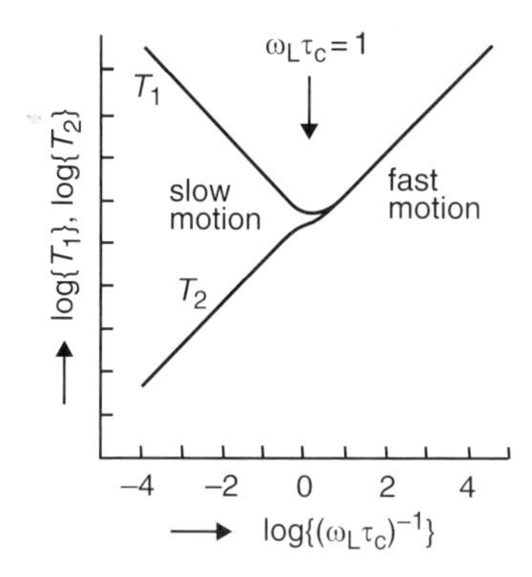

FIG. 3.5.2 Double logarithmic representation of the relaxation times T_1 and T_2 as a function of correlation time τ_c for a fixed Larmor frequency ω_L. Adapted from [Cal2] with permission from Oxford University Press.

On the other side of the T_1 minimum, molecular motion is slow compared to the Larmor period. T_1 and T_2 diverge. This situation is typical for highly viscous liquids and slowed-down or restricted motion in polymers well above the glass transition temperature. Because T_2 is short, the FID decays fast, yet T_1 is long, so that the repetition time for signal averaging is long. Therefore, the *slow motion regime* is experimentally more difficult to access by NMR, and solid-state techniques need to be applied.

3.5.2 Solids and slow motion

When the reorientation is slow compared to the inverse strength of the dipole–dipole interaction, the line-narrowing assumption of the BPP theory is no longer applicable. Explicit motional mechanisms and nuclear spin interactions have to be taken into account in the calculation of the NMR signal decay [Ail1, Spi1].

Dipolar relaxation of protons

For ^{1}H the homonuclear dipole–dipole coupling links spins together. In this case, the expression (3.5.4) for T_1 still remains valid, but the expression for T_2 changes [Abr1, And2]. The fluctuating local fields can be represented as a time-dependent offset $\Omega(t)$ from the Larmor frequency. The normalized FID $s(t)$ is given by the ensemble average

$$s(t) = \overline{\exp\left\{ \mathrm{i} \int_0^t \Omega(t')\,\mathrm{d}t' \right\}}. \tag{3.5.9}$$

This expression can be evaluated under two assumptions. The first one is, that the distribution of $\Omega(t)$ is Gaussian. Then $s(t)$ is Gaussian as well. The second assumption is that the local field fluctuation can be described by a correlation function $a_\Omega(\tau)$ of the frequency offset Ω,

$$\overline{\Omega(t)\Omega(t-\tau)} = \overline{\Omega^2} a_\Omega(t). \tag{3.5.10}$$

Then (3.5.9) simplifies to

$$s(t) = \overline{\exp\left\{-\overline{\Omega^2}\int_0^t (t-\tau)a_\Omega(\tau)\,\mathrm{d}\tau'\right\}}. \tag{3.5.11}$$

In the short-time limit and with an exponential correlation function $a_\Omega(\tau)$ the FID $s(t)$ is approximated by

$$s(t) \approx 1 - \frac{\overline{\Omega^2}t^2}{2} = 1 - \frac{M_2 t^2}{2}, \tag{3.5.12}$$

where M_2 is the second moment of the lineshape function $S(\Omega)$.

Note that the use of moments derives from the days of CW NMR, when Fourier transformation was not standard, and the information about the FID had to be derived from the spectrum. The moments are defined by an expansion of the exponential in the Fourier transform of the FID,

$$s(t) = \frac{1}{2\pi}\int_{-\infty}^{\infty} \mathrm{S}(\Omega)\exp\{\mathrm{i}\Omega \mathrm{t}\}\,\mathrm{d}\Omega = \frac{1}{2\pi}\int_{-\infty}^{\infty} \mathrm{S}(\Omega)\left[1 - \frac{(\Omega \mathrm{t})^2}{2} + \cdots\right]\mathrm{d}\Omega. \tag{3.5.13}$$

For symmetric lineshapes integrals with odd powers in t vanish, and the moments are given by

$$M_{2n} = \frac{\int_{-\infty}^{\infty} S(\Omega)\Omega^{2n}\,\mathrm{d}\Omega}{\int_{-\infty}^{\infty} S(\Omega)\,\mathrm{d}\Omega}. \tag{3.5.14}$$

In the fast motion limit $a_\Omega(\tau)$ decays rapidly, and the integral in (3.5.11) can be approximated by

$$\int_0^t (t-\tau)a_\Omega(\tau)\mathrm{d}\tau' \approx t\int_0^{\infty} a_\Omega(\tau)\,D\tau = t\tau_c, \tag{3.5.15}$$

where the definition (3.2.11) of the correlation time has been used. With the second moment of (3.5.12) the FID then becomes

$$s(t) = \exp\{-M_2\tau_c t\}. \tag{3.5.16}$$

Thus, $M_2\tau_c$ is the relaxation rate $1/T_2$, which has been calculated in (3.5.6) by the BPP theory in the fast motion limit. In the slow motion limit of (3.5.6) only the spectral density (3.5.8a) at frequency zero needs to be considered, and

$$\frac{1}{T_2} = \frac{\mu_0}{4\pi}\gamma^4\hbar^2\frac{3}{2}I(I+1)\frac{J^{(0)}(\omega_\mathrm{L})}{4} = M_2\tau_c \tag{3.5.17}$$

is obtained. In this limit the rate T_2^{-1} of transverse relaxation is proportional to the correlation time τ_c of molecular motion.

Correlation of relaxation with models of molecular structure and dynamics

A way to consider specific slow molecular motions which modulate the signal decay is by means of the exchange matrix $\mathbf{\Pi}$ introduced in Section 3.2.2 for the description of lineshape effects in wideline spectra (cf. eqn. 3.2.14) [Spi1]. Because polymers are rich in protons, considerable effort has been spent to model time-domain NMR signals of polymers in terms of molecular structure and dynamics [Fed1]. Expressions for relaxation times and second moments have been derived, for instance, to characterize the crystallinity in semicrystalline polymers and the crosslink density in elastomers [Göt1], the dependence of the T_2 relaxation time on applied strain for elastomers [Nis1] and gels [Coh1], and the swelling of polymers [Mar1]. Relaxation times are also investigated to characterize molecular motion in confined spaces [Kim1, Kim3], such as in zeolites, porous rocks and on surfaces, where strong dependencies of the relaxation times on pore size and surface properties can be found [Liu1, Liu2]. Moreover, they are analysed in the light of fundamental concepts of molecular motion, as for instance in polymers [Kim2]. Because relaxation times are convenient parameters for generation of image contrast, accurate relations between relaxation times and material properties are highly desirable for image analysis and interpretation. A significant goal for practical interpretation of NMR images is the transformation of relaxation-time parameter images into images displaying the local variation in parameters relevant to chemical engineering (cf. Section 7.1.6).

3.5.3 Spin diffusion

Spin diffusion is a process which describes the migration of magnetization through space. This migration is not the result of particle motion, it is mediated by the dipole–dipole interaction between neighbouring spins. Therefore, spin diffusion is a process characteristic for nuclear magnetization in solid matter. Spatial variations in longitudinal nonequilibrium magnetization, which may arise from the presence of paramagnetic relaxation centres, are attenuated by the spin-diffusion process between abundant nuclei like ^{1}H. Thus spin diffusion is an important mechanism of relaxation in solids. It can also be exploited to probe the size of morphological domains for instance in semicrystalline polymers [Cla1, Sch9, Vee2].

Considering an ensemble of some 10^{20} nuclear spins 1/2 at room temperature, about half of the spins are in the spin-up state and half are in the spin-down state. These states of the spins are distributed at random in space. Only about 10 out of one million spins are unbalanced and form the macroscopic, longitudinal nuclear magnetization. Thus the magnetization is distributed in space on a microscopic scale. Because all spins are coupled by the dipole–dipole interaction, the magnetization can migrate through space by energy-conserving flip-flop transitions of neighbour pairs (Fig. 3.5.3). This space and time-dependent process is well described by Fick's second law, the diffusion equation, where D denotes the spin-diffusion constant,

$$\frac{\partial M_z}{\partial t} = D\nabla^2 M_z. \tag{3.5.18}$$

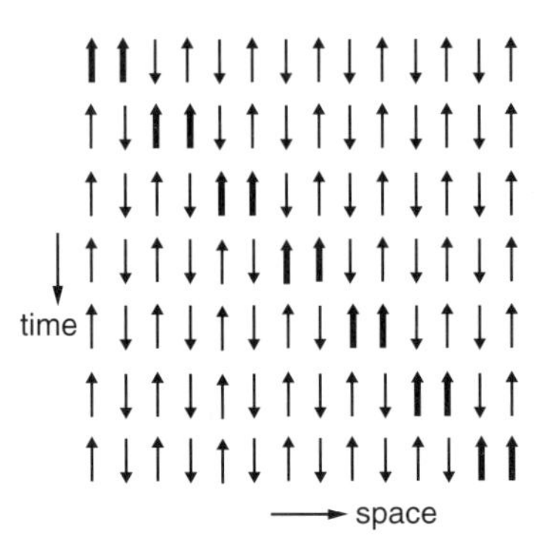

FIG. 3.5.3 [Blü6] Spin diffusion denotes the seemingly random migration of magnetization (bold arrows) through the sample by energy-conserving flip-flop transitions of spin pairs, which, in fact, is not random and can be reversed for solid samples.

Typical values of D for protons in rigid aliphatic polymers are $0.6–0.8 \times 10^{-15}$ m^2/s. Given the spin-diffusion constant, the spin-diffusion equation can be integrated to provide the magnetization change with time for various initial distributions of magnetization and structural models of the sample, for example for magnetized, lamellar structures or spheres of a mobile component in an unmagnetized rigid matrix. Such initial distributions of longitudinal magnetization can be established by the use of magnetization filters (cf. Chapter 7.2.9). Analysis of the time dependence of magnetization transfer from initially magnetized to initially unmagnetized morphological domains provides information about morphological dimensions based on a postulated model of the sample (cf. Section 11.3.2) [Cla1, Dem1, Sch9, Van1, Vee2]. The method is sensitive up to dimensions of the order of 200 nm, depending on the value of the longitudinal relaxation time.

Although the name *spin diffusion* and the description of the phenomenon by the diffusion equation suggests the process of magnetization transfer to be completely irreversible, it is in fact not on a short timescale. In reality the magnetization migrates through the sample along deterministic paths which change only if the network of local dipolar interactions is modified by molecular motions. The spin interaction which mediates the diffusion process is the dipole–dipole interaction. It can be manipulated by pulse sequences, and its sign can be inverted like in the formation of the magic echo (cf. section 3.4). Thus, the magnetization can be made to retrace its paths when the sign of the dipolar interaction is inverted and a *polarization echo* can be observed [Zha1]. If spin diffusion is to be observed between spins with different Larmor frequencies extra energy is necessary to match the energy difference for flipping both spins. This energy can be drawn either from the thermal reservoir of the sample or it can be supplied externally by rf fields or mechanical sample spinning. In the latter case, the spin-diffusion process is accelerated in comparison with the thermally driven process [Col1, Rob1].

4

Transformation, convolution, and correlation

The terms *transformation*, *convolution*, and *correlation* are used over and over again in NMR spectroscopy and imaging in different contexts and sometimes with different meanings. The transformation best known in NMR is the *Fourier transformation* in one and in more dimensions [Bra1]. It is used to generate one- and multi-dimensional spectra from experimental data as well as 1D, 2D, and 3D images. Furthermore, different types of multi-dimensional spectra are explicitly called *correlation spectra* [Ern1]. It is shown below how these are related to nonlinear correlation functions of excitation and response.

Experimental NMR data are typically measured in response to one or more excitation pulses as a function of the time following the last pulse. From a general point of view, spectroscopy can be treated as a particular application of *nonlinear system analysis* [Bog1, Deu1, Mar1, Sch1]. One-, two-, and multi-dimensional *impulse-response functions* are defined within this framework. They characterize the linear and nonlinear properties of the sample (and the measurement apparatus), which is simply referred to as the system. The impulse-response functions determine how the excitation signal is transformed into the response signal. A nonlinear system executes a *nonlinear transformation* of the input function to produce the output function. Here the parameter of the function, for instance the time, is preserved. In comparison to this, the Fourier transformation is a linear transformation of a function, where the parameter itself is changed. For instance, time is converted to frequency. The Fourier transforms of the impulse-response functions are known to the spectroscopist as *spectra*, to the system analyst as *transfer functions*, and to the physicist as dynamic *susceptibilities*.

Impulse-response and transfer functions can be measured not only by *pulse excitation*, but also by excitation with monochromatic, *continuous waves* (CW), and with continuous noise or *stochastic excitation*. In general, the transformation executed by the system can be described by an expansion of the acquired response signal in a series of *convolutions* of the impulse-response functions with different powers of the excitation [Mar1, Sch1]. Given the excitation and response functions, the impulse-response functions can be retrieved by *deconvolution* of the signals. For *white noise excitation*, deconvolution is equivalent to *cross-correlation* [Lee1].

Some basic knowledge about convolution, correlation, and transformation is required for a more general understanding of measurement procedures [Ang1]. In NMR imaging these concepts are needed in particular for the calculation of selective pulses (cf. Fig. 5.3.3), for the description of echo-planar imaging (cf. Section 6.2.8), and in the description of imaging with noise excitation (cf. Section 6.4). In addition, these terms denote fundamental techniques of signal characterization and image processing [Ben1, Bla1, Bog1, Bra1, Deu1, Ern1, Kan1, Pra2]. For instance, the correlation function of the trajectory of molecular motion is a central quantity in the description of molecular dynamics and relaxation processes (cf. Sections 3.2.2 and 3.5), Fourier NMR imaging experiments are performed in the $\boldsymbol{k}$ space, which is the Fourier space of the image (Section 5.4.2), and the $\boldsymbol{q}$-space imaging refers to the acquisition of information in the Fourier conjugate of displacement space (Section 5.4.3). Some relationships relevant in the context of NMR imaging are summarized in the following.

4.1 FOURIER TRANSFORMATION

By *Fourier transformation*, a signal is decomposed into its sine and cosine components [Ang1]. In this way, it is analysed in terms of the amplitude and the phase of harmonic waves. Sine and cosine functions are conveniently combined to form a complex exponential, $\cos \omega t + \mathrm{i} \sin \omega t = \exp\{\mathrm{i}\omega t\}$. The complex amplitudes of these exponentials constitute the spectrum $F(\omega)$ of the signal $f(t)$, where $\omega = 2\pi/T$ is the frequency in units of 2π of an oscillation with time period T. The Fourier transformation and its inverse are defined as

$$F(\omega) = \int_{-\infty}^{\infty} f(t) \exp\{-\mathrm{i}\omega t\}\, \mathrm{d}t, \tag{4.1.1}$$

$$f(t) = \frac{1}{2\pi} \int_{-\infty}^{\infty} F(\omega) \exp\{\mathrm{i}\omega t\}\, \mathrm{d}\omega. \tag{4.1.2}$$

The Fourier transformation above achieves the transition between time and frequency spaces. In this way, for instance, the NMR spectrum $F(\omega)$ is obtained from the FID signal $f(t)$. Similar to time and frequency, a wavenumber $k_x = 2\pi/\lambda$ exists, which is conjugate to the space coordinate x and denotes the frequency in units of 2π of an oscillation in space with wavelength λ (cf. Fig. 2.2.4).

Example 4.1

Fourier transformation of a time-domain signal $f(t)$ decaying with time constant T_2 in an exponential fashion: Such a signal is the NMR *impulse-response function* which can be derived from the Bloch equations (Fig. 4.1.1(b), cf. eqn. 2.2.19),

$$f(t) = f_0 \exp\left\{-\left[\frac{1}{T_2} - \mathrm{i}\Omega_0\right] t\right\}. \tag{4.1.3}$$

Its Fourier transform is obtained by insertion into (4.1.1) and taking care of the fact that the free induction decay(FID) is zero before application of the pulse at $t = 0$,

$$F(\omega) = f_0 \int_0^{\infty} \exp\left\{-\left[\frac{1}{T_2} - \mathrm{i}\,(\Omega_0 - \omega)\right] t\right\} \mathrm{d}t$$

$$= f_0 \frac{(1/T_2) - \mathrm{i}\,(\Omega_0 - \omega)}{(1/T_2)^2 + (\Omega_0 - \omega)^2} = f_0[A(\omega) + \mathrm{i}D(\omega)]. \quad (4.1.4)$$

The real part of this function is proportional to the *absorption signal* $A(\omega)$, and the imaginary part to the *dispersion signal* $D(\omega)$ defined in (2.2.26) (cf. Fig. 2.2.6). The absorption signal is a resonance line with a full width at half height of $\Delta\omega_{1/2} = 2/T_2$. The shorter the T_2 , the faster the time-domain signal $f(t)$ decays to zero and the wider the resonance line $F(\omega)$ becomes.

Example 4.2

Fourier transformation of a *rectangular pulse* (Fig. 4.1.2):

$$f(t) = \begin{cases} 1, & |t| \leq t_\mathrm{p}/2, \\ 0, & |t| > t_\mathrm{p}/2. \end{cases} \quad (4.1.5)$$

The Fourier transform of this function is the *sinc function*,

$$F(\omega) = \int_{-\infty}^{\infty} f(t)\exp\{-\mathrm{i}\omega t\}\, \mathrm{d}t$$

$$= \int_{-t_\mathrm{p}/2}^{t_\mathrm{p}/2} \exp\{-\mathrm{i}\omega t\}\, \mathrm{d}t = 2\frac{\sin\left(\omega t_\mathrm{p}/2\right)}{\omega} = t_\mathrm{p}\mathrm{sinc}\left(\omega t_\mathrm{p}/2\right). \quad (4.1.6)$$

For computation of the Fourier transformation of signals defined on a discrete mesh of equidistant points, a fast algorithm, the *fast Fourier transformation* (FFT) algorithm exists [Bra1]. Its discovery, together with the development of affordable laboratory computers, has accelerated the methodical development of NMR spectroscopy in a unique way [Ern1].

Rules

Fourier transformation is a linear operation, that is, the Fourier transform of a sum of signals is equal to the sum of Fourier transforms, and a scaling factor is preserved in the

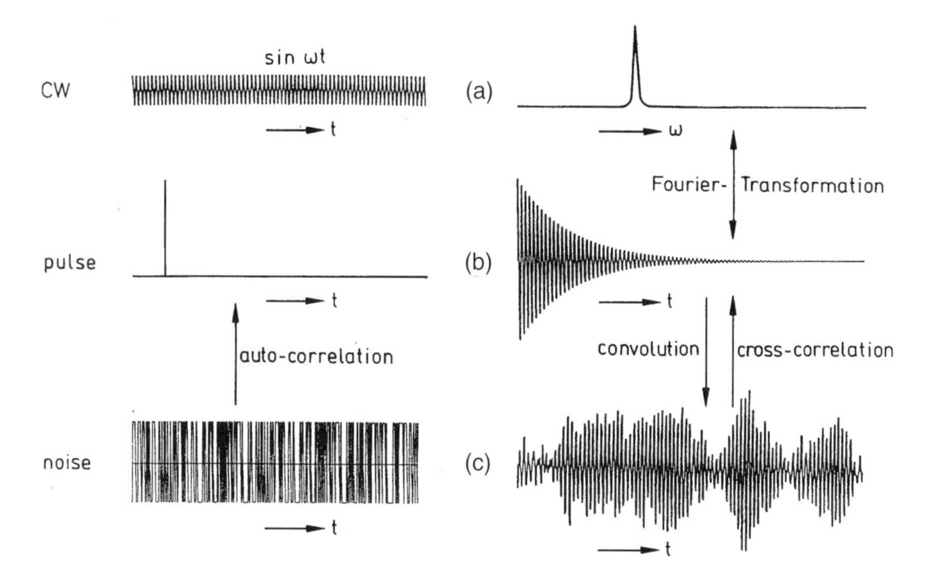

FIG. 4.1.1 Interrelationship between excitation (left) and response (right) in spectroscopy: (a) Excitation with continuous waves (CW excitation) directly produces the spectrum. (b) For pulsed excitation, the spectrum is obtained by Fourier transformation of the impulse response. (c) For stochastic excitation, the impulse response is derived by cross-correlation of excitation and response signals.

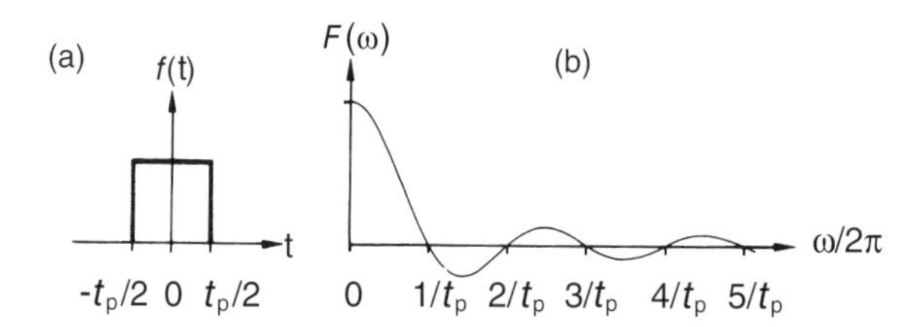

FIG. 4.1.2 The rectangular pulse function (a) and the sinc function (b) form a Fourier pair.

transformation,

$$\int_{-\infty}^{\infty} [f(t) + ag(t)] \exp\{-\mathrm{i}\omega t\}\, \mathrm{d}t = F(\omega) + aG(\omega). \tag{4.1.7}$$

Furthermore, the following rules apply.
Similarity theorem:

$$\int_{-\infty}^{\infty} f(at) \exp\{-\mathrm{i}\omega t\}\, \mathrm{d}t = \frac{1}{a} F\left(\frac{\omega}{a}\right). \tag{4.1.8}$$

Shift theorem:

$$\int_{-\infty}^{\infty} f(t-a)\exp\{-\mathrm{i}\omega t\}\,\mathrm{d}t = \exp\{\mathrm{i}\omega a\}F(\omega). \tag{4.1.9}$$

In analogy to eqns (4.1.1) and (4.1.2), the *2D Fourier transformation* and its inverse are defined as

$$F(\omega_1,\omega_2) = \int_{-\infty}^{\infty}\int_{-\infty}^{\infty} f(t_1,t_2)\exp\{-\mathrm{i}\omega_1 t_1\}\exp\{-\mathrm{i}\omega_2 t_2\}\,\mathrm{d}t_1\,\mathrm{d}t_2, \tag{4.1.10}$$

$$f(t_1,t_2) = \frac{1}{(2\pi)^2}\int_{-\infty}^{\infty}\int_{-\infty}^{\infty} F(\omega_1,\omega_2)\exp\{\mathrm{i}\omega_1 t_1\}\exp\{\mathrm{i}\omega_2 t_2\}\,\mathrm{d}\omega_1\,\mathrm{d}\omega_2. \tag{4.1.11}$$

An n-dimensional Fourier transformation consists in general of n 1D Fourier transformations, each over an independent variable. The number of transformed variables determines the dimensionality of the transform.

4.2 CONVOLUTION

In contrast to the Fourier transformation, *convolution* effects a transformation of the function only, and not of the variable and the function together. Convolution denotes a folding operation (Faltung) of two signals $k_1(t)$ and $x(t)$:

$$y(t) = \int_{-\infty}^{\infty} k_1(\tau)x(t-\tau)\,\mathrm{d}\tau = k_1(t)\otimes x(t). \tag{4.2.1}$$

Here the symbol $\otimes$ is defined to denote the operation of a convolution of two functions. The convolution equation (4.2.1) also describes the response $y(t)$ of a linear time-invariant system to the input signal $x(t)$ (Fig. 4.2.1).

4.2.1 Linear systems

If the input signal is a delta function, $x(t) = \delta(t)$, then the response is given by $k_1(t)$,

$$y(t) = \int_{-\infty}^{\infty} k_1(\tau)\delta(t-\tau)\,\mathrm{d}\tau = k_1(t). \tag{4.2.2}$$

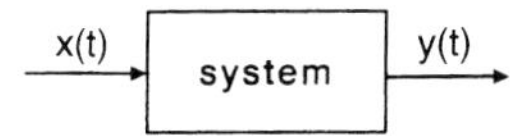

FIG. 4.2.1 A system transforms an input signal $x(t)$ into an output signal $y(t)$. A linear system is described by the linear impulse-response function $k_1(\tau)$. A nonlinear system is described by multi-dimensional impulse-response functions $k_n(\tau_1 \geq \tau_2 \geq \cdots \geq \tau_n)$.

For this reason $k_1(t)$ is also called the *impulse-response function*. For excitation of the *linear response* in NMR, that is, for excitation with small flip-angle pulses, $k_1(t)$ is identical to the FID (Fig. 4.1.1(b)). If the input is a weak *continuous wave* with adjustable frequency ω, then $x(t) = \exp\{i\omega t\}$, and the response is given by the input wave attenuated by the spectrum $K_1(\omega)$ of the impulse-response function $k_1(t)$,

$$y(t) = \exp\{i\omega t\} \int_{-\infty}^{\infty} k_1(\tau) \exp\{-i\omega\tau\} \, d\tau = \exp\{i\omega t\} K_1(\omega). \tag{4.2.3}$$

The input waves can pass the system only at those frequencies where $|K_1(\omega)|$ is large. Therefore, $K_1(\omega)$ is called the *transfer function*, and the system itself can be called a *filter*. In NMR, the system response is measured in a coordinate frame which rotates with the excitation frequency ω. Then the acquired signal is directly given by $K_1(\omega)$, so that the transfer function is the NMR spectrum (Fig. 4.1.1(a)).

For an arbitrary input signal $x(t)$, the response signal $y(t)$ is given by the general convolution (4.2.1) of the input signal with the impulse-response function $k_1(t)$. The impulse response function can also be interpreted as the *memory function* of the system which is centred at time t in the present and points into the past. Then the response signal is the time integral of the input signal weighted by the memory function $k_1(t - \tau)$. This is illustrated in Fig. 4.2.2 for binary stochastic excitation. The corresponding response signal is depicted in Fig. 4.1.1(c).

4.2.2 Nonlinear systems

The convolution defined in (4.2.1) is a linear operation applied to the input function $x(t)$. *Nonlinear systems* transform the input signal into the output signal in a nonlinear fashion. A general *nonlinear transformation* can be described by the *Volterra series*. It forms the basis for the theory of weakly nonlinear and time-invariant systems [Mar1, Sch1] and for general analysis of time series [Kan1, Pri1]. In quantum mechanics, the Volterra series corresponds to time-dependent perturbation theory, and in optics it leads to the definition of *nonlinear susceptibilities* [Blü1].

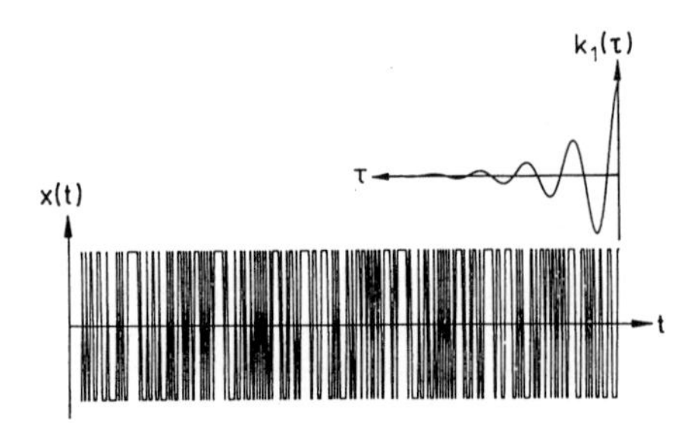

FIG. 4.2.2 [Blü1] Convolution denotes an integration of the input signal $x(t)$, which is weighted by the impulse-response function $k_1(\tau)$.

The Volterra series is an expansion of the response signal into multi-dimensional convolution integrals,

$$y(t) = y_0 + y_1(t) + y_2(t) + y_3(t) + \cdots . \tag{4.2.4}$$

The first few terms of this series are defined as follows:

$$y_0 = k_0, \tag{4.2.5}$$

$$y_1(t) = \int_{-\infty}^{\infty} k_1(\tau)x(t-\tau)\,\mathrm{d}\tau = k_1(t) \otimes x(t), \tag{4.2.6}$$

$$y_2(t) = \int_{-\infty}^{\infty}\int_{-\infty}^{\infty} k_2(\tau_1 \geq \tau_2)\, x(t-\tau_1)\, x(t-\tau_2)\, \mathrm{d}\tau_2\, \mathrm{d}\tau_1, \tag{4.2.7}$$

$$y_3(t) = \int_{-\infty}^{\infty}\int_{-\infty}^{\infty}\int_{-\infty}^{\infty} k_3(\tau_1 \geq \tau_2 \geq \tau_3)\, x(t-\tau_1)\, x(t-\tau_2)\, x(t-\tau_3)\, \mathrm{d}\tau_3\, \mathrm{d}\tau_2\, \mathrm{d}\tau_1. \tag{4.2.8}$$

The linear convolution integral (4.2.6) is the same as in eqn (4.2.1). The kernel $k_n(\tau_1 \geq \tau_2 \geq \cdots \geq \tau_n)$ of the n-dimensional convolution integral is a memory function which correlates n events at times $t - \tau_1, t - \tau_2, \ldots, t - \tau_n$. It is also called the n-dimensional *impulse-response function*, because it describes the response y_n of the nth order convolution to n delta pulses. For example, for $n = 3$,

$$x(t) = \delta(t-\sigma_1) + \delta(t-\sigma_2) + \delta(t-\sigma_3), \tag{4.2.9}$$

and using

$$\sigma_1 > \sigma_2 > \sigma_3 > 0 \tag{4.2.10}$$

one obtains the response

$$y_3(t) = k_3(t-\sigma_1, t-\sigma_2, t-\sigma_3) \tag{4.2.11}$$

from the third-order convolution integral (4.2.8).

Equation (4.2.11) describes the response to three delta pulses separated by $t_1 = \sigma_1 - \sigma_2 > 0$, $t_2 = \sigma_2 - \sigma_3 > 0$, and $t_3 = \sigma_3 > 0$. Writing the multi-pulse response as a function of the pulse separations is the custom in multi-dimensional Fourier NMR [Ern1]. Figure 4.2.3 illustrates the two time conventions used for the *nonlinear impulse response* and in *multi-dimensional NMR spectroscopy* for $n = 3$. Fourier transformation of k_3 over the pulse separations t_i produces the *multi-dimensional correlation spectra* of pulsed Fourier NMR. Fourier transformation over the time delays σ_i produces the *nonlinear transfer functions* known from system theory or the *nonlinear susceptibilities* of optical spectroscopy. The nonlinear susceptibilities and the multi-dimensional impulse-response functions can also be measured with multi-resonance CW excitation, and with stochastic excitation [Blü1].

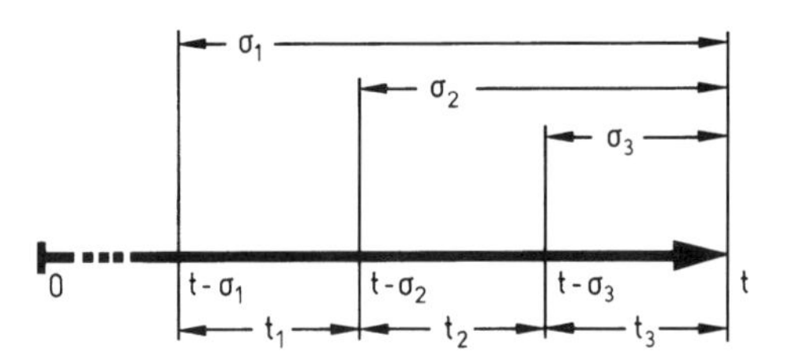

FIG. 4.2.3 [Blül] Time conventions for three-pulse excitation. In 3D correlation spectroscopy, the pulse seperations t_i are used as parameters. In nonlinear system theory, the parameters are the time delays σ_i of the cross-correlation function corresponding to the arguments τ_i of the response kernels.

4.2.3 The convolution theorem

Convolution integrals can readily be calculated in Fourier space. To this end, the Volterra series is Fourier transformed [Bla1, Blü1],

$$Y(\omega) = Y_0 + Y_1(\omega) + Y_2(\omega) + Y_3(\omega) + \cdots, \tag{4.2.12}$$

where the first few terms are given by

$$Y_0 = K_0, \tag{4.2.13}$$

$$Y_1(\omega) = K_1(\omega)X(\omega) = (2\pi)^{-0} \int_{-\infty}^{\infty} K_1(\omega_1)\delta(\omega - \omega_1)X(\omega_1)\,\mathrm{d}\omega_1, \tag{4.2.14}$$

$$\begin{aligned} Y_2(\omega) = (2\pi)^{-1} \int_{-\infty}^{\infty}\int_{-\infty}^{\infty} K_2^{\mathrm{S}}(\omega_1, \omega_2)\,\delta(\omega - \omega_1 - \omega_2) \\ \times X(\omega_1)\,X(\omega_2)\,\mathrm{d}\omega_1\,\mathrm{d}\omega_2, \end{aligned} \tag{4.2.15}$$

$$\begin{aligned} Y_3(\omega) = (2\pi)^{-2} \int_{-\infty}^{\infty}\int_{-\infty}^{\infty}\int_{-\infty}^{\infty} K_3^{\mathrm{S}}(\omega_1, \omega_2, \omega_3)\,\delta(\omega - \omega_1 - \omega_2 - \omega_3) \\ \times X(\omega_1)\,X(\omega_2)\,X(\omega_3)\,\mathrm{d}\omega_1\,\mathrm{d}\omega_2\,\mathrm{d}\omega_3. \end{aligned} \tag{4.2.16}$$

The functions written in capital letters in (4.2.13)–(4.2.16) are the Fourier transforms of the functions written in small letters in (4.2.5)–(4.2.8). The superscript s indicates that the nonlinear transfer functions $K_n^{\mathrm{S}}(\omega_1, \ldots, \omega_n)$ in (4.2.15) and (4.2.16) are the Fourier transforms of impulse-response functions with indistinguishable time arguments, where the causal time order $\tau_1 > \cdots > \tau_n$ is not respected. These transfer functions are invariant against permutation of frequency arguments. Equivalent expressions for the Fourier transforms of impulse-response functions with time-ordered arguments cannot readily be derived.

Equation (4.2.14) is also referred to as the *convolution theorem* [Bla1]. According to it, a linear convolution of the functions $k_1(t)$ and $x(t)$ in the time domain can be evaluated by complex multiplication of the Fourier transforms in the frequency domain and subsequent inverse Fourier transformation of the product back into the time domain. Using the FFT algorithm, this detour into the frequency domain results in considerable savings of computer time. An example from physics illustrating eqn (4.2.12) is the nonlinear optical polarization $Y(\omega)$ in response to excitation by a strong electric field $X(\omega)$, where $K_n^{\mathrm{S}}(\omega_1, \ldots, \omega_n)$ is the *nonlinear optical susceptibility* of order n.

4.3 CORRELATION

The *correlation* of two functions $y(t)$ and $x(t)$ is defined as

$$c_1(\sigma) = \int_{-\infty}^{\infty} y(t)x^*(t-\sigma)\,\mathrm{d}t, \tag{4.3.1}$$

where the symbol * denotes the complex conjugate. This definition is similar to that of the convolution, except that the time axis of the function x is not reversed. In both cases, two functions are multiplied and integrated depending on a relative shift in one of their common variables. Equation (4.3.1) defines a *cross-correlation* if both functions x and y are distinct from one another and an *auto-correlation* if both functions are the same. Correlation is equivalent with *interference* of optical signals [Ern2]. Correlations are used in signal processing and identification, for instance, in order to quantify the common signal components in two similar signals. Therefore, correlation can be considered as the mathematical formulation for a comparison of signals. The magnitude of the correlation function is high for those values of the time shift σ for which both functions are similar.

4.3.1 Linear system analysis

The correlation function corresponds to the memory function, which indicates to which degree values of one function at time t are comparable to values of another function at time $t - \sigma$ before. For statistical signals, the similarity usually decreases rapidly with increasing shift σ. For white noise, all values are independent of the others, and the auto-correlation function is proportional to a delta function. The proportionality factor is the second moment μ_2 of the noise signal.

If $x(t)$ is white noise with a zero mean value, and $y_1(t)$ is the *linear system response* (4.2.6), then the *cross-correlation function* is proportional to the *memory function* $k_1(\sigma)$ (Fig. 4.3.1),

$$\begin{aligned} c_1(\sigma) &= \lim_{T\to\infty} \frac{1}{2T} \int_{-T}^{T} y_1(t)x^*(t-\sigma)\,\mathrm{d}t \\ &= \int_{-\infty}^{\infty} k_1(\tau) \lim_{T\to\infty} \frac{1}{2T} \int_{-T}^{T} x(t-\tau)x^*(t-\sigma)\,\mathrm{d}t\,\mathrm{d}\tau \\ &= \int_{-\infty}^{\infty} k_1(\tau)\mu_2\delta(\tau-\sigma)\,\mathrm{d}\tau = \mu_2 k_1(\sigma). \end{aligned} \tag{4.3.2}$$

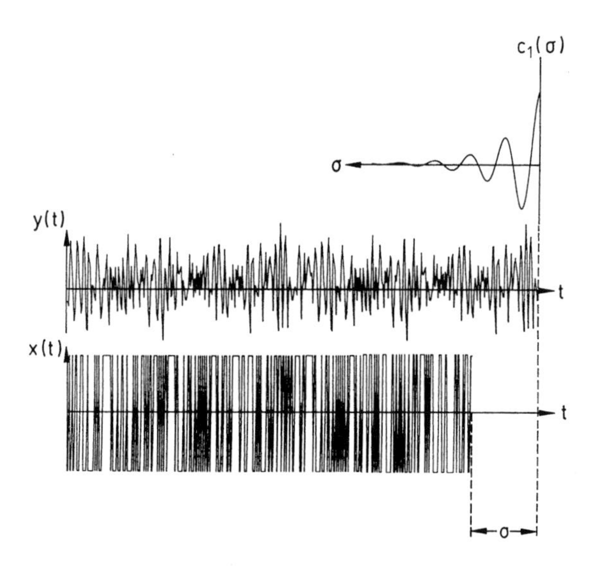

FIG. 4.3.1 [Blül] The cross-correlation function $c_1(\sigma)$ of linear response $y_1(t)$ and stochastic excitation $x(t)$ is proportional to the impulse-response function, and depends on the time delay σ between excitation and response.

Therefore, for measurements with noise excitation, the *linear transfer function* $K_1(\omega)$ (cf. Fig. 4.1.1(a)) is obtained after cross-correlation of excitation and response and subsequent Fourier transformation of the cross-correlation function $c_1(\sigma)$ (cf. Fig. 4.1.1(c)).

4.3.2 Nonlinear cross-correlation

Nonlinear cross-correlation of the system response $y(t)$ (4.2.4) with different powers of a white-noise excitation $x(t)$ yields *multi-dimensional impulse-response functions* $h_n(\sigma_1, \ldots, \sigma_n)$,

$$\begin{aligned} c_n(\sigma_1 > \cdots > \sigma_n) &= \lim_{T\to\infty} \frac{1}{2T} \int_{-T}^{T} y(t)x^*(t-\sigma_1)\cdots x^*(t-\sigma_n)\,\mathrm{d}t \\ &= \mu_2^n h_n(\sigma_1, \ldots, \sigma_n). \end{aligned} \tag{4.3.3}$$

They differ from the kernels $k_n(\tau_1, \ldots, \tau_n)$ of the Volterra series only by a faster signal decay with increasing time arguments [Blü1]. For coinciding time arguments the cross-correlation function is the sum of the n-dimensional impulse-response function h_n with the impulse-response functions h_m of lower orders $m < n$. The stochastic impulse-response functions h_n are the kernels of an expansion of the system response $y(t)$ similar to the Volterra series (4.2.4) but with functionals orthogonalized for white-noise excitation $x(t)$ [Blü1, Mar1, Lee1, Sch1]. This expansion is known by the name *Wiener series*, and the h_n are referred to as *Wiener kernels*.

Fourier transformation of the Wiener kernels $h_n(\sigma_1 > \cdots > \sigma_n)$ over the time delays σ_i produces the *stochastic susceptibilities* $H_n(\omega_1, \ldots, \omega_n)$. Fourier transformation over the delay differences $\sigma_i - \sigma_{i+1} = t_i$ (cf. Fig. 4.2.3) yields *multi-dimensional spectra* similar to correlation spectra measured with n pulse excitation in multi-dimensional Fourier NMR. These stochastic multi-dimensional spectra are the Fourier transforms of multi-dimensional correlation functions and can therefore indeed be called *correlation spectra*. Typical 2D spectra of this kind from *multi-dimensional Fourier NMR* are COSY, NOESY, exchange, and multi-quantum spectra [Ern1].

4.3.3 The correlation theorem

Similar to convolutions, *correlations* can be computed via a detour into the frequency domain. However, for the nonlinear correlation functions $c_n(\sigma_1 > \cdots > \sigma_n)$ this requires lifting the causal time order, just as for calculation of the nonlinear convolutions,

$$C_n^S(\omega_1, \ldots, \omega_n) = \int_0^\infty \cdots \int_0^\infty c_n(\sigma_1, \ldots, \sigma_n) \exp\{-i\omega_1\sigma_1\} \cdots \exp\{-i\omega_n\sigma_n\} \, d\sigma_1 \cdots d\sigma_n. \tag{4.3.4}$$

The expressions derived in this way for the Fourier transforms of the first, second, and third order correlation functions are:

$$C_1(\omega) = Y(\omega) X^*(\omega), \tag{4.3.5}$$

$$C_2^S(\omega_1, \omega_2) = Y(\omega_1 + \omega_2) X^*(\omega_1) X^*(\omega_2), \tag{4.3.6}$$

$$C_3^S(\omega_1, \omega_2, \omega_3) = Y(\omega_1 + \omega_2 + \omega_3) X^*(\omega_1) X^*(\omega_2) X^*(\omega_3). \tag{4.3.7}$$

The superscript s in (4.3.6) and (4.3.7) indicates that the correlation spectra are invariant against permutation of frequency arguments. Numerical evaluation of correlation functions by this detour into the frequency domain can result in considerable savings of computer time.

Equation (4.3.5) is the Fourier transform of the linear correlation (4.3.1). This relationship is referred to as the *correlation theorem*. If $x(t)$ is white noise, then ensemble averages have to be incorporated into equations (4.3.5)–(4.3.7), because the power spectrum of white noise is again white noise, but with a variance as large as its mean value [Ben1]. When the linear correlation theorem (4.3.5) is applied to the same functions, then $C_1(\omega)$ is the power spectrum of $Y(\omega) = X(\omega)$. Conversely, $C_1(\omega)$ is then the Fourier transform of the auto-correlation function of $y(t) = x(t)$ (cf. eqn 4.3.1).

4.4 FURTHER TRANSFORMATIONS

In the context of image and information processing a number of transformations related to the Fourier transformation are useful [Ang1]. In the following some of them are briefly reviewed without complete listings of their properties. More detailed information can be found in the literature cited.

4.4.1 Laplace transformation

The *Laplace transformation* [Spi1] is obtained from the definition (4.1.1) of the Fourier transformation by introduction of a complex variable p instead of a purely imaginary variable $\mathrm{i}\omega$,

$$F_{\mathrm{L}}(p) = \int_{-\infty}^{\infty} f(t) \exp\{-pt\} \, \mathrm{d}t. \tag{4.4.1}$$

If the real part of p is zero, the Laplace Transformation becomes the Fourier transformation. The inversion formula of (4.4.1) consequently resembles the inverse Fourier transformation (4.1.2)

$$f(t) = \frac{1}{2\pi i} \int_{c-\mathrm{i}\infty}^{c+\mathrm{i}\infty} F_{\mathrm{L}}(p) \exp\{pt\} \, \mathrm{d}p, \tag{4.4.2}$$

and c is a suitably chosen constant.

Laplace transformations are mainly used in signal analysis of electrical circuits for mathematical convenience. Differential and integral equations can often be reduced to nonlinear algebraic equations of the complex variable p in the transform domain. Many of the properties of the Fourier transformation can be taken over simply by substituting ω by p . Particularly useful are the Laplace transforms $\mathbf{L}\{\,\}$ for differentiation and for integration. They can be expressed in terms of the transform $F_{\mathrm{L}}(p)$ of a function $f(t)$ by

$$\mathbf{L}\left\{\frac{\mathrm{d}}{\mathrm{d}t} f(t)\right\} = pF_{\mathrm{L}}(p), \tag{4.4.3}$$

and

$$\mathbf{L}\left\{\int f(t)\mathrm{d}t\right\} = \frac{F_{\mathrm{L}}(p)}{p}. \tag{4.4.4}$$

While mathematical insight is gained by use of Laplace transformations, Fourier transformation is used for gaining physical insight in terms of spectra. Theorems for Laplace transforms and the transforms of common functions are tabulated in the literature [Spi1].

4.4.2 Hankel transformation

For 2D systems with circular symmetry, the relevant information is in the radial direction. Therefore, instead of calculating its 2D Fourier transform by (4.1.10), evaluation of a *radial 1D Fourier transformation* is sufficient. In magnetic resonance imaging [Maj1] such a situation is encountered, for instance, in aging studies of cylindrical objects, and in MAS NMR imaging, where axial projections or cross-sections need to be evaluated.

For circular symmetry,

$$F(x, y) = F_r(r), \quad \text{where} \quad r^2 = x^2 + y^2. \tag{4.4.5}$$

The *Hankel transformation* of order 0 is obtained from (4.4.5) by 2D Fourier transformation and subsequent introduction of polar coordinates defined by $x + \mathrm{i}y = r \exp\{\mathrm{i}\varphi\}$ and $k_x + \mathrm{i}k_y = k \exp\{\mathrm{i}\theta\}$,

$$\begin{aligned}
&\int_{-\infty}^{\infty}\int_{-\infty}^{\infty} F(x, y) \exp\left\{-\mathrm{i}(k_x x + k_y y)\right\} \mathrm{d}x\, \mathrm{d}y \\
&= \int_0^{\infty}\int_0^{2\pi} F_r(r) \exp\{-\mathrm{i}kr \cos(\varphi - \theta)\}\, r\, \mathrm{d}r\, \mathrm{d}\varphi \\
&= \int_0^{\infty} F_r(r) \left[\int_0^{2\pi} \exp\{-\mathrm{i}kr \cos\varphi\}\, \mathrm{d}\varphi\right] r\, \mathrm{d}r \\
&= 2\pi \int_0^{\infty} F_r(r) J_0(kr) r\, \mathrm{d}r = p(k).
\end{aligned} \tag{4.4.6}$$

J_0 is the *Bessel function* of the first kind of order 0, and the Hankel transformation of order zero and its inverse are given by

$$p(k) = 2\pi \int_0^{\infty} F_r(r) J_0(kr) r\, \mathrm{d}r, \tag{4.4.7a}$$

$$F_r(r) = 2\pi \int_0^{\infty} p(k) J_0(kr) k\, \mathrm{d}k. \tag{4.4.7b}$$

If objects with symmetries in higher dimensions are reduced to one dimension in a similar fashion, Hankel transforms in terms of higher-order Bessel functions result [Bra1].

4.4.3 Abel transformation

The *Abel transformation* relates the radial information $F_r(r)$ of a circular object to the *projection* $P(x)$ (Fig. 4.4.1). In NMR imaging, the projection $P(x)$ is obtained by Fourier transformation of the FID signal measured in a constant magnetic field gradient G, and the radial information $F_r(r)$ is the inverse Hankel transform of the FID [Maj1].

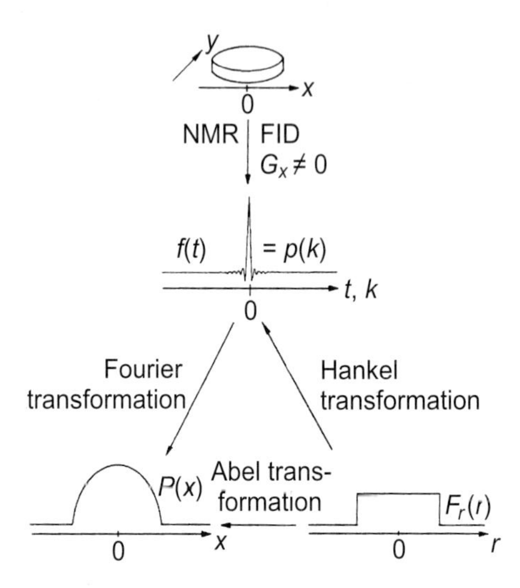

FIG. 4.4.1 [Maji] Relationship between Fourier, Hankel, and Abel transformations in NMR imaging of a 2D object with circular symmetry. The functions for negative values of k, x and r are obtained by mirror symmetry and not as a result of the transformations.

The Abel transformation is defined as

$$P(x) = 2 \int_x^\infty \frac{F_r(r)}{\sqrt{(r^2 - x^2)}} r \, dr. \tag{4.4.8a}$$

After substitution of variables $X = x^2$ and $R = r^2$ the transformation (4.4.8a) can be written as a convolution (4.2.1), which is readily evaluated in the Fourier transform domain by use of the convolution theorem (cf. Section 4.2.3) [Bra1]. The *inverse Abel transformation* is given by [Maj1]

$$F_r(r) = \int_r^\infty \frac{dP(x)/dx}{\sqrt{(x^2 - r^2)}} dx. \tag{4.4.8b}$$

For evaluation of *radial NMR images* $F_r(r)$ of circular objects, processing of the FID in two steps by Fourier transformation and subsequent inverse Abel transformation is preferred over straight forward Hankel transformation, because established phase correction, baseline correction, and filter routines can be used in calculation of the projections $P(x)$ as intermediate results [Maj1]. As an alternative to Hankel and Abel transformations, the *back-projection* technique (cf. Section 6.1) can be applied for radial evaluation of circular objects, using copies of just one projection for input. As opposed to the inverse Abel transformation, however, this provides the radial information with nonuniform spatial resolution.

4.4.4 *z* transformation

In digital signal analysis the values of functions are usually given at equidistant intervals. In NMR this interval is the dwell time or the sampling interval Δt of the measured response. With the time variable being

$$t = n\Delta t, \tag{4.4.9}$$

the discrete time-domain signal

$$\begin{aligned} f(t) =& f(0)\delta(t - 0\Delta t) + f(1)\delta(t - 1\Delta t) \\ &+ f(2)\delta(t - 2\Delta t) + \cdots + f(n)\delta(t - n\Delta t) \end{aligned} \tag{4.4.10}$$

can be written as a collection of samples numbered by n,

$$\{f(0), f(1), f(2), \ldots, f(n)\}. \tag{4.4.11}$$

Its *z transform* is defined by the polynomial

$$F_z(z) = f(0)z^{-0} + f(1)z^{-1} + f(2)z^{-2} + \cdots + f(n)z^{-n}. \tag{4.4.12}$$

The function $f(t)$ is real and is taken to be zero for $t < 0$. The variable z is taken to be complex.

Simple product operations in the z transform domain apply for calculating the system response and the impulse response analogous to the expressions for convolution (4.2.14) and correlation (4.3.5) in the Fourier domain. The z transform of the result is readily found by expansion into a polynomial in z^{-1}, and the coefficients determine the result on the equidistant time grid [Bra1].

A close relationship exists between the z transform $F_z(z)$ and the *Laplace transform*. The Laplace transform $F_L(p)$ of (4.4.10) is given by

$$F_L(p) = \int_{-\infty}^{\infty} f(t)\exp\{-pt\}\,dt = \sum_{n=0}^{\infty} f(n)\exp\{-np\}. \tag{4.4.13}$$

Using $z = \exp\{p\}$, one finds the definition of the z transform of $f(t)$,

$$F_L(p) = \sum_{n=0}^{\infty} f(n)z^{-n}. \tag{4.4.14}$$

Theorems for z transforms and the transforms of common signals are tabulated in the literature [Bra1].

4.4.5 Hadamard transformation

Because in digital signal processing functions are represented as arrays of numbers over discrete variables (cf. eqn 4.4.10), the linear operations of transformation (cf. e.g.

eqn 4.1.1), convolution (4.2.1), and correlation (4.3.1) can be written as multiplications of a vector by a square matrix. The dimension of the matrix is determined by the number of values of the discrete variable. In contrast to the Fourier transformation, the *Hadamard transformation* is an operation based on rectangular instead of harmonic waves. The frequency corresponds to the number ν of sign changes of the rectangular wave. It is referred to as *sequency*. The transformation is applicable to cyclic functions defined on 2^m discrete values of a variable.

Hadamard transformation is achieved by multiplication of a data vector with the *Hadamard matrix* $\mathbf{H}_m$. For $m = 3$ a Hadamard matrix is depicted in Fig. 4.4.2. It is generated by forming the threefold direct product of the matrix

$$\mathbf{H}_1 = \begin{bmatrix} + & + \\ + & - \end{bmatrix} \tag{4.4.15}$$

with itself.

The Hadamard matrix is proportional to its own inverse, $\mathbf{H}_m \mathbf{H}_m = 2^m \mathbf{1}$. For execution of the transformation $y = \mathbf{H}_m \boldsymbol{x}$ of a vector $\boldsymbol{x}$ with the sequency-ordered Hadamard matrix $\mathbf{H}_m$ a fast algorithm is available, similar to the butterfly algorithm of the FFT transformation [Pra1]. However, it only involves subtractions and additions, and no multiplications. Therefore, it is even faster than the FFT algorithm.

4.4.5.1 Hadamard multiplex spectroscopy

The Hadamard transformation is applied, for instance, to signals before transmission, so that transmission errors are smeared out after back transformation [Pra1, Pra2]. Other applications are devoted to multiplexing of signals like the position of the slit in a dispersion spectrometer [Har1], or the frequencies of a multi-frequency pulse in NMR imaging [Bol1, Mül1, Sou1]. The corresponding variable is coded according to the rows of the Hadamard matrix. One value of an integral response is acquired for each row, so that the experiment is executed once for each row. The signal is reconstructed by Hadamard transformation of the data vector. Compared to separate measurements of the response for each of the n values of the variable, $n/2$ values of the variable are interrogated in each step of the Hadamard experiment, and an improvement of the signal-to-noise ratio of $(n/2)^{1/2}$ is gained.

4.4.5.2 Hadamard spectroscopy with m sequences

An interesting application of the Hadamard transformation is its use for determination of the linear impulse-response function with *maximum length binary sequences* or *m*

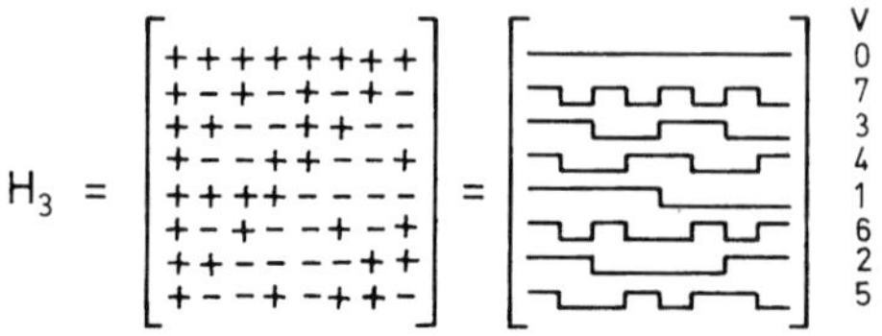

$$\mathrm{H}_3 = \begin{bmatrix} + & + & + & + & + & + & + & + \\ + & - & + & - & + & - & + & - \\ + & + & - & - & + & + & - & - \\ + & - & - & + & + & - & - & + \\ + & + & + & + & - & - & - & - \\ + & - & + & - & - & + & - & + \\ + & + & - & - & - & - & + & + \\ + & - & - & + & - & + & + & - \end{bmatrix} \qquad \begin{matrix} \nu \\ 0 \\ 7 \\ 3 \\ 4 \\ 1 \\ 6 \\ 2 \\ 5 \end{matrix}$$

FIG. 4.4.2 Sequency-ordered Hadamard matrix for $m = 3$. The sequency ν denotes the number of sign changes within one row.

sequences [Zie1, Gol1]. An m sequence is a pseudo-random binary sequence of length $2^m - 1$. Such sequences are generated by *shift registers* with m stages. The input signal of the register is calculated by modulo-2 addition (exclusive or) of the contents of well-defined shift-register stages. Figure 4.4.3(a) depicts an example of such a shift register with $m = 3$. The binary contents of each stage cyclically runs through all values of an m sequence. By (4.2.6) the linear system response $y_1(t)$ to excitation with an m sequence is the convolution of the linear impulse-response function $k_1(t)$ with the m sequence $x(t)$. The impulse-response function $k_1(t)$ can be retrieved from $y_1(t)$ by cross-correlation of $y_1(t)$ with the m sequence $x(t)$. An m sequence is particularly suitable as an input signal for analysis of linear systems, because the variance of its discrete power spectrum is zero, so that formation of an ensemble average can be omitted in evaluation of the cross-correlation in the frequency domain by use of the correlation theorem (4.3.5).

The cyclic, linear cross-correlation of an m sequence with a function y_i, $i = 1, \ldots, 2^{m-1}$, corresponds to multiplication of the vector $y = \{y_i\}$ by the matrix $\mathbf{H}'_m$, the rows of which are formed by the step-wise rotated m sequence (Fig. 4.4.3(b)). Like in a Hadamard matrix, each row of this matrix represents a rectangular wave, but the number of sign changes is the same in each row. The matrix can be converted to a sequency-ordered Hadamard matrix with a different number of sign changes in each row by permutation of rows and columns using the permutation matrices $\mathbf{P}$ and $\mathbf{Q}$, and by adding a row and a column of ones. Therefore, the fast Hadamard algorithm can be employed for linear cross-correlation with m sequences.

Stochastic excitation with m sequences and the use of the Hadamard transformation have been investigated in NMR spectroscopy [Kai1, Zie2] as well as in NMR imaging [Cha1]. Processing of the nonlinear response to m sequences by Hadamard and Fourier transformation results in signal distortions reminiscent of noise, which are caused by the nonlinear parts of the response [Blü2].

4.4.6 Wavelet transformation

It is customary to analyse time-domain NMR signals $f(t)$ in terms of their spectra relating to amplitude and phase of harmonic waves. This decomposition of $f(t)$ is achieved by Fourier transformation. However, the harmonic waves are implicitely assumed to extend

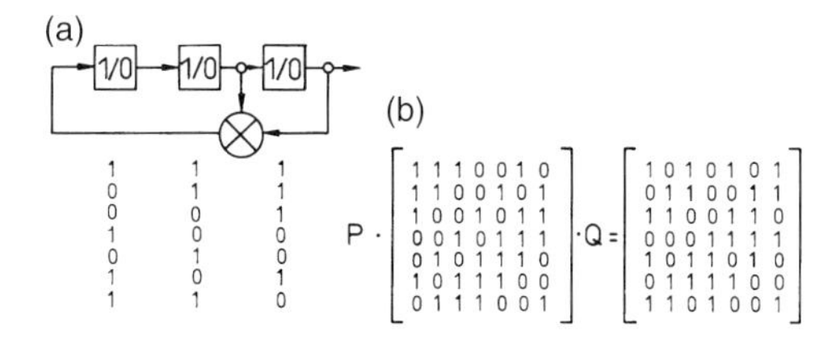

FIG. 4.4.3 (a) Generation of an m sequence of $2^m - 1$ values by a m-stage shift register for $m = 3$. (b) The matrix for cyclic cross-correlation with an m sequence is obtained from the the sequency-ordered Hadamard matrix by transformation with the permutation matrices $\mathbf{P}$ and $\mathbf{Q}$. In replacing '1' by '−' and '0' by '+', the right hand matrix can be converted to $\mathbf{H}_3$ of Fig 4.4.2.

from minus to plus infinity, that is, far beyond the duration of the NMR signal. Therefore, methods of signal analysis, which take account of the limited lifetime of the signal, are being explored [Ang1]. Examples are the *maximum entropy method* [Bar1, Bur1], *linear prediction* with *singular-value decomposition* [Bee1], and the *wavelet transformation* [Ant1, Chu1, Chu2, Coi1]. *Wavelets* are packets of waves with a limited duration. Similar to the use of the Hadamard matrix, wavelets can be used for signal analysis as well as for frequency encoding in imaging and spectroscopy to improve the signal-to noise ratio [Neu1, Sar1, Wea1].

The concept of wavelets is illustrated by example of continuous wavelets [Bar2]. Wavelets are obtained from a single function $w(t)$ by translation and dilatation of the time axis,

$$w_{ab}(t) = \frac{1}{a} w\left(\frac{t-b}{a}\right), \tag{4.4.16}$$

where b defines translation and a defines dilatation of the time axis. Similar to the Hadamard transformation, the *wavelet transformation* of a signal function $f(t)$ is defined by convolution and correlation. For the Hadamard transformation $f(t)$ is cyclically convolved or correlated with an m sequence. For the wavelet transformation, $f(t)$ is correlated with the complex conjugated wavelet,

$$f_{\mathrm{w}}(a,b) = \frac{1}{a}\int w^*\left(\frac{t-b}{a}\right) f(t)\,\mathrm{d}t = \frac{1}{a}\int w\left(\frac{b-t}{a}\right) f(t)\,\mathrm{d}t = \frac{1}{a} w_{ab}(b/a) \otimes f(b). \tag{4.4.17}$$

If the wavelet is symmetric in time, formation of the complex conjugate is equivalent to inversion of the time axis, so that the correlation of f with w^* can also be written as a convolution of f and w (cf. Sections 4.2.3 and 4.3.3). The value of the normalization factor is often chosen in practice to be $1/a$, although different conventions are used mostly in theoretical treatments [Bar2].

To grasp an understanding of the wavelet transform, the convolution (4.4.17) is written with help of the correlation theorem, eqns (4.3.1) and (4.3.5), the similarity theorem (4.1.8), the shift theorem (4.1.9), and the inverse Fourier transformation (4.1.2) as

$$f_{\mathrm{w}}(a,\, b) = \frac{1}{2\pi}\int W_{ab}^*(a\omega) F(\omega) e^{\mathrm{i}\omega b}\mathrm{d}\omega, \tag{4.4.18}$$

where $W_{ab}(a\omega)$ is the Fourier transform of the wavelet (4.4.16), and $F(\omega)$ is the spectrum of the signal function $f(t)$. In this form it becomes clear, that the wavelet $w_{ab}(t)$ acts as a linear filter for the spectrum $F(\omega)$.

If the wavelet satisfies the admissibility condition

$$c_{\mathrm{w}} = 2\pi \int |W_{ab}(\omega)|^2 \frac{1}{|\omega|}\mathrm{d}\omega < \infty, \tag{4.4.19}$$

then the continuous *inverse wavelet transform* is given by

$$f(t) = \frac{1}{c_{\mathrm{w}}}\int\int w_{ab}(t) f_{\mathrm{w}}(a,\, b)\frac{1}{a}\mathrm{d}a\,\mathrm{d}b. \tag{4.4.20}$$

Often the admissibility condition is fulfilled, when the wavelet has no dc component.

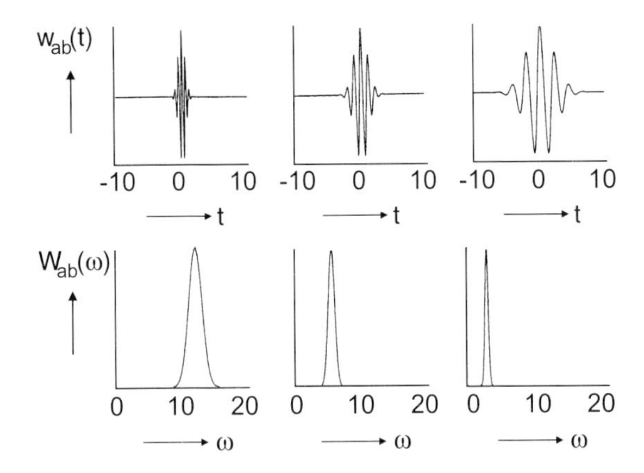

FIG. 4.4.4 [Bar2] Morlet wavelets (top) and their Fourier spectra (bottom) according to eqns. (4.4.21 a and b). Time and frequency are scaled in arbitrary units. Dilatation parameter from left to right: $a = 0.5, 1.0, 2.0$ Widths of $w_{ab}(t) = 3, 6, 12$ (top). Widths of $W_{ab}(\omega) = 3.0, 1.5, 0.75$ and corresponding peak positions at 12, 6, 3 (bottom).

Depending on the type of application, different wavelets can be chosen. In spectroscopy and imaging, signals are often well localized in frequency. In this case the *Morlet wavelet* can be used [Bar2]. The wavelet and its Fourier transform are given by

$$w(t) = \exp\{i\omega_0 t\} \exp\left\{\frac{-t^2}{2\sigma_0^2}\right\} + g(t), \tag{4.4.21a}$$

$$W(\omega) = \sqrt{2\pi}\sigma_0 \exp\left\{\frac{-(\omega - \omega_0)^2\sigma_0^2}{2}\right\} + G(\omega). \tag{4.4.21b}$$

The wavelet is the product of a complex exponential with frequency ω_0 and a Gaussian with variance σ_0. The function $g(t)$ is necessary to enforce the admissibility condition.

The Morlet wavelet can be understood to be a linear bandpass filter, centred at frequency $\omega = \omega_0/a$ with a width of $1/(\sigma_0 a)$. Some Morlet wavelets and their Fourier spectra are illustrated in Fig. 4.4.4. The translation parameter b has been chosen for the wavelet to be centred at time $t = 0$ (top). With increasing dilatation parameter a the wavelet covers larger durations in time (top), and the centre frequency of the filter and the filter bandwidths become smaller (bottom). Thus depending on the dilatation parameter different widths of the spectrum are preserved in the wavelet transform while other signals in other spectral regions are suppressed.

5

Concepts of spatial resolution

Various methods have been proposed for generation of images by NMR. However, only the variants of the Fourier technique are used on a routine basis in *clinical imaging*. The first NMR image was constructed from projections, a technique adapted from X-ray tomography [Lau1], and very quickly, alternative techniques were developed [Bot1, Man1]. It is helpful to classify the imaging techniques by the dimension of the region which produces the acquired signal. Then point, line, plane, and volume techniques are discriminated as illustrated in Fig. 5.0.1 [Bru1]. Point methods [Hin1, Hin2] and some line-scan techniques [Gar1, Hin3, Man2] are simple to execute, because the image can be scanned directly without major computational effort. However, the lower the dimension of the signal-bearing region, the lower is the signal-to-noise ratio of the measured response, because the noise always comes from the entire sample volume. If n volume elements contribute noise to the measured signal, the signal-to-noise ratio is proportional to $n^{1/2}$. To optimize sensitivity, it is advantageous to multiplex the space information during data acquisition, so that the acquired signal derives from as many pixels or voxels simultaneously as possible [Kum1]. Such methods are the topic of Chapter 6. In this chapter, the less sensitive *sensitive-point method* and *line-scan methods* are reviewed (Sections 5.1 and 5.2). Because the requirements for materials imaging differ considerably from those for biomedical imaging, the older methods cannot necessarily be considered obsolete. Sensitivity is not always the prime criterion of choice for a particular method, because the manifold of states covers a much wider range in materials applications than in medical applications. The temperature can be varied and the objects of investigation can be exposed to various conditions like mechanical strain, electric fields, and aggressive environments. In addition, acquisition time and rf power are not necessarily restricted.

A reduction of the space dimensions of the NMR signal [Gar1] can be achieved by the use of *selective excitation* in the presence of field gradients (Section 5.3), for instance, by *selective pulses* [Fre1, Mcd1, War1]. Such pulses are usually much longer than nonselective pulses. In solids, however, slice selection becomes a problem, because a wide linewidth arises from a fast signal decay, and the duration of a selective excitation event must be limited to times shorter than this. Thus, the boundaries of a slice are often ill-defined.

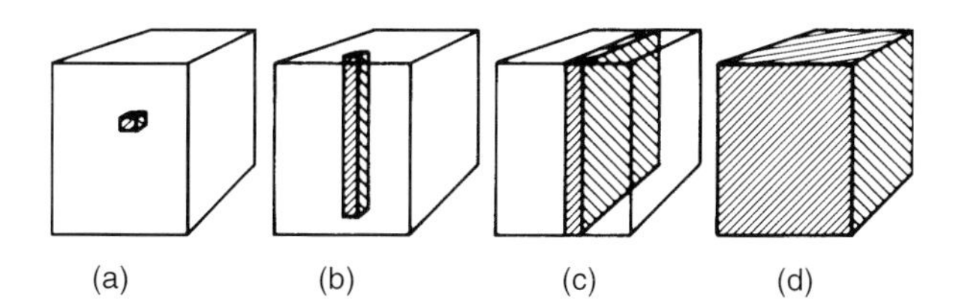

FIG. 5.0.1 [Bru1] Scheme for classification of imaging methods. (a) Point-scan method. (b) Line-scan method. (c) Slice-selective method. (d) Volume method.

The development of sophisticated imaging methods requires an understanding of NMR signals in the presence of time-varying magnetic-field gradients (Section 5.4). Consideration of the *linear response* is sufficient in many cases, but for slice selection, imaging with stochastic excitation, and imaging with 2D spectroscopic resolution the *nonlinear response* may become important [Hou1].

Consideration of the impulse response in a magnetic-field gradient immediately leads to a description and classification of imaging methods in reciprocal space or ***k*** *space* [Man3, Lju1]. This is the space obtained by Fourier transformation over space coordinates $\mathbf{r} = (x, y, z)$. A close relationship exists to the reciprocal space investigated in X-ray and neutron *scattering experiments*. Here the *displacement space* $\mathbf{R} = (\Delta x, \Delta y, \Delta z)$ is investigated. This space can, in fact, be probed by NMR as well as when measuring the effects of particle diffusion [Cal1, Kär1]. The Fourier space conjugate to displacement space is referred to as ***q*** *space*. Because such ***q***-space measurements are associated with particle motion, ***q*** may also be interpreted as the Fourier-conjugate variable to velocity. The ***q*** space can be investigated with spatial resolution for use of diffusion constants and velocity components as contrast criteria in NMR images (cf. Section 7.2.6) [Cal2]. A hierarchy of experiments can be constructed which move up from ***k*** space for encoding of *position* in imaging to ***q*** space for encoding of *velocity*, to other reciprocal spaces for encoding of *acceleration* and higher-order parameters of time-dependent translational motion.

5.1 POINT METHODS

When using *point methods*, individual picture elements are measured sequentially (Fig. 5.0.1(a)). This procedure is time consuming, because the space dimensions are not multiplexed during acquisition. On the other hand, they are simple to execute, tolerant against magnetic field distortions, and can readily be combined with spectroscopic resolution. In the following, the point methods that were significant in the early days of NMR imaging are addressed. Later on, more sophisticated point methods were developed for *volume-selective spectroscopy*. These methods are treated in Section 9.1.

5.1.1 FONAR

The *FONAR* method (field-focussed nuclear magnetic resonance) exploits shaped magnetic fields and rf pulses to produce a resonance aperture the size of a pixel [Dam1].

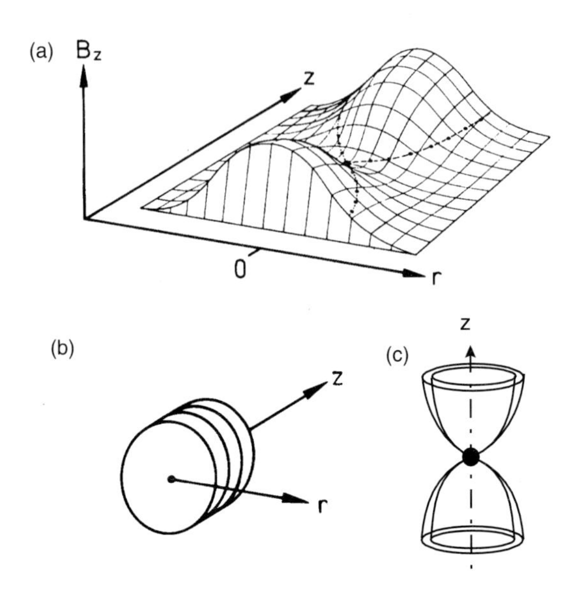

FIG. 5.1.1 [Dam4] Resonance aperture for FONAR imaging. (a) Contour plot of the magnetic field $B_z(z, r)$. (b) Definition of the coordinate system in the magnet. (c) 3D representation of the sensitive region.

By use of this technique, the first human was imaged in 1977 by measuring a cross-section through the chest [Dam2, Dam3]. For image acquisition, the object is moved point by point through the resonance aperture. Typical measurement times are of the order of 1–5 h. For this reason, FONAR is of no interest in medical imaging.

For field focussing, a special superconducting magnet was built by Damadian and his colleagues [Dam4]. It consisted of a short coil with a large diameter which produced the saddle-shaped static magnetic-field profile $B_z(z, r)$ shown in Fig. 5.1.1 [Man1, Mor1],

$$\left.\frac{\partial B_z}{\partial r}\right|_{r=0} = \left.\frac{\partial B_z}{\partial z}\right|_{z=0} = 0. \tag{5.1.1}$$

The *resonance aperture* is given by homogeneous region of the magnetic field at the saddle point. A selective pulse is used to confine the excitation of the magnetization to within the region of the saddle point. However, also the regions sketched in Fig. 5.1.1(c) contribute to the detected signal. Their contributions become smaller for stronger field gradients and more selective excitation.

The pulse response from the homogeneous region is a slowly decaying FID. The FID from the inhomogeneous regions decays fast. The sum of both contributions is acquired. The slowly decaying part is separated from the other and used to assign an intensity value to the pixel. Because the signal for each pixel is the sum of many scans, spin–lattice relaxation can be used for weighting image contrast by varying the repetition time between scans to achieve partial saturation.

In related techniques, the homogeneous magnetic field of standard NMR magnets is degraded, for instance, by a Maxwell coil pair [Tan1], or pulsed $\boldsymbol{B}_0$ field gradients are employed (cf. Section 9.1). The $\boldsymbol{B}_1$ field can also be profiled, for instance, by the use of surface coils (cf. Section 9.2). Either approach is suitable for *volume-selective NMR* (cf. Chapter 9).

5.1.2 The sensitive-point method

The *sensitive-point method* was introduced in 1974 by Hinshaw [Hin1, Hin2]. The z-component B_z of the magnetic field is modulated in a sinusoidal fashion by application of time-dependent field gradients in x-, y-, and z-directions (Fig. 5.1.2),

$$G_i(t) = G_i \cos(\omega_i t), \quad i = x, y, z. \tag{5.1.2}$$

Consequently, the NMR signal becomes time dependent at all points in the sample except at the field-gradient origin. Based on the time invariance of the signal at this point, it can be separated from the signal of the other sample regions by simple addition. For detection, use of the *steady-state free precession* techniques (cf. Section 2.2.1) is suitable. The typical acquisition time for one pixel is one second, so that the simplicity of the method has to be weighted against the duration of image acquisition.

In principle, there is no reason why only the time-independent response should be extracted from the total signal by addition of successive scans with unsynchronized field modulation. Because the time dependence is known for each voxel, the signal can be extracted for any voxel by a lock-in technique, where the reference function is given by the gradient modulation for the particular voxel. This can be executed simultaneously,

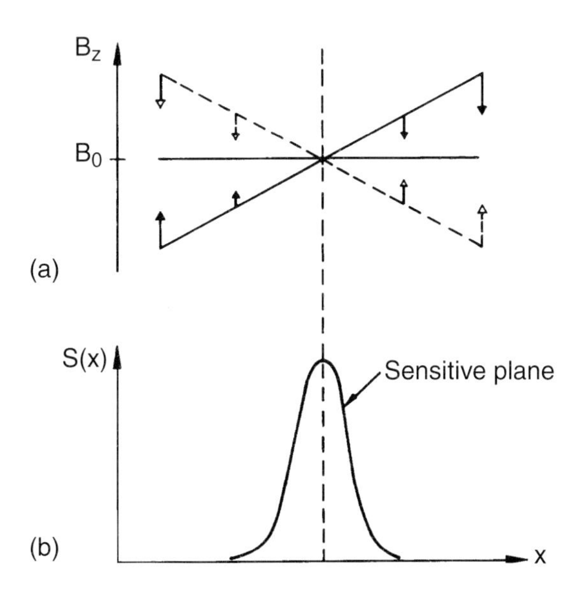

FIG. 5.1.2 [Bot1] Spatial localization with sinusoidal field gradients. The signal from the volume element at the field-gradient origin is independent of the gradient modulation. By averaging the signal acquired in an oscillating gradient the signal of a plane can be selected. (a) Oscillating gradient. (b) Spatial dependence of the selected magnetization.

so that the multiplex advantage is introduced for all three space coordinates, and the point method becomes a volume method [Mac1]. A similar method has been proposed for selection of signal from a slice, a line, or a point in a solid sample to overcome the limitations in spatial resolution posed by large linewidths [Cor2]: signals are repetitively acquired in strong gradients which are incremented from scan to scan. Only the signals from the gradient origin should add coherently, while the other signals are supposed to interfere destructively as a result of the variable frequency shift induced in each voxel by the changing gradient strength.

5.2 LINE METHODS

With *line methods*, the spin density along one line of n voxels in the object is interrogated simultaneously (Fig. 5.0.1(b)) [Man1]. Thus the sensitivity improves by a factor of $n^{1/2}$ compared to point methods.

5.2.1 The multiple sensitive-point method

A line-scan method is obtained by straightforward extension of the sensitive-point method. If only two, instead of three oscillating gradients, are applied (cf. Fig. 5.1.2), the magnetization along a line through the object is time independent. The third space encoding gradient is left static, and the NMR spectrum of the time-independent magnetization components provides a 1D image of the selected line [Hin3, Moo1].

Instead of the SSFP method, pulsed excitation is used [Bot2, Sco1]. However, the lines are broadened for slow modulation frequencies, and sideband artefacts result if the modulation frequency is increased beyond $1/T_2^*$ [Fei1, Man1]. The sidebands appear at multiples of the modulation frequency. If the modulation frequency is increased beyond the spectral width, they can be filtered out. To achieve such high-gradient modulation frequencies, the gradient coils need to be part of a resonant circuit. Modulation frequencies of up to 60 kHz can be achieved on small-gradient systems.

5.2.2 Line-scan methods using selective excitation

Transverse magnetization of a line through the object is prepared by use of selective excitation techniques (cf. Section 5.3) [Fin1, Gar1, Man2, Sut1]. Two gradients are applied for preparation of the selected line and one for read-out of the profile along the line. Pulsed rf excitation is used for recording of the FID. Parameter contrast can be introduced for T_1 by partial saturation resulting from variation of the repetition time, and for the chemical shift by reducing or eliminating the gradient during acquisition of the FID [Man1].

5.3 SELECTIVE EXCITATION

Selective excitation denotes the manipulation of the NMR signal within restricted frequency regions of the magnetization response. It is useful in spectroscopy to select

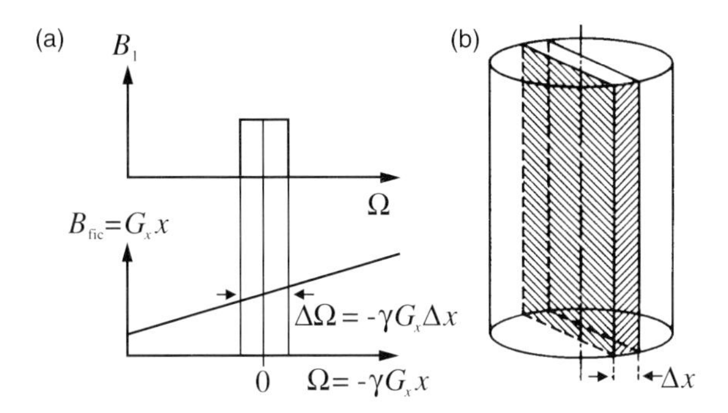

FIG. 5.3.1 [Mor1] Selective excitation of magnetization from a slice of spins.
(a) Radio-frequency field B_1 and magnetic field B_{fic} along the z-axis of the rotating frame as functions the offset frequency Ω in the rotating frame. (b) Localization of the transverse magnetization after excitation of the object with a selective pulse in the presence of a gradient field G_x.

spectral regions of interest and to suppress unwanted signals [Ern1, Fre1]. Its main use in NMR imaging is for slice selection [Gar1, Mor5].

The principle of *slice selection* is illustrated in Fig. 5.3.1 [Mor1]. A magnetic field $G_x x$, which depends linearly on the space coordinate x, is generated by application of the gradient G_x. Then the NMR frequency is proportional to the space coordinate along the gradient direction. Disregarding chemical shift, a slice of transverse magnetization perpendicular to the gradient direction can be excited when starting from z magnetization by restricting the bandwidth of the response to a frequency window centred at the excitation frequency $\omega_{rf} = \omega_0$ or $\Omega = 0$ in the rotating frame. The width $\Delta\Omega$ of this window determines the thickness Δx of the slice. This scheme is useful for preparing transverse magnetization in a slice through the object for use in 2D imaging schemes. At the same time, the z magnetization is reduced within the corresponding frequency window, so that a hole is burnt in the z magnetization profile across the sample.

There are two principal types of magnetization profiles which arise in imaging (Fig. 5.3.2) [Man1]: one generates and the other attenuates magnetization within a given frequency window of the response. Either profile may be applied to transverse and longitudinal magnetization depending on the type of selective pulse used. For slice selection a *selective pulse* which generates a transverse magnetization profile (a) is needed. Such pulses can be produced in a comparatively straightforward fashion. In a first-order approximation, a long rectangular pulse can be used in the presence of a field gradient [Gar1] because its Fourier transform is the sinc function, which exhibits a strong peak in the centre and trailing sidelobes on either side (cf. Fig. 4.1.2). As a rule of thumb, the spectral width of the pulse is given by the inverse length of the pulse.

To reduce the space dimension of the selected transverse magnetization from a slice to a line, a second selective pulse with the excitation profile (b) could be applied in an orthogonal gradient. This would flip the unwanted transverse magnetization back along the z-axis. However, at the same time longitudinal magnetization unaffected by the first pulse would be converted to transverse magnetization. In addition, gradient switching

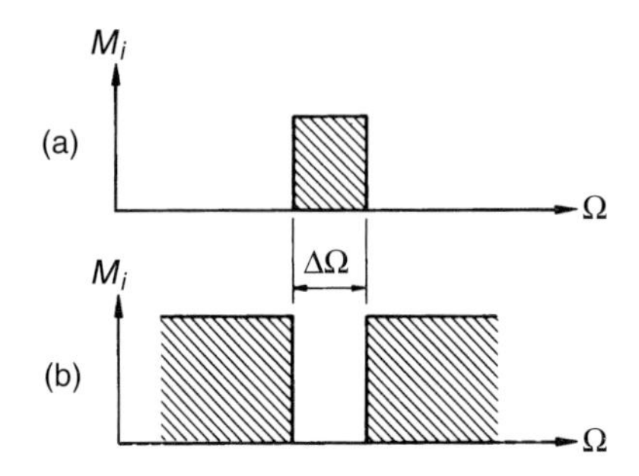

FIG. 5.3.2 Profiles of longitudinal or transverse magnetization M_i after selective excitation. (a) Generation of magnetization within a frequency window of width $\Delta\Omega$ and attenuation outside. (b) Attenuation of magnetization within a frequency window of width $\Delta\Omega$ and generation outside.

takes time, and the transverse magnetization generated by the first pulse will have dephased once the new gradient has stabilized and the second pulse can be applied. Therefore, the order of the pulses is inverted [Man2]. The first pulse is chosen with the excitation profile (b) to saturate longitudinal magnetization outside the frequency region to be selected. The longitudinal magnetization profile generated in this way relaxes on the long timescale of the spin–lattice relaxation time T_1 without the possibility of dephasing. Thus, there is sufficient time for switching to an orthogonal gradient. The second pulse is tailored to convert a line of longitudinal magnetization through the slice to transverse magnetization using the response profile (a) of Fig. 5.3.2 for the transverse magnetization.

Generation of longitudinal magnetization profiles (a) without creation of transverse magnetization is rather difficult, because the magnetization must be saturated or destroyed over large bandwidths and retained within just a narrow window [Cre1, Dod1, Sin1, Sin3]. Limits are imposed by the available rf power and incomplete signal cancellation. To optimize the response profile under the experimental constraints, the rf pulses can be shaped in amplitude and phase [Nil1, War1].

Approaches which are less demanding to the spectrometer hardware use *composite pulses*, that is pulse packages, which consist of a windowless sequence of short, rectangular pulses [Lev1]. Other approaches include dephasing times in between the pulses. The most famous selective excitation pulse sequence of this type is the *DANTE* sequence (delays alternating with nutation for tailored excitation) [Mor2], which is also exploited for selective excitation in solids [Cor1]. Here short relaxation times are of particular concern. To select a slice 0.1 mm thick in a gradient of 10 mT/m, the duration of the selective pulse must be about 20 ms. But in solids, T_2 is larger than this only for very soft materials like a few elastomers. In general, the transverse magnetization of abundant spins like ^{1}H decays under the influence of the strong homonuclear dipole–dipole coupling in about 100 μs to a few ms. Therefore, any excitation used to select transverse ^{1}H magnetization in solid-state NMR should consist of only a short sequence of preferably hard pulses. One approach of achieving this is by the use of *spin-locking* in an rf field of suitably chosen duration and amplitude [Haf2]. A far more favourable situation is met when selecting longitudinal magnetization by saturation and dephasing unwanted components by a pulse with profile (b) in Fig. 5.3.2.

Most types of selective excitation can be modified for simultaneous excitation of n slices or volume elements. Such an approach is advantageous when a limited number of slices or volume elements, but not the entire 3D object, needs to be investigated. By suitable coding of the volume information in n experiments an improvement in signal-to-noise ratio of $n^{1/2}$ can be gained [Bol1, Mül1]. Compared to 3D volume imaging, multi-slice and multi-volume techniques (cf. Section 9.1) suffer from the lack of achieving well-defined boundaries.

Apart from preparation of magnetization in slices and lines, selective excitation can also be used for point selection. Here the objective normally is not to scan an image in a pointwise fashion, but rather to localize a selected volume element to acquire a spectroscopic response from it [Aue1] (cf. Section 9.1).

5.3.1 Excitation and response

The selectivity of the excitation is characterized by the bandwidth of the magnetization response. The response spectrum is determined by the Fourier transform of the selective pulse only in first order. Generally, the NMR response is nonlinear, and *nonlinear system theory* can be applied for its analysis (cf. Section 4.2.2). A model suitable for describing the NMR response in many situations applicable to NMR imaging is given by the *Bloch equations* (cf. Section 2.2.1). They are often relied upon when designing and analysing selective excitation (Fre1).

An instructive approach to the linear and nonlinear response is the perturbative analysis of the Bloch equations in the frequency domain [Hou1]. The *linear response* of the transverse magnetization

$$M^+(t) = M_x(t) + \mathrm{i}M_y(t) \tag{5.3.1}$$

is readily obtained from the Bloch equations (2.2.8) by rewriting them in complex notation in the rotating coordinate frame,

$$\frac{\mathrm{d}M^+}{\mathrm{d}t} + \left[\frac{1}{T_2} - \mathrm{i}\Omega_0\right] M^+(t) = -\mathrm{i}x(t)M_z(t), \tag{5.3.2a}$$

$$\frac{\mathrm{d}M_z(t)}{\mathrm{d}t} = \mathrm{Im}\left\{M^+(t)x^*(t)\right\} - \frac{M_z(t) - M_0}{T_1}. \tag{5.3.2b}$$

Following the nomenclature used in system theory (cf. Section 4.2), the excitation is denoted by $x(t)$. To avoid confusion with the space coordinate x, the time dependence is always explicitly carried along. In general, the excitation is applied in quadrature, that is, in both transverse directions of the rotating frame,

$$x(t) = -\gamma[B_{1x}(t) + \mathrm{i}B_{1y}(t)] = \omega_{1x}(t) + \mathrm{i}\omega_{1y}(t). \tag{5.3.3}$$

In the linear approximation valid for small flip angles, the longitudinal magnetization is unaffected by the rf excitation, and the transverse magnetization is proportional to the first power of the excitation amplitude ω_1. For example, for pulse excitation with flip angle $\alpha = \omega_1 t_p$ one finds for the magnetization at time $t = 0_+$ right after the pulse by expansion of (2.2.19) of the transverse magnetization and by a corresponding treatment of the longitudinal magnetization,

$$M^+(0_+) = M_0 \sin\alpha = M_0 \left[\frac{\alpha}{1!} - \frac{\alpha^3}{3!} + \frac{\alpha^5}{5!} - \cdots \right] \cong \alpha M_0, \tag{5.3.4a}$$

$$M_z(0_+) = M_0 \cos\alpha = M_0 \left[1 - \frac{\alpha^2}{2!} + \frac{\alpha^4}{4!} - \cdots \right] \cong M_0. \tag{5.3.4b}$$

Using this approximation for arbitrarily time-dependent excitation $x(t)$ one obtains a first-order differential equation for the transverse magnetization,

$$\frac{\mathrm{d}M_1^+(t)}{\mathrm{d}t} + \left[\frac{1}{T_2} - \mathrm{i}\Omega_0 \right] M_1^+(t) = -\mathrm{i}x(t) M_0. \tag{5.3.5}$$

It describes the input–output behaviour of a linear system (cf. Section 4.2.1) with input $x(t)$ given by (5.3.3) and output $y_1(t) = M_1^+(t)$. The *transfer function* $K_1(\omega)$ of the system is obtained by Fourier transformation (cf. Section 4.4.1, eqn (4.4.3)),

$$M_1^+(\omega) = Y_1(\omega) = -\frac{\mathrm{i}M_0 X(\omega)}{(1/T_2) - \mathrm{i}(\Omega_0 - \omega)} = K_1(\omega) X(\omega), \tag{5.3.6}$$

and comparison with (4.2.14),

$$K_1(\omega) = -\frac{\mathrm{i}M_0}{(1/T_2) - \mathrm{i}(\Omega_0 - \omega)}. \tag{5.3.7}$$

Thus, the transfer function is the *complex Lorentz function*, and the linear response in the time domain is the inverse Fourier transform of $K_1(\omega)$ (cf. Section 4.1),

$$M_1^+(t) = \frac{1}{2\pi} \int_{-\infty}^{\infty} K_1(\omega) X(\omega) \exp\{\mathrm{i}\omega t\} \, \mathrm{d}\omega. \tag{5.3.8}$$

For a rectangular pulse of length t_p the excitation spectrum is given by the sinc function (cf. Fig. 4.1.2),

$$X(\omega) = -\gamma B_1 \int_{-t_p/2}^{t_p/2} \exp\{-\mathrm{i}\omega t\} \, \mathrm{d}\omega = -\gamma B_1 t_p \frac{\sin(\omega t_p/2)}{\omega t_p/2}. \tag{5.3.9}$$

The relationship between transfer function $K_1(\omega)$, excitation spectrum $X(\omega)$ of the excitation $\omega_1(t)$, and spectrum $Y_1(\omega)$ of the linear system response $M_1^+(t) = y_1(t)$ is illustrated in Fig. 5.3.3 for nonselective (left) and for selective (right) excitation [Hou1]. For nonselective excitation, the spectrum $X(\omega)$ (a) of the rf pulse is broad and for selective excitation it is narrow compared to the widths of the transfer function

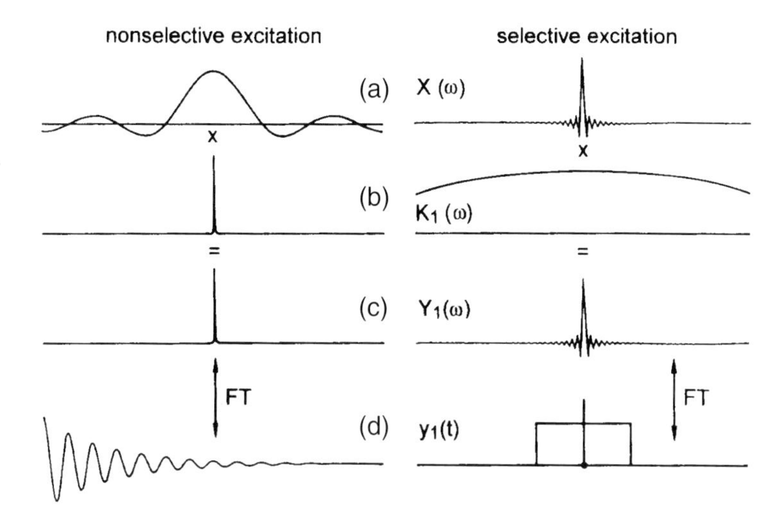

FIG. 5.3.3 [Hou1] Relationships between excitation spectrum $X(\omega)$ (a) of the excitation $\omega_1(t)$, transfer function $K_1(\omega)$ (b), and spectrum $Y_1(\omega)$ (c) of the linear system response $M_1^+(t) = y_1(t)$ (d) for nonselective excitation (left) and selective excitation (right).

or the NMR spectrum $K_1(\omega)$ (b). According to (5.3.6), the spectrum $Y_1(\omega)$ of the linear response is given by the product of these two spectra (c). Therefore, the FID $y_1(t)$ of the entire NMR spectrum is obtained for nonselective excitation (left), while for selective excitation (right), the linear time response is given by the shape of the pulse (d). NMR spectroscopy essentially uses *nonselective excitation*, while NMR imaging employs *selective excitation* in combination with magnetic-field gradients. In the second case, the responses in the time and the frequency domains are copies of the excitation profiles in the same domains. Thus a pulse with a rectangular shape in the time domain is not well suited for selection of the linear response in a slice, because it exhibits strong sidelobes which excite signals from neighbouring regions.

The response profile, however, changes with the excitation flip angle. A 30° pulse can still be considered to excite mainly the linear response. The deviation from linearity is estimated with (5.3.4a) as $(\alpha/\sin\alpha) - 1 = 4.7\%$. To describe the response to a 90° pulse, the first- and third-order terms of the expansion (5.3.4a) are sufficient, while the expansion can no longer be truncated for a 180° pulse. The linear responses, the third-order responses, and the complete responses to a 30° and a 90° x-pulse are illustrated in Fig. 5.3.4 [Hou1]. They have been calculated from the Bloch equations and from an expansion truncated after the first- and the third-order terms. Excellent agreement is obtained for the 30° pulse. The signal is essentially in the y-component ($\mathrm{Im}\{Y(\omega)\}$), while the x-component ($\mathrm{Re}\{Y(\omega)\}$) of the magnetization spectrum is zero (a). If the *nonlinear response* is excited, both components are nonzero. The response spectrum calculated from the first- and third-order response (continuous lines) deviates only slightly from the full response (ticks), whereas the *linear response* (broken line) is far too large. This demonstrates that excitation of the weakly nonlinear response

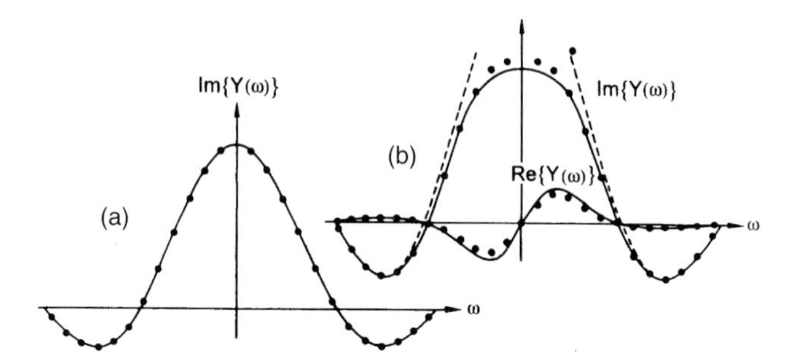

FIG. 5.3.4 [Hou1] Linear (broken lines) and nonlinear (continuous lines) response spectra calculated from the Bloch equations for a 30° pulse (a) and a 90° pulse (b) by the sum of first- and third-order responses. The ticks mark the values obtained for the full, untruncated response.

leads to an attenuation of the signal compared to the unattainable linear response for the same excitation level. Therefore, it is concluded that the *excitation spectrum* of pulses with large excitation power gives only a qualitative estimate for the selectivity of the excitation. For a quantitative description, the response of the spin system described by the Bloch equations or of the density matrix equation of motion (cf. Section 2.2.2) must be evaluated.

5.3.2 Shaped pulses

Pulses with rectangular shapes in the time domain are not well suited for selective excitation, because they exhibit trailing sidelobes in the frequency domain, so that the slice is ill defined and signal from neighbouring sample regions is excited as well. Better selectivity can be achieved with more complicated pulse shapes in the time domain. Here amplitude, phase, and frequency can be modulated [Cap1, Mcd1, War1]. Of course, the more complicated the modulation functions, the more stringent demands are posed to the spectrometer hardware. To judge the selectivity of a 90° pulse, its Fourier transform can be considered a reasonable approximation of the response. However, this approximation should not be used for selective inversion pulses, because of the highly nonlinear response to a 180° pulse (cf. Section 5.3.1).

The sinc pulse

The most simple approach to excitation of a rectangular window of transverse magnetization in the frequency domain is to use the Fourier transform of the frequency profile for modulation of the excitation in the time domain. Now the time and frequency domain profiles are permuted in comparison with a long, selective pulse of rectangular shape in the time domain. The Fourier transform of the rectangle is the *sinc function* or $\sin x/x$ function (cf. Section 4.1) with trailing sidelobes. This function defines the shape of the *sinc pulse*. Because the sidelobes arise in the time domain they need to be truncated in practice (Fig. 5.3.5(a)), leading to wiggles in the centre and near the edges, and to a trapezoidal rather than a rectangular shape of the frequency profile (Fig. 5.3.5(b)) [Cal2]. The phase of the selected magnetization varies linearly with frequency. If it

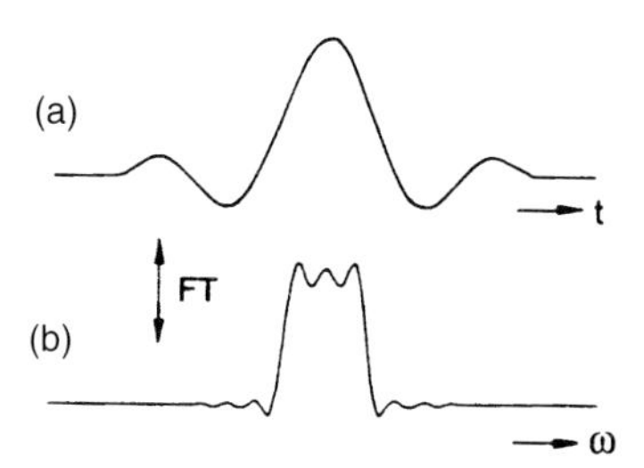

FIG. 5.3.5 [Cal2] The sinc pulse. (a) The amplitude of the rf carrier is modulated by a truncated sinc function. (b) The magnitude of the Fourier transform of the pulse is a rectangular function distorted by wiggles in the centre and near the edges.

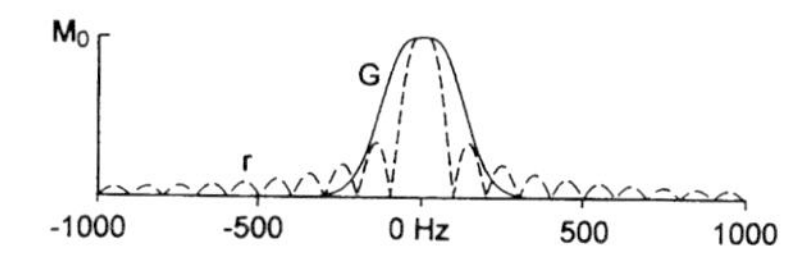

FIG. 5.3.6 Absolute values of the frequency responses to 10 ms long 90° pulses with rectangular (r) and Gaussian (G) shape in the time domain. The response to the Gaussian pulse is negligible beyond an offset of 300 Hz, while that to a rectangular pulse extends to about 1000 Hz. Adapted from [Bau1] with permission from Wiley-VCH.

were constant, the corresponding time-domain modulation function had its maximum at time zero [Man4]. Then it would start with the decay of the sinc pulse and end with its build-up [Cal2]. But such split sinc functions are difficult to generate. In practice, a linear phase shift is taken into account with most selective pulses, because it can be refocused by a *gradient echo*. Depending on the use of the sinc pulse as a selective 90° or a selective 180° pulse, the envelope must be modified accordingly to account for the nonlinear response [Mao1].

The Gauss pulse

Pulses with a Gaussian envelope [Bau1, Sut1] are more easily generated than sinc pulses. Essentially the edges of a rectangular pulse need to be smoothed. The Fourier transform of a Gauss function is again a Gauss function, so that the spectrum of a *Gauss pulse* lacks the wiggles which arise for a rectangular pulse. Apart from an attenuation at the centre, the same shape is essentially obtained for the nonlinear response. Therefore, for a given pulse length the response bandwidth of a Gaussian pulse is shorter than that of a rectangular pulse, leading to better localization (Fig. 5.3.6) [Bau1]. In practice, the tails of the Gaussian envelope are truncated between 1% and 5% of the peak intensity.

The Hermite pulse

A Gauss pulse and a polynomial multiplied by a Gauss pulse satisfy the requirements of rapid fall-off in both the time and the frequency domains. A pulse the shape of which is determined by a product of a Hermitian polynomial and a Gauss function is a *Hermite pulse* [Sil1, War2]. Its width can be more narrow than that of a Gauss pulse. A fairly uniform rotation of z magnetization can be generated if the sum of a zeroth- and a second-order polynomial is multiplied by a Gauss function. Such 90° and 180° Hermite pulses are defined by

$$\omega_1(t) = \omega_1\left[1.0 - 0.667\left(\frac{t}{T}\right)^2\right]\exp\left\{-\left(\frac{t}{T}\right)^2\right\} \quad (90^\circ \text{ pulse}), \tag{5.3.10a}$$

$$\omega_1(t) = \omega_1\left[1.0 - 0.956\left(\frac{t}{T}\right)^2\right]\exp\left\{-\left(\frac{t}{T}\right)^2\right\} \quad (180^\circ \text{ pulse}), \tag{5.3.10b}$$

where T adjusts the width of the curve.

The hyperbolic-secant pulse

Some imaging techniques like multi-slice imaging require selective 180° pulses for refocusing by a Hahn echo, unless stimulated echoes are used. In this case, it is particularly important to avoid trailing edges in the frequency domain, so that the magnetization of neighbouring slices remains undisturbed. Pulses based on continuous phase modulation for population inversion at any desired bandwidth have been proposed to this end [Bau2]. However, these are rather demanding in terms of spectrometer hardware, and composite pulses (cf. Section 5.3.3) consisting of a windowless succession of rectangular pulses with specific duration and phase can be derived from them with little effort. These can be computer optimized for maximum inversion bandwidth.

One remarkable phase- and amplitude-modulated rf pulse is the hyperbolic-secant pulse [Sil1, Sil2]. The B_1 field in the rotating frame field is a complex quantity modulated according to

$$B_{1x}(t) + \mathrm{i}B_{1y}(t) = B_1[\mathrm{sech}(\beta t)]^{1+\mathrm{i}\mu}. \tag{5.3.11}$$

The frequency width $\Delta\nu$ of the selected region is given by

$$\Delta\nu = \frac{\beta\mu}{\pi}. \tag{5.3.12}$$

An excellent 180° pulse results if μ is set to 5. Under such a pulse, magnetization components at frequencies centred at zero in a well-defined frequency band are inverted adiabatically, while components at any other frequency are returned to their initial condition at the end of the pulse. Therefore, such a pulse can be used with inhomogeneous B_1 fields. Adiabatic pulses have been designed for excitation of narrow [Ros1] as well as wide [Hwa1, Kup1, She1] frequency bands at low rf power.

Self-refocusing pulses

All time-symmetric, amplitude-modulated rf pulses introduce a frequency-dependent *phase shift* to the response. This is illustrated in Fig. 5.3.7(a) for the sinc pulse on all three components of the magnetization vector generated by application of the pulse in a magnetic-field gradient [Cal2]. The phase twist can largely be refocused by the formation of a *gradient echo* with a gradient pulse of opposite sign and half the area as the gradient pulse under the selective rf pulse. However, refocusing of magnetization by a

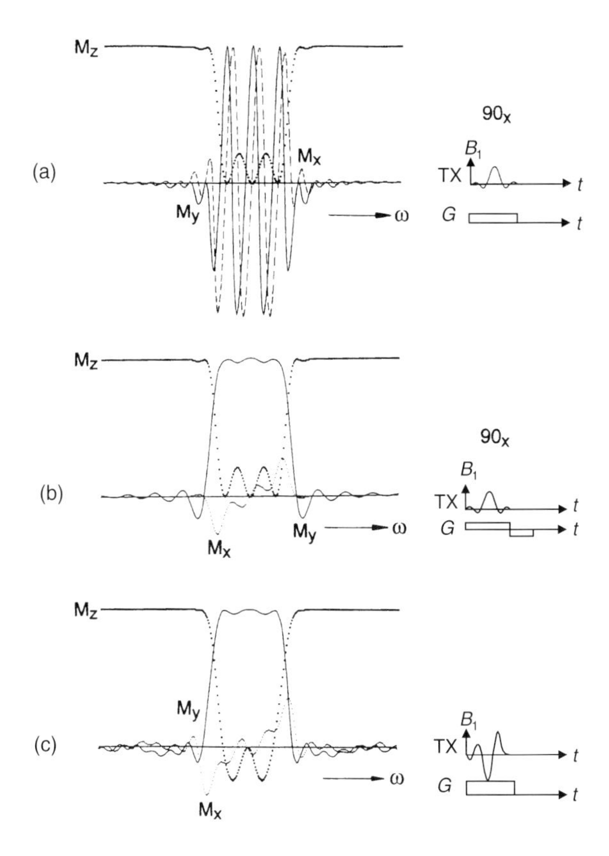

FIG. 5.3.7 Longitudinal and transverse magnetization components after application of selective pulses in a field gradient: (a) sinc pulse without gradient echo, (b) sinc pulse with gradient echo, (c) self-refocusing E-BURP-1 pulse. The Fourier coefficients of the pulse are given by $A_n = +0.23, +0.89, -1.02, -0.25, +0.14, +0.03, +0.04, -0.03, 0.00$ for $n = 0$ to 8, and by $B_n = -0.40, -1.42, +0.74, +0.06, +0.03, -0.04, -0.02, +0.01$ for $n = 1$ to 8 (cf. eqn. (5.3.13)). Adapted from [Cal2] with permission from Oxford University Press.

gradient echo suffers from the shortcoming that the phase twist which the NMR spectrum experiences from the selective pulse is not refocused [Fra1]. Combining the gradient echo with a 180° pulse for generation of a simultaneous Hahn echo can excite extra-slice magnetization and does not refocus the phase twist associated with the homonuclear scalar coupling. For these reasons, pulses are sought which convert z magnetization to in-phase transverse magnetization in a limited frequency window.

A particularly simple pulse with this property is the 270° *Gauss pulse* [Ems1, Ems2]. A whole family of pulses with these properties are the *BURP pulses* (band-selective, uniform-response, pure-phase) [Fre1, Gee1]. Depending on the type of application, they are divided into different classes. Excitation pulses (E-BURP) and inversion pulses (I-BURP) acting on z magnetization are discriminated from general rotation pulses for magnetization in an arbitrary state. Of the latter kind, there are the universal $\pi/2$ rotation pulses (U-BURP), and the π refocusing pulses (RE-BURP).

They have been designed by the method of simulated annealing and further refinement procedures [Gee1]. The pulse shape is represented by a Fourier series expansion,

$$\omega_1(t) = \omega_1 \left\{ A_0 + \sum_{n=1}^{n_{\max}} [A_n \cos(n\omega t) + B_n \sin(n\omega t)] \right\}. \tag{5.3.13}$$

The quality of selectivity, that is, the shape of the edges and the phase of the selected magnetization response improves with increasing $n_{\max}$. Depending on the specifications of the target profile, different BURP pulses are derived. The E-BURP-1 pulse can be optimized for $M_y(\omega) = 1$ for $0 \leq \omega \leq 4\pi/t_p$ and $M_y(\omega) = 0$ for $6\pi/t_p \leq \omega \leq 20\pi/t_p$, where ω is the frequency offset and t_p is the duration of the pulse. This pulse and the magnetization response without gradient echo refocusing are displayed in Fig. 5.3.7(c) [Cal2, Gee1]. The residual transverse magnetization artefacts are smaller than for the refocused sinc pulse. However, precise B_1 amplitude modulation is required, in practice, so that strongly nonlinear rf amplifiers cannot be used.

Magnetization inversion in well-defined frequency regions can also be achieved by *chirp pulses* [Böh1, Böh2, Gar2, Kun1]. Such excitation denotes a fast adiabatic passage through resonance [Sli1] achieved by sweeping the frequency of the B_1 field and keeping its amplitude strong enough for the magnetization to be locked along the direction of the effective field. In this way, populations can be inverted with inhomogeneous B_1 fields, for instance, by using surface coils (cf. Section 9.2.2).

Selection of longitudinal magnetization

The design of selective excitation schemes, which excite large bandwidths with the exception of a narrow region presents a particular challenge. The general excitation profile is shown in Fig. 5.3.2(b). When applied to transverse magnetization, a typical use is for solvent signal suppression in *in vivo NMR* spectroscopy [Zij1]. One of the first attempts to generate such excitation was *tailored excitation* [Tom1]. It has been constructed by Fourier synthesis to match the requested response profile assuming a linear relationship between excitation and response. The resultant excitation was applied in a continuous fashion, but with time-sharing of the coil for interleaved acquisition of the response. This type of excitation requires only low rf power, but frequency profiles with sharp edges could not be obtained so that the technique was not used in practice.

Shaping of the longitudinal magnetization profile by pulses applied prior to the imaging sequence (cf. Fig. 5.3.2(b) and *prepulses* [Sha2] in Section 5.3.3) is most desirable for slice, line, and point selection in imaging and volume-selective spectroscopy. Some techniques of this kind are discussed in Section 5.3.3 in the context of composite pulses. Particularly interesting are those techniques which leave the longitudinal magnetization of the selected sample region untouched in order to avoid phase errors. The resultant nonequilibrium state relaxes with T_1, so that gradient switching times can be longer than T_2^*.

A relatively successful pulse of this kind is the *DIGGER pulse* (Digger is a colloquial term originating from World War I activities of the Australian infantry in digging trenches) [Dod1]. It is obtained by straightforward Fourier transformation of the B_1 frequency profile depicted in Fig. 5.3.8(a). The longitudinal magnetization of the sample

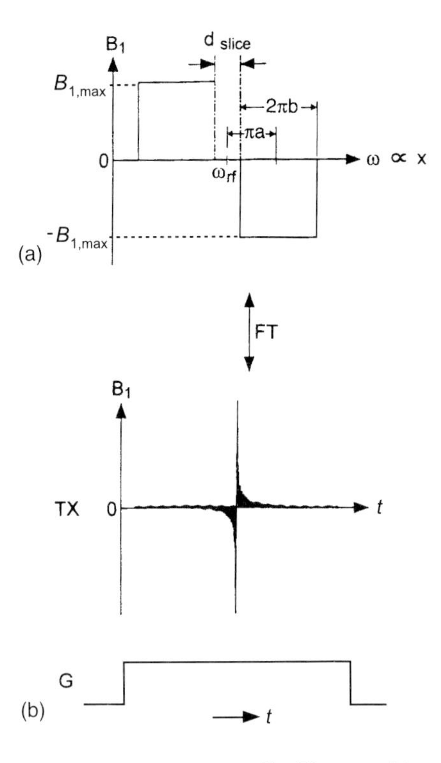

FIG. 5.3.8 The DIGGER pulse. (a) Frequency profile. The quantities a and b are defined in (5.3.14), and d_{slice} is the slice thickness [Dod1]. (b) Radio-frequency amplitude and gradient modulation for slice selection.

regions next to the selected slice is converted to transverse magnetization with opposite phases so as to achieve partial signal cancellation. After the pulse, it rapidly dephases in the presence of the gradient. The corresponding time-domain pulse envelope is given by

$$B_1(t) = -2\mathrm{i}B_{1,\mathrm{max}}\frac{\sin(\pi a t)\sin(\pi b t)}{\pi t}, \tag{5.3.14}$$

where the parameters a and b are used to adjust the pulse shape to the widths of the magnetization slice and the shape of the sample. While profiles with sharp edges can be obtained with such pulses, the rf amplitude still exhibits a pronounced peak in the centre of the time-domain envelope (Fig. 5.3.8(b)). Through repeated application of the pulse, the magnetization profile can be improved, and smaller pulse amplitudes corresponding to less rf power can be used. The technique is particularly suited for samples with long T_1 and short T_2 values, which permit repetitive applications of the pulse and support fast dephasing of transverse magnetization. Thus, it is of interest for slice selection in solids.

With regard to preservation of a selected region of longitudinal magnetization, similar effects can be obtained by saturation pulse trains in rotating gradients [Sin1, Sin2, Sin3] and incoherent cancellation of signals with noise pulses [Ord1, Cre1]. *Noise pulses* are

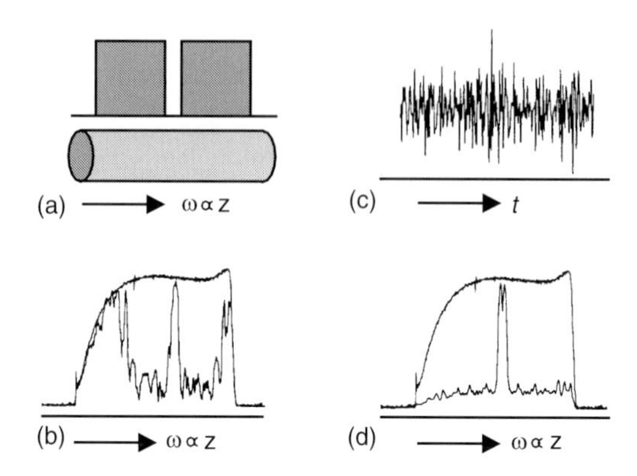

FIG. 5.3.9 [Nil1] SPREAD pulses for saturation of longitudinal magnetization. (a) Saturation profile. Magnetization in the selected slice is to be preserved. (b) Efficiency of the DIGGER pulse. The response to a nonselective pulse applied in a z gradient is shown as well as the same response preceded by a DIGGER pulse. (c) Time-domain profile of SPREAD pulse. (d) Same as (b) but for a SPREAD pulse.

used, for instance, prior to acquisition of zoom images to avoid aliasing of image regions from outside the zoom area [Cre1, Rob1].

The efficiency of noise pulses for suppression of longitudinal magnetization is illustrated in Fig. 5.3.9 by a *SPREAD pulse* (saturation pulses with reduced amplitude distribution) applied to a sample tube filled with water in the presence of a z gradient. [Nil1]. SPREAD pulses are designed with the excitation profile (a) of the DIGGER pulse. The idea is to saturate the longitudinal magnetization outside the selected slice. For saturation, the phase of the rf excitation is of no importance. Therefore, the excitation phase can be different in each frequency window. Exploiting this degree of freedom in the design of the pulse profile, the time-domain signal can be optimized for minimum amplitude, so that the available rf power can be used more efficiently. One among different strategies starts with a random set of phase values for each frequency window. Then, for example, the time-domain signal (c) can be obtained. Application of such a pulse to the simple phantom of a water-filled tube produced the magnetization profile (d), which has been read out by a nonselective 90° pulse after repetitive application of a number of SPREAD pulses with different noise-modulation profiles. This magnetization profile is compared in the same figure to the nonselective response obtained without a SPREAD pulse. The remarkable efficiency of the SPREAD pulse is put into evidence by comparing the magnetization profile (d) with the profile (b) achieved with the DIGGER pulse at the same rf power. A significantly larger bandwidth can be saturated with the SPREAD pulse. It should be noted that SPREAD pulses can be designed for arbitrary profiles of longitudinal magnetization.

5.3.3 Composite pulses

Composite pulses [Lev1, Lev2, Ern1] are sequences of typically hard, rectangular pulses, which are applied in fast succession without delay to replace regular pulses in pulse sequences. They are designed to compensate for instrumental imperfections of the spectrometer and to overcome limitations posed by the physics of conventional pulse excitation. Examples of such features are insensitivity for B_1 inhomogeneity corresponding to flip angle errors, compensation of phase errors, compensation of resonance offset corresponding to an increase of the spectral width, and reduction of the spectral width for selective excitation. Because composite pulses are composed from rectangular pulses, no pulse shaping hardware is required, but the sample is exposed to more rf energy. For this reason, composite pulses are not widely used in medical imaging.

Illustration of the principle

The use of composite pulses is often restricted to rotation of magnetization from a given initial state to a given final state. Universal composite rotations for arbitrary initial orientations are hard to achieve. The principle of this method is illustrated in Fig. 5.3.10 by the magnetization trajectories for a composite 90° pulse, which is compensated for flip angle missets (B_1 inhomogeneity) [Fre2]. A 90° rotation of magnetization initially along the z-axis is achieved by stringing four pulses with nominal flip angles of 90° and 45° together,

$$90^\circ_{-x} = 45^\circ_{+y}\, 90^\circ_{-x}\, 90^\circ_{-y}\, 45^\circ_{-x}. \tag{5.3.15}$$

The trajectories shown in Fig. 5.3.10 are obtained by setting the nominal 45° flip angles to 35°, 40°, 45°, 50°, and 55°, along with proportional changes of the nominal 90° flip angles. All trajectories start at the z-axis and end in close range of the y-axis, despite the angle missets.

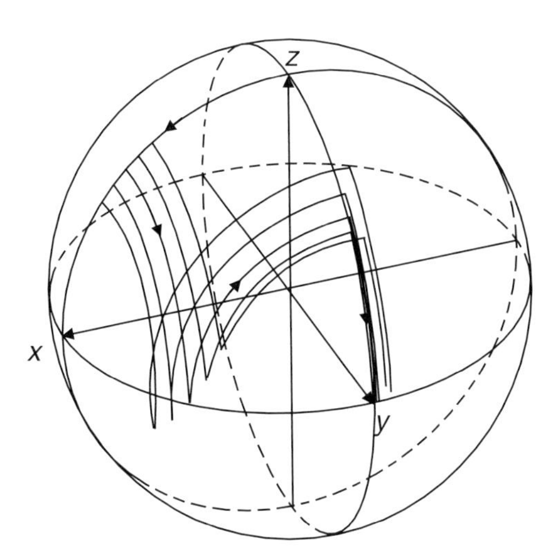

FIG. 5.3.10 [Fre2] Trajectories of magnetization aligned initially along the z-direction for a composite 90°_{-x} pulse. The flip angles of the individual pulses in the composite pulse sandwich have been misset by ±5% and ±10% of their nominal values, respectively. Nevertheless the final magnetization is close to the y-axis in all cases.

Prograde and retrograde pulses

For imaging applications, composite pulses are most suitable when they affect magnetization components over a large frequency range for excitation and refocusing of transverse magnetization in magnetic field gradients (broad-band pulses), or when they affect only a narrow frequency region for slice selective manipulation of magnetization components [Sha1, Wim1]. As an additional constraint, both types of composite pulses are required to avoid a phase twist in the magnetization after the pulse. Such broad- and narrow-band composite 90° pulses have been derived in an iterative fashion. They consist of a series of 90° or 180° pulses of different phases. For example, the *broad-band 90° pulse* (Fig. 5.3.11(a)) [Wim1]

$$B_2 = 180^\circ_{90^\circ}\ 180^\circ_{315^\circ}\ 180^\circ_{45^\circ}\ 180^\circ_{135^\circ}\ 180^\circ_{0^\circ}\ 180^\circ_{225^\circ}\ 180^\circ_{0^\circ}$$
$$180^\circ_{270^\circ}\ 180^\circ_{180^\circ}\ 180^\circ_{315^\circ}\ 180^\circ_{45^\circ}\ 180^\circ_{270^\circ}\ 90^\circ_{0^\circ} \tag{5.3.16}$$

rotates z magnetization in-phase into the transverse plane over a wide range of B_1 field strengths compared to a conventional, hard 90° pulse. Such composite pulses are sometimes called *prograde pulses*. Here the B_1 field strength enters into the strength of the effective field (2.2.30).

Composite 90° pulses, which are optimized for high sensitivity towards B_1 variation are called *narrow-band pulses*. They are also referred to as retrograde pulses.

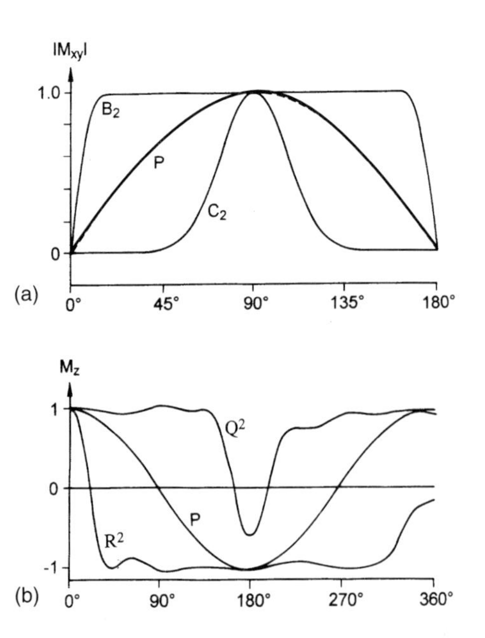

FIG. 5.3.11 The effect of prograde and retrograde composite pulses on the magnetization as a function of the actual flip angle with reference to a nonselective pulse P. (a) 90° pulses B_2 and C_2, respectively [Wim1]. (b) 180° pulses R^2 and Q^2, respectively [Sha1].

One example is the sandwich (Fig. 5.3.11(a))

$$C_2 = 180^\circ_{209^\circ} 180^\circ_{0^\circ} 180^\circ_{209^\circ} 180^\circ_{0^\circ} 180^\circ_{151^\circ}$$
$$180^\circ_{75.5^\circ} 180^\circ_{151^\circ} 180^\circ_{0^\circ} 180^\circ_{104.5^\circ} 180^\circ_{225.5^\circ} 90^\circ_{0^\circ}. \tag{5.3.17}$$

These composite pulses can be used in liquid- and solid-state multi-pulse experiments such as the *INEPT* experiment (insensitive nuclei enhanced by polarization transfer) and the Jeener–Broekaert echo experiment [Wim1].

Originally, the concept of prograde and retrograde composite pulses was developed for *inversion pulses* [Sha1]. For instance, a prograde inversion pulse, is obtained for (Fig. 5.3.11(b)) [Sha1]

$$R^2 = 180^\circ_{0^\circ}\ 180^\circ_{90^\circ}\ 180^\circ_{0^\circ}\ 180^\circ_{90^\circ}\ 180^\circ_{180^\circ}$$
$$180^\circ_{90^\circ}\ 180^\circ_{0^\circ}\ 180^\circ_{90^\circ}\ 180^\circ_{0^\circ}, \tag{5.3.18}$$

and a retrograde inversion pulse is given by (Fig. 5.3.11(b))

$$Q^2 = 180^\circ_{0^\circ}\ 180^\circ_{270^\circ}\ 180^\circ_{180^\circ}\ 180^\circ_{90^\circ}\ 180^\circ_{180^\circ}$$
$$180^\circ_{270}\ 180^\circ_{180^\circ}\ 180^\circ_{90^\circ}\ 180^\circ_{0^\circ}. \tag{5.3.19}$$

The effect of these pulses on the magnetization as a function of the actual flip angle is depicted in Fig. 5.3.11 with reference to excitation by a simple rectangular pulse. The actual flip angle plotted in the figure denotes the flip angle used instead of a 90° pulse in (5.3.16) and (5.3.17), and a 180° pulse in (5.3.18) and (5.3.19).

Prograde and retrograde composite pulses can also be designed for arbitrary flip angles using rectangular pulses [Wim1], as well as *Gaussian pulse cascades* [Ems3]. Composite pulses which act on z magnetization can be used to filter the initial magnetization prior to space encoding. Such *prepulses* have been designed to compensate resonance offset effects without compromising sensitivity to variation in B_1, for instance, for localization by use of the B_1 gradients of surface coils [Sha2]. Furthermore, a composite pulse has been worked out for chemical-shift insensitive slice selection [Wil1]. It is composed of shaped pulses and hard 180° refocusing pulses which are applied at the zero crossings of alternating magnetic-field gradients.

5.3.4 Pulse sequences

Shaped and composite pulses are devoid of free precession periods between periods of rf excitation. *Pulse sequences* combine both free precession and rf pulses. Clearly, given the manifold of shaped and composite pulses, there is an infinite number of pulse sequences which can be constructed from them. A variety of such sequences is used for volume localization. This topic is discussed in Section 10.2. Here two methods relevant to the acquisition of NMR images are reviewed. These are the DANTE technique for single-shot localization, and the Hadamard technique as an example of multi-shot localization.

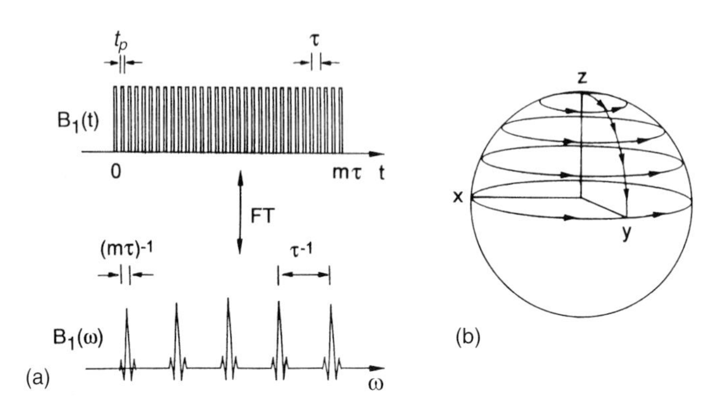

FIG. 5.3.12 Selective excitation by the DANTE sequence. (a) The DANTE sequence and the excitation spectrum. Adapted from [Cal2] with permission from Oxford University Press. (b) Illustration of the origin of the sideband response [Fre1]. The sideband response occurs when the offset from resonance is such as to allow a complete 2π rotation about the z-axis in the interval τ between the pulses. The overall effect is similar to that of the centreband response.

The DANTE sequence

The *DANTE sequence* [Mor2] consists of a sequence of small flip-angle pulses with intermittent delays (Fig. 5.3.12(a)). The spectrum of such an excitation exhibits a strong centreband and many sidebands. This excitation comb is adjusted in such a way that only the desired components of the response are excited. Such pulse trains can be readily generated without sophisticated hardware using standard nonlinear rf amplifiers. The DANTE sequence is applied in high-resolution liquid-state NMR and also in spectroscopic imaging [Hal1] for selective excitation and suppression of signal components.

Each of the n short pulses of amplitude ω_1 and length t_P of the DANTE sequence is followed by an evolution time of duration τ. The flip angle of one pulse is determined by the condition $n\omega_1 t_P = \pi/2$. The effect of the pulse sequence can be readily understood in the vector picture of NMR. At exact resonance, the magnetization moves in a smooth arc from the $+z$-axis to the $+y$-axis. The selectivity is introduced by the precession of the off-resonance magnetization components in the intervals between the pulses. Only those magnetization components which are offset by multiples of $2\pi/\tau$ from the carrier frequency have returned to their starting positions during the time τ after the pulse and before the next pulse (Fig. 5.3.12(b)) [Fre1]. These components experience the same net rotation by the rf pulses as the on-resonance magnetization component does. If the frequency dependence of the magnetization is dominated by the action of a magnetic-field gradient, the DANTE sequence can be used for slice selection in liquid-like samples (cf. Fig. 7.2.16) [Mor2]. A variant of it has been designed for slice selection by use of the residual B_1 gradient along the axial direction of saddle shaped coils, which are part of many standard microimaging probes [Maf1].

Multi-slice excitation

If the region of interest in the sample is not well defined, it is advantageous to acquire more than one slice through the object. Effective use of instrument time is achieved by interleaved acquisition of different slices: during the repetition time for magnetization recovery of one slice, the magnetization of one or more different slices can be measured. [Cro1]. An efficient realization of *multi-slice imaging* is through use of stimulated echoes (cf. Section 6.2.5) [Haa1].

These approaches make effective use of the repetition time T_R, but they are not applicable for samples with short T_R, and whenever the data of one slice is acquired, the noise originating from the entire sample is also acquired. These disadvantages can be overcome by simultaneous multi-slice imaging [Mül2, Sou1]. This technique was pioneered for simultaneous multi-slice line scanning [Cho1]. Several slices are selectively excited at the same time by a *multi-frequency selective pulse* [Mül1]. The slice selective pulses for a set of image planes are added and subtracted according to the signs given by the different rows of a sequence-ordered Hadamard matrix (cf. Section 4.4.5). A 2D image is acquired for each combination of selective pulses corresponding to each row of the Hadamard matrix. For this reason, the method is a multi-shot technique. Hadamard transformation of the resultant set of images unscrambles the multi-frequency encoded

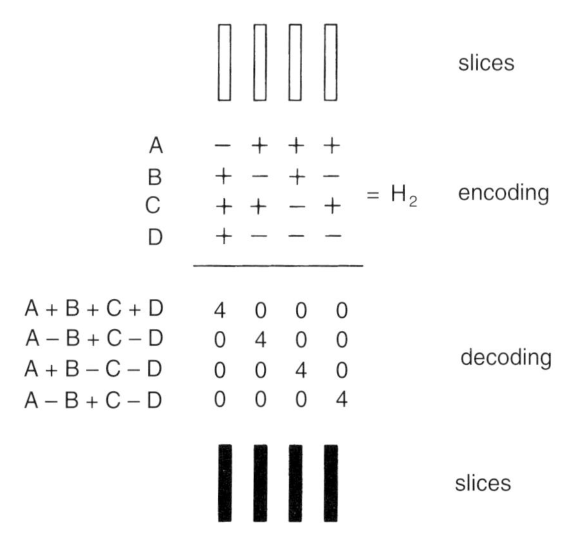

FIG. 5.3.13 Hadamard encoding and decoding for simultaneous four-slice imaging. The encoding is based on four experiments, A–D. In each experiment, all four slices are excited by a multi-frequency selective pulse. Its phase composition is determined by the rows of the Hadamard matrix $\mathbf{H}_2$. The image response is the sum of responses for each individual, frequency selective part of the pulse. Thus, addition and subtraction of the responses to the four experiments separates the information for each slice. This operation is equivalent to Hadamard transformation of the set of image responses. Adapted from [Mül2] with permission from Wiley-Liss. Inc., a division of John-Wiley & Sons, Inc.

images into images of separate slices. This scheme is illustrated in Fig. 5.3.13 [Mül2]. In contrast to successive measurements of n selective slices, the same time has now been spent for simultaneous measurement of n slices. Thus, the slice selection has been multiplexed, giving a signal-to-noise advantage of $n^{1/2}$. This scheme can readily be extended to multi-line and multi-volume selection for linescan and volume-selective spectroscopy (cf. Section 9.1) [Bol1, Haf1, Goe1, Goe2, Mül3].

5.3.5 Solid-state techniques

Solids are characterized by long spin–lattice relaxation times T_1, which are often longer than 1 s even for ^{1}H NMR, and by short spin–spin relaxation times T_2, which are typically shorter than 1 ms. The FID of rigid solids generally decays in times shorter than the duration of a slice-selective pulse. Therefore, techniques are in need by which slices can be selected in short times, and which artificially slow down the decay of the transverse magnetization amplitude or avoid the generation of transverse magnetization altogether. This can be achieved through the use of spin-locking, sequences of hard pulses, and saturation and dephasing of unwanted longitudinal magnetization. A trivial, but time-consuming solution is to measure a 3D image, for instance, by the back-projection method, and compute slices through it at any angle [Cot1].

Selective saturation of z magnetization

By the use of a selective pulse with a wide excitation profile showing a hole at the frequency of the slice (cf. Fig. 5.3.2(b)), the longitudinal magnetization outside the slice in question can be saturated or converted to transverse magnetization for subsequent dephasing. The use of such pulses is favoured by the large differences in T_1 and T_2 encountered in solids. In particular, there will be no phase twist and no intensity loss by dephasing of the selected magnetization. For good results, the pulse can be applied in a repetitive fashion. Pulses of this type are the DIGGER [Dod1] and the noise pulses [Cre1, Ord1, Rob1] mentioned in Section 5.3.2. Problems with wide-band saturation requiring high rf power for the large signal bandwidths of solids can be avoided by the SPREAD family of pulses (cf. Fig. 5.3.9) [Nil1].

Spin-lock slice selection

The most widely studied slice-selection technique uses spin-locking in a magnetic field gradient (Fig. 5.3.14) [Haf2]. It is applicable not only to solids, but also to liquids [Rom2, Win1] (cf. Section 9.1.1). The spin-lock induced slice excitation (*SLISE*) consists of a hard 90° pulse followed by a period where the transverse magnetization is locked in the

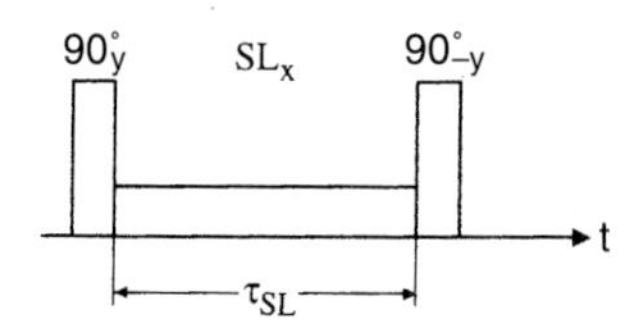

FIG. 5.3.14 [Haf2] Principle of selective excitation by spin-locking.

direction of the B_1 field. During the lock pulse, the locked magnetization relaxes with the relaxation time $T_{1\rho}$ in the rotating frame. The other magnetization dephases with a time constant T_2^*, which is characteristic for the strength of the field gradient. If the gradient is strong enough, T_2^* is short compared to $T_{1\rho}$. Provided that the length τ_{SL} of the spin-lock period is longer than T_2^* and shorter than $T_{1\rho}$,

$$T_2^* < \tau_{SL} < T_{1\rho}, \tag{5.3.20}$$

the magnetization which survives the spin-lock period is the magnetization of the selected slice. Given a gradient G_x in x-direction, the slice profile is determined by the transverse component [Dem1]

$$M_{SL}(x) = M_0 \cos^2\left[\arctan \frac{G_x x}{B_{1SL}}\right] \tag{5.3.21}$$

with a width at half-height of

$$\Delta x_{1/2} = 2\frac{B_{1SL}}{G_x}. \tag{5.3.22}$$

The slice thickness is proportional to the amplitude B_{1SL} of the spin-lock pulse and is inversely proportional to the strength of the field gradient. Therefore, the thickness can be varied by changing B_{1SL} without changing the shape of a pulse, and it is independent of the length of the pulse, which can be adjusted to yield $T_{1\rho}$-weighted image contrast. However, to ensure the spin-lock effect, the lock field must be larger than the local fields in the sample, which arise, for instance, from the homonuclear dipole–dipole coupling among ^{1}H.

The transverse magnetization selected by this SLISE excitation can be employed for subsequent gradient switching and imaging [Cot1] or a second nonselective 90° pulse is appended for storage as longitudinal magnetization (Fig. 5.3.14) [Cot1, Haf2]. Application of three such spin-lock sandwiches can be employed to select a volume element of longitudinal magnetization for localized spectroscopy (LOSY) (cf. Section 9.1.2) [Haf2, Rom1]. The method has been demonstrated to work on a phantom yielding localized ^{1}H spectra of hexamethylbenzene and polyethylene with about 3 mm spatial resolution [Haf2] as well with cross-polarization for localized ^{13}C spectroscopy [Haf3].

Sensitive slice selection

A rather simple technique related to the sensitive point method (cf. Section 5.1.2) has been proposed for imaging and slice selection and demonstrated on a PMMA phantom [Cor2]: FID signals are acquired in static-field gradients which are stepped through a range of values with increasing scan number. The acquired signals are added. In the signal sum only those parts of the signal add coherently for which the applied magnetic field is constant for all scans. The other signal contributions are averaged out. Given one gradient direction only, the signal from the magnetization of a slice is selected in this way.

With this technique, the spatial resolution is not limited by the requirement that B_1 must be larger than the largest frequency offset. In fact, selective excitation can improve the spatial selectivity. The gradients can be increased to large values, and slices down to tenth of millimeters in width can be obtained. Given enough time for data acquisition, the method can also be used for point and line scanning in imaging similar to the sensitive point (Section 5.1.2) and the sensitive line (Section 5.2.1) methods.

Multi-pulse slice selection

Multi-pulse sequences are used for line narrowing, for instance, by selective averaging of the homonuclear dipole–dipole interaction (cf. Section 3.3.4). In their presence, the length of the FID is effectively prolonged, so that time-extended selective excitation can be applied. Different schemes based on the MREV8 sequence have been designed for use with [Car1] and without [Cor1] magic-angle spinning (CRAMPS).

A solid-state variant of the DANTE sequence (Fig. 5.3.15) is obtained by replacing the rf pulses and the free precession periods of the original sequence by line-narrowing multi-pulse sequences [Car1, Cor1, Hep1, Hep2]. Such DANTE sequences can be used for selective excitation in solid-state spectroscopy (cf. Fig. 7.2.8) and for slice selection in solid-state imaging (Fig. 5.3.16).

The original *DANTE sequence* consists of a string of m small flip-angle pulses which are separated by free precession intervals. In these intervals, the magnetization precesses about the z-axis with a frequency which is determined by the resonance offset from the rf excitation. For slice selection, the dependence of this rotation frequency on the chemical shift needs to be replaced by a dependence on the applied gradient field. To achieve this for rigid solids, the dominant dipole–dipole interaction is overcome by replacing each segment consisting of a pulse followed by a free precession interval in the original sequence by a combination of two MREV8-based multi-pulse sequences (Fig. 5.3.15)

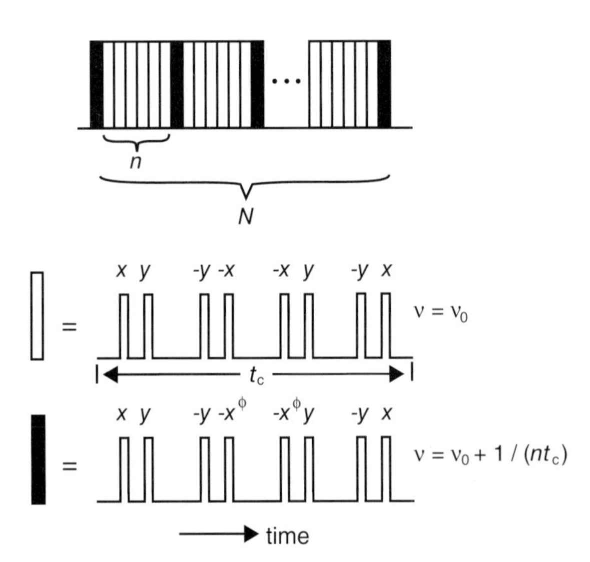

FIG. 5.3.15 [Hep2] Dipolar decoupled DANTE sequence for selective excitation for abundant nuclei in solids. The sequence is composed of a series of phase-toggled MREV8 cycles separated by n normal MREV8 cycles.

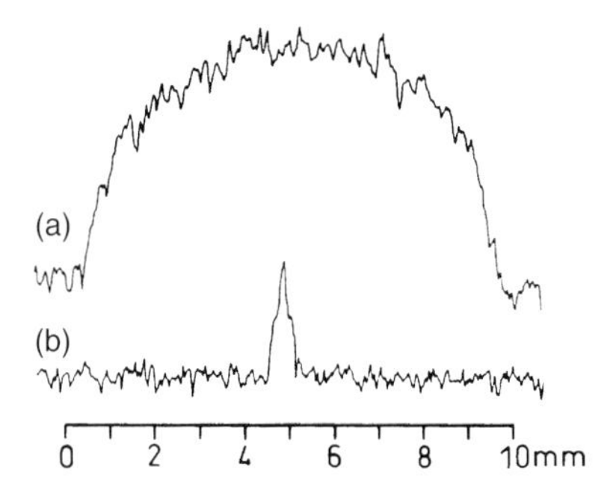

FIG. 5.3.16 [Cor1] An example of DANTE slice selection for a cylinder of ferrocene. (a) Projection without slice selection. (b) Projection after slice selection. The duration of the slice-selection sequence was 3 ms. Gradient pulses 4 μs long were applied giving a mean gradient strength of 55 mT/m.

[Cor1]. One sequence effects a precession around the z-axis which is independent of the chemical shift for all spins. Magnetic-field gradients are applied in intervals between the rf pulses, so that the precession frequency becomes dependent on the space coordinate. This requires gradient pulses which are just a few microseconds long, but it eliminates excess line broadening induced by the offset dependence of the multi-pulse sequence. The other sequence effects a rotation around the y-axis which is identical for all spins. For example, n MREV8 cycles are applied during the free precession intervals of the original DANTE sequence, and frequency-shifted MREV8 sequences with a phase toggle φ on some of the MREV8 pulses are used instead of the small flip-angle pulses of the original sequence (Fig. 5.3.15(a)). The selectivity towards position and width of the slice is determined by the choice of n, the number N of DANTE periods, and by the frequency of the MREV8 cycles. Second averaging by phase modulation of the rf pulses is used to cancel effects of chemical shift in addition to the dipole–dipole interaction.

The effectiveness of slice selection by the dipolar-decoupled DANTE sequence is illustrated in Fig. 5.3.16 with 1D images of a ferrocene cylinder without (a) and with (b) slice selection [Cor1]. The same MREV8 sequence used for inducing the gradient dependent z rotation in the DANTE sequence has been used for detection of the spatial response after slice selection. The technique is based solely on the use of hard pulses. The slice-selection process does not interfere with the decoupling efficiency of the MREV8 sequence, and distortions from chemical shift and susceptibility differences are absent.

5.4 MAGNETIZATION IN FIELD GRADIENTS

Most imaging techniques can be understood within the vector model of the *Bloch equations* (cf. Section 2.2.1). For this reason, the magnetization response calculated from the Bloch equations is investigated for arbitrary rf input and arbitrarily time-dependent magnetic-field gradients. In particular, the response which depends linearly on the rf excitation is of interest not only for imaging the spin density $M_0(\boldsymbol{r})$, but also for *spectroscopic imaging*, that is, for obtaining 1D NMR spectra with spatial resolution. For *spectroscopic 2D imaging*, the nonlinear response must be considered. But in this case the Bloch equations are no longer adequate, because they do not include the spin–spin interactions, which are of interest in 2D spectroscopy. However, significant insight can be gained from a phenomenological treatment of the nonlinear response (cf. Section 5.4.1).

In pulsed NMR spectroscopy, the signal $s(t)$ is measured in the time domain. The NMR spectrum $S(\omega)$ is obtained by Fourier transformation of $s(t)$. In the limit of the Bloch equations spectral dispersion is neglected, and the spectrum of an FID signal acquired in the presence of a time-independent field gradient describes the *projection* $P(x)$ of the *spin density* $M_0(\mathbf{r})$ onto a space coordinate x parallel to the direction of the applied gradient (cf. Fig. 1.1.5) [Gab1]. The Fourier transform of a projection is a *cross-section* in Fourier space or $\mathbf{k}$ space, where the *wave vector* $\mathbf{k} = (k_x, k_y, k_z)$ is the Fourier-conjugate variable to the space vector $\mathbf{r} = (x, y, z)$ (cf. eqn (1.2.5) and Fig. 2.2.4). The variable $\mathbf{k}$ depends on the gradient modulation. It is scanned in NMR imaging experiments. The image of the spin density is, therefore, obtained from the experimental data by Fourier transformation after extrapolation of the $\mathbf{k}$-space data onto a rectangular grid. Different imaging experiments are distinguished in the way how $\mathbf{k}$ space is being sampled (cf. Section 5.4.2) [Man3, Lju1].

Because the NMR response is detected in a phase-sensitive fashion, an image can be obtained by straightforward Fourier transformation. Diffraction methods like light, X-ray, and neutron scattering permit only the detection of signal intensities. As a result of the lack of phase information, the auto-correlation of the respective image and not the image itself is obtained. This auto-correlation function is a function of the *displacement vector* $\mathbf{R} = \Delta\mathbf{r}$. It is called the *Patterson function* [Chi1]. If the phase information is discarded, the spatial resolution of NMR can be increased to resolve averaged structures smaller than 10 μm. For this purpose, studies of translational self-diffusion are performed by observation of the echo attenuation in magnetic-field gradients [Cal1, Kär1, Sti1]. Here the data are acquired in $\mathbf{q}$ space, which is Fourier conjugate to the *displacement space*, and spatial information similar to that of the Patterson function can be gained from NMR self-diffusion studies (cf. Section 5.4.3). However, $\mathbf{q}$ space can also be understood as the Fourier-conjugate space of *velocity*. This can be seen from a description of the translational diffusion experiment in the formalism of 2D exchange NMR for analysis of slow molecular reorientation in solids (cf. Section 3.2.2). In fact, a hierarchy of exchange experiments can be constructed in *pulsed-field-gradient NMR* (PFG NMR), from which the distributions of the coefficients like the initial position $\mathbf{r}$, the initial velocity $\mathbf{v}$, the initial acceleration $\mathbf{a}$, etc., of a Taylor series expansion of the time-dependent space vector $\mathbf{r}(t)$ (cf. eqn (1.2.4)) can be obtained (cf. Section 5.4.4).

5.4.1 Linear and nonlinear responses

Given a magnetic field $\mathbf{B} = (B_{1x}(t), B_{1y}(t), -\Omega_0/\gamma)$ and a gradient vector $\mathbf{G} = \nabla B_z = -(\partial/\partial x, \partial/\partial y, \partial/\partial z)\Omega_{0r}/\gamma$, the NMR frequency is a linear function of position $\mathbf{r} = (x, y, z)$. In the rotating coordinate frame, this frequency is given by (cf. eqn (2.1.17))

$$\Omega_0(\mathbf{r}) = (\omega_0 - \omega_{\mathrm{rf}}) - \gamma\mathbf{G}\mathbf{r} = \Omega_0 - \gamma\mathbf{G}\mathbf{r}, \tag{5.4.1}$$

where ω_0 is the resonance frequency in the laboratory frame, and the chemical shift is neglected for the time being. Following the nomenclature used in system theory (cf. Section 4.2), the excitation is denoted with $x(t)$. In the rotating coordinate frame, the complex transverse magnetization response $M^+(t)$ (5.3.1) is determined by the Bloch equations (5.3.2), where the NMR frequency Ω_0 is replaced by the space-dependent expression (5.4.1),

$$\frac{\mathrm{d}M^+(t)}{\mathrm{d}t} + \left\{\frac{1}{T_2} - \mathrm{i}[\Omega_0 - \gamma \boldsymbol{G}(t)\boldsymbol{r}]\right\} M^+(t) = -\mathrm{i}x(t)M_z(t), \tag{5.4.2a}$$

$$\frac{\mathrm{d}M_z(t)}{\mathrm{d}t} = +\mathrm{Im}\left\{M^+(t)x^*(t)\right\} - \frac{M_z(t) - M_0}{T_1}, \tag{5.4.2b}$$

where M_0 is the thermodynamic equilibrium magnetization.

Linear response

Equations (5.4.2) cannot be solved in integral form unless they are linearized by setting $M_z(t) = M_0$. The *linear response* is obtained for small excitation amplitudes (cf. eqns (5.3.4)). In this case, the solution becomes

$$M_1^+(t) = -\mathrm{i}M_0 \int_0^t \exp\left\{\int_{t-\tau}^{t} \left[\frac{1}{T_2} - \mathrm{i}(\Omega_0 - \gamma \boldsymbol{G}(t')\boldsymbol{r})\right] \mathrm{d}t'\right\} x(t-\tau)\,\mathrm{d}\tau, \tag{5.4.3}$$

which is verified by insertion into (5.4.2). For time-independent gradients, this is the Fourier transform of (5.3.6). The expression for the linear response can be simplified by use of the wave vector (2.2.23),

$$\boldsymbol{k}(t, t-\tau) = -\gamma \int_{t-\tau}^{t} \boldsymbol{G}(t')\,\mathrm{d}t', \tag{5.4.4}$$

yielding

$$M_1^+(t) = -\mathrm{i}M_0 \int_0^t \exp\left\{-\left(\frac{1}{T_2} - \mathrm{i}\Omega_0\right)\tau + \mathrm{i}\boldsymbol{k}(t, t-\tau)\boldsymbol{r}\right\} x(t-\tau)\,\mathrm{d}\tau. \tag{5.4.5}$$

So far the magnetization has been treated for a single NMR frequency Ω_0 without consideration of the chemical shift and a space-dependent spin density. The amplitude M_0 of the thermal equilibrium magnetization is the sum over all magnetization components which differ in space and frequency,

$$M_0 = \iint M_0(\boldsymbol{r}, \Omega_\mathrm{L})\,\mathrm{d}\Omega_\mathrm{L}\,\mathrm{d}\boldsymbol{r}, \tag{5.4.6}$$

where $M_0(\boldsymbol{r}, \Omega_\mathrm{L})$ is the spin density which describes the spatial distribution of spins and their resonance frequencies $\Omega_\mathrm{L} = \Omega_0 - \sigma\omega_0$, and σ is the chemical shift (cf. Section 3.1.1). Parameters other than space and resonance frequency may also be included.

For an inhomogeneous sample, the linear transverse magnetization response (5.4.5) is rewritten with the help of (5.4.6) and using $M_1^+(t) = y_1(t)$,

$$y_1(t) = -\mathrm{i} \int_0^t \iint M_0(\boldsymbol{r}, \Omega_\mathrm{L}) \exp\left\{-\left(\frac{1}{T_2} - \mathrm{i}\Omega_\mathrm{L}\right)\tau\right\} \mathrm{d}\Omega_\mathrm{L} \times \exp\{\mathrm{i}\boldsymbol{k}(t, t-\tau)\boldsymbol{r}\}\, \mathrm{d}\boldsymbol{r}\, x(t-\tau)\, \mathrm{d}\tau. \quad (5.4.7)$$

This equation is fundamental for the explanation of many imaging methods. It describes the linear part of the complex transverse magnetization response in the presence of arbitrary magnetic-field gradient modulation $G(t)$ and quadrature rf excitation $x(t)$. When the excitation is a delta pulse $x(t) = \delta(t)$, the response $y_1(t)$ is given by the kernel (cf. Section 4.2.1) of the outer integral in (5.4.7),

$$y_1(t) = k_1(t) = -\mathrm{i} \iint M_0(\boldsymbol{r}, \Omega_\mathrm{L}) \exp\left\{-\left(\frac{1}{T_2} - \mathrm{i}\Omega_\mathrm{L}\right)t\right\} \mathrm{d}\Omega_\mathrm{L} \exp\{\mathrm{i}\boldsymbol{k}(t, 0)\boldsymbol{r}\}\, \mathrm{d}\boldsymbol{r}$$
$$= \int M_0(\boldsymbol{r}) s(t, \boldsymbol{r}) \exp\{\mathrm{i}\boldsymbol{k}(t, 0)\boldsymbol{r}\}\, \mathrm{d}\boldsymbol{r}. \quad (5.4.8)$$

In the last step, the frequency dependence of the spin density and the exponential Bloch decay have been replaced by a normalized spectroscopic FID $s(t, \boldsymbol{r})$. This is advantageous when arbitrary spectra $S(\omega, \boldsymbol{r})$ need to be considered. Thus, the impulse response measured in the presence of magnetic-field gradients, maps the Fourier transform of the spin density $M_0(\boldsymbol{r})$, as long as the signal decay introduced by the FID can be neglected.

Nonlinear response

With increasing amplitude of the excitation $x(t)$, the response eventually becomes nonlinear. In the case of single-pulse excitation, the *nonlinear response* becomes significant for flip angles larger than 30° (cf. eqns (5.3.4)). The first nonlinear contribution to the transverse magnetization arises in third order [Ern1]. There the spin density depends on three resonance frequencies, Ω_L, Ω_M, and Ω_N, which correspond to neighbouring single quantum transitions in the energy level diagram with coherence orders $p = \Delta m = \pm 1$ [Blü1]. The third-order response is a nonlinear convolution integral (cf. eqn (4.2.8)),

$$y_3(t) = \int_{-\infty}^{\infty}\int_{-\infty}^{\infty}\int_{-\infty}^{\infty} k_3(\tau_1 \geq \tau_2 \geq \tau_3) x(t-\tau_1) x(t-\tau_2) x(t-\tau_3)\, \mathrm{d}\tau_3\, \mathrm{d}\tau_2\, \mathrm{d}\tau_1. \quad (5.4.9)$$

Its kernel provides information about resonance connectivities, which are established through spin–spin couplings and by chemical exchange and cross-relaxation [Blü2]. It can readily be evaluated by quantum mechanical calculations in Liouville space. From a phenomenological point of view it can be written as the product of three successive single pulse responses at frequencies Ω_L, Ω_M, and Ω_N [Blü3],

$$
\begin{aligned}
k_3(\tau_1 \geq \tau_2 \geq \tau_3) = \mathrm{i} \iiiint & M_0(\boldsymbol{r}, \Omega_L, \Omega_M, \Omega_N) \\
& \times \exp\left\{-\left[\frac{1}{T_2} - \mathrm{i}\Omega_L\right](\tau_1 - \tau_2) - \mathrm{i}\boldsymbol{k}(t - \tau_2, t - \tau_1)\boldsymbol{r}\right\} \\
& \times \exp\left\{-\left[\frac{1}{T_{LM}} - \mathrm{i}(\Omega_L + \Omega_M)\right](\tau_2 - \tau_3)\right. \\
& \qquad \left. - \mathrm{i} p_{LM} \boldsymbol{k}(t - \tau_3, t - \tau_2)\boldsymbol{r}\right\} \\
& \times \exp\left\{-\left[\frac{1}{T_2} - \mathrm{i}(\Omega_L + \Omega_M + \Omega_N)\right]\tau_3 + \mathrm{i}\boldsymbol{k}(t, t - \tau_3)\boldsymbol{r}\right\} \\
& \times \mathrm{d}\Omega_L \, \mathrm{d}\Omega_M \, \mathrm{d}\Omega_N \, \mathrm{d}\boldsymbol{r}.
\end{aligned}
\tag{5.4.10}
$$

In this expression p_{LM} denotes the coherence order of the two-quantum transition at the sum frequency $\Omega_L + \Omega_M$, which may be either -2, 0, or $+2$, and T_{LM} is a spin–spin relaxation time for $|\Omega_L| + |\Omega_M| \neq 0$ and a spin–lattice relaxation time for $\Omega_L + \Omega_M = 0$. The kernel may be interpreted with the help of the time scheme of Fig. 4.2.3, when replacing σ_i by τ_i . Each of the exponentials in (5.4.10) describes the evolution of the spin system in one of the three successive time intervals $\tau_i - \tau_{i+1}$ in the presence of time-varying gradients. Nonlinear responses of orders higher than three can be constructed in an analogous fashion. Such nonlinear response functionals become significant if *multi-dimensional correlation spectra* are to be measured with spatial resolution in time-varying gradients.

5.4.2 Position: *k* space

The linear impulse response measured in the presence of field gradients provides the information about the spin density (cf. eqn (5.4.8)). However, the spin density is weighted by functions which depend on parameters of the spin system, here the spin–spin relaxation time T_2, and the resonance frequencies Ω_L observed in the NMR spectrum. If the nonlinear impulse response were evaluated, other NMR parameters would have to be considered as well, in particular, the spin–lattice relaxation time T_1. In fact the value of NMR as an imaging method derives from the existence of these parameters for weighting and for generation of image contrast. Nevertheless, the appearance of the NMR FID in (5.4.8) degrades the spatial resolution of the image, and for understanding the fundamentals of space encoding it is helpful to neglect relaxation and a distribution of resonance frequencies altogether for the time being. In this limit, the wave vector $\boldsymbol{k}$ is the experimental variable to probe position and the change of position. Therefore, the wave vector serves to scan an image in reciprocal space, and to map parameters of translational molecular motion [Cal2, Can2, Sod1].

The projection–cross-section theorem

Neglecting the frequency spread in (5.4.8) by setting $s(t, \boldsymbol{r}) = 1$, the linear impulse response assumes the form of a scattered wave (cf. eqn (2.2.22)) [Man4, Man5],

$$y_1(t) = \int M_0(\boldsymbol{r}) \exp\{\mathrm{i}\boldsymbol{k}(t, 0)\boldsymbol{r}\}\, \mathrm{d}\boldsymbol{r}$$

$$= \iiint M_0(x, y, z) \exp\{\mathrm{i}k_x(t, 0)x\} \exp\{\mathrm{i}k_y(t, 0)y\}$$

$$\times \exp\{\mathrm{i}k_z(t, 0)z\}\, \mathrm{d}x\, \mathrm{d}y\, \mathrm{d}z, \tag{5.4.11}$$

where the wave vector with magnitude $|\boldsymbol{k}| = 2\pi/\lambda$ is defined by (5.4.4), and λ is the wavelength (cf. Fig. 2.2.4). The spin density can be derived from (5.4.11) simply by inverse Fourier transformation with respect to $\boldsymbol{k}$. Thus, to measure an image by NMR, the gradients must be varied in such a way that the complete $\boldsymbol{k}$ space is scanned in the course of the experiment. In this regard, $\boldsymbol{k}$ can be understood to encode *position* in NMR.

The FID (5.4.11) acquired in the presence of a time-invariant gradient defines the signal along a straight line in $\boldsymbol{k}$ space. For example, if the gradient is applied in x-direction, the FID is given by

$$y_1(t) = \iiint M_0(x, y, z)\, \mathrm{d}y\, \mathrm{d}z \exp\{-\mathrm{i}\gamma G_x x t\}\, \mathrm{d}x$$

$$= \int P(x) \exp\{\mathrm{i}k_x x\}\, \mathrm{d}x = p(k_x). \tag{5.4.12}$$

For this FID $k_y = k_z = 0$, so that $t \propto k_x$. Consequently, $y_1(t)$ in (5.4.12) defines a *cross-section* $p(k_x)$ in $\boldsymbol{k}$ space along the k_x-axis. This cross-section is the Fourier transform of $P(x)$, and $P(x)$ is a 1D *projection* of the 3D distribution $M_0(x, y, z)$ of longitudinal magnetization onto the x-axis corresponding to the integral of $M_0(x, y, z)$ over the space dimensions y and z (cf. Fig. 1.1.5). In the mathematical sense, a projection is just the integral over a function of many variables, so that fewer variables remain. Fourier transformation over t leads to

$$Y_1(\omega) = \int y_1(t) \exp\{-\mathrm{i}\omega t\}\, \mathrm{d}t = \int P(x) \int \exp\{-\mathrm{i}(\omega + \gamma G_x x)t\}\, \mathrm{d}t\, \mathrm{d}x$$

$$= \frac{1}{2\pi} \int P(x)\delta(\omega t + \gamma G_x x t)\, \mathrm{d}x = \frac{1}{2\pi} P\left(-\frac{\omega}{\gamma G_x}\right). \tag{5.4.13}$$

Here the delta function introduces the proportionality between frequency and space,

$$\omega = -\gamma G_x x. \tag{5.4.14}$$

Equation (5.4.13) states that the NMR spectrum $Y_1(\omega)$ acquired in the presence of a magnetic-field gradient G produces a projection $P(-\omega/(\gamma G_x))$ of the spin density $M_0(x, y, z)$. In comparison to that, the FID (5.4.12) provides a trace through $\boldsymbol{k}$ space. This relationship is known as the *projection–cross-section theorem*.

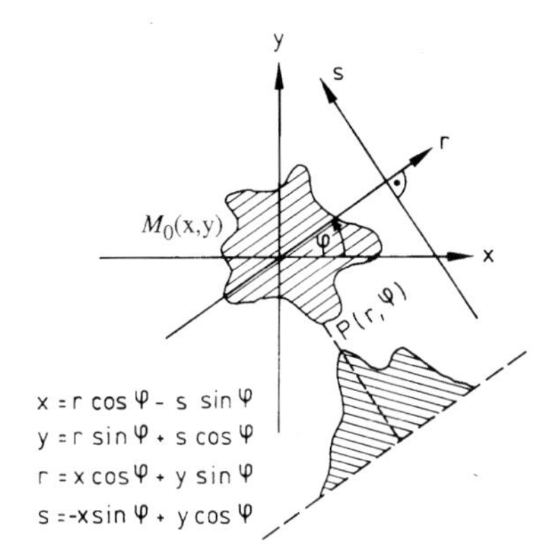

FIG. 5.4.1 Relationship between the Cartesian laboratory coordinate frame with axes x and y and the rotated Cartesian frame with axes r and s. The field gradient is applied parallel to r at an angle φ defining the variables of the projection $P(r, \varphi)$ of the object.

From a set of projections acquired under different angles φ in polar coordinates, a spin density map is reconstructed in *back-projection imaging* (cf. Section 6.1). For a 2D spin density $M_0(x, y)$ the projection onto an axis r which is at an angle φ with respect to the x-axis follows by integration over the space variable s orthogonal to r (Fig. 5.4.1),

$$\begin{aligned} y_1(t) &= \iint M_0(x, y) \exp\{\mathrm{i}k_x x\} \exp\{\mathrm{i}k_y y\} \, \mathrm{d}x \, \mathrm{d}y \\ &= \iint M_0(r, s) \exp\{\mathrm{i}k_s s\} \, \mathrm{d}s \exp\{\mathrm{i}kr\} \, \mathrm{d}r \\ &= \int P(r, \varphi) \exp\{\mathrm{i}kr\} \, \mathrm{d}r = p(k, \varphi) \end{aligned} \qquad (5.4.15)$$

Note that in Fig. 5.4.1 and eqn (5.4.15) s is a space variable and not a normalized FID. The gradient is applied in the direction of r so that $k_s = 0$ and the projection P is obtained by integration over s. Therefore, for arbitrary angles φ, the FID $y_1(t)$ provides the Fourier transform $p(k, \varphi)$ of the projection $P(r, \varphi)$ onto the gradient direction.

Scanning of ***k*** *space*

Different imaging techniques are distinguished in the way how $\boldsymbol{k}$ space is scanned. Some possible ***k****-space trajectories* are depicted in Fig. 5.4.2 for 2D $\boldsymbol{k}$ space [Lju1]. Ideally, the trajectories are evenly spaced. However, variable spacings may be admitted for the benefit of other advantages. Even spiral paths are conceivable [Man4, Vla1]. In a time-invariant gradient, $k = |\boldsymbol{k}|$ is proportional to the acquisition time t by (5.4.4). For large values of k, the FID decays rapidly, but as long as the magnetization is not irreversibly relaxed, it can be refocused to form echoes. Echoes build up following a time inversion, so that in the echo maximum the origin of $\boldsymbol{k}$ space is approached. They can be generated by gradient inversion or by 180° pulses (cf. Section 3.4), but usually the field gradient

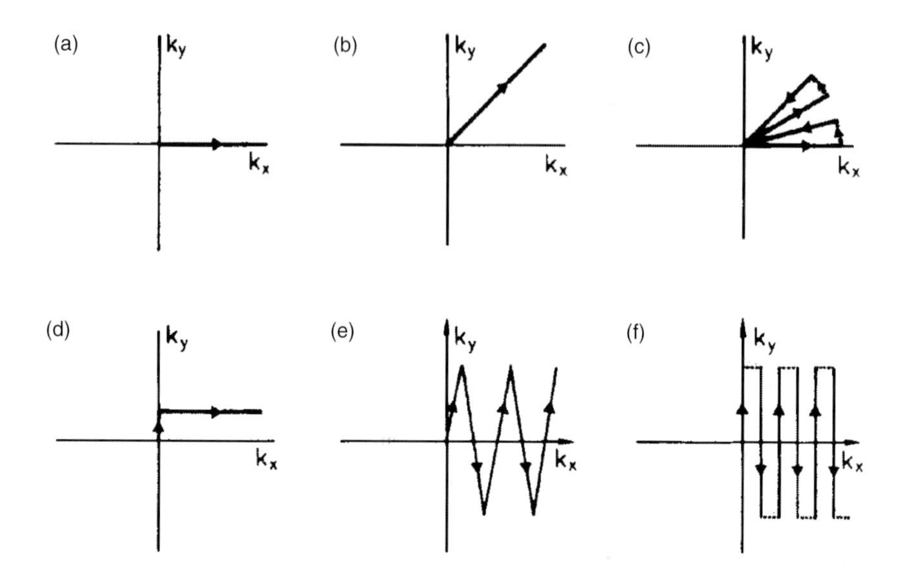

FIG. 5.4.2 [Lju1] Trajectories in $\boldsymbol{k}$ space. (a) Line-scan method. (b) Back-projection method. (c) Modified back-projection method. (d) Fourier imaging. (e) Echo-planar imaging. (f) Modified echo-planar imaging.

is slightly changed in order to avoid scanning the same trajectory in the reverse sense. A 180° pulse generates the complex conjugate signal. Thus the sign of $\boldsymbol{k}$ is inverted and a factor of -1 must be considered in the signal phase in case the phases of the two generating rf pulses are the same.

The $\boldsymbol{k}$-space trajectories depicted in Fig. 5.4.2 are for imaging methods treated in Chapter 6, except for (a), the line-scan method (cf. Section 5.2.2). In the latter case, the FID is acquired from magnetization forming a line through the object in a constant gradient in the direction of the line. For the back-projection method (b) a constant gradient is applied in a direction φ specified in cylindrical coordinates (Section 6.1). The FID decays without being refocused by an echo. This may be expensive in terms of acquisition time, if recycle delays need to be accommodated between scans. Therefore, the magnetization is refocused in a modified version (c), but in a way that a neighbouring trajectory is scanned. In Fourier imaging (d) orthogonal directions are explored in $\boldsymbol{k}$ space by an evolution under one gradient, here G_y, followed by the detection in another gradient, here G_x (cf. Section 6.2.1). The experiment then needs to be repeated for different values of k_y. In echo-planar imaging (e) the magnetization is refocused several times in the presence of a weak, constant x gradient, so that the entire $\boldsymbol{k}$ space can be scanned during the length of a single FID (cf. Section 6.2.7). This is the fastest way of acquiring an image. However, the trajectory follows a zigzag path. A more favourable coverage of $\boldsymbol{k}$ space is achieved if the weak x gradient is pulsed (f).

For all methods, which exploit multiple refocusing of the FID, subsequent echoes are attenuated to different degrees by relaxation, which effects the image contrast. When scanning $\boldsymbol{k}$ space, arbitrary shapes can be excited in 3D space [Pau1, Ser1, Ser2, Ser3].

This approach to the scanning of $\boldsymbol{k}$ space is a generalization of the concept of slice selection (cf. Section 5.3) [Vla1].

The effect of the NMR spectrum

The description of imaging experiments in reciprocal space is not restricted to $\boldsymbol{k}$ space, the Fourier conjugate space of physical space. The modification of the spin density by other parameters like resonance frequencies, coupling constants, relaxation times, etc., can be treated in a similar fashion [Mül4]. For the frequency-dependent spin density, the Fourier transformation with respect to Ω_L is already explicitly included in (5.4.7). Introduction of a T_2-dependent density would require the inclusion of another integration over T_2 in (5.4.7) and lead to a Laplace transformation (cf. Section 4.4.1).

The frequency dependence of the spin density is exploited in *spectroscopic imaging* (cf. Section 6.2.4) to derive spectroscopic NMR parameter contrast from another dimension attached to the imaging experiment. Yet when measuring time is at a premium, a spectroscopic dimension cannot be acquired. The frequency dependence will then lead to a degradation of the spatial resolution. If, however, time-independent gradients are applied, and if the spectrum does not vary over the sample, that is, if $s(t, \boldsymbol{r}) = s(t)$, then the Fourier transform of the FID (5.4.8) is given by the convolution (cf. eqn (4.2.1)) of the spin density with the NMR spectrum $S(\omega)$,

$$\begin{aligned} Y_1(\omega) &= \int M_0(\boldsymbol{r}) \int s(t) \exp\{-\mathrm{i}(\omega + \gamma \boldsymbol{G}\boldsymbol{r})t\} \,\mathrm{d}t \,\mathrm{d}\boldsymbol{r} \\ &= \int M_0(\boldsymbol{r}) S(\omega + \gamma \boldsymbol{G}\boldsymbol{r}) \,\mathrm{d}\boldsymbol{r} = M_0(\omega) \otimes S(\omega). \end{aligned} \tag{5.4.16}$$

The spin density can be separated from the spectrum $S(\omega)$ by deconvolution (cf. Section 4.2.3), provided that the spectrum itself is known [Cor3].

Resolution and the point-spread function

Throughout most of the NMR literature as well as in parts of this book, *spin density* and *NMR image* are used synonymously. However, often the actual image is a map of some NMR parameter, or of a function of several parameters, like the spin density and the relaxation times. In addition to this, the actual image $F(\boldsymbol{r})$ determined in an experiment maps the true image only with the spatial resolution and the parameter resolution provided by the spectrometer and the particular imaging technique employed. If all NMR parameters like the chemical shift Ω_L and the relaxation times T_1 and T_2, which are resolved in the measurement, are collected in the parameter vector $\boldsymbol{p}$, the parameter-weighted NMR image can be written as a convolution of the parameter-dependent spin density $M_0(\boldsymbol{r}, \boldsymbol{p})$ with the *localization function* or *point-spread function* $H(\boldsymbol{r}, \boldsymbol{p})$ [Kie1, Kie2, Mcf1, Met1],

$$F(\boldsymbol{r}) = \int M_0(\boldsymbol{r}, \boldsymbol{p}) H(\boldsymbol{r} - \boldsymbol{r}', \boldsymbol{p} - \boldsymbol{p}') \,\mathrm{d}\boldsymbol{r} \,\mathrm{d}\boldsymbol{p}. \tag{5.4.17}$$

Equation (5.4.17) is a generalized form of (5.4.16), where the localization function is determined by the NMR spectrum $S(\Omega)$ and is limited to just the space coordinates $\boldsymbol{r}$.

Thus, in the case of one resonance dominating the NMR spectrum, the *spatial resolution* achievable by measurement in time-invariant gradients is determined by the linewidth (cf. Fig. 2.3.10) unless deconvolution techniques are applied.

Factors which ultimately limit the spatial resolution [Cal4] are translational diffusion in liquids [Ahn1, Hys1] and distortions of the local magnetic field by susceptibility differences [Cal3, Lud1]. The best spatial resolution is obtained for liquids in confined spaces like plant cells with values in the order of $(5\ \mu\text{m})^{-1}$ [Cof1] and in viscous liquids. With increasing resolution the number of spins decreases. So does the sensitivity, and the acquisition time subsequently increases. Therefore, the available acquisition time imposes a practical limit on the achievable spatial resolution [Ecc1].

The limitation of the resolution by signal decay in the absence of a gradient applies only for *frequency encoding* of the space information, that is for the space dimension which is acquired directly (cf. Section 2.3.4). For indirect detection of the space information by *phase encoding* with the spin-warp technique, the spatial resolution is unrestricted in theory. It is determined just by the range of gradient values scanned in the experiment. Apodization techniques can be used for giving the voxel the desired shape (cf. Section 6.2.4) [Mar1].

5.4.3 Velocity: q space

Magnetization that is acquired in $\boldsymbol{k}$ space can be Fourier transformed to a projection or an image in physical space with a one-to-one correspondence of image and space coordinates $\boldsymbol{r}$. Here the achievable spatial resolution is of the order of $10\ \mu\text{m}$ in one dimension. Nevertheless, by application of magnetic-field gradients the spatial resolution can be increased by one to two orders of magnitude if molecular transport by translational diffusion is measured in a manifold of similar structures. The price paid for increased spatial resolution is a loss of spatial information. Instead of actual space coordinates, the signal intensity is obtained in *displacement space*, that is in a space which measures net distances $\boldsymbol{R} = \boldsymbol{r}_2 - \boldsymbol{r}_1$ which magnetization travels along the gradient directions for a given time. Thus the spatial information refers to average features of similar structures. This explains the gain in spatial resolution, because the signal-to-noise ratio is improved by considering structures averaged over the entire sample in place of individual structures in the sample.

The associated loss of information is a consequence of the measuring schemes. They are most frequently based on combinations of the Hahn echo or the stimulated echo with *constant-field gradients* [Cal2, Dem2, Gei1, Hah1, Kim1, Kim2] or *pulsed-field gradients* [Kär1, Kim2, Mcc1, Ste1]. In the context of imaging PFG NMR [Cal1, Cal2, Cal5, Cal14, Sti1] is the more important variant, because gradient switching allows for more freedom in the design of pulse sequences, and the method can be used for generation of contrast [Cal2]. The displacement information essentially derives from the attenuation of echoes with gradients, where, in the case of diffusion, information about the precession phase due to different positions in the sample is lacking.

Because the measurement of displacement takes some time Δ, the measured quantity can also be considered to be displacement over time, that is average velocity $\boldsymbol{v} = \boldsymbol{R}/\Delta$ during the time lag Δ. The Fourier-conjugate variable to displacement $\boldsymbol{R}$ has been given the symbol $\boldsymbol{q}$. This pair of variables referring to displacement space corresponds to the pair of variables position $\boldsymbol{r}$ and wave vector $\boldsymbol{k}$ for physical space. Similarly, the Fourier conjugate variable to $\boldsymbol{v}$ can be introduced as $\boldsymbol{q}_v = \boldsymbol{q}\Delta$ which differs from $\boldsymbol{q}$ only by the scaling factor Δ.

Physical space and displacement space

In NMR experiments with gradient echoes, the lack of phase information due to position in the sample is analogous to the lack of phase information in many scattering experiments [Bar1, Man3]. As a consequence, the resulting diffraction pattern cannot be readily inverted to a real-space image of the scattering centres, and only the relative distances of the scattering centres can be derived, for example, the lattice vectors in X-ray analysis of crystal structures.

The analogy between diffraction signals from scattering experiments and the loss of phase information in NMR imaging is illustrated in Fig. 5.4.3 [Blü4]. The magnitude square $|s(k_x, k_y)|^2$ of the NMR signal $s(k_x, k_y)$ measured in $\boldsymbol{k}$ space from a more or less regular array of glass capillaries in water forms a familiar diffraction pattern. The separation between diffraction peaks provides inverse distances between scattering centres in the object. The Fourier transform of the diffraction pattern is the auto-correlation function $a(\Delta x, \Delta y) = a(\boldsymbol{R})$ of the image $S(x, y)$. The auto-correlation function is known as the *Patterson function* in X-ray scattering.

In one dimension this relationship is expressed by the following calculation: The signal $s(k)$ measured in an imaging experiment is expressed in terms of the associated image $S(r)$ by inverse Fourier transformation,

$$s(k) = \frac{1}{2\pi} \int_{-\infty}^{\infty} S(r_2) \exp\{\mathrm{i}kr_2\} \, \mathrm{d}r_2. \tag{5.4.18}$$

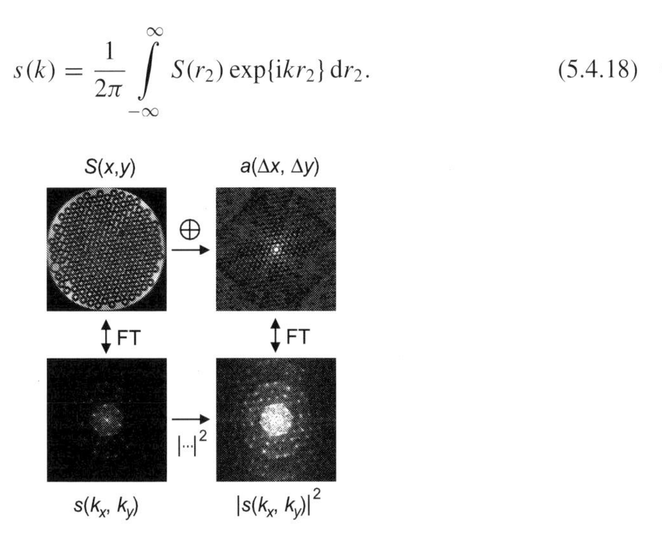

FIG. 5.4.3 [Blü4] Imaging and diffraction by example of 50 similar glass capillaries immersed in water. The image $S(x, y)$ of the spin density derives from the acquired signal $s(k_x, k_y)$ by 2D Fourier transformation. Formation of the magnitude square $|s(k_x, k_y)|^2$ discards the signal phase (unidirectional arrows). 2D Fourier transformation of the squared magnitude signal yields the Patterson function which is given by the auto-correlation function $a(\Delta x, \Delta x) = a(\boldsymbol{R})$ of the spin density. Here the symbol $\oplus$ is used to denote auto-correlation.

Formation of the magnitude square of $s(k)$ eventually introduces the displacement R. For the purpose of conformity with subsequent calculations, the following approach is taken in computing the magnitude square of $s(k)$,

$$|s(k)|^2 = \frac{1}{(2\pi)^2} \int_{-\infty}^{\infty} \int_{-\infty}^{\infty} S(r_2) S^*(r_1) \exp\{\mathrm{i}kr_2\} \exp\{-\mathrm{i}kr_1\} \, \mathrm{d}r_2 \, \mathrm{d}r_1. \tag{5.4.19}$$

For further evaluation the wavenumber k is replaced by

$$k = k_2 = -k_1, \tag{5.4.20}$$

and the following transformation of variables is introduced,

$$\begin{aligned} R' &= \frac{r_2 - r_1}{\sqrt{2}} = \frac{R}{\sqrt{2}}, \quad r' = \frac{r_2 + r_1}{\sqrt{2}} = r\sqrt{2}, \quad r_1 = \frac{r' - R'}{\sqrt{2}}, \quad r_2 = \frac{r' + R'}{\sqrt{2}}, \\ q' &= \frac{k_2 - k_1}{\sqrt{2}} = q\sqrt{2}, \quad k' = \frac{k_2 + k_1}{\sqrt{2}} = \frac{k}{\sqrt{2}}, \quad k_1 = \frac{k' - q'}{\sqrt{2}}, \quad k_2 = \frac{k' + q'}{\sqrt{2}}. \end{aligned} \tag{5.4.21}$$

With these substitutions the magnitude square of the measured signal $s(k)$ can be expressed as

$$\begin{aligned} |s(k)|^2 &= \frac{1}{(2\pi)^2} \int_{-\infty}^{\infty} \int_{-\infty}^{\infty} \int_{-\infty}^{\infty} S(r_2) S^*(r_1) \exp\{\mathrm{i}k_2 r_2\} \\ &\quad \times \exp\{\mathrm{i}k_1 r_1\} \delta(k_1 + k_2) \, \mathrm{d}k_2 \, \mathrm{d}r_2 \, \mathrm{d}r_1 \\ &= \frac{1}{(2\pi)^2} \int_{-\infty}^{\infty} \int_{-\infty}^{\infty} \int_{-\infty}^{\infty} S\left(\frac{r' + R'}{\sqrt{2}}\right) S^*\left(\frac{r' - R'}{\sqrt{2}}\right) \\ &\quad \times \exp\left\{\mathrm{i}\frac{k}{\sqrt{2}} r'\right\} \delta(k) \, \mathrm{d}k \, \mathrm{d}r' \, \exp\{\mathrm{i}q'R'\} \, \mathrm{d}R' \\ &= \frac{1}{(2\pi)^2} \int_{-\infty}^{\infty} \int_{-\infty}^{\infty} S\left(\frac{r' + R'}{\sqrt{2}}\right) S^*\left(\frac{r' - R'}{\sqrt{2}}\right) \mathrm{d}r' \, \exp\{\mathrm{i}q'R'\} \, \mathrm{d}R' \\ &= \frac{1}{(2\pi)^2} \int_{-\infty}^{\infty} \int_{-\infty}^{\infty} S(r) S^*(r - R) \, \mathrm{d}r \, \exp\{\mathrm{i}qR\} \, \mathrm{d}R = f(q) \end{aligned} \tag{5.4.22}$$

Through the delta function in the first step of (5.4.22) the relation (5.4.20) is taken care of. In the second step, the displacement and position variables R and r, respectively, are introduced. In the final result, the inner integral is identified as the auto-correlation function $a(R)$ (cf. eqn (4.3.1)) of the 1D image $S(r)$, so that $|s(k)|^2 = f(q)$ is given by the Fourier transform of the auto-correlation function of the image,

$$|s(k)|^2 = f(q) = \frac{1}{(2\pi)^2} \int_{-\infty}^{\infty} a(R) \exp\{iqR\} \, dR. \tag{5.4.23}$$

The quantity $f(q)$ is the diffraction pattern, and $a(R)$ is the Patterson function (cf. Fig. 5.4.3). This interpretation relates to the description of PFG NMR in terms of *probability densities* or *average propagators*. In fact, the condition (5.4.20) defines a cross-section in the 2D $\boldsymbol{k}$ space spanned by k_1 and k_2, so that by the *projection–cross-section* theorem (5.4.12) the Patterson function can be interpreted as a projection of the corresponding signal $S(r_1)S^*(r_2)$ onto the subdiagonal in the space defined by r_1 and r_2.

PFG NMR for translational motion

The use of PFG NMR for measurement of translational motion is illustrated in Fig. 5.4.4 [Cal2]. Narrow gradient pulses of amplitude G and width δ are inserted into the dephasing and rephasing intervals of an echo sequence. The first gradient pulse defines the initial pitch of the magnetization in terms of the initial wave vector

$$\boldsymbol{k}_1 = -\gamma \int_0^{\delta} \boldsymbol{G}(t) \, dt. \tag{5.4.24}$$

The second gradient pulse unwinds this pitch, provided the initial magnetization vectors have not changed position into a region where the gradient causes a different magnetic field to be present. If a magnetization component has migrated during the time interval Δ from its initial position $\boldsymbol{r}_1$ to its final position $\boldsymbol{r}_2$ into a region of different field strength, it has accumulated the phase difference

$$\Phi = \boldsymbol{k}_2\boldsymbol{r}_2 + \boldsymbol{k}_1\boldsymbol{r}_1 = \boldsymbol{k}_2(\boldsymbol{r}_2 - \boldsymbol{r}_1) = \boldsymbol{q}\boldsymbol{R}, \tag{5.4.25}$$

where $\boldsymbol{k}_2 = -\boldsymbol{k}_1$ has been used as required by the experiment. Consequently, the phase difference is defined by the product of the displacement vector $\boldsymbol{R}$ and its Fourier-conjugate wave vector $\boldsymbol{q}$. The validity of (5.4.25) rests upon the *narrow gradient-pulse approximation* [Cal2].

Each group of molecules which diffuses with a different velocity accumulates a different value of the phase difference (5.4.25). The contribution of one group of molecules to the transverse magnetization is proportional to $\exp\{i\Phi\}$. The contribution from all groups of molecules is given by the statistical average $\langle \exp\{i\Phi\} \rangle$, so that the echo amplitude a_Δ can be written as

$$\begin{aligned} a_\Delta(\boldsymbol{k}_1, \Delta, \boldsymbol{k}_2) &= \langle \exp\{i\boldsymbol{k}_1\boldsymbol{r}_1\} \exp\{i\boldsymbol{k}_2\boldsymbol{r}_2\} \rangle \\ &= \iint \exp\{i\boldsymbol{k}_1\boldsymbol{r}_1\} \exp\{i\boldsymbol{k}_2\boldsymbol{r}_2\} W(\boldsymbol{r}_1, \boldsymbol{r}_2; \Delta) \, d\boldsymbol{r}_1 \, d\boldsymbol{r}_2. \end{aligned} \tag{5.4.26}$$

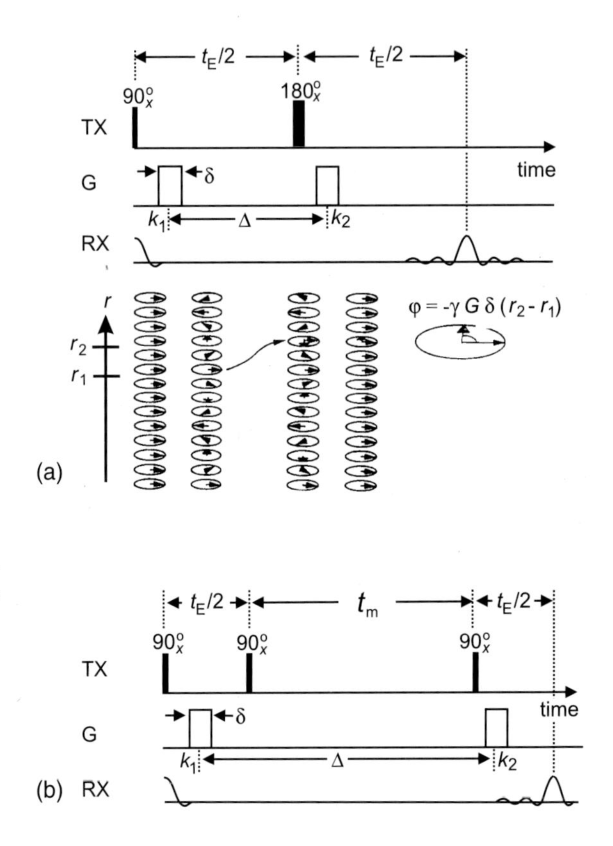

FIG. 5.4.4 Pulsed-field gradient NMR. (a) The Hahn echo is attenuated by translational diffusion during the time interval Δ between two short gradient pulses applied in the dephasing and in the rephasing periods of the echo (top). By use of the gradient pulses initial position $\boldsymbol{r}_1$ and final position $\boldsymbol{r}_2$ of the magnetization are labelled to identify migrating magnetization components (bottom). (b) The sensitivity of the method towards slower processes can be increased if the stimulated echo is used in place of the Hahn echo. Adapted from [Cal2] with permission from Oxford University Press.

In the second line, the statistical average has been reformulated in terms of a *combined probability density*

$$W(\boldsymbol{r}_1, \boldsymbol{r}_2; \Delta) = M_0(\boldsymbol{r}_1) P(\boldsymbol{r}_1 \,|\, \boldsymbol{r}_2, \Delta) \tag{5.4.27}$$

to find magnetization at position $\boldsymbol{r}_1$ at the time of the first short gradient pulse, and at position $\boldsymbol{r}_2$ when the second short gradient pulse is applied after the time Δ has elapsed. It can be written as the product of the spin density $M_0(\boldsymbol{r}_1)$, which is the initial distribution of magnetization at the time of the first gradient pulse and the *conditional probability density* $P(\boldsymbol{r}_1 \,|\, \boldsymbol{r}_2, \Delta)$ to find a given magnetization component at position $\boldsymbol{r}_1$ at the beginning of the molecular transport time Δ and at position $\boldsymbol{r}_2$ at the end. Therefore, the experiments of Fig. 5.4.4 can be understood as *position-change experiments* (Fig. 5.4.5). Then, the

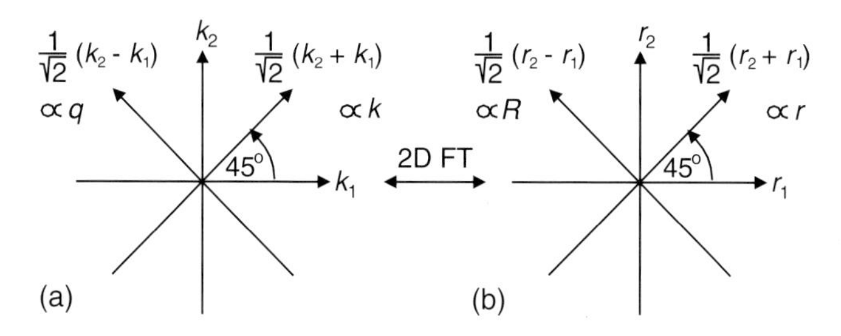

FIG. 5.4.5 Definition of q space in terms of position-change NMR. (a) Initial and final positions are encoded by k_1 and k_2 in the narrow gradient-pulse approximation. The transformation to a coordinate system where the difference wave number q defines one of the axes corresponds to a right-handed 45° rotation of the coordinate system (cf. eqn (5.4.21)). The perpendicular variable is proportional to the wave number k which encodes position. (b) 2D Fourier transformation of such a 2D position-change data set produces the displacement coordinate R in a coordinate frame rotated by 45° on one axis and the space coordinate r on the other axis.

formal description is completely analogous to the description of *rotational exchange in solid-state NMR* [Sch1] by eqn (3.2.12), where $\boldsymbol{k}_1$ and $\boldsymbol{k}_2$ correspond to t_1 and t_2, and $\boldsymbol{r}_1$ and $\boldsymbol{r}_2$ to ω_1 and ω_2, respectively.

With reference to the (k_1, k_2) coordinate system of a position-change experiment, a coordinate frame rotated by 45° can be introduced, where the difference wavenumber q is obtained along one axis and the wavenumber k along the other (Fig. 5.4.5). Correspondingly, after 2D Fourier transformation, displacement R and position r label orthogonal axes in the coordinate frame rotated by 45° with respect to the (r_1, r_2) coordinate system of the position-change experiment. Because the two moments of position encoding are separated by a delay Δ (cf. Fig. 5.4.4), R/Δ can be introduced as one of the axes in the position-change experiment of Fig. 5.4.5 instead of R. Corresponding to v, the Fourier-conjugate variable $q_v = q\Delta$ can be introduced, so that instead of (5.4.25)

$$\Phi = q_v v \tag{5.4.28}$$

can be written for the signal phase. Then the position-change experiment can be read as a *velocity–position correlation experiment*. Once q_v has been identified as the Fourier-conjugate variable of velocity, a *velocity-change experiment* can be designed in a similar way as the position-change experiment [Cal10, Sey1]. In a frame rotated by 45°, the position-change experiment becomes an *acceleration–velocity correlation* experiment (cf. Section 5.4.5).

It should be noted that the elementary PFG experiments of Fig. 5.4.4 is usually not performed in a 2D fashion because the constraint $\boldsymbol{k}_2 = -\boldsymbol{k}_1$ is chosen similar to $t_1 = t_2$ (cf. Fig. 3.2.6(b)) in measuring alignment echo spectra [Spi1] in place of 2D exchange spectra. Diffusing molecules migrate in positive and negative directions with equal probability. Therefore, the statistical average of the signals from all magnetization components with different phases Φ does not exhibit a net phase change, and the echo is attenuated but not shifted in phase. Coherent motion, on the other hand, leads to a modulation of the echo phase.

Velocity imaging

Instead of the 2D phase encoding experiment in (k_1, k_2) space which derives from Fig. 5.4.5 by stepping through the amplitudes of both gradient pulses independent of each other, equivalent experiments are performed in the (k, q) space obtained from the (k_1, k_1) space by 45° rotation as indicated in Fig. 5.4.5(a) and suitable rescaling of axes. Then the space information is encoded in frequency during data acquisition, and the velocity information is encoded indirectly in the signal phase. This saves time compared to 2D phase encoding, and the resolution can independently be optimized along both axes. A suitable pulse sequence and results from blood flow in a capillary are depicted in Fig. 5.4.6 [Chw1, Sey3]. The refocusing pulse of the Hahn echo sequence has been made frequency selective to excite only a given slice in z-direction. This minimizes signal distortions from inflow and outflow effects through the resonator during the experiment. Further gradient modulations are introduced to minimize phase distortions

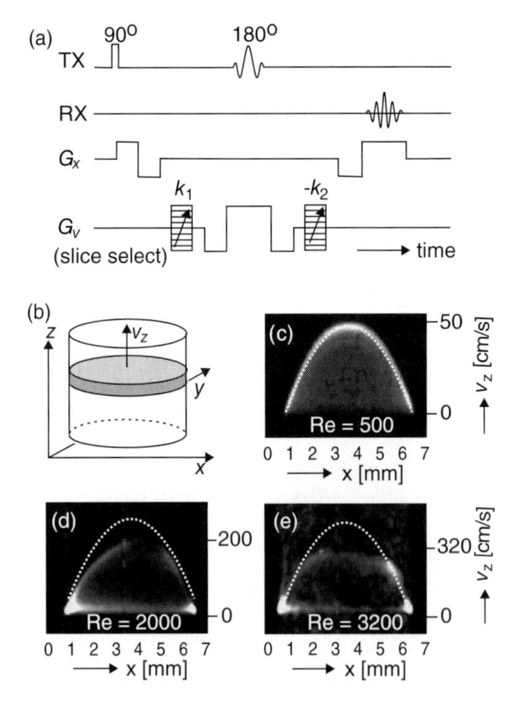

FIG. 5.4.6 [Chw1] Imaging of velocity distributions. The PFG experiment of Fig. 5.4.4(a) is modified to acquire data in the presence of a gradient G_x to gain spatial resolution in x-direction. (a) Pulse sequence. The refocusing pulse of the Hahn echo excitation has been made selective to select a slice of spins flowing in z-direction. The pulsed-field gradients are stepped in the same direction for phase encoding of velocity. Additional gradient modulations are incorporated to reduce phase distortions from flow during slice selection and in the echo maximum.
(b) Schematic drawing of the sample geometry. Blood flows through a capillary in z-direction.
(c)–(e) 1D velocity images. With increasing volume flow rate the Reynolds number *Re* increases. Signal loss at high velocities indicates the Maguss effect, i.e. an increased concentration of erythrocytes in the centre.

from coherent flow during slice selection and in the echo maximum (cf. Section 7.2.6). At high Reynolds numbers *Re* the images show a signal decrease in the centre due to a concentration increase of erythrocytes. Because only the x-direction is resolved in these measurements, the signal amplitude corresponds to the projection of the spin density over the y-direction and the slice thickness in z-direction. These data show that the PFG experiment for measuring translational diffusion essentially corresponds to a phase-encoding method for measuring 1D *velocity images* in a frame rotated by 45° in (k_1, k_2) space or the conjugate (r_1, r_2) space. Flow imaging is of interest in biomedicine, chemical engineering, and in geophysics for analysis of fluid transport in rocks [Dij1].

The average propagator

The average propagator is introduced to describe the echo attenuation in the displacement experiments of Fig. 5.4.4 in terms of the particle displacement R along the gradient direction during the gradient-pulse delay Δ. To this end (5.4.26) is transformed to the rotated coordinate frame of Fig. 5.4.5 with axes defined in (5.4.21). By the same derivation used for (5.4.22) one arrives at

$$a_\Delta(k', q', \Delta) = \int\int W(r', R'; \Delta) \exp\{ik'r'\} \exp\{iq'R'\} \, dr' \, dR'. \tag{5.4.29}$$

If the conditional probability density $W(r', R'; \Delta)$ is independent of position r', the inner integral is different from zero only for $r' = 0$. This defines the *displacement propagator* $P(R', \Delta)$,

$$P(R', \Delta) = \int W(r', R'; \Delta) \, dr', \tag{5.4.30}$$

so that (5.4.29) can be rewritten as

$$a_\Delta(q', \Delta) = \int P(R', \Delta) \exp\{iq'R'\} \, dR'. \tag{5.4.31}$$

The displacement propagator is the probability density that a magnetization component starting at any position is displaced by a value in the interval of $[R', R' + dR']$ in a time Δ. For fluids diffusing in small pores, the displacement depends on the starting position. Here the propagator provides information which is averaged over the pore geometry as a result of the integration in (5.4.30). For this reason, $P(R', \Delta)$ is also referred to as the *average propagator* of the diffusion process.

By the projection–cross-section theorem (5.4.12), the integration over space r' corresponds to a slice in (k_1, k_2) space along the subdiagonal where $k_1 = -k_2$. This condition is, in fact, imposed by the choice of gradient signs in Fig. 5.4.4. Given that the displacement is accumulated in a given time Δ, the average propagator can be interpreted as the probability density of average molecular velocities in the sample [Sey3]. The average propagator of coherent flow can be derived from the data in Fig. 5.4.6 by integration over the space coordinate and displaying the image contrast as a function of the velocity.

Experimental aspects

The propagator is determined experimentally by measuring the echo amplitude (5.4.31) as a function of the gradient-pulse amplitude (or the gradient-pulse spacing Δ if time-invariant gradients are used) and by subsequent inverse Fourier transformation of the echo-decay curve. The initial decay of the echo amplitude yields the *mean-square displacement* $\langle R'^2 \rangle$, which can be seen from an expansion of the exponential in (5.4.31) and by writing the Fourier integral over the average propagator in terms of a statistical average $\langle \cdots \rangle$ [Sey1],

$$a_\Delta(q', \Delta) \approx 1 - \frac{1}{2!} q'^2 \langle R'^2 \rangle + \cdots . \tag{5.4.32}$$

Reverting to the unprimed coordinates defined in (5.4.21) the echo amplitude is given for unrestricted Brownian motion with the self-diffusion coefficient D by

$$a_\Delta(q, \Delta) = \exp\{-q^2 D \Delta\}, \tag{5.4.33}$$

with the propagator

$$P(R, \Delta) = \frac{1}{(4\pi D\Delta)^{3/2}} \exp\left\{-\frac{R^2}{4D\Delta}\right\}. \tag{5.4.34}$$

If gradient pulses of finite length δ are employed, instead of (5.4.33) the *Stejskal–Tanner relation* applies for the echo amplitude [Ste1],

$$a_\Delta = \exp\left\{-\gamma^2 G^2 \delta^2 D \left(\Delta - \frac{\delta}{3}\right)\right\}, \tag{5.4.35}$$

and the coefficient D for unrestricted diffusion can be obtained from a plot of the logarithm of a_Δ *versus* $\gamma^2 G^2 \delta^2 (\Delta - \delta/3)$.

The sensitivity of the method towards slow diffusion and flow is increased if the stimulated echo is used instead of the Hahn echo (Fig. 5.4.4(b)). In this case the signal decay is limited by T_1, which can be considerably longer than the limiting relaxation time T_2 of the Hahn echo. Related schemes for measuring flow and diffusion have been designed which apply B_1 field gradients in the rotating frame of reference [Bou1, Can1, Dup1, Kar1]. The basic scheme of Fig. 5.4.4 can be extended to multiple wave-vector schemes [Cor4, Mit4, Sta1, Sta2, Sta3]. Additional information about the pore structure of the sample can be obtained in this way.

An example for 2D *displacement propagators* corresponding to *2D velocity distributions* is given in Fig. 5.4.7 [Göb1]. The basic pulse sequence of Fig. 5.4.4(a) has been extended by a second pair of gradient pulses, so that $\boldsymbol{q}$ space can be scanned in two dimensions. This pulse sequence was used to characterize water flow through a pipe (b) blocked by a pack of glass beads (c) and a pipe blocked by a polymer fibre plug (d). Clearly, the 2D propagators are quite different for the two different obstructions to the water flow. Considerable back flow is observed for the packed beads, and strong lateral dispersion is observed in flow direction for the polymer fibre plug.

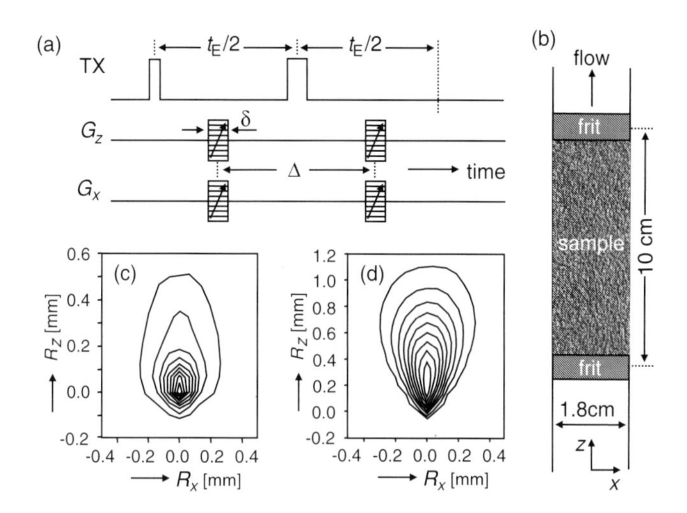

FIG. 5.4.7 [Göb1] 2D displacement propagators corresponding to 2D velocity distributions. (a) The pulse sequence is the same as that for measuring 1D propagators in Fig. 5.4.4(a), except that gradient pulses are applied in two orthogonal directions, so that $\boldsymbol{q}$ space is scanned in two dimensions. (b) Drawing of the sample. Water is pumped through a pipe with circular diameter which contains the sample enclosed between two frits. (c) 2D propagator for 2 mm diameter glass beads at a volume-flow rate of 4.12 ml/s. (d) 2D propagator for a plug made from partially oriented polymer fibres at a volume-flow rate of 8.2 ml/s.

Analogy to neutron scattering

The expression (5.4.31) describing the echo amplitude formally resembles the structure factor of *neutron scattering*, where $\boldsymbol{q}$ is the wave vector of the difference wave between the incoming and the scattered wave [Cal6]. In fact, there is a close analogy between PGSE NMR and neutron scattering [Fle1]. However, time and distance scales of both methods are different. Neutron scattering is sensitive to rms displacements below 5 nm on a timescale not exceeding a few tens of microseconds, whereas PFG NMR is typically applied for probing distances greater than 50 nm on timescales limited by T_2 and T_1, so that $1\,\text{ms} \leq \Delta \leq 10\,\text{s}$.

The incoherent fraction of the inelastic neutron scattering function is given by

$$S_{\text{incoherent}} = \frac{1}{N} \sum_i \langle \exp\{\mathrm{i}\boldsymbol{q}[\boldsymbol{r}_i(t) - \boldsymbol{r}_i(0)]\}\rangle, \tag{5.4.36}$$

where t corresponds to the observation time Δ in NMR and the sum is taken over all scattering centres, for instance, the monomer units of a polymer. On the other hand, the coherent inelastic neutron scattering function and the quasi-elastic light scattering function are described by

$$S_{\text{coherent}} = \frac{1}{N^2} \sum_i \sum_j \langle \exp\{\mathrm{i}\boldsymbol{q}[\boldsymbol{r}_j(t) - \boldsymbol{r}_i(0)]\}\rangle. \tag{5.4.37}$$

The quantity $\boldsymbol{r}_j(t) - \boldsymbol{r}_i(0)$ introduces sensitivity to relative motion of two scattering centres, so that the interpretation of S_{coherent} is more complicated than the self-motion measured by $S_{\text{incoherent}}$ and by PFG NMR.

Short-time limit and surface-to-volume effects

Inhibition of self-diffusion becomes apparent at timescales Δ short compared to d^2/D, where d is the characteristic pore diameter, because a fraction of molecules is always close to the walls, and their diffusion is hindered. This fraction depends on the *surface-to-volume ratio (S/V)* of the sample and the observation time Δ of the diffusion. This simple model leads to a fraction of $(2D\Delta)^{1/2}$ S/V molecules being restricted in their motion. An analytical derivation [Mit1, Mit2, Mit6, Mit7] confirms an approximately linear dependence of the effective diffusion coefficient $D_{\text{eff}}(\Delta)$ on $\Delta^{1/2}$,

$$\frac{D_{\text{eff}}(\Delta)}{D} = 1 - \frac{4}{9\sqrt{\pi}}\frac{S}{V}(D\Delta)^{1/2} + \text{higher-order terms.} \tag{5.4.38}$$

The accuracy of the expression can be improved by taking into account the restricted diffusion occurring during the gradient pulses [For1, Sor1].

Restricted diffusion

For unrestricted Brownian motion the molecules diffuse freely in all directions. The *conditional probability density* $P(r_1 \mid r_2, \Delta)$ and the average propagator $P(R, \Delta)$ then exhibit a Gaussian distribution over molecular displacements R (cf. eqn (5.4.34)) with a width increasing with time. However, when the molecular motion is restricted by some confinement the propagator deviates from the classical Gaussian form [Cal6]. By understanding the influence of restrictions on the form of the propagator or on the echo amplitude in a PFG experiment, information can be obtained not only about the motion of molecules, but also about the boundaries, and hence the pore morphology of the surrounding medium.

In the long-time limit $\Delta \gg d^2/D$ the propagator has a simple relationship to the geometry of closed pores. In this case, the conditional probability density $P(r_1 \mid r_2, \Delta)$ in (5.4.27) is independent of the starting position r_1 so that it reduces to the spin density $M_0(r_2) = M_0(r_1 + R)$ of the pore. In consequence, the average propagator (5.4.30) becomes an *auto-correlation function of the spin density* (cf. eqn (5.4.22)),

$$P(R', \Delta) = \int W(r', \Delta)\,\mathrm{d}r' = \int M_0\left(\frac{r' - R'}{\sqrt{2}}\right) M_0\left(\frac{r' + R'}{\sqrt{2}}\right) \mathrm{d}r', \tag{5.4.39}$$

and by the correlation theorem (4.3.4) the echo attenuation function (5.4.31) is the magnitude square of the Fourier transform of the spin density or pore shape function $M_0(R')$ in analogy to (5.4.22)

$$a_\infty(q') = \iint M_0\left(\frac{r' - R'}{\sqrt{2}}\right) M_0\left(\frac{r' + R'}{\sqrt{2}}\right) \mathrm{d}r' \exp\{\mathrm{i}q'R'\}\,\mathrm{d}R' = |s(q')|^2, \tag{5.4.40}$$

where $s(q')$ is the Fourier transform of $M_0(R')$ in a pore. The spin density cannot be derived from (5.4.40), because the phase information is unavailable in $\boldsymbol{q}$ space. Fourier

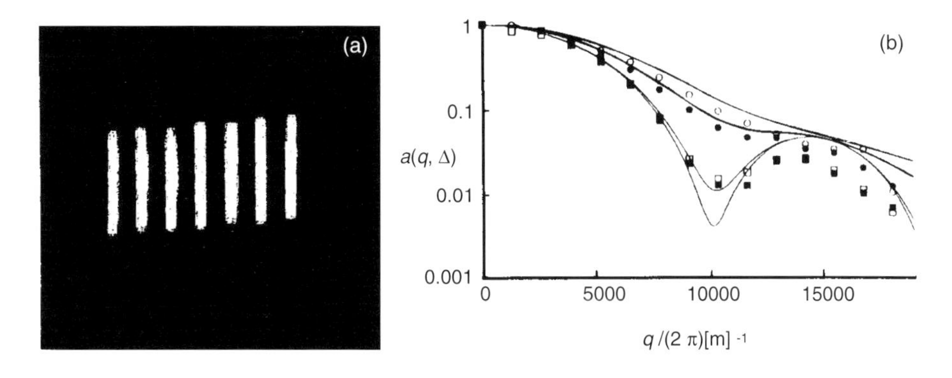

FIG. 5.4.8 (a) NMR image of pentane in seven rectangular microcapillaries forming a 100 μm grid. (b) Echo attenuation curves for different diffusion times Δ of 200 ms (open circles), 300 ms (closed circles), 700 ms, (open squares), and 900 ms (closed squares) corresponding to a range of $\Delta \approx 0.1d^2/D$ to $0.45d^2/D$ with diffraction features at longer diffusion times. Adapted from [Coy1] with permission of the author.

transformation of $a_\infty(q')$ leads to a *Patterson function* (cf. Fig. 5.4.3) which provides information on displacement. This is illustrated in Fig. 5.4.8 by echo attenuation curves for a stack of rectangular microcapillaries filled with pentane [Coy1]. The long-time limit assumed in the derivation of (5.4.40) corresponds to several wall collisions of the confined molecules. Nevertheless, Fig. 5.4.8 illustrates that diffraction effects are already observed for diffusion times $\Delta \approx 0.45\, d^2/D$. When finite-width gradient pulses are considered, translational diffusion has to be taken into account during the pulse duration δ [Ble1, Mit5, Wan1].

The formal similarity of average NMR propagators for restricted translational diffusion and scattering functions is a topic of considerable interest in fundamental science as well as to the characterization of porous media [App1, Ber1, Cal7, Cal8, Cal9, Coy1, Coy2, Fle1, Hür1, Man3, Sen1]. When interconnected pores are of interest, similar information can be derived from the structure of the propagator under flow conditions (cf. Fig. 5.4.7) [Cap2, Sey1, Sey2, Wat1]. However, the analysis is made difficult by wide distributions of pore sizes encountered in realistic materials like polymers [App2], cellulose fibres [Lit1], rocks [Lat1], gypsum [Fil1, Fil2], and other building materials.

Interconnected pores

In many porous structures of practical interest, the pores are interconnected by channels which permit migration of molecules between pores. The long-time limit where each molecule has sampled the entire matrix of interconnected pores is of little interest in diffusion studies. In that case (5.4.40) applies, where $s(q')$ is the spatial Fourier transform of the entire sample. The interesting case is the intermediate scale of diffusion times Δ, where the molecules starting in one pore have moved as far as the first neighbouring pores [Cal8]. If the long-range permeation of the pore matrix by molecules undergoing Brownian motion is describable by some *effective diffusion coefficient* D_{eff}, and the mean pore spacing is denoted by b, the relevant timescale is given by $\Delta \geq b^2/(2D_{\text{eff}})$.

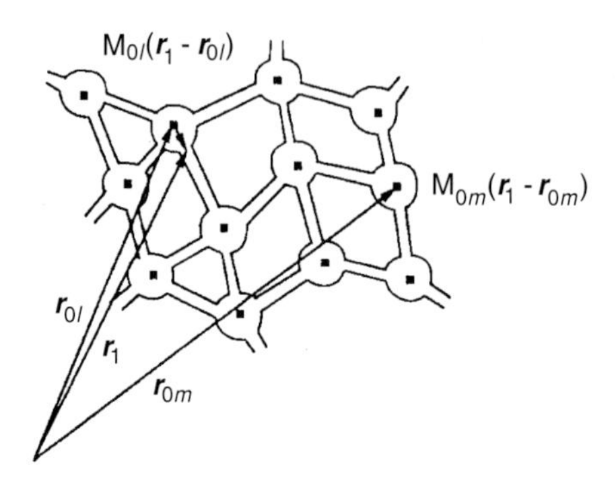

FIG. 5.4.9 Network of identical pores with negligibly narrow connections. Adapted from [Cal6] with permission of the author.

Considering a network of N_p identical pores with densities $M_{0m}(\boldsymbol{r}_1 - \boldsymbol{r}_{0m})$ and negligibly narrow pore connections (Fig. 5.4.9) [Cal6], the connectivity of the pores may be represented by defining a local conditional probability

$$P_m(\boldsymbol{r}_1 \,|\, \boldsymbol{r}_2, \Delta) = \sum_{l=1}^{N_\mathrm{p}} C_{lm} M_{0m}(\boldsymbol{r}_1 - \boldsymbol{r}_{0l}), \tag{5.4.41}$$

where $\boldsymbol{r}_{0l}$ is the vector pointing to the centre of pore l and the matrix element C_{lm} represents the probability that a molecule has migrated from pore l to pore m during the diffusion time Δ. Isolated pores arc represented by diagonal matrix elements and contribute to the echo attenuation according to (5.4.40). For a perfectly periodic structure, nonzero coherences in $a_\infty(\boldsymbol{q})$ arise from (5.4.41), and for irregular structures useful information can be gained over the timescale necessary for the molecules to move just a few pore spacings.

The only connectivities which are of concern are those to pores displaced along the direction of the magnetic-field gradient, that is, along the direction of $\boldsymbol{q}$. For a regular lattice with pore spacing b, C_{lm} has the form of a free diffusion envelope in the long-time limit,

$$C_{lm} = \frac{b}{\sqrt{4\pi D_\mathrm{eff}\Delta}} \exp\left\{-\frac{(l-m)^2 b^2}{4D_\mathrm{eff}\Delta}\right\}. \tag{5.4.42}$$

Assuming that (5.4.42) is also valid for finite distances and times, and representing 1D displacements parallel to the gradient direction by $Z = (z_2 - z_1)$, it can be shown that the echo attenuation function for a lattice of pores in which there is no correlation between lattice spacing and pore size is given by

$$a_\Delta(q) = \langle |s(q)|^2 \rangle \int_{-\infty}^{\infty} C(Z, \Delta) L(Z) \exp\{\mathrm{i} q Z\} \,\mathrm{d}Z, \tag{5.4.43}$$

where $\langle |s(q)|^2 \rangle$ is the *structure factor* of the average pore, $C(Z, \Delta)$ is the diffusion profile equivalent to (5.4.42) for any displacement Z, and $L(Z)$ is the lattice-correlation function which represents the probability that a lattice site is found at displacement Z from the starting site. For a perfectly periodic lattice, $L(Z)$ is equivalent to the lattice array. Equation (5.4.43) is the Fourier transform of a product, which according to the convolution theorem (cf. eqns (4.2.1) and (4.2.14)) can be rewritten as [Cal8]

$$\begin{aligned} a_{\Delta}(q) &= \langle |s(q)|^2 \rangle \mathbf{F}\{C(Z, \Delta)\} \otimes \mathbf{F}\{L(Z)\} \\ &= \langle |s(q)|^2 \rangle \exp\{-q^2 D_{\text{eff}} \Delta\} \otimes \mathbf{F}\{L(Z)\}, \end{aligned} \tag{5.4.44}$$

where $\mathbf{F}\{\cdots\}$ is the Fourier transformation operator and the symbol $\otimes$ represents convolution.

Here $\exp\{-q^2 D_{\text{eff}} \Delta\}$ is equivalent to the free diffusion response (5.4.33). In this form (5.4.44) is reminiscent of X-ray diffraction, in which $\mathbf{F}\{L(Z)\}$ describes the reciprocal lattice, and the diffraction pattern is modulated by the atomic structure factor. However, (5.4.44) requires the validity of the Gaussian form (5.4.42) for C_{lm}, which is hardly fulfilled for the starting pore, where $\boldsymbol{q}$ is large. Therefore, the quantitative values predicted by (5.4.44) will be somewhat inaccurate unless qb is close to a multiple of 2π, that is, unless the wavelength associated with the gradient pulse matches the pore spacing. An improvement of this simple theory is given by the pore hopping approach [Cal9]. Nevertheless, the following qualitative behaviour can be predicted from (5.4.44):

(1) $q \ll b^{-1}$: Long distance scales are probed, so that the convolution $\exp\{-q^2 D_{\text{eff}} \Delta\} \otimes \mathbf{F}\{L(Z)\}$ is dominated by the diffusive width and $a_{\Delta}(q)$ is a Gaussian, yielding D_{eff} on analysis *via* the Stejskal–Tanner relation (5.4.31).

(2) $q \approx b^{-1}$: A peak in $a_{\Delta}(q)$ is observed due to nearest neighbour lattice correlation. The magnitude of this peak depends on the lattice regularity. It decays as Δ increases due to the diffusive convolution.

(3) $q \approx d^{-1}$: The modulation due to the pore structure factor is observed as a local minimum. This modulation can also cause a shift in the peak at $q \approx d^{-1}$ if $d \approx b$.

These conclusions show, that both pore sizes and pore spacings can be determined *via* the structural factors which enter eqn (5.4.44). This is demonstrated, for instance, by experiments with water diffusing around close-packed polymer spheres [Coy1]. Applications are in the analysis of pore-size distributions and tortuosity in porous sedimentary rocks [Lat1, Mit3, Mit7].

Experiments at short distance scales: PFG MASSEY

More recent developments of NMR diffraction methods have increased the spatial resolution to distance scales between 5 and 50 nm, so that the lower NMR limit now reaches the upper resolution limit of neutron scattering. One successful approach utilizes the *time-invariant magnetic-field gradients* of superconducting magnets outside the homogeneous region [Kim1, Dem2]. These gradients can be as large as 80 T/m. Specially designed, superconducting gradient magnets provide even larger gradients of the order of 200 T/m [Fuj1]. In conjunction with the stimulated echo (cf. Fig. 2.2.9(c)), the time $t_E/2$ between the first two 90° pulses defines the duration δ of the gradient induced precession in the transverse plane. Any problems associated with clean switching of large gradient pulses do not arise. After the second pulse half of the transverse magnetization is stored along the z-axis and the residual transverse magnetization dephases. After the third pulse, the stimulated echo is detected in the presence of the magnetic-field gradient. Therefore, spectral information is not accessible. There is also a significant loss of signal for two reasons: first, only a narrow region of spins is excited by the finite width of the rf pulses and the large bandwidth of resonance frequencies in the field gradient. Second, there is a major increase in noise, because the narrow echo needs to be sampled with a large receiver bandwidth. Nevertheless, diffusion coefficients of $2 \times 10^{-15}\, m^2/s$ have been measured in polymer melts with this technique [Fei2], and a value of $3 \times 10^{-16}\, m^2/s$ has been obtained for the plastic molecular crystal of camphene by using a stimulated echo technique with homonuclear multi-pulse decoupling in the dephasing and rephasing periods [Cha1, Fei2, Gei1].

The higher the gradient strengths, the more problematic are sample movements and gradient mismatch in the dephasing and rephasing periods of the echo [Cal5]. Sample movement is critical, in particular, when strong pulsed gradients are used [Cal12]. Their amplitudes can be 40 T/m and higher. These influences lead to variations in the echo phase. Sample movement results in a phase shift common to all spins, while *gradient mismatch* in PFG NMR results in position dependent, local phase shifts. The gradient mismatch between the first and the second gradient pulse can be expressed by a wave vector mismatch Δq. By using the same gradient coils responsible for the gradient mismatch to generate a much smaller read gradient G_{read} the phase shifts can be resolved (Fig. 5.4.10(a)) [Cal5]. The size of the read gradient needs to be in the order of the pulsed-gradient mismatch. The effect of the read gradient is then to enable a coherent superposition of signals at the time $t = -\Delta q / \gamma G_{read}$ either before or after the echo centre, depending on the sign of the mismatch. This is achieved by Fourier transforming the echo acquired in the read gradient to an image. Results of such a transformation are depicted in Fig. 5.4.10(b). If the entire echo is sampled, the signals from successive scans can be coherently added by computing the modulus of the transform despite phase fluctuations from sample movement and pulsed-gradient mismatch.

The method has been given the acronym *MASSEY* for modulus addition using spatially separated echo spectroscopy [Cal5]. Compared to coherent superposition of echoes in the absence of phase fluctuations and a read gradient the method is associated with a loss in signal-to-noise-ratio by a factor of $n^{1/2}$, where n is the number of points under the image obtained by Fourier transformation of the echo. This is a small price to pay for access to displacements on the scale of 10 nm by NMR.

Diffusion-ordered spectroscopy

The PFG method can be combined with a spectroscopic dimension [Joh1, Mor3]. By systematically stepping the gradient pulses and detecting the FID for each gradient value, a 2D data set is obtained, which is Fourier transformed in one dimension to obtain the high-resolution NMR spectrum, and in the other dimension the data are analysed by regularized Laplace transform algorithms like CONTIN [Pro1] for the distribution of diffusion constants.

The method has been given the name *DOSY* for *diffusion-ordered spectroscopy*. It is related to *mobility-ordered spectroscopy* (*MOSY*) for analysis of electrophoretic mobilities [Hei1, Hol1, Joh2, Mor4]. By using shielded gradients with amplitudes larger than

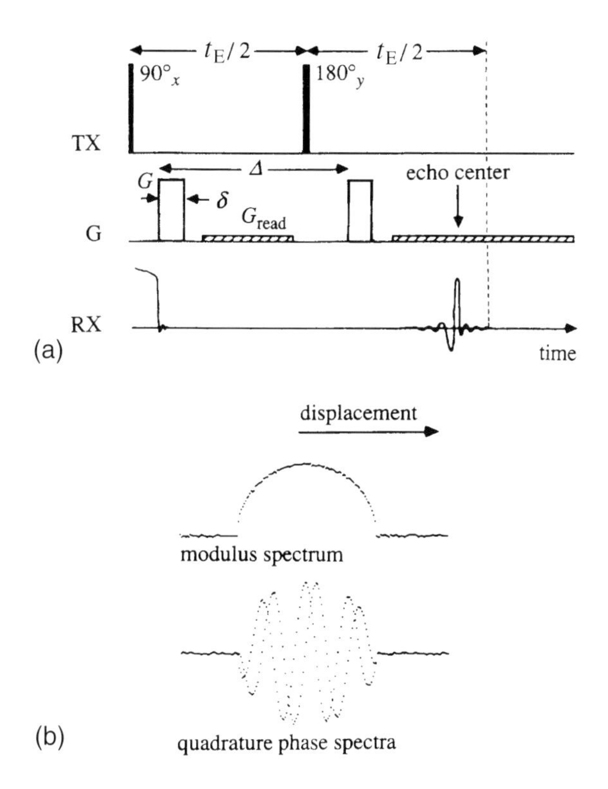

FIG. 5.4.10 [Cal5] (a) PFG-MASSEY pulse sequence with strong gradient pulses G of duration δ for displacement encoding and a weak read gradient G_{read} for coherent addition of signals in the presence of phase jitter from sample movement and gradient mismatch. (b) Simulated Fourier transforms of gradient echoes acquired in the presence of a weak read gradient with a mismatch of the strong gradient pulses. Top: Magnitude image. Bottom: Real part of the phase-sensitive image.

0.3 T/m tracer diffusion coefficients spanning five orders of magnitude with a lower limit of less than 10^{-14} m^2/s can be measured with the DOSY technique.

Frequency analysis of spin motion using modulated gradient NMR

For short diffusion times, the description of PFG experiments for translational diffusion requires consideration of finite gradient pulse widths. This lead to the Stejskal–Tanner correction of the expression (5.4.34) for the echo attenuation in the case of free diffusion. However, a new theory must be developed, if molecular transport is to be probed by several gradient pulses [Cap2]. In general, arbitrary field gradient modulations are possible. For analysis of the signal, it is convenient to compose these waveforms from a series of hypothetical narrow gradient pulses with gradient-free intervals in between [Cal13, Cap2, Cod1, Ste3]. Then, the effect of the gradient can be treated *via* the time dependence of the accumulated magnetization phase [Cal11, Ste2], and a natural description of the translational motion is *via* the *velocity auto-correlation function*, rather than *via* the propagator for displacement. It may be shown that the spectral density of this translational velocity auto-correlation function is probed by the frequency spectrum of the wave vector. The mathematical treatment lends itself to an analogy with the measurement of the rotational dipolar auto-correlation spectral density which is probed by relaxation times, where T_1 samples at frequencies ω_L and $2\omega_L$ (cf. eqn (3.5.4)).

A uniform modulation of the gradient vector leads to a waveform $G(t)$, where interdispersed 180° pulses invert the sign of prior gradients. Provided that one is dealing with molecular spin motion in which sudden local phase changes are avoided, it is possible to show that the phase distribution of the magnetization components exposed to such a motion is Gaussian in character and that the normalized echo attenuation at time t may be written as

$$a(t) = \langle \exp\{\mathrm{i}\theta(t) - R(t)\}\rangle, \qquad (5.4.45)$$

where $i\theta(t)$ is a phase shift arising from molecular flow or drift and $R(t)$ is the relaxation rate or the attenuation factor which arises from random particle migration and is of interest here. The average is taken over all magnetization components, where each component consists of a large enough number of spins for the local density matrix to be evaluated and a small enough number of spins to represent the distribution of boundary conditions in the sample. Without going through a delicate justification why the average can be transferred to the exponent, one obtains upon doing so

$$\theta(t) = -\int_0^t [\boldsymbol{k}(t) - \boldsymbol{k}(t')]\langle \boldsymbol{v}(t')\rangle \, \mathrm{d}t', \qquad (5.4.46)$$

and

$$R(t) = \frac{1}{2}\int_0^t\int_0^t \boldsymbol{k}(t')\langle \boldsymbol{v}(t')\boldsymbol{v}(t'')\rangle \boldsymbol{k}(t'') \, \mathrm{d}t' \, \mathrm{d}t'', \qquad (5.4.47)$$

where $\boldsymbol{k}(t)$ is the wave vector (2.2.23). It is zero at the time of echo rephasing, i.e. in the gradient echo maximum.

Noting that the spectral density of the ensemble-averaged velocity auto-correlation function is the diffusion tensor

$$\mathbf{D}(\omega) = \frac{1}{2}\int_0^\infty \langle \boldsymbol{v}(t)\boldsymbol{v}(0)\rangle \exp\{-\mathrm{i}\omega t\} \, \mathrm{d}t, \qquad (5.4.48)$$

one can rewrite (5.4.47) as

$$R(t) = \frac{1}{\pi}\int_0^\infty \boldsymbol{K}(\omega, t)\mathbf{D}(\omega)\boldsymbol{K}(-\omega, t) \, \mathrm{d}\omega, \qquad (5.4.49)$$

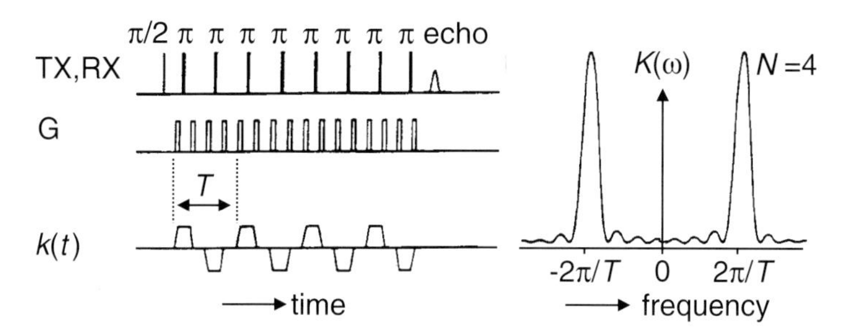

FIG. 5.4.11 [Cal11] Modulated gradient NMR for probing spectral densities of diffusive translational motion. The pulse sequence (left) consists of a CPMG echo train with interdispersed gradient pulses $G(t)$ which produces the time-dependent wave vector $k(t)$. The spectrum $K(\omega)$ of $k(t)$ probes the spectral density of diffusive motion at a single frequency (right).

which simplifies in the case of isotropic diffusion to

$$R(t) = \frac{1}{\pi} \int_0^\infty D(\omega) |\boldsymbol{K}(\omega, t)|^2 \, \mathrm{d}\omega, \tag{5.4.50}$$

where the spectrum of the time-dependent wave vector is given by

$$\boldsymbol{K}(\omega, t) = \int_0^t \boldsymbol{k}(t') \exp\{-\mathrm{i}\omega t'\} \, \mathrm{d}t'. \tag{5.4.51}$$

Different *gradient modulation waveforms* are conceivable. Sinusoidal modulation exhibits a strong contribution at zero frequency. However, a waveform involving a repetitive CPMG train of rf pulses with interdispersed gradient pulses produces a nearly ideal frequency sampling function which samples the diffusion spectrum at a single frequency (Fig. 5.4.11) [Cal11]. Hard 180° pulses are applied in the absence of magnetic-field gradients in order to avoid slice selection effects. With $N = 4$ gradient pulses per period of $\boldsymbol{k}(t)$, a reasonably narrow peak can be achieved. In principle, it is possible to use such sequences to probe *spectral densities of diffusive translational motion* in the frequency range of 10 Hz to 100 kHz. This becomes possible, because rather than using two gradient pulses, for which the attenuation effect disappears as the gradient pulse duration δ is shortened, the repetitive pulse train employs an increasing number of gradient pulses in any time interval t, as the frequency is increased and the period T of $\boldsymbol{k}(t)$ is reduced. Thus, the frequency domain analysis extends the effective timescale of the PGSE experiment to the submillisecond regime.

5.4.4 Acceleration: ε space

The definition of q and the displacement R in Fig. 5.4.5 in terms of a *position-change experiment* performed in 2D k space can be carried one step further: having identified $q_v = q\Delta$ as the Fourier-conjugate variable of average velocity R/Δ in (5.4.28), a

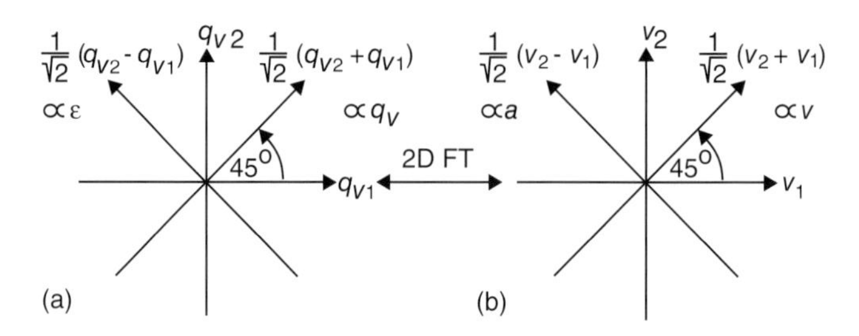

FIG. 5.4.12 Definition of ε space in terms of velocity-change NMR. (a) Initial and final velocities are encoded by q_{v1} and q_{v2} in the narrow gradient-pulse approximation. The transformation to a coordinate system where the difference $q_{v2} - q_{v1}$ defines one of the axes corresponds to a right-handed 45° rotation of the coordinate system (cf. eqn (5.4.52)). The perpendicular variable is proportional to q_v. (b) 2D Fourier transformation of such a 2D velocity-change data set produces the difference velocity $(v_2 - v_1)$ in a coordinate frame rotated by 45° on one axis and velocity v on the other axis.

velocity-change experiment [Cal10, Sey1] can be designed to access acceleration in a similar way as the position change experiment provides access to velocity (Fig. 5.4.12). Based on this principle, a hierarchy of exchange experiments can be designed to measure parameters of translational motion.

Principle of velocity-change NMR

In velocity-change NMR the variables q_{v1} and q_{v2} replace the axes k_1 and k_2 in Fig. 5.4.5, and v_1 and v_2 replace r_1 and r_2 (Fig. 5.4.12). In a coordinate frame rotated by 45° the difference coordinate corresponds to acceleration a and the other to velocity v, so that this exchange experiment can be read as *a velocity–acceleration correlation experiment*. Following the coordinate transformations (5.4.21) for position-change spectroscopy, the following coordinate transformations apply for velocity-change spectroscopy,

$$
\begin{aligned}
a'\Delta_v &= \frac{v_2 - v_1}{\sqrt{2}} = \frac{a\Delta_v}{\sqrt{2}}, \quad v' = \frac{v_2 + v_1}{\sqrt{2}} = v\sqrt{2},\\
v_1 &= \frac{v' - a'\Delta_v}{\sqrt{2}}, \quad v_2 = \frac{v' + a'\Delta_v}{\sqrt{2}},\\
\frac{\varepsilon'}{\Delta_v} &= \frac{q_{v2} - q_{v1}}{\sqrt{2}} = \frac{\varepsilon}{\Delta_v}\sqrt{2}, \quad q_{v'} = \frac{q_{v2} + q_{v1}}{\sqrt{2}} = \frac{q_v}{\sqrt{2}},\\
q_{v1} &= \frac{q_{v'} - \varepsilon'/\Delta_v}{\sqrt{2}}, \quad q_{v2} = \frac{q_{v'} + \varepsilon'/\Delta_v}{\sqrt{2}},
\end{aligned}
\tag{5.4.52}
$$

where Δ_v corresponds to the exchange time between the two moments of velocity encoding. Here a denotes acceleration and ε is the corresponding Fourier-conjugate variable, so that similar to the phase (5.4.28) for constant translational motion, the phase for accelerated motion is given by

$$\Phi = \varepsilon a. \tag{5.4.53}$$

For arbitrary translational motion (cf. eqn (1.2.2))

$$r(t) = r + vt + \tfrac{1}{2}at^2 + \cdots, \tag{5.4.54}$$

the phase of the transverse magnetization is determined by position r, velocity v, acceleration a, and higher-order parameters of motion at time $t = 0$. In one space dimension one obtains

$$\Phi = kr + q_v v + \varepsilon a + \cdots. \tag{5.4.55}$$

Higher-order parameters of motion in the Taylor expansion (5.4.55) can be measured by carrying the hierarchy of exchange experiments to higher orders. For each order, an average propagator can be obtained in terms of a projection of the exchange spectrum onto its subdiagonal. By the *projection–cross-section theorem*, this average propagator

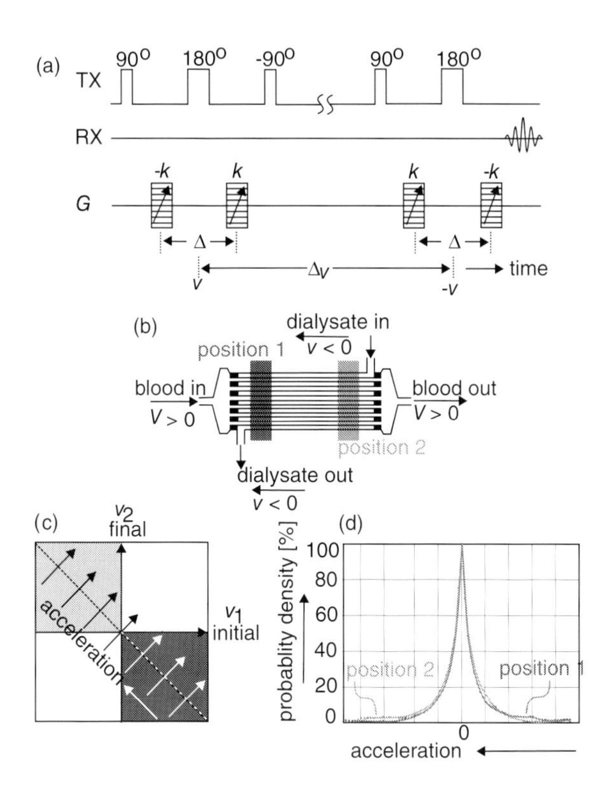

FIG. 5.4.13 Application of velocity-change NMR for characterization of trans-membrane flow in a haemodialyser module from hollow-fibre filter membranes. (a) Pulse sequence for phase encoding of acceleration. (b) Schematic diagram of a hollow-fiber haemodialyser. Molecules crossing the membrane are reversed in their flow velocity. This velocity change is measured in the acceleration distribution. (c) The acceleration distribution is the projection of the velocity exchange spectrum onto its subdiagonal. The experimental data are acquired by measuring a cross-section in 2D $\boldsymbol{q}$ space. (d) Acceleration distributions for positions 1 and 2 [Göb1].

is defined by the subdiagonal slice through the origin in reciprocal space. Because the experiments are carried out in reciprocal space, average propagators are measured much faster than complete exchange experiments.

The acceleration propagator: probability density of acceleration

An example of an acceleration propagator is given in Fig. 5.4.13 by molecular transport across the hollow fibre membranes in a haemodialyser. The experimental pulsed-field-gradient scheme is depicted in (a). One gradient pulse pair is needed to encode initial and one to encode final velocity. In the haemodialyser, blood flows through the fibres in one direction and water washes the fibres in counter flow from the outside. The actual experiment was carried out with water instead of blood, and the pressure difference was adjusted in such a way that at one end of the dialyser water molecules preferably cross the hollow-fibre membrane from the inside to the outside (position 1) and at the other end from the outside to the inside (position 2). Those molecules which cross the membrane change their direction of flow. They are accelerated. Their magnetization can be detected as off-diagonal signal in the 2D velocity-change spectrum (Fig. 5.4.13(c)) but also in the projection onto the subdiagonal. The latter is the *acceleration propagator* in analogy to the *displacement propagator* in position change NMR or 2D $\boldsymbol{q}$ space.

The average acceleration propagator provides the distribution of acceleration or the *probability density of acceleration*. It is obtained by Fourier transformation of the measured slice in reciprocal space and depicted in Fig. 5.4.13(d) for positions 1 and 2 of the haemodialyser [Göb1]. The signal is elevated in the wings of the distribution function at position 1 for negative acceleration and at position 2 for positive acceleration. This proves that water molecules have crossed the membranes at the given operating conditions of the dialyser. Therefore, by measuring the acceleration distribution, the weak signal from the few molecules which have been accelerated is filtered from the signal of the spins with constant flow velocities. The strong peak in the centre of the acceleration distribution is from diffusive molecular acceleration.

6

Basic imaging methods

The *basic imaging methods* of NMR are applicable primarily to imaging of liquids and soft matter, where lines are narrow and spectral dispersion is small compared to the frequency dispersion introduced by application of magnetic-field gradients. The primary incentive for development of these methods comes from applications in medicine and biology, where the dominant signal is most often from water [And1, Cal1, Cer1, Kim1, Mor1, Vla1, Weh1, Weh2].

Practically, all of these methods are designed to scan $\boldsymbol{k}$ *space* as completely as possible, so that the image is obtained by Fourier transformation of the $\boldsymbol{k}$-space signal (cf. Section 5.4.3) [Ern1, Man5, Man6, Mor2]. *Backprojection* methods (Section 6.1) scan $\boldsymbol{k}$ space in cylindrical or spherical coordinates, and most other methods provide a $\boldsymbol{k}$-space signal in Cartesian coordinates. Exceptions are fast imaging methods, most notably *spiral imaging*, where $\boldsymbol{k}$ space is sampled in the shape of a spiral [Duy1, Duy2, Mey1, Tak1, Vla1]. Most backprojection methods exploit *frequency encoding* of the spatial information, so that without spectroscopic resolution, the spatial resolution is limited by the linewidth or the spectrum acquired in the absence of magnetic-field gradients. *Phase encoding* is an indirect way of inscribing the spatial information into the initial conditions of the measured signal (cf. Fig. 2.2.5). It is similar to the indirect detection of a second dimension in 2D NMR spectroscopy [Ern1]. In phase encoding, the spatial resolution is not limited by the shape of the NMR spectrum, but only by the signal-to-noise ratio.

Many imaging methods use combinations of phase and frequency encoding (cf. Fig. 2.2.5 and Section 6.2). Pure phase encoding is usually employed in *spectroscopic imaging* (Section 6.2.4), where phase is encoded in an evolution period and the FID leading to the NMR spectrum is acquired directly without gradients applied. *Single-point imaging* refers to imaging with pure phase encoding and no frequency encoding. This technique and backprojection imaging are the only methods addressed in this chapter which are suitable also for imaging of solids (cf. Chapter 8). For samples with good signal-to-noise ratios and long relaxation times, fast imaging methods have been developed (Section 6.2.8). Here the relaxation delay is kept short by avoiding large flip-angle pulses and using gradient instead of Hahn or stimulated echoes. With such techniques, imaging times of less than 100 ms can be achieved. Basically, all of these methods can also be performed with magnetic-field gradients not in the $\boldsymbol{B}_0$ field but in the $\boldsymbol{B}_1$ field. Methods of this kind are referred to as *rotating-frame imaging* (Section 6.3). The gradient scheme of the backprojection method can also be combined with stochastic instead of pulsed rf excitation, leading to one form of *stochastic imaging*, but more complicated gradient modulations can be considered as well (Section 6.4).

6.1 RECONSTRUCTION FROM PROJECTIONS

Images are readily reconstructed from a set of projections acquired under different angles of the field gradient with respect to the object [Lau1]. The mathematical procedure is quite analogous to the one used in X-ray tomography [Her1]. For historical reasons, *reconstruction from projections* is also called *backprojection*. The principle of the technique is first explained for 2D objects corresponding to a projection along one space axis or to slices which have been prepared by a selective pulse in a field gradient. Then, a third space dimension is introduced into the reconstruction algorithm [Lai1, Lai2], and methods to obtain spectroscopic resolution by frequency encoding are treated [Cor1, Lau2]. A different way of unravelling projections can be pursued for objects with circular symmetry. Here a radial image can be derived from a projection by use of suitable transformations [Maj1].

Sampling of **k** *space*

Projections in the context of NMR imaging are NMR spectra measured in the presence of a magnetic-field gradient. They are given as the integral of the spin density over the space coordinates perpendicular to the gradient direction. According to the projection cross-section theorem (cf. eqn (5.4.12)), the Fourier transform of a projection is defined on a line in $\boldsymbol{k}$ space. The time variable of the FID is proportional to k and the direction of the trace in $\boldsymbol{k}$ space is defined by the directional angle φ of the gradient $\boldsymbol{G}$. Therefore, the signal can be sampled to cover a relevant portion of $\boldsymbol{k}$ space centred at $k = 0$ by simply rotating the gradient by angular increments $\Delta\varphi$ and acquiring an FID for each gradient direction (Fig. 6.1.1(a), cf. Fig. 5.4.2(b)). The rotation is achieved by stepping G_y in a sinusoidal and G_x in a cosinusoidal fashion, so that $\varphi = \arctan\{G_y/G_x\}$ or by rotating the object. By inverse Fourier transformation of the $\boldsymbol{k}$-space signal the image is obtained.

The method is simple to implement and can be readily combined with multi-pulse excitation for solid-state imaging and with noise excitation for rf-power reduction. However, unless a single small flip-angle pulse is used for excitation, the method is slow, and the image quality strongly depends on the degree of linearity of the gradient fields, so that it is no longer used for medical applications. The characteristics of this approach are the following:

(1) The data are sampled on a discrete mesh in $\boldsymbol{k}$ space, which is defined on cylindrical or spherical coordinates. Discrete Fourier transformation requires Cartesian coordinates.
(2) The point density in $\boldsymbol{k}$ space is higher in the centre than farther out.
(3) The spatial resolution is limited by the shape of the NMR spectrum $S(\omega)$ in the absence of field gradients (cf. eqn (5.4.16)).

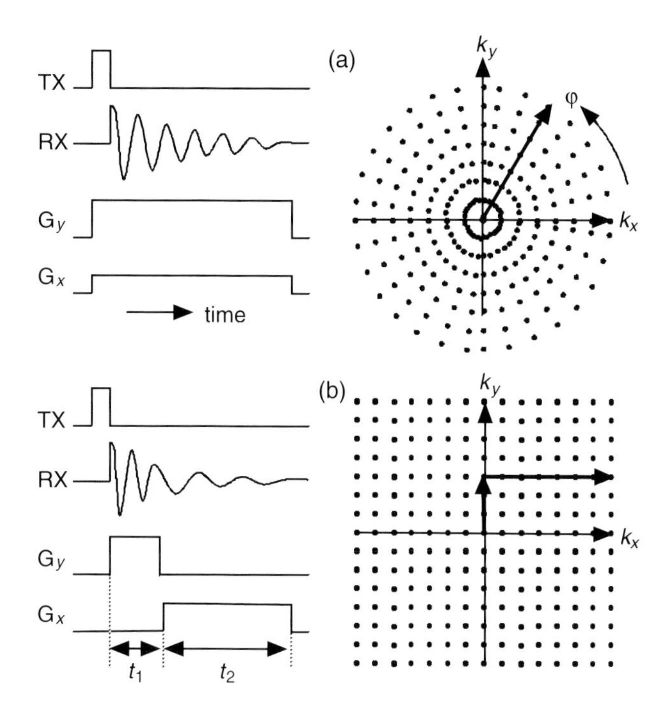

FIG. 6.1.1 Sampling schemes in 2D $\boldsymbol{k}$ space. (a) Cylindrical coordinates. They are used in conjunction with reconstruction from projections. The angle of the field-gradient direction with respect to the x-axis is given by $\Phi = \arctan\{G_y/G_x\}$. (b) Cartesian coordinates. They are used in Fourier imaging (cf. Section 6.2). For rectangular gradient pulse shapes $k_y = -\gamma G_x t_2$. Such sampling schemes are applicable to a slice which can be selected when the rf pulse is applied selectively in the presence of a gradient G_z. Adapted from [Blü7] with permission from Springer-Verlag.

The last point is a problem inherent to all methods which encode space in the NMR frequency. It can be alleviated by introducing spectroscopic resolution. For frequency encoding, this requires measurement of a set of projections for different gradient strengths at each gradient direction (see below) [Cor1, Lau2]. The second point leads to better signal-to-noise ratios at large space coordinates because of the inverse proportionality between $\boldsymbol{k}$ and $\boldsymbol{r}$.

The first point has not been clearly recognized or appreciated in the early days of the method, so terms like *radon transformation* and *filtered backprojection* were introduced [Her1, Man1]. In practical realizations of image reconstruction from projections, however, numerical *filters* must, indeed, be used [Her1].

Backprojection

Use of the expression *backprojection* instead of *reconstruction from projections* is historical. Given a sufficient number of projections of an object acquired at different angles, the shape of the object can indeed be reconstructed with recognizable features, if the projections are just summed over the image plane in the directions over which

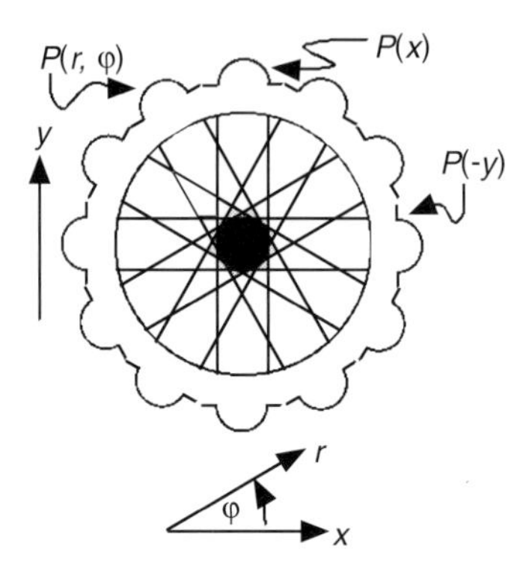

FIG. 6.1.2 [Mor1] Construction of an image from a circular disc by backprojection. The projections taken at different angles φ are added along the directions which have been integrated over in obtaining the projections. The grey values represent image intensities. The backprojection image resembles a disc with star-like distortions.

the integrals were formed in obtaining the projections. However, the image features are convolved with a star-like localization function. This is illustrated in two dimensions in Fig. 6.1.2 for projections of a circular disc [Mor1].

If the magnetic-field gradient is applied in direction r in polar coordinates, which forms an angle φ with the x-axis of a Cartesian reference frame, the projection $P(r, \varphi)$ is obtained according to (5.4.15) by integration of the spin density $M_0(r, s)$ along direction s, which is perpendicular to r (cf. Fig. 5.4.1),

$$P(r, \varphi) = \int M_0(r, s)\, \mathrm{d}s, \tag{6.1.1}$$

where

$$r = x \cos\varphi + y \sin\varphi \quad \text{and} \quad s = -x \sin\varphi + y \cos\varphi \tag{6.1.2a}$$

and

$$x = r \cos\varphi - s \sin\varphi \quad \text{and} \quad y = r \sin\varphi + s \cos\varphi. \tag{6.1.2b}$$

The backprojection approach uses straightforward addition of the projections acquired at different angles φ (cf. Fig. 6.1.2). By taking φ as a continuous instead of a discrete variable for simplicity, the backprojection image M_0' is obtained by integration over φ,

$$M_0'(x, y) = \int_0^\pi P(r, \varphi)\, \mathrm{d}\varphi = \int_0^\pi P(x \cos\varphi + y \sin\varphi) \mathrm{d}\varphi. \tag{6.1.3}$$

This means that each point of a given projection is added to all pixels in direction s perpendicular to r. The simple sum image of the circular disc obtained in this way demonstrates that all projections produce a desired signal increase in the centre at the position of the disc, but image is distorted in a way reminiscent of the rays of a star. These distortions can be eliminated by filtering the projections prior to summation. This approach has lead to the terminology of *filtered backprojection*. The ideal filter corresponds to nothing but the proper scaling routine which is required when changing coordinate systems from cylindrical ones, where the projections are obtained, to Cartesian ones, where the image is reconstructed. Though intuitively appealing, the proper way of obtaining an image from a set of projections is not by backprojection but by reconstruction from projections. Today, the term *backprojection* is used synonymously with *reconstruction from projections*.

Reconstruction from projections

A linear filter performs a convolution of the input function with the Fourier transform of the filter transfer function. According to the *convolution theorem* (cf. Section 4.2.3) application of a filter in one domain corresponds to multiplication of the Fourier transform of the function to be filtered with the filter-transfer function. To filter a backprojection image, eqn (6.1.3) is Fourier transformed,

$$\begin{aligned} p(k, \varphi) &= \int P(r, \varphi) \exp\{ikr\} \, \mathrm{d}r = \int\int M_0(r, s) \, \mathrm{d}s \exp\{ikr\} \, \mathrm{d}r \\ &= \int\int M_0(x, y) \exp\{\mathrm{i}xk \cos\varphi\} \exp\{\mathrm{i}yk \sin\varphi\} \, \mathrm{d}x \, \mathrm{d}y \\ &= p(k\cos\varphi, \, k\sin\varphi) = p(k_x, \, k_y). \end{aligned} \tag{6.1.4}$$

Here the space variables r and s in the Cartesian coordinate frame of the projection have been replaced by the Cartesian coordinates x and y in the laboratory frame (cf. Fig. 5.4.1), and φ is the rotation angle between both frames. This equation is another formulation of the *projection cross-section theorem* (cf. eqn (5.4.12)), which states that the Fourier transform $p(k, \varphi)$ of a projection $P(r, \varphi)$ is defined on a line $p(k\cos\varphi, \, k\sin\varphi)$ at an angle φ through the origin of $\boldsymbol{k}$ space.

According to the projection cross-section theorem $p(k, \, \varphi)$ is given by the FID acquired in the presence of a time-invariant and spatially constant field gradient. If FIDs are acquired for all angles φ of the gradient orientation, the $\boldsymbol{k}$-space signal $s(k\cos\varphi, k\sin\varphi)$ is obtained for all angles φ and all relevant values of k. The spin density M_0 is then obtained by inverse 2D Fourier transformation of the $\boldsymbol{k}$-space signal,

$$\begin{aligned} M_0(x, y) &= (2\pi)^{-2} \int\int p(k_x, \, k_y) \exp\{-\mathrm{i}k_x x\} \exp\{-\mathrm{i}k_y y\} \, \mathrm{d}k_x \, \mathrm{d}k_y \\ &= (2\pi)^{-2} \int\int p(k, \, \varphi) \exp\{-\mathrm{i}kr\} |\boldsymbol{k}| \, \mathrm{d}k \, \mathrm{d}\varphi. \end{aligned} \tag{6.1.5}$$

Because cylinder coordinates are used, the scaling factor $|\boldsymbol{k}| = k$ appears in the integral. Apart from this and the prefactor $(2\pi)^{-2}$, eqn (6.1.5) and the backprojection formula (6.1.3) are identical.

Multiplication of the FID $p(k, \, \varphi)$ by $|\boldsymbol{k}|$ prior to Fourier transformation for use by the backprojection method (6.1.3) can be interpreted in terms of filtering the projection $P(r, \, \varphi)$ by a filter the transfer function of which is given by $f(k) = |\boldsymbol{k}|$. For this reason, calculation of the image by proper transformation of the raw data from cylindrical to

Cartesian coordinates according to (6.1.5) is sometimes referred to as *filtered back-projection*. In theory, the FID is multiplied by a linear ramp, so the large signal in the beginning is strongly attenuated, and the small signal towards the end with low signal-to-noise ratio is strongly amplified. To avoid unnecessary contamination by noise, often filter functions $f(k)$ different from the ideal one are used, which approach zero for large values of k [Man1, Her1].

The effect of different filters on the spatial resolution is illustrated in Fig. 6.1.3 for 2D images corresponding to 2D projections of 50 capillaries filled with water to different heights [Nil1]. Each image is reconstructed from 128 1D projections by use of the filter transfer functions $f(k)$ sketched below each image. The theoretical filter function is given in (a), and image (c) corresponds to classical backprojection without a filter. In either case the resolution is bad. The best spatial resolution is obtained with triangular filter functions (bottom). Here the symmetric triangular filter function (e) yields the best results. It should be noted that calculation of images by s is often replaced today by explicit interpolation of the $\boldsymbol{k}$-space data acquired in cylindrical coordinates to Cartesian coordinates and subsequent 2D Fourier transformation.

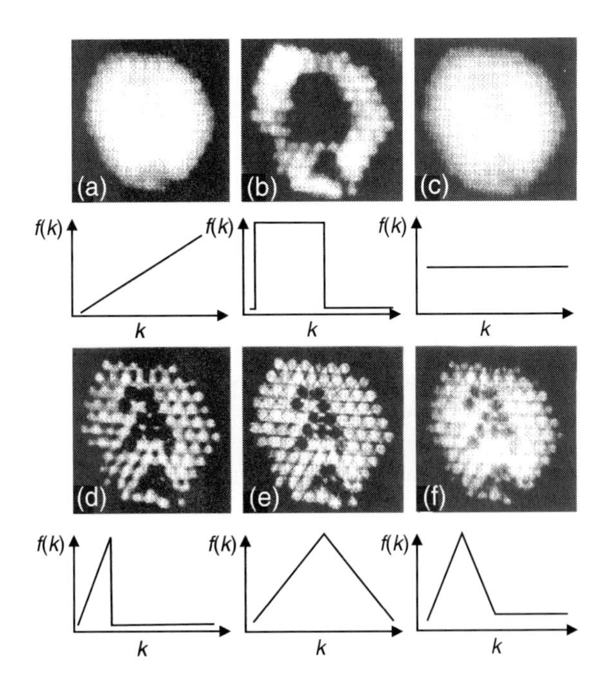

FIG. 6.1.3 [Nil4] 2D filtered backprojection images providing a 2D projection through a set of 50 capillaries filled with water to different heights. The images were calculated from 128 projections acquired with the Hadamard imaging technique (cf. Section 6.4). The filter transfer functions $f(k)$ used are sketched below each image. The best spatial resolution is obtained with a symmetric triangular filter function (e). The theoretically correct filter function for reconstruction from projections (a) and lack of a filter function in the old backprojection method (c) give rather low spatial resolution.

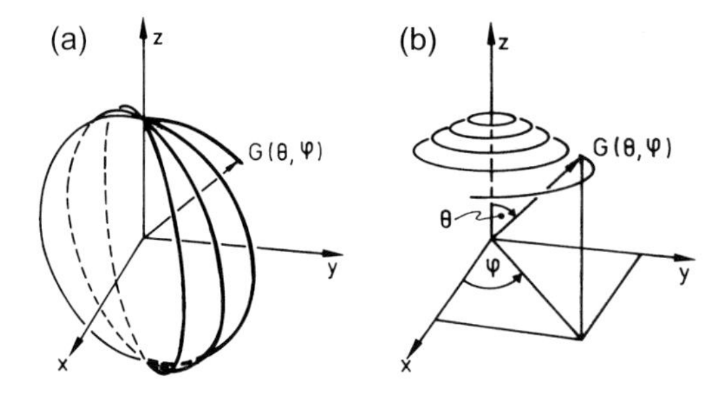

FIG. 6.1.4 Gradient paths for 3D reconstruction from projections. Only half a hemisphere is covered by the gradient paths, because signal for negative gradient values can be acquired by time inversion in echo techniques. (a) 3D space can be covered by a set of 2D projections, so that the 2D algorithm can be applied in two steps. (b) Optimization of the point density in 3D $\boldsymbol{k}$ space requires an integral approach to 3D reconstruction from projections. Adapted from [Lai1] with permission from Institute of Physics.

Reconstruction from projections in three dimensions

Not only 2D but also 3D spin densities can be reconstructed from projections. To this end projections need to be acquired in all three space dimensions. Two possible schemes are illustrated in Fig. 6.1.4 [Lai1]. Reconstruction of the 3D spin density can proceed either via the intermediate reconstruction of a set of 2D spin-density projections (a) or in a direct manner (b). In the first case the reconstruction algorithm is applied twice, once for the variable θ and once for the variable φ, and the resultant ideal filter for the projections is not given by $|\boldsymbol{k}|$ but by $|\boldsymbol{k}|^2$ [Lai1, Lai2]. When using single-pulse excitation with low flip angles the imaging time can be rather short [Haf2, Wuy1]. For a $(128)^3$ point image, less than 1 min was reported for liquid and for solid samples [Haf2].

Spectroscopic resolution by frequency encoding

The method of *reconstruction from projections* can be combined with *spectroscopic resolution*, so that an NMR spectrum can be assigned to each voxel. The spectroscopic information can be separated from the spatial information if a set of projections is acquired with different gradient strengths for each gradient orientation. Two approaches [Cor1, Lau1] are considered below and illustrated for the case of a 1D spin density with space coordinate r.

Given the single-pulse response $y_1(t)$ of the Bloch equations from (5.4.8), it is clear that both spectroscopic and spatial information are encoded along the time axis,

$$\begin{aligned} y_1(t) &= -\mathrm{i}\int\int M_0(r,\Omega_\mathrm{L})\exp\left\{-\frac{t}{T_2}\right\}\exp\{\mathrm{i}\Omega_\mathrm{L}t\}\,\mathrm{d}\Omega_\mathrm{L}\,\exp\{-\mathrm{i}\gamma Grt\}\,\mathrm{d}r \\ &= -\mathrm{i}p(k,\varphi). \end{aligned} \tag{6.1.6}$$

But by variation of the gradient strength G, the time axes for spectroscopic and for spatial evolution can be separated. By dividing the time axis t into two parts,

$$t = t_0 + t_\mathrm{G}, \quad \text{where } t_0 = t\cos\zeta,\ \ t_\mathrm{G} = t\sin\zeta,\ \text{ and } t = (t_0^2 + t_\mathrm{G}^2)^{1/2}, \tag{6.1.7}$$

the time axis t appears at the angle ζ in the Cartesian time frame spanned by the spectroscopic time axis t_0 and the spatial time axis t_G. Using (6.1.7) and the relation $\Omega_G t = -\gamma G r t = kr$ the single-pulse response (6.1.6) is rewritten,

$$y_1(t) = -\mathrm{i} \int \int M_0(\Omega_G, \Omega_L) \exp\left\{-\frac{t}{T_2}\right\} \times \exp\left\{\frac{\mathrm{i}\Omega_L t_0}{\cos\zeta}\right\} \mathrm{d}\Omega_L \exp\left\{\frac{-\mathrm{i}\gamma G r t_G}{\sin\zeta}\right\} \frac{1}{\gamma G} \mathrm{d}\Omega_G. \tag{6.1.8}$$

2D Fourier transformation with respect to $t_0/\cos\zeta$ and $t_G/\sin\zeta$ would produce directly the T_2-weighted spin density $M_0(\Omega_G, \Omega_L)\exp\{-t/T_2\}$. But the only accessible quantity is the Fourier transform of the single-pulse response,

$$Y_1(\omega) = \int y_1(t) \exp\{-\mathrm{i}\omega t\}\,\mathrm{d}t. \tag{6.1.9}$$

According to the *projection cross-section theorem*, (6.1.4) is the Fourier transform of a slice which is a projection in Fourier space. Equation (6.1.9) can, therefore, be understood as a projection of a signal in $\Omega_L\Omega_G$ space onto the frequency coordinate ω. This is illustrated in Fig. 6.1.5 [Cor1], where the frequency Ω_G corresponds to the space coordinate r and the frequency Ω_L to the spectroscopic axis. The projection angle $\zeta = \arctan(-\gamma G r_{max}/\Omega_{L,max})$ can be adjusted *via* the gradient strength G between zero and a maximum value ζ_{max}. Thus, the method of *reconstruction from projections* can be used for evaluation of the frequency dependence of the T_2-weighted spin density $M_0(r, \Omega_L)\exp\{-t/T_2\}$ [Lau1]. As a general rule, an $(n+1)$-dimensional algorithm for reconstruction from projections has to be applied for an n-dimensional spin density with 1D spectroscopic resolution. The disadvantage of this method is that the spectroscopic projection angle ζ cannot be varied to assume all values between 0° and 180°, so the missing information must be extrapolated.

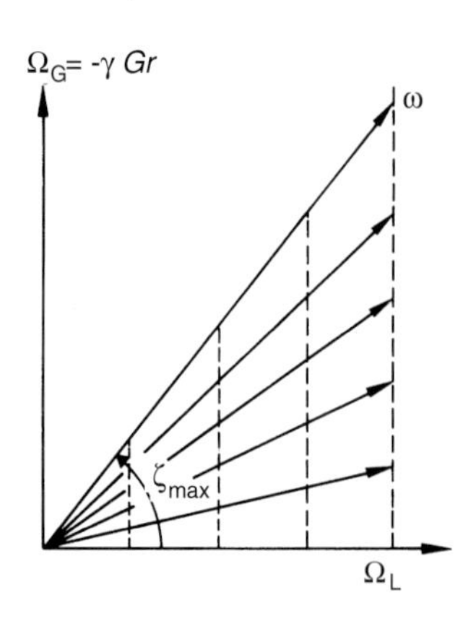

FIG. 6.1.5 [Cor1] The frequency dependence of the T_2-weighted 1D spin density $M_0(r, \Omega_L)\exp\{-t/T_2\}$ can be interpreted as a 2D object in $\Omega_L\Omega_G$ space. The spectra $Y_1(\omega)$ measured in a magnetic-field gradient correspond to projections at an angle ζ through this object.

This disadvantage is avoided by another method [Cor1]. The crucial step is a Fourier transformation of the pulse response (6.1.6) over the field gradients G, which introduces g as the Fourier-conjugate variable of the gradient G,

$$
\begin{aligned}
Y_1(t, g) &= \frac{1}{2\pi} \int y_1(t, G) \exp\{\mathrm{i} g G\} \, \mathrm{d}G \\
&= \mathrm{i} \int \int M_0(r, \Omega_\mathrm{L}) \exp\left\{-\frac{t}{T_2}\right\} \exp\{\mathrm{i}\Omega_\mathrm{L} t\} \, \mathrm{d}\Omega_\mathrm{L} \, \delta(g - \gamma r t) \, \mathrm{d}r. \qquad (6.1.10)
\end{aligned}
$$

For evaluation of (6.1.10) the definition of the *delta function*

$$
\delta(g - \gamma r t) = \frac{1}{2\pi} \int \exp\{\mathrm{i} g G\} \exp\{-\mathrm{i}\gamma G r t\} \, \mathrm{d}G \qquad (6.1.11)
$$

has been used.

After Fourier transformation of a matrix of FID signals (6.1.6) over the gradient values G, the signals belonging to a fixed space coordinate r lie on a straight line through the origin at $g = 0$ with the slope $\mathrm{d}g/\mathrm{d}t = \gamma r$ (Fig. 6.1.6). The reason for this is the appearance of the delta function in (6.1.10). For separation of space and frequency variables, the signal is interpolated in tg space along these lines with different slopes and stored as $Y_1(t, r)$. Fourier transformation over t directly leads to the spin density with spectroscopic resolution. The method can be readily extended to reconstruction from projections with two and three space coordinates as well as for spectroscopic resolution with two and more frequency coordinates.

Radial imaging

In materials science applications of NMR imaging the geometry of the object can often be chosen at liberty. For example, cylindrical samples can be used for studies of porosity by diffusion and flow in rocks and catalyst pellets [Maj2], and investigations of deterioration in polymers and elastomers. Furthermore, many biological samples show radial or close

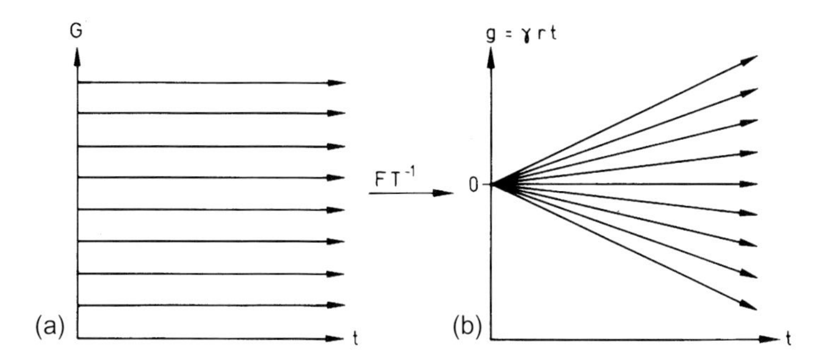

FIG. 6.1.6 Principle of spectroscopic resolution by the Fourier method in combination with reconstruction of the spin density from projections. (a) Arrangement of measured signals $y_1(t, G)$ in tG space before Fourier transformation over G. (b) Interpolation scheme in tg space after Fourier transformation over G for separation of space and frequency coordinates.

to radial symmetry. Examples are arteries and plant stems [Mei1]. In this case, imaging with 1D spatial resolution is often sufficient for sample characterization. According to the projection cross-section theorem, a *radial image* relates to a tangential projection, so that considerable signal-to-noise improvement can be achieved in contrast selection by displaying particular spectroscopic components or other NMR parameters in the image [Mei1].

Different approaches can be taken to obtain radial images. Radial field gradients can be applied by the use of dedicated hardware [Hak1, Lee1, Lee2]. Alternatively, a 2D image can be reconstructed from one projection by the backprojection technique, and a radial cross-section can be taken through it. The most direct way to access the radial image from a projection consists in computing the inverse *Hankel transformation* (cf. Section 4.4.2) of the FID measured in Cartesian $\mathbf{k}$ space (cf. Fig. 4.4.1) [Maj1]. But in practice, the equivalent route *via* Fourier transformation of the FID and subsequent inverse *Abel transformation* (cf. Section 4.4.3) is preferred because established phase and baseline correction routines can be used in the calculation of the projection as an intermediate result.

6.2 FOURIER IMAGING

The imaging techniques used most frequently in medical diagnostics and most often explored for applications to soft materials are based on the scheme shown in Fig. 6.1.1(b), which is directly related to multi-dimensional Fourier NMR spectroscopy [Ern1, Ern2]. The time axis is divided into two intervals, one evolution time t_1 and a detection time t_2. The experiment is started by an rf pulse inducing a free-induction decay signal. During each of the following time intervals a different gradient is applied. Just like in *2D spectroscopy*, the phase of the signal acquired during the detection time is modulated by the signal phase accumulated in the evolution time. The evolution time is incremented independently from scan to scan, while the data are stored as rows in a 2D data array. The idea is that the accumulated signal phase is primarily a result of the applied gradient. A 2D Fourier transformation then immediately produces a 2D image of the object. The scheme is readily extended to three dimensions by incorporating a second evolution time for the third gradient. This original scheme of Fourier imaging has been optimized to suit different needs. Nevertheless, the fundamental idea of *phase encoding* during data acquisition for image construction by simple Fourier transformation in *Cartesian coordinates* proved to be the key to the most successful imaging methods applied to liquids and soft matter [Kum1, Kum2].

6.2.1 The spin-warp technique

Spin-warp imaging denotes the most common form of Fourier imaging with data acquisition in *Cartesian* ***k****-space coordinates* [Ede1]. Instead of a variable evolution time t_1 for *phase encoding*, the gradient amplitude is stepped during t_1 in order to halt phase evolution from spin interactions other than with the applied space-encoding gradient.

During the detection time the signal is recorded with *frequency encoding* of the spatial information (cf. Section 2.3.4).

Phase encoding

The phase acquired during the evolution period t_l in the presence of the applied gradient G_y (cf. Fig. 6.1.1(b)) is given by

$$\phi_y(y, t_1) = -\gamma y \int_0^{t_1} G_y(t)\,\mathrm{d}t = k_y(t_1)y, \tag{6.2.1}$$

where the definition (2.2.23) of the component k_y of the wave vector $\boldsymbol{k}$ or the wave number in y-direction has been used. Because the phase is proportional to y for a given time integral of the gradient pulse, the signal phase is a direct measure of space. If the gradient amplitude is constant and t_1 is increased, the phase is expected to grow linearly with t_1 for each magnetization component in proportion to its space coordinate y, and Fourier transformation over k_y of phase-encoded magnetization $M_+(k_y) = M_0(y)\exp\{\mathrm{i}k_y y\}$ in a voxel at position y provides the spin density $M_0(y)$ directly.

However, this scheme neglects all other spin interactions which influence the NMR frequency. In liquid samples, these are the chemical shift, the indirect coupling, and signal attenuation as a result of relaxation and dephasing in magnetic field inhomogeneities other than the applied gradients. The phase evolution of the detected signal under the influence of these interactions progresses as the evolution time increases. In solids, other interactions dominate. These are homo- and heteronuclear dipole–dipole couplings, chemical-shift anisotropy, and possibly the quadrupole interaction. The influence of all of these effects on the phase of the detected signal evolution can be eliminated, if instead of the time t_1 in (6.2.1), the gradient strength is varied to modulate the phase and t_1 is kept at a fixed value. Then the phase evolution only depends on the applied gradient strength, while a constant phase term is contributed by the evolution of the magnetization for a time t_1 in the presence of all the other interactions. This variant of Fourier imaging is called *spin-warp imaging* [Ede1], because the signal phase can be imagined to warp around the gradient direction with increasing gradient amplitude (cf. Fig. 2.2.4). Consequently, the gradients applied during the evolution periods are called *phase-encoding gradients*. The one applied during the detection period is called the *read gradient*. It is used for *frequency encoding* (cf. Section 2.3.4) of the space information.

Basic Fourier imaging

In most cases, the image of a 2D slice needs to be acquired instead of a complete 3D image. Then the magnetization of the slice needs to be selected from the total magnetization through suitable preparation of the magnetization at the start of the actual imaging sequence. This can be done, for instance, by application of a selective pulse in the presence of a gradient perpendicular to the plane of the slice (cf. Chapter 5.3). A 2D Fourier imaging sequence for acquisition of a single slice then consists of the following steps [Weh1]:

(1) Excitation of the nuclear magnetization by a selective 90° pulse in the presence of a gradient G_z for *slice selection*.

(2) Application of a *phase-encoding gradient* G_y for a fixed time interval.
(3) Application of the *read gradient* G_x and acquisition of n data points.
(4) Incrementation of the phase-encoding gradient and repetition of steps 1–3 for $m - 1$ times.
(5) 2D Fourier transformation of the data matrix of size $n \times m$ for computation of the image.

The use of echoes

In a magnetic-field gradient, the magnetization components dephase. As a result of this, the FID decays faster than in the absence of a gradient. The stronger the gradient, the faster the dephasing. Therefore, any unnecessary dephasing must be avoided or refocused by an *echo*. Refocusing is achieved either by an rf pulse or by gradient manipulation. Echoes generated by rf pulses can be the Hahn echo (cf. Fig. 2.2.9) for nuclei with interactions linear in the spin operator or the solid (cf. Fig. 3.2.6) and magic (cf. Fig. 3.4.3) echoes for nuclei with bilinear and multiple spin interactions, respectively. Echoes generated by gradient manipulation are called *gradient echoes* (cf. Section 2.2.1). A gradient echo appears whenever the accumulated magnetization phase (6.2.1) becomes zero as a result of gradient modulation. In the most simple case of gradients with constant magnitude, the gradient sign is inverted after a time $t_E/2$. Then the gradient echo follows after a time of duration t_E has elapsed. If gradients of different amplitudes are employed, the timing of the echo follows from (6.2.1) with $\phi_y = 0$. The refocusing gradient is often called the *time-reversal gradient*. Gradient echoes are applied, for instance, to cancel the signal loss as a result of dephasing in a gradient during application of a selective pulse for definition of a slice. If the gradient echo is used in combination with a Hahn echo, the magnetization phase is changed in sign by the 180° pulse, so that the time-reversal gradient needs to have the same sign as the dephasing gradient. *Hahn echoes* are used to refocus dephasing from residual magnetic field inhomogeneities.

Aliasing of spatial information

During detection, the *Nyquist sampling theorem* must be observed to avoid signal *aliasing* by *frequency folding* (cf. Section 2.3.4): an object of diameter $x_{\max} = 10\,\text{mm}$ in a maximum field gradient of $G_{\max} = 0.5\,\text{T/m}$ produces a spread of the ^{1}H signal over 200 kHz. The *Nyquist frequency* f_N must be adjusted by proper setting of the sampling interval Δt to accommodate the spectral width $S_w = 200\,\text{kHz}$. In quadrature detection the Nyquist frequency is half the spectral width, so that for simultaneous acquisition of one complex magnetization value every Δt, the sampling rate is limited by

$$\frac{S_w}{2} = f_N = \frac{1}{2\Delta t} > \frac{\gamma\, G_{\max}\, x_{\max}}{2\pi}. \tag{6.2.2}$$

The *acquisition time* T of the signal is then determined by the number n_{data} of discrete frequency components corresponding to the number of pixels in this dimension,

$$T = n_{\text{data}}\, \Delta t < \frac{n_{\text{data}}\, \pi}{\gamma\, G_{\max}\, x_{\max}}. \tag{6.2.3}$$

In medical imaging, T is of the order of 5 ms and is kept constant. For a different field of view, $x_{\max}$, the gradient strength $G_{\max}$ is adjusted rather than the acquisition time.

The Nyquist theorem also applies for the indirectly detected signal, which evolves under variable gradient strengths during a fixed evolution time t_1. Signal aliasing arises if the maximum precession phase during phase encoding evolves by more than 360° for one gradient increment ΔG. For rectangular gradient pulses and an object centred in the gradient origin, aliasing is avoided if

$$\gamma \Delta G (y_{\max}/2) t_1 < 2\pi. \tag{6.2.4}$$

Practical spin-warp imaging

Given these prerequisites, a 2D *spin-warp imaging* sequence is set up in five steps. These are illustrated in Fig. 6.2.1(a)–(e) [Weh1]. The indices assigned to the gradients

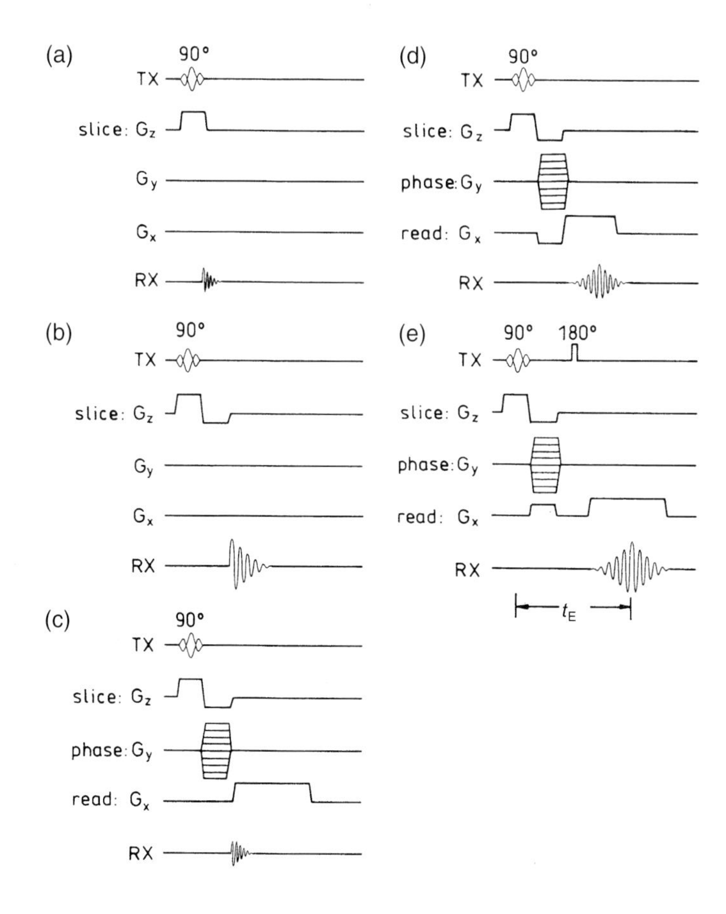

FIG. 6.2.1 The different stages of a 2D spin-warp imaging sequence. (a) Slice selection. (b) Slice selection under a gradient echo. (c) 2D sequence with phase encoding and read gradients. (d) Incorporation of a gradient echo in the read gradient: this is the basis of the FLASH or GRASS method [Utz1]. (e) Combination with a Hahn echo results in the spin-echo imaging sequence. Adapted from [Weh1] with permission from Wiley-VCH.

in the figure denote space directions in an object frame, which may coincide with the laboratory frame or with the frame defined by the gradient coils. For most purposes in the following, the gradient indices refer to a frame which is attached to the object. In general, this object frame appears rotated within the frame defined by the gradient coils, so that the respective gradients are a superposition of gradients generated by the individual gradient coils:

(a) The first step in measuring a 2D image requires the selection of the image plane by *selective excitation* of the magnetization under a gradient in z-direction. The magnetization of the spins in the slice is rotated into the transverse plane by a selective 90° pulse (cf. Section 5.3). For a thin slice, a large gradient and a long selective pulse are required. During this time the magnetization can strongly dephase, leaving only a small fraction of transverse coherence for acquisition.

(b) Because signal dephasing during slice selection is undesired, the transverse magnetization is refocused by a *gradient echo*. The magnetization echo is much stronger than the FID after slice selection without a gradient echo. Its amplitude is attenuated only by T_2 relaxation and inhomogeneities in the static magnetic field $\boldsymbol{B}_0$.

(c) Following slice selection, space information is encoded by application of a *phase-encoding gradient* in y-direction during the evolution time, and a *read gradient* in x-direction for signal acquisition with frequency encoding during the detection time. The FID signal decays much faster in the presence of a field gradient than in the absence of a gradient. However, it now carries information about space. Its Fourier transform is a projection of the spin density perpendicular to the direction of the read gradient.

(d) In practice, the gradients cannot be switched in infinitely short time intervals. The switching time constants are defined by the gradient-pulse shape, the coil inductances and the eddy currents induced in conducting matter in the vicinity of the gradient coils. During the time required for stabilization of the read gradient, a significant part of the signal already dephases before a well-defined signal can be acquired. This *deadtime* strongly depends on the size and the shielding of the gradient coils, and on the strength and the shape of the gradient pulses. It can be handled by inclusion of a *time-reversal gradient* (x gradient) applied simultaneously with the phase-encoding gradient (y gradient) for formation of a gradient echo after the readout gradient (x gradient) has been turned on. The time integral (6.2.1) of the time-reversal gradient has to be adjusted in such a way that the echo maximum appears after the deadtime. Often Gaussian or half-sinc shaped gradient pulses are used for minimization of gradient pulse transients and for optimization of lineshapes, which effectively determine the quality of the image. Apart from the flip angle of the selective pulse, the scheme of Fig. 6.2.1(d) is identical to the methods referred to as *FLASH* (fast low- angle shot [Haa1]) and *GRASS* (gradient recalled acquisition in the steady state [Utz1]), which are used for fast imaging (cf. Section 6.2.7). Formation of an echo provides the opportunity to sample the echo build-up as well as the echo decay. The build-up traces the signal in negative $\boldsymbol{k}$ space, while the decay provides the signal in positive $\boldsymbol{k}$ space.

(e) In addition to the influence of relaxation times the signal intensity is attenuated for the scheme of Fig. 6.2.1(d) by inhomogeneities in the static magnetic field $\boldsymbol{B}_0$ as well as by chemical shift and susceptibility differences. Their influence can be eliminated in the indirectly detected space dimension by formation of a *Hahn echo* on top of the read-gradient echo in x-direction. Because of the 180° refocusing rf pulse, the relative signs of time reversal and read gradients in x-direction are now inverted. The time elapsed from the centre of the selective pulse to the echo maximum is commonly called the *echo time* t_E in magnetic resonance imaging.

Usually the rf pulses applied in an imaging sequence are not perfect. The pulses may differ in shape, phase, and amplitude from the required values. As a result, unwanted transverse magnetization may arise which leads to erroneous signals. This is so, in particular, for the 180° refocusing pulse of the Hahn echo. These transverse magnetization components can be dephased by application of homogeneity-spoil or *crusher gradient* pulses.

Relaxation-time contrast

The intensity of the signal acquired by the spin-warp technique depends on the echo time t_E and on the *repetition time* t_R of successive scans. The echo time introduces T_2 weights to the pixels of the image, resulting in *relaxation-time contrast* in cases where T_2 is a function of space coordinates x, y, and z. Within the validity of the Bloch equations, the weight of the echo at point $\boldsymbol{r} = (x,\ y,\ z)$ is given by

$$a_H(\boldsymbol{r}, t_E) = M_z(\boldsymbol{r}) \exp\left\{-\frac{t_E}{T_2(\boldsymbol{r})}\right\}, \tag{6.2.5}$$

where $M_z(\boldsymbol{r})$ is the initial longitudinal magnetization. Its value is determined by the T_1 weight which is adjustable, for example, *via* partial saturation by choice of the repetition time t_R,

$$M_z(\boldsymbol{r}, t_R) = M_0(\boldsymbol{r}) \left[1 - \exp\left\{-\frac{t_R}{T_1(\boldsymbol{r})}\right\}\right]. \tag{6.2.6}$$

Here T_1 is a function of space coordinates as well, and $M_0(\boldsymbol{r})$ is the thermodynamic equilibrium magnetization giving the spin-density distribution. Thus, the total weight of the signal originating from location $\boldsymbol{r}$ is obtained by insertion of (6.2.6) and (6.2.5),

$$a_H(\boldsymbol{r}, t_R, t_E) = M_0(\boldsymbol{r}) \left[1 - \exp\left\{-\frac{t_R}{T_1(\boldsymbol{r})}\right\}\right] \exp\left\{-\frac{t_E}{T_2(\boldsymbol{r})}\right\}. \tag{6.2.7}$$

Proper choice of t_R and t_E for suitable T_1 and T_2 weights is an important factor for determining image *contrast in medical imaging*. Due to less restrictions in measurement time and the use of rf power, a much wider variety of contrast parameters can be accessed in material applications of NMR imaging (cf. Chapter 7).

6.2.2 Multi-slice imaging

If the region of interest in the sample is not well defined, it is advantageous to acquire more than one slice through the object. Effective use of instrument time is achieved by interleaved acquisition of different slices: during the repetition time t_R for magnetization recovery of one slice, the magnetization of one or more different slices can be measured. In this case not only the 90° pulse but also the 180° pulse must be slice selective.

This approach uses the repetition time t_R efficiently, but it suffers from the shortcoming that the magnetization of just one slice is excited and observed at a time, while the volume of the remaining sample contributes noise and no signal. A way out of this problem consists of the use of *multi-frequency selective rf pulses* [Bol1, Goe1, Goe2, Haf1, Mül1, Mül2, Mül3, Mü4, Sou1]. The *slice selective pulses* for a set of image planes are added according to the values given by the different rows of a frequency-ordered *Hadamard matrix* (cf. Section 4.4.5, Fig. 4.4.2). A 2D image is acquired for each combination of selective pulses corresponding to each row of the Hadamard matrix. *Hadamard transformation* of the resultant set of images unscrambles the multi-frequency encoded images into images of separate slices. In contrast to successive measurements of n selective slices, now the n-fold time has effectively been spent on measurement of each slice in the same total acquisition time. The slice selection has been multiplexed, giving a signal-to-noise advantage of $n^{1/2}$. This gain, however, does not apply in comparison with interleaved multiple slice acquisition, because individual Hadamard scans need to be separated by the repetition time t_R. The Hadamard technique for space encoding is also used for spectroscopic imaging, where typically only a small number of volume elements is selected (cf. Section 10.1.2).

6.2.3 3D imaging

The slice-selective 2D imaging scheme is readily extended to three space dimensions following the original Fourier imaging scheme of Fig. 6.2.1(e): the selective pulse for generation of transverse magnetization is replaced by a nonselective pulse which excites the whole sample. Now two instead of one gradient are stepped independently during the evolution time (Fig. 6.2.2). This increases the duration of the experiment. The

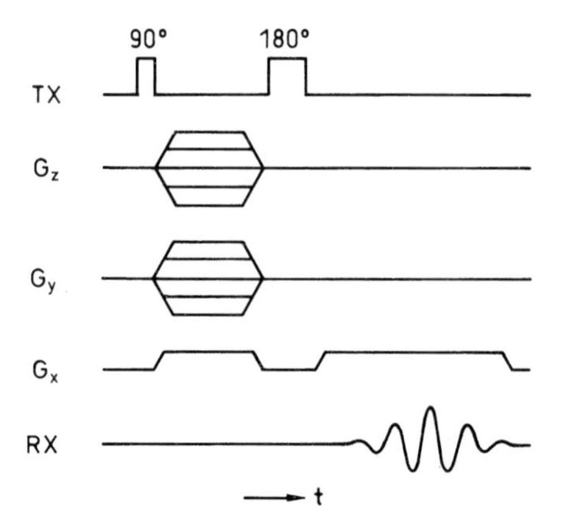

FIG. 6.2.2 Pulse sequence for 3D spin-warp imaging.

experimental time can be reduced, however, if the initial rf pulse is made somewhat selective for restriction of the field of view in the z-direction. In this way, the number of gradient increments in z-direction can be kept small. The 3D image is obtained by straightforward 3D Fourier transformation of the experimental data set. As compared to multiple slice selective imaging, the slices are well defined in *3D imaging*. Signal leakage between neighbouring slices resulting from rf pulse envelope defects is avoided leading to better resolution in the third-space dimension.

6.2.4 Spectroscopic imaging

Spectroscopic imaging denotes imaging with acquisition of an NMR spectrum for each voxel. This increases the dimensionality of the imaging experiment by one for 1D spectroscopic imaging and is expensive in terms of measurement time. The benefit though is access to all spectroscopic parameters for definition of images with contrast from different spectroscopic parameters (cf. Section 7.3) [Bro1, Dec1].

Principle

To decouple spatial and spectroscopic evolutions, the integral in (5.4.16) needs to be unravelled to obtain a product instead of a convolution of spatial and spectroscopic dimensions. The associated phase evolution of transverse magnetization can be decoupled in a 2D fashion by suitable gradient modulation during the experiment (Fig. 6.2.3). Usually, *phase encoding* is employed to inscribe the spatial information into the acquired signal during a fixed evolution time t_1, where only $\boldsymbol{G}(t)$ is varied to step through $\boldsymbol{k}$ space. Then the phase evolution from all spectroscopic interactions is constant for all gradient values. Moreover, this phase evolution can be refocused by Hahn, solid, and magic echoes to extend the usable duration of the phase-encoding time t_1. Subsequently, the

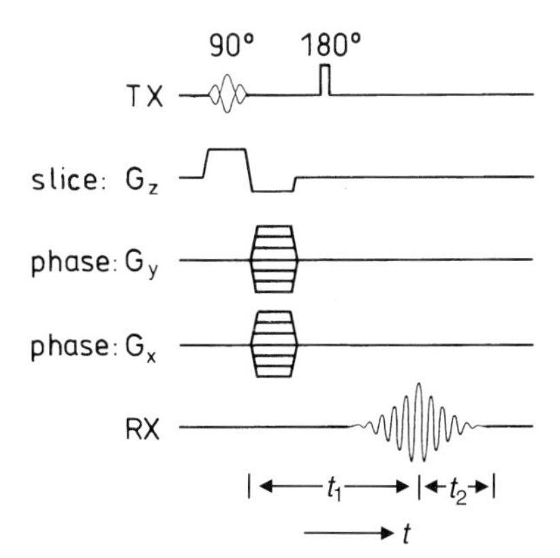

FIG. 6.2.3 [Weh1] Principle of spectroscopic Fourier imaging with slice selection. To separate spatial and spectroscopic responses, the spectroscopic evolution must be constant during the space-encoding period t_1. The spectroscopic signal is acquired during the detection time t_2 in the absence of a gradient.

gradients are turned off and the magnetization evolves only under the spectroscopic interactions for a time t_2. During this time, the signal $f(\boldsymbol{k}(t_1), t_2)$ can be acquired,

$$f(\boldsymbol{k}(t_1), t_2) = \int_V M_0(\boldsymbol{r}) \exp\{\mathrm{i}\boldsymbol{k}(t_1)\boldsymbol{r}\} s(t_2, \boldsymbol{r})\, \mathrm{d}\boldsymbol{r}. \tag{6.2.8}$$

Because the initial phase of the FID $s(t_2, \boldsymbol{r})$ in a voxel at position $\boldsymbol{r}$ is determined by the gradient modulation only, spatial and spectroscopic variables become separated into a product by Fourier transformation over $\boldsymbol{k}$,

$$\begin{aligned}\int f(\boldsymbol{k}(t_1), t_2) \exp\{-\mathrm{i}\boldsymbol{k}\boldsymbol{r}'\}\, \mathrm{d}\boldsymbol{k} &= 2\pi \int_V M_0(\boldsymbol{r}) \delta(\boldsymbol{r} - \boldsymbol{r}') s(t_2, \boldsymbol{r})\, \mathrm{d}\boldsymbol{r} \\ &= 2\pi\, M_0(\boldsymbol{r}') s(t_2, \boldsymbol{r}'). \end{aligned} \tag{6.2.9}$$

After further Fourier transformation over t_2, the product of the spin density and the NMR spectrum is obtained for each space coordinate $\boldsymbol{r}'$ in contrast to (5.4.16). Note, however, that only the phase evolution of $s(t_1, \boldsymbol{r})$ can be refocused during t_1 and not the irreversible signal decay. This is particularly important for solids with molecular motion where, for example, in ^{2}H NMR, the signal decay during the solid-echo time is used specifically to probe the motion (cf. Section 3.2.2) [Jel1, Spi1].

In principle, the spatial information can be inscribed into the detected signal by *frequency encoding* during the detection time and the spectroscopic phase evolution can be detected indirectly [Man7, Sep1]. For good spectroscopic resolution, the number of data points in the frequency dimension is considerably larger than in either of the space dimensions, for example, 1 k versus 256. To keep the number of scans low, the spectroscopic dimension is preferably acquired directly during the detection time, while the space information is phase encoded during the evolution time. In addition, phase encoding of the spatial information improves the spatial resolution and eliminates susceptibility artefacts. But hybrid techniques based on combinations of phase and frequency encoding have been developed as well [Coc1, For1] and are useful in accelerating the acquisition of *spectroscopic imaging* data [Jak1, Poh1]. Spectroscopic imaging of a slice is similar to 3D imaging, because a complete 3D data set has to be acquired. The slice is defined either by a selective excitation pulse (Fig. 6.2.3) or a selective refocusing pulse (cf. Fig. 7.3.1) of the echo. Spectroscopic 3D imaging requires incrementing three gradients independently during t_1, resulting in a 4D data set [Bro1, Mau1].

Shaping of the spatial response

In spectroscopic imaging, the spatial dimensions are usually phase encoded for indirect detection. The shape of the spatial response function is determined by the Fourier transform of the signal in the gradient encoded dimension. Often data are acquired with equal weights for all gradient steps within a finite range of discrete values, resulting in a rectangular profile in gradient space (Fig. 6.2.4(a)) [Mar1]. Thus, the spatial response assumes the shape of a sinc function (cf. Section 4.1), and after zero-filling, the pixel boundaries are ill defined by the wiggles introduced from limiting the values of the gradient to a finite range. The artefact resulting from truncation of the NMR signal is called the *Gibbs artefact* [Par1]. Better defined boundaries are obtained by improved $\boldsymbol{k}$-space sampling [Mar3, Mar4, Pon1], *shaping of the spatial response* [Hen1, Hod1, Mar1, Web1], and data processing schemes which avoid the Fourier transformation of finite-length data sets [Wea1]. Shaping of the response is done by *filtering*, corresponding to *apodization* or multiplication of the data in gradient space with a suitable weighting function. This procedure is completely analogous to the technique of exponential weighting mentioned in Section 2.3.4 for processing of the free-induction decay. Various weighting functions can be used. One function with excellent shape and good sensitivity is the *Hanning function* [Ern1] (Fig. 6.2.4(b)).

Basically, all spectroscopic imaging techniques are used in practice in conjunction with signal averaging. Because space encoding is done indirectly, and signal averaging is employed in most cases, this type of shaping can be obtained without a loss of sensitivity by suitable choice and addition of signals. The spatial position of the response function can be selected by using the *shift property of the Fourier transformation* (cf. eqn (4.1.9)) (Fig. 6.2.4(c)). This is accomplished by altering the overall phase of each acquired NMR signal as a function of the applied encoding gradient. After Fourier transformation, this results in a spatial shift of the response function. With this method, the shape of the spatial response can be selected during data acquisition, and the exact position can be determined during data processing. A selective Fourier transform method has been developed to achieve this localization of a response for all locations [Mar1] instead of just for a single location.

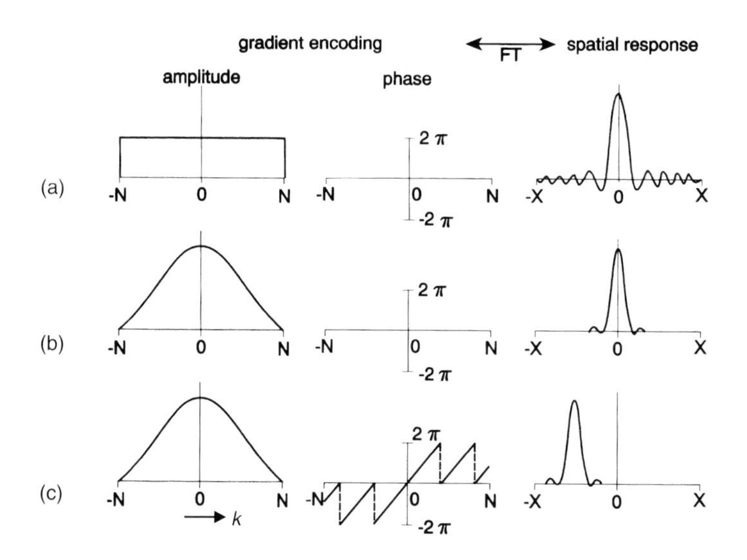

FIG. 6.2.4 [Mar1] Shaping and shifting of the spatial response by suitable addition of space-encoded signals. Left: Weights of the acquired scans as a function of the number N of gradient steps. Middle: The impressed phase shift. Right: Form and position of the spatial response obtained by Fourier transformation. (a) Conventional approach to signal averaging without weighting the number of acquired scans. (b) Weighting of the acquired data with a Hanning window improves the shape of the spatial response. (c) Impressing a gradient dependent phase shift on the acquired data shifts the location of the spatial response function.

6.2.5 Stimulated-echo imaging

The popular *spin-echo imaging* scheme (Fig. 6.2.1(e)) requires execution of a 180° pulse for formation of the Hahn echo. This sequence provides the maximum signal without phase distortions for image construction. However, a 180° pulse also requires considerable rf power, in particular, when it is applied to large diameter coils. As an alternative to Hahn echoes, *stimulated echoes* can be used for imaging [Bur1, Fin1, Fra1]. They are excited by three instead of two rf pulses (cf. Section 2.2.1). Imaging schemes based on stimulated echoes are also referred to as *STEAM* (stimulated-echo acquisition mode) images [Fra1].

As a consequence of having three time periods for the stimulated echo instead of two for the Hahn echo, a considerable variety of imaging schemes can be designed by assigning different functions to the individual pulses. Either one, two, or all three pulses can be made selective, and the selective pulses can be combined with suitable gradients. Thus, imaging sequences which provide more information than those based on Hahn echoes can be designed (cf. Section 7.2.4).

Principle

For optimum signal strength in liquid spin-$\frac{1}{2}$ systems, all pulses are 90° pulses. In this case, the stimulated-echo sequence can be written as

$$90°-t_1-90°-t_{\mathrm{m}}-90°-t_2. \tag{6.2.10}$$

The first pulse generates transverse magnetization, which dephases in an applied magnetic-field gradient. Half of the dephased magnetization is converted to longitudinal magnetization by the second pulse. The remaining transverse magnetization forms a Hahn echo after the second pulse at time $t_{\mathrm{m}} = t_1$. The third pulse recalls the longitudinal magnetization as measurable transverse magnetization, however, with the initial phases determined by the evolution during the time t_1 between the first two pulses. A stimulated echo forms at time $t_2 = t_1$. If t_{m} is long, so that longitudinal magnetization builds up, then an FID will be observed also after the third pulse.

The characteristics of *stimulated-echo imaging* are:

(1) The rf power is reduced by a factor of four as compared to that of the Hahn-echo sequence for a given pulse width.
(2) Longitudinal magnetization exists during t_{m} after the second pulse. It relaxes with T_1, which often is larger than T_2^*, so that t_{m} can be used for gradient switching.
(3) T_1 weighting can readily be introduced by a variation of t_{m}.
(4) The echo amplitude is only half of what it is for the Hahn echo when optimum flip angles are used.

T_1 and T_2 contrast

Some STEAM sequences are summarized in Fig. 6.2.5. Refocusing of the slice-selection gradient by gradient inversion may take place immediately after (a) or before (b) the selective pulse, or by using a gradient of the same sign (c). In schemes (b) and (c), the defocusing gradient is applied before the combined use of the slice-selection gradient and the selective rf pulse. This cannot be done in spin-echo imaging.

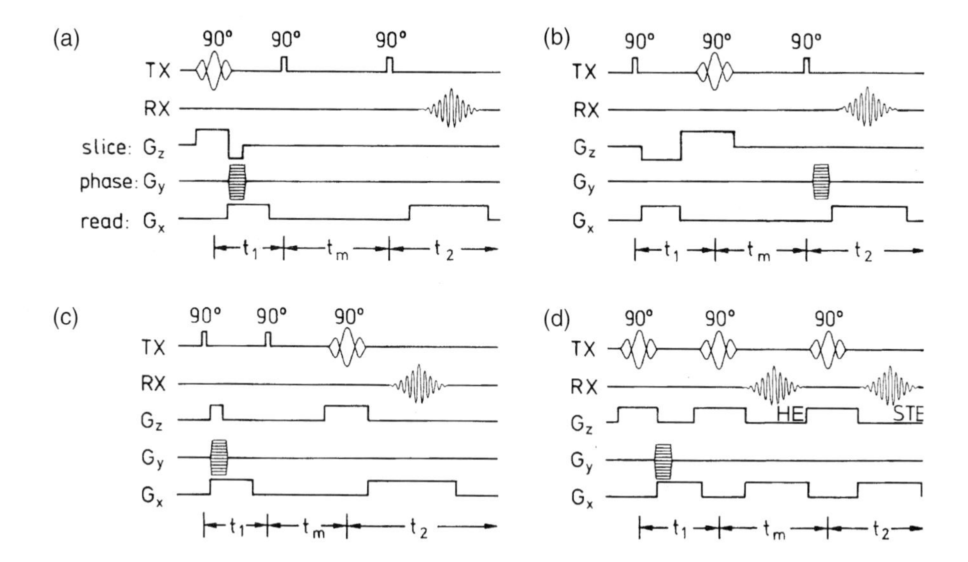

FIG. 6.2.5 [FRA1] Timing diagrams of basic STEAM imaging sequences. Schemes (a)–(c) use a single selective pulse in the first, second and third position, respectively. Sequence (d) uses three slice-selective rf pulses for observation of both the Hahn echo (HE) and the stimulated echo (STE).

If the phase-encoding gradient is applied before the second pulse, both the Hahn echo and the stimulated echo are affected by it (Fig. 6.2.5(d)), and Hahn-echo and stimulated-echo images can be acquired simultaneously. While the contrast in *Hahn-echo images* is determined by T_2 for long repetition times t_R, the *contrast in stimulated-echo imaging* is also determined by T_1 as a result of T_1 relaxation of the longitudinal magnetization during t_m. In fact, a set of T_1-weighted images can be acquired in the time needed for a single image, if the third pulse is broken down into a sequence of small flip-angle pulses (Fig. 6.2.6), where the sum of all flip angles is given by 90° [Haa2]. The amplitudes of the stimulated-echo signals then decay for each pixel with the T_1 relaxation time as a function of the effective mixing time t_m. Neglecting T_2 relaxation,

$$M_x(t_2 = t_1, \boldsymbol{r}) = M_x(t_1, \boldsymbol{r}) \exp\left\{\frac{-t_m}{T_1(\boldsymbol{r})}\right\}. \tag{6.2.11}$$

A fit of (6.2.11) to the pixel intensities in a sequence of T_1-weighted images yields an effective T_1 value for each pixel at position $\boldsymbol{r}$. All T_1 values taken together produce a T_1 image (cf. Chapter 7). The assumption for the use of (6.2.11) is that all T_1-weighted images have been obtained from the same z magnetization. But because the z magnetization is decreased by each of the preceding small flip-angle pulses, the pulse flip angles α_i have to be increased according to [Man2, Ste1]

$$\sin\alpha_i = \tan\alpha_{i-1}. \tag{6.2.12}$$

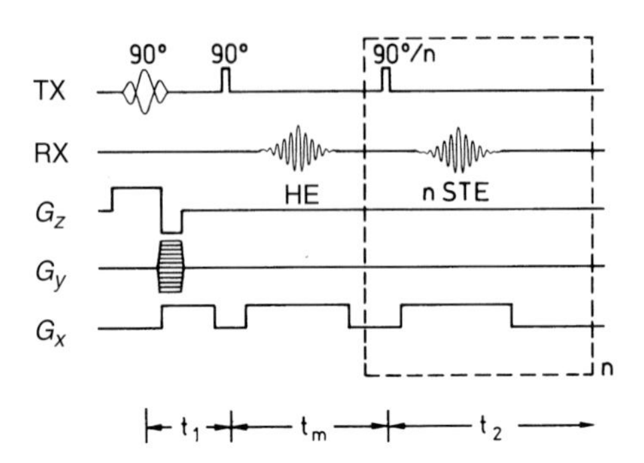

FIG. 6.2.6 [Haa2] Timing diagram for T_1 imaging by the STEAM method. The first rf pulse is used for slice selection. The last pulse is split into n pulses α_i. Using this sequence a T_2-weighted image and a set of T_1-weighted images can be acquired.

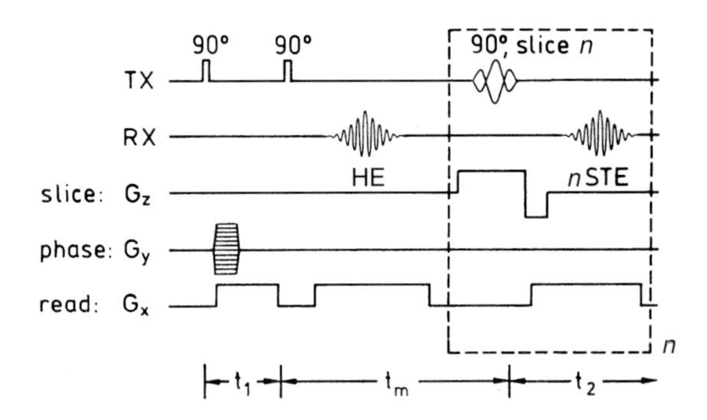

FIG. 6.2.7 [Fra1] Timing diagram for multi-slice imaging by the STEAM method. The third rf pulse is used for slice selection. It is repeated with different centre frequencies for acquisition of different slices.

Multi-slice imaging

A closely related technique can be used for *multi-slice imaging* (Fig. 6.2.7) [Fra1]. The scheme of Fig. 6.2.5(c) is appended by further slice-selective 90° pulses with different centre frequencies, so that the magnetization of other slices is selected [Fra1]. In this way, the otherwise necessary recycle delay can effectively be used for acquisition of additional slices. However, the contrast in each slice is affected by a different T_1 weight, because t_m is different for each slice. The technique can readily be adapted to *line-scan imaging* by applying successive slice-selective pulses in orthogonal gradients [Fin1].

Simultaneous acquisition of images with different spatial resolution

If the phase-encoding gradient is applied right after the second pulse, it only acts on the Hahn echo. A second phase-encoding gradient can then be used independently right after the third pulse acting on the stimulated echo (Fig. 6.2.8) [Fra1], so that the resolution of the phase-encoded space dimension can be set in one image independent of that in the

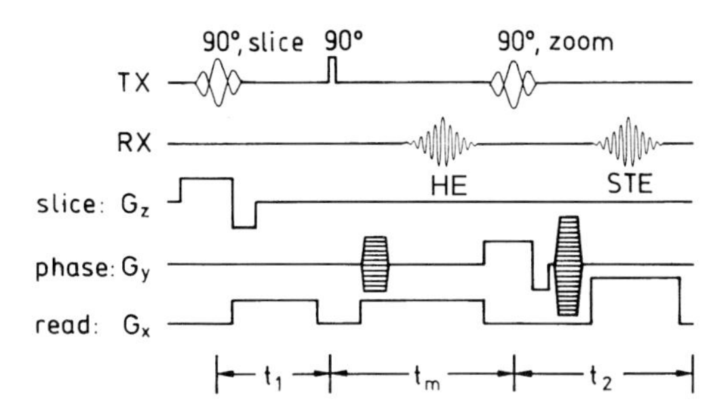

FIG. 6.2.8 [Fra1] Timing diagram for acquisition of two images with independent spatial resolution in the phase-encoded dimension by the STEAM method.

other image. In particular, an overview with a large *field of view* can be acquired with the spin echo and a *zoom image* with the stimulated echo.

6.2.6 Imaging with CPMG echoes

The flexibility in pulse sequence design gained by using three pulses in STEAM imaging as compared to spin-echo imaging can also be accomplished in part by using a *double Hahn echo* or even a *CPMG train of echoes* (cf. Fig. 2.2.10(b)) [Hau1, Hen3, Mor2]. Then the data acquired from successive echoes are affected by increasing T_2 weights and the J modulation from coupled spins [Sta1]. This can be exploited for generation of contrast. In combination with repetitive gradient, switching images can be acquired in a single shot [Hen2].

Relaxometric imaging

The pulse sequence for *relaxometric imaging* is depicted in Fig. 6.2.9(a) [Mor2]. To reduce interference effects from signals outside the slice of interest, all pulses are made slice selective. Repetitive refocusing of the phase-encoding step in k_y-direction has the effect that different $\boldsymbol{k}$-space regions are sampled by odd end even echoes. For each echo an image can be acquired with a different T_2 weight. From a signal analysis in a given pixel of the set of images, relaxation and amplitude parameters can be extracted by a numerical fit, for example, of a mono-exponential function $M_z(x, y, z) \exp\{-nt_{\mathrm{E}}/T_2(x, y, z)\}$, where n counts the echoes and t_{E} is the echo time corresponding to the separation of echoes.

Under realistic conditions, imperfect 180° pulses lead to secondary echoes and other artefacts which contaminate the measured signal, and only the *double Hahn-echo sequence* is used in practice [Mor2]. From the images of the first and the second Hahn echo, parameter images can be readily calculated: in case of mono-exponential relaxation, the ratio of the second to the first image eliminates the spin density, so that the signal amplitude in the image is defined by $\exp\{-t_{\mathrm{E}}/T_2(x, y, z)\}$, while the ratio of the square of the first image to the second image produces an image of the spin density $M_z(x, y, z)$, where $M_z(x, y, z) = M_0(x, y, z)$ in the absence of T_1 weights (cf. Section 7.2.1) [Hep1].

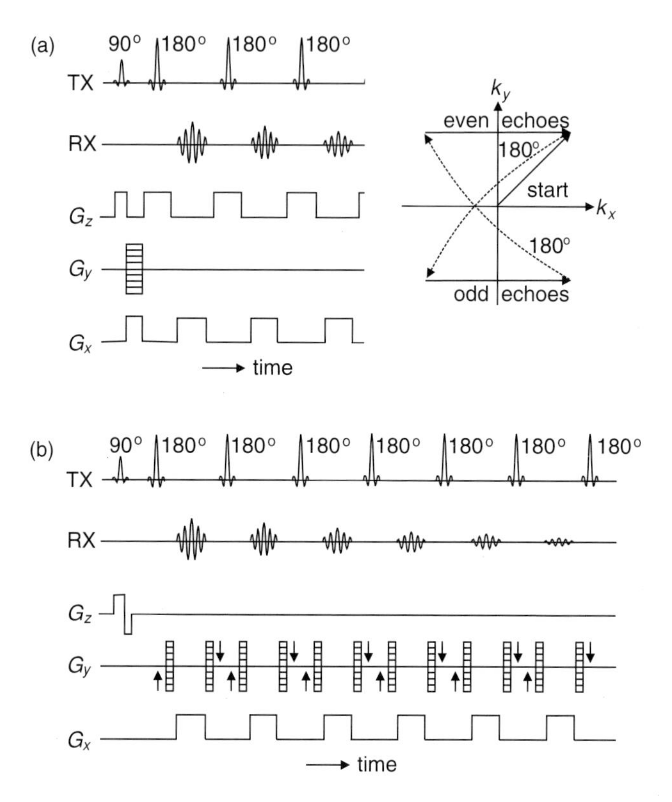

FIG. 6.2.9 [Mor2] Imaging with CPMG echoes. (a) Pulse sequence and $\boldsymbol{k}$-space trajectories for relaxometric imaging. (b) The RARE sequence for rapid imaging. This sequence is also referred to as turbo spin-echo imaging.

RARE: rapid acquisition with relaxation enhancement

By changing the phase encoding scheme of the relaxometric imaging sequence, each echo can be encoded with a different value of k_y in order to scan a complete image with one train of CPMG echoes (Fig. 6.2.9(b)) [Hen2]. Following the acquisition of one echo, the phase evolution along k_y due to the first-gradient pulse is cancelled by formation of a gradient echo after application of a second-gradient pulse with opposite amplitude. Then the next echo can be phase encoded with a different value of k_y. Because each echo carries a different T_2 weight, the sequence has been given the acronym *RARE* for rapid acquisition with relaxation enhancement. The technique is also known by the names *turbo spin-echo imaging* and *fast spin-echo imaging* [Mor2].

Each echo is associated with a specific value of k_y. By judicious choice of the acquisition order of k_y, images with different contrast features can be generated: the signal for small values of $|\boldsymbol{k}|$ defines the signal at large space coordinates after Fourier transformation. Signal at small $|\boldsymbol{k}|$ values is, particularly, important for image contrast, while that at large $|\boldsymbol{k}|$ values is important for the spatial resolution. To obtain an image with strong T_2 *contrast*, signal for small $|\boldsymbol{k}|$ values should be acquired late in the CPMG-echo train. *Vice versa*, images from data where small $|\mathbf{k}|$ values have been acquired early in the CPMG-echo train provide *spin-density contrast*, or T_1 *contrast* in case of short repetition times [Mor2, Mul1]. For small enough samples, the refocusing pulses can be applied while the read gradient G_x is on. Furthermore, the y gradient may be pulsed to achieve small increments in k_y after every second echo. Then an echo-planar imaging sequence is obtained which uses Hahn instead of gradient echoes (cf. Fig. 6.2.14(b) below; the function of G_x and G_y is interchanged in both figures) [Hen6, Rof1]. By incorporating gradient-pulse pairs sandwiching every second refocusing pulse, a fast sequence is obtained for velocity imaging [Hen6].

6.2.7 Gradient-echo imaging

During the generation of a Hahn echo, the magnetization is inverted and a recycle delay t_R of the order of $5T_1$ is required for the magnetization to relax back to thermal equilibrium. If instead of a Hahn echo, only a single small flip-angle pulse is used for excitation, the magnetization is not appreciably perturbed from thermal equilibrium. In fact, the flip angle can be optimized for maximum signal-to-noise ratio, given the recycle delay t_R. The optimum flip angle is the *Ernst angle* (cf. eqn (2.2.35)) [Ern1],

$$\alpha_E = \cos^{-1}\left(\exp\left\{-\frac{t_R}{T_1}\right\}\right). \tag{6.2.13}$$

Choosing the shortest possible excitation pulse separation and setting the flip angle close to the Ernst angle is the idea underlying the fast imaging schemes referred to as *FLASH imaging* [Haa1] or *GRASS imaging* [Utz1].

Typical flip angles are in the order of 15° for the slice-selective excitation pulses. The intensity of the FID after such a pulse corresponds to 25% (sin 15°) of the intensity after a 90° pulse. However, more than 96% (cos 15°) of the longitudinal magnetization are preserved, enabling fast repetition rates. The method can be employed in combination with the backprojection imaging scheme and with spin-warp imaging by phase encoding of spatial information.

2D and 3D FLASH imaging

The timing diagram for slice-selective *2D FLASH imaging* with phase encoding is depicted in Fig. 6.2.10. The scheme is repeated n times, where n is the number of phase-encoding steps. The free-induction decay is detected in the form of a *gradient echo* after time reversal by the read gradient G_x. The repetition time is determined by the duration of the slice selection pulse and the data acquisition time. Compared to spin-warp imaging employing a Hahn echo, it is reduced by a factor of the order of 100, that is from about 1 s to 10–20 ms for each scan.

After about 10–20 cycles of the experiment, the magnetization reaches a steady state, where it is in equilibrium with relaxation and selective excitation pulses. Conditions

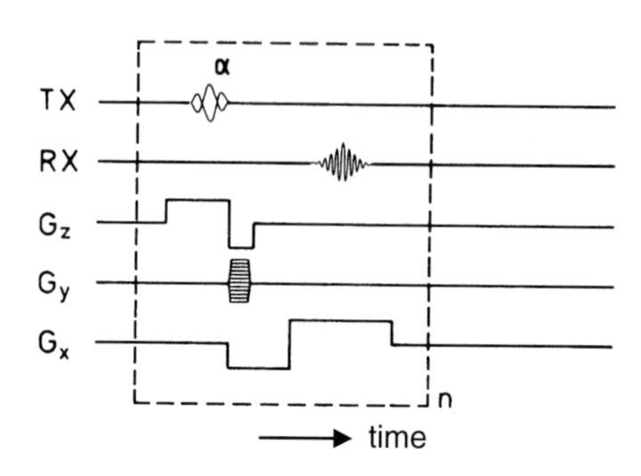

FIG. 6.2.10 [Haa1] Timing diagram of the 2D FLASH sequence for fast imaging. The method employs slice-selective excitation pulses of the order of 15°.

similar to this, but using larger flip-angle pulses (90–120°), are the basis of the *SSFP* technique (cf. Fig. 2.2.10). In this case, the relaxation-time contrast is determined by the ratio of T_1/T_2. For FLASH imaging, the contrast tends to be dominated by T_2 [Weh1]. T_1 weighting can be enhanced by insertion of a 180° pulse prior to the actual imaging sequence [Haa1].

In *3D FLASH imaging* the selective low flip-angle pulse is replaced by a nonselective pulse, and instead of one phase-encoding gradient two are used [Fra2]. Typical imaging times on a medical imager are 2 s for a $(128)^2$ pixel 2D image and 4 min for a $(128)^3$ voxel image. Thus 2D images can be taken of slowly moving objects [Fra3]. In addition to fast imaging times, the rf power exposure of the object is significantly reduced as a result of the use of small flip-angle pulses. The FLASH sequence is one of the standard imaging sequences in medical imaging.

Spectroscopic FLASH imaging

The method has been extended for inclusion of a spectroscopic dimension (*SPLASH* imaging: spectroscopic low angle shot imaging) [Haa3]. For each time-domain data point of the spectroscopic dimension a separate FLASH experiment is performed, where the position of the echo is scanned (Fig. 6.2.11). The *gradient echo* refocusses only dephasing of the transverse magnetization resulting from the presence of the gradients, but not its evolution governed by other spin interactions like chemical shift and J coupling. Therefore, the echo maxima are independent of the applied gradients and modulated by the internal spin interactions. A total of n FLASH experiments needs to be performed to obtain n data values, which can be Fourier transformed to obtain the spectrum. This time may be shortened if multiple gradient refocusing is used. However, the method is highly sensitive to magnetic field inhomogeneities, and the beginning of the FID is always lost for the minimum time it takes to generate the first gradient echo. Therefore, the technique is applicable to the acquisition of magnitude spectra with narrow lines only.

6.2.8 Ultra-fast imaging

Ultra-fast imaging techniques are capable of acquiring a complete image in a time shorter than the spin–lattice relaxation time T_1. They can be used for imaging of dynamic

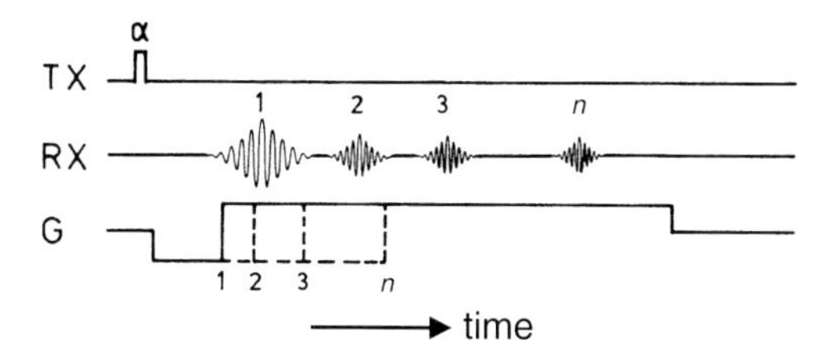

FIG. 6.2.11 [Haa3] Timing diagram for the 1D spectroscopic FLASH experiment. A total of n experiments is performed for different positions of the gradient echo in order to scan the chemical shift dimension.

processes which proceed on comparatively fast time scales, i.e. with time constants in the order of 0.1 s and less. Several approaches have been developed for medical use. Important methods are the *snapshot-FLASH*, the *FAST*, and the *echo-planar imaging* (*EPI*) techniques. They require fast switching of gradients, but they differ in terms of the rf power requirements and subsequently the signal strength or signal-to-noise ratio [Nor1]. Fast gradient switching is avoided in *DANTE-based imaging* sequences, where a train of small flip-angle pulses is applied in a constant field gradient, and the spin response is refocused in a Hahn echo for data collection [Cho2]. Ultra-fast imaging methods are suited for imaging of objects with motions on the time scale of one-tenth of a second, but *parameter images* (cf. Section 7.1.4) can also be measured which demand the acquisition of several *parameter-weighted images* within a limited total measuring time [Ste2, Tur1, Wor1]. Examples are in *diffusion-tensor imaging* of humans [Bas1, Bas2], imaging of *chemical waves* in chemical reactors (cf. Section 10.1.1) [Arm1], and imaging of diffusion and transport of fluids [Cor2, Gat1, Gui1, Kos1, Man9, Mer1].

The snapshot-FLASH technique

The *snapshot-FLASH* technique is a fast version of the FLASH experiment (Fig. 6.2.10). Only as many samples of the transverse magnetization are acquired as needed for the required image resolution. The remaining transverse magnetization is spoiled by *crusher gradients*, so that the next scan can follow immediately. The flip angle of the selective pulse is optimized for maximum transverse magnetization in the experiment. Because the whole imaging experiment lasts less than T_1, the excitation flip angle is no longer given by the Ernst angle [Haa4].

Neglecting T_1 relaxation during the experiment, the total transverse magnetization is given by the sum of the amplitudes M_{x0} of the individual FID signals. For a snapshot-FLASH image of N lines, this is

$$\sum_{n=1}^{N} M_{x0}(n) = M_0 \sin\alpha \sum_{n=1}^{N} \cos^{n-1}\alpha. \tag{6.2.14}$$

From this equation, the optimum flip angle α_{opt} can be obtained as a function of the number N of lines in the image. For example, $\alpha_{\mathrm{opt}} = 16^\circ, 11.3^\circ, 8^\circ$ for $N = 32, 64, 128$, respectively. After each pulse, the remaining z magnetization is lower than before. Consequently, the transverse magnetization generated by the next pulse is reduced. This can be alleviated by scaling successive flip angles according to (6.2.12). The method can be extended to include T_1 and T_2 weighting as well as chemical shift information by use of an appropriate preparation of the initial magnetization before the start of the actual imaging experiment (cf. Section 7.1) [Haa4, Haa5, Jiv1, Sin1].

As in FLASH imaging, each line of the image is acquired with a new rf excitation pulse. Therefore, the method shows little sensitivity to the effective relaxation time T_2^*. In particular, it can be used in cases where T_2^* is short. The echo-planar imaging method below requires sufficiently long values of T_2^*, because the whole image must be acquired in a time shorter than this. For large values of T_2^*, the snapshot-FLASH method is inferior to *echo-planar imaging* (see below) as long as the tails of the FID signals are spoiled and not acquired, but for short values of T_2^* it is superior [Nor1].

The FAST and CE-FAST techniques

The original FLASH technique does not lead to maximum transverse magnetization. The optimum signal is obtained when the FID following a given pulse is refocused by the next pulse to form an echo. In the snapshot-FLASH technique, however, the transverse magnetization dephases in a gradient and is deliberately spoiled before the next rf pulse. The *FAST* (Fourier acquired steady state) sequence is an extension of the FLASH sequence, which meets the phase coherence requirements of the *SSFP* technique [Gyn1, Gyn2]. The effect of the phase-encoding gradient preceding signal collection is cancelled exactly by a second-gradient pulse with opposite sign (Fig. 6.2.12(a)).

Different contrast mechanisms are accessed by a clever modification of the FAST method. If adjacent pairs of rf pulses are identified as the pulses generating a Hahn echo, then the FID following the first pulse can be understood to be refocused by the second pulse to form a spin echo. In the echo maximum the next pulse is applied, so that the echo time t_E corresponds to twice the repetition time t_R. By reordering the timing of

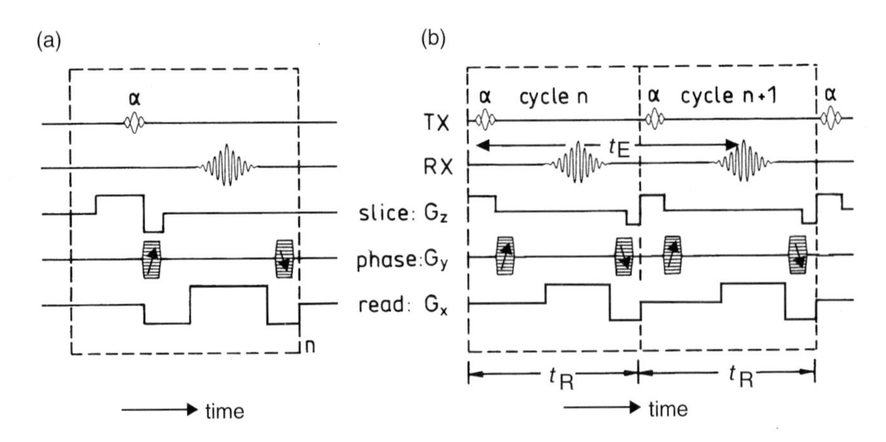

FIG. 6.2.12 [Gyn1] Timing diagrams of (a) the FAST and (b) the CE-FAST sequence. The dephasing of the magnetization as a result of phase encoding is refocused by application of a second-gradient pulse of opposite sign. Using the FAST sequence, the FID of the SSFP signal is acquired. With the CE-FAST sequence the echo is acquired, leading to different image contrast.

the gradients in the FAST sequence, each gradient echo can be generated in such a way that the FID following the rf pulse just before the echo does not contribute to the signal (Fig. 6.2.12(b)). This is achieved by refocusing the magnetization dephased during slice selection at the end of the same cycle. The echo of the read gradient forms in cycle $n + 1$ at a time when the integral gradient matches that of cycle n. The phase-encoding gradient in cycle n has no net effect, so that the spin-echo signal is encoded only by the phase-encoding gradient in cycle $n + 1$. In this way an echo signal is prepared from the SSFP response, which shows significant enhancement in T_2 *contrast*. The method is known as the *CE-FAST* (contrast enhanced Fourier acquired steady state) technique [Gyn1]. It is in fact, a time-reversed version of the FAST sequence. The SSFP technique has been explored for a number of different fast imaging schemes [Haw1, Jon1, Pat1, Zur1].

Echo-planar imaging

In *echo-planar imaging* (*EPI*) the information of an entire 2D or 3D image is encoded into a single FID [Man1, Man5, Man6, Man8, Ste2]. The acquisition times for an image are typically shorter than 100 ms for viscous samples, but strong gradients ($G = 40$ mT/m) have to be switched rapidly ($dB/dt = 10$ T/s). The principle of the method is explained with the help of Fig. 6.2.13 [Man4]: the acquired signal can be interpreted in a way, as if it came from discrete points on a lattice in the 2D image plane. For such a spin density, the image can be reconstructed either from several projections in the conventional way (cf. Section 6.1) or from an appropriately chosen single projection. The angle φ of such a projection has to be set in such a way that all image points are visible from that direction. This approach corresponds to echo-planar imaging [Man1], which can be extended to

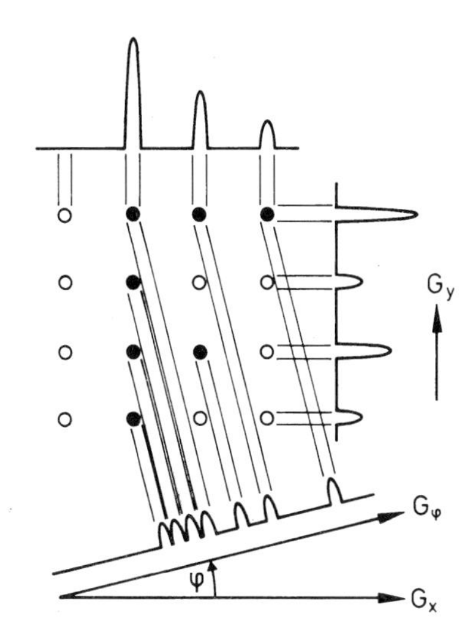

FIG. 6.2.13 [Man4] Principle of echo-planar imaging. Representation of the letter F by a discrete spin density and projections for three gradient directions.

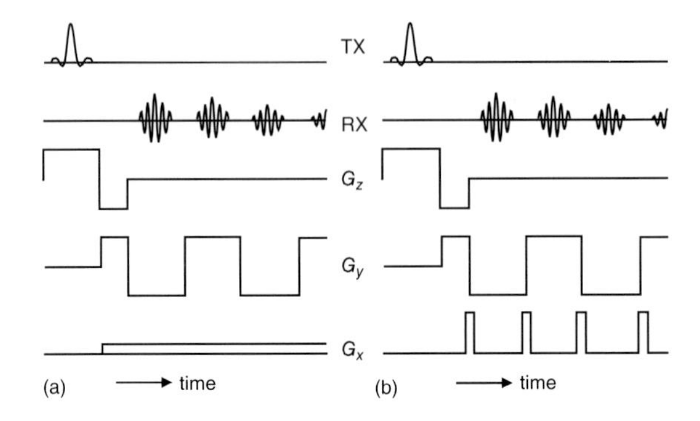

FIG. 6.2.14 [Man6] Pulse sequences for two variants of 2D echo-planar imaging. (a) Continuous x gradient. (b) Pulsed x gradient. The associated $\boldsymbol{k}$-space trajectories are depicted in Fig. 5.4.2(e) and (f).

three space dimensions (*EVI*, *echo-volumnar imaging*) [Man3] as well as to include spectroscopic resolution (*EPSM*, *echo-planar shift mapping*) [Man2].

Figure 6.2.14 depicts the sequence of events for rf pulses, field gradients and receiver signals for *slice-selective 2D EPI* [Man6]. The experiment starts with a selective rf pulse in the presence of a z gradient for slice selection. In the presence of a weak x gradient, a sequence of *gradient echoes* is generated by alternating, strong y gradients (Fig. 6.2.14(a)). The magnetization decay is determined by relaxation and by dephasing under the weak x gradient and other magnetic field inhomogeneities. The 1D Fourier transform of the echo train provides the 2D image.

The relationship between measured signal and image information is explained with the help of Fig. 6.2.15 [Mor1]. As a consequence of the alternating y gradient, the measured signal is given by a string of gradient echoes, which are reminiscent of a series of CPMG echoes (cf. Fig. 3.4.1) or *rotational echoes* in *MAS NMR* (cf. Fig. 3.3.7(a)). Neglecting the effect of the weak x gradient for the time being, the measured signal $s(t)$ is a convolution $\otimes$ of the echo-shape function $e(t)$ with a comb $f(t)$ of delta functions,

$$f(t) = \sum_i \delta(t - t_i), \tag{6.2.15}$$

$$s(t) = \int e(t) f(t - \tau)\, \mathrm{d}\tau = e(t) \otimes f(t). \tag{6.2.16}$$

According to the *convolution theorem* (4.2.14), the Fourier transform $S(\omega)$ of the echo train is given by the product of the Fourier transform $E(\omega)$ of an echo $e(t)$ and the Fourier transform $F(\omega)$ of the delta-comb function $f(t)$,

$$S(\omega) = E(\omega) F(\omega). \tag{6.2.17}$$

Because the Fourier transform of a delta comb is a delta comb again, $S(\omega)$ corresponds to a stick spectrum similar to a MAS spectrum with rotational sidebands (cf. Fig. 3.3.7(c)).

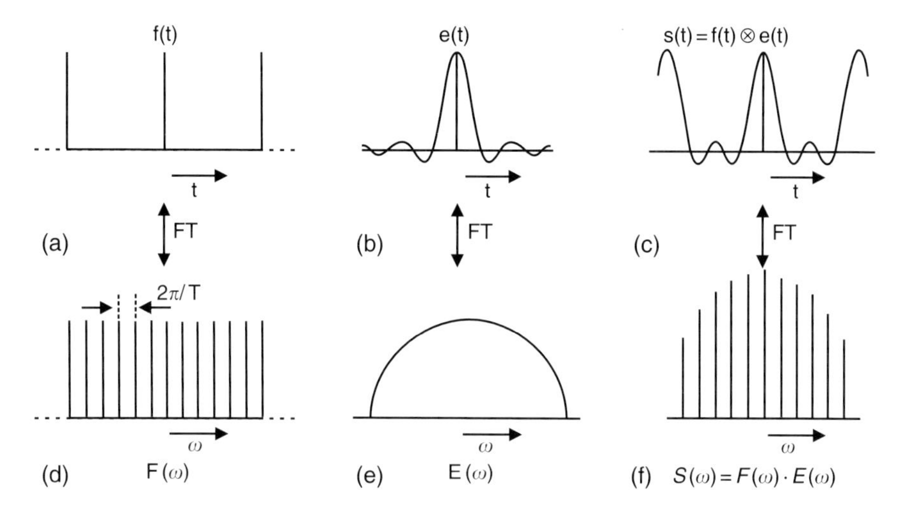

FIG. 6.2.15 Illustration of the Fourier transform of an echo train. (a) Delta comb $f(t)$. (b) Echo signal $e(t)$. (c) Convolution of the delta comb and the echo signal. (d) Fourier transform $F(\omega)$ of the delta comb. (e) Fourier transform $E(\omega)$ of the echo signal. (f) The product of $F(\omega)$ and $E(\omega)$ is the Fourier transform of the convolution of delta comb and echo signal. Adapted from [Mor1] with permission of Oxford University Press.

The echo $e(t)$ describes the evolution of the spin system under the gradient G_y, and its Fourier transform $E(\omega)$ provides a *projection* of the spin density on to the y-direction. The stick spectrum $S(\omega)$ is a discrete representation of this projection, whereby the stick amplitude is not given by the amplitude of the projection at the corresponding frequency, but by the integral over a frequency interval $2\pi/T$, which defines the separation of delta spikes in the frequency domain. Here T is the duration of the signal in the time domain. This appearance is a consequence of the signal periodicity in the time domain, which is completely analogous to MAS NMR. Similar to MAS spectra, the stick-like projection is much more sensitive in terms of signal-to-noise ratio than a continuous projection is.

Apart from relaxation effects, differences among echoes are generated by the phase evolution in the weak gradient G_x. Its magnitude is selected in such a way that $G_x < G_y/n$, where n is the number of pixels in one dimension. The x gradient introduces a decay envelope $h(t)$ of the string of echoes, so that the signal $s'(t)$ acquired in the presence of the x gradient is given by the product of (6.2.16) with $h(t)$,

$$s'(t) = h(t) \int e(t) f(t-\tau)\,\mathrm{d}t = h(t)[e(t) \otimes f(t)]. \qquad (6.2.18)$$

The Fourier transform $S'(\omega)$ of $s'(t)$, is then given by the *convolution* of $S(\omega)$ with the Fourier transform $H(\omega)$ of $h(t)$,

$$S'(\omega) = \int E(\omega')F(\omega')H(\omega-\omega')\,\mathrm{d}\omega'. \qquad (6.2.19)$$

Here $H(\omega)$ provides a projection of the spin density onto the x-axis. The complete image is obtained by simple Fourier transformation of a sequence of gradient echoes and subsequent sorting of line profiles. This procedure is equivalent to derivation of the 2D spectrum in Fig. 3.4.2 from a simple 1D ^{2}H MAS spectrum. It should be noted that even and odd echoes must be Fourier transformed separately because they are acquired in different sum gradients $G_x + G_y$ and $G_x - G_y$ [Yan1].

The $\boldsymbol{k}$-space trajectory of the EPI method described in Fig. 6.2.14(a) is depicted in Fig. 5.4.2(e). Because the gradient G_x is turned on continuously, the different sections of the trajectories are not parallel to the Cartesian coordinate axes. The coverage of $\boldsymbol{k}$ space is improved (cf. Fig. 5.4.2(f)) by pulsing the x gradient (Fig. 6.2.14(b)) or by radial scanning of $\boldsymbol{k}$ space [Sil1]. In either case, the disadvantage is that the decay of echo amplitudes by T_2 relaxation leads to a loss of image quality. Positive and negative k values can be discriminated in y-direction by scanning the gradient echo build-up and decay. By incorporating a gradient echo also with the x gradient, positive and negative values of k can be discriminated in the x-direction as well. Modifications of the basic technique are being explored in NMR microscopy [Man1, Pet1]. Because gradient echoes are used, the technique is sensitive to magnetic field distortions and susceptibility artefacts. Multi-shot variants have been designed to cope with these shortcomings on the expense of measurement speed [Hen4, Hen5].

DANTE-based imaging

Echo-planar imaging, in particular, also the FAST techniques, require fast gradient switching and in turn pose high demands on spectrometer hardware. A less demanding class of techniques exploits refocusing of the response to a DANTE-type excitation in combination with magnetic-field gradients [Cho2, Dor1, Low1].

The basic principle is depicted in Fig. 6.2.16(a) [Cho2]. The rf excitation is a train of small flip-angle pulses forming a DANTE sequence, i.e., the sum of flip angles is 90° (cf. Section 5.3.4). They are applied while the read gradient G_x is tuned on. The response to the DANTE excitation is refocused in a Hahn echo, which itself consists of a train of echoes. This echo train is detected in the presence of the read gradient G_x

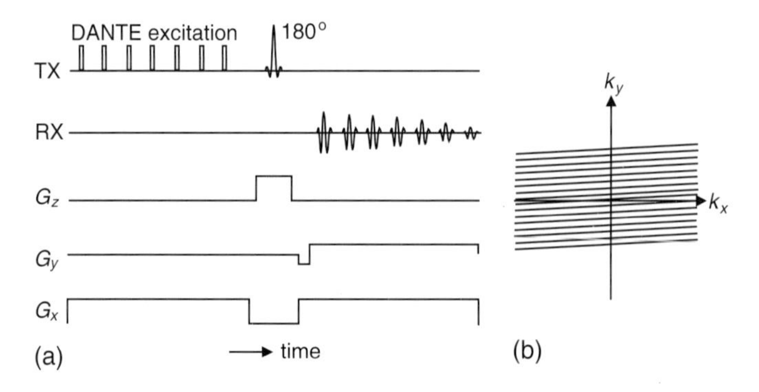

FIG. 6.2.16 [Cho2] 2D DANTE-imaging. (a) Timing of signals. The rf excitation is a DANTE pulse train applied in a read gradient G_x. The Hahn echo of the response is detected while both the phase-encoding gradient G_y and the read gradient G_x are turned on. The slice-selection gradient is G_z. (b) Traces in $\boldsymbol{k}$ space.

as well as in a weak phase-encoding gradient G_y. In this way $\boldsymbol{k}$ space is scanned in a Cartesian but tilted frame (Fig. 6.2.16(b)). The 180° refocusing pulse is slice selective for 2D imaging, but the scheme can readily be extended to three dimensions [Cho2]. A shortcoming of the method is that the DANTE pulse train excites only a magnetization grid in the sample. Consequently, variants of the basic scheme explore better utilization of the available sample magnetization for signal-to-noise improvement [Cho2, Gel1].

6.3 IMAGING IN THE ROTATING FRAME

The magnetic-field gradients required for spatial resolution in imaging do not necessarily have to be applied in the laboratory coordinate frame (LCF) of the $\boldsymbol{B}_0$ field. Spatial resolution can also be achieved by gradients in the $\boldsymbol{B}_1$ field [Hou1]. These gradients oscillate with the frequency ω_{rf} of the rf excitation. In a coordinate frame, rotating with frequency ω_{rf} (RCF) these gradients are time invariant. Methods of imaging based on the use of magnetic-field gradients in the rf field are called *rotating-frame imaging* methods [Can1, Sty1].

In Fourier imaging field gradients have to be switched rapidly. Because the interaction energy of gradients in the RCF with the sample is low, rf gradients can be switched rapidly. Therefore, the use of rotating-frame imaging can be of advantage in imaging in biomedicine as well as in material science [Maf1].

Experimental setup

In contrast to the LCF imaging methods treated above, in rotating-frame imaging the gradients are applied in the transverse plane with the components B_x and B_y of the total magnetic field $\boldsymbol{B} = \boldsymbol{B}_0 + \boldsymbol{B}_1$. The components of interest of the magnetic field-gradient tensor are not $G_{zx} = G_x$ and $G_{zy} = G_y$ as before but rather the components G_{xx} and G_{yy}. They are generated by asymmetric *Helmholtz coils* and *saddle coils* with different numbers of windings in each half or simply by a surface coil. Two coils are needed, one for the x-direction and one for the y-direction of the LCF. For 3D imaging, the z gradient $G_{zz} = G_z$ is identical to the one used in laboratory-frame imaging. The gradients are applied either simultaneously or successively. The magnetization is detected by another rf coil with a homogeneous field, for example, by a saddle coil. A simple coil arrangement for RCF imaging with 1D spatial resolution, suitable for use in a vertical-bore superconducting magnet, is depicted in Fig. 6.3.1 [Can2]. 2D spatial resolution can be achieved by mechanical sample rotation around the z-axis in angular increments for acquisition of projections at different angles and subsequent image *reconstruction from projections*. A scheme for 3D spatial resolution could make use of the laboratory-frame z gradient.

Basic scheme: amplitude modulation

The basic timing sequence for rotating-frame imaging with 2D spatial resolution in y- and z-directions is illustrated in Fig. 6.3.2 [Hou1]. In z-direction the gradient is applied in the LCF. In y-direction it is applied in the RCF. The components of the magnetic field

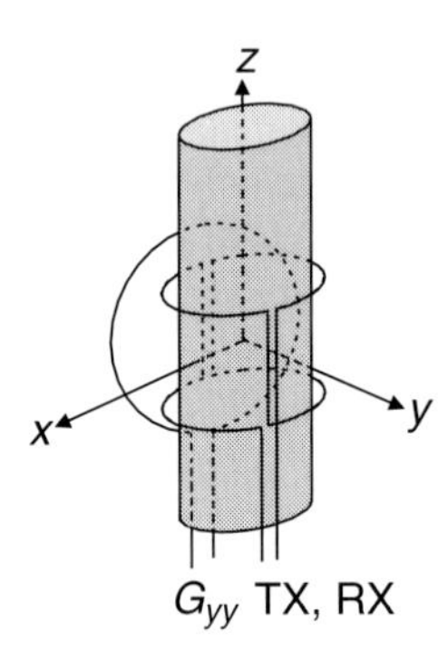

FIG. 6.3.1 [Can2] Coil arrangement for 1D rotating-frame imaging in a vertical-bore superconducting magnet. The magnetic rf field gradient is generated by a surface coil. Homogeneous excitation and detection is achieved with a saddle coil.

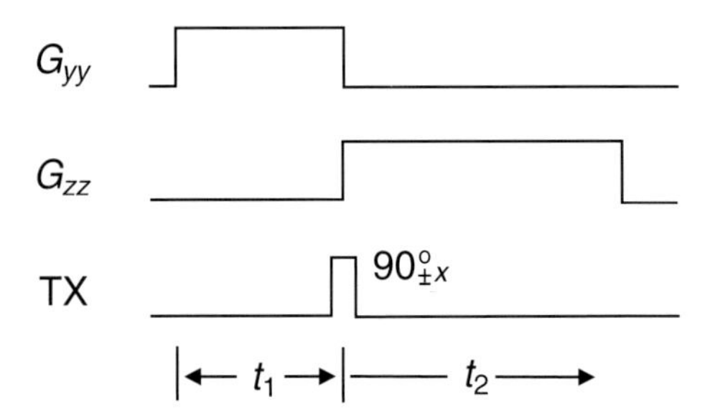

FIG. 6.3.2 [Hou1] Pulse sequence for 2D Fourier imaging in the rotating frame.

in these directions can be written as

$$B_z = B_0 + G_{zz}z, \tag{6.3.1a}$$

$$B_y = B_{1y} + G_{yy}y. \tag{6.3.1b}$$

An rf pulse of length t_1 and average amplitude B_{1y} in the gradient coil rotates the magnetization in the sample by the space-dependent angle

$$\theta = -\gamma(B_{1y} + G_{yy}y)t_1. \tag{6.3.2}$$

The simplest way to obtain a 2D image of the spin density consists of acquiring FID signals for different pulse lengths t_1. The measured signal $s(t_1, t_2)$ can then be expressed as

$$s(t_1, t_2) = M_0(y, z)\sin\{\gamma(B_{1y} + G_{yy}y)t_1\}\exp\{-\mathrm{i}\gamma(B_0 + G_{zz}z)t_2\}. \tag{6.3.3}$$

The image is obtained directly by 2D Fourier transformation over t_1 and t_2, whereby a phase twist can be avoided as a consequence of amplitude modulation. For 1D rotating-frame imaging the field-gradient rf pulse of variable length t_1 is broken down into a series of short pulses for acquisition of one data point within each delay between the pulses (cf. Fig. 6.3.4) [Can1, Met1]. The imaging time can be reduced by flipping the magnetization back after data acquisition with an rf field-gradient pulse in direction opposite to the excitation pulses [Rau1].

Improved scheme: phase modulation

The rotating-frame imaging method of Fig. 6.3.2 is less sensitive by a factor of $2^{1/2}$ with respect to techniques based on *phase modulation*, which also contain the factor $M_0 \cos\theta$. Furthermore, the magnetization has to be completely relaxed back into the z-direction before excitation and data acquisition can be repeated, because after a scan, the starting value for T_1 relaxation of the magnetization is different for different positions within the sample. Both disadvantages can be circumvented if the magnetization, which is located in the xz plane after a y pulse of length t_1, is rotated into the xy plane by a 90°_x pulse (Fig. 6.3.3) [Hou2]. Then the phase of the measured signal is modulated by the spatial information along y, but also the absorptive and dispersive signal contributions are no longer separable, similar to the phase twist in 2D spectroscopy [Ern1] unless two data sets are acquired, which in the case of rotating-frame imaging requires the use of a 90°_x and a 90°_{-x} pulse,

$$s_1(t_1, t_2) = M_0(y, z) \exp\{-\mathrm{i}\gamma(B_{1y} + G_{yy}y)t_1\} \exp\{-\mathrm{i}\gamma(B_0 + G_{zz}z)t_2\}, \qquad (6.3.4a)$$

$$s_2(t_1, t_2) = M_0(y, z) \exp\{+\mathrm{i}\gamma(B_{1y} + G_{yy}y)t_1\} \exp\{-\mathrm{i}\gamma(B_0 + G_{zz}z)t_2\}. \qquad (6.3.4b)$$

The sum and the difference of both signals provide two signals with pure amplitude modulation of the complex signal contribution which depend on the detection time

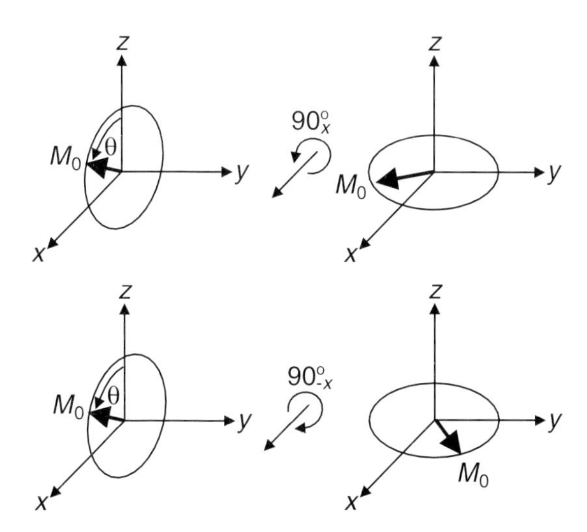

FIG. 6.3.3 [Hou2] Vector diagrams of magnetization illustrating the principle of phase modulation in rotating-frame imaging for a selected volume element. Following an initial y pulse of length t_1 with variable flip angle θ, the magnetization is placed into the xz plane (left). A subsequent $90^\circ_{\pm x}$ pulse rotates the magnetization into the xy plane for acquisition (right). For derivation of a phase-sensitive image two data sets need to be acquired, one with a 90°_x (top) and the other with a 90°_{-x} pulse (bottom).

t_2. The amplitude-modulation factor is given by the sine and the cosine of the signal contribution which depends on the evolution time t_1. In this way phase-sensitive images can be measured. At the same time the relaxation problem has been solved, because the initial position of the magnetization for T_1 relaxation is always in the xy plane, independent of the voxel location in the sample and the evolution time t_1. Now the magnetization in all voxels is equivalent for T_1 relaxation, and the repetition time can be used for adjusting T_1 *contrast* by *saturation recovery* (cf. Section 7.2.1).

Variations of the method

Different types of contrast are readily introduced to rotating-frame imaging by magnetization filters prior to the imaging sequence (cf. Chapter 7). In this case $M_0(y, z)$ in (5.3.3) and (5.3.4) has to be replaced by the weighted spin density $M_z(y, z)$. For *slice-selection* shaped pulses, composite pulses, pulse sequences, and phase cycles have been developed which are effective in B_1 gradients [Bou1, Can1, Hed1, Maf2, Sha1]. Such pulses can also be employed as chemical-shift filters for chemical-shift selective imaging [Bou2, Bow1, Can3, Val1]. For *spectroscopic imaging*, the scheme of Fig. 6.3.2 can be used but with signal detection in the absence of a field gradient (cf.Section 6.2.4) [Cox1]. Because of short space-encoding times, susceptibility effects can be less pronounced than in laboratory-frame imaging [Rau2].

B_1 gradients can also be employed to study translational motion [Dup1, Hum2, Gat2, Kim1, Kim2, Mis1, Mis2, Sim1]. For this purpose, a magnetization grid is generated in the sample by applying rf field-gradient pulses of variable length, and the change of the grid from molecular transport is followed by imaging the grid with the same inhomgeneous B_1 field as that used for establishing the grid [Kim2, Sim1]. The advantage of this approach is that by imaging the grid, variations in the B_1 gradient manifest themselves by variations in the grid spacing, so that nonconstant gradients can be employed, and rf coils can be optimized for maximum gradient strength.

The technique is known as *magnetization-grid rotating-frame imaging* (*MAGROFI*) [Kim1, Kim2]. A refined pulse sequence is depicted in Fig. 6.3.4 [Sim1]. With the rf field-gradient pulses applied during the preparation period, a magnetization grid is established during time t_1'. Diffusion is probed by changing the grid spacing in successive experiments. To this end, the gradient pulse length is incremented in steps Δt_1. During the subsequent delay, Δ the grid is washed out by diffusion. This process is probed by rotating-frame imaging using the same field gradients for space encoding as for formation of the grid. For direct acquisition of the space information, the space-encoding field gradient-pulse of length t_2 is broken down into a series of n small pulses of duration Δt_2 which are separated by detection windows. It may be helpful to adjust the grid spacing before detection to a suitably large value by applying another field-gradient pulse of duration t_1 but with opposite phase. The MAGROFI technique lends itself to investigations of diffusion by the use of *toroid resonators* [Ger1, Rat1], because the radial B_1 field gradient in such a resonator changes with r^{-3}, so that a whole range of grid spacings can be probed in a single shot [Woe1].

In combination with *multi-pulse excitation*, rotating-frame imaging can also be used for *imaging of solids* [Cho1]. A *time-suspension multi-pulse sequence* (cf. Table 3.1) is applied to the sample by use of an rf coil with a homogeneous $\boldsymbol{B}_1$ field so that both

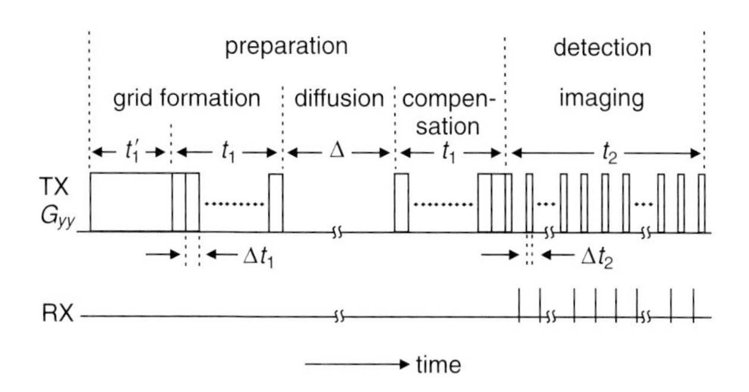

FIG. 6.3.4 [Sim1] Pulse sequence for imaging of diffusion in the rotating frame. Diffusion encoding is achieved by varying a magnetization grid in the rotating frame in a preparation period characterized by t_1 and Δ prior to space encoding. Space is encoded during detection which is characterized by B_1 gradient pulses of length Δt_2. which are interrupted by free-precession periods for signal acquisition.

linear and bilinear spin interactions are made ineffective, and linewidths in the range of a few to 100 Hz are obtained. Synchronized with the time-suspension sequence, a second multi-pulse sequence is applied *via* the $\boldsymbol{B}_1$-gradient coil, which rotates the magnetization by an angle depending on the flip angle of the rf pulses. As a result of the $\boldsymbol{B}_1$ gradient, this rotation angle depends on position, so that space encoding is achieved together with line narrowing in solids.

Applications

Rotating-frame imaging has been applied to investigations of stems with spatial resolution better than 5 μm [Hum1], and to solvent ingress into polymers [Val2, Maf3]. As an example, Fig. 6.3.5 shows slice-selective 2D images of toluene penetrating poly(vinyl chloride) (PVC) and of *n*-pentane in high-density poly(ethylene) (HDPE) together with penetration profiles extracted from cross-sections through the corresponding images [Val2]. The images were acquired after immersing a rod of the polymer material into the solvent for 30 h in each case. Significant differences in the penetration profiles were observed. A steep profile was obtained for PVC, a glassy polymer and a smooth profile for HDPE, a semicrystalline polymer above the glass-transition temperature. These profiles indicate case II diffusion for PVC and case I diffusion for HDPE (cf. Section 10.1.3) [Val2]. The rf gradient of the saddle coil (Fig. 6.3.1) was used to select a vertical slice through the sample. A set of 1D projections was acquired for different sample orientations, and the image was reconstructed from projections. The B_1 gradient pulses had an amplitude of 20 mT/m and a duration of typically 25 μs.

6.4 IMAGING WITH NOISE EXCITATION

Stochastic excitation in NMR denotes excitation of the spin system with noise [Ern3, Kai1], which is, in general, white within the spectral region of the linear response. White means that the average of the power spectral density is independent of frequency. This is automatically fulfilled for a delta pulse. For noise this is valid only if successive values

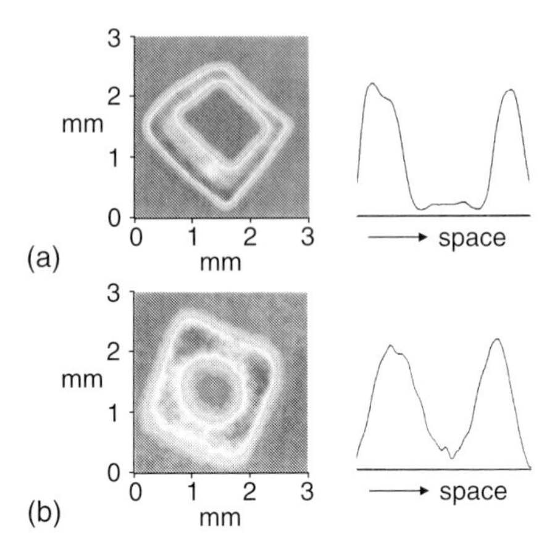

FIG. 6.3.5 Rotating-frame imaging of solvent penetration into rectangular polymer rods. (a) Toluene in PVC. (b) n-heptane in HDPE. Slice-selective 2D images obtained by reconstruction from projections are shown on the left and characteristic penetration profiles extracted from cross-sections through the images on the right. The grey scale is nonuniform. Dark signal next to bright signal inside the rectangular contours of the polymer materials indicates high signal intensity as shown in the penetration profiles. Adapted from [Val2]. Copyright 1995 American Chemical Society.

of the noise signal are linearly independent of each other. For nonlinear stochastic NMR these values have to be nonlinearly independent of each other [Blü1], so that higher order auto-correlation functions are zero except at the origin (cf. Section 4.3). In contrast to pulsed excitation, *noise excitation* is continuous or quasi-continuous. This means, that the excitation energy is not applied to the spin system within a short time t_{p} such as the length of a 90° pulse but over the entire duration of the energy relaxation time T_1. Therefore, the excitation power for noise excitation is smaller by a factor of T_1/t_{p} for noise excitation than for excitation with 90° pulses. A conservative value for T_1/t_{p} is 10^3. This tremendous power advantage is of interest for whole-body imaging of large objects [Hah1], because the excitation power scales with the sample volume, that is, with the third power of the diameter.

Principle

The starting point for a formal description of imaging with noise excitation is the *linear response* (5.4.7) of a spin system in arbitrarily time-dependent magnetic-field gradients with arbitrary rf excitation $x(t)$. Note that $x(t)$ denotes the excitation signal in agreement with the nomenclature established in nonlinear system analysis [Mar2, Sch1, Vol1, Wie1]. To distinguish the excitation $x(t)$ from the space coordinate x, the time argument is always carried along with the excitation.

The spin system contains four input signals. These are the rf excitation $x(t)$ and the three gradient modulation signals G_x, G_y, and G_z. The four input signals can be combined into one excitation function

$$p_1(\mathbf{r}, t, \tau) = \exp\{\mathrm{i}\mathbf{k}(t, t-\tau)\mathbf{r}\} f(t) x(t-\tau), \tag{6.4.1}$$

where the gradient wave vector $\mathbf{k}(t, t-\sigma)$ is defined in (5.4.4), and $f(t)$ is a function suitably chosen for data evaluation by cross-correlation (see below). For calculation of the system response (5.4.7) $f(t)$ is unity. By considering (6.4.1) as the excitation signal in imaging with noise excitation, the analysis is reduced from that of a four-input linear system to that of a one-input linear system.

Cross-correlation of the system response $y_1(t)$ defined in (5.4.7) with $p_1^*(\mathbf{r}', t, \rho)$ over a finite time T produces the correlation function

$$c(\mathbf{r}', \sigma) = \frac{1}{T} \int_0^T y_1(t) p_1^*(\mathbf{r}', t, \sigma) \, \mathrm{d}t, \tag{6.4.2}$$

which not only contains the time displacement σ as a cross-correlation parameter (cf. Section 4.3) but also the space coordinates x', y', and z' (Fig. 6.4.1) [Blü2]. By forming an ensemble average $\langle \cdots \rangle$ of $x(t-\tau)x(t-\sigma)$ in the evaluation of (6.4.2) the *systematic noise* from finite correlation times T is eliminated [Roo1]. For normalized white noise, the ensemble average is given by a delta function $\langle x(t-\tau)x(t-\sigma) \rangle = \delta(\tau - \sigma)$, so that (6.4.2) reduces to

$$c'(\mathbf{r}', \sigma) = \langle c(\mathbf{r}', \sigma) \rangle = \int \int M_0(\mathbf{r}, \Omega_\mathrm{L}) b_1(\mathbf{r} - \mathbf{r}', \Omega_\mathrm{L}, \sigma) \, \mathrm{d}\Omega_\mathrm{L} \, \mathrm{d}\mathbf{r}, \tag{6.4.3}$$

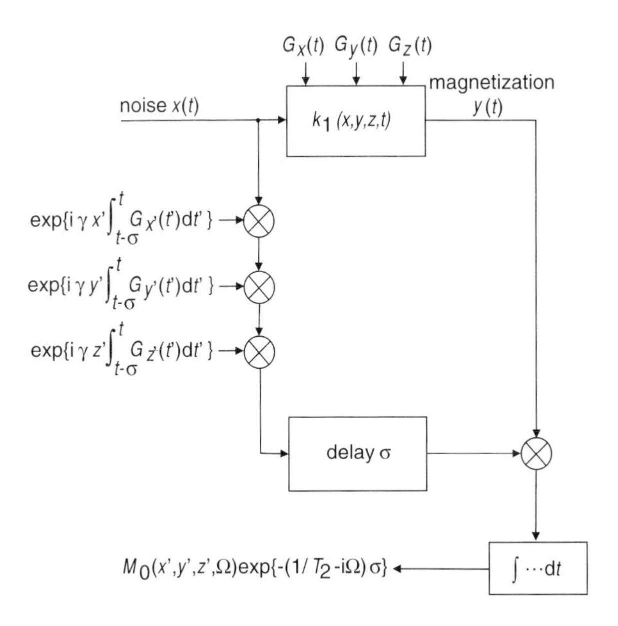

FIG. 6.4.1 [Blü1] In imaging with noise excitation the spin density convolved by a localization function can be retrieved by linear cross-correlation of the system response with a function of the space coordinate $\mathbf{r}' = (x', y', z')$ and the time delay σ. Fourier transformation over σ introduces spectroscopic resolution.

where

$$b_1(\boldsymbol{r}, \Omega_L, \sigma) = -i\frac{1}{T}\int_0^T \exp\{i\boldsymbol{k}(t, t-\sigma)\boldsymbol{r}\} f^*(t)\, dt\, \exp\left\{\left(-\frac{1}{T_2} - i\Omega_L\right)\sigma\right\} \quad (6.4.4)$$

is the *localization function* or *point-spread function* which determines the resolution of the spin density $M_0(\boldsymbol{r}, \Omega_L)$ in the image $c'(\boldsymbol{r}', \sigma)$ [Roo1].

In special cases, the localization function is proportional to a product of delta functions of the space coordinates [Blü2, Blü3]. When the chemical shift Ω_L is neglected, the cross-correlation function (6.4.3) represents the spin density $M_0(\boldsymbol{r}')$ weighted by $\exp\{-\sigma/T_2(\boldsymbol{r})\}$. This case applies in good approximation to imaging of the water signal in biological tissue. The general case, however, is described by introduction of the localization function $b_1(\boldsymbol{r}, \Omega_L, \sigma)$. The quality of spatial resolution can be adjusted by suitable choice of the function $f(t)$ during cross-correlation [Roo1]. The achievable spatial resolution depends on the kind of gradient modulation. For stochastic rf excitation, *stochastic gradient modulation* [Blü2] and *oscillating gradients* have been discussed [Blü3, Roo1]. The signal-to-noise ratio can be improved if images computed for different delays σ are added. On the other hand, the value of σ determines the T_2 *contrast*. In this context, it is interesting to note that the T_2 weight is determined during data evaluation and not during data acquisition. Also, from one set of acquired data overview images can be computed with reduced resolution as well as high-resolution images. On the other hand, T_1 *contrast* is determined during the experiment by adjusting the rf power for partial saturation of the spin response.

1D spectroscopic resolution

Spectroscopic resolution is obtained by systematic variation of σ in the cross-correlation function (6.4.3) and subsequent Fourier transformation over σ. In this way, resonance frequency Ω_L and T_2 relaxation time or spectroscopic lineshape become part of a localization function $b_1'(\boldsymbol{r}-\boldsymbol{r}', \Omega_L - \Omega_L')$, which now determines not only the spatial but also the spectroscopic resolution,

$$c''(\boldsymbol{r}', \Omega_L') = \int\int M_0(\boldsymbol{r}, \Omega_L) b_1'(\boldsymbol{r}-\boldsymbol{r}', \Omega_L - \Omega_L')\, d\Omega_L\, d\boldsymbol{r}. \quad (6.4.5)$$

The validity of (6.4.5) has been verified experimentally by a spectroscopic 1D image of a phantom from water and oil (Fig. 6.4.2) [Roo1]. The image was computed from 64 k of magnetization data acquired in an oscillating gradient in response to binary noise excitation. The gradient strength was 4 mT/m, and the modulation frequency was in the range of 1–5 kHz. The data acquisition rate was less than 100 kHz. When using oscillating gradients it is important that the oscillation frequency ω_G is larger than the signal bandwidth. For example, for an object with a diameter of 10 cm in a 4 T magnetic field, and for gradient amplitudes of 10 mT/m diameter, a spatial resolution of 0.9 cm can be achieved for ^1H with $\omega_G/2\pi = 1.7$ kHz, but only 9.3 cm can be obtained for ^{13}C at $\omega_G/2\pi = 4.5$ kHz [Roo1].

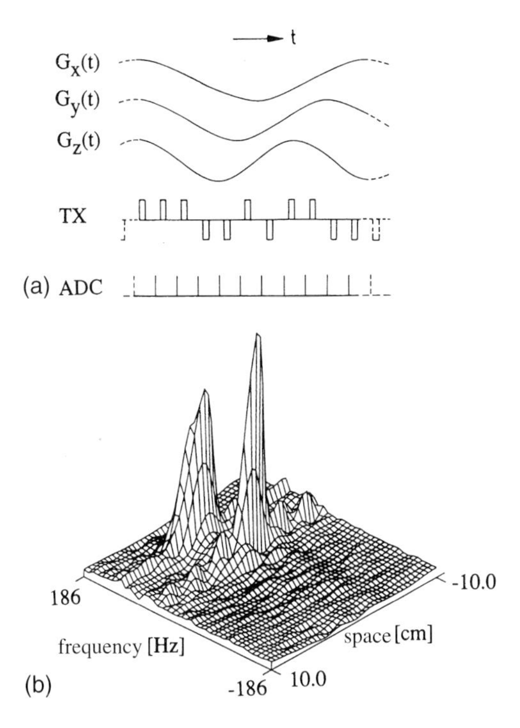

FIG. 6.4.2 (a) Timing diagram for imaging with noise excitation and oscillating gradients. The excitation consists of an uninterrupted string of rf pulses which are modulated randomly in phase and amplitude. Before each pulse a magnetization value is acquired [Blü3]. (b) Spectroscopic 1D image of a 10 cm large phantom from oil and water. The image has been computed by cross-correlation of 64 k data of magnetization acquired in response to binary noise excitation [Roo1].

2D spectroscopic resolution

For computation of 2D spectroscopic resolution at coordinates $\mathbf{r}'$ with stochastic rf excitation the *nonlinear response* (5.4.10) is cross-correlated with the function $p_3^*(\mathbf{r}', t, \sigma_1 > \sigma_2 > \sigma_3)$, which depends on three time variables σ_1, σ_2, and σ_3 as well as on the wave vectors in three successive time intervals [Blü4],

$$p_3(\mathbf{r}, t, \sigma_1 > \sigma_2 > \sigma_3) = \exp\{\mathrm{i}\mathbf{k}(t - \sigma_2, t - \sigma_1)\mathbf{r}\} \exp\{\mathrm{i}p_{\mathrm{LM}}\mathbf{k}(t - \sigma_3, t - \sigma_2)\mathbf{r}\} \\ \times \exp\{\mathrm{i}\mathbf{k}(t, t - \sigma_3)\mathbf{r}\} x(t - \sigma_1) x(t - \sigma_2) x(t - \sigma_3), \quad (6.4.6)$$

where $|p_{\mathrm{LM}}| = 0$ and 2 for zero- and double-quantum coherences, respectively (cf. eqn (5.4.10)). As a result one obtains a cross-correlation function which depends on position as well as on three time coordinates,

$$c_3(\mathbf{r}', \sigma_1 > \sigma_2 > \sigma_3) = \int\int\int M_0(\mathbf{r}, \Omega_{\mathrm{L}}, \Omega_{\mathrm{M}}, \Omega_{\mathrm{N}}) \\ \times b_3(\mathbf{r} - \mathbf{r}', \sigma_1 > \sigma_2 > \sigma_3)\, \mathrm{d}\Omega_{\mathrm{L}}\, \mathrm{d}\Omega_{\mathrm{M}}\, \mathrm{d}\Omega_{\mathrm{N}}\, \mathrm{d}\mathbf{r}, \quad (6.4.7)$$

where b_3 is the third-order localization function,

$$\begin{aligned} &b_3(\boldsymbol{r}-\boldsymbol{r}', \sigma_1 > \sigma_2 > \sigma_3) \\ &= \mathrm{i}\frac{1}{T}\int_0^T \exp\left\{-\left[\frac{1}{T_2} - \mathrm{i}\Omega_\mathrm{L}\right](\sigma_1-\sigma_2) + \mathrm{i}\boldsymbol{k}(t-\sigma_2, t-\sigma_1)\boldsymbol{r}\right\} \\ &\times \exp\left\{-\left[\frac{1}{T_\mathrm{LM}} - \mathrm{i}(\Omega_\mathrm{L}+\Omega_\mathrm{M})\right](\sigma_2-\sigma_3) + \mathrm{i}p_\mathrm{LM}\boldsymbol{k}(t-\sigma_3, t-\sigma_2)\boldsymbol{r}\right\} \\ &\times \exp\left\{-\left[\frac{1}{T_2} - \mathrm{i}(\Omega_\mathrm{L}+\Omega_\mathrm{M}+\Omega_\mathrm{N})\right]\sigma_3 + \mathrm{i}\boldsymbol{k}(t, t-\sigma_3)\boldsymbol{r}\right\} \mathrm{d}t. \end{aligned} \tag{6.4.8}$$

In comparison with the 1D localization function (6.4.4), no processing function $f(t)$ has been introduced for simplicity. Variation of the space vector $\boldsymbol{r}'$ in (6.4.7) and of the time delays σ_1, σ_2, and σ_3 results in a spatially resolved 3D interferogram similar to a spatially resolved response to three generic rf excitation pulses. Depending on the choice of time delays and variables for Fourier transformation, different 1D, 2D, and 3D pulse experiments can be mimicked (Fig. 6.4.3) [Blü1, Blü4, Paf1].

Reconstruction from projections

Numerical execution of cross-correlations on a conventional computer is demanding in time. The processing time could be significantly reduced by use of dedicated processors for parallel computing, but computation in the time domain can be avoided altogether if time-invariant field gradients are employed. Projections of the spin density are obtained by Fourier transformation of the 1D (4.3.2) and 3D (eqn (4.3.3) with $n = 3$) cross-correlation functions. With the *correlation theorem* (cf. Section 4.3.3),

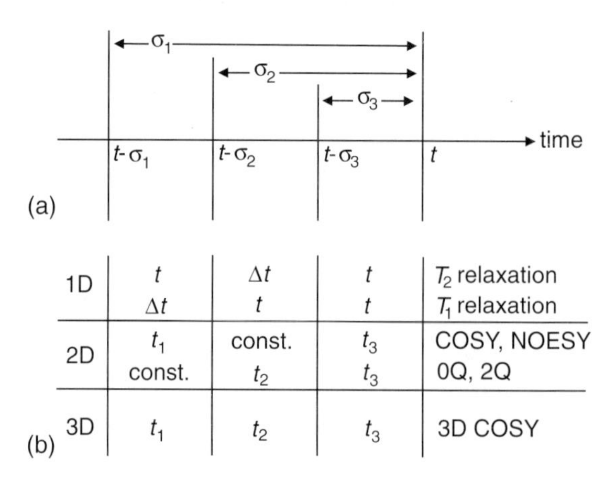

FIG. 6.4.3 [Blü4] Interpretation and choice of time delays in the third-order cross-correlation function. (a) The time dependence of the cross-correlation function corresponds to that of a response to three rf pulses. (b) Depending on the choice of the time delays different 1D, 2D, and 3D pulse experiments can be mimicked.

these computations can conveniently be executed in the frequency domain. By use of (4.3.5) and (4.3.7) the Fourier transforms of the first- and third-order impulse response functions, eqns (5.4.8) and (5.4.10), respectively, can be computed:

$$H_1(\omega) = \frac{\langle Y(\omega)X^*(\omega)\rangle}{\langle |X(\omega)|^2\rangle} \tag{6.4.9}$$

and

$$H_3(\omega_1, \omega_2, \omega_3) = \frac{\langle Y_3(\omega_1 + \omega_2 + \omega_3)X^*(\omega_1)X^*(\omega_2)X^*(\omega_3)\rangle}{\langle |X(\omega_1)|^2|X(\omega_2)|^2|X(\omega_3)|^2\rangle}, \tag{6.4.10}$$

where $Y_3(\omega) = Y(\omega) - H_1(\omega)X(\omega)$. $X(\omega)$ and $Y(\omega)$ are the spectra of the stochastic excitation and the corresponding response (cf. eqns (4.2.4), (5.4.7), and (5.4.9)). Here the Fourier transforms of the impulse response functions are given the symbol H_n because cross-correlation of a nonlinear response introduces saturation features in the spectra similar to those observed in CW NMR as opposed to the K_n obtained by Fourier transformation of pulsed excitation [Blü1]. An ensemble average $\langle \cdots \rangle$ over many products of sample spectra needs to be formed in order to reduce systematic noise (cf. Section 4.3.3).

Images of the spin density can readily be generated from (6.4.9) through *reconstruction from projections* (cf. Section 6.1), if $H_1(\omega)$ is obtained for different orientations of the gradient vector $\boldsymbol{G}$. An example is given in Fig. 6.4.4 by an image of four sample tubes filled with water. The image has been calculated by reconstruction from 64 projections. Each projection was computed from 16 k of response data and 16 k of excitation data. The acquisition time was below 10 min. By variation of the gradient strength, images can be generated with spectroscopic resolution (cf. Section 4.1) [Jan1]. For 2D spectroscopic resolution the algorithm has to be adapted to two spectroscopic dimensions and applied to (6.4.10), for example, with $\omega_1 = -\omega_2$ [Blü4].

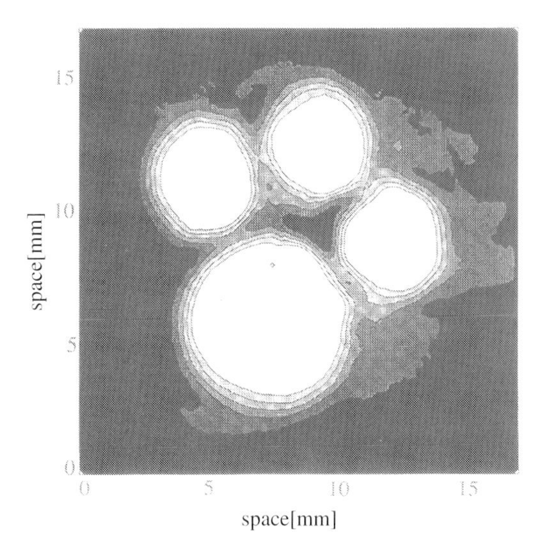

FIG. 6.4.4 [Jan1] Stochastic 2D ^{1}H NMR image of four sample tubes filled with water.

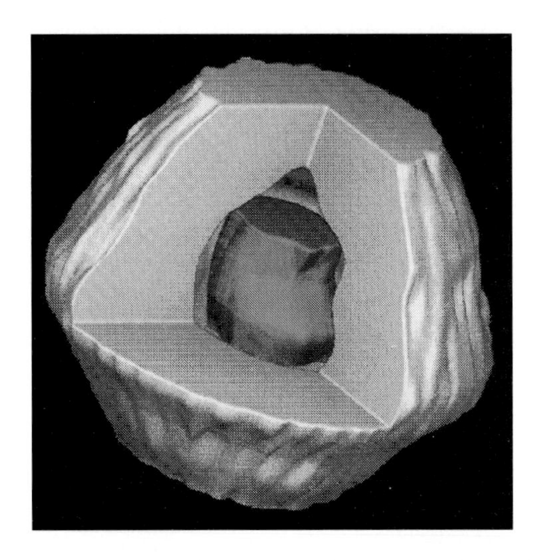

FIG. 6.4.5 [Blü6] Surface-rendered 3D image of an olive computed from 12 slice-selective Hadamard images. The mean excitation flip angle was 1.1°. The slice thickness is 1.6 mm, and the field of view is $4 \times 4\,\text{cm}^2$.

Hadamard imaging with slice selection

One disadvantage of the use of random noise excitation in spectroscopy and imaging is the unavoidable presence of systematic noise as a consequence of the variance of the power spectral density of the excitation [Blü5]. It is reduced but not eliminated by formation of the ensemble averages in (6.4.9) and (6.4.10). Spectroscopy and imaging without systematic noise can be conducted with *pseudo-random excitation*, for example, with excitation by *maximum-length binary sequences* also called *m sequences* (cf. Section 4.4.5) [Kai2, Zie1]. In this case, linear cross-correlation of the response with the excitation corresponds to *Hadamard transformation* of the response. The Hadamard transform of the *linear response* to excitation by *m* sequences is an interferogram free of systematic noise. The advantage of low excitation power also applies to this approach, but also the excitation power must be kept low to ensure a linear response, so that T_1 contrast by partial saturation cannot be easily introduced. The use of excitation with *m* sequences in time-invariant field gradients together with Hadamard transformation and image reconstruction from projections is known as *Hadamard imaging* [Ber1, Blü6, Blü8, Nil1, Nil2, Nil4]. In practice, phase cycling is employed for reduction of instrumental artefacts.

The theoretical treatment of noise excitation demands a dynamic equilibrium of excitation and response. For the linear response this equilibrium is attained with the time constant T_2. For the nonlinear response, the time constant is T_1. When changing the gradient orientation a new equilibrium state has to be assumed, but for small changes the deviation from dynamic equilibrium is small, so that data acquisition can continue. The same reasoning applies to a short break in the *m*-sequence excitation [Gre1]. Such a break can be used for *slice selection* through application of a soft pulse. Most soft pulses, however, are generated with large signal amplitudes in the time domain, so that the rf power advantage of noise excitation is sacrificed unless saturation pulses with randomized phases for different frequency components are used. For example, *SPREAD* pulses can be employed (cf. Fig. 5.4.9) [Nil3], by which the signal from volume elements outside the selected slice can efficiently be suppressed with low excitation power. The same approach can be used for generation of T_1 contrast. Then m-sequence excitation and saturation pulses are interleaved for slice-selective Hadamard imaging.

The validity of this approach has been verified by multi-slice acquisition of a 3D image of an olive (Fig. 6.4.5) [Blü6]. The surface-rendered 3D image has been constructed from 12 slice-selective 2D images, each of which was constructed from 64 projection. Typical imaging times for one slice are 1–2 min. The excitation power was below 1 W for the m-sequence excitation and in the order of 5 W for the slice-selective low-power pulse. In the image, signals are observed from the flesh of the olive as well as from the interior of the pit. Acquisition times are comparable to those of the *FLASH* technique (cf. Section 6.2.7), because only the linear response is excited, and recycle times are not governed by T_1.

7

Contrast

Compared to other imaging methods, the unique feature of NMR imaging is the abundance of parameters that can be exploited for image *contrast* [Blü1, Blü4, Blü6, Blü10, Blü11, Cal1, Cal2, Cor4, Haa1, Kim2, Xia1]. These parameters are mostly molecular in nature and are linked to the chemical and physical properties of the sample. Examples of *molecular chemical parameters* include the chemical shift and the indirect spin–spin coupling. *Molecular physical parameters* are lineshapes, relaxation times, the self-diffusion coefficient, and the strength of the dipole–dipole interaction. The last of these is the fundamental quantity by which distances can be probed either on a molecular level by the dipole–dipole coupling tensor or on a mesoscopic level by spin diffusion. *Nonmolecular contrast parameters* are the spin density, differences in magnetic susceptibility and local magnetic fields inscribed by electric currents in the sample. Parameters of molecular translational motion by diffusion and flow can be probed to elucidate macroscopic heterogeneities. Furthermore, the image contrast can be manipulated by adding contrast agents to the sample and by imaging different nuclear species.

However, these NMR parameters are useful only if they serve to detect inhomogeneities that are invisible to other, less expensive imaging techniques. For interpretation of NMR images, it is necessary to link these parameters to *material properties*. This is a general topic of NMR in materials science and is not particular to imaging itself. Material properties can be divided into those that describe a *state of matter* and those that describe the *change of matter*. The state parameters include the correlation times of molecular motion, the degree of molecular orientation, the shear modulus, the viscosity, the cross-link density of elastomers, the distribution and agglomeration of filler particles, the pore-size distribution, the temperature distribution, and the thermal conductivity. Examples of parameters attributed to the change of matter are (1) the characteristic times of physical ageing which describe phase transitions in microcrystals; (2) kinetic constants of chemical reactions describing chemical ageing by thermo-oxidative processes, chain scission in polymers upon heating, the cross-linking kinetics during rubber vulcanization, and curing times for thermosetting polymer materials; and (3) the progress of fluid ingress into polymers.

7.1 IMAGE CONTRAST

The *contrast* $\Delta M_z/|M_{z,\max}|$ is defined as the relative difference in image intensities M of neighbouring structures i and j [Man1],

$$\frac{\Delta M_z}{|M_{z,\max}|} = \frac{M_z(\boldsymbol{r}_i) - M_z(\boldsymbol{r}_j)}{|M_{z,\max}|}, \tag{7.1.1}$$

where $M_{z,\max}$ is the maximum of $M_z(\boldsymbol{r}_i)$ and $M_z(\boldsymbol{r}_j)$, and $\boldsymbol{r}_i$ and $\boldsymbol{r}_j$ are the space coordinates of the pixels i and j under consideration.

7.1.1 Optimization of contrast

To exploit the manifold of contrast features, it is useful to optimize the imaging experiment for the generation of maximum contrast. A generic scheme applicable to Fourier imaging is illustrated in Fig. 7.1.1. Three time periods are distinguished, which are reminiscent of 2D NMR [Ern1]: a *filter preparation period, a space-encoding period* and a *spectroscopic detection period*. During the filter period, the initial magnetization to be imaged is manipulated by rf excitation, so that only part of the magnetization survives. This may be, for instance, the magnetization of molecules with high molecular mobility or with weak dipole–dipole interactions between abundant spins. Many magnetization filters do not require homogeneous magnetic fields, so gradients need not to be turned off during the filter preparation period. During the space-encoding period the gradients $\boldsymbol{G}$ are on for slice selection and space encoding. At this stage the spatial resolution is determined. During the spectroscopy period a complete FID is acquired from the magnetization of the molecules selected in the filter period with the spatial resolution inscribed during the space-encoding period. The FID transforms into a spectrum for further analysis of chemical shifts, lineshapes, moments of the lineshape, and other parameters. For image contrast other than through the spin density, the space-encoding

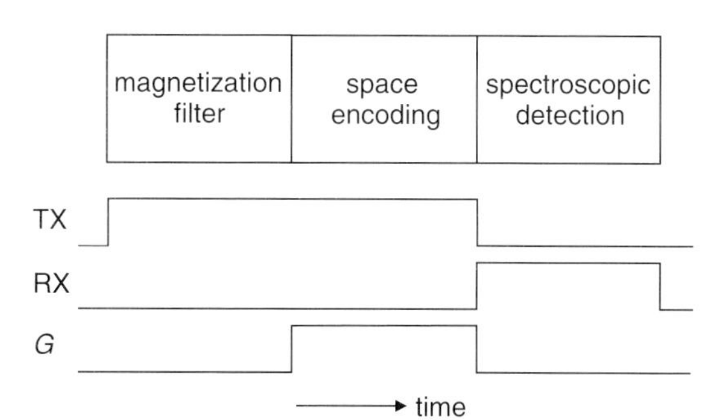

FIG. 7.1.1 Generation of contrast: the initial magnetization is filtered by rf excitation (TX) before space encoding by gradients $\boldsymbol{G}$. In the absence of gradients the free-induction decay is detected (RX) for the extraction of spectroscopic information.

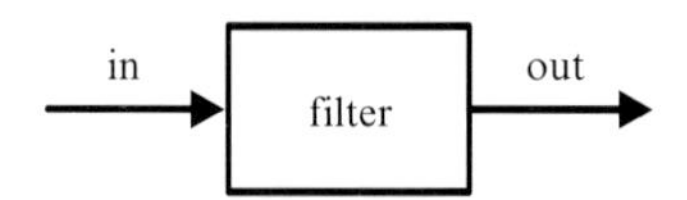

FIG. 7.1.2 A magnetization filter blocks part of the incoming magnetization from passing through the filter.

period needs to be combined with either a magnetization filter, or spectroscopic data acquisition, or both.

7.1.2 Magnetization filters: parameter weights

If thermodynamic equilibrium magnetization is mapped by NMR imaging, the image displays the *spin-density distribution* $M_0(x, y, z) = M_0(\mathbf{r})$ of the nucleus under investigation. For abundant nuclei, this distribution is often equivalent to the density of the material. In this case, the same information can be acquired readily by other methods such as X-ray tomography. The advantages of NMR as a tool for imaging come into play when the spin density is weighted by some function of NMR parameters, or when an NMR spectrum is measured for each voxel.

Parameter weights are introduced through the use of *magnetization filters* (Fig. 7.1.2). A filter blocks part of the incoming signal from passing through. As a result, there is a gain in selectivity of the output signal at the expense of a deterioration of the signal-to-noise ratio. The selectivity obtained in this way is expressed by a parameter weight $W(\mathbf{p}, \mathbf{r})$ of the spin density, which depends on the vector $\mathbf{p}$ of parameters and on the location $\mathbf{r}$,

$$M_z(\mathbf{p}, \mathbf{r}) = W(\mathbf{p}, \mathbf{r}) M_0(\mathbf{r}). \tag{7.1.2}$$

The mapping of filtered magnetization instead of thermodynamic equilibrium magnetization is a powerful concept [Blü1, Haa1] for generation of image contrast. The most notable of these filters is the T_1 *filter*, which can be realized, for instance, by an aperiodic saturation sequence (cf. Section 7.2.1) or by simply reducing the repetition time between successive scans in the experiment. This leads to partial saturation of the magnetization components with long T_1 and suppression of the corresponding signals in the image. Another well-known filter is the T_2 *filter* (Section 7.2.2) which consists of Hahn echoes. This filter is usually integrated into the space-encoding sequence. Here the magnetization components with short T_2 relaxation times are suppressed in the filter-output signal. Because both relaxation times are determined by molecular motion in different frequency windows, the filtering effect of each of these filters is different.

7.1.3 Transfer functions and mobility filters

The effect a filter exerts on a signal is represented by the *filter transfer function*. For linear systems, the filter output is the product of the transfer function and the filter input (cf. eqn (4.2.14), where $K_1(\omega)$ is the transfer function). The output signal is high for those parameters for which the transfer function is high and *vice versa*. For a qualitative

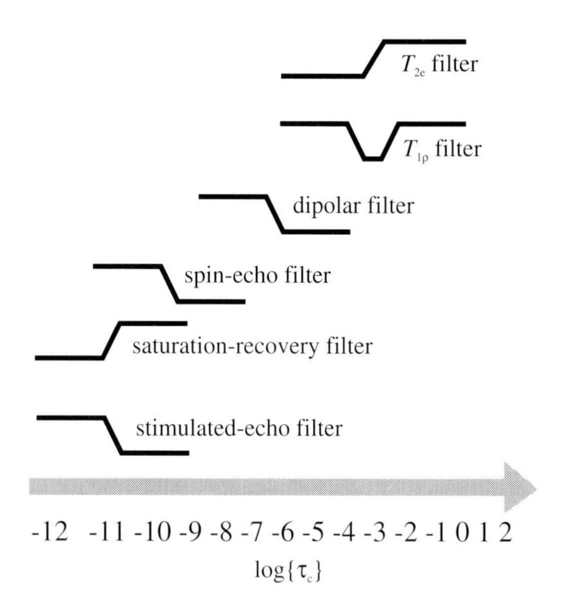

FIG. 7.1.3 [Blü2] NMR-timescale of molecular motion and filter transfer functions of pulse sequences which can be utilized for selecting magnetization according to the timescale of molecular motion. The concept of transfer functions provides an approximative description of the filters. A more detailed description needs to take into account magnetic-field dependences and spectral densities of motion. The transfer functions shown for the saturation recovery and the stimulated-echo filter apply in the fast motion regime.

understanding of the way magnetization filters work, it is helpful to sketch the transfer function. Examples are given in Fig. 7.1.3 [Blü2] with transfer functions for filters, which select magnetization based on the *timescale of molecular motion*. The parameter for discrimination is the characteristic frequency or the *correlation time* τ_c *of molecular motion*.

Depending on the timescale, different pulse sequences can be used. T_1 filters are applicable in the fast and in the slow motion regime. Both the saturation recovery and the stimulated echo can be used for this purpose, the transfer functions of both filters being complementary. The T_2 filter is given by the Hahn or spin echo. Its transfer function is similar to that of the stimulated-echo filter in the fast motion regime, but because T_2 is affected by slower motional processes, the filter cut-off is shifted to longer correlation times. The *dipolar filter* exploits the strength of the dipole–dipole interaction which, in addition to internuclear distances, is determined by molecular reorientation. The $T_{1\rho}$ *filter* is efficient at rather slow motion, because it affects the relaxation at the frequency γB_1 of the effective field in the rotating frame. A similar effect is achieved by the T_{2e} *filter*, which is based on the generation of a solid-echo train and therefore again on the strength of the homonuclear dipole–dipole interaction.

Although the timescale and the principle of *mobility filters* are readily outlined in this way, the validity of the representation is limited, because relaxation times depend on the *spectral densities of molecular motion* at more than one frequency, which is neglected in Fig. 7.1.3. In addition, the spectral densities relevant to NMR of condensed matter

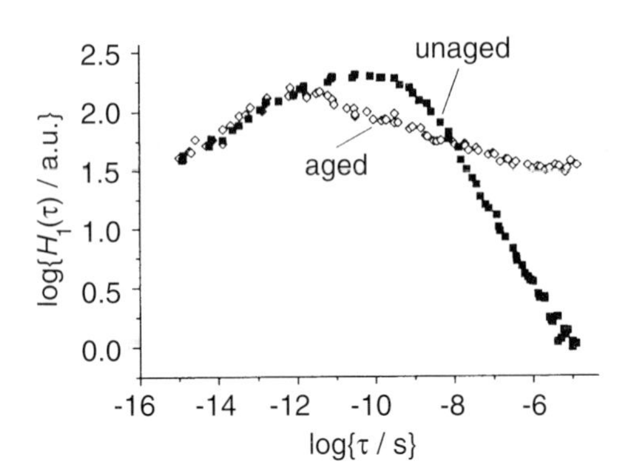

FIG. 7.1.4 Mechanical relaxation strength H_l as a function of the mechanical relaxation time τ for SBR. Filled circles: unaged material. Open circles: material aged in air at 180 °C for 24 h. The timescale is shifted towards shorter times, because the curves are referenced to 25 °C and not to the glass transition temperature. Adapted from [Fül1] with permission from Hüthig Gmbh.

are mainly those referring to fluctuations of magnetic dipolar fields. Thus, they are similar but not identical to the spectral densities of molecular motion relevant for the mechanical properties of materials. The mechanical spectral densities are measured by *dynamic mechanical relaxation spectroscopy* [War1]. An example is shown in Fig. 7.1.4 [Fül1] by a double-logarithmic plot of the mechanical relaxation strengths $H_\mathrm{l}(\tau)$ for two carbon-black filled styrene–butadiene rubber (SBR) samples as a function of the *mechanical relaxation time* τ. This relaxation time corresponds to the *correlation time* τ_c *of molecular motion*. One SBR sample is unaged, the other has experienced extensive thermal–oxidative *ageing*. The relaxation strength for the aged material is lower in the intermediate-motion regime corresponding to T_1 processes and is significantly enhanced in the slow motion regime, which is probed by $T_{1\rho}$. The same trend is, indeed, observed by NMR-relaxation measurements, confirming the close relationship between NMR and mechanical relaxation [Fül1].

7.1.4 Parameter contrast

A magnetization filter consists of modulated rf excitation, for example, from a sequence of nonselective rf pulses with given flip angles and given pulse separations. These parameters of the pulse sequence are adjustable, and they determine the characteristics of the filter, that is, they can be used to tune the filter transfer function. Therefore, they are referred to as *extrinsic contrast parameters*. They must be discriminated from the *intrinsic contrast parameters*, which are the NMR parameters specific of the sample under investigation and are related to the material properties [Man1]. Therefore, the parameter vector $\boldsymbol{p}$ in the weight factor of eqn (7.1.2) is separated into two parts, a vector $\boldsymbol{p}_\mathrm{e}$ of extrinsic parameters and a vector $\boldsymbol{p}_\mathrm{i}(\boldsymbol{r})$ of intrinsic parameters,

$$W(\boldsymbol{p}, \boldsymbol{r}) = W(\boldsymbol{p}_\mathrm{i}(\boldsymbol{r}), \boldsymbol{p}_\mathrm{e}). \qquad (7.1.3)$$

Only the intrinsic parameters depend on position $\boldsymbol{r}$ within the sample.

Systematic variation of the filter pulse sequence by variation of an extrinsic contrast parameter leads to a set of NMR images, where the signal intensity in each image of the set is scaled by a different weight as a function of $\boldsymbol{p}_e$. For each position $\boldsymbol{r}$ in space, the functional dependence of the weight on a given intrinsic parameter can often be approximated by simple expressions, such as an exponential, a Gaussian, or a sum of such functions. By fitting the signal amplitudes of successive images of the set to such an expression, the intrinsic NMR parameters can be obtained as well as the amplitudes of their weight functions.

This procedure is illustrated in Fig. 7.1.5 [Blü1] for the simple case of *spin-lock* (cf. Section 3.3.1) filtered projections of a composite piece of rubber cut from the tread of a car tyre. Such a filter imposes a $T_{1\rho}$ *weight* on the spin density (b). The extrinsic contrast parameter is the duration t_f of the lock field B_1. As it is varied from 0.1 to 10 ms, the $T_{1\rho}$ weight increases and the signal amplitude decreases (a). If $|-\gamma B_1|$ is larger than the frequency offset in the rotating frame, then for each position along the space axis $\boldsymbol{r}$ the signal amplitude decays in an exponential fashion with t_f according to

$$M_z(\boldsymbol{p};\boldsymbol{r}) = W(T_1(\boldsymbol{r}), t_f)M_0(\boldsymbol{r}) = \exp\left\{-\frac{t_f}{T_{1\rho}(\boldsymbol{r})}\right\} M_0(\boldsymbol{r}), \tag{7.1.4}$$

where the decay-time constant is the relaxation time $T_{1\rho}$ in the rotating frame. The collection of decay-time constants as a function of space provides a *parameter image* of the object, where here and in the following parameter means intrinsic contrast parameter in this context. The parameter image obtained for the spin-lock filter is a $T_{1\rho}$ image (c). The amplitudes of the exponential functions correspond to the projection for vanishing filter time t_f and constitute the spin-density image. Other intrinsic parameters can be extracted using other filters.

An example of how much NMR parameter images can differ from photographic images is given in Fig. 7.1.6 [Fül1, Wei4]. A photographic image and a T_2 parameter image of a 23 mm × 31 mm large section of a laminated window pane are compared. The pane consists of two glass sheets separated by a poly(acrylate) layer. For the investigated section, the layer shows ablation from one of the glass sheets, which looks like scratches or streaks in visible light (a). The T_2 *image* (b) looks very different. A slight change of T_2 is observed across the sample and a localized steep change of T_2 somewhat off centre in the upper right-hand section. This points to a significant change in molecular mobility, which indicates a faulty preparation of the poly(acrylate) layer which could be the origin of the ablation. This demonstrates that NMR parameter images give access to novel contrast mechanisms, by which previously invisible inhomogeneities can be localized.

Depending on the NMR parameter chosen, the image contrast can be quite different. The identification of material heterogeneities based on differences in molecular motion is a particular strength of NMR. The importance of slow molecular motion for image contrast is demonstrated in Fig. 7.1.7 with *relaxation-time parameter images* from an *ageing study* of carbon-black filled SBR sheets. The 1D space scale starts at the unaged centre of the rubber sheet at $\Delta r = 0$ and ends at the aged surface at $\Delta r = 700\,\mu$m. Two curves are shown for each of the relaxation times T_1, T_2, $T_{1\rho}$, and T_{2e} as a function of space. The top curve is for unaged material. It varies little with space. The bottom

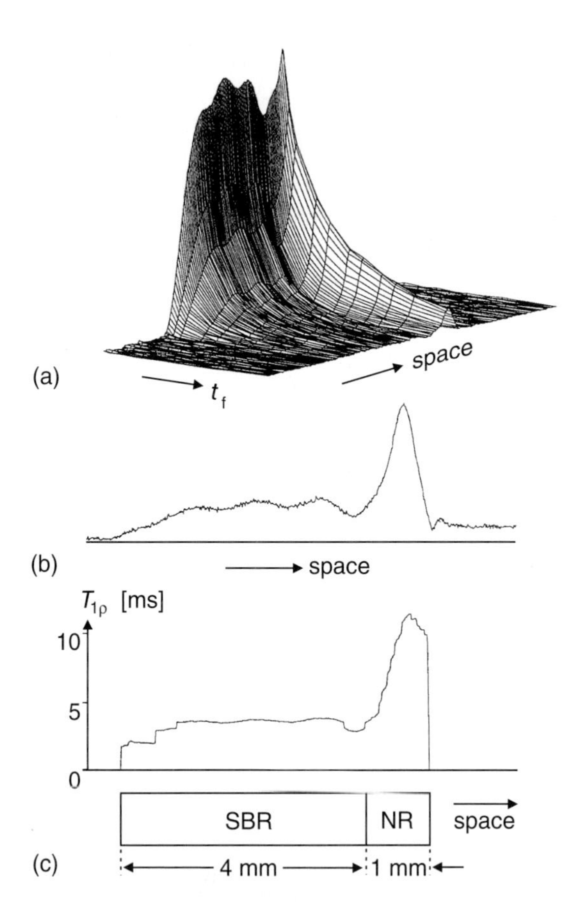

FIG. 7.1.5 Illustration of filter-weighted and of parameter imaging by a stack of 1D images (a) of an elastomer composite of synthetic (SBR) and natural (NR) rubber cut from the tread of a car tyre. A parameter-weighted image is obtained from a set of filter-weighted images by extraction of the filter parameter for each pixel. The applied filter is a spin-lock sequence for $T_{1\rho}$ contrast at a lock-field strength of $|\gamma B_1/2\pi| = 1$ kHz. The extrinsic parameter adjusted before acquisition of each projection (b) is the duration t_f of the lock field. The filter parameter extracted from the signal decay as a function of t_f is the longitudinal relaxation time $T_{1\rho}$ in the rotating frame. The image obtained by displaying the filter parameter as a function of space is the parameter image, in this case a $T_{1\rho}$ image (c). The SBR region (left, weaker signal) is readily discriminated from the NR region (right, stronger signal) in (b) and (c). Adapted from [Blü1] with permission from Elsevier Science.

curve is for the aged material. Both curves overlap in the centre of the sheet at $\Delta r = 0$ and split towards the surface. The splitting is a direct measure of contrast. In agreement with the mechanical relaxation rate of the sample (cf. Fig. 7.1.4), the contrast is largest for those relaxation times which are sensitive to slow molecular motion. T_1 allows no discrimination of aged and unaged material, whereas for the $T_{1\rho}$ and T_{2e} images the contrast is largest.

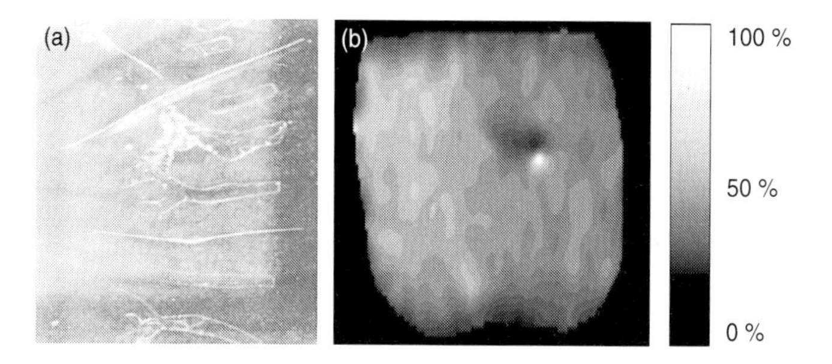

FIG. 7.1.6 [Fül1] Photo (a) and T_2 parameter image (b) of a 23 mm × 31 mm section of a laminated glass window pane illustrating the different contrast features of visible light and the T_2 parameter.

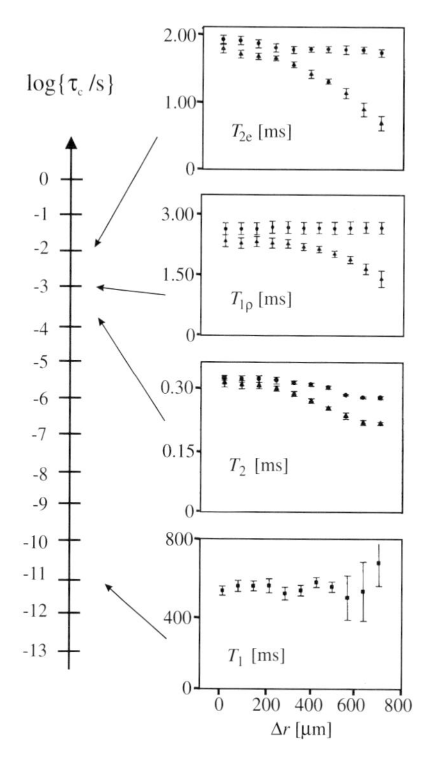

FIG. 7.1.7 Relaxation parameters for contrast in thermal-oxidative aging of SBR. The insets depict the spatial variations of the relaxation times T_1, T_2, $T_{1\rho}$ and T_{2e} for a partially aged (lower curves) and an unaged (upper curves) rubber sheet. For the T_1 data both curves overlap. The space scale starts at the unaged centre of the sheet at $\Delta r = 0$ and it ends at the aged surface at $\Delta r = 700\ \mu$m. The different relaxation times probe the distribution of correlation times of molecular motion which also leads to the mechanical relaxation-time spectrum of Fig. 7.1.4. Adapted from [Fül1] with permission from Hüthig Gmbh.

7.1.5 Contrast parameters

The *contrast parameters* relevant to material characterization through NMR imaging are the intrinsic NMR parameters of the sample. They are referred to as the contrast parameters *per se*. They can be divided into chemical and physical parameters, and into molecular, mesoscopic, microscopic, and macroscopic parameters. A list of NMR parameters for contrast in NMR imaging is compiled in Table 7.1.1.

Chemical contrast parameters are the chemical shielding and the indirect coupling. They relate to the chemical structure and are of molecular nature, although the distribution of the chemical shifts, for instance, also provides information about the physical nature of

Table 7.1.1 NMR parameters for contrast in NMR imaging and associated space scales

Chemical parameters		
	Molecular: 0.1–1 nm	
		Chemical shift
		Indirect coupling
		Chemical shift anisotropy
		Anisotropy of the dipole–dipole interaction
		Multi-quantum coherences
Physical parameters		
	Molecular: 0.1–1 nm	
		Dipole–dipole interaction
		Second moment, fourth moment of lineshape
		Incoherent magnetization transfer: characteristic times for cross-polarization and exchange
	Mesoscopic: 1 nm – 0.1 μm	
		Longitudinal relaxation time T_1
		Transverse relaxation time T_2
		Relaxation time $T_{1\rho}$ in the rotating frame
		Solid-echo decay time T_{2e}
		Spin-diffusion constant
	Microscopic: 0.1–10 μm	
		Molecular self-diffusion constant D
	Macroscopic: 10 μm and larger	
		Spin density
		Differences in magnetic susceptibility
		Velocity and acceleration of flow
		Magnetic fields induced by electric currents
Contrast agents		
	Paramagnetic relaxation agents	
	Magnetite	
	$^{129}Xe, ^{3}He$	
	X nuclei	

the sample, like the degree of molecular orientation. The *physical contrast parameters* are related to different space scales. The dipole–dipole interaction is exerted on a molecular level, yet relaxation as described in terms of correlation functions and spectral densities involves a large number of molecules and, therefore, refers to a mesoscopic space scale somewhere between 1 nm and 0.1 μm. The same argument applies to spin diffusion, a process of magnetization transport between many molecules. Molecular self-diffusion can sample space dimensions in the microscopic regime between 0.1 and 10 μm, whereas the spin density, susceptibility differences, and flow phenomena are usually resolved only with the spatial resolution of the imaging experiments, which for most solid materials is rarely better than macroscopic.

The range of contrast parameters is extended by imaging of another nucleus of the sample. In most cases ^{1}H is being imaged because of its high sensitivity. Imaging of other nuclei, like ^{13}C and ^{17}O, is associated with lower sensitivity, which can be improved partially by isotope enrichment. But isotope enrichment is already a sample manipulation and as such can be considered as a method of introducing a *contrast agent*. Other contrast agents are nuclei or molecules added to the sample, for instance, hyperpolarized xenon [Raf1, Roo1, Sch3, Sch4] or fluorinated spy molecules. Even the addition of small ferromagnetic particles can become a way of manipulating contrast by observing the field distortions introduced in the neighbourhood of their preferential locations.

7.1.6 NMR parameters and material properties

The contrast introduced by spin-density weight of NMR parameters or by imaging specific NMR parameters by themselves already is a powerful tool for detecting otherwise invisible sample inhomogeneities. Yet for quantitative interpretation of images, the NMR parameters must be interpreted in terms of properties used in physics and engineering for characterizing materials. A list of such properties is given in Table 7.1.2. Here parameters that describe the state of matter are discriminated from those that describe the change of matter.

Types of parameters

State parameters are the mechanical moduli such as the shear modulus and the bulk modulus [Par1]. Other parameters are the shear and extensional viscosities [Mar2], the degree of molecular orientation, cross-link density [Fül1, Haf1, Kuh1, Kuh2], and filler-particle distribution of elastomers, the temperature distribution, and correlation times and spectral densities of molecular motion. *Transition parameters* or kinetic parameters characterize the time evolution of systems. They are determined in *in situ* NMR experiments or in stop-and-go type experiments. Examples are characteristic times of physical and chemical ageing [Blü3, Fül1], of chemical reactions such as vulcanization [Fül2] and curing of polymers [Bal1], but also the time dependence of concentration changes in chemical reactions [Arm1], of heat dissipation, and of fluid permeation [Man2, Ran1].

Transition parameters can be determined directly through the analysis of the time evolution of changes in NMR images, whereas state parameters must be related to NMR

Table 7.1.2 Examples of material properties for polymers and elastomers

State parameters
Molecular orientation
Stress, strain
Moduli of shear, compression, and elasticity
Viscosity of shear, bulk, extension
Cross-link density of elastomers
Molecular order parameters
Distribution and agglomeration of filler particles
Pore-size distribution
Temperature distribution
Correlation times of molecular motion
Spectral densities of molecular motion
Transition parameters and kinetic parameters characterizing
Physical ageing
Chemical ageing
Vulcanization and curing processes
Concentration changes in chemical reactions
Thermal heating under load
Fluid permeation, drying

parameters either by theory or by calibration. Theories have been published, for instance, which relate the cross-link density of filled and unfilled elastomers to the parameters of nonexponential relaxation curves of transverse magnetization [Fül3, Kul1, Sim1, Knö1], and the dynamic storage modulus of polymers to the cross-polarization rate [Par1]. Also the relationship between various relaxation parameters and the cross-link density of unfilled rubbers [Coh1, Sot2, Knö1] and gels [Coh1, Coh3], and the relationship between the degree of molecular orientation and the splitting of the deuteron resonance are well understood [Gro1, Sot1]. A simpler approach is based on empirical correlations between NMR parameters and sample properties, which can often be established by calibration of an NMR parameter against a material parameter [Blü5, Dor1, Sch1].

Relaxation in cross-linked elastomers

A simple model of an *elastomer network* is depicted in Fig. 7.1.8. The segmental motion of inter-cross-link chains is fast but anisotropic at temperatures of 100–150 K above the glass transition temperature: The end-to-end vector $\boldsymbol{R}$ of such a chain reorients on a much slower timescale because it appears fixed between seemingly static cross-link points. As a result of the fast but anisotropic motion, the dipolar interaction between spins along the cross-link chains is not averaged to zero, and a *residual dipolar coupling* remains [Coh1, Got1, Lit1].

The simple model of Fig. 7.1.8 is often extended to include dangling chain ends and possibly even a sole part [Kuh1, Men1, Sim1]. Neglecting the sole part, T_2 is assigned to the rapidly moving segments of the cross-link chains, and relaxation of the slowly moving cross-links is described by the associated correlation time τ_s. Based on the

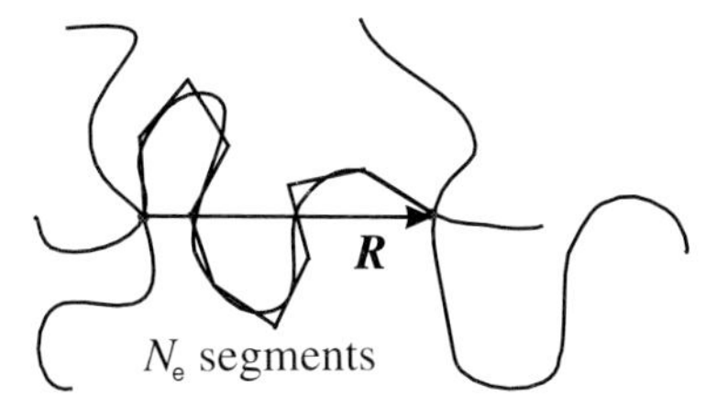

FIG. 7.1.8 A cross-link chain in a rubber network. Following the Kuhn model [Kuh3] it can be decomposed into N_e freely jointed but rigid segments. Their length and their number N_e depends on the temperature and the stiffness of the chemical structure.

Anderson–Weiss model of relaxation [And1] the decay of the Hahn-echo amplitudes of the transverse ^{1}H magnetization as a function of the echo time t_E has been derived [Sim1]:

$$M_x(t_E) = A\exp\left\{-\frac{t_E}{T_2} - qM_2\tau_s^2\left[\exp\left(-\frac{t_E}{\tau_s}\right) + \frac{t_E}{\tau_s} - 1\right]\right\} + B\exp\left\{-\frac{t_E}{T_2}\right\}, \tag{7.1.5}$$

where M_2 is the intramolecular part of the second moment of the rigid-lattice ^{1}H lineshape, and qM_2 is the residual second moment resulting from anisotropic motion of the chains segments. M_2 can be determined at temperatures below the glass transition from transverse relaxation in samples which are swollen in a hydrogen-free solvent like CCl_4. The factor q contains contributions from physical (q'') and from chemical (q') cross-links. Both contributions can be separated through temperature-dependent studies, because the number of effective physical cross-links depends on temperature and the number of chemical cross-links does not. It has been shown that the cross-link density $[M_c]$ is proportional to $(q')^{1/2}$. Given the reduced accuracy of relaxation curves measured with spatial resolution, (7.1.5) has been used for imaging of cross-link densities in an approximated form by expanding the second exponential and truncation after the third term. As a result the signal decay can be described by a sum of a Gaussian and an exponential term,

$$M_x(t_E) = A\exp\left\{-\frac{t_E}{T_2} - qM_2t_E^2\right\} + B\exp\left\{-\frac{t_E}{T_2}\right\}. \tag{7.1.6}$$

From a fit of (7.1.6) to spatially resolved relaxation curves, images of the four parameters A, B, T_2, and qM_2 have been obtained [Haf1, Knö1, Kuh1]. Here $A/(A+B)$ is interpreted as the concentration of cross-linked chains, and $B/(A+B)$ as that of the dangling chains [Fül1].

A crucial point of the four-parameter model is that T_2 is assumed to be independent of cross-link density which is valid only in a first approximation [Fül3]. Moreover, the experimental relaxation signal can be modelled at short and intermediate time scales with good agreement without the inclusion of dangling chains in the model: the free induction decay of one such cross-link chain can be written as a product of an inhomogeneous and a homogeneous decay [Sot1]

$$M_{xR}(t) = M_{0R} \cos(\langle \omega_R \rangle t) \exp\left\{-\frac{t}{T_2}\right\}, \tag{7.1.7}$$

where, as above, T_2 is the time constant of the homogeneous, liquid-like decay. The quantity $\langle \omega_R \rangle$ denotes the average dipolar frequency of the coupling spins on the chain, because as a result of the fast anisotropic motion, the average orientation of the coupling spins is along the *end-to-end vector* $\boldsymbol{R}$. Following the *Kuhn model* [Kuh3] for a random chain of N_e freely jointed, stiff segments of length a the length R of the end-to-end vector is given by $N_e^{1/2} a$, and the *residual dipolar coupling* scales with $N_e^{-1} \Delta$, where $\Delta = \mu_0 \gamma_I \gamma_J \hbar / (4\pi r_{IJ}^3)$ is the angular-independent strength of the dipolar coupling [Coh2, Bre1]. The average distance r_{IJ} of the coupling spins I and J is contained in Δ, and the effective number N_e of fictitious chain segments depends on the chain stiffness, which is determined by the temperature and the chemical structure.

In the fast motion limit, the homogeneous decay is much slower than the inhomogeneous one, and the exponential relaxation factor can be neglected in (7.1.7). Assuming a Gaussian distribution of end-to-end vectors, the magnetization decay assumes the form [Sot2]

$$M_x(t) = \mathrm{Re}\left\{ M_0 \left(\sqrt{1 - \frac{2\mathrm{i}}{3} \frac{\Delta t}{N_e}} \right)^{-1} \left(1 + \frac{\mathrm{i}}{3} \frac{\Delta t}{N_e} \right)^{-1} \right\}. \tag{7.1.8}$$

The two remarkable features of this equation are that a single parameter N_e suffices to describe the shape of the FID, and that the time axis scales with $1/N_e$. As a consequence, relaxation curves from differently cross-linked rubber samples can be mastered onto one curve by rescaling the time axes. This is illustrated in Fig. 7.1.9 [Sot2]. The values of N_e used for scaling of the time axes were obtained by fitting the expression (7.1.8) to the experimental relaxation curves. They are proportional to the maximum vulcameter moments determined during vulcanization [Sot2], the quantity which is used conventionally for measuring cross-link density in test samples. Clearly, N_e comprises contributions from physical and from chemical cross-links, which can be separated by temperature-dependent studies.

The form of the FID (7.1.8) is illustrated in Fig. 7.1.10 by the line of diamonds. In the short-time limit, the curve can be approximated by a Gaussian function (broken line), and in the long-time limit it follows an exponential function (solid line). Therefore, in the long-time limit the cross-link density is expected to be proportional to the relaxation rate of the slow component from a double exponential fit and in the short-time limit it is expected to be proportional to the inverse variance of the Gaussian component from a Gauss-exponential fit. Both cases are, indeed, observed experimentally [Fül3, Fed1, Got1, Sim1]. For accurate measurements, the distribution in chain lengths N_e can be fitted to obtain a molecular-weight distribution of cross-link chains, which can be followed as a function of ageing and degradation processes [San1].

Similar to transverse relaxation *longitudinal relaxation in the rotating frame* is also sensitive to slow molecular motion (cf. Fig. 7.1.3) and is thus a good probe for elastomer

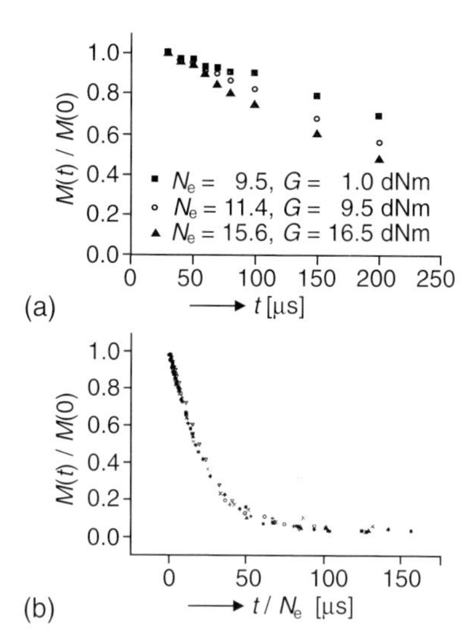

FIG. 7.1.9 ^{13}C-edited ^{1}H transverse relaxation curves of the CH groups in unfilled SBR. (a) Normal time axis. The values of N_e and the vulcameter moment G from vulcanization are given. (b) Master curve with time axis t/N_e. Adapted from [Sot2]. Copyright 1996 American Chemical Society.

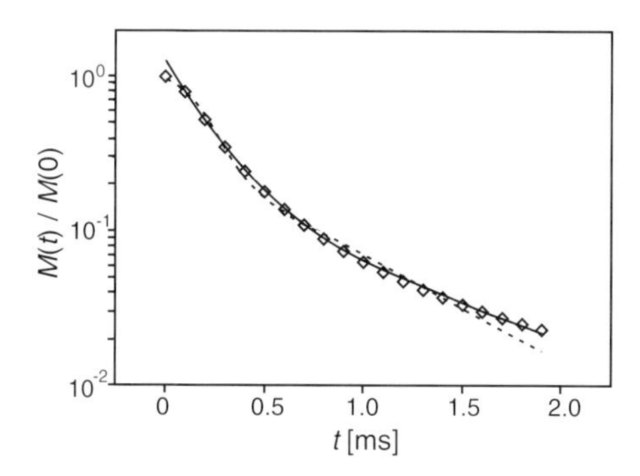

FIG. 7.1.10 The shape of the FID (7.1.7) is given by the line of diamonds. In the short-time limit the curve can be approximated by a Gaussian (broken line) and in the long-time limit it follows an exponential (solid line). Adapted from [Fül4] with permission from Elsevier Science.

properties. For interpretation of the associated relaxation time $T_{1\rho}$, the *defect diffusion model* [Kim1] has been applied [Bar1] to imaging data. In this model, the motion of a chain is described by translational diffusion of structural defects along parts of the chain. It is assumed that the motion of chain defects is 1D and anisotropic, and that physical and chemical cross-links are long-lived barriers to the diffusing defects. Within this model the relaxation rate

$$\frac{1}{T_{1\rho}(2\omega_1)} = \frac{\pi}{\sqrt{2}} M_2 \left[\frac{d}{b}\left(\frac{d}{b} - 1\right)\right]^{-1} \frac{\tau_d^{1-\alpha}}{(2\omega_1)^\alpha}, \tag{7.1.9}$$

is expressed in terms of the ratio of the diffusion length d and the defect width b, the time τ_d for diffusion of the defect from one barrier to the next, and an exponent α. For very small defect concentration along the chain, the diffusion length can be interpreted as the mean length of inter-cross-link chains which is inversely proportional to cross-link density. The diffusion time τ_d is determined by the correlation time of the defect motion which can be derived from the $T_{1\rho}$ minimum as a function of inverse temperature. A typical value of τ_d for unfilled natural rubber is 10^{-7} s at room temperature. The exponent α depends on the shape of the spectral density. It can vary between 0 and 2/3. The ideal value is 1/2. By spatially resolved measurements of $T_{1\rho}$ as a function of the spin-lock field strength ω_1, the experimental data can be evaluated in terms of (7.1.9) and the parameters α and d/b can be displayed as an image [Bar1].

Stress and strain calibration in filled elastomers

An example of *calibration of a material property* against an NMR parameter is given in Fig. 7.1.11 [Blü5], by a plot of T_2 of filled poly(dimethylsiloxane) (PDMS) rubber against strain (a). By use of the stress-strain curve (b) from mechanical measurements, T_2 can be calibrated against stress (c). The experimental values of T_2, strain, and stress are average values over the entire sample. The tensorial properties of stress are neglected. Nevertheless, the procedure can be used to translate the spatially resolved NMR parameter T_2 into values related to the magnitude of local stress for communication of NMR results to scientists and engineers outside the NMR community and for evaluation of NMR images in terms of material properties.

More detailed information about strain in networks can be gained from analysis of lineshapes, in particular the *second moment* M_2. When interchain couplings are neglected, M_2 is related to the elongation ratio Λ according to [War2]

$$M_2 = \frac{1}{3}\left(\Lambda^4 - \frac{2\Lambda}{3} + \frac{2}{3\Lambda^2}\right). \tag{7.1.10}$$

Here the elongation ratio $\Lambda = L/L_0$ is given by the ratio of lengths L and L_0 after and before straining, respectively. Under applied strain, the complicated interplay of changes of angle and mean-square dimension from initial segment orientation and position leads to a second moment which assumes its minimum value near but not exactly at $\Lambda = 1$.

On the other hand, residual anisotropic interactions from anisotropic motion can be studied by investigations of relaxation times [Coh1]. By appropriately superposing the signals obtained from the solid echoes with and without a 90° phase shift, $s_1(t, t_E)$, and $s_2(t, t_E)$, respectively, and the Hahn echo, $s_3(t, t_E)$, where t_E is the echo time, a function $\beta(t, t_E)$ is obtained which for $t = t_E$ is ideally suited to reveal weak dipolar interactions

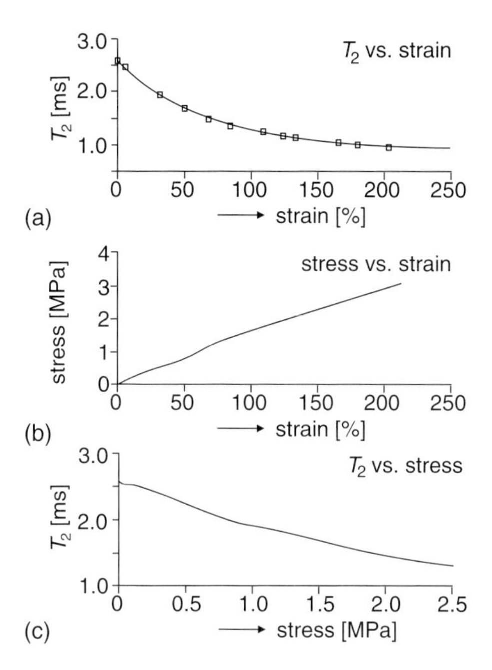

FIG. 7.1.11 [Blü5] Calibration of T_2 values measured for different strains of filled PDMS rubber. (a) Experimental values of T_2 *versus* integral strain. (b) Stress–strain curve from mechanical measurements. (c) The calibration curve of T_2 versus integral stress obtained by combining the data from the NMR measurement (a) and mechanical measurement (b). These curves have been used to calibrate an experimental T_2 image into the stress image Fig. 1.1.7 of a PDMS rubber band with a cut.

of ^{1}H in strained polymers and sheared polymeric fluids [Bal2, Cal3],

$$\beta(t, t_E) = \frac{s_1(t, t_E) - s_2(t, t_E) - s_3(t, t_E)}{2s_3(0, 0)}. \tag{7.1.11}$$

This *beta function* is zero in the absence of dipolar interactions. It vanishes for interactions linear in the spin operator and has nonzero values only for bilinear interactions. Furthermore, it is zero for t_E approaching zero, so that a nonzero signal indicates residual anisotropic interactions, and it is free of signal attenuation by relaxation. The shape of the beta function has been shown to depend strongly on the strain in rubber samples [Cal3].

An alternative to mapping strain by relaxation time imaging of ^{1}H is the observation of the *quadrupole splitting* in deuteron NMR. For small strain, the splitting Δ_Q is linearly proportional to $\Lambda^2 - \Lambda^{-1}$ [Gro1, Sot1] (Fig. 7.1.12). The rubber samples to be investigated need to be prepared by swelling in a solution of deuterated molecules and subsequent drying of the solvent. The deuterated spy molecules follow the network alignment under strain and the *residual quadrupole splitting* that results from their

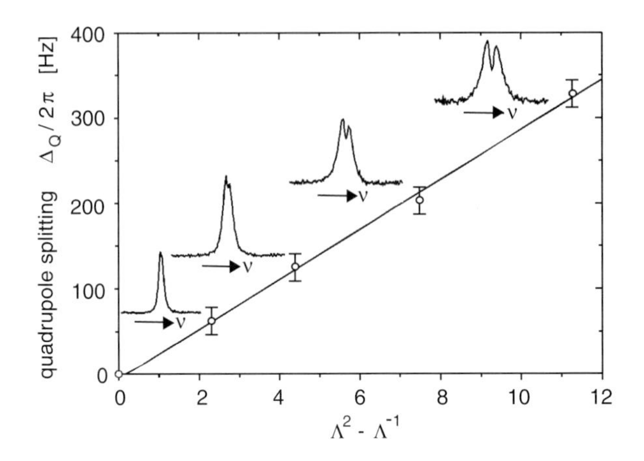

FIG. 7.1.12 Quadrupole splitting of 1,4-tetra-deuterated butadiene oligomers in bands from natural rubber for different strains Λ. Adapted from [Kli2]. Copyright 1997 American Chemical Society.

anisotropic motion is observed either by spectroscopic imaging (cf. Section 7.3) or double-quantum imaging (cf. Section 7.2.10) [Kli1, Kli2].

Temperature mapping

Temperature distributions can be mapped by following various NMR parameters which are sensitive to temperature variations. For medical applications the sensitivity of the self-diffusion coefficient [Hed1, Leb], relaxation times [Dic1, Par2, Sch1], magnetization transfer rates from selective saturation of one chemical group, and of the chemical shift including that of chemical-shift agents have been investigated [Dor1, Lew1, Zuo1]. Often fast methods are needed to follow time-dependent processes. To this end, the spin–lattice relaxation time T_1 [Sch1] and ^{1}H resonance frequency shifts [Zwa1] have been investigated.

For materials applications, the chemical shifts of methanol and ethylene glycol can be monitored in the liquid state to follow temperature [Haw1]. The most sensitive chemical shift is the ^{59}Co resonance of aqueous $Co(CN)_6$ with a sensitivity of ± 0.05 K at 7 T and ± 0.2 K at 2 T [Dor1]. Furthermore, dibromomethane dissolved in a liquid crystal is a temperature sensitive NMR compound [Hed1], and known phase-transition temperatures can be exploited to calibrate the temperature control unit [Haw1]. In *temperature imaging* of fluids, temperature can be determined from the temperature dependence of the self-diffusion coefficient but convective motion may arise in temperature gradients [Hed1]. In the solid state, the longitudinal relaxation time of quadrupolar nuclei like ^{81}Br is a temperature sensitive parameter [Sui1, Sui2]. In elastomers, both T_2 and T_1 depend on temperature (Fig. 7.1.13). In filled SBR, T_2 is the more sensitive parameter with a temperature coefficient of about 30 μs/K [Hau1].

Over small temperature ranges, a linear dependence of the transverse relaxation time T_2 on temperature is observed for technical SBR samples in the vicinity of room temperature (Fig. 7.1.13(a)). The temperature coefficients depend only weakly on the filler contents

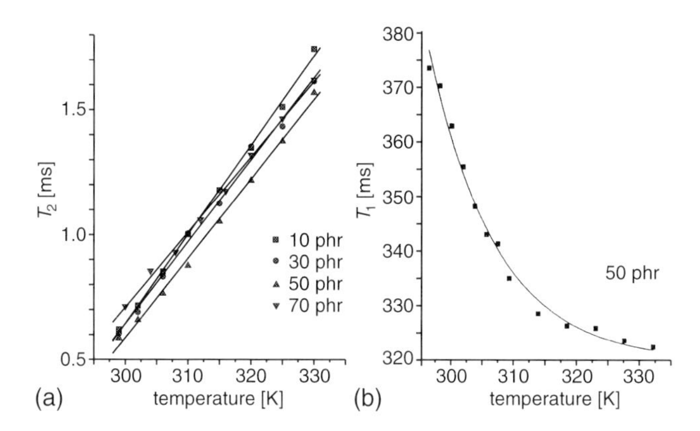

FIG. 7.1.13 Calibration curves of relaxation times *versus* temperature for filled SBR samples. (a) T_2 for samples with different carbon-black content. The straight lines are fits to the experimental data. (b) T_1 for SBR filled with 50 phr carbon black. Adapted from [Hau1] with permission from Wiley-VCH.

from 10 to 70 phr, where the unit phr denotes weight parts per hundred rubber. In fact, temperature appears to affect relaxation times much more than the filler content does. Because transverse relaxation is nonexponential in cross-linked elastomers, the applicable T_2 values refer to the slowly decaying pats of the relaxation curve. This information is more easily accessible in imaging experiments than the initial fast relaxing component. The longitudinal relaxation times T_1 vary in a nonlinear, exponential fashion with temperature for filled SBR and over a smaller relative range of values (Fig. 7.1.13(b)).

Porous media

Analysis of relaxation curves from fluids is of particular interest for the characterization of pores in rocks [Dav1, For1, Kle1] and other *porous materials* such as packs of beads or grains, ceramics, zeolites, foods, colloids, emulsions, gels, suspensions, and is of interest to the retrieval of biochemical and biophysical information [Lab1]. Because longitudinal and transverse relaxation rates can be significantly increased in the vicinity of solid–liquid interfaces, the relaxation behaviour of ^{1}H and other nuclei of fluids confined in porous media can provide important information about porosity, *pore-size distribution*, and pore connectivity [Hür1, Kle1, Kle2].

The majority of rocks conforms with the *fast diffusion* or *surface-limited relaxation regime*. In this regime, the rate-limiting step is relaxation at the surface, not the transport of spins to the surface, for instance by translational diffusion. Thus the spins experience a rapid exchange of environments so that the local fields in each region of a pore are averaged to their mean value. As a consequence, a single exponential relaxation decay is observed for a given pore, and the rate of magnetization decay does not depend on the pore shape but only on the *surface-to-volume ratio*. The time evolution of the magnetization of a sample having a distribution of pores sizes can then be expressed as a sum of exponentials. Assuming constant surface relaxivity, the spectrum or the *distribution of relaxation times* is a direct map of the pore-size distribution. Unfortunately, the inversion of the relaxation curve is ill-conditioned, and an inverse Laplace transformation for the distribution of relaxation times is unstable and often not unique. Thus, sophisticated regularization methods like the *CONTIN algorithm* need to be applied [Dav1, For1, Kle2, Lab1, Pro1] or other forms of numerical analysis of multi-component signals [Pey1, Win1, Win2].

According to the theory of fast exchange [Hal1], the relaxation rate of a single pore is a weighted average of the surface-liquid relaxation rate $1/T_{2s}$ and the relaxation rate $1/T_{2b}$ of the bulk liquid,

$$\frac{1}{T_2} = \frac{V_s}{V}\frac{1}{T_{2s}} + \frac{V_b}{V}\frac{1}{T_{2b}}, \tag{7.1.12}$$

where V_b and V_s are the rapidly exchanging volumes of bulk and surface liquid, and $V = V_s + V_b$. The surface volume can be approximated by the product of a local liquid surface area S and the liquid depth λ. The liquid within this shell close to the pore surface has a short-relaxation time T_{2s}, and the other liquid has a long relaxation time T_{2b}. According to (7.1.12) the initial relaxation rate is then proportional to the average surface-to-volume ratio experienced by diffusing molecules.

This approximation is valid when λ is much less than the linear dimension of the liquid volume, that is for pores with low curvature. Therefore, experimental relaxation rates from a system in the fast exchange limit can provide information on the ratios of the surface-to-bulk volumes of liquid, given the relaxation rates of surface and bulk liquid are known as well as the value for λ. If relaxation values are acquired for different stages of filling, it is possible to determine details of the filling process, like the formation of puddles within the pore structure [All1, Boo1].

The other limiting regime, where the decay of magnetization is controlled by the transport of molecules to the surface, is the slow-diffusion or *diffusion-limited regime*. It arises when the pores are large or relaxation is fast. Then the relaxation rates can be correlated to parameters of the pore structure [Bor1, Bor2]. For instance, the proton T_2 relaxation times of liquids like water or cyclohexane vary according to their location within the pores of porous silica. When eliminating susceptibility effects with echo techniques, in the centres of large pores the relaxation times approach those of the bulk liquid, while the relaxation times of molecules near the surface of the pores are much shorter, often by several orders of magnitude. If the molecules in the different regions of the liquid do not mix appreciably during the NMR experiment, the multi-exponential relaxation decay can be analysed for a set of discrete relaxation rates which can be assigned to the different regions of the liquid within a pore.

7.2 FILTERS

Magnetization filters are used to generate selectivity in spectroscopy [Ern1] and contrast in imaging [Blü1, Blü6, Blü10, Blü11, Cal1, Haa1, Xia1]. *Filters* are given by pulse sequences which in most cases generate longitudinal magnetization which differs from the thermodynamic equilibrium state in some specific way. Following the generic imaging scheme in Fig. 7.1.1, filters are often applied in a preparation period, so that the filtered magnetization is subsequently encoded in space and detected. Many different filters can be conceived and combined to produce parameter weights of spin-density images, or the filter parameters are varied systematically to calculate a parameter image from a set of parameter-weighted spin-density images (cf. Section 7.1.4) [Dor2, Liu1]. In the following sections, different magnetization filters are presented. Though not necessarily complete, this collection of filters is intended to provide insight into the manifold of possibilities to generate contrast in NMR imaging and selectivity in NMR spectroscopy.

The subsequent discussion more or less ignores *spin density* as a quantity which gives contrast readily accessible by other imaging methods as well. Spin density can conveniently be calculated from two T_2-weighted images provided the relaxation is purely exponential: the image $M_1(t_E, \boldsymbol{r})$ is acquired with echo time t_E and the image $M_2(2t_E, \boldsymbol{r})$ with twice the echo time. The ratio $M_1^2(t_E, \boldsymbol{r})/M_2(2t_E, \boldsymbol{r})$ then provides a pure spin-density image $M(\boldsymbol{r})$ (cf. Section 6.2.6) [Hep3].

One of the most common filters is T_1 weighting of the spin density by partial saturation (Section 7.2.1) from reduced repetition times in signal averaging. Other magnetization filters exploit T_2 by using Hahn echoes (Section 7.2.2), the relaxation in the rotating frame (cf. Section 7.2.3), multi-solid echoes and multi-pulse excitation (Section 7.2.7), and combinations of relaxation-time filters (Section 7.2.8). Further filters are generated through chemical-shift selective excitation (Section 7.2.4), multi-quantum coherences (Section 7.2.10), different forms of magnetization exchange and magnetization transfer (Sections 7.2.11 and 7.2.12), and by encoding of molecular transport from diffusion and flow (Section 7.2.6). The particular importance of mobility filters, which are sensitive to molecular motions in different time windows for detection of chemical or physical material change has been addressed in Section 7.1.3.

7.2.1 T_1 filters

T_1 and related contrast can be introduced to an image by different techniques. The most common filtering effect is achieved by manipulation of the recycle delay t_R, leading to progressive saturation as t_R is reduced [Dor2, Liu1]. This method corresponds to the saturation recovery technique for measurement of T_1 [Gar1]. Alternatively, an inversion recovery filter [Vol1], or a stimulated-echo filter may be used to generate T_1 *contrast*. For imaging with T_1 contrast, rf pulses with nonstandard flip angles can be used [Bon1, Kin1].

Partial saturation

Instead of t_R for the recycle delay in *partial saturation* experiments, the generic symbol t_f is used in the following to denote an adjustable filter time. Within the validity of the Bloch equations, the longitudinal magnetization $M_z(t_f, \boldsymbol{r})$ of a pixel corresponding to position $\boldsymbol{r}$, which has been partially saturated, relaxes according to (cf. eqn (2.2.36)),

$$M_z(\boldsymbol{r}, t_f) = M_0(\boldsymbol{r}) \left[1 - \exp\left\{ \frac{-t_f}{T_1(\boldsymbol{r})} \right\} \right], \tag{7.2.1}$$

where M_0 is the longitudinal magnetization in thermodynamic equilibrium. Instead of varying the recycle delay, the magnetization to be imaged can be prepared by a saturation recovery filter [Mar1] (cf. Fig. 2.2.8(a)).

Three different T_1 filters are illustrated in Fig. 7.2.1 in terms of pulse sequences, typical evolution curves of the relevant magnetization components, and the characteristic transfer functions. In the slow motion regime with respect to the T_1 minimum (cf. Fig. 3.5.2), the saturation recovery filter (Fig. 7.2.1(a)) is a low-pass filter which allows magnetization components with short correlation times to pass. Thus, the signal amplitudes from molecular segments undergoing slow motion are attenuated. The longitudinal magnetization is first destroyed by a sequence of 90° pulses [Die1, Mar1]. It is then allowed to build up for a time t_f before the actual imaging experiment starts. The build-up is described by eqn (7.2.1). From a set of images acquired with different values of t_f, a parameter image of T_1 and a spin-density image depicting $M_0(\mathbf{r})$ can be calculated by fitting (7.2.1) to each pixel in a set of images [Liu1].

An application of the *saturation-recovery filter* to the suppression of signal from rigid components in bisphenol-A poly(carbonate) is shown in Fig. 7.2.2 [Han1]. The wideline ^{2}H solid-echo spectrum of the phenyl deuterons exhibits a range of broad and narrow components (a) as a result of a distribution of motional correlation times. The mobile components are characterized by a shorter T_1 than the more rigid components. Consequently the rigid components can be suppressed by partial saturation. After application of the saturation-recovery filter the shape of the wideline spectrum is dominated by the narrow signal in the centre from the mobile ring deuterons (b).

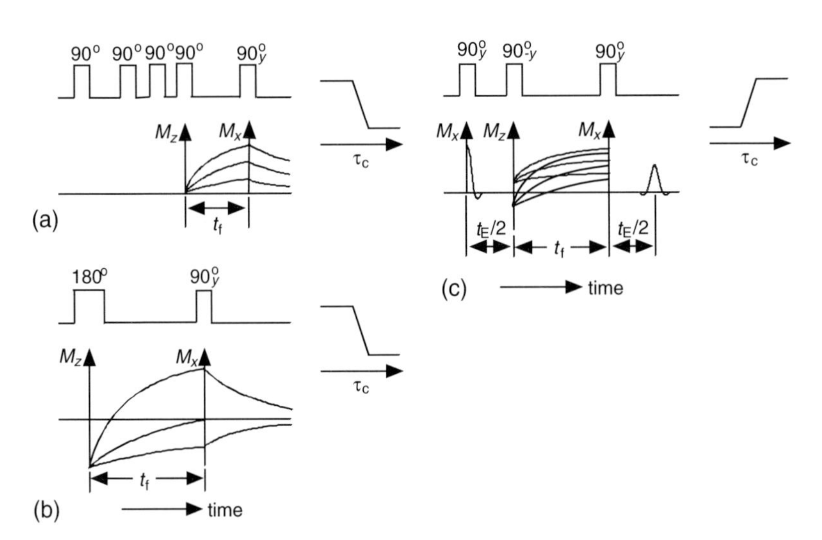

FIG. 7.2.1 Pulse sequences for T_1 and related magnetization filters, typical evolution curves of filtered magnetization components, and schematic filter transfer functions applicable in the slow motion regime. Note that the axes of correlation times start at $\tau_c = \omega_0^{-1}$. (a) Saturation recovery filter. (b) Inversion recovery filter. (c) Stimulated echo filter.

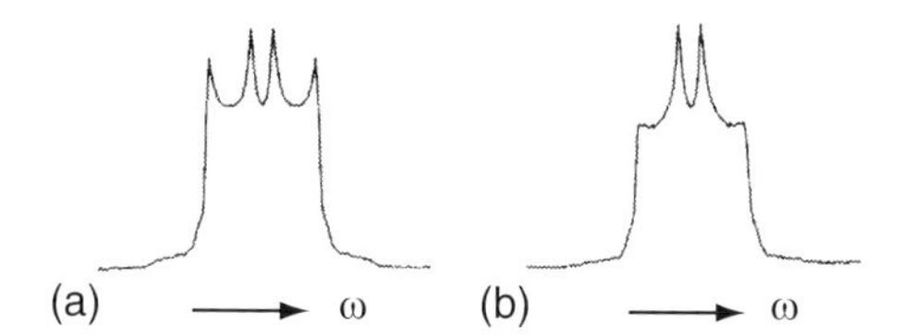

FIG. 7.2.2 [Han1] Solid-echo wideline spectra of the ring deuterons of bisphenol-A poly(carbonate-d_4) at 253 K. The phenyl rings undergo a 180° flip motion with a wide distribution of motional correlation times. (a) Spectrum with signals from fast and slow flipping rings. (b) Application of the saturation-recovery filter suppresses the signals from the slow components.

Inversion recovery

A filtering effect similar to saturation recovery is obtained for the *inversion recovery filter* (Fig. 7.2.1(b), cf. Fig. 2.2.8(b)). The longitudinal magnetization is inverted before recovery during the filter time t_f, so that the contrast range is doubled with respect to the saturation-recovery filter (cf. eqn (2.2.36)) and negative magnetization values are admitted,

$$M_z(\boldsymbol{r}, t_f) = M_0(\boldsymbol{r})\left[1 - 2\exp\left\{-\frac{t_f}{T_1(\boldsymbol{r})}\right\}\right]. \tag{7.2.2}$$

For a quantitative analysis, the recycle delay between successive scans has to be large enough to assure complete build-up of all longitudinal magnetization components in the sample. T_1 *parameter imaging* is therefore slower by this method than by the saturation-recovery method. In the slow motion regime with respect to the T_1 minimum the saturation-recovery and the inversion-recovery filter are both low-pass filters for magnetization components with long correlation times.

The stimulated echo

The inverse filter transfer function is obtained for the *stimulated-echo filter* (Fig. 7.2.1(c), cf. Fig. 2.2.10(c)). It consists of three 90° pulses. The second pulse generates longitudinal magnetization, which is modulated in amplitude by the precession phases accumulated during the evolution time $t_E/2$ between the first two pulses. The filter time t_f is the time between the second and the third pulse. Here the modulated components relax towards thermodynamic equilibrium with the longitudinal relaxation times $T_1(\boldsymbol{r})$, and the memory of the initial two pulses is lost as t_f increases. Therefore, the amplitude of the stimulated echo is given by (cf. eqn (2.2.39))

$$a_s(\boldsymbol{r}, t_E, t_f) = \frac{M_0(\boldsymbol{r})}{2}\exp\left\{-\frac{t_f}{T_1(\boldsymbol{r})}\right\}\exp\left\{-\frac{t_E}{T_2(\boldsymbol{r})}\right\}. \tag{7.2.3}$$

Clearly, to avoid T_2 weighting, t_E must be kept as short as possible. The magnetization components which relax slowly with $T_1(\boldsymbol{r})$, pass the filter while the fast relaxing components are rejected. As they approach their thermodynamic equilibrium values during t_f, they cannot contribute to the formation of the echo.

Filter transfer functions and spectral densities of motion

The filter characteristics symbolized by the transfer functions in Fig. 7.2.1 and in the following figures with mobility filters are schematic only, because, in general, the relaxation times depend on more than one relaxation process acting at different frequencies (cf. Section 3.5). In liquids, for instance, T_1 depends on the spectral densities at the Larmor frequency and twice the Larmor frequency (cf. eqn (3.5.4)). As a result, the shape of the T_1 filter characteristics can be inverted depending on the time regime or the side of the T_1 minimum (cf. Fig. 3.5.2) in which the dominant relaxation mechanisms are active. The transfer functions plotted in Fig. 7.2.1 apply to the *slow motion regime* as opposed to the transfer functions in Fig. 7.1.3, which apply to the *fast motion regime*. T_1 filters can also be designed for heteronuclear systems like ^{1}H and ^{13}C. Here contrast can be introduced by $T_1(^1\text{H})$, $T_1(^{13}\text{C})$, and magnetization transfer weights [Fry1].

7.2.2 T_2 filters

T_2 relaxation causes dephasing of transverse magnetization as a result of time-dependent internal magnetic fields which modulate the resonance frequencies of individual magnetization components in an incoherent fashion (cf. Section 3.5). The resulting signal decay is irreversible. It cannot be refocused by an echo (cf. Section 2.2.1 and Fig. 2.2.10). Consequently, T_2 *filters* are based on the formation of echoes [Hah1] to identify the irreversible signal decay.

The Hahn echo

A *Hahn echo* [Hah1] of the initial FID is generated by application of a 180° refocusing pulse (Fig. 7.2.3(a), cf. Fig. 2.2.10(a)). The pulse is applied at time $t_f/2$, and the echo appears at time t_f, which is identical to the echo time t_E. The pulse spacing and therefore

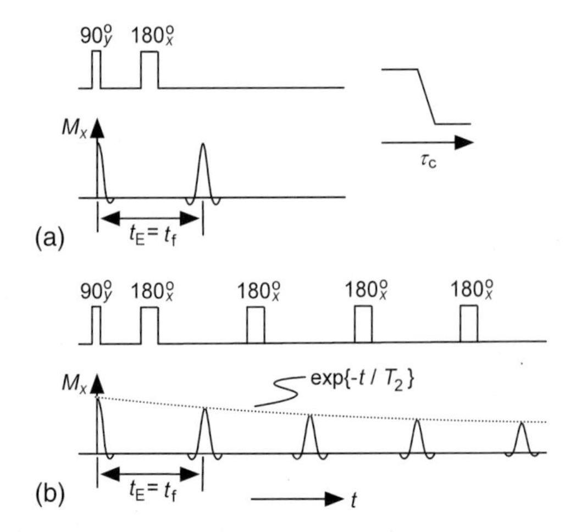

FIG. 7.2.3 Pulse sequences for T_2 filters and schematic filter transfer functions. (a) Hahn-echo filter. (b) CPMG filter.

the timing of the echo determines the resultant T_2 contrast. Disregarding diffusion the echo maximum a_H is attenuated in simple liquids by single-exponential T_2 relaxation only (cf. eqn (2.2.38)),

$$a_H(\boldsymbol{r}, t_f) = M_0(\boldsymbol{r}) \exp\left\{-\frac{t_f}{T_2(\boldsymbol{r})}\right\}. \tag{7.2.4}$$

From a set of images acquired with different values of t_f, a T_2 image and an image of the spin density $M_0(\boldsymbol{r})$ can be calculated [Liu1].

The relaxation decays for extraction of $T_2(\boldsymbol{r})$ are measured indirectly by variation of the filter time t_f. Optimum *sampling strategies* exist for the choice of values of t_f [Jon1, Jon2]. If the magnetization is sampled for only two values of t_f, one sampling point should be placed at the beginning where $t_f = 0$ and one at 1.11 T_2. For larger numbers of points it is best to place some of the sampling points at zero filter time and the rest at some values of time that are proportional to T_2. For a very large number of points, 22% should be placed near $t_f = 0$ and 78% near $t_f = 1.28T_2$ or simply one at 0 and four at $1.3T_2$. This rule applies to single relaxation times. For distributions of relaxation times a compromise has to be found, which is reasonably good over the spread of relaxation times. Clearly, to apply these rules, some prior knowledge about the range of relaxation times is necessary.

In spin-echo imaging (cf. Section 6.2.1) the Hahn echo is already integrated into the imaging sequence. However, to decouple contrast from space encoding the Hahn echo should be used as a filter in preparation of the initial magnetization. T_2 weights as well as T_2 parameter images are particularly important in medical imaging [Ger1, Ger2] but are also important for materials applications, because T_2 is sensitive to slow molecular motion as well as to the anisotropy of motion (cf. Section 7.1.6).

Multi-echo trains

When contrast is to be introduced in liquid-like samples based on large values of T_2, the precision of the simple two-pulse echo method of Fig. 7.2.3(a) suffers from molecular self-diffusion. Instead the Meiboom–Gill modification [Mei1] of the Carr–Purcell sequence [Car1], that is, the *CPMG method* needs to be applied by which signal attenuation from translational motion can be reduced for short echo times (Fig. 7.2.3(b), cf. Fig. 2.2.10(b)). The magnetization is refocused repeatedly by 180° pulses which are 90° out of phase with the initial pulse generating the transverse magnetization. This phase shift makes the method less sensitive to flip-angle errors in the 180° pulses. On the other hand, the multiple spin echoes of the *CPMG sequence* can be combined with gradient modulation for acquisition of several T_2 weighted images (cf. Section 6.2.6), and the echo modulation can be exploited for filtering of magnetization from coupled spins [Nor2].

Multi-echo refocusing in strongly inhomogeneous magnetic fields leads to interference effects from primary and stimulated echoes. As a result a build-up of the initial echo amplitudes is observed before signal decay by relaxation can be detected [Ben1]. Also the signal-to-noise ratio depends on the ratio of rf amplitude and the receiver bandwidth. In the optimum case it is unity [Goe1]. If the effective field is inhomogeneous, pulse phase and magnetization flip angle vary from position to position. Then, effects reminiscent of a spin-lock can be observed [Gut1]. Pulse imperfections can be compensated for by suitable phase cycles of the refocusing pulses [Coh3, Gul1]. One example is the XY16 sequence with phases $(xyxy\ yxyx\ \ -x-y-x-y\ \ -y-x-y-x)_n$ of refocusing 180° pulses [Gul1]. In contrast to the CPMG sequence all three magnetization components are treated equally. The sequence is also useful for generation of T_2 *contrast in solid materials*, because it eliminates chemical-shift effects in the presence of dipolar couplings. At the same time it preserves the full coupling strength, which is desired, because the dipolar couplings are indicative of material properties.

7.2.3 $T_{1\rho}$ filters

Relaxation in the rotating frame denotes relaxation in the effective magnetic field of the rotating frame. Through *spin-locking* of the magnetization along the direction of the effective field, the *energy-relaxation time in the rotating frame*, $T_{1\rho}$, can be measured. The on-resonance condition with zero field along the z-axis is distinguished from the off-resonance conditions with a fictitious field along z.

On-resonance spin lock

Following an initial 90° excitation pulse the transverse magnetization is locked in the xy plane by an rf field which is shifted in phase by 90° with respect to the first pulse (Fig. 7.2.4). In a frame rotating with the rf carrier frequency about the laboratory z-axis, the quantization axis of the locked magnetization is in the direction of the $\boldsymbol{B}_1$ field. Any on-resonance magnetization component off-axis from the lock field precesses about the lock field with frequency $\omega_1 = -\gamma B_1$, and the magnetization reduced towards zero in the rotating frame.

$T_{1\rho}$ exhibits a frequency dependence similar to T_2, with the difference that the frequency in the spectral density $J^{(0)}$ at low field is ω_1 instead of zero (cf. eqns (3.5.6) and (3.5.7)). Therefore, relaxation at frequencies close to ω_1 is particularly efficient, and the filter is a band-block filter for correlation times $\tau_c \approx 2\pi/\omega_1$ of molecular motion (cf. Fig. 7.1.3).

Off-resonance spin lock

Typical values of $\omega_1/(2\pi)$ are 1–100 kHz for ^{1}H given a 1 kW rf amplifier and a 10 mm coil diameter. Therefore, $T_{1\rho}$ relaxation is particularly useful for generating contrast based on differences in slow molecular motion, such as in elastomers and in solids.

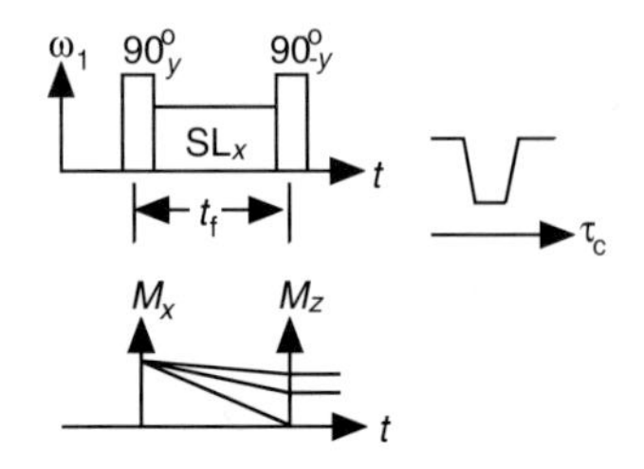

FIG. 7.2.4 [Rom2] Pulse sequence for the $T_{1\rho}$ filter and schematic filter transfer function.

The particular value of ω_1 can be adjusted by the strength B_1/γ of the spin-lock field. The range of effective spin-lock field strengths can be increased if off-resonance techniques are applied. In this case, the effective field $\boldsymbol{B}_{\text{eff}}$ forms an angle θ with respect to the z-axis (cf. Fig. 2.2.7), and the effective relaxation rate $(T_{1\rho,\text{eff}})^{-1}$ is a combination of the longitudinal relaxation rate $(T_1)^{-1}$ and the relaxation rate in the rotating frame $(T_{1\rho})^{-1}$ [Kim2],

$$\frac{1}{T_{1\rho,\text{eff}}} = \frac{1}{T_1}\cos^2\theta + \frac{1}{T_{1\rho}}\sin^2\theta. \tag{7.2.5}$$

For large off-resonance frequencies, the angle θ in (7.2.5) is small enough so that the second 90° pulse in the sequence of Fig. 7.2.4 can be discarded. *Off-resonance spin locking* can also be seen as a low-power alternative to spin locking, so that the technique may be used in medical applications [Fai1].

Spin diffusion during the spin-lock time can be suppressed if the magnetization is locked in the effective field of a multi-pulse sequence for homonuclear dipolar decoupling, such as MREV8 (cf. Table 3.3.1) [Hav1, Sch2]. In this case, the pulse sequence is $(\pi/4)_{-y}$-(MREV8)$_n$-$(\pi/4)_y$, and the $T_{1\rho}$ relaxation of different nuclei in close distance is decoupled. Dipolar decoupling by use of an effective field along the magic angle has been introduced by Lee and Goldburg [Lee2].

$T_{1\rho}$ parameter and dispersion imaging

From a set of images acquired with different spin-lock times (t_f in Fig. 7.2.4) a $T_{1\rho}$ *parameter image* can be calculated for a given strength of the spin-lock field [Rom1]. When applied to elastomers, the value of $T_{1\rho}$ as a function of the spin-lock field strength ω_1 in each pixel can be interpreted in terms of the defect diffusion model (cf. Section 7.1.6) [Bar1, Bar2]. $T_{1\rho}$ contrast can also be introduced in solid-state magic-angle rotating-frame imaging [Del1] (cf. Section 8.6).

The contrast can be enhanced by calculation $T_{1\rho}$ dispersion images [Rom2], where the contrast parameter is extracted from the signal variation with the strength of the spin-lock field at fixed spin-lock times t_f. In the most simple case, a $T_{1\rho}$ *dispersion image* is the difference of two $T_{1\rho}$ images taken at different spin-lock field strengths for the same spin-lock times t_f. Alternatively the difference of two $T_{1\rho}$-weighted images can be analysed. Then the signal amplitude in a given pixel is described by

$$\Delta M_z(\boldsymbol{r}) = M_0(\boldsymbol{r})\left[\exp\left\{-\frac{t_f}{T_{1\rho,\text{eff}}\left(\omega_{\text{eff}}^{(2)}\right)}\right\} - \exp\left\{-\frac{t_f}{T_{1\rho,\text{eff}}\left(\omega_{\text{eff}}^{(1)}\right)}\right\}\right], \tag{7.2.6}$$

where $\omega_{\text{eff}} = -\gamma B_{\text{eff}}$ is the precession frequency about the effective field $\boldsymbol{B}_{\text{eff}}$. For short spin-lock times the exponentials can be approximated linearly so that

$$\Delta M_z(\boldsymbol{r}) = M_0(\boldsymbol{r}) t_f \frac{\Delta T_{1\rho,\text{eff}}}{T_{1\rho,\text{eff}}\left(\omega_{\text{eff}}^{(1)}\right) T_{1\rho,\text{eff}}\left(\omega_{\text{eff}}^{(2)}\right)}, \tag{7.2.7}$$

where $\Delta T_{1\rho,\text{eff}}$ is the difference of the relaxation times at $\omega_{\text{eff}}^{(2)}$ and $\omega_{\text{eff}}^{(1)}$. In such images regions with vanishing $T_{1\rho}$ dispersion are suppressed. Contrast is enhanced in close vicinity of relaxation agents like ^{17}O [Red1] or in morphological regions where the macromolecular composition leads to sufficient spin–lattice relaxation dispersion in the kHz regime [Rom1, Rom3].

7.2.4 Chemical-shift filters

Chemical-shift filters are applied to generate images from signals at selected chemical shifts in the NMR spectrum. In general, two approaches are taken to generate such images. This section covers one approach in dealing with methods to select magnetization at the chemical shifts in question before space encoding. However, also the entire NMR spectrum can be acquired for each volume element and a *chemical-shift selective image* can be generated from the signal intensity at a particular frequency or some other parameter extracted from the spectrum. This second approach is referred to as *spectroscopic imaging* (cf. Section 6.2.4) and is treated in Section 7.3.

Chemical shift difference filters

For biomedical applications, the discrimination of signals from two distinct chemical shifts, in particular from fat and water, is particularly significant. The effect of such a *chemical-shift difference filter* can be achieved by measurement of two data sets with the spin-echo method of space encoding (Fig. 7.2.5(a)) [Dix1, Weh1]. One data set is acquired in the standard mode, where Hahn and gradient echoes are generated to appear at the same time. For the other set, both echoes are separated by a time interval $2\pi/(2\Delta\omega)$, where the frequency difference $\Delta\omega = \omega_A - \omega_B$ is defined by the difference of chemical shifts of water and fat signals. Then, in the maximum of the gradient echo, the signals

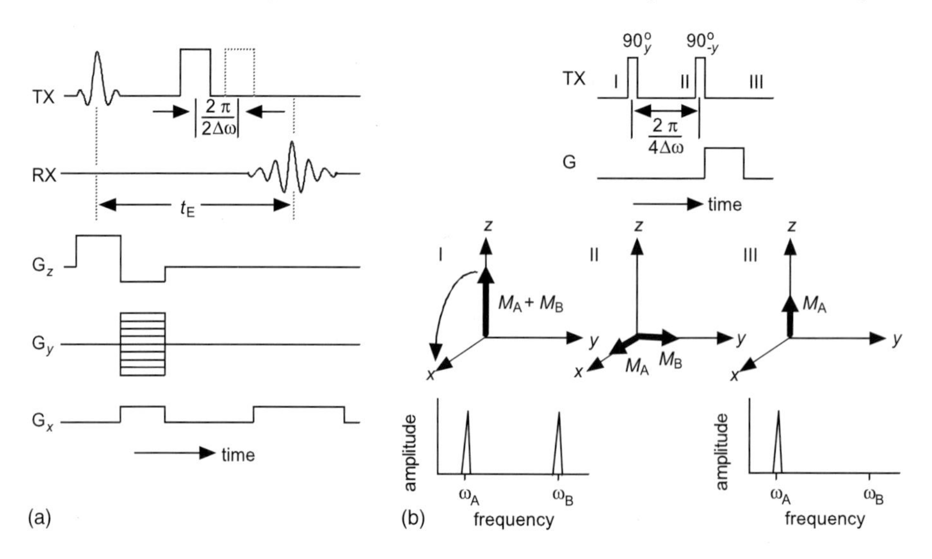

FIG. 7.2.5 Chemical-shift difference filters. (a) Spin-echo method [Dix1, Weh1]. Two data sets are measured. For one set the Hahn and gradient echoes coincide, for the other both echoes are separated by $2\pi/(2\Delta\omega)$. Here $\Delta\omega$ denotes the frequency difference between two lines.
(b) Two-pulse method [Dum1].

of fat and water are both in phase for the first data set and 180° out of phase for the other data set. Addition and subtraction of both data sets produces pure fat and pure water images. This basic scheme has been refined into a single-shot technique [Maj3].

Also, by exploiting the frequency difference $\Delta\omega$ between two resonances, one of the resonances can be suppressed. Two 90° pulses are applied 180° out of phase, a time $2\pi/(4\Delta\omega)$ apart, on resonance with line A to be selected (Fig. 7.2.5(b)) [Dav2, Dum1]. The magnetization from line A does not precess in the rotating frame, while that from line B precesses by 90°, so that it is aligned with the B_1 field at the time the second pulse is applied. Then, the A magnetization is flipped back into the z-direction and the B magnetization dephases possibly with the help of a homogeneity-spoil gradient pulse.

Selective excitation and saturation

If signals at more than two chemical shifts are relevant in the spectrum, a *chemical-shift selective image* of one signal can be acquired by excitation with a *frequency-selective pulse* in the absence of a magnetic-field gradient. The frequency of the pulse can be tuned to the chemical shift of interest. Alternatively the signals of all the other resonances can be suppressed, for instance, by the DIGGER pulse [Dod1] or a coloured-noise pulse [Nil1] (cf. Section 5.3). Dominating signals can be eliminated by established techniques of water suppression [Bra1, Hor1, Xia2]. *Chemical-shift selection* in imaging is often referred to as *CHESS*.

Selective excitation using stimulated echoes

Stimulated-echo imaging or *STEAM imaging* (cf. Section 6.2.5) provides several interesting opportunities for chemical shift-selective excitation [Bur3, Fra2]. As an example Fig. 7.2.6(a) shows the basic pulse sequence for *CHESS-STEAM imaging* and Fig. 7.2.6(b) its extension to simultaneous acquisition of n slice-selective images [Haa2, Haa3]. The first rf pulse is applied in a homogeneous magnetic field. It is

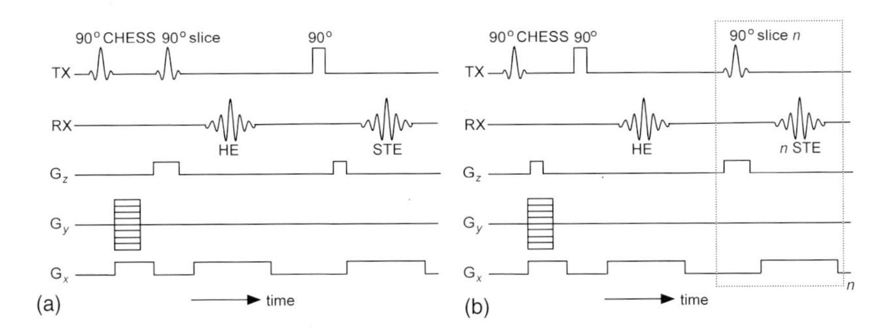

FIG. 7.2.6 [Haa3] Pulse sequences for CHESS-STEAM imaging. The NMR signals are the Hahn or spin echo (HE) and the stimulated echo (STE). (a) Basic sequence for chemical-shift selective measurement of HE and STE images. (b) Sequence for acquisition of n slice-selective images from CHESS stimulated echoes.

frequency-selective and generates transverse magnetization for a narrow range of chemical shifts. This magnetization is subsequently encoded in space. The second rf pulse is also frequency-selective. It is applied in the presence of a gradient and serves to select the slice to be imaged. Only the third pulse can be nonselective. Both the Hahn echo HE and the stimulated echo STE can be imaged, so that T_2- and T_1-weighted CHESS images are acquired in one and the same experiment.

If the functions of the second and the third pulses are interchanged (Fig. 7.2.6(b)), then the third pulse can be repeated for different frequencies to read out stimulated echoes from different slices. The *CHESS multi-slice method* is comparatively insensitive to spatial variations in flip angles from B_1 inhomogeneities and makes efficient use of the available rf power.

The STEAM method may also be employed to acquire individual images from several chemical shifts in one pass (Fig. 7.2.7(a)). The first pulse is selective and is applied in the presence of a gradient for slice selection. The third pulse is also selective but repetitively applied in the absence of a gradient for selection of different chemical shifts. The Hahn echo can then be explored to yield a slice-selective image without chemical-shift selection, and a train of stimulated echoes is acquired where the signals from each echo can be transformed into an image of signals from one particular chemical shift. Finally, instead of the third pulse, the first pulse can be applied repetitively for selection of several chemical shifts. If combined with the multi-slice capabilities of the third pulse, separate images from several chemical shifts and several slices are obtained (Fig. 7.2.7(b)).

Slice selection and chemical-shift effects

When acquiring chemical-shift selective images with slice selection, the influence of the chemical-shift spectrum on the performance of the slice selection in a magnetic-field gradient has to be considered. Only in the limit of strong gradients are the signals of all chemical shifts approximately located in the same slice. Slice selection in weak gradients

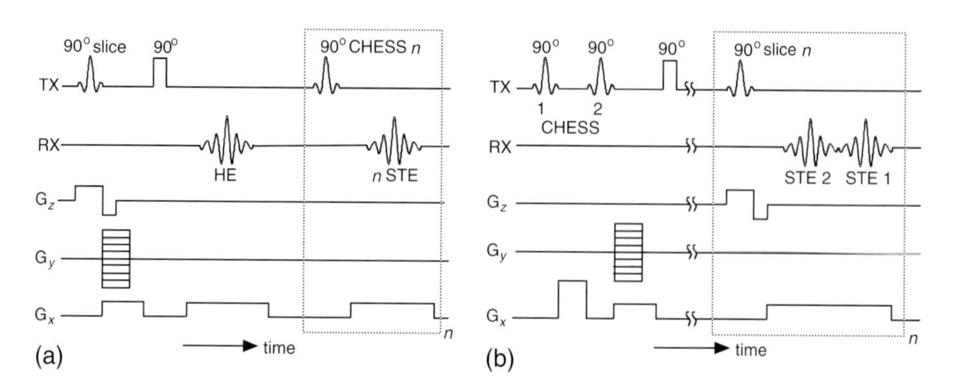

FIG. 7.2.7 [Haa3] Multi-CHESS-STEAM imaging. (a) Simultaneous acquisition of one nonselective HE image and n CHESS-STEAM images from one slice. (b) Multi-slice double CHESS-STEAM method for simultaneous acquisition of two CHESS images at n slices.

leads to chemical-shift selective images from neighbouring slices. These artefacts can be avoided by use of modulated gradients and suitably shaped pulses [Hal2, Vol2].

Frequency-selective excitation in solids

In solid-state imaging, frequency-selective pulses are difficult to apply because of the rapidly decaying FID. Nevertheless, for rare spins, selective excitation may be possible [Dav1], but also saturation sequences which suppress longitudinal magnetization at selected frequencies for durations of the order of the spin–lattice relaxation time T_1 can be applied (cf. Section 5.3.2) [Dod1, Nil1]. For abundant spins, demanding line-narrowing sequences composed of hard pulses must be employed.

One sophisticated pulse sequence, which has been used successfully in *solid-state imaging*, is based on the *DANTE sequence* (cf. Fig. 5.3.15) [Car3, Cor1, Hep1, Hep2]. Decoupling of the homonuclear dipole–dipole interaction between abundant spins like ^{19}F and ^{1}H in solids is achieved by uninterrupted multi-pulse irradiation during the entire length of the DANTE sequence. For example, n MREV8 cycles are applied during the free precession intervals of the original DANTE sequence, and frequency-shifted MREV8 sequences with a phase toggle φ on some of the MREV8 pulses are used instead of the small flip-angle pulses of the original sequence. The selectivity towards the frequency band of interest is determined by the choice of n, the number N of DANTE cycles, and by the frequency of the MREV8 cycles. Further selectivity is achieved by modulating the magnitude of the phase toggle. A Gaussian-shaped modulation results in a Gaussian excitation shape and removes sinc lobes in the excitation spectrum. The effect of several adjustments of the selective excitation sequence on the ^{19}F spectrum of poly(tetrafluorethylene) (PTFE or Teflon) is shown in Figs. 7.2.8(c)–(e), where the centre frequency for shift selection has been set to the edge of the spectrum (b) at about 3 kHz [Hep1].

7.2.5 Susceptibility contrast

The difference of volume *magnetic susceptibility* between two adjacent media gives rise to distortions in the local magnetic field and thus to image artefacts. These artefacts may be detrimental in many applications of imaging in medicine and materials science [Pos1], because they limit the achievable spatial resolution, especially at high fields [Cal14, Cal15], but on the other hand, can provide useful information about material inhomogeneity [Blü5, Cha1, Hwa1].

The susceptibility χ is a dimensionless tensor, which relates the contribution of the magnetic polarization $\boldsymbol{M}_\mathrm{p}$ of matter to the magnetic field $\boldsymbol{H}$ applied *in vacuo*, so that the magnetic induction $\boldsymbol{B}$ in matter can be written as $\boldsymbol{B} = \mu_0(1 + \chi)\boldsymbol{H}$ (cf. Section 2.2.1). Susceptibility induces a frequency shift similar to the chemical shift induced by the magnetic shielding of the nucleus from the surrounding electrons (cf. Section 3.1.1). However, only the susceptibility differences with respect to a reference value are commonly significant similar to the chemical shift with respect to a reference compound. Because of the similarity of chemical shift and frequency shifts from susceptibility differences, conventional selection of slices by selective pulses in applied field gradients is ill-defined. On the other hand, the field distortions are localized on a much larger scale

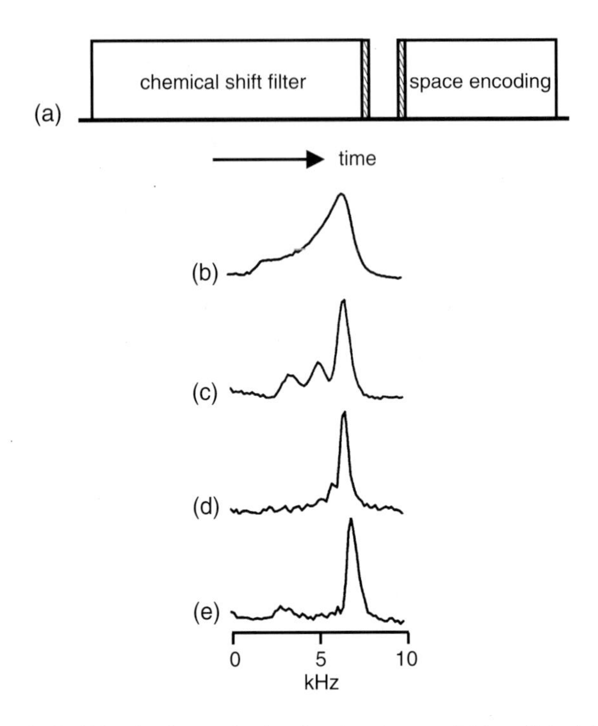

FIG. 7.2.8 Chemical-shift selective excitation for abundant nuclei in solids. (a) Components of the pulse sequence for chemical-shift selective imaging. The chemical-shift filter consists of a dipolar decoupled DANTE sequence [Hep2]. It is composed of a series of phase-toggled MREV8 cycles separated by n standard MREV8 cycles. The shaded pulses form a z filter with long-lived longitudinal magnetization between the pulses for gradient switching. (b) ^{19}F chemical-shift powder spectrum of PTFE [Hep1]. The effect of different adjustments of the DANTE filter is illustrated in (c)–(e). The parameters of the DANTE sequence are defined in Fig. 5.3.15. (c) $n = 3$, $N = 4$, $\varphi = 11.2°$. (d) $n = 3$, $N = 8$, $\varphi = 5.6°$. (d) $n = 3$, $N = 14$. A Gaussian-shaped phase error band with $\varphi = 0°, 0°, 2.8°, 5.6°, 8.4°, 11.2°, 14°, 14°, 11.2°, 8.4°, 5.6°, 2.8°, 0°, 0°$ was used.

than the chemical shift, so that they can give rise to strong gradients in regions large enough to have diffusive attenuation, for instance, for liquids in porous structures [Cal10, Fra1, Hür2, Pos1]. A particular challenging object for imaging in this respect is lung, which consists of tissue, water, and air. Here images can be measured which are free of diffusion attenuation but are weighted with the second moment of the lineshape arising from inhomogeneous broadening by the distribution of susceptibility differences within one pixel [Gan1]: such images are obtained as the quotient of images measured with a time offset between Hahn and gradient echoes and with coincidence of both echoes.

Image distortions from susceptibility mismatch

For a given object immersed in an environment with different magnetic susceptibility the image distortion depends on the shape of the object, that is, on the tensorial properties of

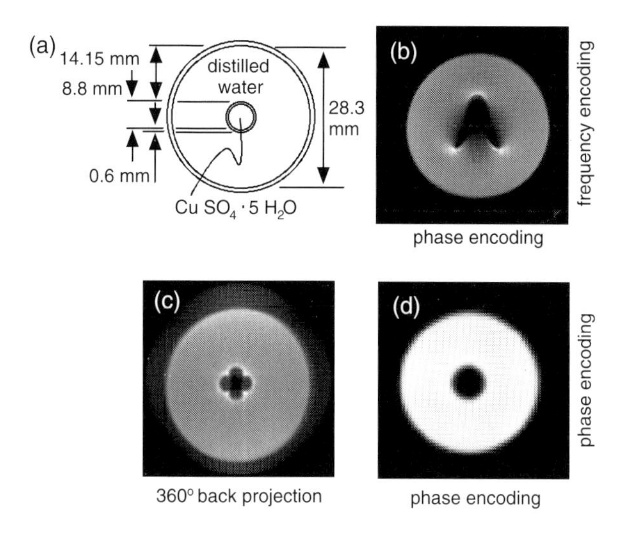

FIG. 7.2.9 [Beu1] Susceptibility effects in NMR imaging. (a) Phantom of two coaxial glass cylinders. The outer one is filled with distilled water, the inner one with a copper sulfate solution. (b) Spin-warp image of the phantom. The core appears dark because of fast signal relaxation. The spearhead image distortion results from the susceptibility difference of glass *versus* water. (c) Image reconstruction from projections acquired at angles spaced equally on a full circle leads to a four-leaf-clover shaped artefact. (d) An undistorted image is obtained by 2D phase encoding.

the *susceptibility difference*. A simple and model object to study susceptibility effects in imaging is that of a cylinder immersed in distilled water (Fig. 7.2.9(a)) [Beu1]. The inner cylinder is a glass tube with a wall 0.6 mm thick. The cylinder is filled with a copper sulfate solution for signal attenuation by T_1 relaxation relative to the distilled water. The glass wall is responsible for field distortions through susceptibility differences.

For 2D imaging by the spin-warp method, a spearhead shape is observed instead of a dark circular spot in the centre (b). The direction of the spear is aligned with the direction of frequency encoding, and its orientation depends on the sign of the susceptibility difference. At the sharp corners of the spear, bright regions are observed because the signal is shifted away from its proper position by the field distortions. With the projection-reconstruction method a four-leaf-clover shape results (c).

The frequency shifts caused by the *susceptibility mismatch* can be resolved with spectroscopic imaging, they can be refocused in both space dimensions by 2D phase encoding, and they can be corrected [Cal10, Wei5]. In the case of pure phase encoding, the circular core of the object is imaged free of distortions (d). This method requires indirect detection in two dimensions and consequently is time consuming. On the other hand, the dephasing of transverse magnetization by static background gradients can repetitively be refocused by a CPMG echo train [Ben3], and to obtain images, gradients can be switched so that their signs alter at the position of each 180° refocusing pulse. Then the dephasing proceeds under the applied gradient only. This scheme can be readily realized with rotating gradients and leads to satisfactory images even in the presence of a ferromagnetic nail near the object [Mil1].

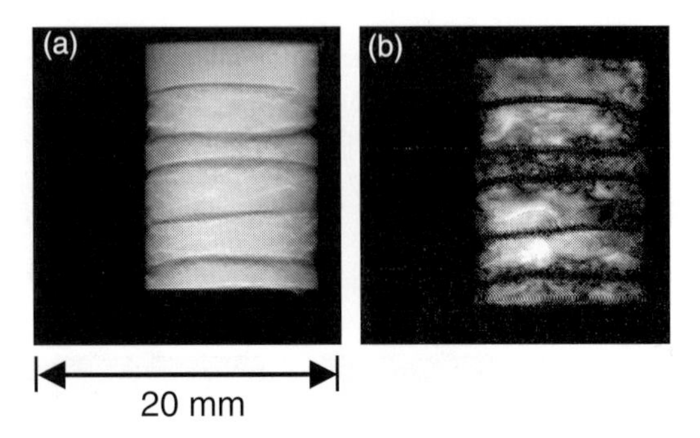

FIG. 7.2.10 Susceptibility contrast in commercial EPDM samples. (a) 2D projection acquired by spin-warp imaging. (b) 2D projection acquired by gradient-echo imaging. The images were measured at a temperature of 363 K. Adapted from [Blü8] with permission from Hüthig Gmbh.

Susceptibility contrast

The distortions of image features by susceptibility differences are not necessarily unwanted. *Susceptibility contrast* can be exploited to emphasize small inhomogeneities which are otherwise barely visible. In that case, it is helpful to avoid a spin echo in the space-encoding part of the imaging sequence altogether and to apply a gradient-echo imaging technique instead [Blü5, Cha1]. In this way, interfaces between two materials, voids, and paramagnetic impurities like clusters of carbon-black can be readily detected in elastomers. A comparison of 2D projections obtained from a stack of six layers of industrial EPDM rubber sheets separated by PTFE layers is shown in Fig. 7.2.10 [Blü8]. The layers appear only slightly inhomogeneous in projection (a) acquired by the spin-warp method, but severe inhomogeneities become apparent in projection (b), which was measured by gradient-echo imaging. These structures are attributed to the folding of thin rubber layers during processing of the material in a rolling mill before vulcanization.

7.2.6 Translational diffusion and transport filters

In addition to information about space, information about molecular transport in terms of displacements from diffusion [Hol3, Pri1, Pri2], velocity [CapP1, Tur2], acceleration [Tli1], etc. can be measured by NMR with magnetic-field gradients (cf. Section 5.4) [Cal1, Can1, Kim2, Pac2, Sod1]. The gradients can be static or pulsed. In either case, echo techniques are employed. *Pulsed-field-gradient NMR* has been pioneered by Stejskal and Tanner [Ste1, Ste2, Tan1, Tan2]. It can be combined with space-encoding schemes to image spatial distributions of displacements, and components of the velocity and acceleration vectors [Cal1, Cal2, Cal4, Cal7, Cap1, Fuk1, Kim2, Pop1, Sey1]. This unique attribute of NMR makes NMR imaging of molecular transport possibly the most interesting tool for analysis of many rheological features such as the non-Newtonian behaviour of fluids and the transport and mixing of fluids in porous media, pipes, filters, mixers, and extruders [Cal16, Cal17, Fuk1].

Phase encoding of molecular transport and gradient moments

The most sensitive method of acquiring information about molecular transport is by use of the phase ϕ of the transverse magnetization, which can be measured with great accuracy in NMR. It is given by the time integral of the resonance frequency $\omega_{0\mathrm{r}}(t)$, which can be expressed in terms of the NMR frequency $\omega_0 = -\gamma B_0$ in homogeneous fields, the frequency offset associated with the gradient vector $\boldsymbol{G}(t)$, and the space vector $\boldsymbol{r}(t)$ (cf. eqn (1.2.4)),

$$\begin{aligned}\phi(t) &= \int_0^t \omega_{0\mathrm{r}}(t')\,\mathrm{d}t' = -\gamma B_0 t - \gamma \int_0^t \boldsymbol{G}(t')\boldsymbol{r}(t')\,\mathrm{d}t' \\ &= -\gamma B_0 t - \gamma \int_0^t \boldsymbol{G}(t')\,\mathrm{d}t'\boldsymbol{r}(0) - \gamma \int_0^t \boldsymbol{G}(t')t'\,\mathrm{d}t'\boldsymbol{v}(0) \\ &\quad - \gamma \int_0^t \boldsymbol{G}(t')t'^2\,\mathrm{d}t'\frac{\boldsymbol{a}(0)}{2} + \cdots \\ &= \phi_0 + \boldsymbol{k}\boldsymbol{r} + \boldsymbol{q}_v\boldsymbol{v} + \boldsymbol{\varepsilon}\boldsymbol{a} + \cdots. \end{aligned} \tag{7.2.8}$$

Here $\boldsymbol{G}$ is a time-dependent, experimental variable, and the space vector $\boldsymbol{r}(t)$ is time dependent because of translational motion of the nuclear spins (cf. Section 1.2). Therefore, the phase ϕ is time-dependent as well. For short times, the final position $\boldsymbol{r}(t)$ of the spins assumed after the time has elapsed, can be approximated by a Taylor series with a finite number of terms (cf. eqn (5.4.54)). These terms are discriminated by the power of the time lag t and involve initial position $\boldsymbol{r}(0)$, velocity $\boldsymbol{v}(0)$, and acceleration $\boldsymbol{a}(0)$ as coefficients to different moments $\boldsymbol{m}_k$ of the time-dependent gradient vector $\boldsymbol{G}(t)$,

$$\boldsymbol{m}_k = \int_0^t \boldsymbol{G}(t')t'^k\,\mathrm{d}(t'). \tag{7.2.9}$$

For convenience, but not by necessity, the moments are usually defined with respect to the gradient $G_i = \partial B_z/\partial x_i$ of the magnetic field. Moments with higher-order spatial $\partial^l B_z/\partial x_i^l$ derivatives of the magnetic field can be considered as well and may define another class of interesting NMR experiments to study translational motion (cf. eqn (1.2.4)).

The second term in the second line of (7.2.8) provides the familiar definition of the $\boldsymbol{k}$ vector for space encoding (cf. eqn (2.2.23)), because

$$\boldsymbol{k} = -\gamma \int_0^t \boldsymbol{G}(t')\,\mathrm{d}t' = -\gamma\boldsymbol{m}_0. \tag{7.2.10}$$

The higher-order *gradient moments* encode parameters of motion like velocity $\boldsymbol{v}$ and acceleration $\boldsymbol{a}$ and are related to the Fourier conjugate variables $\boldsymbol{q}_v$ and $\boldsymbol{\varepsilon}$ of these quantities, so that the signal phase contains contributions from position, velocity, and acceleration (cf. eqn (5.4.55)).

Because the time dependence of the gradient vector is determined by external variables, the gradient moments can be individually made to assume certain values at given times t. In particular, they can be made to vanish. Then, at that time, there is no phase evolution from the associated motional variable, and an echo is obtained. For instance, a *position echo* is generated by the familiar *gradient echo* for which $\boldsymbol{m}_0 = \mathbf{0}$, and a *flow echo* is obtained whenever $\boldsymbol{m}_1$ vanishes. Therefore, in the presence of constant flow, position can be measured accurately if at the time of space encoding, $\boldsymbol{m}_0$ is different from zero and $\boldsymbol{m}_1$ is equal to zero. Conversely, motion from coherent flow and incoherent diffusion can be encoded independent of position if $\boldsymbol{m}_0 = \mathbf{0}$ and higher-order moments are different from zero. Schemes for nulling of gradient moments to eliminate effects of position, velocity, and acceleration on the magnetization phase are referred to as *compensation schemes*.

Illustration of position and velocity compensation

The effects of *position compensation* and *velocity compensation* are readily illustrated by the phase evolution of magnetization from a liquid flowing in an inhomogeneous field under CPMG irradiation (Fig. 7.2.11). Each 180° pulse inverts the sign of the effective field inhomogeneity, for example, the sign of the effective field gradient. The zeroth moment $\boldsymbol{m}_0$ vanishes at every echo time t_E leading to the familiar occurrence of a Hahn echo as an example of phase compensation from position. The first moment $\boldsymbol{m}_1$, however,

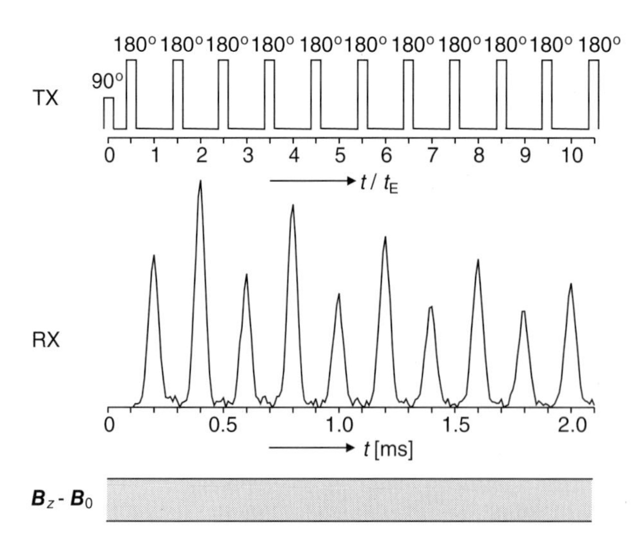

FIG. 7.2.11 Illustration of echo attenuation in a CPMG train by molecular translational motion in an inhomogeneous magnetic field. For constant gradients the inhomogeneous part $\boldsymbol{B}_z - \boldsymbol{B}_0$ of the magnetic field is fully determined by the field gradient $\boldsymbol{G}$ as $\boldsymbol{B}_z - \boldsymbol{B}_0 = \boldsymbol{G}\boldsymbol{r}$. The echoes shown were measured by the NMR-MOUSE (cf. section 9.3.4) for $CuSO_4$-doped water flowing through a tube in z-direction at a volume-flow rate of 2 l/h [Eym1]. The inhomogeneous field of the NMR-MOUSE in z-direction is given by a linear-field gradient, i.e. by $B_z - B_0 = (\partial^2 B_z/\partial z^2) z^2$.

is nonzero for the first echo and all odd echoes, so that the phase evolution associated with velocity is not compensated for. This is achieved for the second echo and all even echoes, where both $\boldsymbol{m}_0$ and $\boldsymbol{m}_1$ vanish, as can be seen by explicit calculation of moments, which, for simplicity, is performed for a linear space dependence expressed by the gradient $\boldsymbol{G}$:

First echo:

$$\boldsymbol{m}_0 = \boldsymbol{G}\left[\left(\frac{t_E}{2}\right)^1 - (0)^1 - (t_E)^1 + \left(\frac{t_E}{2}\right)^1\right] = \boldsymbol{0}, \tag{7.2.11a}$$

$$\boldsymbol{m}_1 = \frac{\boldsymbol{G}}{2}\left[\left(\frac{t_E}{2}\right)^2 - (0)^2 - (t_E)^2 + \left(\frac{t_E}{2}\right)^2\right] = -\boldsymbol{G}\frac{t_E^2}{4}. \tag{7.2.11b}$$

Second echo:

$$\boldsymbol{m}_0 = \boldsymbol{G}\left[\left(\frac{t_E}{2}\right)^1 - (0)^1 - \left(\frac{3t_E}{2}\right)^1 + \left(\frac{t_E}{2}\right)^1 + (2t_E)^1 - \left(\frac{3t_E}{2}\right)^1\right] = \boldsymbol{0}, \tag{7.2.12a}$$

$$\boldsymbol{m}_1 = \frac{\boldsymbol{G}}{2}\left[\left(\frac{t_E}{2}\right)^2 - (0)^2 - \left(\frac{3t_E}{2}\right)^2 + \left(\frac{t_E}{2}\right)^2 + (2t_E)^2 - \left(\frac{3t_E}{2}\right)^2\right] = \boldsymbol{0}. \tag{7.2.12b}$$

As a consequence, every odd echo is attenuated by molecular diffusion and flow and every even echo is not [Car1].

Practical realizations of this principle for *velocity encoding* and *flow compensation* are illustrated in Fig. 7.2.12 [Cap1, Pop1]. If the gradient pulses are applied with equal durations, amplitudes and separations, the sequences (a) produce the familiar gradient and Hahn echoes, which eliminate phase evolution from different positions in the sample. The sequences on the left of the figure are pure gradient-modulation sequences, and those on the right are used in Hahn-echo sequences, where the inversion pulse has the effect of changing the sign of the gradient. The sequences (b) are constructed as two echoes

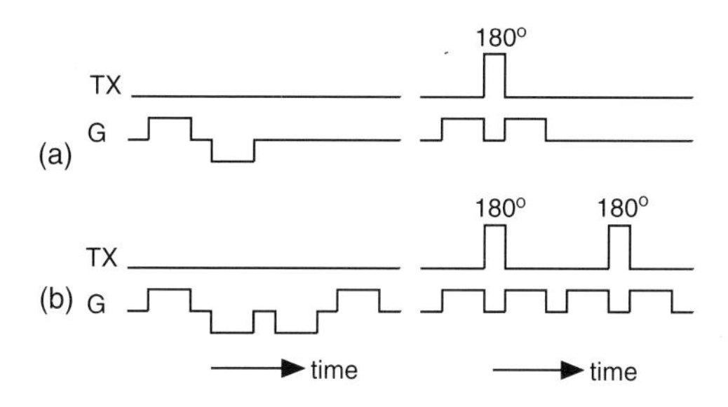

FIG. 7.2.12 [Cap1] Schemes for compensation of gradient moments. Left: gradient waveforms without refocusing pulses. Right: gradient waveforms for use in Hahn-echo sequences.
(a) Nulling of the zeroth order moment $\boldsymbol{m}_0$ (gradient or position echo) for velocity encoding.
(b) Nulling of the zeroth and first-order moments $\boldsymbol{m}_0$ and $\boldsymbol{m}_1$ (position and velocity echoes) for flow compensation in frequency encoding of space and for phase encoding of acceleration.

back to back, so that again the phase evolution associated with the position in the sample is refocused. In addition to that a velocity echo is obtained also, so that $\boldsymbol{m}_1 = \mathbf{0}$, and all magnetization components with constant but different velocities contribute to the echo. In fact the sequence (b) on the left is just a section of a *CPMG sequence* with two echoes and pulsed instead of static gradients (cf. Fig. 7.2.11), for which the moments have explicitly been calculated in (7.2.12). If the amplitudes of one of the two gradient pulse pairs in (b) are changed, a position echo is still obtained, but no longer a velocity echo. In this way velocity can be encoded independent of position.

This principle must be observed in flow imaging [Cal1, Cal4, Cap1, Pop1, Pop2]: during the flow-encoding period the conditions $\boldsymbol{m}_0 = \mathbf{0}$ and $\boldsymbol{m}_1 \neq \mathbf{0}$ and during the space-encoding period $\boldsymbol{m}_0 \neq \mathbf{0}$ and $\boldsymbol{m}_1 = \mathbf{0}$ must be met. More complicated sequences result if phase evolution from accelerated motion is considered as well [Tli1].

Echo attenuation from molecular self-diffusion and flow

Translational diffusion is measured by the echo attenuation in field gradients relative to that in the absence of field gradients (cf. Fig. 7.2.11). For simplicity, the motion of a molecule during the encoding time of a *Hahn echo* is considered in a constant field gradient. By moving from its initial location $\boldsymbol{r}_\mathrm{i}$ with a given strength of the magnetic field to another location $\boldsymbol{r}_\mathrm{f}$ with different field strength, the transverse magnetization associated with the molecule accumulates a phase change $\Delta\phi$, which derives from (7.2.8). In an experiment, the sum of the transverse magnetization components from all molecules is measured. Because diffusion is an incoherent process, the associated phase changes from all molecules in the sample equally cover positive and negative values so that the echo amplitude is attenuated but the echo phase remains unaffected. For flow or acceleration, which arises coherently on a space scale of the voxel resolution, the phase changes as well. Evaluation of the position-dependent term in (7.2.8) produces the phase change in the echo maximum of the transverse magnetization from a single molecule when it is displaced by $\boldsymbol{R} = \boldsymbol{r}(t_\mathrm{E}/2) - \boldsymbol{r}(0) = \boldsymbol{r}_2 - \boldsymbol{r}_1$ in a field gradient,

$$\Delta\phi = \boldsymbol{k}\left(\frac{t_\mathrm{E}}{2}, 0\right)\boldsymbol{r}(0) + \boldsymbol{k}\left(t_\mathrm{E}, \frac{t_\mathrm{E}}{2}\right)\boldsymbol{r}\left(\frac{t_\mathrm{E}}{2}\right) = \boldsymbol{k}(\boldsymbol{r}_2 - \boldsymbol{r}_1) = \boldsymbol{q}\boldsymbol{R}, \tag{7.2.13}$$

where $\boldsymbol{k} = \boldsymbol{k}(t_\mathrm{E}, t_\mathrm{E}/2) = -\boldsymbol{k}(t_\mathrm{E}/2, 0)$. Here the wave vector evolution has been taken to start anew at the beginning of the dephasing and the rephasing times of a spin-echo experiment (rf scheme (a) and gradient scheme (d) in Fig. 7.2.13 below) in analogy to a 2D experiment with an evolution and a detection period. Alternatively the change $\Delta\boldsymbol{k} = \boldsymbol{q}$ in wave vector $\boldsymbol{k}$ accumulated during each of these time intervals can be used in the terminology of a 1D experiment. The same expression is obtained if two equal gradient pulses are applied a time Δ apart, one in the dephasing period and one in the rephasing period (rf scheme (a) and gradient scheme (c) in Fig. 7.2.13, cf. also Fig. 5.4.4 and eqn (5.4.25)).

The normalized echo amplitude is given by the ensemble average $\langle\cdots\rangle$ of all magnetization components with different $\Delta\phi$ (cf. eqn (5.4.26)),

$$a_\Delta = \langle \exp\{\mathrm{i}\Delta\phi\}\rangle = \langle \exp\{-\mathrm{i}\boldsymbol{q}\boldsymbol{r}_2\}\exp\{\mathrm{i}\boldsymbol{q}\boldsymbol{r}_1\}\rangle. \tag{7.2.14}$$

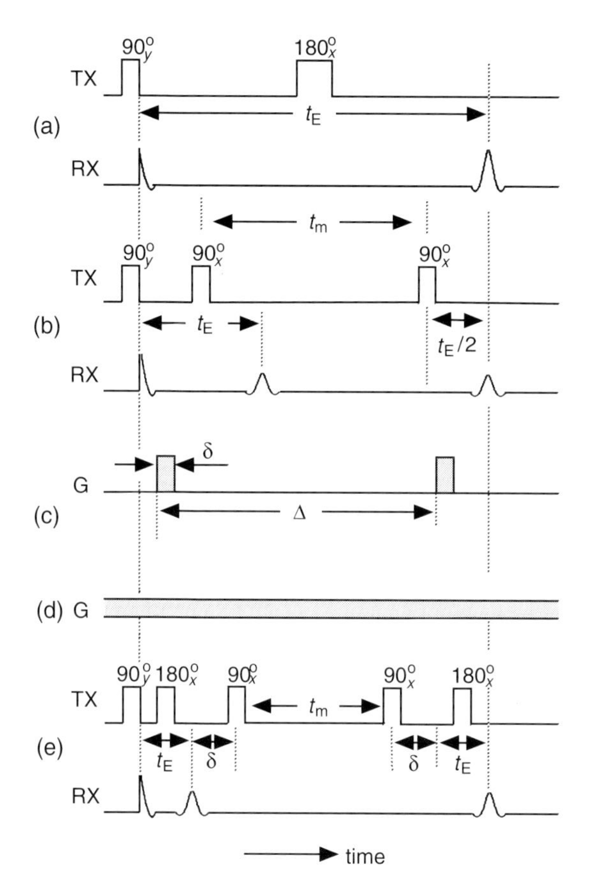

FIG. 7.2.13 [Kim2] Radio-frequency and gradient schemes for measurements of translational diffusion in simple liquids. (a) Hahn echo. (b) Stimulated echo. (c) Scheme for PFG NMR. (d) Static-field gradients. (e) PFG analogon for use with static-field gradients. Note, that all filters end with transverse magnetization.

This factor expresses the *echo attenuation by diffusion* in formal analogy to quasi-elastic incoherent neutron scattering [Fle1]. Depending on the type of experiment, the echo amplitude is attenuated by T_1 and T_2 relaxation as well, so that the total attenuation of the echo amplitude a can be written as

$$a = M_0' a_\Delta a_{T1} a_{T2}, \tag{7.2.15}$$

where M_0' is the echo amplitude in absence of any attenuation. For Hahn echoes, M_0' is given by the thermodynamic equilibrium magnetization M_0, but only half of the magnetization is refocused in the stimulated echo, so that in this case $M_0' = M_0/2$. The echo attenuation factors are collected in Table 7.2.1 for the basic Hahn and stimulated-echo experiments with static (cf. eqns (2.2.40) and (2.2.41)) and pulsed-field gradients [Kim2], where D is the constant of molecular self-diffusion.

Table 7.2.1 Echo attenuation factors for molecular diffusion and mono-exponential relaxation. Experiments and variables are defined in Fig. 7.2.13

rf	Gradient	M_0'	a_Δ	$a_{T1}\, a_{T2}$
(a)	(d)	M_0	$\exp\{-\gamma^2 G^2 D t_E^3/12\}$	$\exp\{-t_E/T_2\}$
(b)	(d)	$M_0/2$	$\exp\{-\gamma^2 G^2 D(t_E/2)^3[2/3 + 2t_m/t_E]\}$	$\exp\{-t_m/T_1\}\exp\{-t_E/T_2\}$
(a)	(c)	M_0	$\exp\{-\gamma^2 G^2 D\delta^3[\Delta/\delta - 1/3]\}$	$\exp\{-t_E/T_2\}$
(b)	(c)	$M_0/2$	$\exp\{-\gamma^2 G^2 D\delta^3[\Delta/\delta - 1/3]\}$	$\exp\{-t_m/T_1\}\exp\{-t_E/T_2\}$

Translational-diffusion filters

Experiments for measurement of translational self-diffusion corresponding to *translational-diffusion filters* are illustrated in Fig. 7.2.13 [Kim2]. *Hahn-echo* experiments (a) are limited in their sensitivity to fast diffusion, because the timescale for formation of the echo is determined by T_2. Accordingly, the attenuation of the echo amplitude in addition to effects from diffusion is given by T_2 only. Slower diffusion processes can be tackled by *stimulated-echo sequences* (b), where the periods of dephasing and rephasing under the influence of the gradient are separated by the mixing time t_m, which can be in the order of T_1. Time-invariant gradients [Car1, Hah1] (d) can provide exceptional strength and stability. Particularly in the stray field of superconducting magnets, values of up to 80 T/m can be obtained [Dem10, Kim5] and dedicated gradient magnets may achieve values in the order of 200 T/m [Fuj1]. Pulsed gradients (c) may also achieve considerable strength [Cal6]. For example, current flow in single wires can provide gradients up to 50 T/m in direct vicinity of the wire [Cal13].

The evolution of transverse magnetization in a magnetic-field gradient establishes a helical *magnetization grid* (cf. Fig. 2.2.4) which is washed out by the diffusion process [Jon3, Kim2]. This interpretation is common to static and pulsed-field gradient methods and bears close analogy to diffusion studies by LASER-induced gratings [Saa1]. The fundamental differences between both types of experiments are that sample vibrations induced from strong currents pulsed in magnetic fields are avoided with static gradients, and that a reduction of the sensitive volume arises from slice selection, which can be eliminated by pulsing the gradients in the absence of rf irradiation.

If the diffusion measurement is performed in the presence of an electric field, the migration of simple ions, charged molecules and charged molecular aggregates can be investigated. A variety of techniques has been investigated for *electrophoretic NMR* [Hei1, Hol1, Joh1] including imaging of electroosmotic flow [Dwu1].

In the absence of residual dipolar and quadrupolar broadening, the echo attenuation from relaxation can be separated from the diffusion attenuation through the use of *relaxation compensated sequences* in static-field gradients [Dem10, Kim6]. One such sequence is given by the stimulated-echo sequence (b) in Fig. 7.2.13, for which both the primary or Hahn echo and the stimulated echo need to be sampled. The ratio of the stimulated to the primary echo eliminates spin density and the transverse relaxation factor a_{T2} in (7.2.15). Diffusion contrast can be introduced by varying t_E and keeping t_m constant or short enough to eliminate signal variation with T_1. The same principle of eliminating relaxation weights is applied in the five-pulse sequence (e), which is used in static-field gradients to emulate a PFG experiment. Within the dephasing and the rephasing periods of a stimulated-echo sequence, the transverse magnetization is refocused by a Hahn echo. Different diffusion weights are obtained for constant relaxation weights by varying δ at the expense of t_E and keeping t_m constant. From a set of *diffusion-weighted images* a *diffusion-parameter image* can be calculated [Gib1, Ler1, Wei5].

Another way to introduce diffusion contrast is to acquire two separate images, one with and one without diffusion gradients for example in a spin-echo diffusion filter (combination of schemes (a) and (c) in Fig. 7.2.13). In the quotient of both images, the relaxation weight is eliminated as well as spin density,

$$\frac{M_0(\boldsymbol{r})\,\exp\{-(\gamma^2 G^2 D(\boldsymbol{r})\delta^3/12)(\Delta/\delta-(1/3))\}\exp\{-t_E/T_2\}}{M_0(\boldsymbol{r})\exp\{-t_E/T_2\}} =$$
$$\exp\left\{-\frac{\gamma^2 G^2 D(\boldsymbol{r})\delta^3}{12}\left(\frac{\Delta}{\delta}-\frac{1}{3}\right)\right\}, \tag{7.2.16}$$

and the contrast is solely defined by a function of the space-dependent diffusion coefficient $D(\boldsymbol{r})$.

The *diffusion filters* employing pulsed-field gradients are readily combined with different space-encoding techniques such as spin-echo imaging [Cal4], back-projection imaging [Cal7], and echo-planar imaging [Tur1], as well as with additional spectroscopic resolution [Wei5]. Although a well-defined diffusion preparation is necessary for quantitative analysis of diffusion data, a diffusion weight can be directly integrated into the space-encoding pulse sequence. Examples are stimulated-echo imaging [Bur3, Mer3] and fast methods based on the steady-state free precession technique (SSFP) [Mer2].

Restricted diffusion

The echo attenuation factors given in Table 7.2.1 are valid for free diffusion only. For *restricted diffusion*, the echo attenuation is reduced because the maximum diffusion distance is limited by spatial confinements. The average geometry of the confinements, pores, or cells within the sensitive volume of the experiment or within a voxel modifies the shape of the echo attenuation curves measured as a function of $\boldsymbol{q}$ and may give rise to diffraction patterns (cf. Fig. 5.4.8) [App1, Bal3, Cal8, Cal9, Cal16, Coy1]. Pore sizes, the pore-size distribution [Gar3, Lat1], and generally, information about the anisotropy of matter can be obtained from analysis of these curves or from diffusion-encoded images [Bah1, Che1, Dou1, Mos1, Swi1]. In the long-diffusion time limit, the Fourier transform of the echo amplitude as a function of the gradient strength is the auto-correlation function of the pore geometry. The simple PFG echo techniques of Fig. 7.2.13 can be extended to include combinations of orthogonal gradient pulses to probe the local geometry in greater detail [Cor2], and generalized gradient forms can be explored [Cal5, Cal11, Cap2] (cf. Section 5.4.3).

The restricted diffusion in local magnetic field variations from susceptibility effects associated with sample inhomogeneities gives rise to signal enhancement and attenuation also known as *edge enhancement* in imaging [Cal10, Des1, Mer1, Nes1, Püt1, Ste3]. For quantitative diffusion measurements the effect of these time-invariant field gradients can

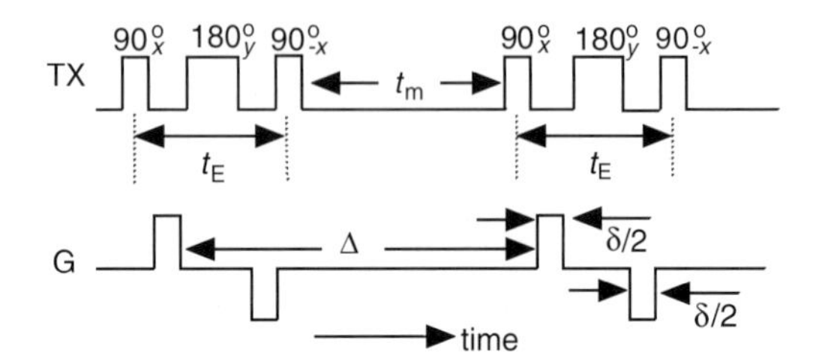

FIG. 7.2.14 [Pac1] Stimulated echo filter for measurement of diffusion in the presence of static background gradients. Note that this filter ends with longitudinal magnetization.

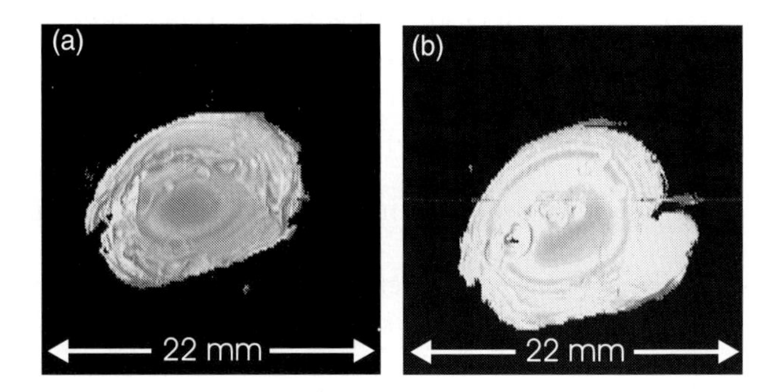

FIG. 7.2.15 [Gie1] Functional diffusion maps according to (7.2.16) of an intervertebal disk from a rabbit acquired *in vitro*. (a) Without compression. (b) With compression. The diffusion gradients have been applied in the direction of applied pressure.

be refocused by incorporation of Hahn echoes into the dephasing and rephasing periods of a stimulated echo (Fig. 7.2.14) [Lia1, Luc1, Pac1].

A manifestation of restricted diffusion is illustrated in Fig. 7.2.15 by two diffusion images according to (7.2.16) of an intervertebral disk from a rabbit without (a) and with (b) compression [Gie1]. The water diffusion is restricted within the cellular structure of the disk, and the cells are deformed by compression. The diffusion gradient has been applied in the direction of the external stress perpendicular to the image plane. The brighter diffusion image of the stressed disk demonstrates that the path lengths available for diffusion with subsequent signal attenuation have been reduced by the stress. By applying the diffusion gradients in a number of different directions, the anisotropy of the confinements can be studied [Weg2]. A so-called *diffusion tensor* can be derived [Mat2]. It can be decomposed according to its transformation properties under symmetry operations, and the elements of the irreducible tensor components are contrast parameters suitable for analysis of tissue properties in medical applications [Bas1, Bas2, Bas3]. This approach is known as *diffusion tensor imaging* [Pap1].

Diffusion measurements in the rotating frame

A spatial magnetization grid similar to the one established by field gradients applied in the laboratory frame can be generated by *field gradients in the rotating frame*, that is by gradients in the $\boldsymbol{B}_1$ field [Kim2]. The loss of the grid structure by molecular self-diffusion and flow phenomena can be observed by rotating-frame imaging methods (cf. Section 6.3) [Kim7, Maf1, Mis1]. No echoes are involved in this magnetization rotating-frame imaging (MAGROFI) technique, and apart from detection, only long-lived longitudinal magnetization is involved [Kim2, Kim7].

Amplitude encoding of flow

Coherent magnetization transport associated with the velocity of *flowing spins* leads to a phase shift in the transverse magnetization according to (7.2.8). This phase shift is the signature of the velocity component parallel to the direction of the pulsed-field gradient, and each component has to be measured in a separate experiment. Flow-imaging techniques faster but less accurate than phase-encoding methods are based on amplitude modulation of the magnetization from flowing spins [Cap1, Fuk1].

In the simplest case of amplitude encoding the continuous replacement of magnetization by flow through the sensitive volume leads to enhancement of the signal from the flowing spins for saturation conditions of the magnetization from stationary spins by short repetition times or saturation pulses [Gar4, Sin1]. Alternatively, a planar slice can selectively be excited and the slice deformation by the different flow velocities within the slice can be imaged [Pop1, Sin2]. This method is called *time-of-flight imaging*. Space encoding in the presence of flow requires *flow compensation* through nulling of the first-order gradient moment $\boldsymbol{m}_1$ to eliminate phase shifts due to motion. This is particularly important for imaging of fast flow [Gui1, Gui2, Kos1].

Flow-sensitive excitation-pulse combinations can also be designed, which exploit the flow-induced phase evolution of the signal during a given pulse delay in the same way as a chemical-shift difference causes a phase evolution during such a pulse delay. The principles explored in water suppression by selective excitation with binomial pulses [Hor1] can be applied for suppression of the signal from stationary spins [Pop1], and velocity-selective rf pulse trains can be designed for the selection of magnetization flowing at particular velocities [Nor1].

A method used to image the *fast flow* of matter tags selected spins in such a way that a spatial grid of longitudinal magnetization is established, and its deformations are followed as a function of time between tagging and imaging [Kim2, Kos1, Ice1]. In combination with fast imaging methods such as FLASH imaging and echo-planar imaging, even turbulent flow [Kue1] can be studied [Gat1, Gui1, Kos1, Kos2, Kos3, Kos4, Rok1].

For tagging of stripes in one dimension, a *DANTE sequence* [Mor1] is applied in the presence of a gradient in such a way that its envelope follows a sinc function (Fig. 7.2.16(a)). Then the shape of each individual stripe is given by the Fourier transform of the envelope, which is a rectangular pulse [Kim2]. In this way, a sharp grid of transverse magnetization can be generated. By applying the same sequence for a second time but in the presence of a gradient in direction orthogonal to the first, a rectangular grid of longitudinal magnetization is obtained.

Figure 7.2.16 gives an example for a *flow-tagging* pulse sequence (a) and an image (b) of stationary water in a tube acquired by a spin-echo sequence following the tagging sequence [Kos3]. By variation of the delay between the tagging pulses and the data

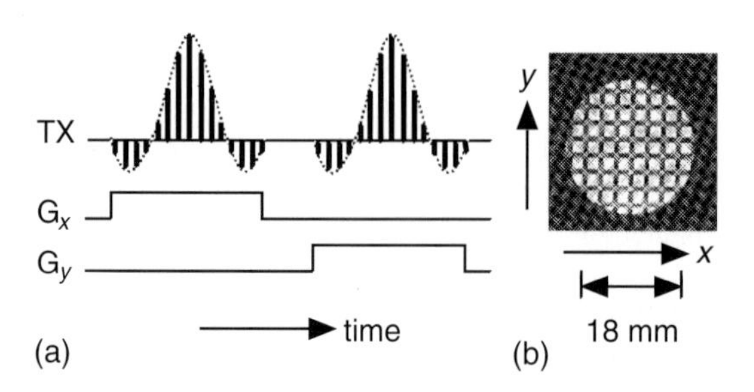

FIG. 7.2.16 [Kos3] 2D tagging of longitudinal magnetization. (a) Sequence of two DANTE pulses applied in orthogonal field gradients. (b) Spin-echo image of water at rest in a tube measured with a delay time of 64 ms between the tagging pulses and the start of data acquisition.

acquisition, any flow of spins can be detected by distortions of the magnetization grid in the image. Through an analysis of the distortions, magnitude and orientation of local velocity vectors can be reconstructed [Kos1].

Phase encoding of flow

Phase encoding of flow [Cal1, Cal12, Hah2, Mor3, Pop1, Pop2, Ste1, Ste2] provides the most accurate measurements of velocities, because even small directional changes of the flow direction can be detected by phase sensitive measurement of transverse magnetization. As a result, velocity vector fields in optically opaque media can be acquired with great accuracy in 3D space.

Phase encoding of flow requires the use of suitable gradient modulation throughout the entire imaging sequence, because the gradient moments $\boldsymbol{m}_i$ have to be manipulated appropriately for phase and frequency encoding of space and *phase encoding of velocity*. As an example, the pulse sequence for *flow-compensated spin-echo imaging* is shown in Fig. 7.2.17. The sequence has been set up to image through-plane flow in z-direction. The phase and frequency gradients for space encoding are flow compensated, so that there is no phase evolution from flow along the directions of these gradients. This means that for the spatial phase encoding gradient $m_{1y} = 0$ and $m_{0y} \neq 0$, and for the spatial frequency encoding $m_{1x} = 0$ and $m_{0x} = 0$ in the echo maximum. Furthermore, $m_{0z} = 0$ is required during phase and frequency encoding of the space information when imaging the xy plane. In addition, flow compensation in z-direction must be active during phase encoding of space, that is $m_{1z} = 0$ is required while G_y is on. However, to detect flow in z-direction, $m_{1z} \neq 0$ is the condition during data acquisition. Clearly, G_z must be switched on for slice selection during the selective pulse. To illustrate these points, the evolution of moments is plotted underneath the pulse sequences for flow-compensated imaging of flow through a plane transverse to the flow direction in Fig. 7.2.17.

The power of the technique is illustrated in Fig. 7.2.18 by two cross-sectional images through a single-screw *extruder* pumping different fluids [Rom4]. The data acquisition had been triggered to the screw phase. No signal is received where the screw blade passes the imaged plane. The screw rotation is responsible for the transverse motion, which is indicated by arrows proportional to the transverse velocity. The axial speed is identified

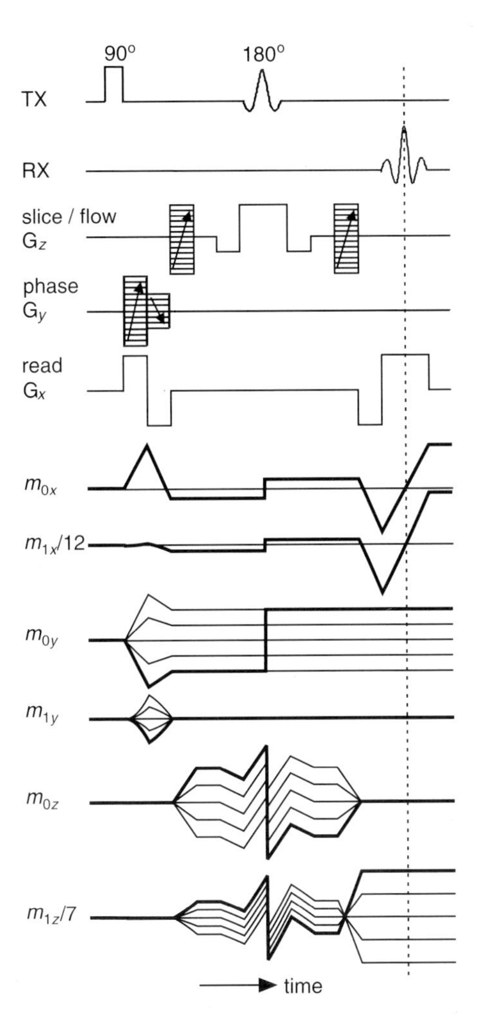

FIG. 7.2.17 Flow imaging by phase encoding. A bipolar flow-encoding gradient is applied for slice selection to image through-plane flow $v_z(x, y)$. Both phase and frequency gradients for space encoding are flow compensated.

by grey-scale values. For water representing a Newtonian fluid, high axial velocity is detected in the centre of the space between the screw and the extruder wall. But for *tomato ketchup* the highest axial velocity is found close to the extruder wall. This is a manifestation of the *wall-slip effect*, which arises in some non-Newtonian fluids. It is responsible for the difficulty in getting small amounts of ketchup in a controlled way out of a glass bottle just by shaking it. In fact, flow imaging is of great importance in studies of fluid rheology [Bri1, Bri2, Bri3, Cal17, Xia3]. For multi-phase flow, the imaging schemes can be extended to include for instance a chemical-shift [Der1, Maj1] or a relaxation filter [Dec1]. In clinical diagnostics, flow imaging is applied for *angiography* [Dum5, Dum6, Dum7, Mor4, Tur2].

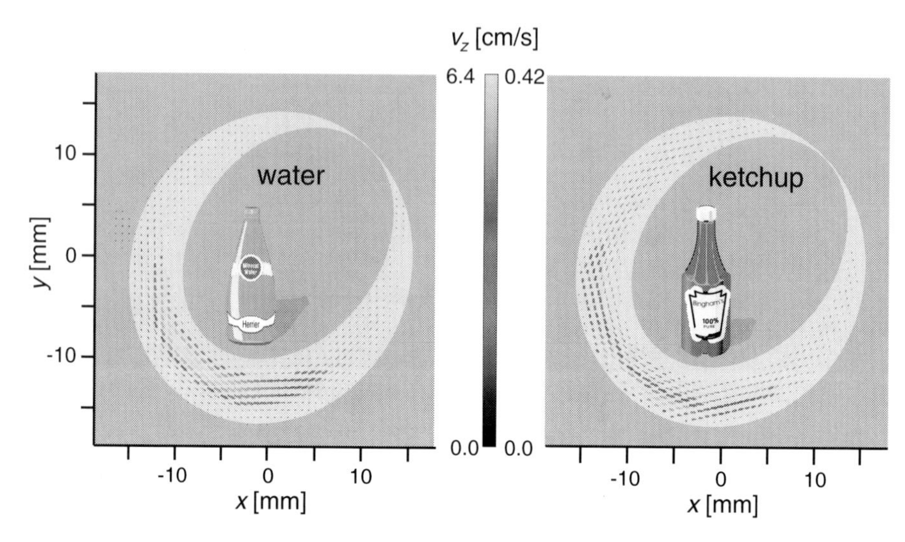

FIG. 7.2.18 [Rom4] Cross-sectional images through an extruder pumping fluids. Blank areas indicate the position where the screw blade passes the imaged slice. Transverse motion is encoded by arrows the length of which is proportional to transverse velocity. Axial speed is encoded as grey-scale values. In the reproduction of the image the arrows appear as short lines. The screw rotation speed was 98 rpm. Left: water image. It provides reference velocities for a Newtonian fluid. The average axial velocity was 8.7 cm/s. Right: tomato ketchup. The velocity profile is typical for a non-Newtonian fluid which exhibits wall slip. The average axial velocity was 13.3 cm/s.

At very fast velocities, the phase-encoding method fails because a very short echo time is required for phase encoding and gradient switching. This time must be shorter than the residence time of the spins in the resonator. Very *slow flow* may also become undetectable because of insufficient phase evolution or strong signal attenuation from transverse relaxation. In that case, SSFP-based methods may be employed, because they rely on phase coherence during T_1, and any disturbance by coherent flow can be detected with high sensitivity [Tys1]. Also, stimulated-echo methods can be applied similar to those used for measuring diffusion [Bou2, Cap4, Dum6, Hol2, Xia4].

7.2.7 Local-field filters

T_1, T_2, and $T_{1\rho}$ filters explore the magnetic fields $\boldsymbol{B}_0$ and $\boldsymbol{B}_1$, which are applied to the object. Other filters are based upon signal change resulting from the presence of local fields arising, for example, from susceptibility differences, chemical shift, and dipolar couplings. Filters probing local dipolar fields modulated by slow molecular motion are highly important for contrast generation in material science applications of ^{1}H NMR imaging. The T_2 filter is sensitive to both relaxation in local fields and in applied fields. In general, T_2 relaxation depends on the magnitude of $\boldsymbol{B}_0$, that is, on motions in the vicinity of the Larmor frequency and twice that frequency (cf. Fig. 7.1.3 and eqn (3.5.6)). For solids and elastomers, motions at the Larmor frequency and at higher frequencies are less significant than slow motions, which average the dipolar interactions only partially. Then the rate of transverse relaxation is directly proportional to the correlation time of the slow molecular motion (cf. eqn (3.5.17)). Therefore, the T_2 filter is a local-field filter when applied to rigid and soft solid matter.

Solid echo and Jeener–Broekaert echo

A collection of *local-field filters* other than the T_2 filter is depicted in Fig. 7.2.19. The solid-echo [Pow1] filter (a) and the alignment or Jeener–Broekaert echo [Jee1] filter (b) require the presence of a bilinear interaction like the *dipole–dipole coupling* of two protons or the *quadrupole coupling* of deuterons (cf. Fig. 3.2.6). Both interactions can be averaged partially by molecular motion depending on the timescale and the anisotropy of the process. Slow processes with correlation times longer than the inverse static coupling strength lead to partial averaging. The motionally unaveraged, residual part of these interactions is refocused by the dipolar and Jeener–Broekaert echoes. Multi-centre dipolar couplings between more than two spins $\frac{1}{2}$ cannot be refocused completely by these echoes, so that the echo amplitudes for abundant protons in rigid solids may be quite low.

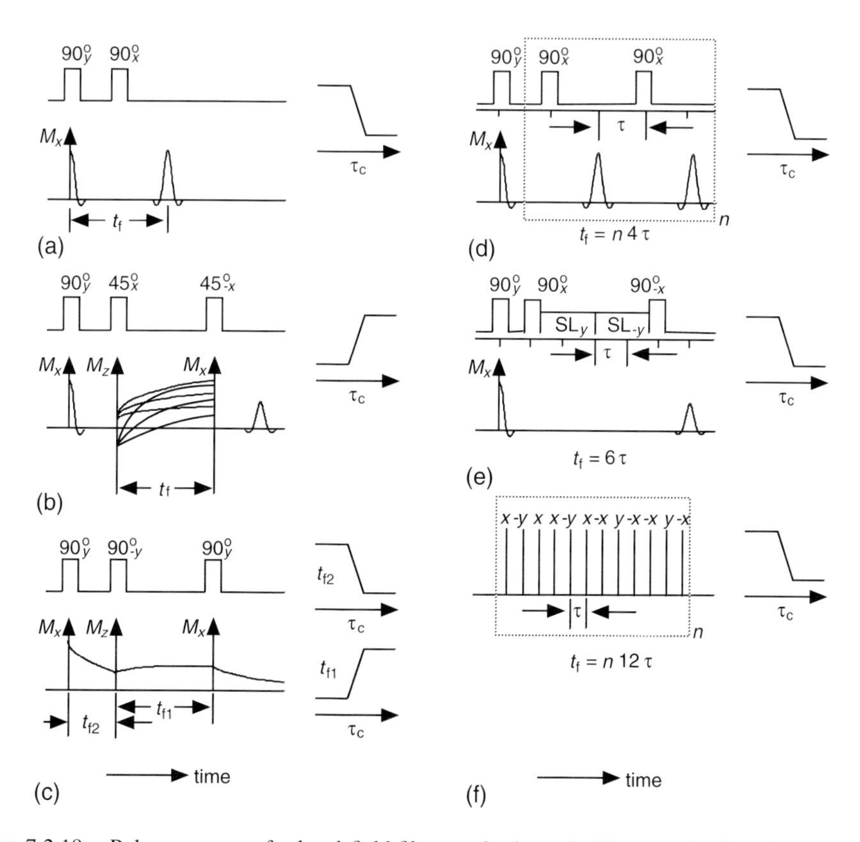

FIG. 7.2.19 Pulse sequences for local-field filters and schematic filter transfer functions. (a) Solid-echo filter. (b) Alignment-echo filter. (c) Goldman–Shen filter. Formally the same pulse sequence is used to generate the stimulated echo (cf. Fig. 7.2.1(c)) for measurement of the dipolar correlation effect. (d) OW4 filter. (e) Magic-echo filter. (f) Dipolar filter.

The *solid-echo* sequence (a) (cf. Fig. 3.2.6(a)) differs from the optimum *Hahn-echo* sequence (cf. Fig. 2.2.9(a)) in the flip angle and possibly the phase of the refocusing pulse, which is required to be in quadrature to the first pulse. The *Jeener–Broekaert sequence* (b) (cf. Fig. 3.2.6(b)) differs from the stimulated-echo excitation (cf. Fig. 2.2.9(c)) in the flip angles and the phases of the second and the third pulses. These phases need to be in quadrature to the phase of the first pulse. The second pulse of the three-pulse sequence generates *dipolar order* or *quadrupolar order*, which denotes an antiparallel alignment of paired spins $\frac{1}{2}$ in a spin-1 system. This ordered state relaxes in the filter time t_f with the dipolar and the quadrupolar relaxation times T_{1D} and T_{1Q}, respectively. Referring to the dipole–dipole interaction among protons, the Jeener–Broekaert sequence is sensitive to ultra-slow molecular motions on the timescale of T_{1D}, which can be in the order of 1 s and longer, whereas the solid-echo sequence is sensitive to motions with correlation times in the slow motion regime, that is between 10 μs and 100 μs.

The stimulated echo and dipolar correlations

A modified stimulated-echo sequence with phases x, $-x$, y (Fig. 7.2.19(c)) can be used to measure ultra-slow reorientations of dipolar couplings in a sensitive way by computing the ratio of the stimulated echo to the primary echo observed in polymer melts and elastomers [Kim3]. Denoting the phase evolution under the orientation-dependent NMR frequency (3.1.23) from the dipole–dipole interaction during the evolution time t_{f2} between the first two pulses by ξ_1, that between the second pulse and the primary echo by ξ_2, and the phase evolution between the last pulse and the stimulated echo by ξ_3, the ratio of amplitudes a_{st} of the stimulated echo to the amplitude a_p of the primary echo is given by [Kim2, Kim3]

$$\frac{a_{st}(2t_{f1} + t_{f2})}{a_p(t_{f2})} = \exp\{-\langle \xi_1 \xi_2 \rangle\} \cosh\langle \xi_1 \xi_2 \rangle, \tag{7.2.17}$$

where $\langle \cdots \rangle$ denotes an ensemble average over coupling spin pairs and thus an orientational correlation function. Because t_{f1} can be as large as the longitudinal relaxation time, extremely slow motions can be probed. For example, the *dipolar correlation effect* is a very sensitive probe for the detection of differences in the cross-link density of elastomers even in inhomogeneous fields (Fig. 7.2.20) [Gut2, Fis2, Zim1]. Another sensitive way of probing residual dipolar couplings is with the *beta function* (cf. Section 7.1.6) [Bal2, Cal3]. This is a sine-correlation function, which has nonzero values only in the presence of dipolar couplings. Alternatively, double-quantum NMR techniques can be applied [Gra1] (cf. Section 7.3.10).

The Goldman–Shen sequence

A sequence seemingly similar to the stimulated echo excitation is the *Goldman–Shen sequence* (Fig. 7.2.19(c)) [Gol1]. It is applied on resonance in ^{1}H NMR and is used for separation of the slow from the fast decaying portions of the solid-state FID following the first pulse. Therefore, t_{f2} is chosen to be considerably shorter in this case than in most cases for generation of a stimulated echo. The components with strong dipolar couplings exhibit a fast decay of the free-induction signal, those with weak dipolar couplings a slow decay. The slowly decaying part of the transverse magnetization is

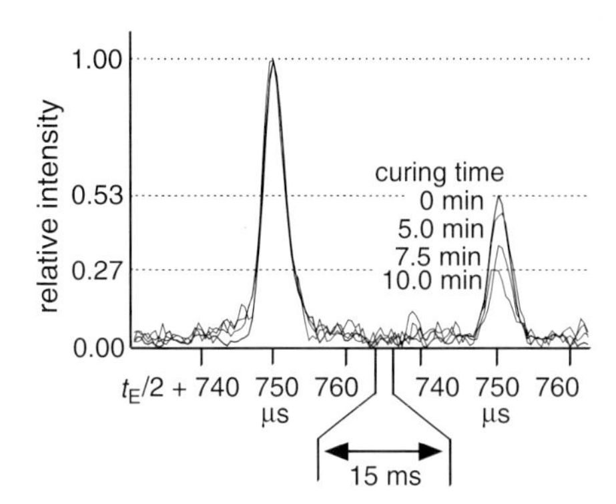

FIG. 7.2.20 [Zim1] Comparison of primary and stimulated echoes for a curing series of carbon-black filled NR measured in inhomogeneous $\boldsymbol{B}_0$ and $\boldsymbol{B}_1$ fields. The amplitudes of the primary echoes have been normalized to 100%. The amplitudes of the stimulated echoes are very sensitive to the change in cross-link density with increasing curing time. $t_{f2} = 15$ ms.

converted to z magnetization by the second pulse following a time t_{f2} after the first one. The z magnetization is recalled as transverse magnetization after a time t_{f1} by the third pulse. The time t_{f2} determines the position of the cut-off correlation time. A large value selects magnetization components with highly mobile molecular segments. Therefore, the first two pulses form the local-field filter. The following delay t_{f1} and the third pulse are used to allow for *spin diffusion* from the regions with high molecular mobility into regions with lower molecular mobility with subsequent spectroscopic detection (cf. Section 7.2.9). Analysis of the change in longitudinal magnetization during the spin-diffusion time gives insight into material morphology on a space scale of 1–200 nm (cf. Section 3.5.3). Therefore, the complete three-pulse sequence can be considered an extension of a local-field filter for detection of spin diffusion.

The OW4 filter

Just like the Hahn echo, the solid echo can also be generated repetitively in a CPMG-like fashion. However, instead of linear interactions, bilinear interactions are refocused (Fig. 7.2.19(d)). With the second and successive echoes of the multi-solid-echo sequence, a multi-pulse line-narrowing effect is observed. In terms of a minimum cycle time, the line-narrowing effect sets in after four times the separation τ of the first two pulses. Therefore, the sequence is referred to in the literature as *OW4 sequence* for Ostroff and Waugh [Ost1] or *MW4 sequence* for Mansfield and Ware [Man3, Man4]. By variation of the pulse spacing τ, the line-narrowing effect is manipulated. For large values of τ, only motionally preaveraged signals can sufficiently be narrowed to pass the pulse sequence, while the signals from broader lines with strong dipole–dipole interactions decay during the pulse sequence.

The line-narrowing effect can be explained in simple terms by the action of an effective lock field of strength $B_{1\text{eff}}$ which changes the transverse relaxation to a relaxation effective in the rotating frame. This relaxation time is referred to as $T_{2\text{e}}$ or *effective relaxation time*. Because $T_{2\text{e}}$ is longer than T_2, the signal is preserved for a longer time, which is equivalent to line narrowing. The signal is observed stroboscopically at the echo maxima in between the refocusing pulses, which is typical for multi-pulse NMR (cf. Section 3.3.4). In fact, extension of the CPMG sequence from liquid-state NMR to multi-solid echoes gave rise to the discovery of multi-pulse solid-state NMR.

Given the flip angle α of the refocusing rf pulses applied every $t_E = 2\tau$ in the OW4 sequence of Fig. 7.2.19(d), the effective strength $B_{1\text{eff}}$ of the $\boldsymbol{B}_1$ field can be varied in terms of its time average, $\gamma B_{1\text{eff}} = \alpha/(2\tau)$ [Man3], and the effective relaxation time becomes [Meh1]

$$\frac{1}{T_{2e}} = M_2 \tau_c \left[1 - \frac{\tau_c}{\tau} \tanh\left(\frac{\tau}{\tau_c}\right)\right], \tag{7.2.18}$$

where τ_c is the correlation time of molecular motion. By adjusting $B_{1\text{eff}}$ *via* the rf power or the cycle time 4τ, magnetization components from spins exposed to different strengths of the local fields can be selected. Those which are exposed to weak local fields pass the filter with little attenuation. To average the dipolar couplings among abundant ^{1}H spins in solids completely, isotropic motions with correlation times $\tau_c < 10^{-5}$ s are necessary. However, slower processes are also sufficient if the partial motional averaging is supplemented by multi-pulse sequences like the OW4 sequence or magic angle spinning. Therefore, the T_{2e} or OW4 sequence forms a band-pass filter for magnetization from segments with short correlation times of molecular motion. The number n of OW4 cycles determines the steepness of the filter transfer function, and the strength of the effective lock field determines the position of the threshold. Because the 90° pulses are spaced far apart compared to the pulse width, the effective fields achievable with the OW4 filter are small compared to a simple $T_{1\rho}$ filter (cf. Section 7.2.3). On the other hand, hard pulses are applied only so that the filter is less demanding on the transmitter phase and amplitude stability than a long spin-lock pulse, and the use of hard pulses avoids selective excitation of particular resonances. Furthermore, spin diffusion is suppressed by the OW4 filter.

The magic echo

The *magic echo* [Rhi1, Sch9, Sch10, Sli1] shown in Fig. 7.2.19(e) (cf. Fig. 3.4.3) is used to refocus magnetization dephased under the influence of the homonuclear dipole–dipole coupling similar to the solid echo. The restriction of the solid echo to spin-1 systems for complete refocusing of magnetization does not apply so that for ^{1}H magnetization in solids with multi-centre dipole–dipole interactions, refocusing of the signal decay by magic echoes is more efficient than by solid echoes. Because the spin evolution under the *dipole–dipole interaction* is reversed during formation of the magic echo, multi-magic-echo trains are being referred to as TREV for time reversal [Tak1].

The *TREV4 sequence* consists of the four pulses from one magic sandwich (Fig. 7.2.19(e)), which are repeated for acquisition of the echoes in between the sandwiches. Because the dipolar interaction is inefficient in the echo maxima, the evolution of the echo amplitudes is determined by the chemical shift and by T_2. Therefore, the TREV4 sequence can be used for measurement of solid-state high-resolution spectra similar to other multi-pulse sequences like WAHUHA or MREV8 (cf. Section 3.3.4) [Hep3, Tak1]. Two magic-echo sandwiches can be combined in one block of eight pulses, and a 180°

pulse can be put into the middle of that block, which in effect changes the phase of the last 90° pulse of the first sandwich from $-x$ to x. This is the *TREV8 sequence*. It forms a *time-suspension sequence*, where chemical shift and homonuclear dipole–dipole interaction are refocused in every second magic echo [Mat1]. The signal evolution under TREV8 irradiation then solely is determined by T_2 relaxation [Hep3]. Extension of the time interval τ which defines the echo time $t_E = 6\tau$ of the magic echo eventually reintroduces dephasing from strong dipolar couplings, so that the TREV sequences can be used as a *dipolar filter* [Dem1] which attenuates magnetization from segments with long correlation times of molecular motion.

Multi-pulse filters for solid matter

Clearly, multi-pulse sequences with averaging properties other than those of the OW4 and magic-echo filters can be used as filters as well (cf. Table 3.3.1). Depending on the preparation pulse preceding the *multi-pulse sequence*, one could observe stroboscopically either an FID-like signal during the pulse sequence or a spin-locked component of the magnetization. For instance, the sequence 45°_x-(MREV8)$_n$-45°_{-x} will lock the magnetization in the effective field of the MREV8 sequence, which is at 45° in the yz plane [Sch2]. Because the MREV8 sequence is used for homonuclear dipolar decoupling, the chemical groups in a solid relax independently of their neighbours with an effective relaxation which is characteristic of the pulse sequence and the chemical group, so that during the *MREV8 spin-lock filter* a thermodynamic nonequilibrium state of longitudinal magnetization is established. This nonequilibrium magnetization equilibrates first, by spin diffusion once the dipolar couplings are no longer suppressed and subsequently by longitudinal relaxation.

The dipolar filter

Through the use of suitable pulse sequences, various filters can be designed with different orientations of the effective field and different averaging properties for individual spin interactions, so that different distributions of longitudinal magnetization can be obtained in the sample. Clearly, the dipole–dipole interaction is the most important one to average in solid materials, but for soft matter resonance offset from chemical shift may become significant and influence the filter performance. In fact, chemical shift differences can be used by themselves for filtering (cf. Section 7.2.4). Therefore, multi-pulse sequences of the time-suspension type which average dipolar and chemical shift interactions are useful for filtering according to molecular mobility in soft matter. Examples of such pulse sequences are WIM24 [Car2], TREV8 (see above), and the *dipolar filter* [Egg1].

For efficient averaging of an interaction by most multi-pulse sequences, the cycle time t_c must be shorter than the inverse of the anisotropy Δ (cf. eqn (3.2.1)) of the interaction. If the cycle time is increased, only the weak couplings which are preaveraged by molecular motion are averaged to their isotropic values. The magnetization components with strong couplings decay during the multi-pulse irradiation. The use of the dipolar filter (Fig. 7.2.19(f)) is based on this principle in the same way as the OW4 and the TREV8 filters are. The steepness of the filter threshold is adjusted by the number n of cycles, typically 10–40. The cut-off correlation time is adjusted *via* the cycle time or the pulse

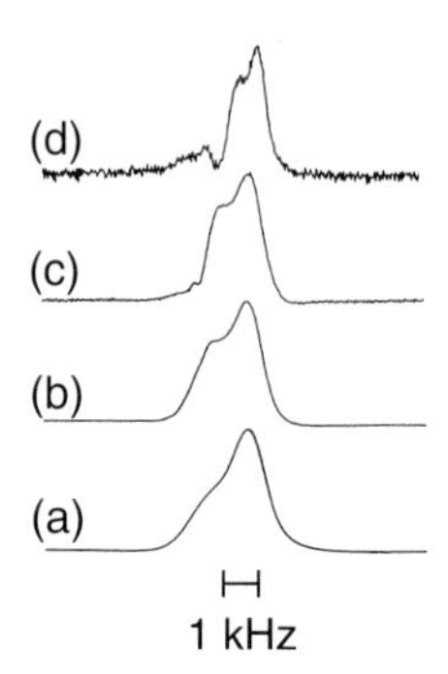

FIG. 7.2.21 Application of the dipolar filter to the ^{1}H magnetization of SBR. (a) No filter. (b) Weak filter: $n = 10$, $\tau = 100\ \mu$s. (c) Medium strong filter: $n = 10$, $\tau = 200\ \mu$s. (d) Strong filter: $n = 20$, $\tau = 400\ \mu$s. Adapted from [Blü1] with permission from Elsevier Science.

spacing τ. Figure 7.2.21 illustrates the action of the dipolar filter on the ^{1}H magnetization of SBR [Blü1]. With increasing cycle time, the resonance lines in the spectrum become narrower, because the signals from the more rigid regions of the sample are eliminated. At the same time, the signal-to-noise ratio decreases.

While short-range dipolar interactions can only be observed in systems with restricted molecular mobility, long-range dipolar interactions can be observed also in fluids. Although the strength of the dipole–dipole coupling decreases with the cube of the distance between the coupling partners, the number of potential coupling partners increases with the cube of the distance. As the distance increases, the internuclear vector is less and less likely to rotate isotropically so that the dipole–dipole interaction is no longer averaged to zero. Therefore, effects of *long-range dipolar interactions* can be observed in liquids, and it was found that the dipolar field experienced by a spin depends on the spatial modulation of magnetization [War3]. By tuning the modulation wavelength, different length scales can be probed in reciprocal space [Rob1].

7.2.8 Combination filters

Different filters can be combined to form *combination filters*. Such filters have been worked out for a variety of cases. For instance, longitudinal and transverse relaxation weights can be introduced simultaneously [Gut1, Pee1, Sez1]. An important class of applications concerns investigations of polymer morphology [Göt1, Göt2, Göt3, Gol1, Sch2, Sch5] and filters which introduce selectivity in such NMR measurements are particularly promising tools for contrast generation in material imaging. For example, magnetization components can be selected from motionally heterogeneous samples which are characterized by hierarchies of relaxation times pertaining to more complicated situations other than simple two-component distributions of mobile and rigid segments [Göt1]. Such hierarchies may be encountered in semicrystalline polymers such as poly(ethylene), where crystalline, amorphous and interfacial regions can be discriminated.

Combinations of T_1 and T_2

The simple T_2 weight (7.2.4) introduced in Hahn-echo measurements is to be modified if partial saturation of the sample magnetization results from repetition times $t_R = t_{f0}$ that are short compared to 5 T_1 [Pop3]. The associated reduction in thermodynamic equilibrium magnetization can be described within the validity of the Bloch equations by replacing $M_0(\boldsymbol{r})$ in (7.2.4) by the saturated magnetization (7.2.1) (cf. eqn (6.2.5)). For perfect pulses one obtains

$$a_H(\boldsymbol{r}, t_R, t_E) = M_0(\boldsymbol{r}) \left[1 - \exp\left\{ -\frac{t_R}{T_1(\boldsymbol{r})} \right\} \right] \exp\left\{ -\frac{t_E}{T_2(\boldsymbol{r})} \right\}. \tag{7.2.19}$$

The echo maxima are weighted by a function of both T_1 and T_2. Similarly, the stimulated echo (Fig, 7.2.1(c)) can be used as a combination filter to introduce T_1 and T_2 weights. The echo time (t_{f2} in Fig. 7.2.19(c)) determines the T_2 weight and the mixing time between the second and the third pulses (t_{f1} in Fig. 7.2.19(c)) the T_1 weight. Note that the filter transfer functions for T_1 contrast by saturation recovery and the stimulated echo are inverted (cf. Fig. 7.2.1(a) and (c)), so that both combination filters introduce different contrasts (cf. eqn (7.2.3)).

T_1 and T_2 *combination filters* are also used for NMR measurements in inhomogeneous B_0 and B_1 fields, where the spatial localization is not necessarily achieved by gradient switching for $\boldsymbol{k}$-space encoding, but by positioning a surface coil [Eid1, Kle2] (cf. Section 9.3). These techniques can, in principle, also be used as combination filters for preparation of magnetization. Figure 7.2.22 depicts two such filters, a *steady-state inversion-recovery filter* (a) [Sez1] and a *saturation recovery filter* (b) [Gut1]. In both cases, the filters start from dynamic equilibrium of the magnetization and the filter parameter t_{f1} determines the T_1 weight of the signal given a fixed value of t_{f0} for the cycle time in (a). The T_2 decays are probed by CPMG trains of n echoes to suppress signal attenuation from diffusion, so that the second filter variable is $t_{f2} = nt_E$. When used in inhomogeneous B_1 fields the 90° pulse becomes a space-dependent pulse of flip angle θ and the 180° pulse becomes a 2θ pulse. A train of echoes acquired with the saturation-recovery sequence (b) is depicted in (c). It shows the T_1 build-up of magnetization with increasing filter time t_{f1} followed by the T_2 decay on the time axis, which is determined by the number n of the CPMG echo. The saturation cycle with filter time t_{f1} of sequence (b) derives from the SSFP sequence (cf. Fig. 2.2.19). This sequence has been extended by 180° pulses for generation of Hahn echoes to permit detection of transverse magnetization in inhomogeneous B_0 fields [Gut1].

Filters which combine measurements of two types of relaxation lead to 2D parameter sets which characterize the evolution of magnetization for each type of relaxation along one axis and correlate both processes. In case of multi-component relaxation such time-domain correlation maps can be analysed for accurate identification of individual components, where discrimination of components is improved by the 2D nature of the available data as compared to the decomposition of 1D relaxation curves [Ole1, Pee1, Weg1].

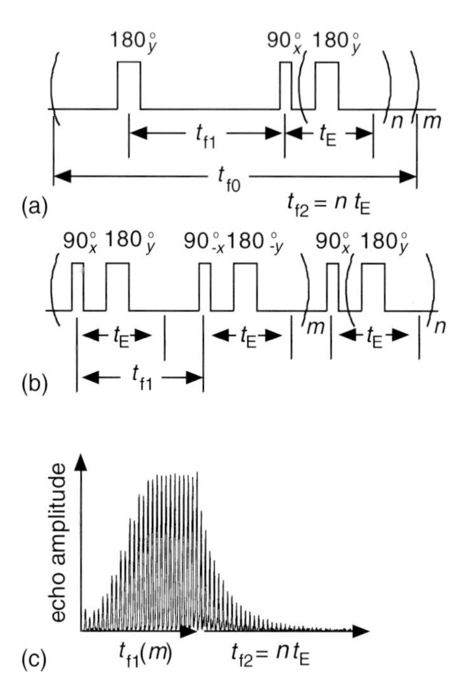

FIG. 7.2.22 [Gut1] Pulse sequences for combined determination of T_1 and T_2 in inhomogeneous B_0 fields. (a) Steady-state inversion recovery filter [Sez1]. (b) Steady-state saturation recovery filter [Gut1]. (c) Train of echoes measured by sequence (b) which shows the magnetization build-up with T_1 and the decay with T_2 of unfilled, cross-linked SBR.

Chemical-shift difference and T_1 relaxation

The scheme of Fig. 7.2.5(b) for selective excitation of one of two lines by two 90° pulses which are 180° out of phase can readily be modified into a combination filter where both lines are excited and the longitudinal magnetization of one line is weighted by the T_1 relaxation [Ger3]. To this end, the rf frequency is set on resonance A as before, but the duration between the two pulses is set to half the inverse frequency difference. Then the second pulse is applied when both magnetization components are 180° out of phase so that they are aligned antiparallel to the z-axis after the second pulse. Depending on the delay following the second pulse and the T_1 relaxation time of the inverted line, different contrast weights can be generated. In particular, the zero crossing of the inverted magnetization can be chosen, so that a pure chemical-shift image of resonance A is obtained.

Combination filters for heterogeneous soft solids

Heterogeneous soft matter, in particular polymer materials, are often characterized by distributions of correlation times of molecular motion. For relaxation studies of polymers, sophisticated filters have been developed which fit the classification of combination filters because they combine T_2 and $T_{1\rho}$ relaxation [Göt1, Göt2, Göt3]. These filters can be used

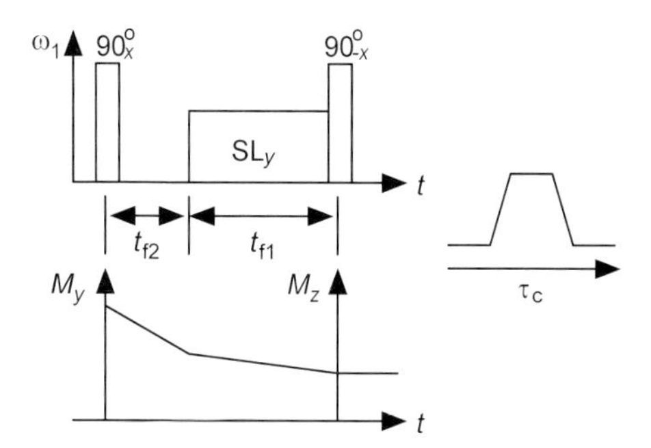

FIG. 7.2.23 [Göt1] Combination filter for selection of magnetization from polymer chains with intermediate mobility.

to discriminate between magnetization from rigid (crystalline), intermediately mobile (interfacial) and mobile (amorphous) polymer chains in polycrystalline materials like poly(ethylene).

The filter in Fig. 7.2.23 exploits a succession of a free induction decay for a time t_{f2} and a $T_{1\rho}$ decay in a *spin-lock field* of amplitude ω_1 for a time t_{f1} [Göt1]. The contribution from the rigid chains decays rapidly during the time t_{f2} after the first pulse. Because $T_{1\rho}$ is longer in poly(ethylene) (PE) for magnetization from more rigid chains than for mobile chains, the magnetization from chains with intermediate mobility is less attenuated during the subsequent spin-lock time than that of the mobile chains. For PE the filter is a band-pass filter for the magnetization from chains with intermediate mobility. In a similar fashion, a spin-lock period can be incorporated in the echo times of a Hahn, a solid, a stimulated, and an alignment echo [Göt2]. For three-pulse echoes, a B_1 field can also be applied during the time between the second and the third pulse [Göt3]. This changes the magnitude and the direction of the field effective for relaxation during this time.

7.2.9 Morphology filters

A particular class of combination filters is obtained when two filters are combined *via* a time for mixing of magnetization components through spin diffusion. From the signal dependence on the spin-diffusion time, information about morphology of heterogeneous polymers can be derived in terms of domain sizes on a scale between 1 and 200 nm [Dem1, Sch5]. Such domains can be discriminated by their chemical structure or their molecular mobility. Chemical domains form in blends of incompatible polymers and in block copolymers. Mobility domains are encountered, for instance, in semicrystalline polymers like poly(ethylene) and poly(propylene) (PP), which consist of layers of rigid crystalline lamellae and mobile amorphous material in between.

The *morphological domain structure* of polymers is determined during processing and often has a significant influence on the properties of polymers as construction materials. By mechanical load, electrical load, or by exposure to elevated temperatures, the morphological structures can be modified. Thus it is important to acquire information of polymer *morphology* and its spatial change as a consequence of processing and handling. *Spin-diffusion NMR* experiments (cf. Section 3.5.3) are one way to achieve this goal.

The generic scheme for *spin-diffusion imaging* is shown in Fig. 7.2.24(a). Spin diffusion is detected by two filters which are separated by a spin-diffusion time. Starting from thermodynamic equilibrium, the longitudinal magnetization is evenly distributed across the sample. The first filter prepares an initial, space-dependent nonequilibrium distribution of longitudinal magnetization by eliminating the magnetization in parts of the sample, for instance in the rigid domains of a polycrystalline polymer. Ideally, a sharp magnetization grating is established by the first filter at the beginning of the spin-diffusion time (Fig. 7.2.24(b), top). With increasing diffusion time, this grating is washed out by magnetization spilling from the magnetized regions into less magnetized regions (middle). For long-diffusion times the total amount of longitudinal magnetization, which at the beginning of the spin-diffusion time was located in the amorphous regions only, is evenly distributed over the entire sample (bottom). In spin-diffusion experiments, images are acquired of the depleting or the recovering part of the longitudinal magnetization for different values of the spin-diffusion time. This part is selected by the second filter, the so-called detection filter. In this way, a diffusion curve is measured for each pixel of the image. The diffusion curves can be analysed in terms of a model of the polymer

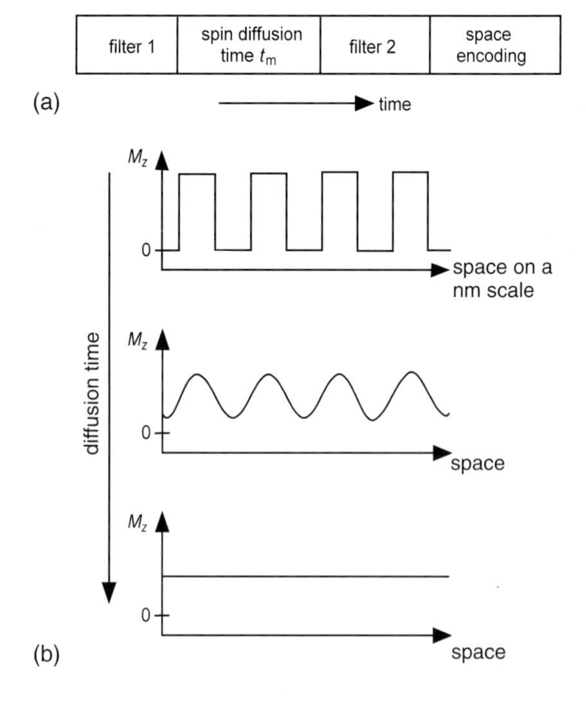

FIG. 7.2.24 (a) Generic scheme for spin-diffusion imaging. Spin diffusion is detected by use of two filters which are separated by a spin-diffusion time. (b) Evolution of longitudinal magnetization during the spin-diffusion time in one dimension.

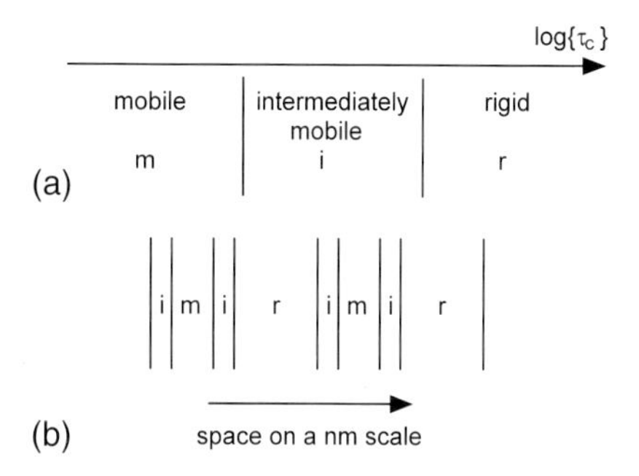

FIG. 7.2.25 (a) Arbitrary mobility timescale of a semicrystalline polymer. Mobile (m), interfacial (i) and rigid (r) segments are discriminated by their correlation times τ_c of molecular motion. (b) 1D domain structure of PE. The mobile domains correspond to the amorphous material, the rigid domains to the crystalline material.

morphology, and the model parameters can be obtained from the analysis [Dem1, Sch5]. Typically these model parameters are the minimum domains sizes. Spin-diffusion imaging has been applied to study the average change of domains sizes with electrical ageing in high-voltage PE cable-insulation material [Wei1, Wei2].

In spin-diffusion studies it is possible to detect not only two but three domain sizes. The third domain can be considered the interface (i) between the other two domains, which can be different chemical species in a polymer blend or rigid crystalline (r) and mobile amorphous (m) material in a semicrystalline polymer. To illustrate this point, a mobility timescale is depicted in Fig. 7.2.25(a) and the simplified 1D domain structure of PE underneath in (b). Rigid crystalline and mobile amorphous materials exhibit motion of chain segments with different correlation times τ_c. The chains at the interface between both domains exhibit intermediate mobility. The exact ranges of correlation times in the individual domains depend on the particular choice of filters. Therefore, the values of domain sizes derived through spin-diffusion NMR also depend on the type of filters used. In particular, the interface is defined solely by the NMR experiment and can only be detected if the filters are properly chosen.

Certain combinations of filters need to be employed to detect an interface. If both major domains can be identified on the chemical-shift scale by the resonances of particular chemical groups, *chemical-shift filters* can be used for the first (preparation) and the second (detection) filter [Sch5, Sch6, Cla1], and an interface can often be detected because the magnetization has to cross the molecule with the selected chemical group to reach the other chemical group. If *mobility filters* are used, more complicated situations can be encountered. The simplest case is selection of the mobile component as the magnetization source with the preparation filter and detection of the slow component as the magnetization sink, or *vice versa*. If either filter avoids the magnetization from segments with intermediate mobility, there will be an initial delay in the diffusion curve before magnetization can be detected at the sinks, because the magnetization has to cross the invisible barrier of molecules with intermediate mobility. In this case, the diffusion curve exhibits a sigmoidal magnetization build-up. No interface will be detectable if one of these filters also selects the magnetization from the interface. However, an interface will become visible if the preparation filter selects magnetization from molecules with high and intermediate mobility and the detection filter detects only the magnetization from the mobile components. Then, in the diffusion time, the magnetization first drains from the interfacial regions into the rigid domains and only later from the mobile domains into the interfacial and rigid domains. In this case, the diffusion curve shows a sigmoidal magnetization decay [Egg1, Wei1, Wei2]. Spin diffusion in homogeneous solids can be probed by monitoring the disappearance of a magnetization grid established by ultra-strong gradient pulses [Zha1].

7.2.10 Multi-quantum filters

Multi-quantum filters [Ern1] generate *multi-quantum* (MQ) *coherences* in a preparation period prior to the space-encoding period. Single-quantum magnetization is eliminated from the image, so that water signals, for example, can efficiently be suppressed in biomedical applications of imaging [Dum2]. When integrating the space-encoding gradients into the multi-quantum filter, methods of *multi-quantum imaging* are obtained in contrast to *multi-quantum-filtered imaging*, where single-quantum coherences are utilized for space encoding. The advantage of applying gradients during multi-quantum evolution times is that the dephasing of coherence proceeds p times faster for MQ coherence of order $p = m_{\mathrm{f}} - m_{\mathrm{i}}$ than for single-quantum coherence. Here m_{i} is the magnetic quantum number of the initial and m_{f} that of the final energy level (cf. Figs. 2.2.11 and 3.2.1). In effect, the gradient strength is multiplied by the *coherence order* [Ern1, Gar2].

Different MQ filters are applied in liquid- and in solid-state imaging. Prerequisite for their use is a total spin quantum number I greater than $\frac{1}{2}$ for the system of spins under consideration. Two or more spins $\frac{1}{2}$ may be considered, which are coupled indirectly by J coupling or directly by dipolar coupling, or quadrupolar nuclei like ^{2}H and ^{23}Na with spins 1 and $\frac{3}{2}$, respectively. In either case the energy-level diagram consists of more than two levels. The multi-quantum frequency of the largest coherence order in solid-state NMR is independent of molecular orientation and thus possesses special properties for space encoding. For example, this would be the double-quantum transition in ^{2}H NMR (cf. Fig. 3.2.1) and the triple-quantum transition in ^{23}Na NMR.

Principle of multi-quantum filtering

Multi-quantum filters consist of a sequence of pulses for preparation of longitudinal magnetization, a MQ evolution time t_1 and a sequence of pulses for reconversion of MQ coherences into longitudinal magnetization (Fig. 7.2.26(a)) [Ern1]. The resultant longitudinal magnetization ($p = 0$ in Fig. 7.2.26(b)) is modulated by the phase evolution of the MQ coherence during t_1 and can be converted for detection into single-quantum magnetization ($p = -1$) by a $\pi/2$ reading pulse. The time evolution of the spin system during generation and reconversion of MQ coherences is described by the *time-evolution*

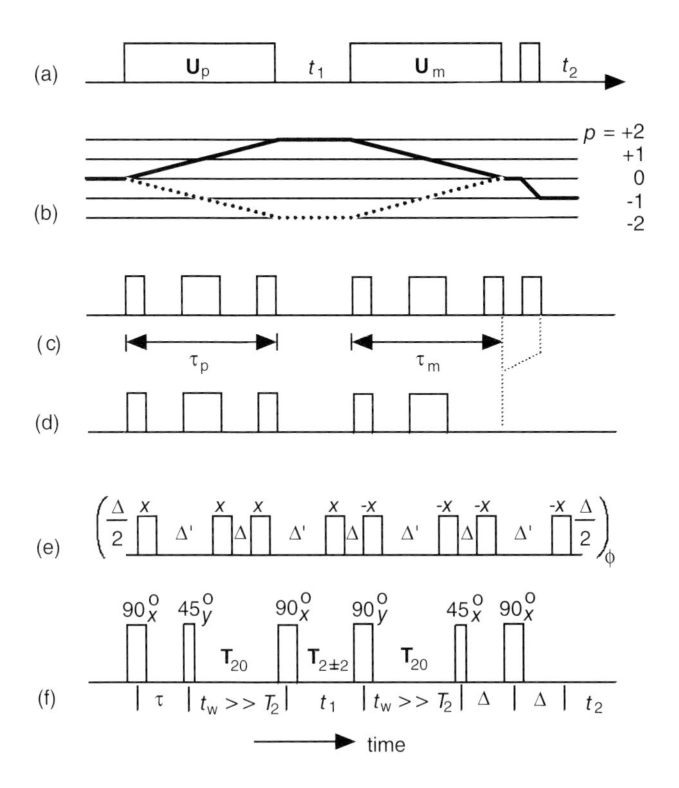

FIG. 7.2.26 Basic schemes for homonuclear multi-quantum NMR [Ern1]. (a) Generic scheme. $\mathbf{U}_{\mathrm{p}}$ and $\mathbf{U}_{\mathrm{m}}$ are evolution operators describing the pulse sequences for generation and reconversion of multi-quantum coherences from and into longitudinal magnetization. t_1 is the multi-quantum evolution time. A $\pi/2$ pulse follows the MQ filter to generate detectable transverse magnetization. (b) Coherence transfer pathways for double-quantum NMR. Individual pathways can be selected by phase cycling. (c) The operators $\mathbf{U}_{\mathrm{p}}$ and $\mathbf{U}_{\mathrm{m}}$ can be realized with nonselective pulses by two $\pi/2$ pulses which sandwich a π pulse to generate an echo for elimination of chemical-shift effects. (d) In practice, the detection pulse is integrated into the reconversion sequence. (e) Pulse sequence for the operator $\mathbf{U}_{\mathrm{p}}$ to generate even-order multi-quantum coherences in a system of strongly dipolar coupled spins [Gar2]. (f) Pulse sequence for generation and reconversion of double-quantum coherence in ^{2}H NMR *via* spin alignment [Gün1].

operators $\mathbf{U}_{\mathrm{p}}$ and $\mathbf{U}_{\mathrm{m}}$, respectively, which are solutions of the density-matrix equation of motion (cf. Section 2.2.2). In high-resolution NMR the read pulse is merged in practice with the reconversion sequence (Figs 7.2.26(c) and (d)). Phase cycling is used to either select a unique pathway 0 to p to -1 or to allow two mirror image pathways simultaneously (0 to $\pm p$ to -1 in Fig. 7.2.26(b)). The pulse sequences shown in Fig. 7.2.26 use nonselective pulses for homonuclear spin systems. Selective pulses can be used as well, and the pulse sequences can be adapted to generate heteronuclear MQ coherences [Ern1].

Basic sequence for preparation of multi-quantum coherences

The basic pulse sequence to generate MQ coherences consists of three pulses,

$$\left(\frac{\pi}{2}\right)_x - \frac{\tau}{2} - (\pi)_x - \frac{\tau}{2} - \left(\frac{\pi}{2}\right)_x. \tag{7.2.20}$$

It can be employed for a wide range of systems involving scalar, dipolar, or quadrupolar couplings. For weak J couplings, the central π pulse removes the effect of chemical shifts without affecting the couplings. By adjusting the time τ to half the inverse line splitting, for example to $1/(2J)$ for a J doublet, antiphase magnetization is created and converted into MQ coherences by the second $\pi/2$ pulse. Only even coherence orders can be excited by the sequence (7.2.20). Odd orders can be excited if the phases of the second and the third pulses are shifted by 90°,

$$\left(\frac{\pi}{2}\right)_x - \frac{\tau}{2} - (\pi)_y - \frac{\tau}{2} - \left(\frac{\pi}{2}\right)_y. \tag{7.2.21}$$

For selection of particular MQ coherences, *phase cycling schemes* can be used which are based on phase shifts φ of the complete pulse sequences (7.2.20) and (7.2.21) [Ern1]: if the change in coherence order effected by the pulse sequence is $\Delta p = p_\mathrm{f} - p_\mathrm{i}$, where i and f denote initial and final coherence orders, the phase of the MQ coherence generated is changed by $\Delta p\,\phi$. This phase shift is carried over to the observable single-quantum magnetization during the detection period, so that by suitable choices of ϕ particular coherence orders can be selected by addition of signals.

The proper sequence for reconversion of MQ coherences derives from the preparation propagator but with a sign change in the exponent. The sign change is achieved by forming a time reversed copy of the preparation pulse sequence corresponding to a copy of the preparation pulse sequence but phase shifted by 90°. In principle a single $\pi/2$ pulse would be sufficient to reconvert the MQ coherence into detectable transverse magnetization. However, an additional π pulse is necessary to generate in-phase multiplets or purely positive lineshapes. Otherwise the integral over the multiplet vanishes and no signal can be detected for small couplings in inhomogeneous fields [Dum2]. For *double-quantum filtering*, different variations of the basic scheme of Fig. 7.2.26(d) have been worked out [Jun1].

Double-quantum filters for rigid solids

Strong dipolar coupling in solids requires schemes different from that in Fig. 7.2.26(d) for excitation of MQ coherences [Gar2, Gra2]. The pulse sequence shown in Fig. 7.2.26(e) has been used to generate even-order MQ coherences among ^{1}H in solid adamantane [Gar2]. It consists of a string of $\pi/2$ and $-\pi/2$ pulses separated by delays Δ and $\Delta' = 2\Delta + t_\mathrm{p}$, where t_p is the duration of a pulse. For fast magic angle spinning, a variety of different multi-pulse schemes have been developed, which partially recouple the dipolar interaction into the spin evolution so that MQ coherences can be generated (see e.g. [Fei1] and references therein).

For well-defined spin-1 systems like deuterons, double-quantum coherence $\mathbf{T}_{2\pm2}$ can be generated *via* intermediate, long-lived spin alignment $\mathbf{T}_{20}$ (Fig. 7.2.26(f)) [Gün1]. The spin-alignment time t_w can be used for switching of gradients in *double-quantum imaging*. Following reconversion, the transverse magnetization is detected *via* a solid echo with echo time 2Δ because of the fast signal decay associated with broad inhomogeneous lines.

Applications

Multi-quantum filtering and MQ imaging are applied in biomedicine for imaging of particular chemical groups [Lei1] as well as for imaging of quadrupolar nuclei like ^{23}Na and ^{39}K with spins $I = \frac{3}{2}$. Multi-quantum coherences, in general, cannot be excited in the limit of fast isotropic motion. For example, extra- and intracellular sodium are sufficiently different in their motional behaviour, so that intracellular sodium gives rise to a *residual quadrupolar interaction*, and double- and triple-quantum signals can be excited, whereas no MQ coherences can be generated in extracellular sodium [Coc1, Coc2, Coc3, Hut1, Kel1, Lyo1, Pay1, Wim1, Wim2].

Furthermore, MQ NMR is used to probe internuclear distances [Gra2, Sch14] and to detect *molecular order* in materials by probing residual anisotropic interactions which are not completely averaged to zero by fast isotropic motion. This concept has been applied to study the residual quadrupolar interaction of molecules in oriented materials like cartilage, tendon, and muscle by MQ NMR of ^{23}Na [Eli1, Hug1, Red2], ^{2}H [Eli2, Seo1, Sha1, Sha2, Shi1], and ^{1}H [Eli1, Eli2, Tso1]. Multi-quantum coherences from long-range dipolar interactions have been proposed as a contrast parameter for bio-medical imaging [War4]. With respect to materials applications, the change in the residual quadrupolar interaction of ^{2}H from deuterated spy molecules incorporated into rubber bands has been studied in *rubber networks* as a function of applied *strain* and filter preparation time τ_p [Kli1, Kli2]. With increasing strain, the double-quantum signal amplitude increases for short preparation times τ_p (Fig. 7.2.27) [Kli1]. For longer preparation times the contrast is inverted. When using the MQ filter (cf. Fig. 7.2.26(c)), the double-quantum signal depends on the quadrupole coupling constant which scales with the elongation ratio $\Lambda = L/L_0$ as $\Lambda^2 - \Lambda^{-1}$ (cf. Fig. 7.1.12).

Residual dipolar couplings in soft matter are smaller than residual quadrupolar couplings. Nevertheless, they can be observed in nonspinning samples under conditions suitable for imaging [Gas1, Sch11, Sch12, Sch13]. In fact, hierarchies of dipolar couplings exist, so that with increasing preparation time τ_p, signals from weaker and weaker couplings can be retrieved. By proper choice of the coherence order and the preparation time, signals from different chemical groups and intergroup couplings can be discriminated [Bay1]. Similar observations apply for clusters of coupled spins in polycrystalline materials [Lac1, Lev3, Mun1, Scr1].

The selectivity of multi-quantum filters is demonstrated in Fig. 7.2.28 by spectra recorded for poly(isoprene) (PI) with the multi-quantum filter from Fig. 7.2.26(d) [Sch11, Sch12, Sch13]. The conventional single-quantum ^{1}H NMR spectrum of PI shows a broad line with little to no chemical-shift resolution. But the double- and triple-quantum filtered spectrum of the material exhibit peaks at different chemical shifts which can be assigned primarily to the signals from CH_2 and CH_3 groups, respectively. At longer

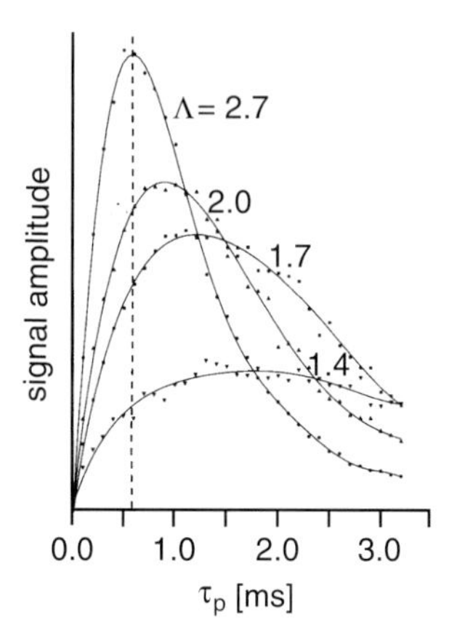

FIG. 7.2.27 [Kli1] Amplitudes of double-quantum coherences from deuterons of 1,4-deuterated butadiene oligomers incorporated into elastic bands of natural rubber as a function of the preparation time τ_p of the MQ filter and the stretching ratio Λ of the rubber bands. The curves drawn serve to guide the eye.

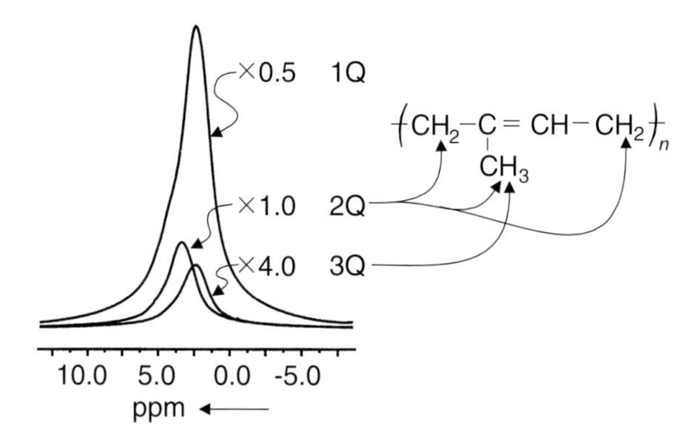

FIG. 7.2.28 [Sch12] Chemical selectivity of multi-quantum filters demonstrated for nonspinning *cis*-1,4 poly(ispoprene) rubber. By multi-quantum filtering the single-quantum (1Q) signal can be decomposed into signals at different chemical shifts. The double-quantum (2Q) signal mainly arises from the CH_2 group with a small contribution from the CH_3 group. The triple-quantum signal (3Q) derives from the CH_3 group. At longer preparation times contributions from intergroup couplings may become important.

preparation times, signal in addition to that from individual chemical groups may arise from intergroup dipolar couplings. Also, the double-quantum signal is dominated from the CH_2 group, but signal contributions from the CH_3 group cannot be excluded.

Depending on the particular pulse sequence used and on the coherence order, the multi-quantum build-up curves can be calculated in the short-time regime for a particular chemical group [Sch12]. From these known expressions, parameter images of the strength of the residual dipolar coupling [Sch13] and the associated dynamic order parameter [Gas1] can be derived.

It should be noted that multi-quantum filters can also be used to formally select zero-quantum coherence. Ignoring chemical shift, no zero-quantum coherences exist in the case of dipolar couplings, so that the filter produces *dipolar-encoded longitudinal magnetization*. This magnetization is the complement of all multi-quantum signals, so that its build-up curve starts at maximum signal and progressively decreases with increasing filter time τ_p. Because the signal is strong, it is particularly suited for good signal-to-noise ratios in contrast generation based on residual dipolar couplings. The filter curves for dipolar-encoded longitudinal magnetization, *double-quantum coherences*, and *triple-quantum coherences* are depicted in Fig. 7.2.29 for three samples of PI with different

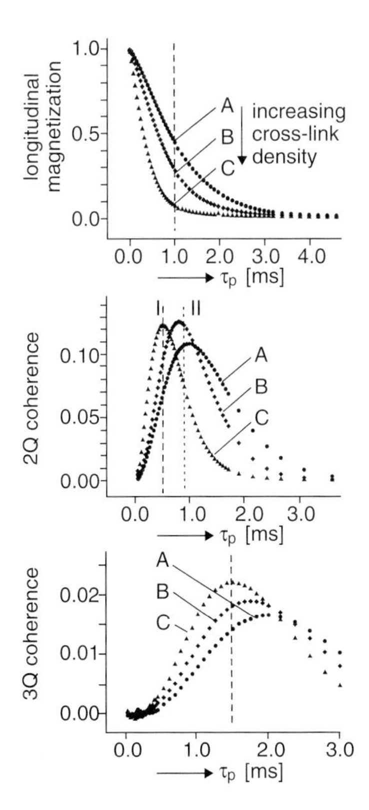

FIG. 7.2.29 [Sch13] Multi-quantum filter curves for *cis*-1,4-poly(isoprene) rubber at different cross-link densities. The cross-link density increases from sample A to C. The dipolar-encoded longitudinal magnetization filter (top) provides the highest signal intensity and is most suitable for generation of contrast in imaging. The double-quantum signals and the triple-quantum signals run through a maximum with increasing filter-preparation time τ_p starting from zero. Depending on the filter time chosen, different contrast is obtained in the images. The broken lines mark the filter-preparation time for optimum contrast.

cross-link density [Sch13]. The maximum signal is obtained for dipolar-encoded longitudinal magnetization. With increasing coherence order, the multi-quantum filtered signal amplitude decreases, and depending on the value of the filter time τ_p, different contrast can be obtained in *multi-quantum filtered images*. For example, for the double-quantum filter, contrast is scrambled with respect to cross-link density in regions I and II. Better contrast is obtained with the triple-quantum filter, but signal amplitudes are very low.

Feasibility and contrast features of multi-quantum filtered imaging are illustrated in Fig. 7.2.30 with images of a phantom composed of three 4 mm diameter PI pieces separated by 1 mm teflon spacers [Sch13]. The PI pieces differ in cross-link density from left to right according to medium, high and low, where medium cross-link density corresponds to sample C and low cross-link density to sample A in Fig. 7.2.29. The filter-preparation time τ_p has been adjusted for maximum contrast in each case. For the double- and triple-quantum filtered images, contrast is inverted compared to the spin-echo image and the image recorded with the dipolar-encoded longitudinal magnetization filter (cf. Fig. 7.2.29). The top image is a conventional spin-echo image for reference. Contrast is similar to that for dipolar-encoded longitudinal magnetization, but the contrast in the image from dipolar-encoded longitudinal magnetization is more sensitive to

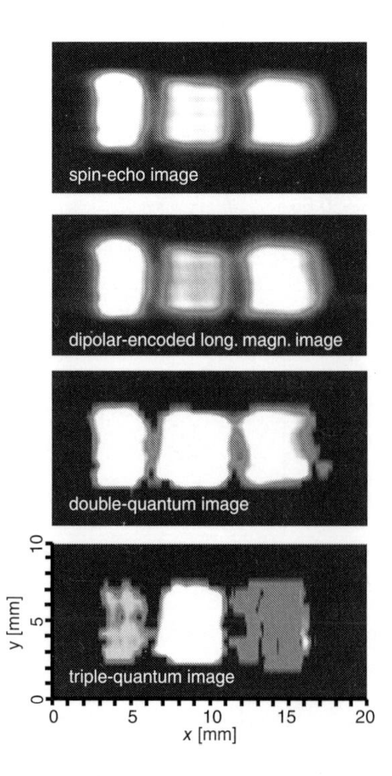

FIG. 7.2.30 [Sch13] Multi-quantum filtered images of a phantom made from PI samples with different cross-link densities. The cross-link density varies from left to right according to medium, high and low. The top image is a conventional spin-echo image for reference.

variations in the residual dipolar couplings than the relaxation weight in a spin-echo image.

7.2.11 Homonuclear magnetization-transfer filters

Magnetization transfer denotes the transfer of coherence and of polarization from one group of spins to another. If the transfer proceeds among nuclei of the same kind, it is called homonuclear and is treated below. Otherwise, the magnetization transfer is a heteronuclear process and is described in Section 7.2.12. Coherent and incoherent transfer must be distinguished from each other. Coherent transfer processes exploit the structure of the energy-level splittings from indirect or direct spin couplings and the manipulation of coherences by pulses, so that information about the magnetization phase is preserved during the transfer. In most cases, the purpose of coherence transfer is to edit the signals according to their multiplet structure. Incoherent magnetization transfer usually only affects *longitudinal magnetization* or the *polarization* in effective fields. It is driven by thermal processes, and individual spin pairs in the magnetization source and the sink flip at different times. There is no net phase associated with the transfer.

Homonuclear coherent magnetization transfer

A collection of filters for *homonuclear coherent magnetization transfer* is depicted in Fig. 7.2.31. All filters start from and end with longitudinal magnetization. Filter (a) represents a general class of filters, which uses selective pulses for excitation and reconversion of coherences in combination with a nonselective mixing period. Such filters can be thought of as frequency-selective realizations of 2D spectroscopy [Brü1, Ern1, Kes1]. A particular realization of a nonselective mixing period is depicted in (b). A short mixing time t_m is sandwiched between two $\pi/2$ pulses. Through the use of suitable phase cycling schemes it can be assured that the filtered magnetization had been transformed to longitudinal magnetization during the mixing time. Such a *z filter* for mixing avoids scrambling of absorptive and dispersive components from coherences evolving during t_1 and t_2.

The filter complementary to filter (a) is shown in (c). Nonselective pulses are used for excitation and reconversion of magnetization, but the mixing of coherences is achieved selectively. A particular realization of this scheme is the *multi-quantum filter* (d) (cf. Fig. 7.2.26(c)), where magnetization is mixed *via* even-order coherence pathways. This filter blocks magnetization from uncoupled spins $\frac{1}{2}$. The same effect is obtained with the filter (e). It is a homonuclear adaptation of the refocused *INEPT method* (insensitive nuclei enhanced by polarization transfer), which has been designed for spectral editing of high-resolution ^{13}C spectra [Ern1, Mor2] according to their multiplet structures from heteronuclear J couplings: just before the central $\pi/2$ pulse is applied in y-direction, the components of a J doublet are aligned in antiphase along the x-axis, if the evolution time is set to $t_1 = 1/(2J)$. Any transverse magnetization from uncoupled spins is refocused along the $-y$-axis as a result of an echo generated by the π pulse in the centre of the t_1 period. Therefore, the $(\pi/2)_y$ pulse in the middle of the sequence does not affect the magnetization of uncoupled spins, but it exchanges the antiparallel magnetization

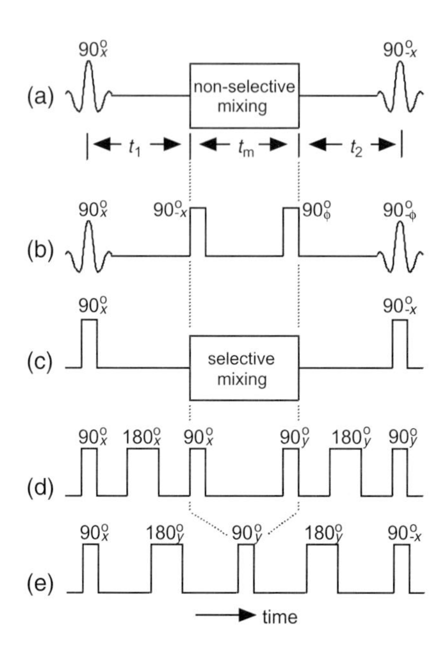

FIG. 7.2.31 Filters for homonuclear coherent magnetization transfer. All filters start from and end with longitudinal magnetization. (a) Selective excitation and reconversion of coherences with a nonselective mixing period. (b) Realization of a nonselective mixing period in a z filter *via* longitudinal magnetization. (c) Nonselective excitation and reconversion of coherences with a selective mixing period. (d) Realization of a selective mixing period by a multi-quantum filter. (e) Selective exchange of transverse magnetization within the multiplets of coupled homonuclear spin pairs by a homonuclear version of the INEPT method.

components of the doublet along x. At the end of the sequence, both the magnetization from uncoupled spins and that from the coupled spin pair are aligned along z. However, if the central $(\pi/2)_y$ pulse is left off, the magnetization from the uncoupled spins points along z, while that of the coupled spin pair points along $-z$. If two data sets are acquired, one with and the other without the middle y pulse, the sum of both data sets contains signals from uncoupled spins only and the difference collects the signals from coupled spins only. Therefore, the filter can be applied for editing signals according to their coupling patterns, for example, to separate signals from fat and water.

Filters based on *coherent magnetization transfer* have been applied in the context of imaging using hard [Dum3, Dum4] as well as soft pulses [Har1]. A *homonuclear magnetization transfer filter* of the INEPT type which has been worked out in the context of volume-selective spectroscopy is the *CYCLPOT* (cyclic polarization transfer) technique [Kim2, Knü1]. The editing principle of this technique is to transfer spin polarization between two spin groups of the molecule in a cyclic way. Integration of the *refocused INEPT sequence* [Bur1] into a spin-echo imaging scheme is illustrated in Fig. 7.2.32 [Dum3]. The reconversion pulse to longitudinal magnetization at the end of the sequence in Fig. 7.2.31(e) has been discarded, and the second π pulse for refocusing

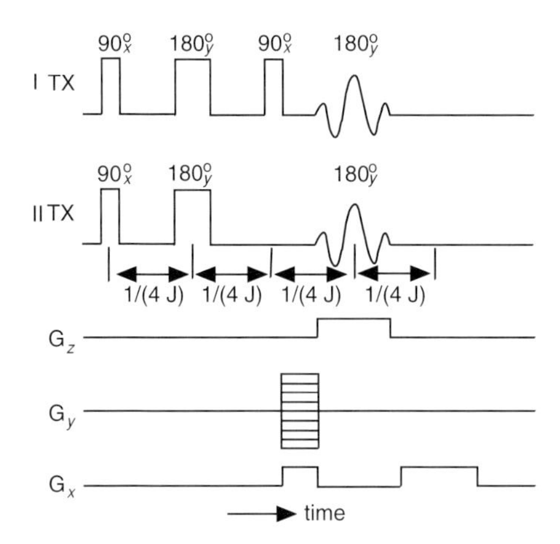

FIG. 7.2.32 [Dum3] Imaging with homonuclear polarization transfer based on the INEPT sequence. Slice selection and space encoding are incorporated into the filter.

of antiphase coherence is selective for definition of a slice. Two images must be acquired, one with and one without the central $(\pi/2)_y$ pulse. For a sample from water and fat, the difference image shows the fat signal only, because fat contains indirectly coupled protons. Because a difference of strong signals is calculated, the method is demanding with regard to accuracy of rf phases, flip angles and dynamic range of the analog-to-digital converter.

In addition to *INEPT-based filters* [Dum3, Knü1], filters with homonuclear polarization transfer *via* multi-quantum coherences (Fig. 7.2.31(d)) have been used in imaging as well [Hur1, Tri1]. These filters exploit the fact that a homonuclear bilinear coupling is not refocused in the peak of the Hahn echo. The same is true for the echo maxima in a CPMG train. For short echo times, the exponential echo decays can be expanded into a series of the first two leading terms. In this limit, the sum of amplitudes from the first and the fourth CPMG echoes and the sum of amplitudes from the second and the third echoes are the same for uncoupled spins. This is not the case for coupled spins. Therefore, the signal from uncoupled spins can be cancelled by forming the difference of the two sums [Web1]. By proper choice of the echo time, the signal from particular multiplets can be optimized for *homonuclear spectral editing*.

Homonuclear incoherent magnetization transfer

Incoherent magnetization transfer refers to transfer of polarization and not to transfer of coherences. Longitudinal magnetization belonging to different chemical groups or morphological entities is transferred by cross-relaxation [Nog1], chemical exchange, and spin diffusion [Bou1]. A seemingly straight forward filter for homonuclear incoherent polarization transfer derives from the chemical exchange experiment and is depicted in Fig. 7.2.31(b). A nonequilibrium state of longitudinal magnetization is prepared by a selective $\pi/2$ pulse followed by a precession period t_1 and a nonselective $\pi/2$ pulse. This nonequilibrium state is allowed to move towards thermodynamic equilibrium by incoherent *magnetization exchange* for some fixed mixing time t_m. The new magnetization state reached can be imaged as such or is read out by another filter with a precession period t_2 and a reconversion pulse which is selective at another frequency. Instead of chemical-shift selective excitation and precession any filter to generate a nonequilibrium state of longitudinal magnetization, for instance selective saturation and inversion recovery filters, can be used for preparation of magnetization at the beginning of the mixing time.

For long mixing times t_m, longitudinal magnetization is transferred by the nuclear Overhauser effect and by chemical exchange [Ern1]. In solids, the magnetization of abundant spins is also transferred through space by *spin diffusion*, and a number of filters for spin diffusion have been reviewed in Section 7.2.9. By selective double irraditation in ^{1}H NMR of liquids, longitudinal magnetization can be transferred in the rotating frame. This method is called *selective Hartmann–Hahn coherence transfer* [Kon1].

Chemical exchange plays an important role in biological systems [Hsi1, Wol1] with enzyme-catalysed reactions but also for other systems where, for example, labile protons are exchanged between some compound and its solvent [Ken1]. With methods which introduce contrast based on homonuclear polarization transfer, images can be generated which show the distribution and rate of the proton exchange. When proton exchange occurs between a low-concentration component with short T_1 relaxation and an abundant solvent with long T_1 relaxation, a substantial increase in sensitivity for the detection of the low-concentration component can be realized. This feature can lead to considerable contrast and signal enhancement in biomedical imaging [Hsi1]. It has also been studied to map tracer diffusion in heterogeneous materials including porous rock [Fis1].

A method often used to exploit this effect is *saturation transfer*. The magnetization from the low-concentration species is attenuated through selective saturation, and the saturation is carried over to the high-concentration solvent by the exchange process. Two images may be acquired for the dominant spin species, one image with selective irradiation of the exchanging partner and the other one without. In the difference image, those regions of the sample appear with enhanced contrast where both exchanging species are in close contact and the exchange actually happens. For proton exchange, the contrast enhancement strongly depends on pH, and for large molecules, the effect may be masked by polarization transfer from cross-relaxation [Bou1].

Because the exchange rate affects the effective longitudinal relaxation time of the observed species, the simple selective saturation method can be extended by appending an inversion-recovery sequence to image the zero crossing of longitudinal magnetization for a given exchange rate and to identify different *exchange rates* based on positive and negative signals in the image [Fis1, Tes1].

7.2.12 Heteronuclear magnetization-transfer filters

Just as in the homonuclear case, *heteronuclear magnetization transfer* is achieved by the transfer of transverse magnetization or, more generally, of coherences, and by transfer of longitudinal magnetization or, more generally, by polarization. Off-diagonal density-matrix elements are transferred in the first case and diagonal density-matrix elements in the second case. Coherence transfer is typical for liquid-like systems which exhibit spectral resolution in the absence of applied gradients, because the coherences need to survive the transfer times. Polarization transfer can be applied in both liquid and solid systems, as long as the transfer is achieved in times shorter than the longitudinal relaxation times. Heteronuclear magnetization transfer schemes are used to access spectroscopic and relaxation parameters of a second nucleus, to exploit the topology of the coupling between both nuclei for spectral editing, and to increase the sensitivity of the measurement. *Heteronuclear filters* are readily used in combination with other filters which introduce selectivity of the NMR measurement on one of the participating nuclear species. One example for such combinations are filters for polarization transfer from ^{1}H to ^{13}C and subsequent relaxation filters for ^{13}C [Fry1]. However, the inverse combination of a ^{1}H relaxation filter succeeded by magnetization transfer to ^{13}C is also useful, for example, in morphological studies of heterogeneous solids [Tek1, Zum1].

Heteronuclear coherent magnetization transfer

In the context of imaging, heteronuclear magnetization transfer is of interest to generate contrast from the properties of a second nucleus or the combined properties of two nuclear species. Depending on the sensitivity of the nuclei involved, either direct or indirect detection schemes of the second nucleus may be advantageous [Ern1]. For *transfer of transverse magnetization*, the sensitivity of the experiment is determined by (1) the amounts of available polarizations which further depend on the gyromagnetic ratio γ_{exc} of the excited nucleus, and, for partial saturation, on the ratio of the repetition time t_{R} to the longitudinal relaxation time T_1, (2) the response of the observed nucleus, which is proportional to the amplitude and the frequency of the observed nucleus and thus to γ_{obs}^2, and (3) the detector noise, which is proportional to $\gamma_{\mathrm{obs}}^{1/2}$. The overall signal-to-noise ratio S/N then is determined by the expression

$$\mathrm{S/N} \propto \gamma_{\mathrm{exc}}\gamma_{\mathrm{obs}}^{3/2}\left[1 - \exp\left\{-\frac{t_{\mathrm{R}}}{T_{1,\mathrm{exc}}}\right\}\right]. \tag{7.2.22}$$

Using this expression to compare excitation of ^{1}H with direct detection of ^{13}C involving a single coherence transfer step from ^{1}H to ^{13}C and with indirect detection of ^{13}C by double transfer from ^{1}H to ^{13}C to ^{1}H shows that indirect detection provides an S/N which is $(\gamma_{1\mathrm{H}}/\gamma_{13\mathrm{C}})^{3/2} \cong 8$ times better than for direct detection.

Different coherence transfer schemes with direct detection are known in liquid-state NMR as the *INEPT* [Mor2] and *DEPT* (distortionless enhancement by polarization transfer) [Ben2, Dod2] methods and their variants [Chi2, Ern1, Hei1]. They can readily be combined with space-encoding sequences to form heteronuclear imaging schemes [Yeu1]. For reasons of sensitivity, mostly indirect detection methods with single or double coherence transfer have been investigated for localized spectroscopy as well as for imaging [Hal5, Kim2, Sil1]. Therefore, only indirect detection schemes are reviewed in the following.

Volume selective schemes are cyclic polarization transfer volume-selective editing (*CYCLPOT-VOSING*), referring to the laboratory frame [Knü2, Knü3] and cyclic cross-polarization localized spectroscopy (*CYCLCROP-LOSY*) referring to the rotating frame [Kös1, Kös2, Kun1, Kun2]. Here the names already indicate an important use of these techniques for signal identification by spectral editing, and also their classification as incoherent polarization transfer methods (see below). The principles for proton detected editing and imaging of heteronuclei like ^{10}B, ^{13}C, ^{14}N, ^{31}P, include echo methods like *spin-echo double resonance* (SEDOR) [Cap5, Cap6, Del2, Del4, Del5, Del6, Kin2, Wat1] for discrimination of signals from linear and bilinear spin interactions, a two-transient subtraction method, single-transient multi-quantum filtering (HYCAT: hydrocarbon tomography) [Knü4, Knü5], and CYCLCROP [Kun3, Spy1].

The *SEDOR* method is based on subtraction of large signals leading to small signal differences. It is sensitive to small signal errors and critically depends on spectrometer adjustments. The *HYCAT method* is illustrated in Fig. 7.2.33(a) [Kim4]. The initial

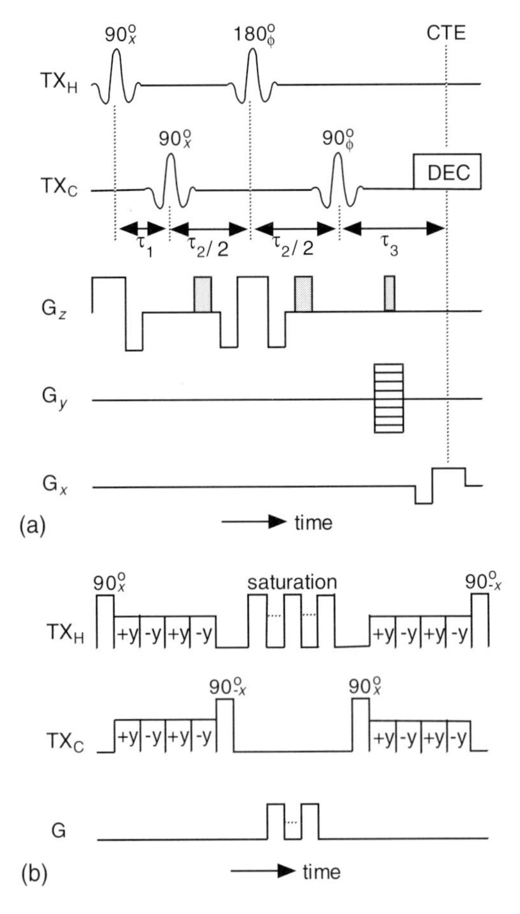

FIG. 7.2.33 [Kim4] Imaging techniques for indirect detection of heteronuclear coherences. (a) HYCAT method. A coherence-transfer echo (CTE) is recorded in the presence of an optional ^{13}C decoupling pulse (DEC). The subscript ϕ indicates an arbitrary rf phase. The grey gradients have the function of selecting the coherence-transfer echo. (b) CYCLCROP filter. It can be used in combination with different space encoding techniques. The gradient pulses are applied to spoil magnetic-field homogeneity.

excitation on the ^{1}H channel is made to be slice selective. The transverse magnetization of a ^{1}H doublet is then allowed to evolve for a time $\tau_1 = 1/(2J_{CH})$, where J_{CH} is the heteronuclear J-coupling constant, so that the coherences of the doublet are in antiphase at the time of conversion into heteronuclear multi-quantum coherences by the following $\pi/2$ pulse in the ^{13}C channel. This pulse may be a soft pulse for chemical-shift selectivity. The slice-selective π pulse applied to ^{1}H in the middle of the multi-quantum evolution period interchanges zero- and double-quantum coherences for formation of a *coherence-transfer echo* (CTE) at a time τ_3 after the end of the multi-quantum evolution period τ_2. The position of the CTE is determined by the time to unwind by single-quantum precession during t_3 the phase evolution acquired during the multi-quantum evolution time τ_2. The echo-formation time is used for encoding the space information in phase and frequency of the echo. To discriminate the CTE from potential homonuclear echoes, the multi-quantum coherences are dephased during τ_2, and the phase transferred to single-quantum coherences is rephased during τ_3 by the grey gradients in Fig. 7.2.30(a). The gradient strengths for dephasing and rephasing are determined by the quotient of the gyromagnetic ratios of the nuclei involved forming the heteronuclear multi-quantum coherences.

Heteronuclear incoherent magnetization transfer

Heteronuclear incoherent magnetization transfer is the transfer of longitudinal magnetization. It can proceed in the laboratory frame and in the rotating frame. The *nuclear Overhauser effect* (NOE) [Nog1] is a manifestation of *polarization transfer in the laboratory frame*. In the extreme narrowing limit saturation of dipolar relaxation of the I doublet of a heteronuclear IS two-spin-$\frac{1}{2}$ system leads to an enhancement of the S-spin polarization by a factor

$$\eta = 1 + \frac{\gamma_I}{2\gamma_S}. \tag{7.2.23}$$

For ^{13}C coupled to ^{1}H the Overhauser enhancement assumes a maximum value of $\eta = 3$, while for ^{15}N coupled to ^{1}H $\eta = -4$. The enhancement factor depends only on the gyromagnetic ratios of the coupled spins, not on their abundance. If additional relaxation mechanisms are involved for the S spins, the enhancement factor is reduced. The build-up of the enhancement proceeds with the longitudinal relaxation rate of the S spins and may be relatively slow, so that extended presaturation periods may be necessary [Ern1].

Polarization transfer in the rotating frame is often referred to simply as *cross-polarization* (CP). It represents a standard technique in solid-state NMR to acquire signals from rare spins like ^{13}C coupled to abundant spins like ^{1}H, because significant sensitivity enhancement can be achieved (cf. Section 3.3.1). Instead of direct observation of the rare spins [Har2, Meh1, Pin1] CP is also used for indirect detection of rare spins *via* high-sensitivity nuclei such as ^{1}H in spectroscopy [Ble1, Man5] as well as in imaging [Nak1]. The signal build-up of the second nucleus depends on the cross-polarization time, which in return depends in the strength of the heteronuclear coupling [Meh1]. For the dipolar ^{1}H–^{13}C coupling in elastomers it has been shown that the CP build-up rates scale with the cross-link density of the elastomers [Fül5]. The effective strength of the heteronuclear dipolar coupling can be manipulated by simultaneous irradiation of both nuclei by multi-pulse sequences like WIM24 (windowless mixing) [Car2] and suitable phase cycling of different lock-pulse intervals. This adds a further feature of selectivity to the heteronuclear cross-polarization techniques by which even heteronuclear J couplings can be probed [Bur2, Ern1, Zil1].

The standard cross-polarization scheme due to Hartmann and Hahn requires the simultaneous application of spin-lock fields to both nuclear species, e. g. ^{1}H and ^{13}C. For optimum polarization transfer, the amplitudes B_{eff} of the fields effective in the rotating frames of each nucleus have to match the *Hartmann–Hahn condition* (cf. Section 3.3.1)

$$\gamma_{\mathrm{C}} B_{\mathrm{effC}} = \gamma_{\mathrm{H}} B_{\mathrm{effH}}, \tag{7.2.24}$$

where γ_{C} and γ_{H} are the gyromagnetic ratios of ^{13}C and ^{1}H. This transfer is not restricted to spins coupled by the dipolar interaction, but may also be applied to J-coupled spins in liquids [Ber1, Chi1, Mau1, Mül1]. In this case it is referred to as *J cross-polarization*. Double J cross-polarization with the CYCLCROP method (see below) has been studied for indirect imaging of ^{13}C in liquids [Kun3], for monitoring the carbohydrate metabolism and transport in plants [Hei2], as well as for imaging of the elastomer distribution in multi-component systems [Spy1, Hei3].

Polarization transfer by the Hartmann–Hahn method is not an adiabatic process, so the transfer efficiency can be improved. In solids this can be done by adiabatic demagnetization and remagnetization in the rotating frame (*ADRF* and *ARRF*, cf. Section 3.3.1) [And2, Gol2, Sli1]: following an initial spin lock of the source magnetization, the lock field is adiabatically reduced to zero. The spin system then always remains very near to equilibrium, and the spin polarization is transferred into the local dipolar fields of the coupling spins. From there, it is recalled by the inverse process performed on the coupling partner by adiabatically ramping up the B_1 field of the sink spins. It should be noted that such cross-polarization schemes can be devised not only by single-quantum coherences but also by multi-quantum transitions. For example, cross-polarization by double-quantum transitions [Dem5, Dem6] has been explored in the context of imaging for editing and for slice and volume selection. In the context of slice selection, single-quantum cross-polarization [Dem4, Haf2] and J cross-polarization [Dem8, Dem9] have been investigated as well, also in combination with other filters to introduce further selectivity [Dem7].

An example for *polarization transfer in the rotating frame* is the *CYCLCROP filter* for indirect detection and editing of rare-spin magnetization by double J cross-polarization (Fig. 7.2.33(b)) [Kun3]. For efficient polarization transfer of a selected chemical group, the effective field has to be turned on for a duration of J^{-1} for an AX, $(J2^{1/2})^{-1}$ for an AX_2 system and $0.62/J$ for an AX_3 system. The efficiency of polarization transfer is 100% for an AX system. For other systems complete polarization transfer cannot be achieved, but the total signal enhancement is still higher than for an AX system. For purposes of spectral editing, the Hartmann–Hahn match is performed off-resonance with field strengths and contact times scaled appropriately to match the magnetic structure of the chemical group. *Hartmann–Hahn mismatch* can be overcome by the *MOIST* (mismatch optimized IS transfer) method [Lev1, Lev2], which consists of simultaneous reversal of the spin-lock fields for intervals short compared to J^{-1}, and by the *PRAWN* (pulsed rotating-frame transfer sequences with windows) method, which consists of a windowed comb of low flip-angle contact and spin-lock pulses [Cha2, Spy1]. Following the polarization transfer from ^{1}H to ^{13}C, the ^{13}C in-phase coherences are stored along the z-axis, and all ^{1}H coherences are destroyed by a sequence of saturation pulses, for instance by ten phase-cycled $\pi/2$ pulses in combination with homogeneity-spoil gradient pulses. Afterwards, the ^{13}C magnetization is transferred back to ^{1}H, and the transverse magnetization obtained after the spin-lock period is stored along z for subsequent space-encoding at a later time. ^{13}C decoupling may be applied during the detection time to collapse the heteronuclear multiplet into one line for improved sensitivity. In combination with magnetic-field gradients during the cross-polarization time, the CYCLCROP scheme and an adiabatic version of it have been used for slice and volume selection [Dem2, Dem3, Kös1].

7.3 SPECTROSCOPIC PARAMETERS

In terms of acquisition time, the use of filters for preparation of nonequilibrium magnetization is a rather economic way to introduce contrast to images, and often the requirements for homogeneity of the $\boldsymbol{B}_0$ field are not demanding. However, many *spectroscopic parameters* are inaccessible by magnetization preparation, and 1D or 2D spectra need to be acquired for each voxel, so that the contrast parameters can be extracted from the spectroscopic response. Then, in addition to the space dimensions required by the image, the spectroscopic dimensions need to be measured as well, and measurement times can become very long. Two approaches are distinguished: *Frequency encoding* and *phase encoding* of the spatial–spectral response. The first is an off-spring of back-projection imaging [Cor3, Lau1], where projections need to be acquired over a range of field-gradient strengths. By extrapolation and transformation of the experimental data, spectroscopic and spatial responses can be disentangled (cf. Section 6.1). This approach is recommended for spectroscopic imaging of widelines. In high-resolution NMR the spectroscopic response covers many data points so that it is usually acquired in the absence of magnetic field gradients after phase encoding of the spatial information [Cox1, Lau3]. In either case, the spectroscopic resolution is limited by the magnetic-field inhomogeneity within a voxel. Imaging methods which acquire spectroscopic, in addition to spatial, dimensions are classified as *spectroscopic imaging* [Haa2] or *chemical-shift imaging* [Bro1].

Different forms of spectroscopic imaging have been elaborated. They primarily differ in the space-encoding schemes. Examples are back-projection imaging [Cor3, Lau1], different versions of echo-based Fourier imaging [Blü9, Coc1, Cox1, Dem11, For2, Gün3, Haa4, Kli1, Lau3, Mau2, Rom5], echo-planar imaging [Man6], multi-quantum imaging [Gün1], multi-pulse imaging [Hep2, Hep3, Hep4], and MAS imaging [Gün2, Sch7, Sun1]. Further selectivity can be introduced by combination with magnetization filters, for instance by partial saturation [Gün1], chemical-shift selective pulses [Hal3], water suppression [Pos2], and spin-diffusion filters [Wei1, Wei2].

For biomedical applications, the definition of image contrast by signal intensities at particular chemical shifts in the high-resolution spectrum of biological tissue is most important. Examples are the formation of images from amino acids and lactate signals in brain, which are indicative of the distribution of healthy tissue and tissue affected by stroke [Hur1, Luy1, Moo1], and proton chemical-shift imaging of human muscle metabolites [Huj1]. Furthermore, spectroscopic imaging is applied to heteronuclei like ^{13}C [Bec1, Ber2, Sze1], ^{23}Na [Koh1], and ^{31}P [Bro2, Koc1]. Sensitivity can be improved by indirect imaging, for example, of ^{13}C *via* heteronuclear multi-quantum coherence as an extension of the HYCAT scheme by variation of τ_2 as a spectroscopic evolution time in Fig. 7.2.30(a) [Lee1]. Further applications concern the distribution of different fluids in rock cores [Hal4, Maj2].

Imaging of heteronuclei as such may be viewed as a form of spectroscopic imaging, when considering the total scale of NMR frequencies from all nuclei. Improvement of the often low signal-to-noise ratio can be achieved by indirect detection and cross-polarization (cf. Section 7.2.12) in addition to isotope enrichment and hyperpolarization (cf. Section 7.4.4). For example, ^{2}H can be incorporated into organic matter like synthetic polymers (cf. Fig. 10.2.3), plants, and animals [Lin1, Mül2, Seo1]. Direct imaging of isotopes like ^{7}Li [Son3], ^{11}B [Lee3], ^{13}C [Fry2, Hal6], ^{14}N [Kin3], ^{15}N [Kin4], ^{23}Na [Sui1], ^{27}Al [Ack1, Bra1, Con1], ^{51}Va [Bra1], ^{63}Cu [Swa1], and other nuclei has been explored in phantom studies (cf. Section 8.1). Imaging investigations of ^{11}B in biological tissue are relevant to radiotherapy of cancer, where boron-rich compounds are localized in tumours and then irradiated with neutrons [Kab1]. Also ^{17}O bears interesting potential in the research of oxygenation in biomedicine [Mat4], and ^{19}F imaging has been used for localization of antitumour drugs in animal studies [Doi1]. ^{23}Na and ^{39}K are of great importance in the chemistry of life, and diverse imaging studies of mainly ^{23}Na are concerned with its role in biochemistry [Coc1, Dai1, Hut1, Kel1, Koh2, Lyo1, Pay1, Wim1, Wim2]. In contrast to ^{129}Xe (cf. Section 7.4.4), ^{131}Xe has a quadrupole moment which has been investigated near the critical point as a surface-sensitive probe in imaging of aerogels [Pav1].

For material imaging, also quantities other than the spin density of a particular isotope, chemical shifts, and associated signal amplitudes are of interest. These may be lineshapes [Blü10, Fry2], signal intensities at a particular frequency of a *wideline powder spectrum* which denotes a particular orientation of the coupling tensor [Hep1, Hep4, Gün1, Gün2, Gün3], and moments of wideline spectra [Mat3, Wei3] which contain information about molecular order and dynamics (cf. Sections 3.2.1 and 3.2.2). For most practical purposes, the analysis of 1D spectra provides sufficient spectroscopic information, but examples of 2D spectroscopic imaging have been demonstrated as well [Sch7].

7.3.1 1D spectroscopy

Principle of spectroscopic imaging by phase encoding of space

The principle of spectroscopic imaging is based on the separation of the spectroscopic response from the spatial response (cf. Section 6.2.4). Because the spatial response depends on the presence of magnetic-field gradients, while the spectroscopic response requires the absence of such gradients, spatial and spectral responses can be separated by processing of a sufficient number of data sets measured with different gradient strengths. Clearly, homo- and heteronuclear variants exist.

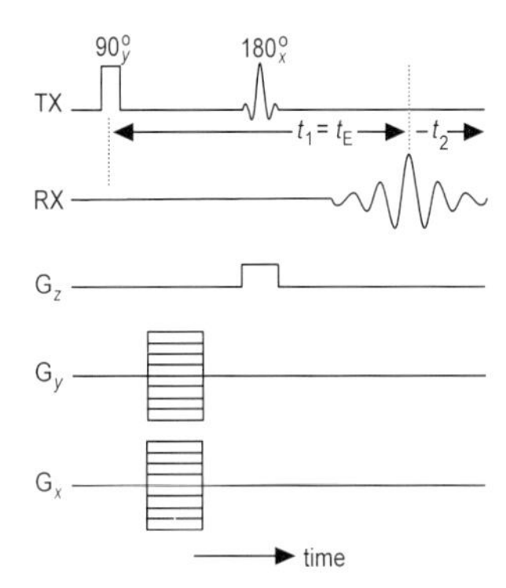

FIG. 7.3.1 Principle of spectroscopic Fourier imaging. To separate spatial and spectroscopic responses, the spectroscopic evolution must be constant during the space encoding period t_1, or it can be refocused by formation of an echo. Space is encoded during the evolution time t_1 and the spectroscopic response is acquired during the detection time t_2.

For spectroscopic imaging, phase encoding is most often the method of choice: the acquisition time is determined by the number of steps in the indirectly detected dimensions. Because the FID usually encompasses more data points than the projection of the object, the FID is detected directly during t_2, and space is phase-encoded during t_1 (Fig. 7.3.1). For definition of the slice, one of the rf pulses is a selective pulse applied in the presence of a gradient (cf. Fig. 6.2.3). The evolution time t_1 is kept constant so that the signal phase is determined only by the gradient modulation and not by the frequency offset in the NMR spectrum. To obtain pure-phase spectra and to prolong the time t_1 available for phase encoding, the spectroscopic phase evolution is refocused to form a Hahn, solid, or magic echo at time t_1. The irreversible signal decay during t_1 can serve as a relaxation filter to generate additional contrast. Instead of an echo [Blü9, Dem11, Rom5], also the orientation-independent double-quantum coherence can be used for space encoding in spectroscopic ^{2}H imaging [Gün1].

Spectroscopic wideline imaging in solids

Apart from mapping spatial distributions of particular chemical structures by high-resolution multi-pulse spectroscopic imaging of ^{1}H in polymers [Hep3], *molecular orientation* is a material property of particular interest, which can be tackled by *spectroscopic wideline imaging in solids*. For imaging of rare spins, cross-polarization and heteronuclear dipolar decoupling may suffice [Sze1], but imaging of abundant nuclei like ^{1}H or ^{19}F requires homonuclear dipolar decoupling during space encoding and acquisition of the spectroscopic response. Then, the sample needs to be small enough to fit into a MAS rotor for MAS imaging [Sch7] or into a 1 cm diameter coil for high-power multi-pulse irradiation [Hep2, Hep4].

In terms of instrumentation the less demanding approach is multi-pulse excitation for averaging of the homonuclear dipolar interaction by trains of repetitive *magic echoes* during space encoding and acquisition of the spectroscopic response. Two such magic-echo

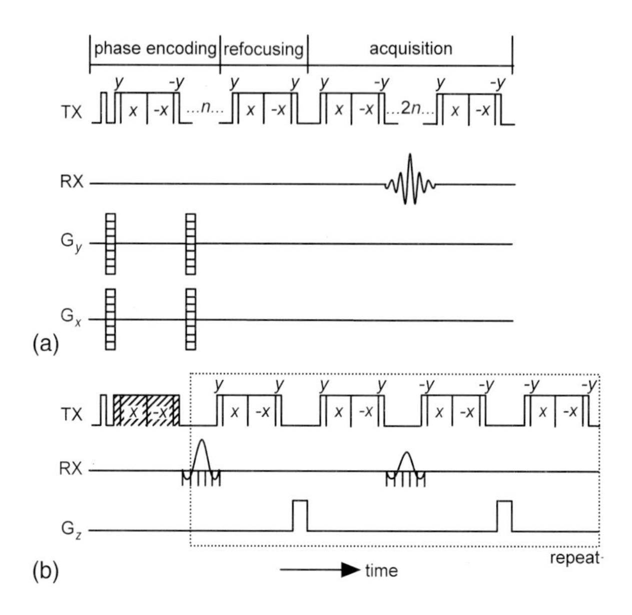

FIG. 7.3.2 Magic-echo pulse sequences for spectroscopic imaging and imaging of moments. (a) Spectroscopic 2D imaging with homonuclear dipolar decoupling [Hep4]. Phase encoding and acquisition periods are defined by trains of successive magic sandwiches which are connected by a magic sandwich combined with a 180° refocusing pulse for formation of a Hahn echo. (b) 1D imaging of moments from dipolar broadened widelines [Mat3]. Gradient pulses are applied for every second echo, while every other echo is acquired to extract spectroscopic moments. The hatched magic-echo sandwich represents a block of four echo cycles [Mat3].

trains have been combined *via* a magic sandwich with a 180° pulse on top to refocus the chemical shift dephasing under the preceding train of magic echoes for acquisition of the orientation-dependent ^{19}F spectrum in strained Teflon sheets (cf. Section 10.3.1) (Fig. 7.3.2(a)) [Hep4]. The first train before the refocusing sandwich is used for space encoding by pulsed-field gradients. The dipolar coupling is suppressed by the magic echo refocusing, but the chemical-shift evolution is unaffected. By adding a 180° pulse to the second 90° pulse of the refocusing sandwich, the sign of that pulse is changed and a chemical-shift echo is invoked during the second magic-echo train which again suppresses the dipole–dipole interaction in the echo maxima. Therefore, a homonuclear, dipolar decoupled ^{19}F spectrum can be derived from the sequence of magic-echo maxima, which forms a chemical-shift echo.

A much simpler situation is encountered when rare nuclei are imaged. Then the homonuclear dipole–dipole interaction is negligible, and orientation-dependent spectra can readily be acquired. For example, a *spectroscopic image of* ^{13}*C* from highly oriented, syndiotactic poly(propylene) has been measured, which reveals the degree of molecular orientation as well as a surface-core layer structure, which often arises in melt casting of polymers [Gün3]. Here, conventional cross-polarization from ^{1}H to ^{13}C has been applied together with high-power decoupling of the heteronuclear

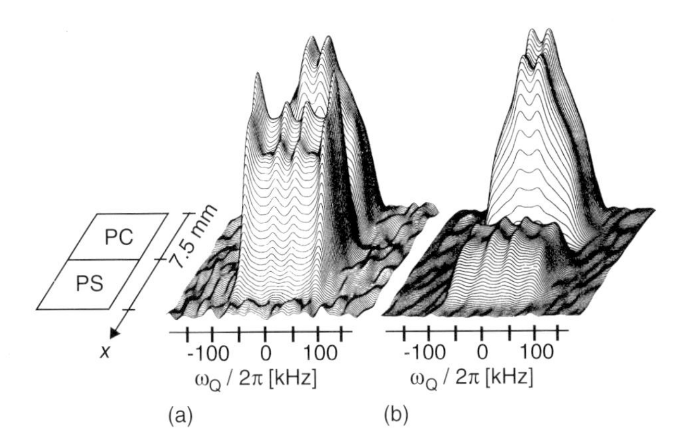

FIG. 7.3.3 Spectroscopic ^{2}H 1D image of a phantom of ring-deuterated PS and PC. The lineshapes reveal different timescales of the phenyl ring flip for both materials. (a) Spectra from Boltzmann magnetization. (b) Spectra with partial saturation from reduced repetition times. The PS signal is attenuated and the shoulders from the more rigid chain segments are partially suppressed. Adapted from [Gün1] with permission from Taylor & Francis.

dipole–dipole interaction during data acquisition. Overlap of broad lines at different chemical shifts is reduced by high molecular orientation which results in comparatively narrow lines.

Overlap of broad resonances can also be avoided by isotope enrichment at selected chemical sites. This approach has been taken in spectroscopic MAS imaging [Gün2] but also in ^{2}H imaging of spy molecules incorporated into elastomer bands. The splitting of the deuteron resonance under strain [Del3] has been mapped by spectroscopic imaging to access the spatial distribution of strain [Kli1]. In rigid polymers the ^{2}H lineshape reveals information about both molecular order and slow molecular dynamics (cf. Sections 3.2.1 and 3.2.2). The orientation-dependent lineshape has been imaged by *spectroscopic ^{2}H imaging* on a phantom of perdeuterated and drawn PE by the double-quantum method of Fig. 7.2.26(f) [Gün1]. Because this technique also involves periods t_w with a spin-alignment state, ultra-slow molecular motions can be probed and the associated ^{2}H lineshapes can be imaged. This is demonstrated in Fig. 7.3.3 for a phantom made of ring-deuterated poly(styrene) (PS) and ring-deuterated poly(carbonate) (PC) [Gün1]. At room temperature, both materials are in the glassy state. They exhibit different lineshapes and longitudinal relaxation times T_1 owing to different correlation times of the phenyl-flip motion. In PS, most of the phenyl rings are rigid on the NMR timescale, while in PC most of the rings perform rapid flips by 180°. The spectra on the left have been measured with a repetition time of 2 s consistent with space encoding of thermodynamic equilibrium magnetization. The spectra on the right were acquired with a repetition time of 0.2 s. The signal of the rigid PS with longer T_1 is partially suppressed, and the lineshapes have changed (cf. Fig. 7.2.2). This demonstrates the combination of spectroscopic imaging with a partial saturation filter for creation of additional contrast. Applications of such imaging methods can be envisioned for studies of distributions and effects of plasticizers in polymer materials [Blü10].

Imaging of spectral moments

The *moments of a lineshape* (cf. eqn (3.5.14)) are sensitive measures of molecular mobility [Abr1]. They pose an alternative to the characterization by relaxation times. For instance, in solid adamantane and hexamethylbenzene, T_2 is similar, but the second moments M_2 differ by more than a factor of two [Mat3]. To obtain the moments it is not necessary to integrate the spectrum but they can be deduced from the initial part of the FID $s(t)$ (cf. eqn (3.5.13)),

$$s(t) = 1 - \frac{M_2 t^2}{2!} + \frac{M_4 t^4}{4!} - \frac{M_6 t^6}{6!} + \cdots, \tag{7.3.1}$$

where a symmetric lineshape centred at ω_0 has been assumed, which is often encountered for abundant spins in solids.

To measure the space-time response for extraction of moments, multi-dimensional schemes typical for spectroscopic imaging can be employed [Wei3]. Most successful are techniques based on *magic echoes*. Because the decay of the wideline is very rapid, the time dimension associated with the wideline spectrum can be incorporated into a train of magic echoes set up for space encoding if every second echo is used for measuring the time-domain response in the absence of a magnetic-field gradient (Fig. 7.3.2(b)) [Mat3]. By writing the successively sampled echo signals into successive rows of a data matrix, the columns of that matrix yield a projection of the object after Fourier transformation and the rows provide the signal decay (7.3.1) for extraction of the moments. This is an economic way for *imaging of moments*. Spatial resolution in more than one dimension is readily obtained by pulsing the gradients in different directions and reconstruction of the shape of the object by the back-projection method. In a way the pulse sequence in Fig. 7.3.2(b) is complementary to the one in Fig. 7.3.2(a) as a narrow spectroscopic signal is sampled in (a) and a broad spectroscopic signal in (b).

7.3.2 Multi-dimensional spectroscopy

In principle, any scheme of multi-dimensional liquid-state or solid-state spectroscopy can be combined with spatial resolution to mD spectroscopic nD imaging. However, measurement times rapidly become excessive, so that meaningful applications are quite rare.

Schemes for the liquid state

The use of *2D spectroscopic imaging* schemes can be justified in cases where the desired spectroscopic information is hidden in complicated 1D spectra which are often encountered in biological systems such as in humans and in plants. In this case, cross-peak intensities of 2D spectra can be converted into image contrast. In biological systems water suppression is a major issue, so that preparation sequences to eliminate the water signal need to be employed or coherence transfer schemes. To reduce acquisition times to a minimum, phase cycling is avoided by use of gradient-selected coherence transfer [Hür2, Knü5, Mau3, Zhu1, Zij1].

The principle of *2D-COSY imaging* has been demonstrated on the hypocotyl of the castor bean, enabling the formation of images showing the distributions of sucrose,

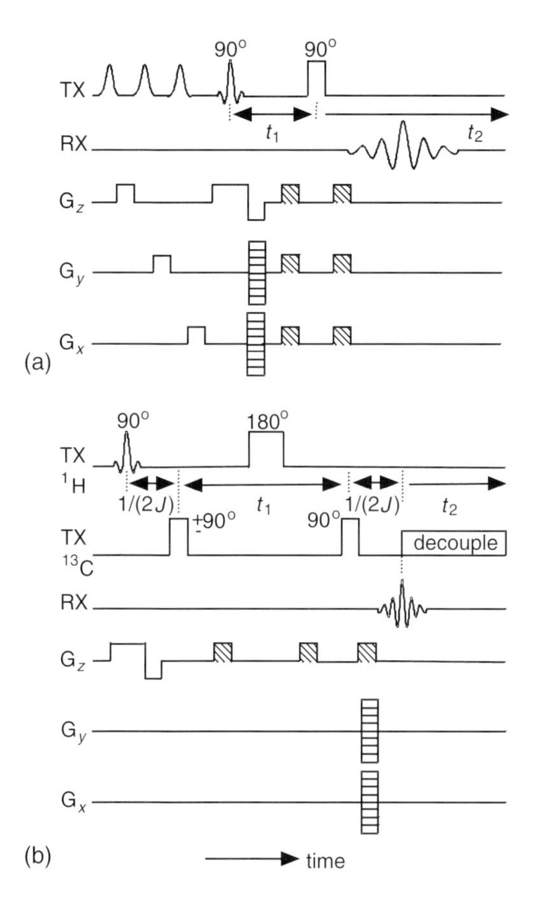

FIG. 7.3.4 Pulse sequences for 2D spectroscopic 2D imaging. (a) COSY spectroscopy [Zie1]. The spectroscopic imaging sequence is preceded by a water suppression sequence of three selective pulses. (b) Heteronuclear multi-quantum coherence-transfer spectroscopy [Zij2]. The hatched gradients are for selection of coherence transfer pathways.

β-glucose, glutamine/glutamate, lysine, and arginine [Zie1]. The pulse sequence is depicted in Fig. 7.3.4(a). The spectroscopic imaging sequence is preceded by three selective pulses with subsequent homogeneity-spoil gradient pulses for suppression of the water signal. The first 90° pulse of the COSY sequence is frequency selective for definition of the slice. The remaining two space coordinates are phase encoded. The coherence-transfer pathways are selected by the hatched gradient pulses right before and after the second, hard 90° pulse.

Efficient suppression of water signals is a side benefit of heteronuclear coherence transfer schemes. The heteronuclear multi-quantum coherence (HMQC) method (Fig. 7.3.4(b)) is a broad-band version of the HYCAT experiment of proton detected ^{13}C imaging (cf. Fig. 7.2.30(a)) [Knü4]. The initial 90° pulse on ^{1}H is used for slice selection. For a heteronuclear AX system, single-quantum proton magnetization is transferred into heteronuclear zero- and double-quantum magnetization by a 90° ^{13}C pulse after a preparation time of duration $(2J_{CH})^{-1}$. The multi-quantum coherence evolves for an evolution time t_1, in the middle of which zero- and double-quantum coherences are interchanged by the 180° pulse on ^{1}H. The second ^{13}C-90° pulse converts the multi-quantum coherence back to single-quantum ^{1}H coherence for detection during t_2. The hatched gradient pulses are adjusted to generate a multi-quantum coherence transfer echo at a time $(2J_{CH})^{-1}$ after the last ^{1}H pulse. Space is encoded on the magnetization phase by gradients applied immediately before detection. The *HMQC imaging* sequence has been used to detect ^{13}C-labelled glucose and its metabolic products in cat brain [Zij2].

Schemes for the solid state

While justifications to apply *2D spectroscopic imaging* can be found for living systems, its true value in materials science needs to be demonstrated. So far, an experiment has been reported for MAS imaging with 1D spatial resolution on a suitably chosen phantom made of two sheets of highly oriented PE sandwiched between two sheets of isotropic PE (Fig. 7.3.5(a)) [Sch7]. The *rotor-synchronized 2D MAS experiment* (cf. Section 3.3.3) [Har3] for characterization of *molecular order* was combined with pulsed field gradients for frequency encoding of spatial and spectral resolution (Fig. 7.3.5(c)) to determine the second moment of the orientational distribution function as a function of space (Fig. 7.3.5(b)).

This 2D experiment is well suited for imaging, because the indirectly detected spectroscopic dimension requires stepping of the spectroscopic evolution time t_1 from an arbitrary but fixed rotor phase in only 16 steps through one rotor cycle t_R. This dimension introduces the information about molecular order into the spectrum. Following t_1, magnetization is cross-polarized from ^{1}H to ^{13}C, and the ^{13}C response is acquired by frequency encoding in field gradients of different strengths, while ^{1}H is being decoupled (DD). After disentangling spectroscopic and spatial responses (cf. Section 6.1) [Cor3] a set of rotor-synchronized 1D ^{13}C-MAS spectra is obtained for each position along the space axis, which is defined by the rotor radius r for the sample investigated. Slow spinning with the rotor frequency ω_R being less than the chemical-shift anisotropy of the line is required for generation of spinning sidebands. Fourier transformation over t_1 produces 2D spinning sideband patterns for each space coordinate.

Analysis of each 2D spectrum (d, e) in terms of *Legendre subspectra* yields the *orientational distribution function* for each point along the space axis. But given the low signal-to-noise ratio and the long measuring times of the method, only the second moment P_2 (b) of the distribution function could be extracted with confidence. It peaks near the centre of the oriented sheet and vanishes for the unoriented material. The spatial resolution is low and the values determined by imaging (stepped line) are smeared out compared to the values expected from prior knowledge of the sample (broken line). This deviation of experimental signal amplitudes from the ones expected from the separately measured spectra of the individual components can formally be analysed in terms of a convolution of the ideal values with a resolution function of the experiment. Assuming a Gaussian resolution function, the smooth shape in Fig. 7.3.5(b) is obtained. From the width of the resolution function the spatial resolution has been determined to be 0.3 mm.

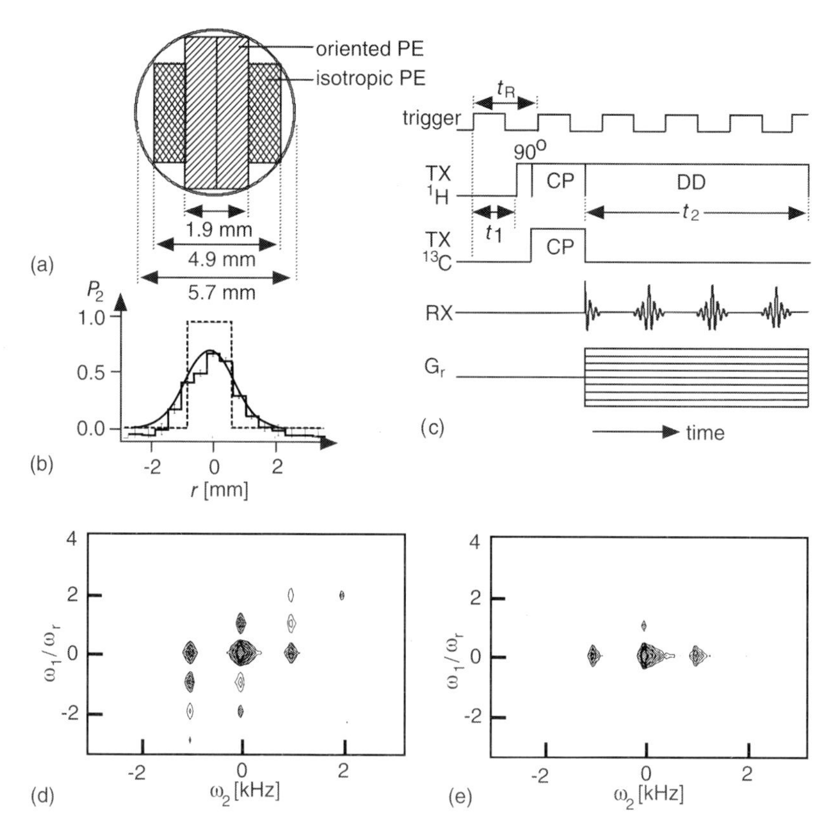

FIG. 7.3.5 [Sch7] 2D spectroscopic 1D MAS imaging of molecular orientation. (a) Phantom made of sheets of highly-oriented and isotropic PE. (b) Second moment P_2 of the orientational distribution function as expected (broken line), as measured (stepped line), and as obtained by convolution of the expected shape with a Gaussian resolution function. (c) Pulse sequence. A CP-MAS experiment is synchronized to the rotor phase, which provides the trigger for the pulse sequence. (d) 2D MAS spectrum of the oriented region extracted from the 3D data set. Positive signals are hashed. (e) 2D MAS spectrum of one of the isotropic regions. Experimental parameters: Rotor frequency $\omega_R/(2\pi)=1$ kHz, length of 90° pulse: 5 μs, CP contact time: 2.5 ms; spectral width in ω_2: 10 kHz, t_1 increment: 125 μs, 64 gradient steps, maximum gradient strength: 75 mT/m.

7.4 SAMPLE MANIPULATION

Manipulation of nuclear magnetization by rf and magnetic-field gradient pulses is one way to introduce contrast or selectivity into NMR imaging. An alternative is physical *sample manipulation*. This may be a destructive or a nondestructive process. Apart from cutting the sample to fit the equipment in material applications of NMR imaging, swelling of solids like polymers is often destructive [Erc1, Erc2] and, therefore, is not considered further. Only more or less reversible or nondestructive sample manipulations are treated. Examples are temperature changes [Car4], the generation of magnetic fields in the sample by application of electric fields [Joy1] and incorporation of magnetic nanoparticles [Kre1], the change of relaxation times from paramagnetic impurities like oxygen [Oga1] or dedicated contrast agents [Lau2], and imaging of noble gases [Kob1, Mil2], in particular, of hyperpolarized ^{129}Xe [Son1] and ^{3}He [Sch4]. Hyperpolarized gases can transfer part of their magnetization to the contact surfaces leading to enhanced ^{1}H and ^{13}C signals [Roo1].

7.4.1 Temperature variation

The choice of *temperature* is much at liberty in material applications of NMR imaging. An increase in temperature enhances the rate of molecular motion. In polymers this reduces the effective strength of the dipole–dipole interaction to the benefit of signal-to-noise ratio and spatial resolution. Changes of T_2 by more than a factor of ten can be observed in thermoplastic polymers and resins by raising the temperature from room temperature to 370 K [Car4, Jez1], and sample heterogeneities undetectable at lower temperatures may show up in images acquired at elevated temperatures. Higher temperatures often are required for *in situ* imaging of polymers and rubbers during standard industrial processing like polymerization [Jez2, Jac1, Gün4, Alb1], injection molding and curing [Jac2], and vulcanization [Fül2].

By going the other way, that is to lower temperatures, interesting effects can be observed in aromatic polymers [Cap3]. When decreasing the temperature from the glass transition temperature on, all nonaromatic polymers as well as any degassed aromatic polymers show a continuous increase of their proton T_1 relaxation time. At low temperatures T_1 is usually larger than 10–20 s. However, due to O_2 molecules selectively adsorbed on aromatic rings, nondegassed aromatic polymers show a marked shortening of the proton relaxation with decreasing temperatures. This effect is maximal at rather low temperatures, where T_1 can be as short as 1 ms. Because the amount of adsorbed O_2 is a function of the chemical nature and the morphology of the polymer, T_1 can be an interesting contrast parameter at temperatures below 200 K.

7.4.2 Magnetic field distortions

The phase of NMR coherences is determined by the active magnetic fields. These fields can be manipulated so that they reveal sample heterogeneities. For example, magnetic particles [Kre1] can be incorporated into the sample at preferential positions, and electric currents can be generated within the sample [Joy1], which identify conducting regions by the associated distortions of the polarizing magnetic field B_0. The resulting type of contrast is similar to that caused by susceptibility effects (cf. Section 7.2.5).

Electric currents

Imaging in the presence of electric currents through the object is known as *current-density imaging* (cf. Section 10.1.4) [Joy1, Sco1, Sco2, Sco3, Sco4, Ser1]. The magnetic field associated with the imposed electric current leads to a frequency shift of the

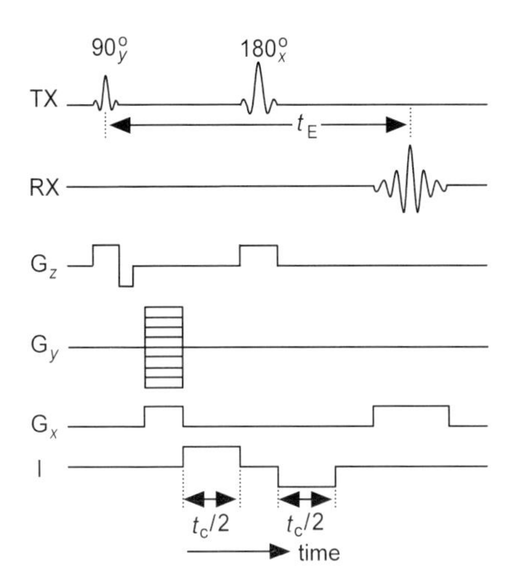

FIG. 7.4.1 [Ser1] Pulse sequence for current-density imaging in the laboratory frame. Electric current pulses I are applied with opposite polarity in the de- and rephasing periods of a standard spin-echo imaging sequence.

resonance, and consequently to a phase shift of the transverse magnetization at the end of the current pulse. From the magnitude of the phase shift the current density and thus the local electrical conductivity can be inferred. Because current is a vector, orientation-dependent measurements are required if magnitude and direction of the local currents need to be known. There are applications in studies of electrotherapy [Joy1], plant physiology [Ser1], monitoring of chemical reactions [Ber3], and investigations of neural firing [Bod1]. Laboratory as well as rotating frame techniques [Sco4] are known.

The basic scheme for current-density imaging in the laboratory frame is depicted in Fig. 7.4.1 [Ser1]. Two current pulses I of total duration t_c are applied with opposite polarity in the de- and rephasing periods of a standard spin-echo imaging sequence so that the local phase shifts induced by each of them add constructively,

$$\varphi = -\gamma B'_z t_c. \tag{7.4.1}$$

Here B'_z is the component of the magnetic field $\boldsymbol{B}'$ associated with the current density $\boldsymbol{j}$, which is parallel to the polarizing magnetic field $\boldsymbol{B}_0$. In 2D current-density imaging, two phase images are needed to construct the current density in the image plane. For example, if the electric current flows in the x'-direction of the sample, then from Ampere's law, the current-density component $j_{x'}$ is defined in the coordinate frame attached to the sample by the field gradients $\partial B'_{z'}/\partial y'$ and $\partial B'_{y'}/\partial z'$ generated by the current,

$$j_{x'} = \frac{1}{\mu_0}\left(\frac{\partial B'_{z'}}{\partial y'} - \frac{\partial B'_{y'}}{\partial z'}\right), \tag{7.4.2}$$

and two perpendicular orientations of the sample, i.e. $z' \| z$ and $y' \| z$ need to be measured to obtain the gradients for computation of $j_{x'}$ by differentiation of the phase images with respect to space. Because the phase is periodic modulo 2π, corrections to the values of the gradients have to be introduced periodically.

Magnetic particles

Magnetic particles have been investigated, for example, as contrast agents in medical imaging, for magnetic-field enhanced sedimentation, and for magneto therapy [Kre1]. Pure iron oxides do not form stable colloidal sols, so that the *nanoparticles* have to be stabilized, e.g. by dextrane, starch, glycosamino glycan, or albumin. The magnetic properties of iron oxides are determined by their crystal structures. αFe_2O_3, commonly known as rust, is paramagnetic, γFe_2O_3 is ferromagnetic, and Fe_3O_4 is ferrimagnetic. The latter compound is also known as *magnetite* and is the most stable iron oxide. It is this compound, which is used for diagnostic purposes. Typical particles have a diameter of 5 nm and contain about 5000 iron atoms. The coatings of the nanoparticles can be functionalized to undergo enzyme-specific reactions, and preferential accumulation in liver and spline has been observed [Col1]. Because tumours do not accumulate such particles, their contrast is enhanced in medical imaging. Applications to materials science have not been reported yet, but preferential accumulation of such particles may be explored, for example, in studies of polymer and elastomer processing. Being at the border between classical and quantum mechanical particles, ferrite nanoparticles exhibit interesting, anisotropic NMR features [Bak1].

7.4.3 Contrast agents

Contrast agents in the conventional sense of medical NMR are paramagnetic compounds which are introduced in low concentration to shorten longitudinal relaxation. Paramagnetic salts [Blo1] like copper sulfate ($CuSO_4$) are used since the early days of NMR to manipulate T_1 relaxation of water in test samples and materials applications. Oxygen gas (O_2) is another effective *relaxation agent*. That is why solutions have to be degassed for T_1 measurements. Similarly, the oxygenation level of blood changes the proton signal from water surrounding the haemoglobin: when diamagnetic oxyhaemoglobin releases its oxygen, the resulting deoxyhaemoglobin is paramagnetic [Oga1, Oga2]. This type of contrast can be utilized to quantify the venous blood oxygenation *in vivo*. The role of oxygen adsorbed in morphologically complex materials like semicrystalline polymers has been addressed in Section 7.4.2 in the context of imaging far below room temperature.

The effect of relaxation agents is based on the electron–nuclear dipole–dipole interaction between the electrons of the relaxation agent and the nearby nuclei to be imaged. Complexes of paramagnetic transition ions and lanthanide ions have received the most attention as potential contrast agents for clinical use [Lau2]. In medicine a nontoxic nature, water solubility and shelf stability of the compounds are essential. The formation of a stable complex is a medical necessity to shield the patient from the toxicity of the transition metal and lanthanide ions. Gadolinium(III), iron(III), and manganese(III) complexes are of great interest, because of their high magnetic moments and relaxation efficiency. The first transition-metal compound administered to humans was Gd(III) diethylenetriaminepentaacetate [$[Gd(DTPA)(H_2O)]^{2-}$]. A further class of contrast agents are microporous metal oxides such as clays and zeolites modified with paramagnetic metal ions or complexes [Bal4, Bre2].

The use of dedicated contrast agents in material imaging appears to be of little importance, as toxicity often is irrelevant and paramagnetic salts can readily be used in imaging studies involving transport and mixing of fluids and pastes. Interesting applications could be seen in target-specific relaxation agents, which, for example, show preferential miscibility with one component in a multi-component mixture.

7.4.4 Noble gases

Noble gases are characterized by chemical inertness and large chemical-shift ranges [Pie1]. For example, for ^{129}Xe the chemical-shift range amounts to 8000 ppm (cf. Table 2.2.1), so that the NMR frequency is extremely sensitive to the environment. In particular, ^{129}Xe has been used extensively to probe confinements in zeolites and porous structures [Bar3, Gre1, Ito1, Jam1, Raf1] as well as morphology and molecular packing in polymers [Mil2, Sch8, Yan1] and elastomers [Ken2, Spe1]. In addition to the chemical shift, the diffusion of xenon in polymers [Jun1, Yan1] and porous materials [Mai1] can be explored to probe morphological features. Sensitivity and Larmor frequency of ^{129}Xe are comparable to ^{13}C, and T_1 relaxation times are long, for example, of the order of 30 s in contact with acrylic acid. Direct imaging of this nucleus under normal pressure is therefore difficult [Pfe1].

^{129}Xe and ^{3}He are spin-$\frac{1}{2}$ nuclei with long relaxation times, that can be hyperpolarized by *optical pumping* and *spin exchange* [Pie1], and they can be transferred to the application of interest while maintaining their spin polarization. Optical pumping enhances the signals by four to five orders of magnitude. *Hyperpolarized gases* can be employed to introduce selectivity in NMR in several ways; by filling empty space [Bru1, Möl1, Son2, Tse1, Tse2], by selective adsorption [Jän1, Son2], by chemical-shift mapping of their environment [Yan1], by introducing diffusion contrast [Son4], and by signal enhancement of ^{1}H at surfaces [Pie2, Roo1]. Signal averaging is complicated by the lack of signal build-up from longitudinal relaxation. Therefore, new spin-polarized gas is introduced for further measurements. A complication in materials applications is the strong increase in the relaxation rate, so that the polarized gas may relax before equilibrating within the sample.

Optical pumping

The possibility of optically pumping spin populations of nuclear energy levels [Tyc1] was first described by Kastler *et al.*, and the importance for NMR had been pointed out already in 1950 [Kas1]. By using left or right circularly polarized light, it is possible to populate either sublevel in a two-state system corresponding to an increase or decrease in the spin temperature. A decrease in spin temperature is referred to as *hyperpolarization*. In practice, hyperpolarization is generated in a two-step process [Hap1]: first the valence electrons of an alkali metal, usually rubidium, are optically pumped with circularly polarized laser light. The angular momentum of the rubidium electrons is then transferred to the nucleus of the noble gas atoms through electron–nuclear dipole–dipole interactions. The nuclei can maintain their hyperpolarized states for more than an hour before interactions with the container walls and other atoms gradually cause the spins to relax into their ground state. To reduce wall relaxation, the container walls need to be treated with a silicon agent. The first use of optical pumping for signal enhancement of physisorbed ^{129}Xe was achieved in 1991 [Raf2]. It opened up a new line of research in NMR spectroscopy and imaging.

Applications

^{129}Xe is considered to be a safe general anaesthetic. It is rapidly transferred in mammals from the lungs to the blood and other tissues, while ^{3}He tends to stay in the gas phase. Therefore, *helium* may be used for imaging void spaces, while *xenon* may be more appropriate for tissue and blood imaging. In fact, xenon retains its hyperpolarization long enough to be injected into blood and to produce an NMR spectrum. Thus, hyperpolarized ^{129}Xe and ^{3}He are of great interest as contrast agents in medical imaging [Alb2, Koe1]. Gradient-echo methods with small flip angles such as FLASH [Haa5] are applied to generate a large number of free induction decays, because repetitive sampling of the residual longitudinal magnetization causes nonrenewable depletion [Alb2].

Imaging of hyperpolarized gases is used to detect voids ranging in size from imperfections in semiconductors all the way to the lung [Möl1, Pie1, Son1, Son2, Tse1, Tse2], and to measure large confinements by diffusion and flow imaging [Bru1, Mai1, Saa2, Sch4]. On one hand, rapid diffusion limits the achievable spatial resolution [Son4]. On the other hand, it replenishes the magnetization destroyed by previous rf pulses. Other applications are the study of materials that are sensitive to water and other liquids like food products, certain polymers, catalysts, and aerogels. In addition to gaseous xenon, solid xenon adsorbed on surfaces can be imaged depending on the relaxation properties of the surface [Son2].

Magnetization transfer

The high polarization of noble gases has been shown to be transferable to low sensitivity nuclei from hyperpolarized ^{129}Xe [Pie1], and the signal of surface ^{1}H spins can be increased by *cross-polarization* from ^{129}Xe frozen on the surface [Dri1, Gae1, Lon1]. The latter approach requires immobilization of xenon and low-field mixing or Hartmann–Hahn frequency match in the rotating frame [Har2]. However, the magnetization of ^{1}H and ^{13}C nuclei in differently modified surfaces of high-surface-area solids, like aerosols, can also be enhanced at high fields in the laboratory frame [Pie2, Roo1]. This is achieved by exposing the surfaces to hyperpolarized ^{129}Xe or ^{3}He gas. In contrast to the cross-polarization experiments the magnetization transfer is established by *cross-relaxation* of surface ^{1}H or ^{13}C spins and mobile gas atoms adsorbed on the surface. This process is called *spin polarization induced nuclear Overhauser effect* or *SPINOE* [Nav1]. The restrictions to use isotopically enriched xenon gas and to work with batches of hyperpolarized gas can be overcome by the production of a continuous flow of hyperpolarized xenon [Haa6], so that applications in materials imaging can be envisioned.

8

Solid-state NMR imaging

NMR imaging of solid matter is made difficult by linewidths, which can be readily larger by a factor of 10^3 to 10^5 compared to those from liquid matter. Consequently, the spatial resolution is degraded by the same factor in frequency encoding, and phase encoding is hampered by rapidly decaying signals and, therefore, by the achievable signal-to-noise ratios. However, despite the resolution and sensitivity problems *imaging of solid matter* remains attractive, because measurement time is less restricted than for many biological samples, so that time-consuming experiments are feasible and a large variety of phenomena can be investigated [Ack3, Blü1, Blü2, Blü3, Blü7, Blü8, Blü9, Bot3, Cal1, Cha1, Cor1, Gar3, Jez1, Jez2, Kim1, Man4, Mil1, Mil6]. In fact, the first publications on NMR imaging in 1972 and 1973 were addressed to both liquid [Lau1, Lau2, Lau3] and solid materials [Man1, Man2]. Clearly, the nondestructiveness and the availability of a multitude of contrast features (cf. Chapter 7) trigger ample curiosity for the use of the method. But because spectroscopic linewidths of 100 kHz and more may need to be excited in ^{1}H NMR imaging of solids, the sample size usually is restricted to 10 mm and less by the availability of suitable rf power amplifiers, which commonly are available in the 1 kW regime, unless surface techniques are used.

The limitations in spatial resolution and sensitivity imposed by the large linewidths can be overcome in different ways. Soft solid matter like elastomer materials often exhibits ^{1}H linewidths of less than 3 kHz, which can still be tackled by Fourier imaging on modern solid-state imaging spectrometers [Blü3], but ^{1}H imaging of rigid matter requires either the use of strong gradients or application of techniques from solid-state NMR spectroscopy for line narrowing to extend the decay of the transverse signal for the benefit of better spatial resolution and sensitivity. The class of imaging methods defined by using large gradients comprises stray-field imaging (Section 8.1), the use of strong oscillating gradients (Section 8.2), imaging by pure phase-encoding (Section 8.3), and multi-quantum imaging (Section 8.4). Imaging by phase encoding includes spectroscopic imaging techniques. The class of imaging methods defined by the use of line-narrowing techniques comprises MAS imaging (Section 8.5), magic-angle rotating-frame imaging (Section 8.6), imaging with multi-pulse line narrowing (Section 8.7), and magic-echo imaging (Section 8.8).

Most of these techniques have been developed on phantoms in search for an optimum approach, and applications to genuine problems are quite rare. From an experimental point of view, phase-encoding methods like *single-point imaging* and *magic-echo imaging* are the most successful approaches to solid-state imaging of abundant nuclei. With a few exceptions, this chapter is devoted primarily to imaging of ^{1}H. The lineshape of such abundant nuclei in solids is dominated by multi-spin dipolar interactions, which, traditionally, are approximated by sums over pairwise interactions [Sli1]. This is why, in this case, the dipolar interaction is still referred to as a *bilinear interaction*, an expression which actually applies only to the quadrupole interaction, because the quadrupole interaction is quadratic in the spin operator. Given the fact, bilinear spin interactions dominate the lineshape in most solid materials, the formal treatment in this chapter uses the quantum-mechanical density-matrix formalism (Section 2.2.1) and the Hamiltonians $\mathbf{H}_\lambda$ dealt with in Section 3.1.

8.1 STRAY-FIELD IMAGING

The development of *stray-field imaging* (STRAFI) represents an important advance in solid-state imaging, because the technique combines high spatial resolution with the ability to image systems with short transverse relaxation times ($T_2 < 100\,\mu$s) [Sam1]. The gradient scheme of the method is simple, because the large static field gradients are utilized. They are present, for example, in the fringe field of all solenoids, where the high field strengths of superconducting solenoids are of particular interest. During the experiment, the sample is physically moved in the fringe field, so that in addition to a strong magnet the equipment needed for STRAFI experiments is a solid-state spectrometer suitable for handling short rf pulses, a controller for driving the sample motion, and a specialized probe.

Typical field gradients are 40 T/m for 4.7 T and 80 T/m for 9.4 T superconducting magnets, respectively [Sam2]. Such gradients are large enough to be the dominant factor in determining the linewidth of the NMR signal. This simplifies the treatment of STRAFI experiments considerably, because all other spin interactions can be neglected. These interactions include the effects from magnetic susceptibility, from homo- and heteronuclear dipole–dipole interactions, from chemical shielding, and in some cases from first- and second-order quadrupole interactions. Also, the gradient strength dominates the electron–nuclear interactions, whether dipolar in nature or arising from the Fermi contact interaction.

Within the fringe field a plane can be found, which exhibits a constant field gradient. This is the *STRAFI plane* used for generation of the NMR signal (Fig. 8.1.1) [Mal3]. The sample is moved through this plane in different orientations, and projections of the object are acquired point by point by moving the sample in steps. Alternatively, the STRAFI plane can be shifted by field-sweep coils which are either part of the superconducting magnet or mounted externally [Mal3]. 2D and 3D images can be reconstructed from projections (cf. Section 6.1) by combining sample rotation and translation or field sweep with data acquisition. Not the repetition time but mechanical complexity of the

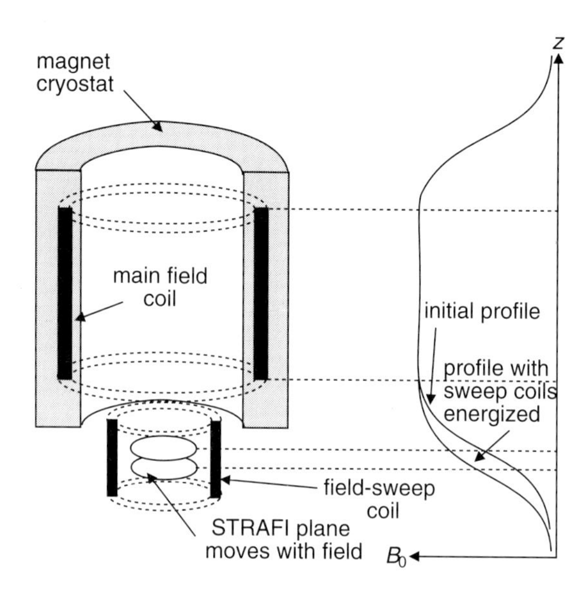

FIG. 8.1.1 [Mal3] Principle of STRAFI imaging. The STRAFI plane is defined by a region of constant field gradient in the fringe field outside the magnet. The plane can be shifted through the sample by energizing sweep coils, or the sample is moved through the STRAFI plane for pointwise acquisition of projections.

corresponding probe limits the speed of measurements. Because individual planes of the sample are excited separately, there is no need for a recycle delay.

Another important aspect of the STRAFI method is the large resonance-field range which results from the large gradient. This means that a pulse centred at a given frequency can, for instance, excite protons in a sensitive slice at one position and fluorine nuclei in a second slice at a position at higher magnetic field [Ran1]. The distance between the two slices is governed by the difference in the gyromagnetic ratio of the two nuclear species and shape of the time-invariant magnetic field. When working with only one gradient strength, the method cannot be combined with spectroscopic resolution, so that the parameters accessible for contrast exclude the spectroscopic dimension.

Principle

In the following, a collection of spins with a heterogeneous spin density and different spin interactions λ is considered. The spin interactions are described by the Hamiltonians $\mathbf{H}_\lambda$ (cf. Section 3.1). For simplicity, only the case of the sample with heterogeneities along the z-axis is considered in terms of the normalized spin density $M_0(z)$, where the integral (5.4.6) over the distribution of thermodynamic equilibrium magnetization is set to $M_0 = 1$. In terms of quantum mechanics, $M_0(z)$ is then given by the expectation value (cf. eqn (2.2.51)) of the angular momentum operator $\mathbf{I}_z$ in z-direction

$$M_0(z) = \mathrm{Sp}\,\{\mathbf{I}_z \boldsymbol{\rho}_0(z)\}\,, \tag{8.1.1}$$

where $\boldsymbol{\rho}_0(z)$ is the space-dependent thermodynamic-equilibrium density matrix (2.2.58).

If the field gradient is applied along the z-direction, the Hamiltonian for the spin interaction λ is given by

$$\mathbf{H}(z) = -\gamma G_z z \mathbf{I}_z + \mathbf{H}_\lambda(z). \tag{8.1.2}$$

This expression applies in the rotating frame where $\omega_0 = \omega_{\rm rf}$ so the fictitious field (cf. Fig. 2.2.7) is zero. If the gradient strength is larger than the local fields corresponding to the spin interactions λ, the total Hamiltonian is dominated by the Zeeman interaction with field gradient, i.e.,

$$\mathbf{H}(z) \approx -\gamma G_z z \mathbf{I}_z. \tag{8.1.3}$$

This is the *high-gradient approximation* [Ran2]. It corresponds to a truncation of all spin interactions including moderate quadrupole interactions.

The spatial distribution of the transverse magnetization after application of a hard 90°_x pulse of duration $t_{\rm p}$ can be evaluated from the density operator

$$\boldsymbol{\rho}\left(z, t_{\rm p}\right) = \exp\left\{ i\gamma \left[B_1 \mathbf{I}_x + G_z z \mathbf{I}_z\right] t_{\rm p}\right\} \boldsymbol{\rho}_0(z) \exp\left\{-i\gamma \left[B_1 \mathbf{I}_x + G_z z \mathbf{I}_z\right] t_{\rm p}\right\}, \tag{8.1.4}$$

where B_1 is the strength of the applied rf field. In addition to the sample-inherent space dependence of the initial density matrix $\boldsymbol{\rho}_0(z)$, a further space dependence is introduced in $\boldsymbol{\rho}(z, t_{\rm p})$ by application of the rf pulse in the presence of the gradient G_z. It can be exploited in two ways: first, in a sufficiently strong gradient the rf pulse is selective so that the centre frequency of the pulse defines the position of a slice through the sample, and second, by rotating the sample the gradient direction relative to the sample is under control of the experiment. The gradient direction and the position of the slice are external variables and can be used to investigate the space dependence of a density matrix $\boldsymbol{\rho}_0(x, y, z)$ which may depend on three space dimensions. Sample reorientation is not considered further, so that the treatment continues with slice selection and one space dimension along z.

Equation (8.1.4) is rewritten in a reference frame oriented along the local effective field, which changes direction with the resonance offset from the rf excitation frequency (cf. Fig. 2.2.7),

$$\begin{aligned}\boldsymbol{\rho}\left(z, t_{\rm p}\right) = & \exp\left\{i\theta(z)\mathbf{I}_y\right\} \exp\left\{i\omega_{\rm eff}(z) t_{\rm p} \mathbf{I}_z\right\} \exp\left\{-i\theta(z)\mathbf{I}_y\right\} \boldsymbol{\rho}_0(z) \\ & \times \exp\left\{i\theta(z)\mathbf{I}_y\right\} \exp\left\{-i\omega_{\rm eff}(z) t_{\rm p} \mathbf{I}_z\right\} \exp\left\{-i\theta(z)\mathbf{I}_y\right\},\end{aligned} \tag{8.1.5}$$

where the space-dependent tilt angle $\theta(z)$ of the effective field with respect to the z-axis of the rotating frame is given by

$$\tan\theta(z) = \frac{B_1}{G_z z}, \tag{8.1.6}$$

and $\omega_{\rm eff}(z)$ is defined as (cf. eqn (2.2.32))

$$\omega_{\rm eff}(z) = -\gamma\sqrt{B_1^2 + (G_z z)^2}. \tag{8.1.7}$$

In the presence of a strong field gradient the pulse flip angle depends on the space coordinate z, so that the rectangular rf pulse is a *selective pulse* and excites a slice through

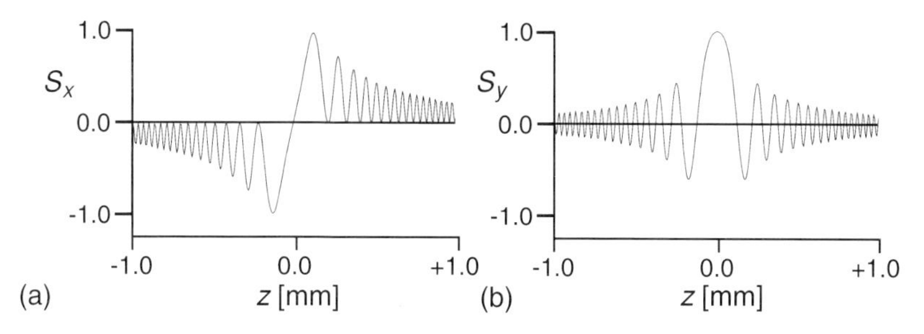

FIG. 8.1.2 Excitation by delta pulses in a strong field gradient G_z as a function of the space coordinate z. Normalized signals $S_x(z)$ (a) and $S_y(z)$ (b) from an rf pulse with flip angle $\pi/2$ at $z = 0$. For the simulation, the gradient strength has been set to $G_z = 20$ T/m, and the amplitude of the rf field was $\gamma B_1 = 100$ kHz. The width of the excited slice is of the order of 100 μm.

the sample. The normalized complex NMR signal $S_x(z) + iS_y(z)$ from a slice centred at $z = 0$ is evaluated from eqns (8.1.5)–(8.1.7) as

$$S_x(z) = \frac{G_z z/B_1}{1 + (G_z z/B_1)^2}\left[1 - \cos\left(\frac{\pi}{2}\left(1 + (G_z z/B_1)^2\right)^{1/2}\right)\right], \tag{8.1.8a}$$

and

$$S_y(z) = \frac{1}{\left(1 + (G_z z/B_1)^2\right)^{1/2}} \sin\left(\frac{\pi}{2}\left(1 + (G_z z/B_1)^2\right)^{1/2}\right). \tag{8.1.8b}$$

In eqns (8.1.8) the flip angle of the rf pulse is taken to be $\pi/2$ at $z = 0$.

Slice profiles generated by infinitesimally short delta pulses and simulated by eqns (8.1.8) are illustrated in Fig. 8.1.2 (cf. Fig. 5.3.6). They are independent of the gyromagnetic ratio. The shape of the slice is not rectangular and shows strong spatial modulation. The slice definition can be improved by use of *shaped pulses* (cf. Section 5.3.2) at the expense of short pulse durations. But short pulses are essential for interrogation of rapidly decaying signals, so that the slice shape is hard to improve in stray-field imaging. The width of the slice is inversely proportional to the field-gradient strength and the duration of the rf pulse [Ben1]. For ^{1}H, a slice width of 40 μm in a 50 T/m gradient requires a 10 μs pulse, equivalent to an excitation bandwidth of 100 kHz. Moreover, absorptive and dispersive components are present and lead to strong nonlinear phase distortions of the NMR signal. For accurate selection of a flat slice, it is important that the magnetic field is uniform in the plane orthogonal to the gradient direction. Because the magnetic field is axially symmetric, this requires that $\partial^2 B_z/\partial x^2 = \partial^2 B_z/\partial y^2 = 0$. It is this equation which defines the STRAFI plane.

Practical considerations

It is difficult to detect a free induction decay following a short rf pulse in a strong fringe-field gradient because the effective relaxation time is very short and can be easily less than the spectrometer dead time. This drawback can be overcome by refocusing the signal in Hahn or solid echoes, where a signal-to-noise advantage and access to relaxation-time contrast is obtained from generating echo trains by repetitive refocusing [Bai1, Ben1]. Trains of Hahn echoes are generated by the CPMG method according to 90°_x–τ–(180°_y–τ-echo–τ)$_n$ (cf. Figs 3.4.1 and 7.2.3) [Car1, Mei1], and trains of solid echoes by 90°_x–τ–(90°_y–τ-echo–τ)$_n$ (OW4 sequence, cf. Fig. 7.2.19(d)) [Man3, Ost1]. In practice, the intensity of each echo in the train is recorded.

The OW4 sequence is applied preferentially, because in addition to overcoming spectrometer dead-time problems associated with short transverse relaxation and providing a convenient way to enable signal averaging, all the pulses are of the same length so that the width of the selected slice is not reduced by refocusing pulses of smaller bandwidth. Multi-echo trains show an increase in the amplitudes of the first echoes before the transverse relaxation decay can be observed [Ben1]. This is explained by interference of primary and secondary or stimulated echoes. In case of pulse-width modulation this effect is stronger for the CPMG than for the OW4 sequence. Furthermore, the solid echo partially refocuses dipolar broadening and the sequence as a whole leads to extended echo trains from partial line narrowing in the limit of short echo times $t_E = 2\tau$ (cf. Section 7.2.7). In this limit, the effective transverse relaxation time T_{2e} determines the signal decay, and images derived from different echoes show different relaxation weights. Lastly, such a train of pulses can be conveniently used to accurately set the 90° pulse flip angle [Ben1].

The STRAFI experiment can be performed without serious penalty under continuous sample movement without stopping the sample for each slice [Ben1]. In this case, the magnetic field experienced by a given nucleus is changing during the pulse sequence at a rate $G_z v$, where v is the magnitude of the sample velocity in the gradient direction. Two critical velocities can be defined [Ben1]: for sample motions slower than the first critical velocity, useful spatial information relating to density is encoded, but the information about spin–spin relaxation is heavily obscured. For motions slower than the second, it is possible to extract relaxation times, although the echoes are still modulated by the sample motion. Typically, the sample is physically moved during a STRAFI experiment within the rf coil [Iwa1]. However, surface coils can be used as well together with the sample, providing a sensitivity advantage for the analysis of planar films and layers [Glo1]. Then the coil selectively excites a small central region of a larger sample.

The performance of the STRAFI method is demonstrated in Fig. 8.1.3 by a proton image of a solid composite fabricated from the thermoplastic polymer poly(phenylene sulphide) (PPS) and carbon fibres [Iwa1]. The composite was fabricated from layers of pure PPS alternating with layers of carbon-fibre containing PPS. This sample presents a challenge to NMR imaging, because the proton density is relatively low, the multi-echo relaxation is fast, and the sample is highly filled with conductive carbon fibres which reduce the signal-to-noise ratio significantly by lowering the quality factor of the probe. In addition, the fibres induce strong magnetic susceptibility effects, which limit the effectiveness of other imaging methods with frequency encoding using lower gradient strengths. Despite these difficulties, the STRAFI image presented in Fig. 8.1.3 clearly shows the various pure PPS layers (bright) against a lower-intensity background (dark) of adjacent graphite-containing layers with less PPS. The image contrast is largely due to

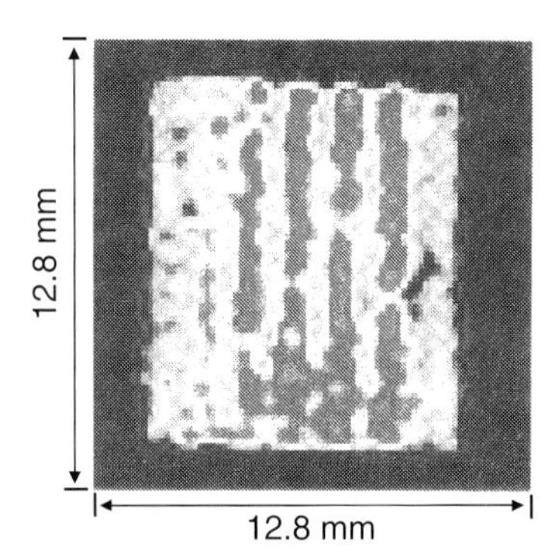

FIG. 8.1.3 STRAFI image of a PPS/carbon-fibre composite. The image was obtained from 100 profiles. Each profile was measured with 16 echoes and 32 acquisitions. The spatial resolution is 100 μm. Adapted from [Iwa1] with permission from Elsevier Science.

differences in proton density, because all PPS layers exhibit basically the same relaxation characteristics.

A number of applications of the STRAFI method have been reported. They include solvent ingress into polymers [Kin1, Mcd1, Per1, Ran1], overcoming of susceptibility broadening [Bla1, Kin1, Kin2, Nun1, Ran2, Ran3], for example, in samples with paramagnetic impurities, and investigations of drying of films [Hug1]. Potential use of the STRAFI method in medicine is likely to be associated with biomaterials. By example of the dentistry and the visualisation of teeth [Bau1], it has been shown that stray-field imaging is well able to map the rigid enamel, dentine, and cement. The possibility to image *quadrupolar nuclei* like ^{7}Li, ^{11}B, ^{23}Na, ^{27}Al, ^{51}V, ^{59}Co, ^{65}Cu, and ^{115}In has also been explored [Bod1, Ran3, Ran4]. Quadrupolar effects for nuclei with half-integral spin do not prevent the use of STRAFI imaging, because the central transition is not broadened in first order and can be observed.

8.2 IMAGING WITH OSCILLATING GRADIENTS

In all NMR imaging techniques it is advantageous to apply the rf pulses in the absence of large gradients and to observe the signal in the presence of a gradient. To switch large gradients reliably and have them to settle in times short compared to T_2 of a typical solid is difficult and is severely hampered by inductive effects. To switch the gradients within T_2^*, the decay time in the presence of the gradient is even more stringent. A reliable way to achieve a known and rapidly varying, time-dependent gradient is to use sinusoidal variation. The use of such a scheme of *imaging with oscillating gradients* has been shown to have a number of advantages [Cot1, Mil2, Str1]:

1. The power requirement for the large gradients can be reduced by making the gradient coils part of a tuned circuit.

2. The gradient can be applied with sufficient strength to dominate the homogeneous dipolar broadening in soft solids.
3. The rf pulses can be applied at a zero crossing of the sinusoid, so that the rf power requirement is simply that normally associated with solid-state NMR.
4. The refocusing property of the oscillating gradient can be exploited for generation of gradient echoes.

In the strong-gradient limit the NMR spectrum is dominated by the applied gradients. This feature is a common goal for both imaging with oscillating gradients and stray-field imaging (cf. Section 8.1), so that images can be recorded with reasonable spatial resolution [Dau1, Mal1]. Although oscillating gradients can be made quite strong in resonant mode, the maximum achievable gradient strength is still significantly smaller than that available in stray fields of superconducting magnets. Hence, imaging with oscillating gradients is suitable for medical imaging [Sta1] and for imaging of soft solids with linewidths of up to 3 kHz [Cod1, Dau1, Mal1] including imaging of rare nuclei [Mil7] and fluids in porous media [Att1] unless multi-pulse line narrowing is used as well [Mcd3, Mil2].

Principle

The basic scheme of imaging with a sinusoidally oscillating gradient and some simple variants are illustrated in Fig. 8.2.1. Given the sine dependence (a) of the gradient, the 90° pulse or a small flip-angle read pulse can be applied either in the gradient maximum (b) or in the gradient zero crossing (c) [Sta1]. In the first case both halves of $\mathbf{k}$ space are covered, in the latter case only one half is sampled. Also, the number of gradient echoes for scheme c is only half of that as for scheme b. However, the advantage of scheme c is that strong gradients can be used at moderate rf pulse powers. Contrast can be introduced by magnetization filters synchronized to the gradient phase [Att1]. For example, T_1 contrast is obtained by an inversion recovery sequence (d) and T_2 contrast by a CPMG sequence. The basic scheme can be further modified to accommodate spectroscopic resolution [Sta1], and translational diffusion weights can be realized by the use of additional bursts of gradient oscillations similar to the use of pulsed-field gradients (cf. Section 5.4.3) [Mal2].

By admitting the second harmonic at three times the gradient modulation frequency, the advantages of schemes (b) and (c) are combined (Fig. 8.2.2(a)) [Cod1]. Here the read gradient for frequency-encoding is modulated in amplitude according to

$$G_x(t) = G_{0x}\,[\sin(\omega_G t) - 3\sin(3\omega_G t)]\,, \tag{8.2.1}$$

where ω_G is the fundamental angular frequency of the oscillating gradient. The gradient form consists of a series of lobes with areas alternating between $-A$ and $2A$. The 90° pulse at $t = 0$ is synchronized to the read gradient to occur at the zero crossing between the lobes of areas $2A$ and $-A$, so that a gradient echo arises half way through the next lobe of area $+2A$ with the echo time $t_E = \pi/(2\omega_G)$. The gradient echo is sampled at equally spaced points $t = n_x \Delta t$ during the read window, which covers the gradient lobe of area $2A$ (cf. Fig. 8.2.2(a)).

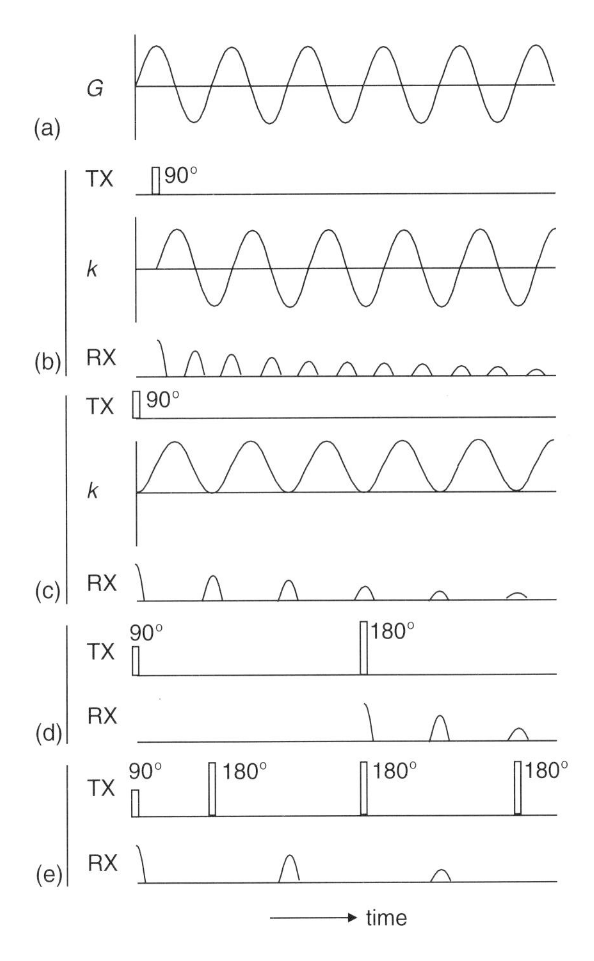

FIG. 8.2.1 Basic schemes of imaging with a sinusoidal oscillating gradient. (a) Simple sinusoidal gradient oscillation. (b) Application of the 90° pulse in the gradient maximum leads to coverage of positive and negative $\boldsymbol{k}$ space, and the gradient-echo spacing is half the modulation period. (c) Application of the 90° pulse at a gradient zero crossing covers one half of $\boldsymbol{k}$ space and the echo spacing is one modulation period [Sta1]. (d) T_1 contrast can be introduced, for example, by an inversion recovery filter synchronized to the gradient period. (e) T_2 contrast can be obtained by a gradient synchronized CPMG sequence [Att1].

For 3D imaging this pulse sequence is repeated in the presence of two phase-encoding gradients oscillating at frequency $2\omega_G$,

$$G_y\left(t, n_y\right) = \frac{2n_y}{N_y} G_{0y} \sin\left(2\omega_G t\right), \tag{8.2.2a}$$

$$G_z\left(t, n_z\right) = \frac{2n_z}{N_z} G_{0z} \sin\left(2\omega_G t\right), \tag{8.2.2b}$$

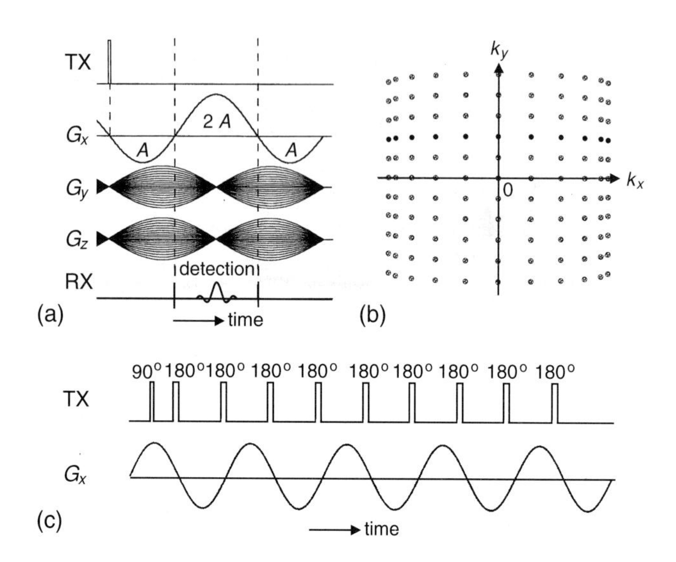

FIG. 8.2.2 Imaging with oscillating gradients. (a) Gradient and rf scheme. The gradient echo centre appears in the centre of the lobe with area $2A$ of the frequency encoding gradient G_x [Cod1]. (b) Sampling points in 2D $\boldsymbol{k}$ space. (c) CPMG variant for refocusing of chemical shift and background gradients [Mil3].

where $-N_y/2 \leq n_y < N_y/2$ and $-N_z/2 \leq n_z < N_z/2$. The corresponding $\boldsymbol{k}$-space trajectories are derived from eqns (8.2.1) and (8.2.2) by integration according to (2.2.23).

For two dimensions the coverage of $\boldsymbol{k}$ space is illustrated in Fig. 8.2.2(b) [Cod1]. Because the refocusing lobe in the read gradient G_x is twice as large as the defocusing lobe, the read trajectory covers both sides of $\boldsymbol{k}$ space. As a consequence of the sinusoidal gradient modulation, the data points are not spaced evenly in $\boldsymbol{k}$ space, so that a linearization algorithm [Mal1] needs to be applied. But even without that, the sampling is fairly uniform near the origin. Therefore, straightforward Fourier transformation of the sampled data will produce pictures with recognizable and only slightly blurred features. Equal spatial resolution in all three dimensions of the image can be obtained by ensuring that the maximum $\boldsymbol{k}$ space excursion is the same in all three directions. This can be accomplished if the gradient amplitudes are adjusted to fulfil the condition $8G_x/3^{3/2} = G_y = G_z$ [Cod1].

Multi-pulse schemes

The basic imaging scheme of Fig. 8.2.1(a) can be extended to *multi-pulse schemes* for repetitive refocusing of particular spin interactions. The effects of linear interactions like inhomogeneities in the static magnetic field $\boldsymbol{B}_0$, sample-inherent background gradients, and chemical shielding are eliminated by the use of a CPMG sequence (cf. Fig. 3.4.1), where the 180° refocusing pulses are applied at the gradient zero crossings (Fig. 8.2.2(c)) [Mil2, Mil3]. This basic scheme has been extended for imaging of rare nuclei to include cross-polarization and heteronuclear dipolar decoupling [Mil7].

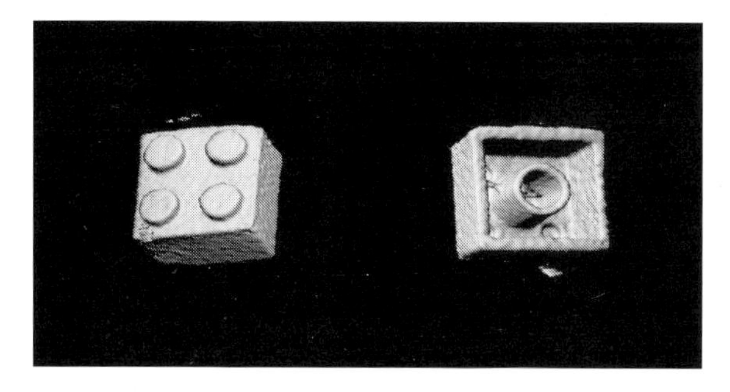

FIG. 8.2.3 [Cod1] 3D surface-rendered images of a Lego® brick measured with oscillating gradients. The dimensions of the brick are approximately $15 \times 15 \times 11\,\text{mm}^3$.

For solid-state imaging of abundant nuclei with spin $\frac{1}{2}$ the homonuclear dipole–dipole interaction must be averaged to zero as well. In principle, the *CPMG sequence* and a pulse sequence like *MREV8* (cf. Table 3.3.1) can be interleaved, but in practice this approach is not feasible, because all pulses can no longer be applied at nonzero gradient strength. As a consequence, pulse-width effects become important, and special pulse sequences of the time-suspension type need to be developed. They work either in the presence of magnetic field gradients [Mil2] or use equal pulse spacings so that the pulses can be applied at the gradient zero crossings [Mcd3, Str1]. Imaging with line narrowing is discussed below in Section 8.6.

Example

The technique is illustrated in Fig. 8.2.3 by surface-rendered 3D images of a Lego® brick [Cod1, Mal1]. The transverse relaxation curve of the polymer material displays two significant components: 90% of the transverse magnetization relaxes with $T_{2\text{f}} = 760\,\mu\text{s}$, and 10% relaxes with $T_{2\text{s}} = 6\,\text{ms}$. Through imaging at an echo time of $128\,\mu\text{s}$, it was possible to view both of the components present in the sample. The data were collected on a 64^3 point 3D grid in 28 min. The peak gradient strength in the frequency-encoding direction was 1.2 T/m, and the gradient oscillation frequency was $\omega_\text{G} = 2\pi\,2\,\text{kHz}$. For image processing, the raw data were linearized in $\boldsymbol{k}$ space prior to Fourier transformation.

8.3 IMAGING WITH PURE PHASE ENCODING

The method of encoding spatial information in the phase of the transverse magnetization by variation of the gradient amplitude in a fixed evolution time has been introduced in Section 2.2.1 (cf. Fig. 2.2.5) and was also addressed in Section 6.2.1 in the context of spin-warp imaging. Usually, all space dimensions are phase encoded in spectroscopic Fourier imaging (cf. Section 6.2.4). The advantage of *phase encoding* in a fixed evolution time is that the evolution of transverse magnetization is constant for all spin interactions other than the interaction with the magnetic-field gradient. Therefore, the spatial resolution is no longer limited by the width of the NMR spectrum in the absence of a gradient, which is the case in frequency encoding. Therefore, even in rigid solids, high spatial resolution can be obtained [Emi1, Emi2]. In solid-state imaging, the method is also referred to as *constant-time imaging* [Cor2, Gra1, Mcd2] or *single-point imaging* [Bal1, Bog1]. It provides superior signal-to-noise ratios when the gradient-switching time is longer than T_2 compared to imaging techniques with frequency encoding [Gra1]. In the following, the general principle of the method is discussed as well as its extension to *spectroscopic imaging of solids*.

Principle

The general timing scheme for *imaging with pure phase encoding* is shown in Fig. 8.3.1(a) [Cor2]. It corresponds to that of spectroscopic imaging (cf. Fig. 6.2.3). Slice selection is usually avoided in solid-state imaging, because most samples are smaller than the slice thickness which can be achieved by conventional use of selective pulses, so that the free-induction decay is excited by a nonselective pulse. The flip angle α is kept small to enable rapid repetition. During the evolution time following the pulse single-quantum coherences evolve in the presence of field gradients. The gradients are pulsed with variable amplitude, and the pulse duration after the rf pulse is kept short in order to obtain a strong signal for detection after the gradient pulse has been switched off. Only one data point needs to be acquired for each gradient value. This is why the

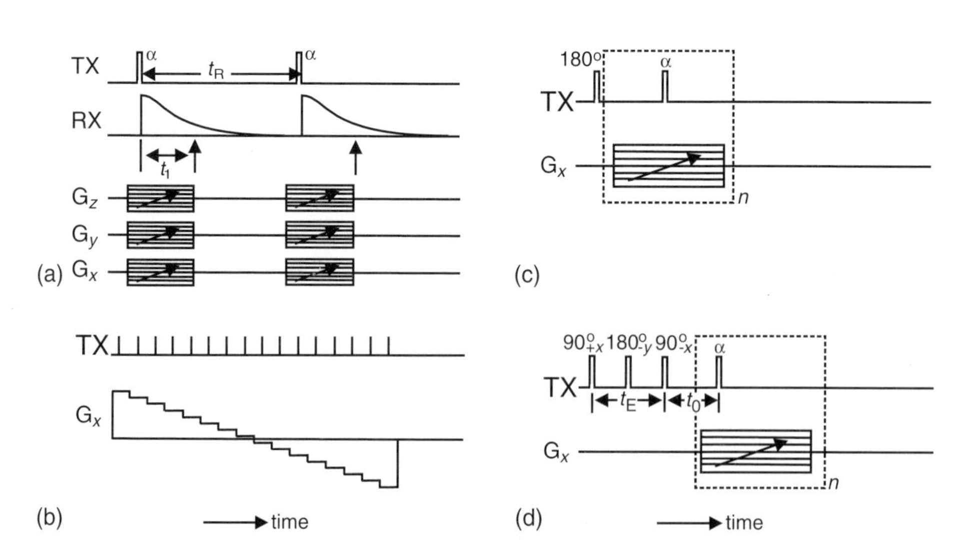

FIG. 8.3.1 Imaging with pure phase encoding. Detection pulses are used with small flip angles α. (a) 3D single-point imaging [Cor2]. The gradients consist of short pulses with variable amplitude. Right after each gradient pulse (arrow) one data point is acquired. (b) Single-point ramped imaging (SPRITE) offers improved data acquisition time efficiency. (c) T_1 contrast can be introduced by an inversion recovery filter. (d) T_2 contrast can be introduced by a Hahn-echo filter, where additional T_1 contrast can be achieved by adjusting the delay t_0 [Pra1].

method is called single-point imaging. Better signal-to-noise ratios can be achieved by sampling and averaging several data points of the FID after each pulse [Pra2]. Because the phase-encoding time t_1 is constant, the method is also called constant-time imaging. Successful performance of the experiment requires fast switching of strong gradient pulses.

The high demands on the gradient switching times [Bal2] are relaxed when ramping the gradient instead of applying short pulses (Fig. 8.3.1(b)) [Bal1, Bal3, Bog1]. This method is called *single-point ramped imaging with T_1 enhancement* or *SPRITE* for repetition times shorter than $5T_1$. The savings in gradient switching time translates into improved efficiency of the data acquisition time. Further contrast is introduced into single-point and SPRITE imaging by the use of filters for longitudinal magnetization (cf. Chapter 7). Two such filters are depicted in Fig. 8.3.1, an inversion-recovery filter c for T_1 weighting and a Hahn-echo filter d for T_2 weighting. In the latter, the delay t_0 can be varied to introduce an additional T_1 weight [Pra1]. In soft matter, mainly spin-density contrast is obtained for short sampling times t_1 (cf. Fig. 8.3.1(a)), while for long sampling times a chemical-shift weight is introduced, unless the signal is refocused by a Hahn echo (cf. Fig. 8.3.1(d)).

The different information obtainable from a chemical-shift weight and a relaxation-parameter image is illustrated in Fig. 8.3.2 [Pra5]: In covulcanization of different rubber sheets, for example, sheets from SBR and NR, an interface may arise depending on the materials and the conditions of vulcanization (Fig. 8.3.2(a)). A sufficiently long acquisition delay t_1 without chemical-shift refocusing introduces the chemical-shift

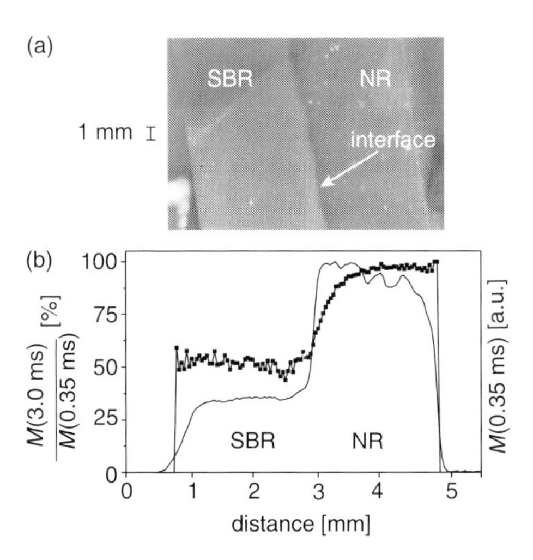

FIG. 8.3.2 Images from the interface between unfilled, covulcanized sheets of SBR and NR. (a) Photo showing the interface (arrow). (b) 1D profiles acquired by single-point imaging with the acquisition delayed by t_1. The interface between both materials assumes a different width for a chemical-shift weighted profile (rugged curve, right scale) and a relaxation parameter weighted profile (smooth curve, left scale). The corresponding t_1 delays are indicated in the figure. Adapted from [Pra5] with permission from Wiley-VCH.

modulation from the different lineshapes of SBR and NR and thus chemical contrast into the acquired image data. If a Hahn echo is formed at the beginning of data acquisition, this modulation is absent and the contrast arises only from spin density and transverse relaxation. When the ratio of two such images is formed, spin density is eliminated, and image contrast derives mainly from relaxation, that is, from the physics of molecular motion. The 1D projections in Fig. 8.3.2(b) across the interface between both materials reveal that the chemical interface marking the transition from SBR to NR is narrow, while the physical interface marking the change in modulus is much wider.

Single-point imaging is particularly suited for imaging of samples with short T_2^*. Examples are solid polymers [Axe1] and water in cementitious materials [Bey1, Bog1] including catalysts [Pra1]. Unless contrast filters [Mas1] other than partial saturation are employed, SPRITE imaging is extremely fast and has been applied to samples with relaxation times of down to T_2^*=100 μs. Examples are industrial objects from poly(ethylene), plexiglas or poly(methylmethacrylate), poly(vinylchloride) [Ken1], fluids in porous media [Pra4], and gases with short relaxation times [Pra3].

Theory

In the high-field approximation the space-dependent thermodynamic-equilibrium density matrix $\boldsymbol{\rho}_0(\boldsymbol{r})$ is proportional to $\mathbf{I}_z(\boldsymbol{r})$ (cf. eqn (2.2.58)). A 90°_y pulse converts this density matrix into $\boldsymbol{\rho}\left(0^+, \boldsymbol{r}\right) \propto \mathbf{I}_x(\boldsymbol{r}) = \left[\mathbf{I}_+(\boldsymbol{r}) + \mathbf{I}_-(\boldsymbol{r})\right]/2$, where $\mathbf{I}_x$ is the x-component of the spin operator, $\mathbf{I}_+ = \mathbf{I}_x + \mathrm{i}\mathbf{I}_y$, and $\mathbf{I}_- = \mathbf{I}_x - \mathrm{i}\mathbf{I}_y$. In the subsequent free-precession period the density matrix evolves for a time t_1 under the spin Hamiltonian $\mathbf{H}_\lambda(\boldsymbol{r})$ (cf. eqn (3.1.1)) and under the influence of the applied gradient, which introduces a phase evolution given by $\boldsymbol{kr} = -\gamma \boldsymbol{Gr} t_1$, where $\boldsymbol{k}$ is the experimental variable,

$$\boldsymbol{\rho}(\boldsymbol{k}, \boldsymbol{r}) = \exp\{-\mathrm{i}\mathbf{H}_\lambda(\boldsymbol{r})t_1\} \exp\{\mathrm{i}\boldsymbol{kr}\mathbf{I}_z\}\, \boldsymbol{\rho}\left(0^+, \boldsymbol{r}\right) \exp\{-\mathrm{i}\boldsymbol{kr}\mathbf{I}_z\} \exp\{\mathrm{i}\mathbf{H}_\lambda(\boldsymbol{r})t_1\}. \quad (8.3.1)$$

The different spin interactions λ active during the evolution period can be the chemical shielding, the dipolar, and the quadrupolar interactions. All of these interactions commute with the $\mathbf{I}_z$ operator.

For given values of field gradient and evolution time, i.e. for a given $\boldsymbol{k}$ vector, the total NMR signal $s(\boldsymbol{k})$ is determined by the integral of the density matrix over the sample volume,

$$s(\boldsymbol{k}) = \mathrm{Sp}\left\{\boldsymbol{\rho}_\mathrm{s}(\boldsymbol{k}) \left\{\mathbf{I}_x - \mathrm{i}\mathbf{I}_y\right\}\right\} = \mathrm{Sp}\left\{\boldsymbol{\rho}_\mathrm{s}(\boldsymbol{k})\mathbf{I}_+^*\right\}, \quad (8.3.2)$$

where

$$\boldsymbol{\rho}_\mathrm{s}(\boldsymbol{k}) = \int_V \boldsymbol{\rho}(\boldsymbol{k}, \boldsymbol{r})\, \mathrm{d}\boldsymbol{r}. \quad (8.3.3)$$

Equation (8.3.2) is the quantum-mechanical formulation of (5.4.8). The prefactor −i is missing in (8.3.2), because the equation has been derived from the pulse response and not from a perturbation expansion (cf. Section 5.4.1). The similarity of both equations becomes more obvious by writing (8.3.2) in Liouville space, where the density matrix elements are arranged in vector form (cf. Section 2.2.2). Then,

$$s(\mathbf{k}) = \mathrm{Sp}\left\{\left[\int_V \exp\left\{-\mathrm{i}\,\mathbf{H}_\lambda^{\mathrm{L}}(\mathbf{r})\,t_1\right\}\exp\left\{\mathrm{i}\,\mathbf{k}\,\mathbf{r}\,\mathbf{I}_z^{\mathrm{L}}\right\}\boldsymbol{\rho}^{\mathrm{L}}\left(0^+,\mathbf{r}\right)\,\mathrm{d}\mathbf{r}\right]^{\mathrm{H}}\mathbf{I}_+^*\right\}. \qquad (8.3.4)$$

The operators which carry the superscript L are written in Liouville-space notation. The most important simplification arising from this notation is that a transformation $\mathbf{O}\boldsymbol{\rho}\mathbf{O}^{-1}$ in Hilbert space can be written as $\mathbf{O}^{\mathrm{L}}\boldsymbol{\rho}^{\mathrm{L}}$ in Liouville space. Here $\exp\left\{\mathrm{i}\,\mathbf{k}\,\mathbf{r}\,\mathbf{I}_z^{\mathrm{L}}\right\}$ describes the phase evolution in the applied gradients, and $\exp\left\{-\mathrm{i}\,\mathbf{H}_\lambda^{\mathrm{L}}(\mathbf{r})\,t_1\right\}$ describes the evolution under the internal Hamiltonians of the sample, corresponding to the free-induction signal. Relaxation is neglected in (8.3.4).

The superscript H on the square bracket in (8.3.4) indicates that the Liouville expression inside the bracket needs to be transformed back into Hilbert space before multiplication with the Hilbert-space operator $\mathbf{I}_+^* = \mathbf{I}_x - \mathrm{i}\mathbf{I}_y = \mathbf{I}_-$ and formation of the trace. Inverse Fourier transformation over $\mathbf{k}$ produces an image of the spin density. Given the form of the density matrix after the initial pulse, the spin density corresponds to $\mathbf{I}_+^{\mathrm{L}}(\mathbf{r})$. It is weighted by the phase evolution under the internal Hamiltonians $\mathbf{H}_\lambda(\mathbf{r})$ during the space-encoding time t_1,

$$S\left(\mathbf{r}',t_1\right) \propto \mathrm{Sp}\left\{\int_V \left[\exp\left\{-\mathrm{i}\,\mathbf{H}_\lambda^{\mathrm{L}}(\mathbf{r})\,t_1\right\}\mathbf{I}_+^{\mathrm{L}}(\mathbf{r})\right]^{\mathrm{H}}\mathbf{I}_-\,\delta\left(\mathbf{r}-\mathbf{r}'\right)\,\mathrm{d}\mathbf{r}\right\}$$

$$= \mathrm{Sp}\left\{\left[\exp\left\{-\mathrm{i}\,\mathbf{H}_\lambda^{\mathrm{L}}(\mathbf{r})\,t_1\right\}\mathbf{I}_+^{\mathrm{L}}(\mathbf{r})\right]^{\mathrm{H}}\mathbf{I}_-\right\}. \qquad (8.3.5)$$

Spectroscopic imaging

The phase evolution from the internal Hamiltonians $\mathbf{H}_\lambda$ leads to a signal loss, which may be refocused by generation of suitable echoes, for example, the Hahn echo [Gün1], the rotational echo [Cor3, Cor4], the solid-echo [Rom1, Dem1], the alignment echo [Blü4, Gün2], and the magic echo [Dem2, Haf1, Haf3]. The time inversion pulse for echo formation must be applied in the centre of the gradient evolution time t_1. Then, image contrast is defined by relaxation and spin density only and not by the distribution of resonance frequencies. Therefore, ultra-fast gradient switching is less stringent, and the signal can be sampled in the absence of the gradient. In this way, the sensitivity can be improved substantially and the gradient-free detection time can be used for acquisition of the spectroscopic wideline response.

Spectroscopic imaging of solids is most rewarding if the spectrum contains information about molecular ordering or chemical structure. For soft solids or highly ordered rigid solids, this information may be found in the spectra of abundant nuclei like ^{1}H and ^{19}F [Hep1]. For rare nuclei like ^{13}C, the homonuclear dipole–dipole interaction is negligible, and in the presence of heteronuclear decoupling the chemical-shift interaction determines the spectrum [Sze1]. This is an orientation-dependent interaction [Mar1] which is linear in the spin operator. Hence, it can be refocused by a Hahn echo.

The feasibility of spectroscopic imaging of ^{13}C in natural abundance has been demonstrated by an investigation of a drawn tensile bar from syndiotactic poly(propylene)

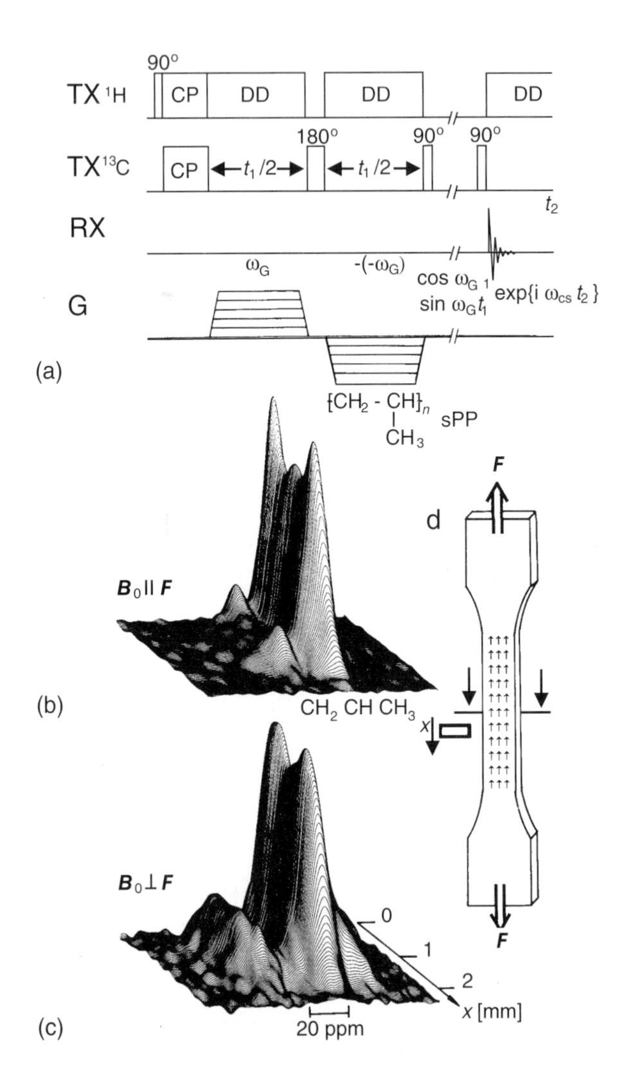

FIG. 8.3.3 Spectroscopic imaging of ^{13}C in natural abundance. (a) Pulse sequence. The chemical shift is refocused by a 180° pulse. The gradient is reversed to preserve space encoding. The ^{1}H decoupled ^{13}C spectrum is recorded in a 2D fashion by stepping the gradients in the indirectly detected dimension. Adapted from [Gün1]. Copyright 1992 American Chemical Society. (b) 1D spectroscopic ^{13}C image of an injection-moulded, drawn tensile bar of syndiotactic poly(propylene) (sPP) obtained with the pulse sequence (a). The draw axis is parallel to $\boldsymbol{B}_0$. On the spectroscopic axis methylene and methine resonances can be clearly distinguished. From their lineshapes the orientations were calculated from the surface and core parts of the sample. Different material densities can be recognized from both regions by the signal amplitudes along the space axis. (c) Spectroscopic image with the draw axis perpendicular to $\boldsymbol{B}_0$. (d) Sketch of the sample. A small piece was cut out from the centre and oriented with the drawing direction parallel (b) and perpendicular (c) to $\boldsymbol{B}_0$. Adapted from [Blü5] with permission from Wiley-VCH.

[Blü5, Gün1]. The pulse sequence (Fig. 8.3.3(a)) involves cross-polarization, heteronuclear dipolar CW decoupling, and refocusing of the chemical shift evolution of ^{13}C during the space encoding time t_1. To minimize artefacts from gradient switching a z filter is incorporated between space encoding and detection of the spectroscopic response which evolves with frequency ω_{CS} during t_2 in the absence of a gradient. By appropriate rf phase cycling of the z-filter pulses, a cosine and a sine data set of the signal are recorded, which can be combined to provide the spectroscopic response with modulation of the phase $\omega_G t_1 = -\gamma G x t_1$ of the spatial information in x-direction. A phase-sensitive spectroscopic image is derived from the experimental data set by 2D Fourier transformation [Ern1].

Using this pulse scheme, a rectangular sample has been imaged which had been cut out from the centre of a tensile bar fabricated from syndiotactic poly(propylene) by injection moulding (Fig. 8.3.3(b)). The tensile bar had been drawn to 500% so that the polymer chains were highly oriented. The spectroscopic dimension reveals the CH and CH_2 signals. Due to high mobility of the CH_3 groups, the corresponding signal is suppressed, because it cannot be enhanced through cross-polarization. A signal variation is observed along the space axis indicating higher material density for the surfaces than for the core. Such a skin-core layer effect is not unknown for injection-moulded polymer products. Analysis of the lineshapes provides information about the orientational distribution function of the chemical shift tensors of CH and CH_2 groups. The variance of the orientational distribution function is larger for the core than for the surface layers.

Constant-time imaging of ^1H in solids has been applied in combination with MAS for line narrowing [Cor3, Cor4]. The images were recorded for a dynamically stressed poly(isoprene) phantom and of poly(butadiene) in two poly(butadiene)/poly(styrene) blends. Spectroscopic MAS imaging has also been tested on deuterated polymers to probe differences in molecular mobility from the ^2H lineshape of the rotary-echo envelope [Gün3]. *Spectroscopic imaging of* ^2H is particularly simple as the quadrupolar interaction dominates the homo- and heteronuclear dipole–dipole interactions, so that decoupling is unnecessary, but isotope enrichment is required. The advantage of spectroscopic ^2H wideline imaging is that information about molecular order and mobility can be derived in a very accurate way through analysis of 1D wideline spectra (cf. Section 3.2). This has been demonstrated on phantoms by spectroscopic ^2H double-quantum imaging (cf. Section 7.3.1) [Blü5, Gün2].

8.4 MULTI-QUANTUM IMAGING

For both frequency and phase encoding methods, the spatial resolution is determined by the gradient strength. Hence, methods by which the effective gradient strength can be increased are of considerable interest. Other than improving the gradient hardware, the enhanced phase evolution of *multi-quantum coherences* can be exploited: if single-quantum coherence dephases with frequency Ω in the rotating frame, then multi-quantum coherences of coherence order p evolves with frequency $p\Omega$ [Ern1, Mun1]. Double-quantum coherence, for example, evolves with frequency 2Ω. The influence of any variation in the effective magnetic field on the precession frequency, for example that introduced by a magnetic field gradient, is amplified by the factor p. In the case of imaging this means, that the effect of the field gradient is amplified by the coherence order. Moreover, multi-quantum coherences in solids can be less sensitive to certain spin interactions, leading to spectra with substantially less line broadening or improved spatial resolution without the application of line-narrowing techniques.

The use of *multi-quantum imaging* inherently includes a multi-quantum contrast filter in the sequence. Such *multi-quantum filters* (cf. Section 7.2.10) can also be combined with single-quantum space encoding techniques. They are used to discriminate coupled spins $\frac{1}{2}$ from uncoupled spins in liquids and signals from ordered materials from those of disordered materials.

Principle

Formally, multi-quantum coherences of order p are described by irreducible tensor operators $\mathbf{T}_{qp}$ (cf. Table 3.1.2 for coupled spins $\frac{1}{2}$ [Ern1]. The *coherence order* is described by $p = m_\mathrm{f} - m_\mathrm{i}$ (cf. Fig. 2.2.11), where m_f and m_i are the final and initial magnetic quantum numbers of a transition. For double-quantum coherence, for example, $p = \pm 2$. The *total spin coherence* q corresponds to the maximum order possible. In this case, $p = q$ so that maximum coherence order is described by $\mathbf{T}_{qq}$.

If $\mathbf{T}_{qp}$ evolves under a field gradient applied along the z-direction parallel to the static magnetic field, the evolution of spin coherences under the Zeeman interaction leads to changes described by the transformation

$$\exp\{\mathrm{i}\boldsymbol{k}\boldsymbol{r}\mathbf{I}_z\}\mathbf{T}_{qp}\exp\{-\mathrm{i}\boldsymbol{k}\boldsymbol{r}\mathbf{I}_z\} = \mathbf{T}_{qp}\exp\{\mathrm{i}p\boldsymbol{k}\boldsymbol{r}\}. \tag{8.4.1}$$

This equation demonstrates that the phase of the multi-quantum coherence depends on position $\boldsymbol{r}$, and that the effect of the gradient strength $\boldsymbol{G}$ in the wave vector $\boldsymbol{k} = -\gamma \int \boldsymbol{G}(t)\,\mathrm{d}t$ can be expressed as $p\boldsymbol{G}$. Clearly, excitation of the total spin coherence for which the maximum value of p is reached is desirable. Unfortunately, the intensity of the NMR signal decreases strongly with increasing p. For dipolar multi-spin systems, the signal intensity exhibits an approximately Gaussian dependence on p [Mun1]. Hence, for solid-state imaging the most promising approach is to excite double-quantum coherences from deuterons or protons.

Methods

In NMR, multi-quantum coherences can be excited by just two pulses [Ern1, Mun1] but for rigid samples multi-pulse sequences are more efficient (cf. Fig. 7.2.26) [Bau2, Mun1]. Because the receiver coil in the NMR experiment corresponds to a magnetic dipolar detector, only dipolar single-quantum coherence can be detected directly and not multi-polar multi-quantum coherences. However, the latter can be detected indirectly by methods of 2D NMR spectroscopy [Ern1].

The general scheme of multi-quantum imaging is illustrated in Fig. 8.4.1. In an initial preparation period (not shown) the spins are polarized by T_1 relaxation and contrast can be introduced by using some type of magnetization filter. Starting with initial z magnetization a multi-quantum excitation pulse sequence with propagator $\mathbf{U}_\mathrm{p}$ is applied during time τ_p. Because this sequence is said to prepare the multi-quantum coherences, it carries the index p and is also called the preparation sequence [Ern1]. The multi-quantum

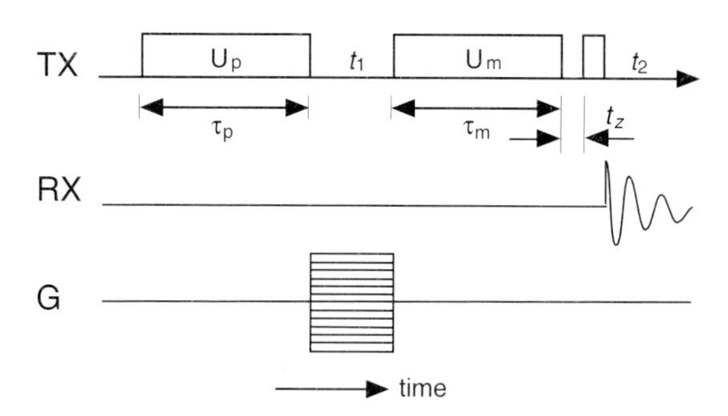

FIG. 8.4.1 General scheme for excitation and indirect detection of multi-quantum signals. The evolution operators for generation and reconversion of multi-quantum coherences are denoted by $\mathbf{U}_p$ and $\mathbf{U}_m$, respectively. The space-encoding field gradient is applied during the evolution period t_1 to modulate the precession phases of the multi-quantum coherences.

coherences subsequently evolve for a time t_1 under the effect of phase-encoding gradients. In order to detect these coherences, a conversion into single-quantum coherences is necessary. This is achieved by converting the multi-quantum coherences into longitudinal magnetization with a pulse sequence with propagator $\mathbf{U}_m$ and then converting longitudinal magnetization into transverse magnetization by a simple 90° pulse. The pulse sequence for reconversion of multi-quantum coherences of order p into longitudinal magnetization can be a time-reversed copy of the preparation sequence. It can also be realized by the pulse sequence for generation of the multi-quantum coherences, but with all the pulse phases shifted by $\pi/2$. Because the reconversion sequence and subsequent generation of observable transverse magnetization by a read pulse can be understood as a process of mixing multi-quantum coherences into single-quantum coherences, the index m is used for mixing [Ern1].

Different pulse sequences for homonuclear multi-quantum NMR are depicted in Fig. 7.2.26. In many cases, an appended 90° read pulse cancels the last pulse of the reconversion sequence (cf. Fig. 7.2.26(d)). The resulting single-quantum signal is modulated in phase or amplitude by the multi-quantum coherences prevalent during the evolution time t_1 and is detected during the detection time t_2. The different coherence orders can be separated by use of phase cycling schemes [Ern1, Bod2]. The most simple pulse sequence for multi-quantum NMR consists of three 90° pulses [Ern1, Mun1]. This pulse sequence is obtained from that shown in Fig. 7.2.26(d) by omitting the 180° pulses for refocusing the phase evolution from chemical shift dispersion and other linear spin interactions. A somewhat more sophisticated sequence consisting of five 90° pulses is obtained from the one shown in Fig. 7.2.26(c) by omitting the 180° refocusing pulses. In addition to the three-pulse sequence, two further 90° pulses are appended to obtain a z filter by which the coherences are channelled through longitudinal magnetization during time t_z to eliminate unwanted signals by dephasing and phase cycling.

Double-quantum imaging

The coherence pathways (cf. Fig. 7.2.26(b)) for multi-quantum pulse sequences can be obtained with the help of the transformation algebra of the irreducible tensor operators

$\mathbf{T}_{qp}$ (cf. Table 3.1.2) [Bow1, Sch1, Spi2]. The case of a rigid, isolated pair of spins $\frac{1}{2}$ is simple to treat. Such a pair is equivalent to a spin-1 nucleus like a deuteron with a quadrupole interaction [Sli1]. The coherence pathways generated by the five-pulse sequence lead to the following transformations of tensor operators:

$$
\begin{aligned}
&\left[\mathbf{T}_{10}\right]_{\text{contrast}} - \\
&\left[90^\circ_x - \mathbf{T}^{\text{s}}_{11} - \tau_{\text{p}} - \left(\mathbf{T}^{\text{s}}_{11} \text{ and } \mathbf{T}^{\text{a}}_{21}\right) - 90^\circ_x - \left(\mathbf{T}_{10} \text{ and } \mathbf{T}^{\text{a}}_{22}\right)\right]_{\text{preparation}} - \\
&\left[\mathbf{T}^{\text{s}}_{22} \text{ and } \mathbf{T}^{\text{a}}_{22}\right]_{\text{evolution}} - \\
&\left[90^\circ_y - \left(\mathbf{T}^{\text{a}}_{11}, \mathbf{T}_{20}, \mathbf{T}^{\text{s}}_{21}, \text{ and } \mathbf{T}^{\text{s}}_{22}\right) - \tau_{\text{m}}\right. \\
&\quad \left. - \left(\mathbf{T}^{\text{a}}_{11}, \mathbf{T}^{\text{s}}_{21}, \text{ and } \mathbf{T}^{\text{s}}_{22}\right) - 90^\circ_y - \mathbf{T}_{10}\right]_{\text{mixing}} - \\
&\left[t_z - 90^\circ_\phi - \left(\mathbf{T}^{\text{s}}_{1\pm1} \text{ and } \mathbf{T}^{\text{a}}_{1\pm1}\right)\right]_{\text{detection}}.
\end{aligned}
\tag{8.4.2}
$$

Here τ_{p} and τ_{m} are the durations of the preparation and mixing periods (Fig. 8.4.1). Only the double-quantum coherences $\mathbf{T}_{22}$ existing during the evolution period t_1 are of interest in this experiment. The symmetric and antisymmetric irreducible tensor operators are defined by

$$\mathbf{T}^{\text{s}}_{pq} = \frac{-\text{i}}{\sqrt{2}}\left[\mathbf{T}_{pq} + \mathbf{T}_{p-q}\right] \quad \text{and} \quad \mathbf{T}^{\text{a}}_{pq} = \frac{1}{\sqrt{2}}\left[\mathbf{T}_{pq} - \mathbf{T}_{p-q}\right]. \tag{8.4.3}$$

By using the density-matrix formalism for calculation of the detected signal, the evolution of the double-quantum part $\boldsymbol{\rho}_{2\text{Q}}$ of the density matrix during the double-quantum space-encoding period t_1 is needed as an intermediate result [Gün2, Got1],

$$\boldsymbol{\rho}_{2\text{Q}}\left(\tau_{\text{p}} + t_1\right) \propto \left[\mathbf{T}_{22}\exp\{\text{i}\,2\boldsymbol{k}\boldsymbol{r}t_1\} - \mathbf{T}_{2-2}\exp\{-\text{i}\,2\boldsymbol{k}\boldsymbol{r}t_1\}\right]\sin\left(\sqrt{3/2}\Delta\tau_{\text{p}}\right), \tag{8.4.4}$$

where Δ is the anisotropy of the dipolar or quadrupolar interaction. The factor 2 in the exponent indicates that the efficiency of gradients for phase encoding has doubled. Choosing equal durations of the double-quantum preparation and mixing periods, $\tau_{\text{p}} = \tau_{\text{m}} = \tau$, the complex signal of one voxel detected during the detection time t_2 is modulated by the factor

$$s\,(\tau, t_1, t_2 = 0) \propto \exp\{\text{i}\,2\boldsymbol{k}\boldsymbol{r}t_1\}\sin^2\left(\sqrt{3/2}\Delta\tau\right). \tag{8.4.5}$$

The square of the sine term arises because the term is introduced once by the double-quantum preparation propagator $\mathbf{U}_{\text{p}}$ and once by the mixing propagator $\mathbf{U}_{\text{m}}$. The proportionality constant contains the spin density.

Applications

Multi-quantum imaging has first been explored on dipolar coupled protons in adamantane (Fig. 8.4.2) [Gar2]. Together with the anisotropy Δ of the bilinear spin interaction, the duration τ_{p} of the multi-quantum preparation period determines the order to which the

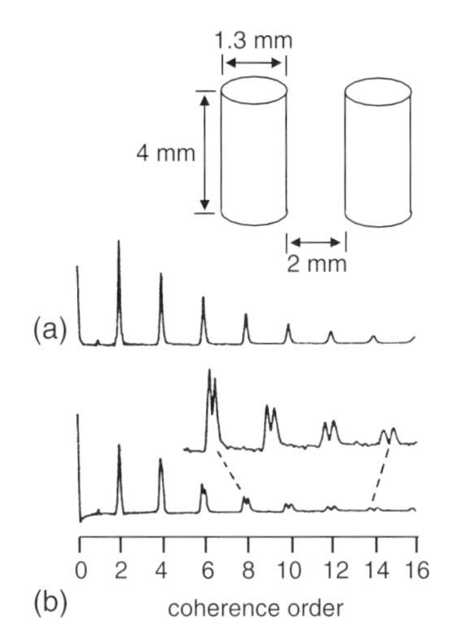

FIG. 8.4.2 Proton multi-quantum spectra (bottom) of an adamantane phantom (top) without (a) and with (b) application of a static gradient of 48 mT/m. Evolution time and phase of the preparation pulse sequence were incremented in 32 steps of 0.1 μs and $2\pi/32$, respectively. Only even-order multi-quantum coherences were detected. The signals from orders 8 through 14 are also displayed on an expanded scale. The multi-quantum spectra demonstrate the increase in the spatial resolution of the two cylinders with increasing coherence order p. Adapted from [Gar1] with permission from the American Physical Society.

coherences build up. At grain boundaries, the homonuclear dipole–dipole interaction among ^{1}H is often reduced, so that the maximum achievable coherence order may be a parameter characteristic of the material under investigation. This is the idea underlying the method of *spin counting* [Mun1].

The pulse sequence used for generation of multi-quantum coherences in adamantane, a strongly dipolar coupled spin system, is shown in Fig. 7.2.26(e). The multi-pulse sequence with an eight-pulse cycle [Bau2] has been used to pump multiple-quantum coherences of orders up to $p = 20$ in a 396 μs long preparation period (cf. Fig. 8.4.2). The contribution of a given coherence order can be separated from the sum of all by manipulation of phases of the preparation pulse, because the phase of different spin-coherence orders is proportional to $p\omega_0\tau_p$ [Bod2]. Spectroscopic information can be recorded in the detection period without an applied gradient, because the space information is detected indirectly (cf. Fig. 8.4.1). Although the signal intensity decreases for higher coherence orders, the amplification of the gradient strength is clearly demonstrated in Fig. 8.4.2 for coherence orders as high as 14.

Spectroscopic double-quantum imaging has been explored in deuteron NMR [Blü6, Gün1, Gün2, Gün4]. Deuteron spectra of solid samples exhibit a spectral width up to approximately 250 kHz due to the quadrupole interaction. Given such linewidths, frequency encoding of the space information leads to unacceptably low resolution. For example, minimum resolvable distances can be as low as 0.6–8 cm. The use of stronger gradients broadens the spectra to an extent, that nonselective excitation can no longer be achieved owing to limited rf power and receiver bandwidth. These problems can be alleviated through phase encoding and the use of the maximum spin coherence, that is the double-quantum coherence in deuteron NMR. The double-quantum transitions of $I = 1$ nuclei are independent of the quadrupole coupling tensor in the magnetic field [Mun1]. Thus, they live long and correspond to a narrow line (cf. Fig. 3.2.1), which can favourably be exploited for space encoding. As an extra benefit, the dephasing speed in an applied gradient is amplified by a factor of $p = 2$. While the space information is detected indirectly *via* phase modulation of single-quantum coherences, the spectroscopic information is detected directly. Therefore, each point in space is characterized by a deuteron wideline spectrum, which provides information about molecular order and slow molecular reorientation [Gün2].

Instead of the five-pulse sequence, the pulse sequence of Fig. 7.2.26(f) can be applied, which includes long-lived spin alignment states $\mathbf{T}_{20}$ [Spi3] in the preparation and mixing propagators of the double-quantum coherences $\mathbf{T}_{2\pm2}$. During the alignment periods, the gradients can readily be switched on and off without detrimental effects of magnetization dephasing [Gün2]. An application of the ^{2}H double-quantum imaging for the detection of molecular mobility is depicted in Fig. 7.3.3, where differences in slow molecular mobility can be detected from difference in the deuteron wideline spectra for each position in space. An application to the detection of molecular order is depicted in Fig. 8.4.3. The spectroscopic image (b) of a phantom (a) made from three pieces of highly oriented poly(ethylene) shows different lineshapes depending on the orientation angle of the order axes in the PE samples with respect to the direction of the magnetic field $\boldsymbol{B}_0$. The projection (c) of the spectra onto the space axis provides a spin-density image. The high potential for obtaining detailed information with this method of spectroscopic imaging is offset by the fact, that the sample needs to be deuterated and that long spin–lattice relaxation times may require long recycle delays, so that acquisition times can be as long as one day.

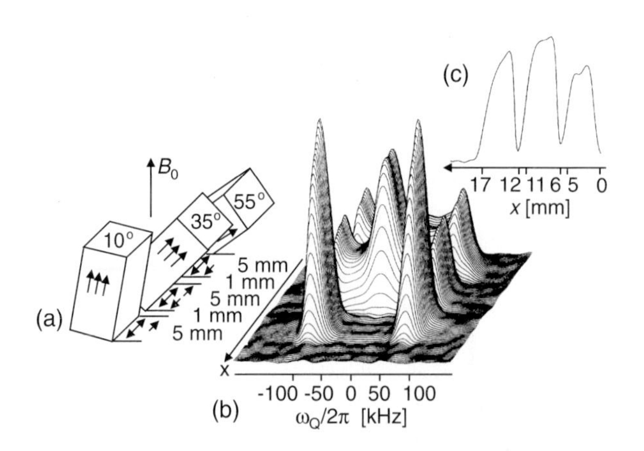

FIG. 8.4.3 Spectroscopic ^{2}H double-quantum imaging of molecular order. (a) Phantom made of three parts of highly oriented, perdeuterated poly(ethylene). The direction of macroscopic molecular order is oriented differently with respect to the magnetic field for each part. (b) Spectroscopic double-quantum image. The ^{2}H lineshape is different for each of the three parts of the sample. (c) Projection onto the space axis. Adapted from [Gün1]. Copyright 1999 American Chemical Society.

Double-quantum ^{2}H images have also been measured using the simple three-pulse sequence extended by one 180° pulse in the preparation period to refocus the effects of chemical shift and magnetic field inhomogeneities (Fig. 8.4.4(a)) [Kli1, Kli2, Sha1]. In an application to *imaging of strain* in stretched rubber bands with a cut, (Fig. 8.4.4(c)) the double-quantum coherences are used for space encoding and for eliminating signals from unstrained regions of the sample. Deuterated poly(butadiene) oligomers were incorporated into elastic rubber bands as spy molecules which sense molecular orientation by deformation of their environment. The quadrupole splitting corresponding to the anisotropy Δ of the deuteron resonance is a measure of the local strain (cf. Fig. 7.1.12). It can be mapped by spectroscopic imaging [Kli1] as well by the amplitude of double-quantum coherences. The amplitude of the double-quantum signal is defined by the double-quantum build-up curve (cf. Fig. 7.2.27), which, for the five-pulse sequence, is described by eqn (8.4.5). The double-quantum sequence also serves as a filter. Unstrained sample regions do not exhibit a double-quantum splitting of the spy molecules, so that signal is observed only from strained regions of the sample and signal from unstrained regions is suppressed. In Fig. 8.4.4(c) the highest strain is observed in the middle of the sample, where the quadrupole splittings are of the order of 350 Hz. The strain decreases towards the sides. Compared to a spin-echo image in Fig. 8.4.4(b) the contrast is inverted. In the spin-echo image, high signal amplitude arises in the unstrained regions, where T_2 is long. Nevertheless, the double-quantum image in Fig. 8.4.4(c) and the spin-echo image in Fig. 8.4.4(b) are both functions of spin density and transverse relaxation. Relaxation and spin-density weights can be eliminated through division of the double-quantum image by

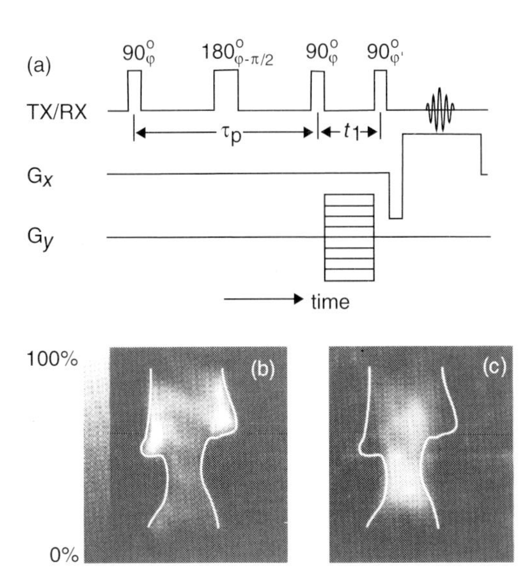

FIG. 8.4.4 [Kli1] ^{2}H double-quantum imaging of deuterated spy molecules in strained rubber bands with a cut on each side. (a) Pulse sequence. A 180° pulse in the preparation period refocusses the chemical shift and off-resonance interactions. (b) Spin-echo image. (c) ^{2}H double-quantum image. The contrast is determined by the double-quantum coherence amplitude weighted by the spin density.

a corresponding spin-echo image acquired with the same echo time (cf. Fig. 10.2.3). Then the contrast is determined only by the sine function in (8.5.5). A fit of the sine function to the signal values in each pixel of a set of normalized double-quantum images provides a *double-quantum parameter image*, which reveals the double-quantum splitting Δ as a function of space [Kli2].

In general, quadrupole nuclei like ^2H in polymers and ^{27}Al in ceramics [Ack2, Con2] are promising candidates for this concept of spectroscopic multi-quantum imaging, because their spectra can provide information about the local electron symmetry and, therefore, about molecular mobility and orientation. Instead of the multi-quantum coherences of quadrupolar nuclei, multi-quantum transitions of the dipolar-coupled abundant spins can be used for the same purpose. The feasibility of this approach has been demonstrated by ^{19}F images of glassy samples [Mil5].

8.5 MAS IMAGING

Given the relationship $1/\Delta x = G_x/\Delta\omega_{1/2}$ among spatial resolution $1/\Delta x$, gradient strength G_x, and linewidth $\Delta\omega_{1/2}$ at half-height, it is clear that the spatial resolution can be increased either by increasing the gradient strength or by reducing the linewidth. Increasing the gradient strength requires larger receiver bandwidths to accommodate the spread in resonance frequencies. Therefore, more noise is introduced into the resultant image. The noise content is proportional to the square root of the receiver bandwidth, and hence, the time to acquire an image of given quality in both resolution and sensitivity, depends on the linewidth. For this reason, line narrowing techniques are of interest in solid-state imaging.

Line broadening in solid-state NMR arises from spin interactions which can be described in first order by coupling tensors of rank two (cf. Section 3.1) [Hae1, Meh1, Sch1]. The spin interactions are either linear or bilinear in the spin operator. Linear interactions are the Zeeman interaction, the chemical shielding, and the interaction with the rf field. Bilinear interactions are the J coupling, the dipole–dipole coupling, and the quadrupolar interaction. In isotropic materials like powders, glasses, and undrawn polymers, wide lines are observed as a result of an isotropic orientational distribution of coupling tensors.

Narrowing of lines broadened by tensorial interactions of rank two can be achieved in a number of ways. Unless the sample is macroscopically oriented by crystallization or application of external fields, the time average of either the space part or the spin part of the interaction Hamiltonian must be modified through mechanical sample reorientation or through rf pulses. *Magic-angle spinning* (MAS) manipulates the space part of the interaction, while spin-locking at the magic angle and *multi-pulse excitation* manipulate the spin part.

In the following, *MAS imaging* is treated. The technique has been proposed by Wind and Yannoni [Win1] and implemented first by Cory, de Boer, and Veeman [Cor5, Cor6]. Conceptually, it is one of the simplest methods to eliminate the anisotropic spin interactions, but in most cases it is necessary that the coupling strength measured by the linewidth of the static NMR spectrum does not exceed the MAS frequency. Therefore, the

^{13}C chemical shift anisotropy can be readily eliminated in this way, but the homonuclear dipole–dipole coupling among ^{1}H can only be averaged sufficiently for soft solids by MAS alone, because spinning speeds are limited by the use of rotors large enough to accommodate heterogeneous samples. If the anisotropy of the interaction is smaller than the spinning speed, then spinning sidebands are observed (cf. Fig. 3.3.7).

Different experiments have been proposed for MAS imaging. In the approach studied most extensively, the gradients are rotating synchronously with the spinner, so that the gradients appear stationary in the sample frame [Vee1]. As an alternative to synchronized gradient-rotation, gradient pulses synchronized to the rotor phase can be used [Ack1] or B_1 gradients can be employed [Mal4]. Such techniques promise access to high-resolution *spectroscopic imaging of* ^{13}C, but long measurement times, the requirements for technically demanding hardware [Sch2, Vee1], and small and balanced samples limit the applicability of the method to selected cases of interest. Furthermore, centrifugal forces can cause sample deformation, fracture, and separation of materials. Nevertheless, once the hardware is available, most imaging techniques can be performed with MAS.

Summary of MAS for dipole–dipole coupling

The principle of sample spinning has been described in Section 3.3.3. As a result of sample spinning, the interaction Hamiltonian and thus the resonance frequency (cf. eqn (3.3.6)) becomes time dependent. For the simple case of a pair ij of dipolar coupled spins $\frac{1}{2}$, equivalent to a quadrupolar nucleus spin 1 like ^{2}H with, the time-dependent spin Hamiltonian is given by [Meh1, Sch1]

$$\mathbf{H}_{\mathrm{D}}(t) = \sum_{m=-2}^{2} \left(-\sqrt{6}\right) D^{ij} d^{(2)}_{-m0}(\psi)\, D^{(2)}_{0-m}\left(\Gamma^{ij}\right) \exp\{\mathrm{i} m \omega_{\mathrm{R}} t\}\, \mathbf{T}^{ij}_{20}, \tag{8.5.1}$$

where $D^{ij} = (\mu_0/4\pi)\gamma^2\hbar/r^3_{ij}$ depends on the internuclear distance r_{ij}, and $\mathbf{T}^{ij}_{20}$ is the irreducible spin tensor operator of the spin pair ij (cf. Table 3.1.2). ψ is the orientation angle of the rotor axis relative to the magnetic field $\boldsymbol{B}_0$. Γ^{ij} denotes the set of Euler angles $(\alpha^{ij}, \beta^{ij}, \gamma^{ij})$ which relate the principal axes of the coupling tensor to the spinner frame (cf. Fig. 3.3.6). The notation $D^{(2)}_{mm'}(\Omega) = \exp\{-\mathrm{i}m\xi\}\, d^{(2)}_{mm'}(\psi) \exp\left\{-\mathrm{i}m'\xi\right\}$ used for the Wigner rotation matrices (cf. Table 3.1.3) [Spi2] connecting the spinner frame to the laboratory frame (cf. Fig. 3.3.6) is defined in [Sch1] (cf. Table 3.1.3). The initial azimuthal angle ξ_0 of the spinner axis with respect to the magnetic field $\boldsymbol{B}_0$ is of no relevance for isotropic samples, and $\xi(t) = \xi_0 + \omega_{\mathrm{R}} t$, where ω_{R} is the sample-rotation frequency (cf. eqn (3.3.4)). For sample rotation at the magic angle $\psi_{\mathrm{m}} = 54.7^\circ$ the reduced Wigner matrix $d^{(2)}_{00}(\psi_{\mathrm{m}}) = \left(3\cos^2\psi_{\mathrm{m}} - 1\right)/2 = 0$ vanishes, and the strength of the dipolar coupling is reduced independent of the rotation frequency.

Following a 90°_y pulse, the NMR signal for an ensemble of dipolar coupled spin-$\frac{1}{2}$ pairs is given by

$$s_x(t) \propto \mathrm{Sp}\left\{\exp\left\{-\mathrm{i}\int_0^t \mathbf{H}_{\mathrm{D}}(t')\mathrm{d}t'\right\} \boldsymbol{\rho}(0_+) \exp\left\{\mathrm{i}\int_0^t \mathbf{H}_{\mathrm{D}}(t')\mathrm{d}t'\right\} \mathbf{I}_x\right\}, \tag{8.5.2}$$

where the initial density operator after the preparation pulse is given by $\boldsymbol{\rho}(0_+) \propto \mathbf{I}_x$. Similar to (3.3.8), the spin coupling is modulated by the rotor frequency ω_R and its first harmonic $2\omega_R$. Neglecting relaxation, this modulation leads to complete refocusing of the dephased magnetization after one rotor period t_R, so that $s_x(nt_R) = s_x(t_R)$, and *rotational echoes* are detected in the time-domain signal (cf. Fig. 3.3.7(a)). Fourier transformation of the rotational echo train leads to a spinning-sideband pattern (cf. Fig. 3.3.7(c)), the envelope of which is given by the Fourier transform of the decay of a rotational echo (cf. Fig. 3.3.7(d)). Stroboscopic detection of the echo maxima and subsequent Fourier transformation leads to a single narrow centreband in the spectrum and no sidebands.

In the presence of *multi-spin dipolar interactions*, the *spin response under MAS* is more complicated. The dipolar Hamiltonian does not commute with itself at different moments in time, so that the simple expression (8.5.2) for the detected signal is not longer valid. Nevertheless, a train of decaying rotational echoes is observed, depending on the strength of the dipolar couplings in the dipolar network (Fig. 8.5.1). The width of the centreband varies in inverse proportion with ω_R^2. In first approximation, the same behaviour is observed for the intensities of the spinning sidebands (Fig. 8.5.1(b)). When the rotor

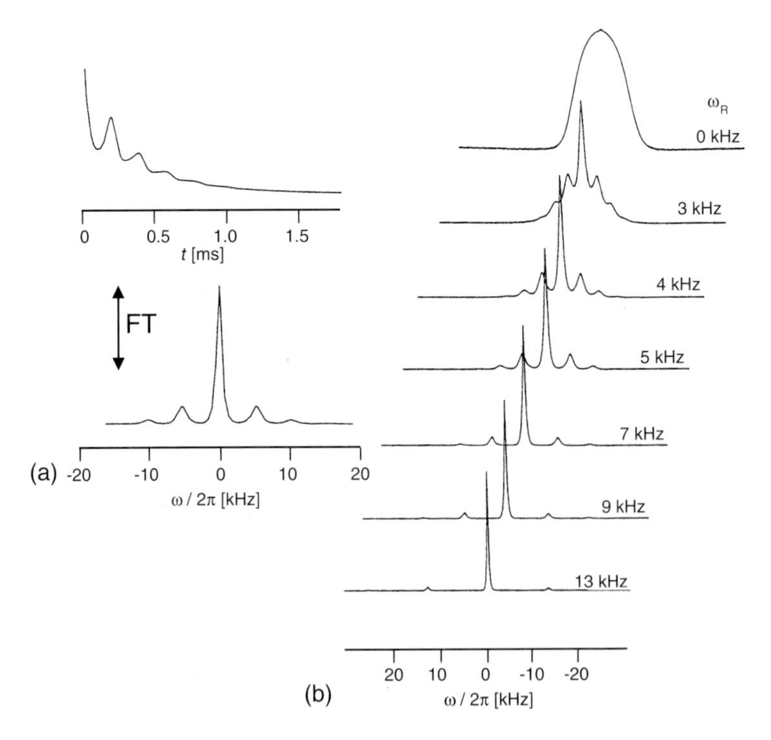

FIG. 8.5.1 ^{1}H MAS NMR spectroscopy of adamantane. (a) The train of the rotational echoes (top) transforms into a spinning-sideband pattern (bottom). (b) With increasing spinning speed $\omega_R/(2\pi)$ the intensity of the sidebands decreases and the centreband narrows. Adapted from [Fil1] with permission from the author.

frequency becomes larger than the dipolar coupling the complex dipolar topology is reduced to spin pairs with scaled dipolar couplings [Fil1].

Compared to the case of dipolar coupled multi-spin systems, the spinning-sideband patterns due to chemical shielding anisotropy can be described in a simpler manner (cf. Section 3.3.3) [Meh1]. Spinning sidebands are observed even at rotor frequencies small compared to the chemical-shift anisotropy. In the fast spinning regime the intensity of spinning-sidebands scales with the inverse of the rotor frequency [Meh1], and in the limit of infinitely high spinning speeds the anisotropic NMR spectrum is coherently averaged to a narrow line positioned at the frequency of the isotropic chemical shift.

The effect of the field-gradient tensor

In MAS imaging it may be important to take into account the tensorial character of the *magnetic-field gradient* [Sch3]. The tensorial properties of the field gradient (cf. Section 2.1.9) cannot be neglected if the gradient coils are oriented along the spinning axis at the magic angle with respect to the magnetic field $\boldsymbol{B}_0$. Such an arrangement is convenient for small dimensions of the gradient coils corresponding to low gradient currents [Vee1]. Denoting the coordinate system of the gradient coils by (x', y', z'), the z' gradient is generated by a pair of Maxwell coils centred at the rotor axis z', and the x' and y' gradients are generated by Golay coils [Bot1]. Projections of the spin density along the perpendicular directions (x', y', z') of the gradient-coil system cannot be generated by applying currents to just one of the corresponding coils. This is illustrated in Fig. 8.5.2, where such projections are shown for a static MAS rotor filled with water [Sch3]. Because

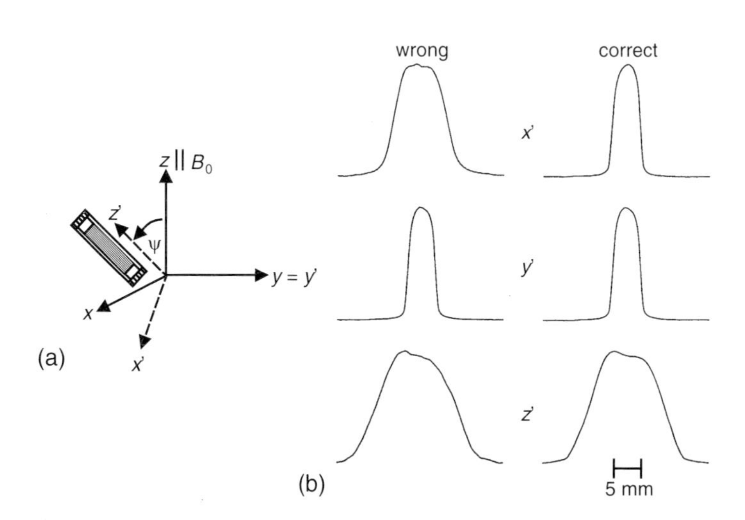

FIG. 8.5.2 [Sch3] (a) Orientation of the gradient coordinate system (x', y', z') with respect to the magnetic field $\boldsymbol{B}_0$. (b) Projections of a static rotor filled with water produced with the field gradients applied along the three orthogonal directions in the gradient coordinate frame. Application of currents to the tilted gradient coils results in distorted x' and y' projections (*wrong*, left). Undistorted projections are obtained by applying well-defined currents to all three coils for each projection (*correct*, right).

of the cylindrical symmetry of the sample, the projections onto the directions orthogonal to the rotor axis should be identical.

In general, the gradient of the magnetic field $\boldsymbol{B} = (B_x, B_y, B_z)^{\mathrm{t}}$ is a second rank tensor $\mathbf{G}$. In most imaging experiments the z-axis of the Cartesian laboratory coordinate frame is parallel to $\boldsymbol{B}$, so that the *gradient tensor* can be approximated with good accuracy by a *gradient vector* $\boldsymbol{G} = (G_x, G_y, G_z)$, where, for instance, $G_x = G_{zx} = \partial B_z/\partial x$ (cf. Section 2.1.2). Because in MAS imaging the z'-axis of the gradient system is oriented along the magic angle, the field-gradient components in the laboratory frame (x, y, z) are related to the gradient components $G_{i'j'}$ generated in the gradient frame (x', y', z') by (cf. Fig. 8.5.2(a)) [Sch3]

$$\begin{aligned} G_x &= -G_{x'x'} \sin\psi + G_{x'z'} \cos\psi, \\ G_y &= -G_{y'x'} \sin\psi + G_{y'z'} \cos\psi, \\ G_z &= -G_{z'x'} \sin\psi + G_{z'z'} \cos\psi. \end{aligned} \tag{8.5.3}$$

Six gradient components of the gradient tensor in the gradient frame contribute to the gradient vector in the laboratory frame. This is so only because the y'-axis of the gradient frame is aligned with the y-axis of the laboratory frame. Otherwise all nine gradient components would have to be considered.

At the magic angle $\psi_{\mathrm{m}} = 54.7°$. Moreover, Maxwell's equations require

$$G_{i'j'} = G_{j'i'} \quad \text{and} \quad G_{x'x'} + G_{y'y'} + G_{z'z'} = 0, \tag{8.5.4}$$

and the symmetry of the Maxwell coil pair around the z'-axis leads to $G_{x'x'} = G_{y'y'}$. With these simplifications (8.5.3) is rewritten

$$G_x = \frac{1}{\sqrt{3}}\left(\frac{G_{z'z'}}{\sqrt{2}} + G_{z'x'}\right), \quad G_y = \frac{1}{\sqrt{3}} G_{z'y'}, \quad G_z = \frac{1}{\sqrt{3}}\left(\frac{-G_{z'x'}}{\sqrt{2}} + G_{z'z'}\right). \tag{8.5.5}$$

Application of currents to the gradient coils generates the tensor elements $G_{z'x'}$, $G_{z'y'}$, and $G_{z'z'}$. From (8.5.5) it is clear that only $G_{z'y'}$ corresponds to a field gradient along the y-axis. The other two gradient components in the gradient frame contribute to the gradients along both the x- and the z-axes. Therefore, to generate gradients along one of these axes proper adjustment of the currents which generate $G_{z'x'}$ and $G_{z'z'}$ is needed. For example, to remove the z dependence $G_{z'z'} = G_{z'x'} 2^{1/2}$ and to remove the x dependence $G_{z'x'} = -G_{z'z'}/2^{1/2}$ must be fulfilled. Based on the last condition projections in the z-direction can be obtained by application of currents to the z' and x' gradient coils. In this way, imaging of the static objects can be achieved with gradients inclined under the magic angle with respect to the static field. This is demonstrated in Fig. 8.5.2(b), where the projections of the water-filled rotor are depicted. On the left side the results are shown when applying currents of equal amplitude to just one of the gradient coils. Clearly, the transverse projections across the rotor are different, demonstrating that this approach is wrong. On the right side projections are shown, with the currents suitably adjusted to satisfy the conditions discussed above.

Rotating gradients

In MAS imaging the spin density to be imaged rotates with angular frequency ω_R about the z'-axis of the gradient coordinate frame. In the frame with axes x'', y'', and z'', which rotates around the $z' = z''$-axis, the gradient components are given by

$$\begin{aligned} G_x &= \frac{1}{\sqrt{3}}\left[\left(\frac{1}{\sqrt{2}}G_{z'z'} + G_{z'x'}\right)\cos(\omega_R t) + G_{z'y'}\sin(\omega_R t)\right], \\ G_y &= \frac{1}{\sqrt{3}}\left[\left(\frac{1}{\sqrt{2}}G_{z'z'} + G_{z'x'}\right)\sin(\omega_R t) + G_{z'y'}\cos(\omega_R t)\right], \\ G_z &= \frac{1}{\sqrt{3}}\left(\frac{-1}{\sqrt{2}}G_{z'x'} + G_{z'z'}\right). \end{aligned} \tag{8.5.6}$$

The gradients appear stationary in the rotating sample frame, if the gradient coils are driven with currents alternating at the rotor frequency. For imaging of $x''y''$ cross-sections, the scaling and time dependence of the gradients must be adjusted to [Sch3]

$$\begin{aligned} G_{z'x'} &= \frac{1}{2}G_0\cos(\omega_R t + \varphi), \\ G_{z'y'} &= G_0\sin(\omega_R t + \varphi), \\ G_{z'z'} &= \frac{1}{\sqrt{2}}G_0\cos(\omega_R t + \varphi), \end{aligned} \tag{8.5.7}$$

where φ denotes the relative phase between the gradient direction and the x''-axis of the sample frame, and G_0 is the reference amplitude of the rotating gradients.

Inserting eqn (8.5.7) into eqn (8.5.6) yields the field gradient components in the sample frame,

$$\begin{aligned} G_{x''} &= \frac{1}{\sqrt{3}}G_0\cos\varphi, \\ G_{y''} &= \frac{1}{\sqrt{3}}G_0\sin\varphi, \\ G_{z''} &= 0. \end{aligned} \tag{8.5.8}$$

The gradient is static in the sample frame so that the rotating spin density leads to a time-independent signal. The phase φ is varied in small steps for imaging with reconstruction from projections, or it is switched by multiples of $\pi/2$ for Fourier imaging methods.

If the condition (8.5.7) is not fulfilled, error signals are introduced which lead to spinning-sideband images. These images can be shifted outside the centreband image if the rotor period is set twice the spectral width of the image. They can also be eliminated by small gradient misadjustments [Cor1, Cor7, Vee1] or averaged to zero in multiple accumulations with random rotor phases [Cor5].

Because the dipolar interaction is independent from the magnetic field strength B_0, the dipolar interaction is averaged out by MAS, even if the spins rotate in an inhomogeneous field, because the dipole–dipole interaction is independent from B_0. However, the chemical shift is proportional to B_0, so that an extra line broadening is introduced, but as long as the maximum field difference caused by the gradient is smaller than B_0 this broadening can be neglected. Consequently, the line-narrowing capability of MAS is effective in magnetic-field gradients as long as the change in field a spin experiences during MAS is much smaller than B_0.

For phase encoding in a 2D fashion, the signal detected is given by

$$s(t_1, t_2) \propto \cos\left\{ \int_0^{t_2} \omega_\lambda \, \mathrm{d}t \right\} \exp\{\mathrm{i}\gamma \boldsymbol{G} \boldsymbol{r} t_1\}, \tag{8.5.9}$$

where λ represents for different anisotropic interactions. In the first dimension t_1 is chosen to be an integral multiple of rotor periods, so that the magnetization is modulated only by $\exp\{\mathrm{i}\gamma \boldsymbol{G} \boldsymbol{r} t_1\}$. During the detection period t_2 MAS spectra can be accessed. To convert phase to amplitude modulation, two 2D data sets differing in the gradient signs need to be measured and processed [Jan1].

Applications

Frequency- as well as phase-encoding techniques have been applied to generate MAS images of nuclei like ^{1}H [Bus1, Bus2, Bus3, Cor1, Cor4, Cor5, Cor6, Cor7, Cor8, Sun1, Vee1], ^{2}H [Gün3], and ^{13}C [Sch2, Sch4, Sch5, Sch6, Sun2]. For rare nuclei like ^{13}C the signal can be enhanced through cross-polarization and the detection is done in the presence of high-power heteronuclear decoupling.

The proton dipolar couplings in soft solids can be easily averaged by MAS and hence MAS imaging has been applied to map the poly(butadiene) fraction of two poly(butadiene)–poly(styrene) blends (Fig. 8.5.3) [Cor4]. Single-point imaging (cf. Section 8.2) was used to map the signal decay as a function of the gradient amplitude for a fixed gradient time t_1, equal to a multiple of the rotor period. A full set of 1D projections suitable for reconstruction from projections was acquired by varying the phase φ of the gradient signal in equal steps over 180°. The molecular mobility of the poly(butadiene) fraction is high enough for MAS at 5 kHz to average the residual dipolar couplings to zero. This is not so for the protons in poly(styrene). Therefore, the signal of poly(styrene) is suppressed and the image contrast is formed by the protons in poly(butadiene) only. Differences in the phase separation of the blend components can be identified in the two images. They result from different preparation techniques of the samples.

Spectroscopic *MAS imaging of deuterons* with phase encoding of the space dimension has been used to average the proton–deuteron dipolar couplings in order to obtain angular-dependent deuteron wideline spectra with high spatial resolution [Gün3]. In this 2D experiment, the quadrupole splitting in the deuteron spectrum has been separated from the effect of the field gradients through phase encoding of the space dimension according to (8.5.9). Consequently, the spatial resolution was not determined by the large linewidth of a deuteron powder spectrum, but by the loss of signal during the evolution period

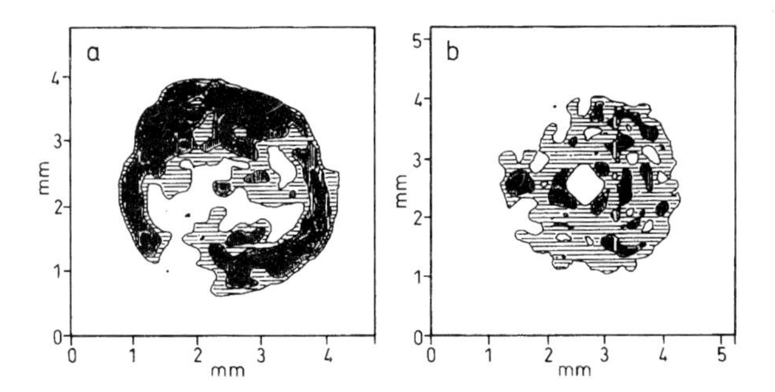

FIG. 8.5.3 Proton MAS images of solid discs (750 μm thick and 3.5 mm diameter) of poly(butadiene)–poly(styrene) blends. (a) Mechanical blend of both components. (b) Blend cast from solution in toluene. The image contrast is caused by the differences in the strengths of the dipole–dipole couplings among ^{1}H for the two polymers. The signals from poly(styrene) are filtered out (white) so only poly(butadiene) protons contribute to the image (dark). The spatial resolution is better than 50 μm at a spinning frequency of 5 kHz. Adapted from [Cor4]. Copyright 1989 American Chemical Society.

caused by decay of coherences under the influence of the heteronuclear ^{1}H–^{2}H dipole–dipole couplings. These couplings can be attenuated by MAS resulting in significantly improved spatial resolution of better than 100 μm.

MAS imaging of ^{13}C bears promise for mapping rigid polymers for which current proton-based techniques suffer from a lack of spatial resolution. The smaller magnetogyric ratio of ^{13}C leads to a loss in spatial resolution which, however, is more than compensated for by much narrower lines obtained in ^{13}C NMR by high-power proton decoupling in conjunction with MAS. Moreover, even in ^{13}C enriched samples the homonuclear dipolar couplings among ^{13}C are weak enough to be averaged by spinning frequencies of a few kilohertz. The main advantage of MAS imaging is the access to spatially resolved ^{13}C-NMR spectra which can provide detailed molecular information, extending the range of contrast parameters accessible to ^{1}H. These parameters can be exploited to map molecular properties like chemical structure, order, and dynamics [Spi1], which, in return, determine many macroscopic properties of the sample such as the viscoelastic parameters.

In the case of ^{13}C, at low spinning frequencies, spinning sidebands appear which degrade the resolution image. These can be removed using the TOSS sequence [Dix1] prior to space encoding [Sch4, Sch6]. The fact that spinning sidebands can be manipulated, even if some loss in signal intensity is present, implies that there is no necessity for high spinning speeds and that it is possible to use larger diameter rotors to accommodate a larger variety of samples.

The large chemical shift range of ^{13}C has been utilized to differentiate chemical structures [Sch6]. Spinning sidebands from the ^{13}C chemical shielding anisotropy have been analysed to characterize the variation of the polymer-chain orientation in an oriented sample as a function of space in a 2D spectroscopic MAS imaging experiment (cf. Fig. 7.3.5) [Sch4] by introducing rotor synchronization of the data acquisition (cf. Fig. 3.3.8) [Har1].

8.6 MAGIC-ANGLE ROTATING-FRAME IMAGING

Line narrowing through dipolar decoupling and reduction of chemical shift anisotropy can be achieved by operating on the spatial and on the spin part of the interaction Hamiltonian. Molecular motion and MAS operate on the spatial part (cf. Section 3.3.3). Multi-pulse and other rf excitation methods operate on the spin part (cf. Section 3.3.4). One particular rf line-narrowing method is the *magic-angle rotating-frame method* (MARF) [Del1, Del2, Del3, Del4, Del5, Del6, Del7, Del8, Del9, Del10, Lug1]. It is based on the Lee–Goldburg technique [Lee1] of line narrowing by spin-locking at the magic angle [Ats1, Mef1]. Here, the resonance offset of the rf excitation is adjusted in such a way that the effective field is at the magic angle, so that the average spin orientation is in that direction with respect to the applied magnetic field $\boldsymbol{B}_0$.

In the Lee–Goldburg approach the magic-angle condition is fulfilled only while the rf field is turned on. For multi-pulse excitation the magic-angle condition may be fulfilled at the time average. In fact, line-narrowing multi-pulse sequences have been designed which produce a frequency shift of the narrowed line position that is proportional to the flip angle of certain pulses in the sequence. Given the line-narrowing capability and the linear relationship between line position and rf field amplitude, a coil for pulsed rf gradients could be used to achieve spatial discrimination of nuclear spins by observing their forced precession frequency in analogy to liquid-state *rotating-frame imaging* (cf. Section 6.3) [Cho1].

MARF imaging

If the angle between $\boldsymbol{B}_0$ and the effective field is denoted by θ in the *Lee–Goldburg experiment*, interactions linear in the spin operator, for example the chemical shielding, are scaled by $\cos\theta$. Bilinear interactions like the dipole–dipole interaction scale with the second Legendre polynomial $P_2(\cos\theta) = (3\cos^2\theta - 1)/2$ in first order. P_2 vanishes at the magic angle $\theta_{\mathrm{m}} = 54.7^\circ$ (cf. Fig. 2.2.7). The second-order contributions are nonsecular relative to the effective Zeeman field. If the Zeeman field is large enough, they periodically average the anisotropic interactions to zero with period $2\pi/\omega_{\mathrm{eff}}$, where $\omega_{\mathrm{eff}} = -\gamma B_{\mathrm{eff}}$ is the angular frequency of the effective field B_{eff}, and narrow liquid-like NMR lines can be observed.

This simple approach to line narrowing was applied to imaging of dipolar solids with direct and an indirect detection [Mar2]. It is known as *MARF imaging* in the literature. The advantage of this method is an improved dipolar decoupling efficiency for solids compared to MAS, because the modulation frequencies of the Hamiltonian can be higher by one order of magnitude. Furthermore, the sample remains static. However, the chemical-shift anisotropy is only scaled, and the method is less efficient for materials with strong dipolar couplings than multi-pulse methods are. A low signal-to-noise ratio is intrinsic to this method. Also, the accuracy of the magic-angle condition strongly depends on the stability of $\boldsymbol{B}_0$ and $\boldsymbol{B}_1$ field gradients over the sample. This is not easy to achieve in large sample volumes, especially for the required $\boldsymbol{B}_1$ gradient. The MARF imaging methods can be combined with contrast filters [Del9, Del10].

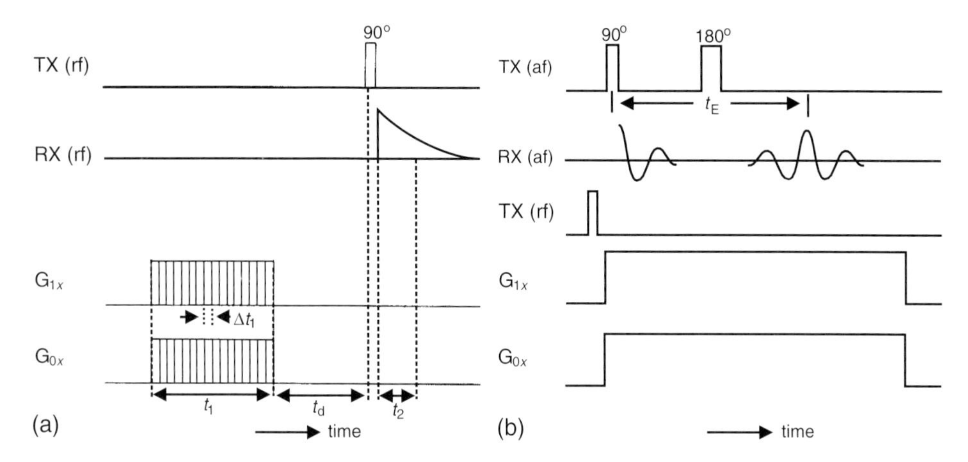

FIG. 8.6.1 [Mar2] Timing diagrams for MARF imaging. (a) Indirect detection of one data point at time t_2 in the absence of gradients. During t_1 the coherences evolve in the tilted rotating frame. During the delay t_d the transverse magnetization components are eliminated by dephasing. (b) Direct detection of transverse magnetization in the tilted doubly rotating frame. Here the precession frequency is in the audio frequency regime.

Principle

The excitation schemes for MARF imaging (Fig. 8.6.1) are based on the spin evolution in the effective field $\boldsymbol{B}_{\mathrm{eff}}$ [Sli1], which, for the purpose of imaging, is generated by a combination of a static field gradient G_{0x} and a rf field gradient G_{1x}. Both gradients are needed in order to produce an effective field with linearly varying amplitude but constant orientation. Following Fig. 2.2.7 and eqn (2.2.32), the precession frequency around the effective field is given by $\omega_{\mathrm{eff}} = [\omega_1(x)^2 + \Omega(x)^2]^{1/2}$, where the resonance offset frequency $\Omega(x) = -\gamma G_{0x} x$ is defined by a time invariant gradient G_{0x} in the polarizing magnetic field $\boldsymbol{B}_0$, and the amplitude of the rf field is spatially modulated by a gradient G_{1x} according to $\omega_1(x) = -\gamma G_{1x} x$. The angle between the effective field and the fictive field along the laboratory z-axis is defined according to (2.2.31) as $\tan\theta(x) = B_1(x)/B_{\mathrm{fic}}(x) = G_{1x}/G_{0x}$. It is independent of space and is defined by the ratio of the gradient values of G_{0x} and G_{1x}. At the magic angle $\tan\theta_{\mathrm{m}} = 2^{1/2}$, so that the static and the rf gradients need to be scaled in amplitude to $G_{1x}/G_{0x} = 2^{1/2}$ in order for the effective field to be at the magic angle at all points in the chosen space direction of the laboratory frame. This condition limits the resolution of the method because of the limiting values of the rf field gradient which can be achieved experimentally. To assure a finite lock field at all positions, the sample is placed outside the origin $x = 0$ of the coordinate system.

In the presence of dipolar couplings the total spin Hamiltonian in the rotating reference frame is given by

$$\mathbf{H}(x) = -\omega_1(x)\mathbf{I}_x - \Omega(x)\mathbf{I}_z + \mathbf{H}_D^{(0)}(x), \tag{8.6.1}$$

where the secular part $\mathbf{H}_D^{(0)}$ of the dipolar Hamiltonian depends on space for an inhomogeneous sample. At time t_1 just after application of the field gradient pulses (Fig. 8.6.1(a)) the density matrix is given by

$$\boldsymbol{\rho}(t_1, x) = \exp\{-\mathrm{i}\mathbf{H}(x)t_1\}\,\boldsymbol{\rho}_0(x)\exp\{\mathrm{i}\mathbf{H}(x)t_1\}. \tag{8.6.2}$$

Here the initial density matrix $\boldsymbol{\rho}_0(x)$ corresponds to the spin density (cf. eqn (8.1.1)).

Evaluation of the density matrix is facilitated by transformation of the spin Hamiltonian into the rotating frame, the z'-axis of which is tilted at the magic angle (*tilted rotating frame*),

$$\exp\{\mathrm{i}\psi_\mathrm{m}\mathbf{I}_\mathrm{y}\}\,\mathbf{H}(x)\exp\{-\mathrm{i}\psi_\mathrm{m}\mathbf{I}_\mathrm{y}\} = -\omega_\mathrm{eff}(x)\mathbf{I}_{z'} + \mathbf{H}_\mathrm{D,ns}(x) \approx -\omega_\mathrm{eff}(x)\mathbf{I}_{z'}, \tag{8.6.3}$$

where $\mathbf{I}_y$ is the spin angular momentum operator along the y-direction in the rotating frame, and $\mathbf{H}_\mathrm{D,ns}(x)$ is the nonsecular part of the dipolar Hamiltonian in the tilted rotating frame. If the strength ω_eff of the effective field is larger than the dipolar couplings the nonsecular Hamiltonian $\mathbf{H}_\mathrm{D,ns}$ can be truncated in the tilted rotating frame. Therefore, the line-narrowing efficiency may be space dependent and can lead to nonuniform image resolution.

MARF imaging with indirect detection

In *MARF imaging with indirect detection* [Del7, Del8, Del9, Del10, Del11] the signal is detected in the rotating frame after the application of a 90° rf detection pulse following the evolution for a variable time t_1 in the spin-lock field at the magic angle. The FID can be transformed into a spectrum, or single-point detection is used if the spectroscopic information is unnecessary (Fig. 8.6.1(a)). For amplitude modulation of the signal, a dephasing delay t_d is introduced before detection [Del9]. Contrast by T_2 and $T_{1\rho}$ weights can be incorporated into the sequence [Del9, Del10], and a method for slice selection has been proposed [Del11]. Different projections of the sample can be recorded with gradient rotation and with sample rotation. MARF imaging with indirect detection is conceptually simple, yet time consuming in the experiment.

MARF imaging with direct detection

An alternative to MARF imaging with indirect detection is *MARF imaging with direct detection* (Fig. 8.6.1(b)) [Del1, Del2, Del3, Del4, Del5, Del6, Lug1]. The longitudinal magnetization is first rotated into the direction of the magic angle to align it with the effective field. Then, a spin-echo experiment is performed, and transverse magnetization is generated in the tilted rotating frame. It precesses around the z'-axis with frequency ω_eff. This frequency is in the audio-frequency (af) range. Therefore, the echo excitation pulses are applied with that frequency using a separate coil. For the purpose of decoupling this coil from the rf coils, it is aligned along the z-axis of the laboratory frame. The same af coil is also used for detection. In principle, all experiments usually performed in the rotating frame by conventional rf excitation can also be performed in the tilted rotating frame with af excitation. However, problems may arise with mechanical oscillations of the af coil [Del5]. Because transverse magnetization is generated in the rotating frame by the scheme of Fig. 8.6.1(b), the echo time t_E introduces a parameter weight from $T_{2\rho}$, the *transverse relaxation time in the rotating frame*.

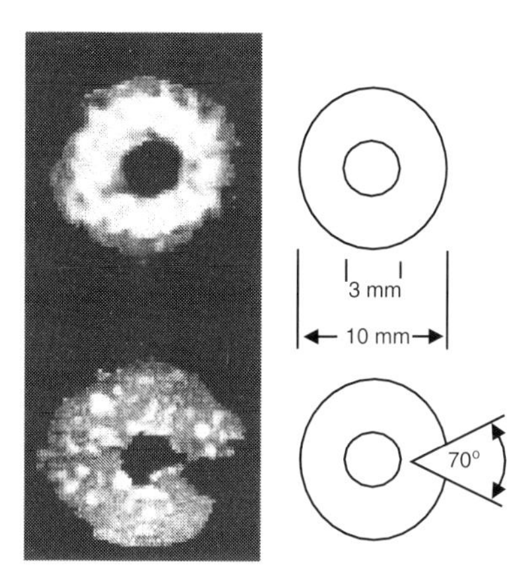

FIG. 8.6.2 2D MARF images of compressed polycrystalline adamantane. Left: MARF images obtained by reconstruction from projections. Right: shapes of the objects and dimensions. Adapted from [Mar2] with permission from the author.

Because of the demanding apparatus, experiments have been performed without switching the gradient direction, and 2D images were obtained through reconstruction from projections measured for different sample orientations. An example demonstrating the feasibility of this sophisticated method is given in Fig. 8.6.2 by two images of phantoms made from polycrystalline adamantane [Mar2]. The images were constructed from 16 1D projections. Clearly, the shape of the objects (right) can be recognized in the images.

8.7 IMAGING WITH MULTI-PULSE LINE NARROWING

The dipolar broadening is the main contribution to the spectral linewidth of ^{1}H in rigid materials. In many cases it is too large to be averaged by MAS, even for the high spinning speeds that are available today. Line narrowing must, therefore, be achieved with rf irradiation by coherent averaging techniques alone or in combination with MAS [Ger1, Hae1, Meh1]. In the presence of multiple-pulse excitation the space-encoding gradients should be introduced in such a fashion that they do not influence the line-narrowing efficiency of rf irradiation.

Line narrowing by multiple-pulse techniques (cf. Section 3.3.4) is a promising approach to solid-state imaging with good spatial resolution. However, high demands on experimental set-up and equipment hamper the use of the technique. Moreover, for line narrowing by multiple-pulse techniques the data points of the free-induction decay are acquired between pulses (cf. Fig. 3.3.10), and the efficiency depends on the quality of the rf pulses [Hae1, Meh1]. Radio-frequency pulses with short rise and fall times require a low quality factor of the rf coil which, at the same time, reduces the sensitivity and therefore limits the spatial resolution.

8.7.1 Principle

In the following, some aspects of multi-pulse excitation are reviewed which are relevant for space encoding. The effects of time-invariant and pulsed gradients are addressed as well as methods to increase the line-narrowing efficiency through *second averaging* and *CRAMPS* (cf. Section 3.3.4).

Multi-pulse excitation in time-invariant field gradients

In strong field gradients, pulses are frequency selective. This has been illustrated in Fig. 8.1.2 for a rectangular pulse. The 90° flip angle of the pulse is valid only in a narrow slice around zero offset frequency. Many multi-pulse sequences use solid echoes for refocusing of the dipole–dipole interaction. However, when applied off resonance, already the two-pulse sequence for generation of a solid echo (cf. Fig. 3.2.6(a)) exhibits complex behaviour [Dem1]. The spatial selectivity is enhanced compared to that produced by just a single rf pulse (cf. Fig. 9.2.2). This effect is even stronger for sequences of many pulses applied in the presence of a field gradient. Consequently, the efficiency of dipolar decoupling is localized and affects the image quality.

To consider the efficiency of dipolar decoupling in more detail, a *WAHUHA* pulse sequence is taken as an example [Hae1]. One pulse cycle (cf. Table 3.3.1) leads to the time evolution operator (cf. eqn (3.3.22))

$$\mathsf{U}(6\tau) = \mathsf{L}^z\mathsf{P}_{-x}\mathsf{L}^z\mathsf{P}_{+y}\mathsf{L}^z\mathsf{L}^z\mathsf{P}_{-y}\mathsf{L}^z\mathsf{P}_{+x}\mathsf{L}^z, \tag{8.7.1}$$

where P_{+x} represents a 90° pulse about the $+x$-axis and L^z the time evolution of the free induction for a time interval τ. On the time average the spin system rotates about the main diagonal of a cube in the laboratory frame ($\boldsymbol{a} = (1, 1, 1)$ in Table 3.3.1), which is oriented at the magic angle with respect to the applied magnetic field $\boldsymbol{B}_0$. The Hamiltonian H is switched between the different states of the *toggling coordinate frame* (cf. eqn (3.3.24)), which are obtained by the transformations

$$\mathsf{H}^x = \mathsf{P}_{+y}\mathsf{H}^z\mathsf{P}_{+y}^{-1} \quad \text{and} \quad \mathsf{H}^y = \mathsf{P}_{-x}\mathsf{H}^z\mathsf{P}_{-x}^{-1}. \tag{8.7.2}$$

For the WAHUHA pulse sequence the average Hamiltonian is given by

$$\overline{\mathsf{H}} = \tfrac{1}{3}\left(\mathsf{H}^x + \mathsf{H}^y + \mathsf{H}^z\right). \tag{8.7.3}$$

For the truncated or secular dipolar Hamiltonian H_D (3.3.26) the average Hamiltonian (8.7.3) vanishes, i.e.

$$\mathsf{H}_\mathrm{D}^x + \mathsf{H}_\mathrm{D}^y + \mathsf{H}_\mathrm{D}^z = 0. \tag{8.7.4}$$

This equation is sometimes called the *magic-angle condition*, because when it is fulfilled, the spins precess about the cube diagonal which is at the magic angle with respect to $\boldsymbol{B}_0$. Therefore, the WAHUHA pulse sequence decouples the homonuclear dipole–dipole interaction in first order (cf. Section 3.4.3).

In the presence of a field gradient, the Hamiltonians must be evaluated in a space-dependent toggling rotating frame,

$$\mathbf{H}^{x} = \exp\{\mathrm{i}\theta(z)\mathbf{I}_x\}\exp\{\mathrm{i}\omega_{\mathrm{eff}}(z)t_{\mathrm{p}}\mathbf{I}_z\}\exp\{-\mathrm{i}\theta(z)\mathbf{I}_x\}\mathbf{H}^{z} \times \exp\{\mathrm{i}\theta(z)\mathbf{I}_x\}\exp\{-\mathrm{i}\omega_{\mathrm{eff}}(z)t_{\mathrm{p}}\mathbf{I}_z\}\exp\{-\mathrm{i}\theta(z)\mathbf{I}_x\}, \quad (8.7.5a)$$

and

$$\mathbf{H}^{y} = \exp\left\{\mathrm{i}\theta(z)\mathbf{I}_y\right\}\exp\left\{\mathrm{i}\omega_{\mathrm{eff}}(z)t_{\mathrm{p}}\mathbf{I}_z\right\}\exp\left\{-\mathrm{i}\theta(z)\mathbf{I}_y\right\}\mathbf{H}^{z} \times \exp\left\{\mathrm{i}\theta(z)\mathbf{I}_y\right\}\exp\{-\mathrm{i}\omega_{\mathrm{eff}}(z)t_{\mathrm{p}}\mathbf{I}_z\}\exp\left\{-\mathrm{i}\theta(z)\mathbf{I}_y\right\}, \quad (8.7.5b)$$

where $\theta(z)$ is the angle between the local effective field of magnitude $B_{\mathrm{eff}}(z) = -\omega_{\mathrm{eff}}(z)/\gamma$ and the applied magnetic field $\boldsymbol{B}_0$. As a result of the transformations (8.7.5) the secular part of the dipolar Hamiltonian transforms in such a way that terms describing both single- and double-quantum contributions are present. The presence of these terms demonstrates that the magic-angle condition (8.7.4) is no longer fulfilled for all z-coordinates. The coherent averaging pulse sequence thus exhibits a dipolar decoupling efficiency, which depends on the position z within the sample. This dramatically degrades the image resolution for large field gradients.

Second averaging

In the context of multi-pulse solid-state NMR, criteria have been established which relate the frequency offset to the line-narrowing efficiency [Gar1, Meh1, Mil4, Pin1]. Enhanced line-narrowing efficiency can be achieved off-resonance by *second averaging* of the Hamiltonian [Meh1, Cor12]. Second averaging is effective if the offset frequency Ω is much smaller than the inverse of the cycle time t_{c} of the pulse sequence. For example, if $\Omega/2\pi = 0.1/t_{\mathrm{c}}$, and if $t_{\mathrm{c}} = 20$ µs, then $\Omega/2\pi$=5 kHz which corresponds to a sample size of 1 cm in a modest field gradient of about 0.01 T/m (1 G/cm) for ^{1}H. In addition to this the value of the field gradient effective for spatial resolution is reduced by the multi-pulse sequence in the same way as the chemical shift is scaled (cf. Table 3.3.1). For the WAHUHA sequence, for instance, the scaling factor R_σ is $3^{-1/2}$. In this case the spatial resolution is of the order of $(0.1\ \mathrm{mm})^{-1}$ for a linewidth of 30 Hz. Obviously, this resolution is considerably lower than in liquid-state imaging [Vee1].

Pulsed-field gradients

A remedy to the offset problem is to apply the rf pulses in the absence of a field gradient. Then *pulsed-field gradients* need to be applied. However, the shorter the succession of rf pulses in the pulse sequence, the better the dipolar decoupling efficiency, so that ultra-fast gradient switching is necessary. Gradient pulses with durations of the order of 2–5 µs are required in practice. This demand presents a considerable technical challenge, especially for strong pulsed-field gradients, which may also need to be stepped in well-defined intervals for phase encoding. Longer gradient pulses can be employed if the pulse sequence is based on magic echoes instead of solid echoes (cf. Section 8.8). In this case inter-pulse spacing can be extended to durations of the order of T_2.

Time suspension

Multi-pulse sequences of the *time-suspension* type are particularly useful for imaging of solids, because both the bilinear and linear spin interactions are scaled to zero [Cho1, Cor9, Cor10, Man2]. When eliminating the linear spin interactions, line broadening from chemical shift, susceptibility, and heteronuclear dipolar couplings is suppressed in addition to line broadening from the bilinear homonuclear dipole–dipole interaction. Apart from relaxation, no evolution of the spin system is observed, because all spin interactions are eliminated. The NMR spectrum reduces to just one line with a width of some 10 Hz, so that the spatial resolution is increased significantly. Typical linewidths are 3.5 Hz for adamantane and 66 Hz for high-density poly(ethylene) [Cor9].

Because multi-pulse sequences require stroboscopic data sampling, time appears to be suspended only at the sampling point in the cycle. Time-suspension pulse sequences can be improved by incorporating the principle of second-averaging for the linear spin interactions. A coherent spin rotation can be produced by toggling the phases of the subcycles [Cor9, Cor10]. Through skilful combination of such multiple-pulse sequences with time-dependent magnetic field gradients, the gradient-induced spin evolution, and therefore, space encoding can be preserved without destroying the line-narrowing benefits of the cycle.

CRAMPS

Conventional multi-pulse methods alone cannot be used to separate the contributions from the isotropic and anisotropic chemical shielding. Either both are removed or both are preserved. However, the chemical shift anisotropy can be removed from spectra of abundant spins in a straightforward way by *CRAMPS*, that is, by applying rf pulses and MAS simultaneously (cf. Section 3.3.4) [Bur1, Haf2, Sch7, Tay1]. Because fast spinning of samples with spatial heterogeneities is difficult to achieve, only the case of slow sample spinning is relevant for imaging. If the MAS rotor frequency is low, the modulation of the spin interaction by the sample rotation is ineffective during one pulse cycle. In this case, dipolar decoupling by one pulse cycle is achieved under quasi-static conditions, and interferences between rf pulses and sample rotation are largely avoided. Then, the dipolar couplings are removed by the multiple-pulse sequence, whereas the chemical shift anisotropy is dealt with by the relatively slow MAS. The isotropic chemical shift information is preserved and can be exploited for spectroscopic imaging. In the fast MAS regime, the rotor period is short compared to the cycle time of the pulse sequence. Here, the pulse modulation of the space and spin parts of the Hamiltonian becomes important and the pulses need to be synchronized to the rotor period [Dem3]. One example are multi-pulse sequences based on magic echoes (cf. Section 8.8).

8.7.2 Methods

Imaging with multi-pulse line narrowing has been realized in different variants. These include time-invariant gradients, oscillating gradients, pulsed gradients, and CRAMPS. Each of these variants is addressed in the following.

Time-invariant gradients

The simplest realization of solid-state imaging with multiple-pulse excitation for line narrowing uses the *MREV8* pulse sequence (cf. Table 3.3.1) in the presence of a time-invariant field gradient [Bot2, Die1, Chi1, Sin1]. MREV8 reduces the homonuclear dipole–dipole interaction to zero in first order and scales the chemical shift as well as the gradient strength by a factor of $R_\sigma = 3^{1/2}/2$. If the gradient is sufficiently strong to dominate the frequency shifts caused by magnetic shielding and magnetic susceptibility, images of solids can be obtained with good spatial resolution.

Suitable imaging schemes are illustrated in Fig. 8.7.1. 2D and 3D images can be reconstructed from projections (a) and measured by Fourier schemes (b) [Chi1]. In the first case, slow gradient switching can be afforded so that the gradient settling time does not interfere with the pulse sequence. In the second case, time intervals have to be implemented for phase encoding and gradient switching prior to detection. The gradient-switching period can be realized preferably by storing the magnetization along the z-axis [Sze1] and by spin-locking [Chi1]. This time delay introduces T_1 and $T_{1\rho}$ contrast, respectively. Optimum spatial resolution requires a careful balance between gradient strength and multi-pulse cycle time.

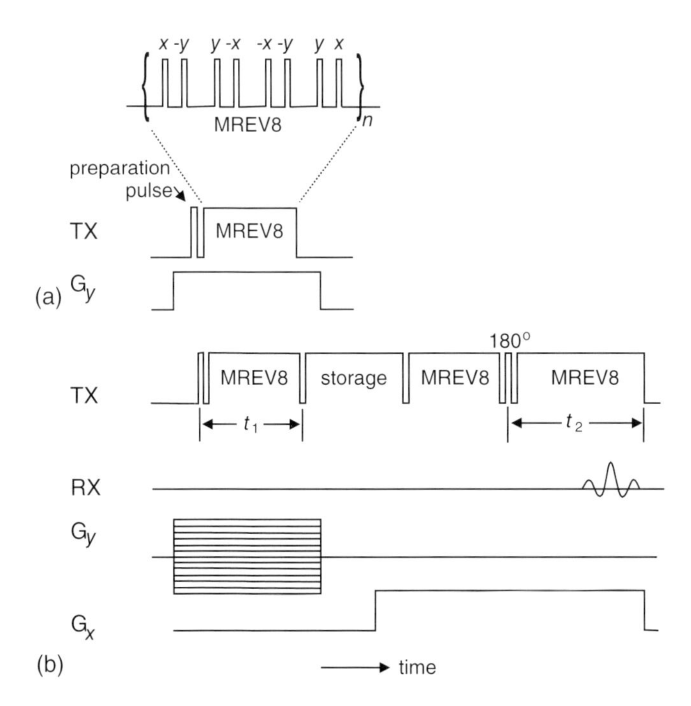

FIG. 8.7.1 [Chi1] Imaging with multi-pulse line narrowing in the presence of time-invariant magnetic field gradients. (a) Frequency encoding. (b) Combination of phase and frequency encoding with intermediate storage of the magnetization components for gradient switching. The 180° pulse generates a Hahn echo during t_2 which coincides with an echo from magnetization dephasing during t_1. During the storage period magnetization is stored along $\boldsymbol{B}_0$ or $\boldsymbol{B}_1$. This period can be exploited to introduce contrast.

Oscillating gradients

An improvement in dipolar decoupling efficiency and image resolution is obtained when the main magnetic field gradients are applied periodically in the windows of the pulse sequence (cf. Section 8.2). One way is to use *oscillating gradients* of constant amplitude for frequency encoding, the other is to use variable amplitude for phase encoding [Con1, Cot1, Mcd2, Mcd3, Mcd4, Mcd5, Mil2, Mil3, Mil4]. The main advantage of the method is related to an improvement in the gradient strength, if the gradient circuit is driven in resonance mode and if it is tuned to the oscillation frequency. The oscillating gradient periodically crosses zero amplitude. These zero crossings can be used for application of the rf pulses to avoid off-resonance effects.

Pulsed gradients

Comparatively good spatial resolution can be obtained by combining *time-suspension line narrowing* [Cor9, Cor10, Mat1, Wei1] with *second averaging* through phase shifting and *pulsed-field gradients*. These improvements in performance of pulse sequences and image quality are gained at the expense of demanding gradient accessories [Con1], which must be capable of producing short gradient pulses in the windows of the multiple-pulse cycle [Mil4, Wei1].

A suitable time-suspension pulse sequence that utilizes second averaging is the 48-pulse sequence of Fig. 8.7.2 [Cor9]. The rf pulses are applied in the absence of the gradients, and by removing resonance offset, the variation of the spatial resolution over the sample is eliminated. The cycle can be applied either with oscillating or with pulsed gradients. The superior line-narrowing efficiency of this sequence leads to high-resolution images such as that of a poly(acetal) (Delrin) phantom shown in Fig. 8.7.3 [Cor12].

The longer the pulse cycle, the larger the sampling interval and the smaller the size of the object that can be investigated without frequency folding. The spectral width of the acquired data can be enlarged by *oversampling*, that is, by collecting more than one data point per cycle [Cor10, Cor11, Cor12]. This feature becomes particularly evident, when considering, that the *spatial resolution* Δx^{-1} is proportional to the inverse linewidth divided by the length l of the object. Because the inverse linewidths is proportional to the number n_{data} of data points which can be sampled during a time of length T_2, the spatial resolution is proportional to [Cor10, Cor11]

$$\Delta x^{-1} \propto n_{\text{data}}/l. \tag{8.7.6}$$

It is obvious that n_{data} should be maximized to achieve the optimum resolution.

In choosing a pulse sequence for imaging, improvement in the decoupling efficiency with longer, more highly compensated cycles must be weighted against relaxation of the signal over longer cycle times. The best cycle is that with the largest n_{data}. However, sampling the signal many times within one pulse cycle may reintroduce the spin interactions, which are averaged with the periodicity of the cycle time. By suitable data treatment the time-suspension 48-pulse sequence (Fig. 8.7.2) can be sampled eight times per cycle, while the WAHUHA cycle only admits one sampling point [Cor9]. In the imaging experiments with magic-echo line narrowing each echo maximum can be sampled stroboscopically [Wei1] instead of every second echo [Mat1]. As a result, the

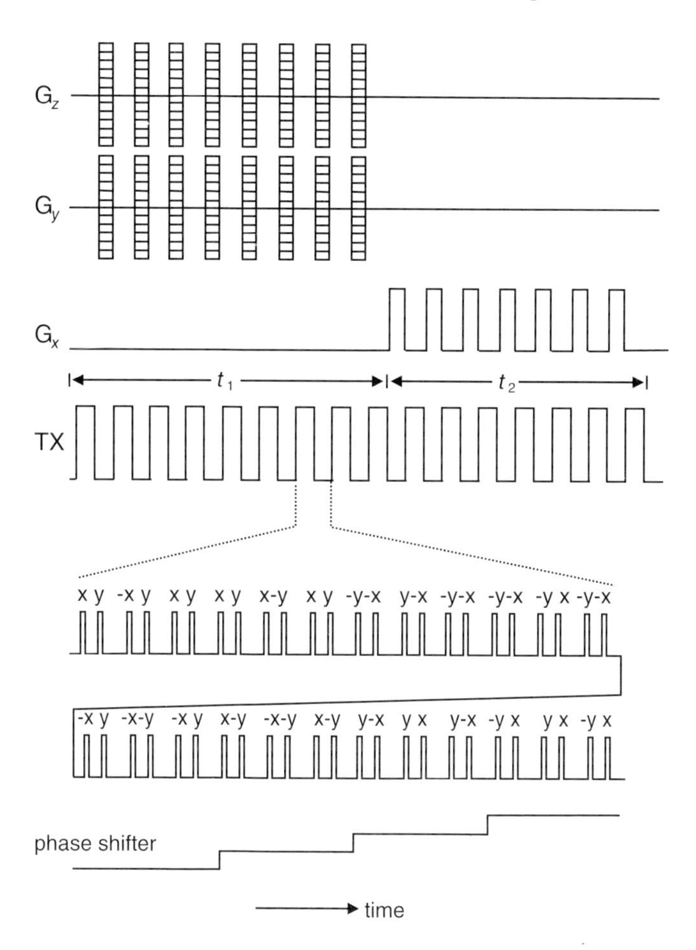

FIG. 8.7.2 The 48-pulse sequence used for imaging with phase and frequency encoding. For second averaging the phases of the rf pulses are shifted in small increments (phase shifter). The signal is observed stroboscopically during t_2. Adapted from [Cor13] with permission from Wiley-VCH.

spectral width is doubled, so that stronger field gradients can be employed for better spatial resolution.

Line narrowing with multiple-pulse techniques can also be implemented for imaging with pulsed $\boldsymbol{B}_1$ gradients [Cho1]. This approach provides fast switching times, but gradients are low and additional rf coils have to be used.

Multi-pulse sequences based on magic echoes

Multi-pulse sequences for line narrowing based on *magic echoes* represent another possibility to realize time-suspension. Such sequences are less difficult to implement than pulse sequences based on solid echoes and consequently, appear to be more popular for use in imaging. Therefore, they are treated separately in Section 8.8.

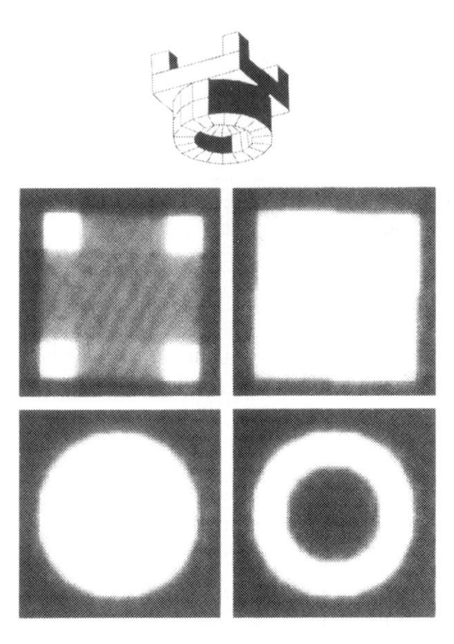

FIG. 8.7.3 Proton images (bottom) of a phantom made from poly(acetal) (Delrin) acquired with the time-suspension pulse sequence of Fig. 8.7.2. The field of view is $(5.9\,\text{mm})^2$. The sample (top) has been imaged in three dimensions at 400 MHz with a voxel size of $92 \times 92 \times 625(\mu\text{m})^3$. Adapted from [Cor12] with permission from Elsevier Science.

CRAMPS

The main purpose of line-narrowing through *CRAMPS* is to improve the spatial resolution by reducing the homonuclear dipole–dipole interaction and the chemical shift anisotropy [Bus2, Sun1]. Spectroscopic images acquired under MAS (cf. Section 8.5) with line narrowing by pulse sequences based on *magic echoes* (TREV8, cf. Section 8.8) have been reported [Bus1]. By using magic echoes the line-narrowing efficiency is sustained for large offset, unlike other multi-pulse sequences based on solid echoes [Bus3]. Phase encoding of spatial information along the rotor axis has been achieved on simple phantoms through application of pulsed-field gradients during an evolution period [Bus3]. In this case, the spatial information is detected indirectly, and the spectroscopic information is detected directly.

8.8 IMAGING WITH MAGIC ECHOES

Magic echoes are generated from transverse magnetization by using the magic sandwich (cf. Section 3.4) [Dem4, Pin2, Rhi1, Sch8]. Unlike a 90° refocusing pulse in solid echo, the magic sandwich is capable of refocusing magnetization, which decays under the dipole–dipole interaction not only between just two spins $\frac{1}{2}$ but also among many spins. An outstanding feature of magic echoes is that single-quantum coherences can be refocused even after delays as long as the transverse relaxation times, so that gradients can be switched on and off during these delays. This significantly improves the space-encoding potential with rigid solids, and *magic-echo imaging* can be considered one of the more successful solid-state imaging techniques. For this reason *imaging with magic echoes* is discussed as a separate topic. Time-suspension multi-pulse sequences based on magic echoes have been used in imaging [Mat1, Mat2, Mat3, Mat6, Mat7, Tak1] just as multi-pulse sequences based on solid echoes (cf. Section 8.7). Such sequences are suitable for frequency encoding of space information. Nevertheless, magic echoes can be used also for phase encoding [Dem2, Haf1, Haf3]. Both aspects of using magic echoes in imaging are treated below.

8.8.1 Multi-pulse sequences from magic echoes

Multi-pulse sequences can be constructed from magic sandwiches in two ways. If magic sandwiches are chained together in a straightforward way, the signal decay from dipole–dipole interactions is refocused, but magnetization dephasing caused by linear spin interactions continues. For example, chemical shift is preserved although scaled down, but homonuclear dipolar decoupling is effective. However, if the magic echo is combined with a Hahn echo, a time-suspension sequence is obtained so that the signal is cleared of spectroscopic information and highly suitable for space encoding [Mat1, Tak1].

The magic echo

In the following, application of a symmetrical magic-sandwich pulse sequence to a nonspinning system of abundant spins with chemical shielding and homonuclear dipolar interactions is considered. The magic sandwich is composed of two 90°_y and 90°_{-y} sandwich pulses which enclose two spin-lock pulses with phases x and $-x$, respectively (cf. Fig. 3.4.3). This magic sandwich of duration 4τ is applied in the centre of the pulse cycle of duration $t_c = 6\tau$. If the phase of the last sandwich pulse is chosen to be $-y$, a magic echo of dipolar or quadrupolar type is generated [Kim1], whereas for the phase $+y$, a *mixed echo* is produced, which refocuses coherence loss from dipolar as well as linear spin interactions like resonance offset, susceptibility variation, and chemical shift [Mat1]. The phase alternation of the spin-lock pulses is not important for the spin evolution under homogeneous spin interactions. It merely serves the formation of a rotary echo [Sli1] to reduce the influence of dephasing as a result of rf field inhomogeneities.

During the spin-lock period of duration 4τ the secular dipolar Hamiltonian $\mathbf{H}_d^{(0)}$ is scaled by the factor $-1/2$ (cf. Section 3.4) so that the Hamiltonian averaged over the echo time $t_E = 6\tau$ vanishes [Rhi1],

$$\overline{\mathbf{H}}_D^{(0)} = \left[\mathbf{H}_D^{(0)}\tau - \mathbf{H}_D^{(0)}4\tau/2 + \mathbf{H}_D^{(0)}\tau\right] \Big/ (6\tau) = 0. \tag{8.8.1}$$

The factor $-1/2$ is obtained from the second Legendre polynomial $P_2 = \frac{1}{2}\left(3\cos^2\theta - 1\right)$ for the angle $\theta = 90^\circ$, which the quantization axis of the spins forms with the $\boldsymbol{B}_0$ field during spin locking. As a result of self-compensation of the dipolar interaction in the different evolution intervals, the initially excited coherences are completely refocused under the magic echo. Another echo arises after half the echo time during the spin-lock

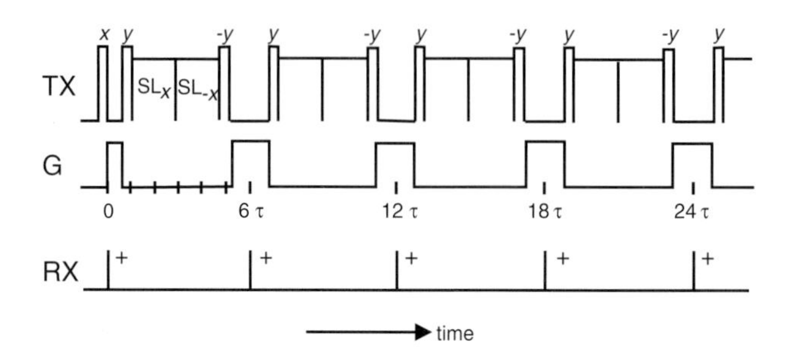

FIG. 8.8.1 [Haf3] Basic scheme for magic-echo frequency encoding in one space dimension. All narrow rf pulses are 90° pulses.

period. This is called the *rotary echo*, but it is usually not observed because of difficulties in switching transmitter and receiver rapidly enough.

Frequency encoding

Frequency-encoding *magic-echo imaging* methods [Mat2, Mat3, Mat4, Mat5, Mat6, Mat7, Wei1, Wei2] use homonuclear line-narrowing techniques based on the stroboscopic observation of the multiple magic echoes [Mat1, Tak1]. Space encoding is performed by gradient pulses of constant strength applied in the windows of the magic-echo multi-pulse sequence.

In order to demonstrate the salient features of the technique, the pulse sequence of Fig. 8.8.1 is considered. If $\overline{\mathbf{H}}^{(0)}$ denotes the average Hamiltonian of order zero, the density operator describing the FID, which is detected stroboscopically at times $t = nt_c$, is given by

$$\boldsymbol{\rho}(nt_c) = \exp\left\{-\mathrm{i}\overline{\mathbf{H}}^{(0)} nt_c\right\} \boldsymbol{\rho}(0) \exp\left\{+\mathrm{i}\overline{\mathbf{H}}^{(0)} nt_c\right\}, \tag{8.8.2}$$

where n is an integer and $t_c = 6\tau$ is the cycle time. Because the dipolar couplings are eliminated in the echo maxima, the average Hamiltonian $\overline{\mathbf{H}}^{(0)}$ is given by

$$\overline{\mathbf{H}}^{(0)} = \left[-\gamma G_z z \mathbf{I}_z + \mathbf{H}_\sigma\right]/3. \tag{8.8.3}$$

Here $\mathbf{H}_\sigma$ is the Hamiltonian of the chemical-shielding interaction, and the field gradient is applied in the z-direction. From the average Hamiltonian it is clear, that the space-encoded NMR signal also depends on the chemical-shielding interaction which affects the image resolution. All linear spin interactions including the effect of the field gradient G_z are scaled by a factor of 1/3. This reduction in gradient efficiency has to be compensated for by higher gradient amplitude or longer encoding time.

The *magic-echo frequency-encoding method* of Fig. 8.8.1 has been demonstrated on a phantom of adamantane and hexamethylbenzene [Mat1]. The spatial resolution achieved was better than $(100\,\mu\mathrm{m})^{-1}$. For both materials the ^{1}H linewidth at half-height is about 15 kHz without homonuclear decoupling.

Time-suspension multi-pulse sequences

The original magic-echo sandwich (cf. Fig. 3.4.3) only refocuses the evolution from the homonuclear dipole–dipole interaction. A time-suspension sequence is obtained by adding a 180° refocusing pulse to the last 90° pulse of the magic sandwich. Then, the phases of both 90° pulses become equal and the evolution from linear spin interactions is refocused as well [Mat1]. The resultant time reversal sequence is also referred to as TREV4, because the refocusing sandwich consists of four pulses [Tak1]. Two of these refocusing sandwiches form the TREV8 sequence. Such sequences have been applied for imaging of abundant spins in solids [Mat1, Mat2, Mat3, Tak1, Wei1].

In order to preserve the effect of the gradients on the evolution of magnetization in a time-suspension sequence, the gradient pulses must be applied with alternating polarity in successive free-evolution windows of the magic-echo sequence. In this case, the evolution of the density operator can be expressed similar to (8.8.2),

$$\boldsymbol{\rho}(n\,t_c) = \exp\{\mathrm{i}\gamma G_z z\,\mathbf{I}_z n\,t_c/3\}\boldsymbol{\rho}(0)\exp\{-\mathrm{i}\gamma G_z z\,\mathbf{I}_z n\,t_c/3\}, \tag{8.8.4}$$

and the normalized components of transverse magnetization can be evaluated as

$$s_x(n\,t_c, z) = \mathrm{Sp}\{\mathbf{I}_x\boldsymbol{\rho}(nt_c)\}\,/\mathrm{Sp}\{\mathbf{I}_z\boldsymbol{\rho}(0)\} = \sin\{\gamma G_z z\,n\,t_c/3\}\,, \tag{8.8.5a}$$

$$s_y(n\,t_c, z) = \mathrm{Sp}\big\{\mathbf{I}_y\boldsymbol{\rho}(nt_c)\big\}\,/\mathrm{Sp}\{\mathbf{I}_z\boldsymbol{\rho}(0)\} = \cos\{\gamma G_z z\,n\,t_c/3\}\,, \tag{8.8.5b}$$

where $\boldsymbol{\rho}(0)$ is the initial density operator. It is evident, that the acquired signals oscillate with a frequency determined by the gradients strength G_z and the space coordinate z. The maximum time available for the gradient to be effective is 2τ, which is the total duration of the free-evolution periods. Averaged over the full cycle time of 6τ, the effective gradient strength is $G_z/3$.

Example of a magic-echo time-suspension multi-pulse sequence

A rather favourable magic-echo multi-pulse technique [Wei2] consists of a time-suspension version (Fig. 8.8.2(a)) [Mat4, Tak1] of the original frequency-encoding sequence (Fig. 8.8.1) [Mat1]. Because of the combination with a Hahn echo, the gradient sign must be inverted in every second window. Therefore, the data should be acquired in every second free-evolution window, resulting in a reduced spectral width [Mat3]. This problem can be overcome through oversampling [Wei1]. In this case data are acquired in every window of the sequence, but the receiver phase must be switched as well (RX in Fig. 8.8.2(a)). To suppress a zero-frequency artefact [Mat1], the sequence incorporates phase cycling of magic sandwich pairs, which shifts the artefact to the Nyquist frequency, where it can readily be cut off [Wei1]. The sequence is capable of handling homonuclear dipole–dipole couplings of up to 50 kHz, and a typical value for the achievable spatial resolution is $\Delta x^{-1} = (0.9\,\mathrm{mm})^{-1}$. For example, with $\tau = 6.6\,\mu\mathrm{s}$, the linewidth of ultra-high molecular-weight poly(ethylene) is reduced from 31 kHz to 258 Hz [Wei1].

Spectroscopic imaging with homonuclear dipolar decoupling

For *spectroscopic imaging with magic echoes* the space encoding sequence can be appended by an MREV8 sequence for frequency encoding of the spectrum in the absence

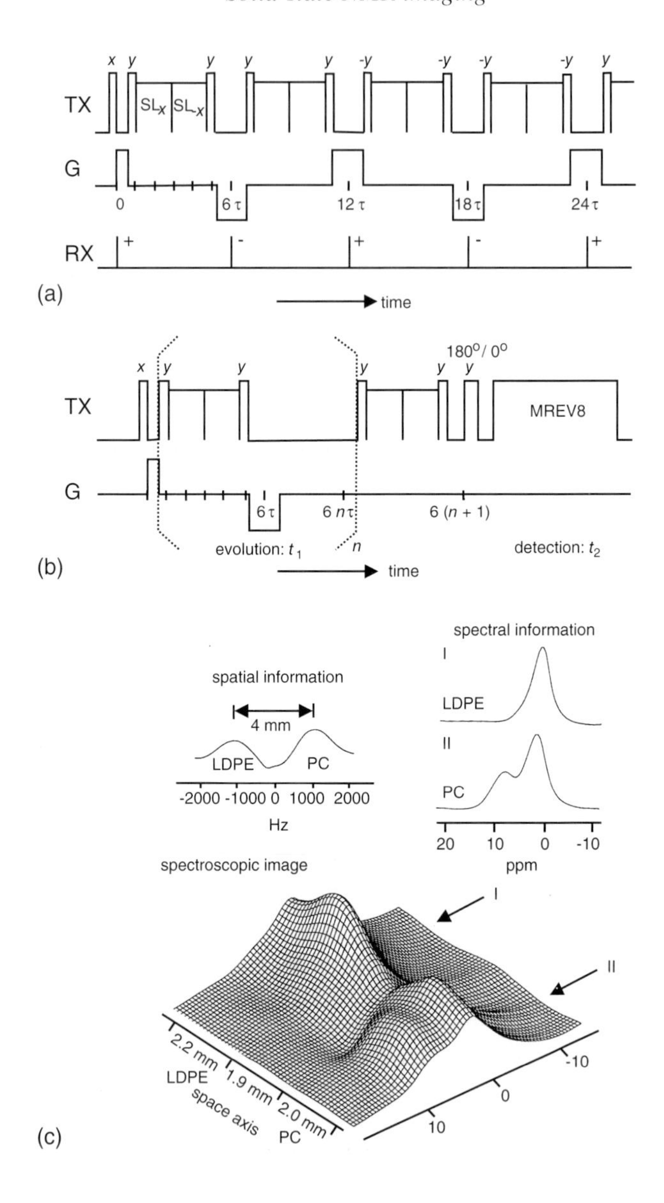

FIG. 8.8.2 (a) Magic-echo time-suspension pulse sequence for 1D imaging with data sampling every t_c, where $t_c = 6\tau$ is the cycle time. The field gradient G is applied only in the windows of the pulse sequence. It is inverted in successive windows to account for a change in the sign of the magic echo after each magic sandwich. All narrow rf pulses are 90° pulses. (b) Magic-echo pulse sequence for spectroscopic 1D imaging. Magic echoes are used for space encoding during the evolution time, and MREV8 multi-pulse irradiation is used for line narrowing during detection of the spectroscopic response. (c) Spectroscopic image of a phantom obtained by scheme (b). The phantom consisted of a 2.2 mm thick sheet of LDPE, a 1.9 mm thick sheet of PTFE, and a 2.0 mm thick sheet of PC. The left inset shows the spin-density profile extracted from the 2D data set. The right inset depicts independently phased cross-sections through the 2D plot taken at the maximum values the LDPE and PC signals. Adapted from [Wei1] with permission from Elsevier Science.

of a gradient (Fig. 8.8.2(b)). By incrementing the evolution time t_1 the space information is phase encoded in the spectroscopic signal detected during t_2. Two data sets need to be acquired for phase sensitive detection of the space information, one with positive and one with negative phase evolution during t_1. This sign change is achieved by incorporating a 0° or a 180° pulse before the detection period t_2.

An application of the sequence of Fig. 8.8.2(b) to a phantom from layers of poly(carbonate) (PC, 2 mm thick), poly(tetrafluoroethylene) (PTFE or Teflon, 1.9 mm thick), and low-density poly(ethylene) (LDPE, 2.2 mm thick) is demonstrated in Fig. 8.8.2(c)) [Wei1]. The gradient strength was 40 mT/m and $\tau = 16\,\mu$s. The two components are clearly separated with a spatial resolution of approximately $(0.1\text{ mm})^{-1}$. The ^{1}H spectra exhibit lines at the correct chemical shift values known from liquid-state measurements (LDPE: 1.25 ppm; PC: 1.7 and 7.2 ppm, corresponding to the phenyl and methyl protons in the sample). Spectroscopic images of abundant nuclei in solids have also been obtained in an experiment, which employs magic-echo phase encoding of space and homonuclear decoupling through formation of a train of magic echoes for spectroscopic detection in the absence of gradients (cf. Fig. 7.3.2(a)) [Hep2].

8.8.2 Magic-echo phase encoding

Pure phase encoding of the space information is a powerful approach to imaging with high spatial resolution (cf. Section 8.3). The effect of bilinear spin interactions can be reduced with the help of suitable echo techniques, which maintain the information content of the linear spin interactions, because the latter are often of primary interest in the solid-state spectrum. For example, imaging techniques using solid echoes [Dem1, Mcd2], Jeener–Broekaert echoes [Dem1, Rom1], and magic echoes [Bar1, Dem2, Haf1, Haf3, Kim1] have been demonstrated. For imaging of abundant spins *magic-echo phase-encoding solid-state imaging* (MEPSI) is a powerful approach, because long echo times can be employed, but hardware for generation of suitably short gradient pulses must be available.

Magic-echo phase-encoding solid-state imaging

For phase encoding of the space information in magic-echo imaging the amplitude of the gradient pulses applied during the echo time is varied in a systematic fashion to cover $\mathbf{k}$ space. The scheme is illustrated in Fig. 8.8.3 [Bar1, Haf3] together with slice selection by the SLISE technique (cf. Section 5.3.5) and with spectroscopic detection. The basic principle of *imaging by magic-echo phase encoding* is the same as for spin-echo imaging (cf. Fig. 6.2.3) and imaging with pure phase encoding (cf. Section 8.3): because the echo time and the gradient-pulse time are kept constant, the only evolution of the spin system is due to the variable gradient amplitude apart from a signal loss by transverse relaxation.

To grasp the principle of the method, chemical shielding is neglected and resonant rf irradiation is considered. Then, essentially the same arguments leading to (8.8.5) describing the signal under frequency encoding also apply to phase encoding. The difference is that for frequency encoding the variable is the number n of magic sandwiches, and for phase encoding the variable is the gradient amplitude G_z while $n = 1$.

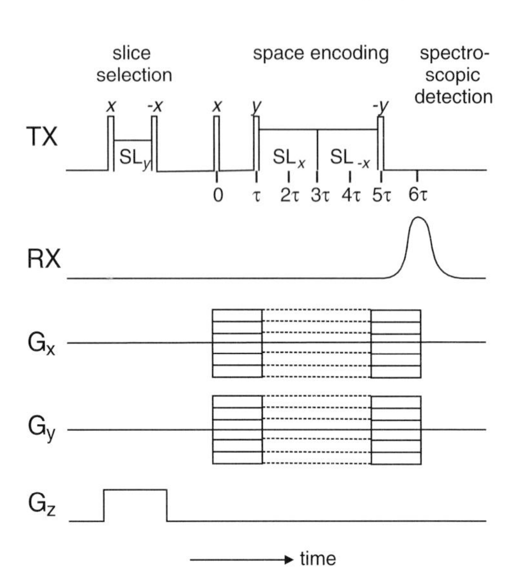

FIG. 8.8.3 [Haf3] Scheme for magic-echo phase encoding with slice selection by the SLISE technique and spectroscopic detection. The strength of the field gradient is varied in constant increments. The gradients may be left on during the magic sandwich. The amplitude of the magic-echo is acquired as a function of the gradient amplitudes. The durations of all time intervals are kept constant. All narrow rf pulses are 90° pulses.

If the gradients cannot be pulsed sufficiently rapid, they can be left on during the magic sandwich at the expense of minor image artefacts (broken lines in Fig. 8.8.3). The gradient build-up and ring-down times can be accomodated in intervals of the pulse sequence where z magnetization prevails. This is the case for gradient build-up between slice selection and the 90°_x pulse at the beginning of the magic-echo sequence. A similar z storage interval can be incorporated into the pulse sequence for gradient turn-off in order to acquire the spectroscopic response free of distortions from gradient switching [Haf3].

The *MEPSI method* can be modified for space encoding by rf gradients [Dem2]. For the same gradient values the spatial resolution is expected to be better for the rotating-frame technique than for the original MEPSI technique in the laboratory frame, because the usable space encoding time is 4τ instead of 2τ. Moreover, $\boldsymbol{B}_1$ gradients can be switched considerably faster with negligible rise and fall times.

Application of MEPSI to 3D imaging

A 3D MEPSI image can be recorded using a scheme slightly modified from that shown in Fig. 8.8.3. The slice-selection sequence is replaced by a third phase-encoding gradient [Haf4, Haf5]. Figure 8.8.4(b) shows the surface-rendered 3D image of a cylindrical phantom sample of two hexamethylbenzene (HMB) parts separated by a teflon spacer (Fig. 8.8.4(a)). The spatial resolution is about $(300\,\mu\text{m})^{-3}$, so that the sample shape could be reproduced quite well.

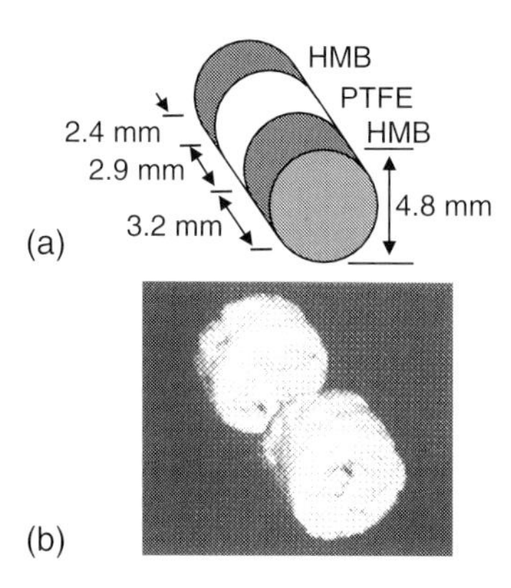

FIG. 8.8.4 3D imaging by magic-echo phase encoding. (a) Schematic representation of a test sample consisting of two compartments filled with hexamethylbenzene powder separated by a PTFE spacer which is free of protons. (b) Surface-rendered 3D ^{1}H image of the object. The voxel resolution is about (300 μm)$^{-3}$. Adapted from [Haf3] with permission from Elsevier Science.

Because the sequence solely relies on phase-encoding, the measuring times tend to be relatively long. Therefore, methods were proposed to accelerate image rendering [Haf1, Haf5]. On the other hand, the spectroscopic information of the magic-echo decay can be acquired when the space information is encoded in the signal phase (cf. Section 8.3). Then additional information for generation of contrast becomes accessible. Because magic echoes can be generated in the complete absence of any molecular motional averaging, imaging with magic echoes appears suitable for application to glassy and crystalline solids. In the other extreme, magic-echo techniques have the potential to provide spatial information, irrespective of the rigidity of the material, as long as a residual dipolar interaction is present, so that also soft matter like elastomers is suited to magic-echo imaging [Mat3].

9

Spatially resolved NMR

Spatially resolved NMR denotes the measurement of spectra, relaxation times, and other NMR parameters from within a small region of a large object [Rud1, Rud2, Rud3, Cer1]. It can be considered the limiting case of imaging, where NMR information is acquired for each volume element. In spatially resolved NMR just one or in some cases a few volume elements are considered (Fig. 9.0.1). The methodical challenge is posed by the selection of the signal from the volume element in question [Bry1]. This can be done in two major ways. One approach exploits selective excitation in combination with gradients for preparation of the initial magnetization in the volume element in question, which is then subject to further investigation by NMR. Because the use of gradients is inherent to this approach, localization techniques based on this principle are referred to as *gradient methods* (Fig. 9.0.1(a),(b)) [Aue1]. Alternatively, the volume upon which the excitation acts and from which the NMR signal is received can be reduced by shaping the transmitter and receiver coils, respectively. Consequently, the second approach encompasses *surface-coil techniques* (Fig. 9.0.1(c)) [Ack1, Ben1].

The sensitive volume, which can be investigated by a circular surface coil in a homogeneous polarizing field, is approximately given by a half sphere with the coil radius. As the coil radius is typically small compared to the largest dimension of the object, the use of surface-coil techniques is restricted to volume elements near the sample surface.

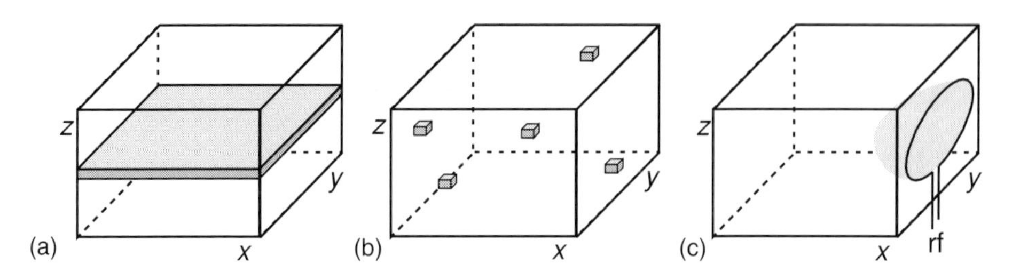

FIG. 9.0.1 Different approaches to spatial resolution in NMR. (a) NMR imaging. Contiguous volume elements on a plane are interrogated. (b) Spatially resolved NMR by gradient methods. Individual volume elements are interrogated with volume selection by pulsed-field gradients. Data are acquired for a single volume element or a set of elements. (c) Volume selection by local rf excitation and detection with surface coils.

Both approaches, gradient- and surface-coil techniques, have been developed in the biomedical field, where spectroscopic investigations of the brain are of particular interest [Luy1]. Indeed, most applications of spatially resolved NMR appear to be in *in vivo* magnetic resonance spectroscopy [Hau1, Rud1, Rud2, Rud3, See1]. However, in materials science spatially resolved NMR is of interest as well, in particular to process and quality control, where spatially selective determination of material properties is needed to quantify the quality of a product [Gib1, Gla1, Gla2]. If this is done on the production line, corrective action can be taken in case of deviation from a standard range of values, so that spatially resolved NMR can be used to interface with a process by interactive feedback control. Therefore, NMR can be used for both *quality control* and *process control*. In this respect, *single-sided NMR*, where the polarizing magnetic field is provided by the stray field of a magnet and the rf field is generated by a surface coil, bears considerable potential [Eid1, Mat2].

9.1 GRADIENT METHODS

Gradient methods utilize magnetic-field gradients and selective excitation for localization of the volume element of interest. Three orthogonal slices can be selected by successive application of slice-selection pulses in mutually orthogonal gradients. Their common intersection defines the volume element (Fig. 9.1.1) [Ako1]. In this case, the excitation is split into three time periods, each of which is used for selection of a slice. This provides considerable flexibility in the design of localization methods, but also of imaging methods, which has been demonstrated in Section 6.2.5 by the STEAM methods [Fra1]. The various techniques for *volume selection* differ in the manner in which the selective pulses are applied and combined with the magnetic-field gradients. They are of particular interest for *volume-selective spectroscopy* [Aue1], including polarization transfer (cf. Sections 7.2.11 and 7.2.12) [Aue3] and solvent suppression [Luy2], and for determination of other NMR parameters such as relaxation times, diffusion constants, and velocities [Kla1, Rom1, Yan1]. In principle, they can be used for point-scan imaging

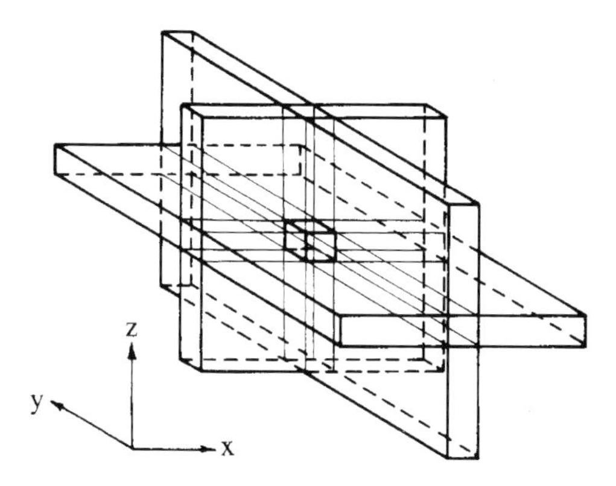

FIG. 9.1.1 [Ako1] Gradient methods define a volume element as the intersection of three slices.

(Section 5.1) as well, and the excitation can be optimized for arbitrary shapes of the sensitive volume [Hod1].

Examples for shaped pulses which select longitudinal magnetization [Sin1] are the DIGGER pulse and SPREAD pulses (Section 5.3.2) [Dod1, Nil1]. Techniques based on such pulses benefit from long spin–lattice relaxation times and short effective spin–spin relaxation times, and they can be used for slice selection in solid-state imaging. However, the elimination of magnetization by dephasing or saturation in large frequency windows is still a challenge in practice. Therefore, approaches which tolerate the creation of transverse magnetization for short time periods are often preferred for samples with sufficiently long spin–spin relaxation times, despite a phase error which may be introduced in this way.

The volume-selection techniques can be classified into single- and multi-shot methods. By single-shot techniques (Section 9.1.1) the magnetization of the volume element in question is selected in a single shot. This gives access to fast shimming on localized regions. Multi-shot techniques (Section 9.1.2) require addition and subtraction of signals acquired in several scans to isolate the signal in question from the selected volume. Such methods are demanding on the dynamic range of the AD converter, but provide other advantages like tolerance of the slice shape against flip-angle missets and high signal sensitivity.

9.1.1 Single-shot techniques

Single-shot methods permit fast acquisition of volume-localized NMR data. They are preferred for localized shimming. The techniques can be discriminated according to those which store the selected magnetization along the z-direction for gradient switching, and those where gradients are switched in the presence of transverse magnetization. Realizations of the former are the VSE, SPACE, SPARS, and LOSY methods. Realizations of the latter are the PRESS, STEAM, VOSY, and VOISINER techniques, as well as forms of localized 2D spectroscopy.

Volume-selective excitation (VSE)

An elegant approach to *volume-selective excitation* (*VSE*) has been designed by Aue (Fig. 9.1.2) [Aue2]. The excitation sequence consists of three *composite pulses*. Each composite pulse is applied in a different field gradient orthogonal to the others and reduces the space dimension of the selected magnetization by one. The composite pulse itself consists of two selective 45° pulses with a hard 90° pulse in the middle. The selected magnetization is aligned parallel to the z-direction after each step, so that gradients can be switched in the time between the pulses. The z magnetization of a slice is selected after application of one such composite pulse. The magnetization of the other volume elements remains in the transverse plane, where it dephases in the presence of the field gradients. Similarly, a line is selected by two and a volume element by three such composite pulses. The remaining longitudinal magnetization can, subsequently, be investigated by any spectroscopic technique. For example, the technique can be applied for definition of the sensitive volume in surface-coil spectroscopy [Mül5].

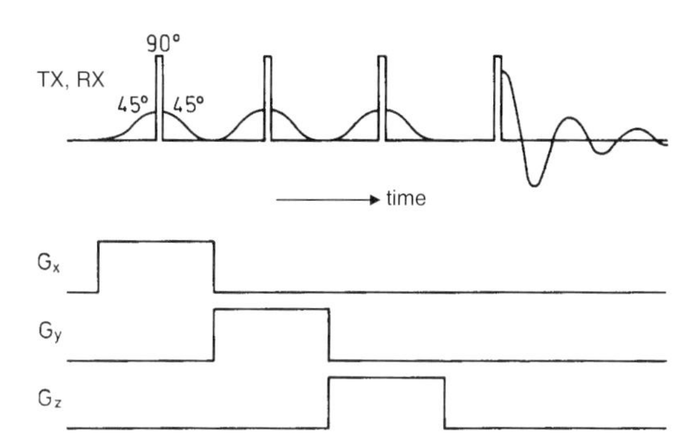

FIG. 9.1.2 [Aue2] Radio-frequency field gradients and rf pulses for volume-selective excitation. Each composite 45°–90°–45° pulse is applied in a different field gradient, so that the space dimension of the magnetization to be investigated is reduced by one in each step.

Phase cycling of the excitation pulses helps to suppress artefacts [Mül1]. Polarization transfer techniques can be linked to the VSE sequence for localized spectroscopy of heteronuclei, like the DEPT sequence for ^{13}C observation in liquids [Aue3, Wat2]. The problems with this particular approach are (1) the lack of refocusing for the magnetization which dephases during the composite pulses, and (2) the use of a nonselective pulse, which may require excessive rf power for localization of narrow regions within large objects.

SPACE and spatially resolved spectroscopy (SPARS)

The dephasing of transverse magnetization during the composite pulses of the VSE sequence (Fig. 9.1.2) can be eliminated by formation of a Hahn echo, so that the selective and the nonselective 90° pulses are applied in the dephasing and in the rephasing periods, respectively. The Hahn echo also refocuses magnetization dephasing from chemical-shift dispersion. Depending on whether the selective 90° pulse is applied first or last, the resulting excitation has been given the name SPACE (not intended to be an acronym, Fig. 9.1.3) [Dod2] and *SPARS* (*spatially resolved spectroscopy*) [Luy2], respectively. But the incorporation of a nonselective 180° pulse aggravates the rf power problem and introduces significant off-resonance effects. Better phase and off-resonance properties are obtained by use of self-refocusing 270° Gaussian pulses in combination with hard 90° pulses [Ems1].

Lock-pulse selective spectroscopy (LOSY)

A *slice-selection* technique which is, particularly, simple to implement exploits locking of spins in a magnetic-field gradient [Rom1, Win1]. The same idea is used for slice selection and localized spectroscopy in solids (cf. Section 5.3.5) [Haf2]. Transverse magnetization is spin locked in a magnetic-field gradient. Depending on the strengths of the lock field and the gradient, the magnetization is locked only within a given frequency window centred at the rf carrier frequency. It relaxes with $T_{1\rho}$, and this time is required to be longer than the effective transverse relaxation time T_2^* in order for the method to

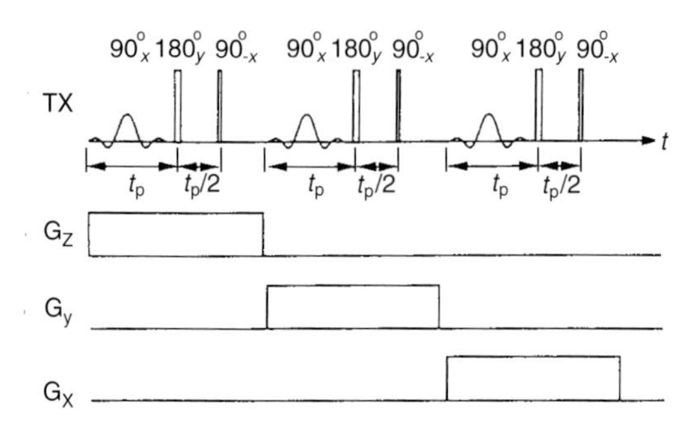

FIG. 9.1.3 [Cal2] Pulse and gradient scheme for selective preservation of z magnetization in a narrow frequency window by the SPACE technique.

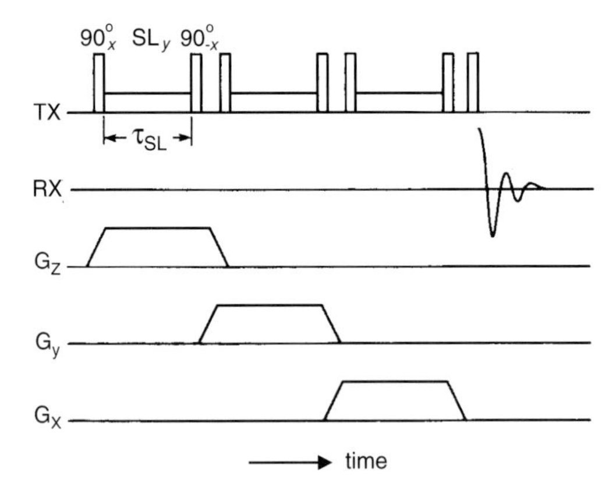

FIG. 9.1.4 [Rom2] Pulse and gradient scheme for lock-pulse selective spectroscopy. The slice selection is based on selective spin locking.

work. But T_2^* can be shortened by increasing the gradient strength. For magnetization components with large offset frequencies, the spin-lock condition is not fulfilled, so that these components dephase.

The *spin-lock induced slice excitation* (*SLISE*) consists of a hard 90° pulse followed by a spin-lock period of length τ_{SL}. By augmenting this sequence by another 90° pulse, longitudinal magnetization is obtained for the slice (cf. Fig. 5.3.14). The slice thickness can be varied by changing the lock-field amplitude B_{1SL} without changing the shape of a pulse (cf. Section 5.3.5). It is independent of the length of the pulse, which can be adjusted to yield $T_{1\rho}$-weighted contrast. Volume-selective excitation for localized spectroscopy is obtained when three such spin-lock sandwiches are applied in orthogonal magnetic-field gradients (Fig. 9.1.4). This type of spatially resolved spectroscopy is called *lock-pulse selective spectroscopy* (*LOSY*) [Rom2].

Point-resolved spectroscopy (PRESS)

In the *point-resolved spectroscopy* (*PRESS*) method [Ord1], volume selection is achieved by double refocusing of the transverse magnetization to form Hahn echoes (Fig. 9.1.5) [Ako1]. Each pulse is selective to the magnetization in a slice orthogonal to the others. The first pulse is applied in the presence of a z gradient. It selects the magnetization from a slice parallel to the xy plane. The transverse magnetization of a line from this plane is refocused by the first 180° pulse applied in the presence of a y gradient to form an echo after the echo time t_E has elapsed. The second 180° pulse is applied in the presence of an x gradient, refocusing the magnetization of the first echo in a second echo after time $2t_E$, but only from the volume element at the intersection of the three planes (cf. Fig. 9.1.1). The magnetization from outside the plane selected by the first pulse of the sequence remains as longitudinal magnetization. The magnetization from this plane but outside the selected volume is transverse magnetization, which dephases rapidly, because it is not refocused. The method is suitable only for samples with long transverse relaxation times T_2, because the selected magnetization exists as transverse magnetization for the duration of $2t_E$. For short relaxation times, short pulse spacings must be chosen, which may result in spurious signals induced by insufficient pulse turn-off and gradient stabilization after switching. These can be eliminated to some degree by phase cycling [Hen1]. The technique has been applied, for instance, to spectroscopic studies of the human brain [Hen2, Moo1], of muscles [Kre1], and of tumours [Sta1]. Based on the J modulation of the Hahn echoes, the method supports spectroscopic editing in high-resolution NMR [Sta1].

Stimulated echo acquisition mode (STEAM)

The PRESS method provides the capability of refocusing the complete magnetization of the selected volume element. Yet 180° pulses are needed, which are more difficult

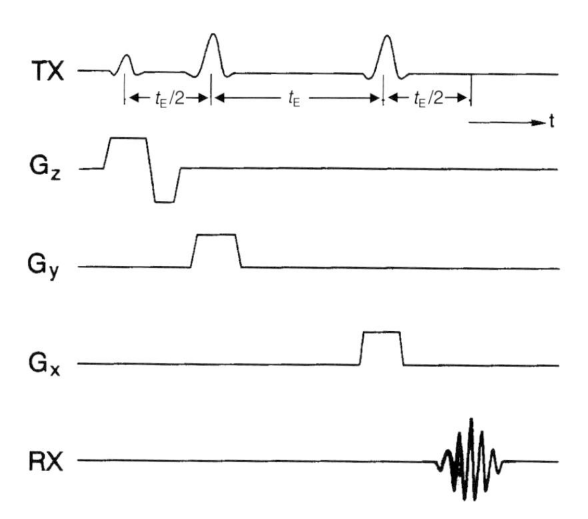

FIG. 9.1.5 [Ako1] Timing of rf and gradient signals for the PRESS method. Transverse magnetization of a volume element is selected by formation of a double Hahn echo with selective pulses in the presence of orthogonal field gradients.

to achieve in practice, and the method is unsuitable for samples with short T_2. These disadvantages are alleviated if the stimulated echo is used instead of two double Hahn echoes (Fig. 9.1.6), leading to volume selection in the stimulated echo acquisition mode (*STEAM*) principle (cf. Section 6.2.5) [Fra1, Fra2]. This method is also referred to as *VEST* (selected-volume excitation using stimulated echoes) [Gra1]. The same timing of rf and gradient signals is followed as for the PRESS method, except that all pulses are now selective 90° pulses, and that the time interval between the second and the third pulse can be extended beyond the echo time t_E, as long as it is not much longer than the spin–lattice relaxation time T_1. During this time, half of the magnetization localized in the line defined by the second pulse exists as long-lived longitudinal magnetization, of which the part localized in the selected volume element is converted to transverse magnetization by the third pulse to form the stimulated echo.

The STEAM method can be combined with image guidance for localized spectroscopy and imaging [Bru1, Fra2, Fra3], with signal suppression techniques [Fra3, Fra4, Zij1], and with diverse magnetization filters including a spin-lock period following the first pulse for volume-selective determination of the spin–lattice relaxation time $T_{1\rho}$ in the rotating frame [Rom1]. Regions as small as 1 ml can be localized in liquids with the technique [Fra4, Zij1]. For good spectral resolution, shimming of the magnetic field in the selected volume is essential [Zij2]. The method has been shown to be more sensitive to influences from motion [Gyn1], diffusion, and multi-quantum excitation than the PRESS method [Moo1]. It refocuses only half of the total magnetization of the selected volume element. The other half constitutes the antiecho, which can be explored for localized spectroscopy as well [Zhu1]. However, the method is more efficient in terms of rf power, because it uses 90° pulses only, and signal decay from transverse relaxation arises only for a period of length t_E instead of $2t_E$ for the PRESS method.

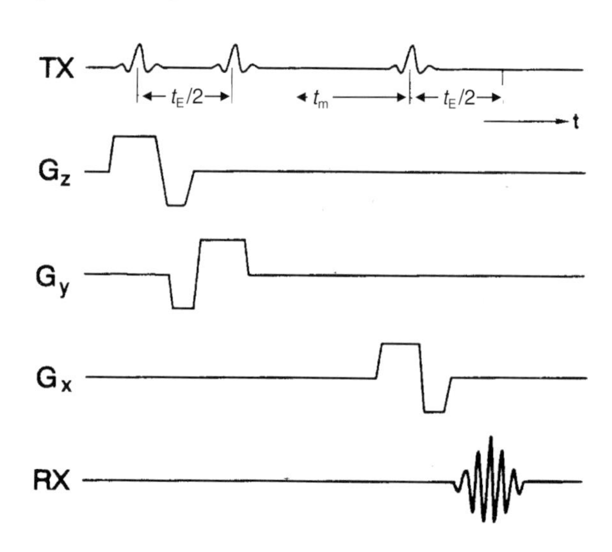

FIG. 9.1.6 [Fra1] Timing of rf and gradient signals for the STEAM method. Transverse magnetization of a volume element is selected by formation of a stimulated echo with selective pulses in the presence of orthogonal field gradients.

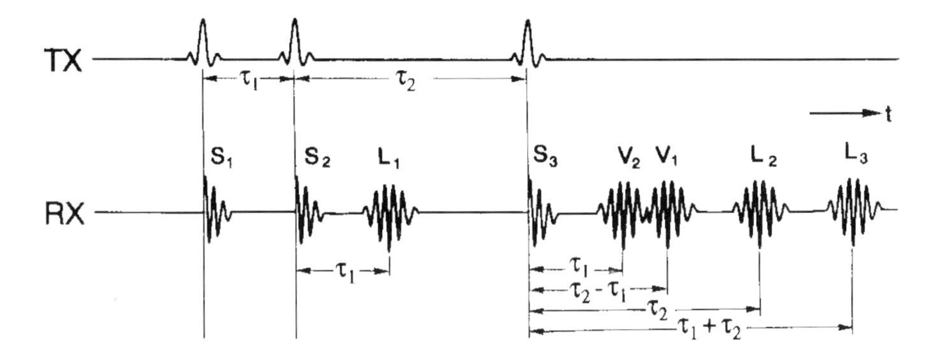

FIG. 9.1.7 [Kim1] Magnetization response to three rf pulses which are applied in the presence of orthogonal magnetic-field gradients. The gradient signals are not shown. S_i: slice-selective signals. L_i: line-selective signals. V_i: volume-selective signals.

Volume-selective spectroscopy (VOSY)

Volume-selective spectroscopy (*VOSY*) denotes a generalized view of the *PRESS* and *STEAM* methods [Kim1]. Given the imperfections in $\boldsymbol{B}_1$ homogeneity, accurate 90° and 180° pulses are hard to achieve in large inhomogeneous objects. Therefore, it is helpful to consider the general response to three selective pulses of arbitrary flip angle [Hah1] as illustrated in Fig. 9.1.7 [Kim1]. After each pulse a free-induction signal S_i is observed, which arises from the magnetization of a slice. The Hahn echoes L_i in response to two pulses are obtained from magnetization corresponding to lines through the sample. Two signals V_i are observed which arise from the response to all three pulses, and thus from magnetization of the selected volume. The echo V_1 is the refocused Hahn echo L_1, and the echo V_2 is the stimulated echo. Exploitation of these signals for spatially resolved spectroscopy is equivalent to using the PRESS and the STEAM methods, respectively, where the flip angles have been optimized for maximum signal strength. With the help of appropriate gradients, the desired signals can be enhanced and the undesired signals can be eliminated [Kim1]. This is done with so-called *crusher gradients* or *spoiler gradients*.

The different ways of using the VOSY principle can be combined with various spectroscopic experiments including suppression, filtering, and transfer of magnetization. This philosophy has been followed systematically primarily for liquid-like samples and biomedical applications [Kim5]. Examples are solvent suppression by saturation [Kim2] and magnetization transfer [Knü1], sequential [Knü2] and simultaneous [Knü3] spectral editing [Kim3], indirect [Knü4] and direct [Knü5] detection of X nuclei with editing [Knü6], multi-quantum filtering [Knü7], and relaxometry [Kim4]. This manifold of techniques illustrates the universality and importance of the three-pulse approach for selection of magnetization from a specific volume element in the sample.

The VOISINER technique

The volume of interest by selective inversion, excitation and refocusing (*VOISINER*) technique [Bri1] is a hybrid volume localization method (Fig. 9.1.8). It consists of three selective excitation modules. The first one is an inversion module like VSE, SPACE, or

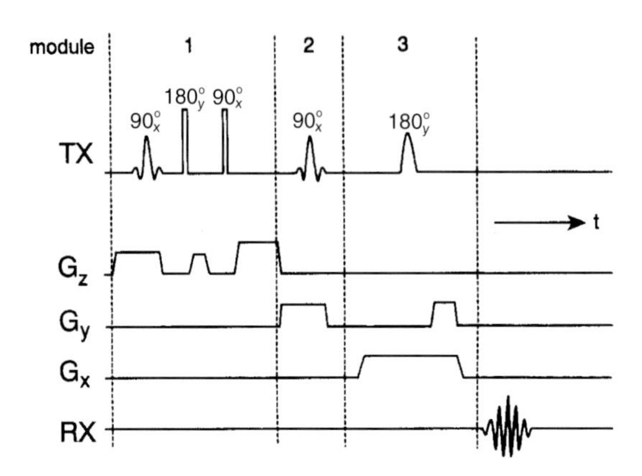

FIG. 9.1.8 [Bri2] Schematic illustration of the VOISINER pulse sequence for volume-selective excitation. It consists of three excitation modules, a SPACE module, a selective 90° pulse, and a selective 180° pulse.

SPARS for selection of longitudinal magnetization in a slice. It is followed by a selective 90° pulse for generation of transverse magnetization in a line, of which the magnetization within one volume element is refocused by a selective 180° pulse. This scheme differs from others in the magnetization states after each selective excitation module. It can be used in a universal fashion [Bri2] similar to the VOSY technique for measurement of local spectra, relaxation times [Bri3], reaction rates, and diffusion constants.

Spatially resolved 2D spectroscopy

Volume selection is particularly useful for spatially *localized 2D spectroscopy*, because acquisition times for 2D spectroscopic imaging would be prohibitively long. Localized 2D spectroscopy is of interest, for instance, to detect information hidden beneath overlapping signals in 1D spectra [Alo1, Ber1, Coh1, Sot1] as well as to follow reaction kinetics [Bal1].

Pulse sequences for 2D correlated spectroscopy inherently consist of more than one pulse [Ern1]. In combination with magnetic-field gradients these can be exploited for spatial localization. Pulse and gradient sequences for 2D spatial localization of *COSY* and *NOESY* responses are depicted in Fig. 9.1.9 [Bla1]. To accommodate gradient switching times, data acquisition starts in the echo maximum after the last pulse. Therefore, the technique is suitable for samples with long T_2 values. The NOESY sequence can, of course, also be used for studies of localized magnetization exchange. For 3D spatial localization the COSY sequence (a) can be extended by a Hahn echo, and in the NOESY sequence (b) the remaining hard 90° pulse can be made selective. These and other spatially localized 2D experiments may prove useful for investigations of liquids in heterogeneous phases, such as multi-layer structures with liquid–liquid, permeable solid–liquid, and permeable solid–solid interfaces. Clearly, the principle of spatially resolved 2D spectroscopy can be generalized to various forms of spatially resolved multi-dimensional spectroscopy [Ryn1].

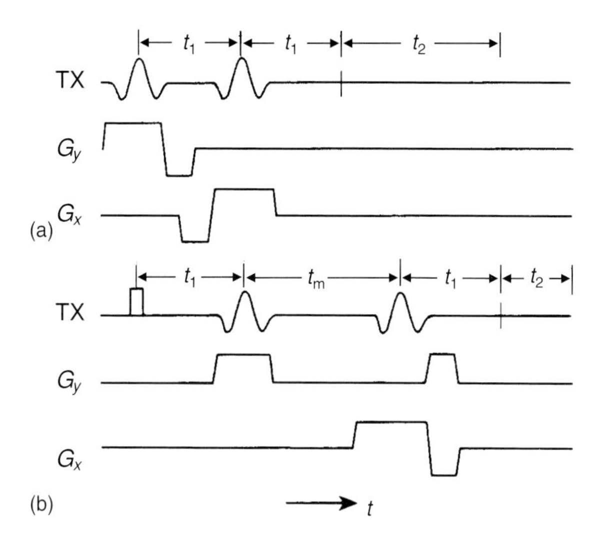

FIG. 9.1.9 [Bla1] Timing diagrams for rf and gradients applicable to spatially resolved 2D spectroscopy. (a) Spatially localized COSY. (b) Spatially localized NOESY.

9.1.2 Multi-shot techniques

Multi-shot localization techniques require acquisition of a set of data, from which the localized information is extracted. Given that signal averaging is employed in most cases, this method offers additional freedom in data manipulation during processing, but a large dynamic range of the AD converter may be required, and shimming on a localized volume element is more difficult than with single-shot techniques. Two methods are reviewed below. The spatial selectivity of the ISIS method is independent from the accuracy of the flip angle, and the HSI method provides optimum signal sensitivity by multi-volume localization.

Image-selected in vivo *spectroscopy (ISIS)*

The most popular multi-shot technique for spatially resolved spectroscopy is the method of *image-selected in vivo spectroscopy* (*ISIS*) [Ord2]. It is based on selective inversion of the magnetization within orthogonal slices using selective 180° pulses applied in orthogonal gradients (Fig. 9.1.10) [Ako1]. The spectrum of a plane is selected by subtracting two data sets. One of them, obtained with just one nonselective 90° read pulse, contains the signal from the entire sample within the coil (Fig. 9.1.10(a)). The other contains a slice-selective 180° pulse before the nonselective read pulse (Fig. 9.1.10(b)). The difference of the signals acquired with both excitations results from magnetization of only the slice selected by the 180° pulse. The advantage of isolating the slice signal this way is that an imperfect 180° pulse does not affect the selectivity of the method. Instead, only the signal intensity is affected, which scales with $(1 - \cos\alpha)$, where α is the actual flip angle of the selective pulse.

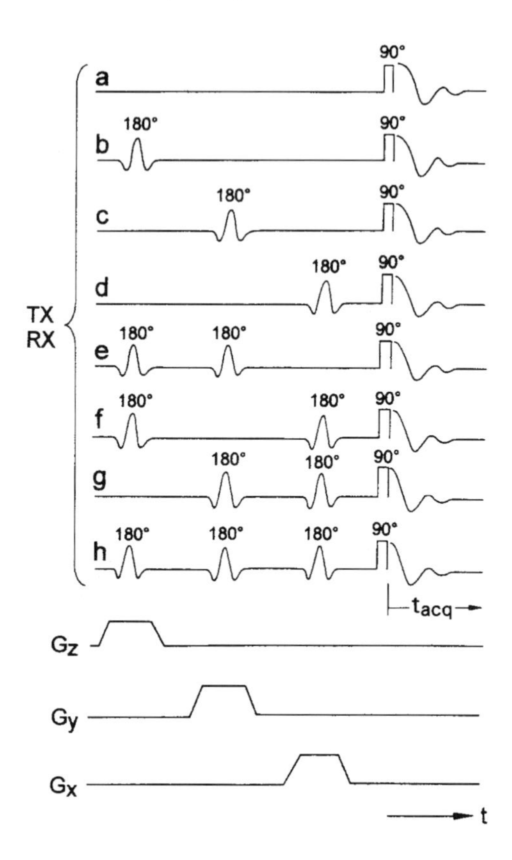

FIG. 9.1.10 [Ako1] Spatially resolved spectroscopy by the ISIS method. The signal from the selected volume element is obtained by the combination of eight signals, $a - b - c - d + e + f + g - h$.

Selection of a cube is achieved in a similar way, except that eight acquisitions must be combined with different gradients and different numbers of the selective 180° pulses. Then the signal strength scales with $(1 - \cos\alpha)^3$, so that the sensitivity strongly depends on the accuracy of the flip angle α. This limitation can be overcome by use of composite inversion pulses. Because selective excitation by the ISIS method does not involve transverse magnetization, it is applicable to samples with short transverse relaxation times and is insensitive to transient field gradient distortions resulting from switching. Like other forms of volume selection, the ISIS method can be combined with different filter and spectroscopy techniques [Tho1].

Hadamard spectroscopic imaging (HSI)

Hadamard spectroscopic imaging (HSI) is a technique to obtain localized spectroscopic information from n regions of interest in n scans [Bol1, Haf1, Goe1, Goe2, Goe4, Mül4]. It is a straightforward extension of the multi-frequency selective-pulse technique

[Mül2] used for simultaneous acquisition of m selected slices [Mül2, Mül3, Sou1] (cf. Section 5.3.4). There, multi-slice selective pulses are composed of sums and differences of regular selective pulses centred at different frequencies. The signs in the summation are determined by the rows of a *Hadamard matrix* (cf. Section 4.4.5). The same procedure is applied in the other space dimensions. For a 2D HSI experiment two successive pulses, the responses of which correspond to the rows of a Hadamard matrix, are applied successively in orthogonal gradients. For each multi-slice selective Hadamard encoded pulse in one dimension, all pulses corresponding to the rows of a Hadamard matrix must be applied in the second dimension. Therefore, the number of experiments to perform increases in a multiplicative fashion with the number of volume elements in each dimension. Given that more than n signals need to be averaged, the Hadamard technique provides the spectroscopic information of n volume elements in the same time as other volume-selective techniques do for just one volume element. Hadamard spectroscopic imaging of n volume elements provides a signal-to-noise advantage of $n^{1/2}$ over techniques by which one voxel is selected at a time. A practical value of n is 4^3.

Hadamard frequency encoding can be applied to longitudinal and to transverse magnetization [Goe1, Goe4]. A flip angle of 180° for the selective pulses can be used for space encoding in one, two, and three dimensions, corresponding to encoding in the longitudinal direction. A flip angle of 90° is applicable for encoding one and three dimensions in the transverse plane and two dimensions in the longitudinal direction. Longitudinal and transverse encoding can also be combined [Goe4]. Longitudinal encoding is less sensitive to rf field inhomogeneity and can be used for surface-coil spectroscopy. Special, chemical-shift insensitive pulses have been designed as building blocks for the multi-slice selective 90° pulses [Goe1]. Multi-frequency inversion pulses have been synthesized from modified hyperbolic secant pulses for use with inhomogeneous B_1 fields from surface coils [Goe2].

The measured responses to the combinations of multi-frequency selective pulse excitation can be unscrambled for each volume element by transformation with a super-Hadamard matrix. The dimension of this matrix equals the product of the dimensions of the Hadamard matrices used for encoding each space axis.

9.2 SURFACE-COIL METHODS

Localization of the sensitive volume is achieved by the use of *surface coils* without the necessity of manipulating magnetic-field gradients [Ack1, Bos1]. The great advantage is a large filling factor, so that the signal-to-noise ratio is high close to the coil, if the surface coil is used as for signal reception. For volume elements away from the coil the signal-to-noise ratio drops, because the sensitivity of the antenna drops. The receiver sensitivity is described by the same space-dependent $\boldsymbol{B}_1$ field which is generated when the antenna is used as a transmitter. If surface coils are used for excitation, the inhomogeneous $\boldsymbol{B}_1$ field gives rise to a variation of the excitation flip angle throughout the volume of interest. Considerable effort has been devoted to cope with the $\boldsymbol{B}_1$ field inhomogeneity. Special excitation pulses have been developed that are less sensitive to variations in $\boldsymbol{B}_1$ [Gar1].

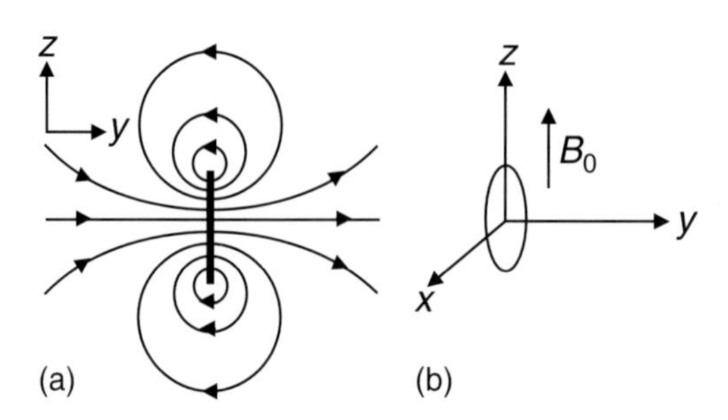

FIG. 9.2.1 [Bos1] A circular surface coil with radius r. (a) Magnetic field lines. (b) Optimum orientation of a surface coil with respect to the magnetic field $\boldsymbol{B}_0$. $\boldsymbol{B}_0$ points along z. The surface coil lies in the xz plane. The coil is centred at $x = y = z = 0$.

But the $\boldsymbol{B}_1$ variation with distance from the coil centre can also be exploited for spatial localization by rotating-frame methods (cf. Section 6.3) [Bot1, Sty1]. If surface coils are used for excitation and detection the space dependence of signal excitation and signal detection enters into the acquired signal in a multiplicative fashion.

A simple surface coil consists of a flat circular loop of wire. In the plane of the coil, the magnetic field lines produced by current in the coil are parallel to the coil axis. Beyond the coil plane the field lines diverge, wrap around the coil and reconverge (Fig. 9.2.1(a)). The best efficiency of the coil is achieved when it is oriented with its coil axis perpendicular to the polarizing magnetic field $\boldsymbol{B}_0$ (Fig. 9.2.1(b)). In this case, the magnetic field lines are nearly all perpendicular to $\boldsymbol{B}_0$ for regions near the coil, where the flux density is greatest. The signal-bearing volume accessible by the surface coils depends on the experiment and the sample. If the surface coil is used for excitation and detection, the signal comes from a volume section in the sample underneath the coil which approximates the shape of a hemisphere with the coil radius in case of a single excitation pulse. This is demonstrated in Fig. 9.2.2(b) with a gradient-echo image of the phantom with water-filled slots shown in (a). Excitation with two pulses flattens the accessible volume as seen in Fig. 9.2.2(c) which shows a spin-echo image.

9.2.1 Theory of surface coils

This treatment of surface coils is divided into the use of surface coils as transmitters only, as receivers only, as transmitters and receivers combined, and on experimental aspects relating to the dependence of the interrogated volume on pulse-sequence parameters [Bos1, Bos2].

Surface coils for reception

A magnetic dipole oscillating with frequency ω_0 at a distance $\boldsymbol{R}$ away from a loop of wire induces a time-varying voltage in the loop. The voltage amplitude is directly related to the strength of the magnetic field $\boldsymbol{B}_1$ that would be produced at the point of the oscillating dipole by the unit dc current in the loop [Hou1, Hou2]. This fact is known as the *principle of reciprocity* [Hou1]. According to it, the oscillating signal voltage

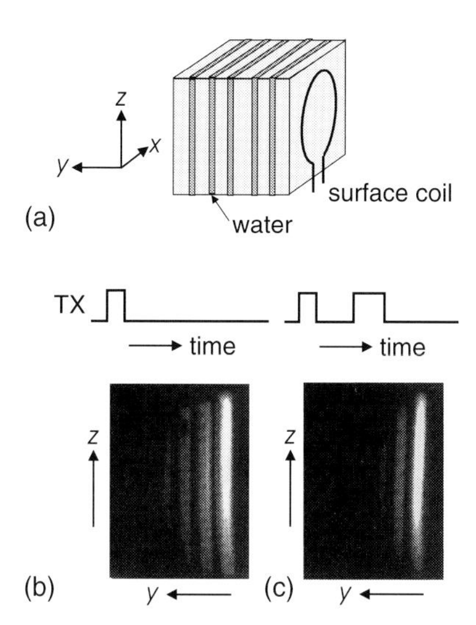

FIG. 9.2.2 Interrogated volumes for a circular surface coil used for excitation and detection. (a) Phantom from slots filled with water. (b) Single-pulse excitation by gradient-echo imaging. (c) Double-pulse excitation by spin-echo imaging.

induced by a precessing magnetic dipole moment $\boldsymbol{M}$ at position $\boldsymbol{R}$ relative to the coil is given by

$$s(t) = -\frac{\partial}{\partial t}[\boldsymbol{B}_1(\boldsymbol{R})\,\boldsymbol{M}(\boldsymbol{R}, t)], \tag{9.2.1}$$

where $\boldsymbol{B}_1$ is the magnetic field produced by the unit current through the receiver coil. For transverse magnetization, only the components of $\boldsymbol{B}_1$ and $\boldsymbol{M}$ transverse to $\boldsymbol{B}_0$ are responsible for the induced signal. With the usual convention that $\boldsymbol{B}_0$ points along the z-direction, (9.2.1) simplifies to

$$s(t, \boldsymbol{R}) = -B_{1xy}(\boldsymbol{R})\, M_{xy}(\boldsymbol{R})\, \omega_0 \sin\{\omega_0 t + \varphi(\boldsymbol{R})\}, \tag{9.2.2}$$

where $B_{1xy}(\boldsymbol{R})$ and $M_{xy}(\boldsymbol{R})$ are the magnitudes of the transverse components of $\boldsymbol{B}_1$ and $\boldsymbol{M}$, and φ is the phase angle between $B_{1xy}(\boldsymbol{R})$ and $M_{xy}(\boldsymbol{R})$. It is not to be confused with the receiver phase ϕ (cf. Section 2.3.4). The total signal is given by the integral of (9.2.2) over $\boldsymbol{R}$. Therefore, the signal depends on how B_{1xy}, M_{xy}, and φ vary with $\boldsymbol{R}$.

The spatial distribution of B_{1xy} completely describes the receptivity pattern of the surface coil. This distribution can be obtained for any coil geometry from the *Biot–Savart law* (2.3.8) [Let1]. Field distributions simulated for a one-turn receiver coil are shown in Fig. 9.2.3 [Bos1]. Contours of constant field are shown for the xy plane at $z = 0$ (Fig. 9.2.3(a)) and for the yz plane at $x = 0$ (Fig. 9.2.3(b)). High values of B_{1xy} and thus high signal sensitivity is found near the coil wires. With increasing distance y from the coil the transverse magnetic field falls off rapidly. The shape of the B_{1xy} distribution in the xy plane is distinctly different from that in the yz plane. This results from the fact that two transverse components of $\boldsymbol{B}_1$ exist in the xy plane but only one in the yz plane. In

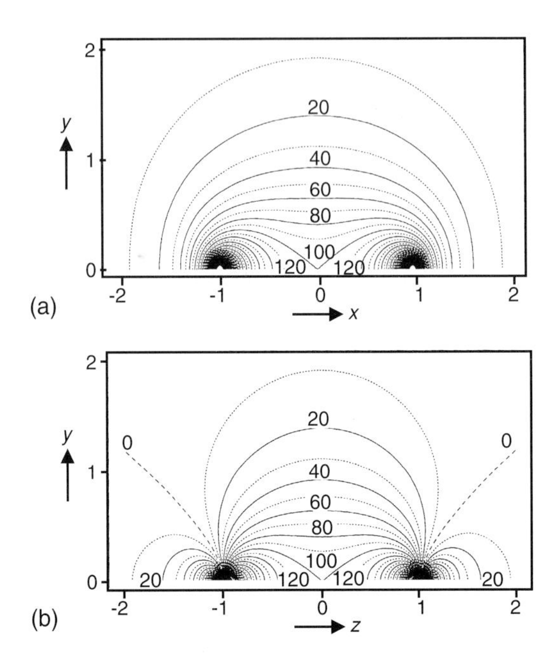

FIG. 9.2.3 [Bos1] Distributions of the magnitude of B_1 simulated for a circular single-turn receiver coil. The coil lies in the xz plane. The y-axis denotes depth into the sample (cf. Fig. 9.2.1). The axes are labelled in units of the coil radius. The magnetic field intensity is normalized to 100 in the centre of the coil. (a) Contour lines of constant B_{1xy} in the xy plane at $z = 0$. (b) Contour lines of constant B_{1xy} in the yz plane at $x = 0$.

addition, the closed loops of the $\boldsymbol{B}_1$ field lines in the yz plane (cf. Fig. 9.2.1(a)) produce components parallel to $\boldsymbol{B}_0$, so that no signal can be detected at these positions at all. The locations of zero B_{1y} field form nodal curves in the yz plane, which radiate outward from the coil wires (dashed lines in Fig. 9.2.3(b)). Beyond these curves, the transverse component of the field is reversed in direction.

Surface coils for excitation

The field distribution B_{1xy} of the surface coil defining the received signal (9.2.2) is given by the distributions depicted in Fig. 9.2.3 in the laboratory frame. It can be understood to be produced by a hypothetical dc current through the coil. The rf field for excitation is produced by an rf current through the coil. In a frame rotating at the rf excitation frequency ω_{rf}, the excitation profile is given by the same field distribution with the possible differences of a constant amplitude factor and a constant phase offset between laboratory and rotating frames for all positions $\boldsymbol{R}$ [Ack2]. Therefore, for a given surface coil, excitation and reception patterns are identical.

Excitation near resonance converts longitudinal magnetization M_z into the transverse magnetization M_{xy} according to

$$M_{xy} = M_z \sin \alpha, \tag{9.2.3}$$

where α is the flip angle;

$$\alpha(\boldsymbol{R}) = -\gamma t_{\mathrm{p}} B_{1xy}(\boldsymbol{R}). \tag{9.2.4}$$

The flip angle is related to the pulse width t_{p} and on the field distribution $B_{1xy}(\boldsymbol{R})$. Because the latter is proportional to the current I through the coil, the flip angle varies with the distance $\boldsymbol{R}$ of the volume element of interest from the centre of the coil. In return the distribution of flip angles produces a dependence of the excited transverse magnetization on the space coordinates,

$$M_{xy}(\boldsymbol{R}) \propto \sin\alpha(\boldsymbol{R}). \tag{9.2.5}$$

If the spins at the centre of the coil at $x = y = z = 0$ receive a 180° pulse, then the spins on the coil axis at 0.77 radii into the sample experience a 90° pulse. The regions of the sample which receive pulses of the same flip angle are curved following the B_{1xy} field lines of Fig. 9.2.3(a). Near the coil wires B_{1xy} and with it the flip angle increase sharply, so that by (9.2.5) the excited signal rapidly changes sign with distance $\boldsymbol{R}$. Therefore, given one pulse length t_{p}, several distinct regions of the sample are excited to produce maximum signal.

Surface coils for excitation and reception

When the same surface coil is used for excitation and for reception, $B_{1xy}(\boldsymbol{R})$ enters into the detected signal from the transmitter side as well as from the receiver side. As a result the sensitive volume changes in size compared to excitation or reception only. For a homogeneous sample excited by a single pulse, the distribution of signal amplitude is described by

$$s(\boldsymbol{R}) \propto B_{1xy}(\boldsymbol{R}) \sin\alpha(\boldsymbol{R}). \tag{9.2.6}$$

Signal amplitude maps calculated for a homogeneous sample are depicted in Fig. 9.2.4 [Bos1] for a 90° excitation pulse in the coil centre. In comparison with the corresponding plots of Fig. 9.2.3 for single-mode operation of the surface coil, the lines of constant signal are contracted closer to the origin. The regions of significant intensity extend approximately one coil radius from the centre of the coil into the sample (cf. Fig. 9.2.2(b)). Although the signal falls off more rapidly into the sample, the advantage of using the same coil for excitation and for detection is that the phase $\varphi(\boldsymbol{R})$ in (9.2.2) between excitation and receiver reference is the same for all points in space, i.e., it is independent of $\boldsymbol{R}$, so that the signals from all points in space add coherently.

If a separate coil is used for excitation, the phase φ depends on space $\boldsymbol{R}$. This is also the case, if a homogeneous $\boldsymbol{B}_1$ field is used for excitation, which is the case in many applications of clinical tomography with surface coils. Then the direction of the magnetic field from the surface coil is not constant over the volume of detection, so that the phase relationship between $\boldsymbol{B}_1$ and the transverse magnetization M_{xy} depends on the position $\boldsymbol{R}$ in the sample relative to the origin of the surface coil. This lack of phase coherence leads to partial cancellation of the detected signal. But this effect is offset by the advantage of being able to produce an optimum excitation flip angle over the entire sample by using

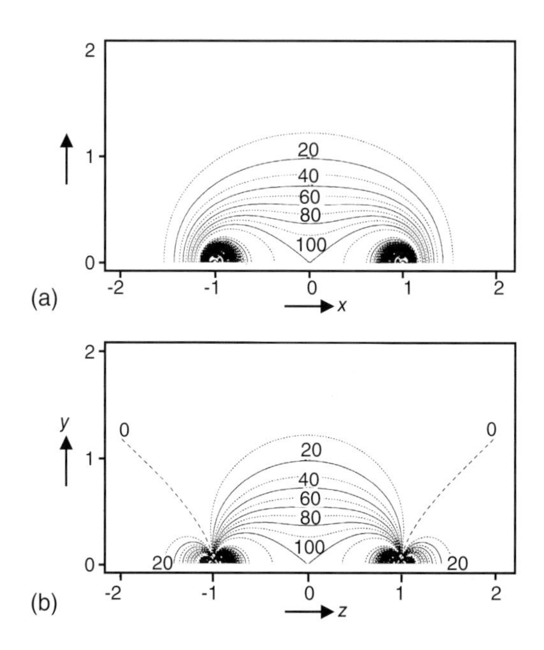

FIG. 9.2.4 [Bos1] Distributions of signal magnitudes simulated for single-pulse excitation and detection by a circular single-turn coil. The coil lies in the xz plane. The y-axis denotes depth into the sample (cf. Fig. 9.2.1). The axes are labelled in units of the coil radius. The signal magnitude is normalized to 100 for a 90° flip angle in the centre of the coil. (a) Contour lines of constant signal magnitudes in the xy plane at $z = 0$. (b) Contour lines of constant signal magnitudes in the yz plane at $x = 0$.

a homogeneous $\boldsymbol{B}_1$ field for excitation. A further benefit of this is the increased depth y into the sample which can be interrogated. This can readily be seen by comparison of Figs 9.2.3 and 9.2.4. Increased sensitivity at larger depth can also be achived by using coaxial coplanar surface coils with different diameters, where the larger one is used for excitation and the smaller one for reception [Fro1]. This also reduces the variation of $\varphi(\boldsymbol{R})$, but at the expense of homogeneous excitation.

Dependence of the sensitive volume on experimental parameters

Shape and location of the *sensitive volume* are determined by the inhomogeneous distribution of B_{1xy} and by the nonlinear relationship between B_{1xy} and the induced transverse magnetization [Dec1, Eve1, Haa1]. For simple systems of spins $\frac{1}{2}$, this relationship is expressed by the Bloch equations (2.2.8). Resonance offset, relaxation times, and details of the pulse sequence are important parameters. The influence of the pulse sequence on the size of the sensitive volume is illustrated in Fig. 9.2.2, which demonstrates that for spin-echo excitation the sensitive volume is smaller than for single-pulse excitation.

Under repetitive pulsing the transverse magnetization in the steady state is given by the SSFP formula (2.2.34),

$$M_{xy} = M_0 \frac{1 - \exp\{-t_R/T_1\}}{1 - \cos\alpha \exp\{-t_R/T_1\}} \sin\alpha = M_z \sin\alpha. \tag{9.2.7}$$

For surface-coil excitation, the flip angle α and, consequently, the transverse magnetization M_{xy} are functions of position $\boldsymbol{R}$ relative to the coil origin. If $t_R/T_1 < 5$, then also M_z is a function of α and, consequently, of position. Moreover, the spin density M_0 can be a function of $\boldsymbol{R}$. The signal induced in the receiver coil can be written by inserting (9.2.7) into (9.2.2):

$$s(t, \boldsymbol{R}) = -\omega_0 B_{1xy}(\boldsymbol{R}) M_0 \frac{1 - \exp\{-t_R/T_1\}}{1 - \cos\alpha(\boldsymbol{R}) \exp\{-t_R/T_1\}} \sin\alpha(\boldsymbol{R}) \sin\{\omega_0 t + \varphi(\boldsymbol{R})\}. \tag{9.2.8}$$

Therefore, for *surface-coil excitation and detection*, the distribution of signal amplitudes in the steady state depends (a) on the excitation in terms of the ratio t_R/T_1 and in terms of $B_{1xy}(\boldsymbol{R})$ contained in the factor $\sin\alpha(\boldsymbol{R})$ and (b) on the detection *via* $B_{1xy}(\boldsymbol{R})$ of the receiver coil.

When reducing t_R relative to T_1 for homogeneous excitation the signal amplitude per pulse decreases from partial saturation equally for all positions $\boldsymbol{R}$ in the sample. This is not the case for surface-coil excitation. Here, the $\cos\alpha(\boldsymbol{R})$ term in (9.2.8) introduces a spatial dependence, not considering a possible spatial dependence of $s(t, \boldsymbol{R})$ from sample heterogeneity in $T_1(\boldsymbol{R})$. As a consequence of the space dependence of the excitation flip angle $\alpha(\boldsymbol{R})$, the transverse magnetization (9.2.7) is saturated more for larger flip angles. Because these arise close to the coil, magnetization from volume elements farther away from the coil are saturated less. Consequently, the shape of the sensitive volume changes and larger depths can be interrogated with rapid pulsing [Dec1]. This is illustrated in Fig. 9.2.5 [Bos1], by signal profiles along the y-axis which originates in the centre of the surface coil and points into the sample. With decreasing t_R/T_1 the overall signal amplitude per pulse decreases, but at the same time the curves become flatter, so that the loss of relative signal with depth is less pronounced. Because many signals can be accumulated in a given time under rapid-pulsing conditions, there is a net gain in signal: rapid pulsing with $t_R/T_1 = 0.05$ results in a gain of about 130% in total signal-to-noise ratio compared to slow pulsing conditions [Eve1]. This is more than the 60% gain achieved with a homogeneous coil under rapid pulsing conditions [Bec1]. The difference is attributed to the increase in the sensitive volume under rapid pulsing with surface coils [Bos1].

As a result of the strong space dependence of $B_{1xy}$for surface coils, their sensitivity is much higher for regions near the surface than that of large-volume coils with homogeneous fields. However, due to the rapid fall-off of B_{1xy} with depth into the sample, large-volume coils outperform surface coils for regions deeper within the sample [Bos1, Har1, Hay1].

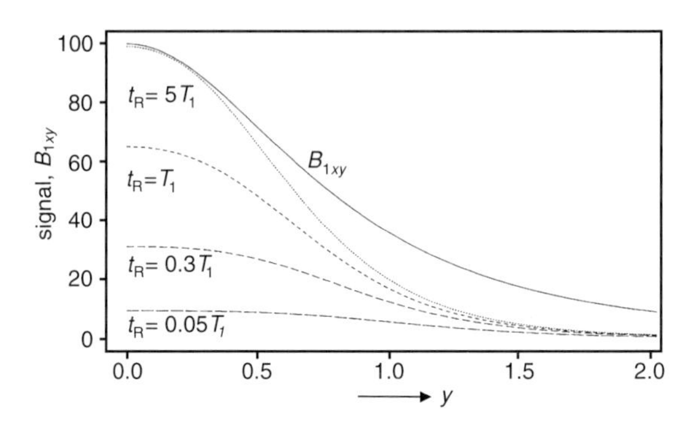

FIG. 9.2.5 [Bos1] Comparison of signal amplitudes for volume elements placed along the y-axis and different repetition times t_R. The y-axis which starts in the centre of a single-turn circular surface coil and points into the sample. The axes are labelled in units of the coil radius. The curves are normalized relative to an amplitude of 100 at $y = 0$ for the response to a single pulse applied to the sample in thermal equilibrium. For reference the axial space dependence of B_{1xy} is also given.

9.2.2 Working with inhomogeneous B_1 fields

The *inhomogeneous* $\boldsymbol{B}_1$ *fields* generated by surface coils provide advantages and disadvantages in practical applications. The main advantage is the high sensitivity towards signals from regions near the coil. The main disadvantages are the difficulties associated with localizing signals from well-defined volume elements deeper within the sample, and the variation of the excitation flip angle with space [Bos1]. There are remedies to both. In fact, the space dependence of $\boldsymbol{B}_1$ can be exploited for space encoding by rotating-frame methods, and parameter images, for example, relaxation time images, can be obtained without contrast distortions despite the space dependence of B_{1xy}.

Localization techniques with $\boldsymbol{B}_0$ *gradients for use with surface coils*

The large intensity of B_{1xy} near the surface makes it difficult to detect signals from deeper volume elements without contamination by signals from lower volume elements. The inherent volume localization of the surface coil can be improved by using gradients in $\boldsymbol{B}_0$ or in $\boldsymbol{B}_1$ to define a specific sample volume within the total volume accessible by the surface coil. The respective techniques are mainly used for localized spectroscopy in the biomedical context of *in vivo NMR* [Bec2, Bot2, Hen4, Mül5, Sau1, Slo1, Web1]. They have been addressed in Section 9.1.

Localization using inhomogeneous $\boldsymbol{B}_1$ *fields*

The spatial derivative $\mathrm{d}\boldsymbol{B}_1(\boldsymbol{R})/\mathrm{d}\boldsymbol{R}$ of the space-dependent $\boldsymbol{B}_1$ field defines the $\boldsymbol{B}_1$ field-gradient tensor $\mathbf{G}_1(\boldsymbol{R})$ similar to the field-gradient tensor of the $\boldsymbol{B}_0$ field (cf. eqn (2.1.9)). Because the variation of $\boldsymbol{B}_1$ is nonlinear in $\boldsymbol{R}$, $\mathbf{G}_1$ is a function of position $\boldsymbol{R}$, so that images derived with $\boldsymbol{B}_1$ gradients from surface coils are defined on nonlinear space axes. These image distortions can be corrected by post-processing the experimental data. Images using $\boldsymbol{B}_1$ gradients are obtained by *rotating-frame techniques* [Cox1, Haa2, Hou3] and related techniques [Gar2, Met1, Pec1] based on the acquisition of a series of data for which the excitation pulse width is incremented (cf. Section 6.3).

Spatial selectivity can be enhanced and phase distortions from off-resonance effects be reduced through the use of composite pulses where the sensitivity depends on the excitation flip angle of the individual pulses (cf. Fig. 5.3.10) [Sha1, Sha2, Sha3, Tyc1]. Similarly, phase cycling can be employed to introduce sensitivity towards the excitation flip angle. This concept is used by the family of *DEPTH pulses* [Ben2, Ben3, Ben4, Ben5, Tyc1]. These sequences are designed to be highly efficient only for a narrow band of flip angles. Because the flip angle can be chosen at will by adjusting the rf amplitude or the pulse width, DEPTH pulses can be used to focus on specific, narrow regions within the sensitive volume. Cascading DEPTH pulses further increases selectivity.

An example of a DEPTH pulse is the simple *echo sequence* $\alpha_{\pm x}$–τ–$2\alpha_{\pm y}$. The spatial discrimination along the depth axis achieved by this pulse is compared to that of a single pulse in Fig. 9.2.6 [Bos1] for a surface coil used for both excitation and detection of the signal. The α pulse has been scaled to produce a 180° flip angle at the coil centre. It is seen that the signal bearing region is more restricted for the DEPTH pulse. An extension of this principle is Fourier series windowing [Gar2, Gar3, Met1, Pec1], where the volume section of interest is narrowed based on the principles of rotating-frame imaging [Hou3].

A more detailed analysis shows that additional signals from regions close to the coil wires cannot be excluded by the simple DEPTH pulses. However, the selectivity can be improved further with the help of additional $\boldsymbol{B}_0$ gradients [Ben3, Bot2, Che1, Che2, Che3], by variation of the pulse lengths and by addition of low flip-angle preparation pulses [Ben3, Ben6, Bou1, Sha4]. Nevertheless, long pulses and severe off-resonance effects can limit the use of such pulse techniques which exploit the $\boldsymbol{B}_1$ gradient for achieving spatial selectivity [Bos1]. In that case two coaxial surface coils can be used, where typically the larger one acts as a transmitter and the smaller one as the receiver [Ben7, Che4, Rat1]. Depending on the particular demands of the experiment, various coil geometries have been designed [Hyd1], including quadrature detection coils [Che5], phased coil arrays for lateral extension of the sensitive region [Roe1], and doubly tuned

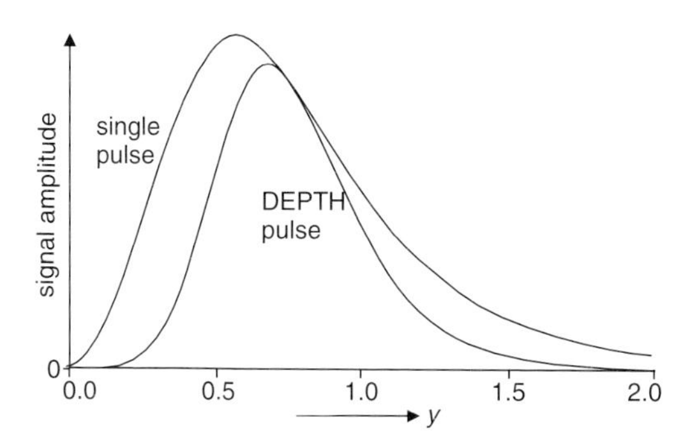

FIG. 9.2.6 [Bos1] Theoretical signal amplitudes from a single-turn circular surface coil used for excitation and detection. The flip angle is scaled to 180° at the coil centre. The y-axis denotes depth into the sample. It is scaled in units of the coil radius. For a single pulse the signal is proportional to $B_{1xy} \sin \alpha(y)$, and for the simple DEPTH pulse it is proportional to $B_{1xy} \sin^3 \alpha(y)$.

coils [Gon1, Gri1, Mit1] for heteronuclear investigations with polarization transfer [Ben8]. When using separate excitation and detection coils, particular attention must be drawn to the decoupling of both antennas [Bos1, Haa3].

Homogeneous rf excitation for inhomogeneous $\boldsymbol{B}_1$ fields

Inhomogeneous $\boldsymbol{B}_1$ fields are advantageous in cases, where spatial resolution is in demand, but they are disadvantageous whenever homogeneous excitation is required. Special techniques have been developed to extend the volume of homogeneous excitation by reducing the spatial flip-angle dependence of surface coils. To this end *shaped pulses* (cf. Section 5.3.2) and *composite pulses* (cf. Section 5.3.3) [Bau1, Kem1, Sha1, Sha2, Sha3, Sha4, Sha5, Woe1] have been developed [Bos1]. The use of field-insensitive pulses in the single-coil mode increases the signal-to-noise ratio by more uniform excitation, and the excitation and reception fields exhibit a constant phase relationship within the homogeneously excited volume [Joh1].

Shaped pulses with reduced sensitivity towards $\boldsymbol{B}_1$ inhomogeneity have been developed based on the *fast adiabatic passage* of continuous wave (CW) NMR [Sli1]. They are referred to as *adiabatic pulses* [Gar1] and can be used for exitation of large sensitive volumes [Gar1, Gar4, Gra2, Gra3, Hen3, Hwa1, She1, Zwe1], slice selection [Gra4, She2] inluding water suppression [Gra5], and double resonance experiments [Cap1, Ram1]. In CW NMR an inversion of the spin polarization can be achieved for all resonance offsets by a rapid sweep through the resonance. This is the effect of a 180° pulse on longitudinal magnetization. Because the $\boldsymbol{B}_1$ field amplitude is of little importance in this experiment, the fast adiabatic passage is equivalent to a $\boldsymbol{B}_1$ insensitive 180° pulse. The effect of a $\boldsymbol{B}_1$ insensitive 90° pulse on longitudinal magnetization can be obtained by a fast adiabatic half passage.

The fast adiabatic passage can be understood by considering the effective field $\boldsymbol{B}_{\text{eff}}$ in the rotating frame as a function of frequency offset (Fig. 9.2.7; cf. Fig. 2.2.7). The effective field is the vector sum of the fictitious field $\boldsymbol{B}_{\text{fic}}$ along the z-direction and the $\boldsymbol{B}_1$ field along the y-direction in the rotating frame, $\boldsymbol{B}_{\text{eff}} = \boldsymbol{B}_{\text{fic}} + \boldsymbol{B}_1$. The size of the fictitious field is defined in terms of resonance offset Ω by $\Omega = -\gamma \boldsymbol{B}_{\text{fic}}$. Far off resonance the contribution of the rf field $\boldsymbol{B}_1$ to the effective field is insignificant, because Ω is large. Then the spin polarization $\boldsymbol{M}$ is essentially aligned along z. By changing the frequency ω_{rf} of the $\boldsymbol{B}_1$ field in the laboratory frame, the frequency offset $\Omega = \omega_0 - \omega_{\text{rf}}$ can be reduced. Then the relative contribution of $\boldsymbol{B}_1$ in the vector sum defining $\boldsymbol{B}_{\text{fic}}$ changes so that not only the magnitude of $\boldsymbol{B}_{\text{fic}}$ but also the direction of $\boldsymbol{B}_{\text{fic}}$ are affected. On resonance, $\Omega = 0$, so that $\boldsymbol{B}_{\text{fic}} = \boldsymbol{B}_1$, and the direction of the fictitious field has been changed by 90°. The trick is to change the direction of $\boldsymbol{B}_{\text{fic}}$ slowly enough, so that the magnetization can follow the orientation of the fictitious field, but also fast enough so that the magnetization is not lost by relaxation. This first condition gives rise to the attribute 'adiabatic' and the second to the attribute 'fast' for the passage through resonance. The adiabatic condition is fulfilled, whenever the rate of change ω of the orientation of $\boldsymbol{B}_{\text{eff}}$ is much smaller than the NMR frequency ω_0 in the laboratory frame [Sli1]. When the passage is not stopped on resonance at $\Omega = 0$ but continued through resonance, the transverse magnetization which exists on resonance continues to follow the fictitious field and finally ends up with its orientation along the $-z$-direction.

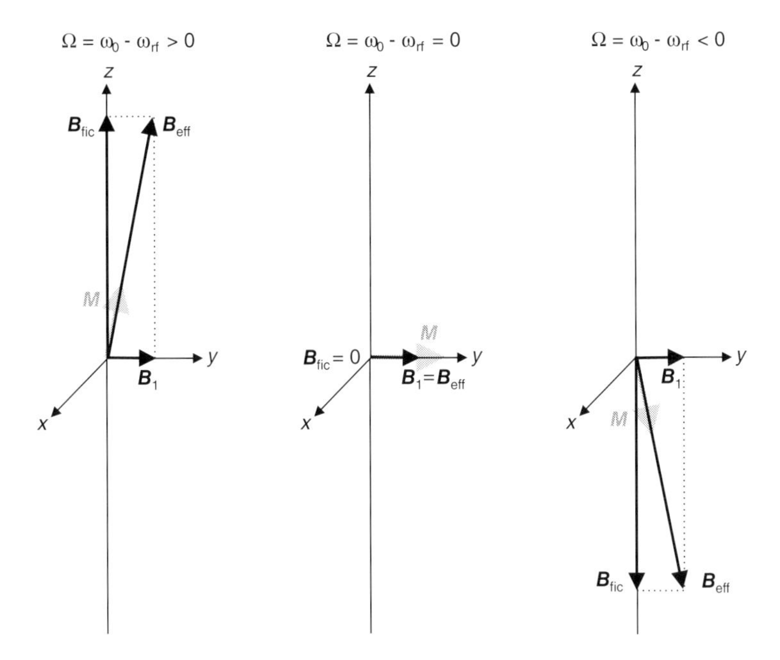

FIG. 9.2.7 Illustration of the fast adiabatic passage through resonance. The magnetization $\boldsymbol{M}$ follows the direction of the effective field $\boldsymbol{B}_{\text{eff}}$. The effective field in the rotating frame is the vector sum of the fictitious field $\boldsymbol{B}_{\text{fic}}$ and rf excitation field $\boldsymbol{B}_1$. Both fields are applied in orthogonal directions. Because the fictitious field is proportional to the resonance offset Ω, the magnitude of the fictitious field and thus the direction of the effective field can be changed by adjusting the resonance offset frequency Ω.

Parameter imaging in inhomogeneous $\boldsymbol{B}_1$ fields

In NMR images acquired with surface coils the spatial dependence of B_{1xy} introduces an intensity loss with distance from the coil (Fig. 9.2.2). This signal attenuation appears like a loss in spin density, but derives from the fact that B_{1xy} is part of the image contrast. By calculating true *parameter images*, such an intensity loss in surface-coil images can be avoided at the expense of a loss in signal-to-noise ratio with increasing distance from the coil centre. In fact, even concentrations can be quantified by surface-coil methods [Bos2].

Longitudinal relaxation times can be determined despite inhomogeneous $\boldsymbol{B}_1$ fields by the *saturation-recovery* and the *inversion-recovery* methods as long as three conditions are met [Bos2, Dec1, Eve2, Mat1, San1]: (1) The number of spins should remain invariant over the course of the experiment. This is important in flow experiments. (2) The magnetization of each volume element must be prepared consistently in the same initial nonequilibrium condition prior to relaxation into the thermodynamic equilibrium state during a variable recovery period. (3) Each volume element must consistently be interrogated at the end of the recovery period. It is not necessary that all volume elements experience the same preparation and interrogation conditions as long as they are the same for all recovery periods. Phase cycling of the excitation pulses and the receiver phase, as well as pulsed-field gradients can be used to reduce unwanted magnetization from imperfect pulses.

Transverse relaxation is determined by the Hahn-echo and CPMG sequences (cf. Section 7.2.2) also in inhomogeneous $\boldsymbol{B}_1$ fields [Blü1, Pop1], where already the simple Hahn echo provides enhanced localization according to the *DEPTH effect* (see above) [Bos2, Wux1]. As an example for surface-coil parameter imaging, the acquisition of a T_2 parameter image of a phantom from three sheets of poly(urethane) rubber is considered (Fig. 9.2.8(a)) [Blü1]. 1D spatial resolution along the depth axis is obtained through application of a $\boldsymbol{B}_0$ field gradient along y. A Hahn-echo type sequence (Fig. 9.2.8(b)) with two pulses of flip angles α and 2α is applied to the sample *via* the surface coil and the echo is detected for different echo times $t_E = 2\tau$. A stack of the depth projections (Fig. 9.2.8(c)) acquired in this way shows an approximately exponential signal decay with τ from which T_2 can be derived for each position y (d). In the T_2 parameter image the three sheets of rubber can clearly be distinguished based on differences in T_2. Within each sheet, T_2 is constant, but the signal-to-noise ratio deteriorates with increasing depth y.

Despite the inhomogeneous $\boldsymbol{B}_1$ field of the surface coil, *spin-lock experiments* can also be performed to interrogate $T_{1\rho}$ relaxation (cf. Sections 3.5 and 7.2.3) [Blü2]. If τ_c denotes the correlation time of molecular motion then, from eqn (3.5.7) $T_{1\rho}$ can be expressed in the weak collision limit as

$$T_{1\rho}(\boldsymbol{R}) = T_2(\boldsymbol{R})[1 + \gamma^2 B_{1xy}^2(\boldsymbol{R})\tau_c^2]. \qquad (9.2.9)$$

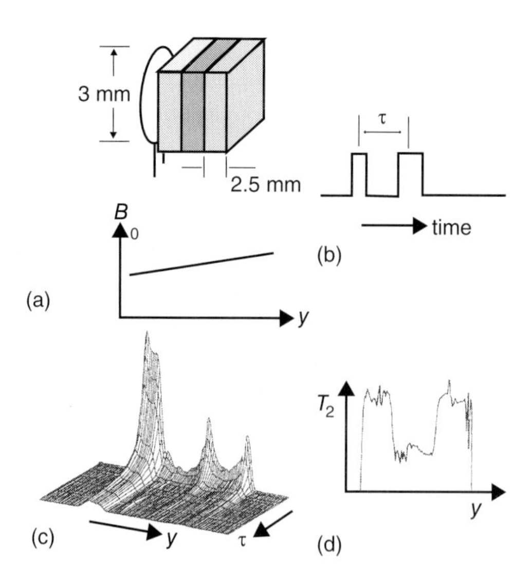

FIG. 9.2.8 T_2 parameter imaging with a surface coil. (a) The phantom consists of three sheets from poly(urethane) rubber. A magnetic-field gradient was applied perpendicular to the surface coil for space encoding along the depth axis y. (b) A two-pulse sequence with flip angles α and 2α, respectively, was used for excitation. (c) Stack of 1D images for different echo times $t_E = 2\tau$. (d) T_2 parameter image extracted from the signal decays in c.

Here T_2 enters as the limit of $T_{1\rho}$ for vanishing lock-field strength B_{1xy}. Therefore, given a T_2 parameter image and a map of $B_{1xy}^2(\boldsymbol{R})$, the *correlation time* τ_c of molecular motion can be mapped in the weak collision limit.

Typical solid-state NMR experiments are more difficult to achieve with surface-coil excitation. Nevertheless, multi-quantum surface-coil NMR has been applied in investigations of biological tissue [Lyo1], and MREV8 multi-pulse excitation has been studied on a phantom of adamantane in an imaging experiment to exploit the $\boldsymbol{B}_1$ gradient for spatial resolution [Mil1]. The low sensitivity of the MREV8 sequence to rf amplitude results in low-spatial resolution, which is interpreted in terms of a large sensitive volume for this sequence. Higher-spatial resolution can be achieved with $\boldsymbol{B}_1$ sensitive pulses sequences which make use of the magic angle [Mil1].

9.3 DEDICATED NMR

NMR spectroscopy is usually applied in batch processes: a sample is drawn, prepared for the NMR measurement, positioned in the magnet, measured, and discarded. These steps as well as data analysis can proceed with varying degrees of automation. Sample preparation often involves dissolution and purification of the compound in question for high-resolution liquid-state NMR or the sizing of solid samples to fit into a spinner for magic-angle spinning in high-resolution solid-state NMR [Bov1]. A step beyond this scenario is the analysis of samples not particularly prepared for NMR which are passed by an *NMR sensor* for analysis. The NMR sensor may be a conventional high-resolution liquid-state NMR spectrometer equipped with a continuous-flow [Bay1] or a stopped-flow probe [Wat1], an NMR imager with a horizontal-bore magnet which accommodates a *conveyor belt* with the samples [Arm1, Bur1, Mue1, Rol1], or a dedicated NMR device with locally applied coils [Ack1, Cor1], which sample only a selected part of the total magnetization from the object. The latter approach is routinely applied for studies of humidity and salt transport in building materials [Bli1, Kop1, Kru1, Pel1]. Alternatively, the object can be left at rest and the NMR sensor can be moved. Such *mobile NMR devices* [Pow1] have been developed for analysis of oil wells [Bro1, Coo1, Kle1, Kle2], water reservoirs [Gol1, Shu1], Antarctic sea ice [Cal1], water in soil [Rol1], polymer materials [Blü3, Eid1, Gut1, Mat2], and plant growth and life cycles [Ash1]. Depending on their area of use, different design principles have been implemented [Pow1]. Examples are robust tube-shaped spectrometers including magnets which can be lowered into *bore holes* [Bro1, Coo1, Kle1, Sez1], endoscope coils [Hak1, Sch1], *single-sided NMR* devices, where the polarizing $\boldsymbol{B}_0$ and the rf $\boldsymbol{B}_1$ field are applied to the object from one side [Eid1, Mat1, Rol1], or just surface coils [Ack1, Bos1] and short solenoidal coils [Cor1] for localized detection. Use of a polarizing magnetic field can be abandoned in *earth-field NMR* [Ben1, Cal4, Pow1], where the rf coil may be up to 100 m in diameter [Gol1, Shu1], or the polarizing field can be supplied by the stray field of a magnet in the vicinity of the rf coil (cf. Section 8.1) [Kim5, Kim6, Mcd1]. Similar ideas can be followed with other magnetic resonance techniques such as electron spin resonance (ESR) [Ike1, Yam1] and nuclear quadrupole resonance (NQR) [Bue1, Gar5, Gar6, Gre1, Gre2].

A particularly interesting development in this context is the use of NMR for *process and quality control* [Pow1]. Here, nondestructiveness is a key demand. It implies the lack of invasive sample preparation. Also the measurements often take place in a harsh environment where high fields and homogeneous fields are difficult to attain. As a consequence, high-resolution NMR spectroscopy must be discarded in most cases in favour of a *relaxation analysis* [Kim5]. For this reason, this type of NMR is sometimes referred to as *low-resolution NMR*. Nevertheless, low-resolution NMR provides information about molecular mobility in solids and liquids through relaxation measurements and about molecular transport of fluids by diffusion, flow, and acceleration in microconfined environments through judicious use of magnetic-field gradients (cf. Section 7.2.6). Applications of low-resolution NMR are in the food industry [Ban1, Bar1, Bar2, Bot3, End1], the paper and pulp industry [Arg1], and in the analysis of building [Bli1, Kru1, Pel1] and polymer materials [Blü3, Gut1, Zim1].

Low-resolution NMR devices can readily be incorporated into production processes to quantify the quality of a product or the performance of a process [Gib1, Gla1, GLA2]. In case of deviation from reference values, the NMR signal can be used to trigger adjustments of the production process [Pow1]. However, NMR imaging [Blü4, Blü5, Blü7, Cal2, Kim5] is also explored for *on-line detection* in industrial applications, for example, in the food industry [Mcc1]. In such applications real-time imaging systems are needed, which can be realized, for example, with a personal computer [Hai1, Kos1]. NMR imaging of sample batteries on trays is being explored for simultaneous analysis of multiple samples [Eva1, Hal1]. These approaches are referred to as *process tomography* [Dya1, Gib1]. Applications of NMR in the fields of process control and nondestructive detection are old [Mat2, Pow1], but rare in the literature. Nevertheless, more and more small- and medium-size enterprises focus on dedicated applications of magnetic resonance to industrial processes [Bru2, Fox1, Igc1, Qua1, Swr1, Usi1]. This trend is accompanied by an increasing use of technology driven applications of NMR in chemical engineering and in industry [Gla2, Gla3].

9.3.1 Experimental aspects

Dedicated NMR often requires special instrumentation and adapted measurement techniques. A general treatment of the experimental aspects is complicated by the diversity of solutions. Nevertheless, some general developments are the following: low-field instruments are preferably employed for applications in *process control* and for mobile use like in geophysical exploration [Ash1, Bro1, Cal4, Coo1, Gri2, Kle1, Rud4, Sez1]. Special surface coils have been developed for investigations of large [Cor1] as well as of small volumes [Pec2, Ols1]. A remarkable result of this work are microcoils for analysis of nanolitre volumes [Web1]. Single-sided application of the polarizing magnetic field and the rf field is experimentally diverse and may appear simple [Eid1, Mat1, Rol1]. NMR experiments in such inhomogeneous $\boldsymbol{B}_0$ and $\boldsymbol{B}_1$ fields are most complicated and demand the use of large spectral widths and of Hahn echoes, possibly in combination with other types of echoes for refocusing of magnetic field inhomogeneities [Blü3, Gut1, Gut2, Zim1]. In fact, the conventional NMR in homogeneous fields can be considered as a limiting case of *NMR in inhomogeneous fields*.

Low-field spectrometers

Low-field spectrometers [Cal3, Cap2, Gri2, Rud4, Saa2] operate at frequencies below 10 MHz. They need to be designed for small signal amplitudes because the NMR frequency is low. Therefore, it is advantageous to employ high-sensitivity coils with high-quality factors Q (cf. Section 2.3.3). At low frequency and high Q, square pulses are problematic to apply, because the falling edge of the pulse has a strong influence on the receiver deadtime. The deadtime is determined, the time it takes for the excitation energy to be removed from the coil. It increases with higher values of Q. Therefore, a Q switch to toggle between low Q for excitation and high Q for detection is a critical element of the spectrometer [Gri2].

Figure 9.3.1 illustrates the signal flow of a low-field spectrometer [Gri2]. The leading edge of the rf pulse is overdriven until the $\mathbf{B}_1$ amplitude has increased to its nominal value. The pulse amplitude is then reduced to a lower value until completion of the pulse. Already before the end of the pulse, the Q switch is toggled, so that high Q is available for reception at the end of the Q-switch pulse. The duplexer signal actuates the transmit–receive switch (cf. Fig. 2.3.5). A low duplexer signal indicates high impedance of the receiver to the coil for protection of the receiver during the rf pulse. The receiver impedance is switched back to the 50 Ω state while the Q switch is still operating. This helps damp ringing of elements in the transmit–receive switch as a result of switching the impedance. The delays between successive rf pulses are adjusted to integral multiples of Larmor periods, so that the transmitter gate is synchronized with the carrier period. The receiver gate is turned on just for detection of the echo [Gri2]. At ultra-low frequencies down to 200 Hz, signal detection through nuclear induction is less sensitive than the use of superconducting quantum interference devices (SQUID) [Aug1]. The development of *SQUID* technology for magnetic resonance shows great potential for interesting applications in medical imaging as well as in process control by NMR and NQR.

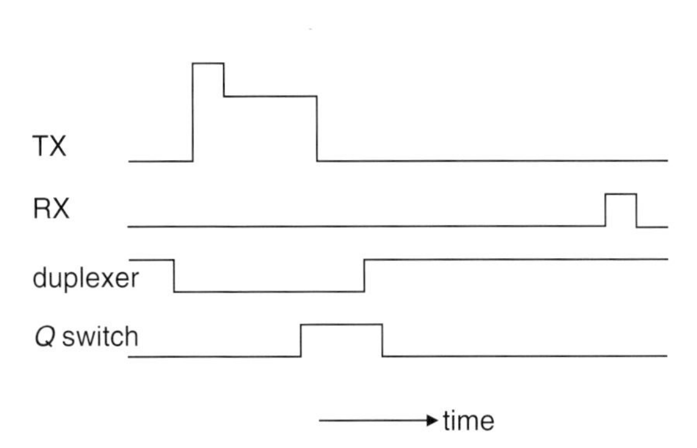

FIG. 9.3.1 [Gri2] Timing diagram for operation of a low-field spectrometer. The input pulse for the transmitter is overdriven until the $\mathbf{B}_1$ magnetic field has reached its maximum value. It then falls to a maintenance level until completion of the pulse. The Q switch signal toggles the rf coil between low and high values of the quality factor. The duplexer signal switches between transmit (low) and receive (high) operation of the coil. The receiver gate is turned on just for detection of the signal.

Dedicated coils

Radio-frequency coils or more generally rf antennas are designed to excite and receive the NMR signal. Depending on the type of application the coil dimensions range from extremely large to extremely small. Large coils of 100 m diameter are used in the earth magnetic field for *hydrogeological surveys* down to a depth of 100 m [Gol1, Shu1]. Antennas for objects of up to 1 m in size are used for identification of mines and for detection of explosives and narcotics in luggage security checks on airports by ^{14}N NQR [Bue1, Gar5, Gar6, Gre1, Gre2, Gre3, Gre4, Sui2, Sui3]. In some cases, the objects are placed on a conveyor belt and passed through the coil. In a way, the conveyor belt with the objects can be considered as an extended sample, where measurements are executed at selected positions, while the sample is moved through the spectrometer. If those positions are closely spaced, a 1D image of the object is obtained by pointwise measurements. With sufficiently small coils dimensions, this approach to 1D NMR imaging can be applied even in NMR with multi-pulse line narrowing [Cor1].

Very small coils are used for analysis of small sample volumes, because the sensitivity of a coil increases with the inverse diameter of the coil. *Microcoils* are coils with a volume of less than 1 μl [Ols1, Pec2, Web1]. Applications are explored with magic-angle spinning for structural analysis in combinatorial chemistry, on-line coupling of microseparation techniques with NMR, and microimaging. Even planar microcoils with diameters of 200 μm and less have been reported (Fig. 9.3.2(a)) [Web1]. Gallium arsenide is a suitable substrate for such coils, because preamplifiers and mircocoils can be combined on a single device [Pec3]. With such a coil a sample volume of 0.8 nl has been detected. For reduction of susceptibility-induced line broadening, sample and coil need to be immersed in a liquid fluorocarbon [Web1]. In addition to increasing the sensitivity in the measurement of small sample volumes, the use of microcoils reduces the requirements on magnetic field homogeneity, because the applied magnetic field needs to be homogeneous over smaller volumes than for conventional coils.

Surface coils (cf. Section 9.2) [Ack1, Bos1] are applied to the object from one side so that the associated $\boldsymbol{B}_1$ field is inhomogeneous within the sensitive volume. Apart from symmetry, *semitoroidal coils* offer particular advantages in spatially resolved NMR. A semitoroidal coil is a solenoid which is bent into a semicircular arc, and the rf field

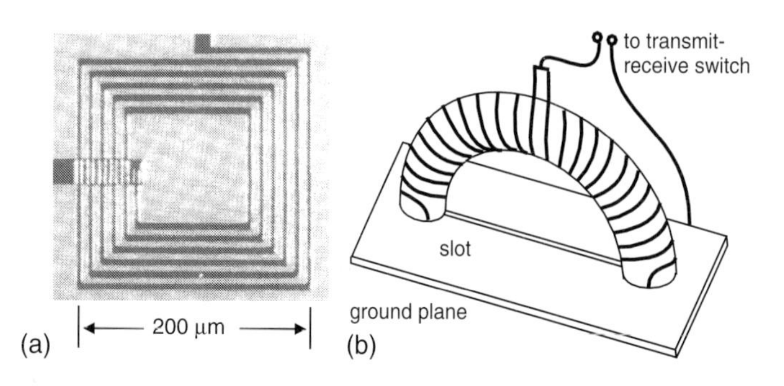

FIG. 9.3.2 Special surface coils. (a) Electron micrograph of a planar microcoil. Adapted from [Web1] with permission from Elsevier Science. (b) Drawing of a semitoroidal coil [Ass1].

emanating from both ends is used (Fig. 9.3.2(b)) [Ass1]. Such a coil produces an rf field which penetrates into the object to a depth which is of the order of the diameter of the semitoroidal coil. The wire size and the spacing can be chosen to make the Q factor much higher for a semitoroid than for a planar coil at a given frequency. Furthermore, the coil can be used in a way that the rf electric field is virtually eliminated at the sample, so that the dielectric properties of the sample are less influential in changing the Q factor and the frequency of the circuit. To this end, the coil is fed in the centre, and the coil ends, which are closest to the sample, are grounded in a conducting plane (Fig. 9.3.2(b)). A slot in the plane between the two holes for the coil ends prevents the formation of induced currents around the holes which would reduce the field from the main coil.

Other interesting geometries of surface coils are also based on the cylindrical coil: if the windings of the solenoid are not perpendicular to the coil axis but tilted, the main field component generated by the coil is no longer parallel to the coil axis, and a component perpendicular to the coil axis is generated, the magnitude of which depends on the tilt angle [Jeo1]. Alternatively, the stray field of the solenoid can be used as such for excitation and reception by using the solenoid as an endoscope [Hak1, Sch1] or by just placing it on the object with its axis parallel to the surface [Glo1].

An interesting alternative to the semitoroid is obtained if two coaxial circular surface coils with different diameters are operated together with opposing fields [Rat1]. This type of coil produces a region of relatively homogeneous field outside the coil volume. Applications to *in vivo* imaging, noninvasive monitoring of reactions, and production line monitoring have been suggested. For bore-hole investigations half coaxial rf sensors have been designed [Sez1]. Their sensitive volumes exhibit noncircular sensitive regions which are extended along the axis of the half coaxial sensor. Such deviations from circular symmetry are advantageous when sample and sensor are in uniaxial relative motion.

Techniques in low-resolution NMR

The experimental techniques in *low-resolution NMR* often are rather simple. The reason for this is that many of the techniques developed so far require homogeneous polarizing fields $\boldsymbol{B}_0$ and homogeneous excitation fields $\boldsymbol{B}_1$. The number of applicable techniques rapidly decreases if either $\boldsymbol{B}_0$ or $\boldsymbol{B}_1$ is inhomogeneous. In inhomogeneous $\boldsymbol{B}_0$ fields, spectroscopic resolution of interactions linear in the spin operator, like the chemical shift, can no longer be achieved by simple means, because the magnetic field inhomogeneity is linear in the spin operator as well and must be eliminated by formation of Hahn echoes for detection of transverse magnetization. But based on the different properties of echoes for linear and bilinear spin interactions the molecular information contained in bilinear spin interactions can still be detected in principle. In the limit of hard pulses most experiments may be restricted to homogeneous $\boldsymbol{B}_1$ fields, and for reasons of good signal-to-noise ratio, many pulses need to be applied for generation of multi-echo trains. Because pulse imperfections can be compensated usually only for small deviations from ideality, sophisticated NMR techniques appear to be excluded for applications like on-line monitoring in process and quality control, which would benefit from single-sided application of both $\boldsymbol{B}_0$ and $\boldsymbol{B}_1$. On the other hand, much less effort has been devoted to the development of NMR methods suitable for use in inhomogeneous $\boldsymbol{B}_0$ and $\boldsymbol{B}_1$ fields, and it can be expected that significant advance in NMR methodology can be achieved, for example, by the use of shaped adiabatic pulses [Gar1] in echo sequences to cope with inhomogeneities in both, $\boldsymbol{B}_0$ and $\boldsymbol{B}_1$.

In weakly inhomogeneous $\boldsymbol{B}_0$ fields the single-pulse response already contains significant information about heterogeneity of food and polymer samples, and the observed signal shape can be correlated with parameters like oil and water content, additive concentrations, processing conditions, degree of polymerization, solid content, liquid content, viscosity, cross-link density, etc. [Ban1, Ber2, Bru3, Bar1, Bar2, Dem1, Her1, Hie1, Kha1, Mon1, Rut1]. As an example, Fig. 9.3.3 depicts the free- induction decay signals of sun- and oven-dried peanut kernels [Ban1, Bar1]. The solid and semisolid carbohydrates have the shortest T_2, and their signal decays within the first 20 μs of the FID. The oil has the longest T_2, and its signal dominates the FID at times beyond 150 μs. Therefore, the window between 25 and 150 μs is left to observe the contributions from water. Drying at 105 °C removes more water from the peanuts than does drying in the sun. For characterization of the drying process, it is sufficient to extract the *y* intercept of the slow relaxation component after normalization of the signal [Ban1].

In strongly inhomogeneous $\boldsymbol{B}_0$ fields all rf pulses are selective (cf. Fig. 5.3.3). Such a situation is encountered in *fringe-field NMR*, which has been developed to exploit the high-field gradients of superconducting magnets for measuring slow diffusion [Kim5, Kim6], and for improving the spatial resolution in solid-state imaging by *STRAFI* (cf. Section 8.1) [Mcd1]. Because of the selective character of the excitation, pulse-width and pulse-amplitude modulation are no longer equivalent in sequences which employ pulses with different flip angles. In multi-echo trains, echo interference effects can be observed, which modulate the echo amplitude in the initial part of the echo envelope [Ben9, Ben10, Gut2], while repetitive acquisition of single echoes leads to simple decay functions. The decay of a train of CPMG echoes measured in inhomogeneous $\boldsymbol{B}_0$ and $\boldsymbol{B}_1$ fields is illustrated in Fig. 9.3.4(a) for unfilled cross-linked SBR [Eid1]. A fit of an exponential function to the echo envelope yields the transverse relaxation time T_2.

In inhomogeneous $\boldsymbol{B}_0$ and $\boldsymbol{B}_1$ fields the flip angle may be relatively well defined if the $\boldsymbol{B}_0$ gradient is strong. Then the $\boldsymbol{B}_1$ field is quite homogeneous over the sensitive

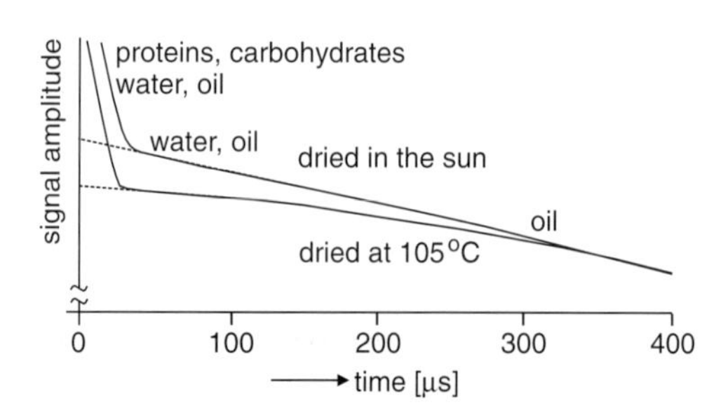

FIG. 9.3.3 [Ban1] Free-induction decays for peanut kernels dried in the sun and at 105 °C. Based on the water signal visible in the time window between 20 and 150 μs both drying processes can be discriminated.

volume (Fig. 9.3.4(b)) [Gut3]. Depending on the rf frequency chosen, different sample regions are probed in strongly inhomogeneous $\boldsymbol{B}_0$ fields. Hahn and solid echoes can be discriminated for solid samples (Fig. 9.3.4(c)), because the dipolar interaction is unaffected by the Hahn echo. But the multi-pulse versions of Hahn and solid echo often lead to similar echo envelopes, an effect which may be attributed to different sensitive volumes for both cases. In either case a so-called *spin-lock effect* can be observed in elastomer samples, which denotes a decrease of the effective transverse relaxation time with increasing pulse separation in the multi-echo sequence (Fig. 9.3.4(d)) [Gut1]. One possible interpretation of this effect is that for short echo times the signal decay is determined by $T_{1\rho}$-type relaxation and for long echo times by T_2-type relaxation. In addition to such simple echo techniques, *stimulated echoes* can also be generated in

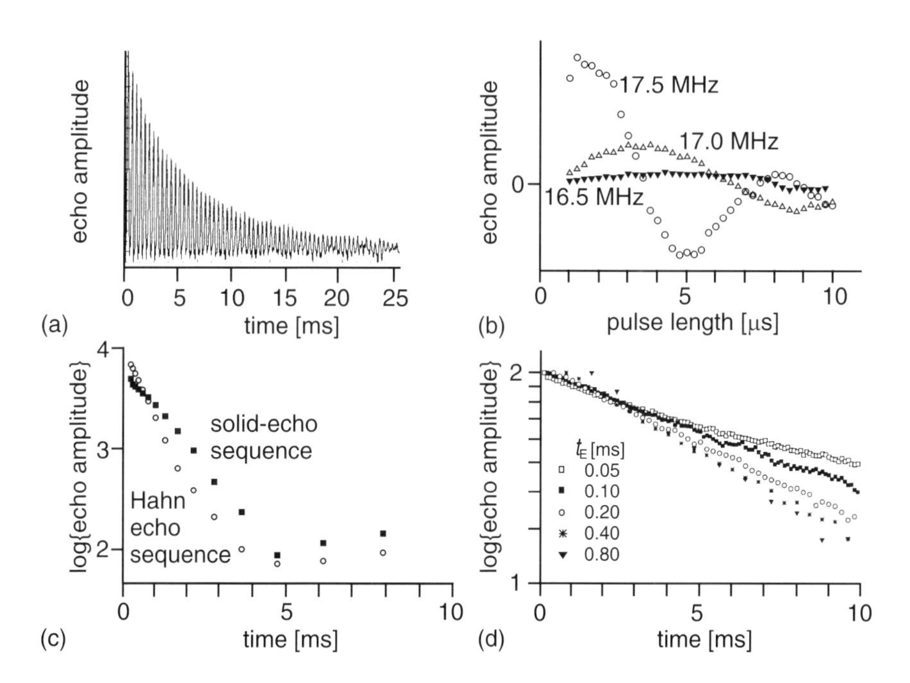

FIG. 9.3.4 Excitation with a surface coil of 9 mm diameter in a $\boldsymbol{B}_0$ gradient of the order of 10 T/m by the NMR-MOUSE: (a) Series train of CPMG echoes from a carbon-black filled SBR section of an intact car tyre with a steel belt. A fit of the echo envelope with an exponential decay function yields a transverse relaxation time T_2 [Eid1]. (b) Variation of the pulse duration in an $\alpha - t_E/2 - 2\alpha - t_E/2$ pulse sequence for different rf frequencies. For each frequency maxima and minima are observed which define the nominal 90° and 180° pulse widths. With decreasing rf frequency the distance of the sensitive volume from the rf coil increases. A frequency of 17.5 MHz correspond to depths of 0–0.5 mm, 16 MHz to 0.5–1.0 mm, and 16.5 MHz to 1.0–1.5 mm into the sample [Gut3]. (c) Hahn- and solid-echo envelopes for a sample of carbon-black filled cross-linked SBR. The Hahn-echo decay is faster because of residual dipolar couplings which are partially refocused by the solid-echo [Gut3]. (d) Multi-echo excitation. A so-called spin-lock effect manifests itself in a dependence of the effective transverse relaxation time T_{2e} on the echo time t_E of the multi-echo sequence [Gut3].

inhomogeneous fields and steady-state excitation sequences like *SSFP* can be employed to improve the signal-to-noise ratio. The disadvantage of steady-state sequences is a possible loss in contrast for discrimination of material properties, which are associated with molecular mobility. However, better signal-to-noise ratio results from exploitation of the Ernst angle (cf. Fig. 7.2.22) as well as from an increase in the size of the sensitive volume for bulk samples (cf. Fig. 9.2.5) [Gut1, Sez2].

In systems which contain liquids, molecular self-diffusion can be studied with inhomogeneous $\boldsymbol{B}_0$ fields (cf. Sections 5.4.3 and 7.2.6) [Kim6] but also with inhomogeneous $\boldsymbol{B}_1$ fields [Goe3, Mcc2, Wud1]. This is, particularly, interesting for restricted diffusion within the droplets of emulsions and the pores of rocks. The former case is of interest in low-resolution NMR for product development and quality analysis of dairy products and creams in combination with pulsed-field gradients [Bal2, Bot3, End1, Fou1], and the latter is a fundamental subject of study in context of oil exploration [Kle2]. If the field inhomogeneities over the sample assume the size of the underlying homogeneous field, the simple exponential expressions of the signal decay which hold for small field inhomogeneities are no longer valid [Ste1].

9.3.2 On-line coupling of high-resolution NMR

On-line coupling of high-resolution NMR has been explored extensively for chemical identification following separation processes in analytical chemistry [Alb1, Web1]. Although the NMR chemical shift is a most powerful contrast parameter for discrimination of different chemical species, the major disadvantage of NMR is its low sensitivity with regard to sample volume compared to other detection methods. But because compounds eluted from chromatographic columns are usually concentrated in small volumes, microcoils for nanolitre sample volumes are being explored for their sensitivity advantage compared to conventional coils with low-filling factor [Beh1, Wun1]. The commercial availability of NMR *flow cells* facilitates the use of on-line ^{1}H [Mai1] and ^{13}C NMR [Alb4] for observation of chemical reactions.

Liquid chromatography

The major advantage of *on-line NMR detection* as opposed to off-line detection for *liquid chromatography* (LC) are improved chromatographic resolution, consistent response, and rapid data acquisition. The disadvantages include poorer sensitivity due to limited measurement time and a flow-rate dependence of the NMR linewidth [Web1]. Following the first stopped flow design in 1978 [Wat1], an on-line system had been developed within a year [Bay1], and the techniques have been improved continuously since then [Web1].

The standard *LC-NMR* detector is a saddle coil which is wound on the flow cell with a diameter of 2–4 mm and a length of 12–14 mm (Fig. 9.3.5) [Web1]. The cell consists of a glass tube. Inflow and outflow are provided by tubings from PTFE with an inner diameter of typically 0.25 mm. The associated *chromatography column* is positioned outside the magnet. Because of time requirements 2D-NMR spectra can be acquired only in stopped flow mode, while in continuous flow mode 1D-NMR spectra are acquired as a function of the elution time.

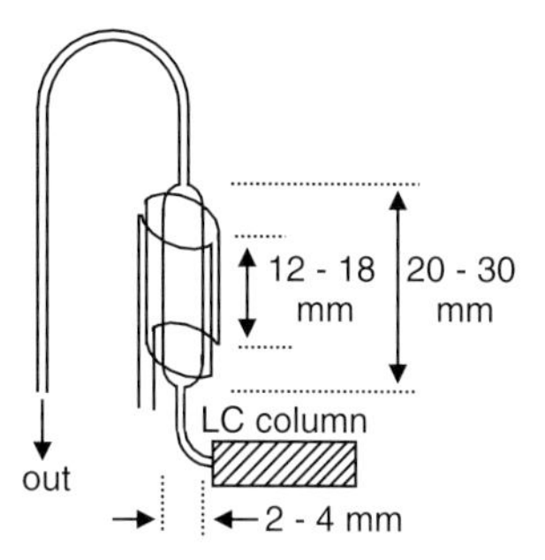

FIG. 9.3.5 [Web1] Schematic drawing of an LC-NMR flow cell. A saddle coil is wound on the cell. The cell is centred in the magnet and the chromatography column is positioned outside.

In continuous flow mode the relaxation times are affected by the *flow rate* and may be expressed in terms of an effective residence time τ within the flow cell,

$$\frac{1}{T_1} = \frac{1}{T_{1,\text{static}}} + \frac{1}{\tau} \tag{9.3.1}$$

and

$$\frac{1}{T_2} = \frac{1}{T_{2,\text{static}}} + \frac{1}{\tau}. \tag{9.3.2}$$

Increasing the flow rate decreases the effective longitudinal relaxation time, which allows shorter repetition delays t_R. However, at the same time the effective transverse relaxation rate increases, leading to line broadening.

As an example of a continuous flow chromatogram Fig. 9.3.6 shows a stack of NMR spectra (right) as a function of elution time for five vitamin A acetate isomers on a cyanopropyl column in *n*-heptane recorded at a flow rate of 0.2 ml/min [Alb1]. The first two peaks somewhat overlap in the ultraviolet (UV) detected chromatogram (left), but in the NMR spectra (right) the peaks are well separated due to different chemical shifts. Forty-eight scans were added for each trace, so that the time resolution was 62 s.

Solvent suppression is a major issue in NMR identification of *high-pressure liquid chromatography* (HPLC) eluates [Alb2]. One way to solve this problem is by using proton-free solvents in *superfluid chromatography* (SFC). Separations in supercritical CO_2 are only feasible at temperatures higher than 31.3 °C and pressures higher than 72.9 bar. Therefore, for coupling SFC and high-resolution NMR a high-pressure probe is needed. In such a probe the glass sample cell is replaced by a sapphire cell, and the PTFE tubing is replaced by titanium tubing. In contrast to T_2, T_1 increases two to three orders in magnitude because of lower viscosity in the supercritical state as compared to the liquid state. Pressure stability is critical because the chemical shifts are pressure dependent [Alb1, Alb3].

Electrophoresis

Electrophoretic separation is based on mobility differences of charged molecules within an applied electric field. The flow rate depends on the product of electrophoretic mobility

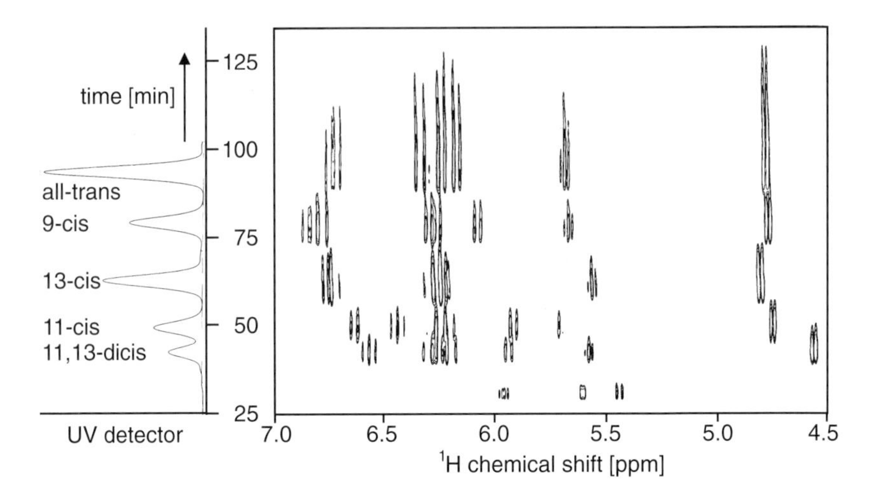

FIG. 9.3.6 HPLC chromatograms of five vitamin A acetate isomers. Left: UV chromatogram in *n*-heptane. Right: ^{1}H NMR spectra of the olefinic region recorded at 400 MHz. Adapted from [Alb1] with permission from Elsevier Science.

and applied field strength. Larger voltages typically lead to better separation of ions. The first coupling of electrophoretic separation with high-resolution NMR has been achieved in 1972 [Pac1], and the technique has been developed since [Hol1, Hol2, Joh2, Saa1]. The major difficulty with the technique is the use of electric fields in the range of 30 V/cm, which may lead to sample heating from resistive loss with subsequent disruption of the separation process by the formation of bubbles [Web1]. This problem can be alleviated by increasing the surface area for better heat dissipation. Because the ratio of surface area to volume is inversely proportional to the cell diameter capillaries are being used for electrophoretic investigations. An NMR detection cell for *capillary electrophoresis* (CE) has been built with a microcoil wound around a fused silica capillary with 75 μm diameter, creating a detection volume of 5 nl. [Wun2, Ste2]. With it linewidths of less than 7 Hz were obtained at 300 MHz.

9.3.3 Geophysical exploration

Dedicated NMR equipment is used for geophysical exploration in different fields: the free precession in the earth magnetic field is measured for mapping groundwater reservoirs with large diameter surface coils [Gol1, Shu1] and for analysis of Antarctic sea ice [Cal1, Cal5]. *Earth-field NMR* [Pac2] has been explored rather early for oil-well analysis [Bro1] but later abandoned in favour of low-field *inside-out NMR*. Here the polarizing field is generated by permanent magnets, which are lowered into the bore hole together with the spectrometer hardware [Bur2, Coo1, Jac1, Kle1, Mar1].

Groundwater exploration

The detection of fresh *groundwater* is a challenging task in geophysics, because the electrical conductivity of fresh groundwater-bearing aquifers does not differ much from that of dry lithological units. For this reason earth-field NMR techniques have been studied to facilitate the search for groundwater and other hydrocarbons in hydrogeological surveys [Gol1, Shu1]. Surface coil and hardware used for such purpose have been called the *NMR hydroscope* [Shu1].

The principle of the NMR hydroscope is illustrated in Fig. 9.3.7(a) [Shu1]. A single-turn transmitter coil with a diameter between 50 and 500 m is laid out on the earth's surface and driven by *mobile NMR spectrometer* hardware inside a truck. The coil radius determines the maximum depth accessible by the technique. Either the same coil is used for signal reception or a much smaller coil, for example, a multi-turn air coil of 1 m [Gol1]. Depending on the geographic location the Larmor frequency of water in the earth magnetic field is in the range of 2–2.7 kHz. As a consequence of the high homogeneity of the earth magnetic field, signal decays are long, so that single-pulse excitation is suitable for excitation and subsequent measurement. Due to the skin effect of conducting earth formations, the exploration depth for a given coil radius is determined by the time interval after the excitation pulse has been turned off. As time increases the excitation current density migrates to greater depths. In order to make deeper explorations it is necessary to record the signal for a longer time. Typical times vary in the range of several microseconds for an exploration of depths of 5–10 m to tens of milliseconds for exploration depths of a few 100 m.

The signal induced in a surface coil used for transmitter and receiver by a voxel at position $\boldsymbol{r}$ following a pulse of flip angle $\alpha(\boldsymbol{r})$ is given by eqn (9.2.6). The expression needs to be modified to accommodate effects of conductivity [Sui1, Wei1]. Figure 9.3.7(b) demonstrates for a 100 m diameter coil that with increasing depth of a horizontal water-bearing layer the signal amplitude decreases and the amplitude maximum is shifted towards higher flip angles [Shu1]. In reality, the earth's surface consists of many, and diverse, layers with different conductivity and water content, so that signal interpretation can be ambiguous. Because of the open loop geometry the method is susceptible to noise from power lines and other electromagnetic contamination. Therefore, applications are restricted to areas with low population density.

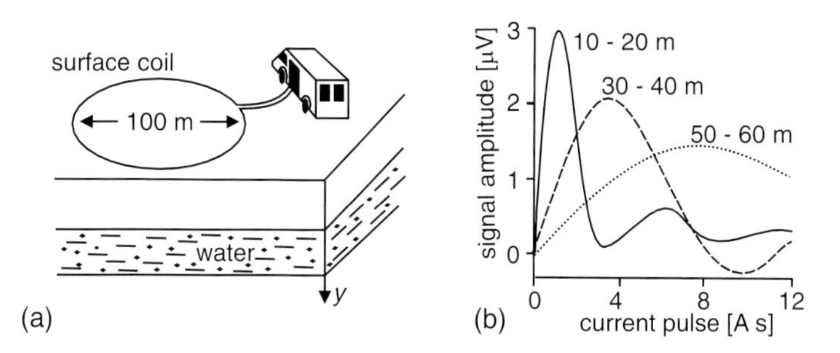

FIG. 9.3.7 Water exploration by surface-coil NMR in the earth magnetic field.
(a) Principle of the NMR hydroscope. (b) Signal amplitude for a 100 m diameter surface coil and a horizontal 10 m thick water layer located at the depth indicated. The surface coil is positioned at the origin. Instead of the flip angle the current-pulse intensity is given in A s. Adapted from [Shu1] with permission from the Society of Exploration Geophysicists.

Earth field NMR with conventional size coils

Conventional-sized coils may be understood to possess diameters of the order of 1 m and less, so that power demands can be satisfied with conventional rf amplifier technology. Such coils were explored already in the fifties for NMR in the earth magnetic field in the context of *oil-well characterization* [Bro1] and for mapping the magnetic field of the earth by the NMR frequency of a test sample even from outer space [Pow1]. A strong current is run through the coil for a few seconds so that nuclear polarization of the object under study can build up in the generated magnetic field $\boldsymbol{B}_0$. The current is then turned off and the precession of magnetization in the earth magnetic field is observed with the same coil.

The same principle has been applied for studies of *sea ice* in Antarctica, except that different coils were used for polarization, excitation, and reception (Fig. 9.3.8) [Cal1, Cal3]. In Antarctica the magnetic field lines of the earth are essentially perpendicular to the earth's surface and most noise sources from civilization are absent, so that long free-induction decay signals can be observed with signal-to-noise ratios much better than in more populated areas. The sensitivity of the measurement is improved by a polarizing coil, which does not have to produce a highly homogeneous magnetic field. If the polarizing coil is parallel to the receiver coil and perpendicular to the earth magnetic field (Fig. 9.3.8(c)), the nuclear magnetization, which has built up during the polarization-current pulse, immediately precesses about the earth magnetic field after the current pulse has been turned off nonadiabatically. In practice, nonadiabatic switching is difficult, an adiabatic removal of the polarizing field is preferred (Fig. 9.3.8). It leaves the nuclear magnetization $\boldsymbol{M}$ parallel to the earth field, so that conventional NMR studies can be conducted, for example, a one-pulse experiment (Fig. 9.3.8(a)) and a spin-echo

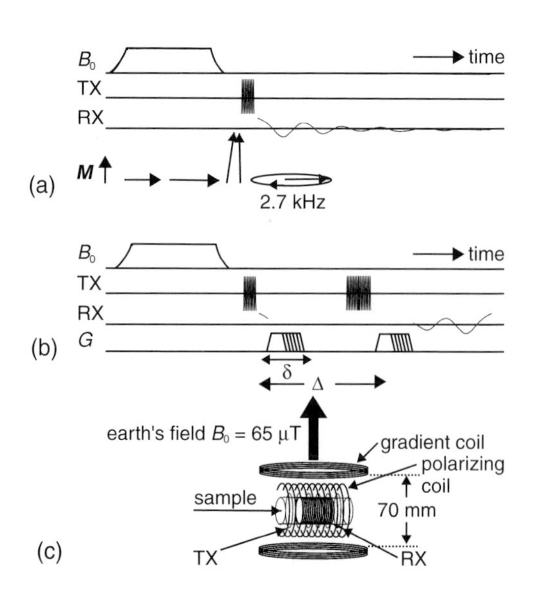

FIG. 9.3.8 [Cal1] NMR in the earth's magnetic field. The pulse sequences were used for studies of Antarctic sea ice. (a) Single-pulse excitation. (b) Pulsed field-gradient NMR. (c) Schematic drawing of the probe showing the coil arrangement and the sample.

experiment with pulsed-field gradients for investigations of restricted diffusion of brine in sea ice (Fig. 9.3.8(b)) [Cal4].

NMR in the earth magnetic field is also of interest in biomedical studies [Ben1] including imaging, because relaxation rates $1/T_1$ of biological tissues exhibit strong dispersion at low frequencies [Meh1, Moh1, Pla1, Ste3]. When using earth field NMR as a *magnetometer* to map the magnitude and the direction of the magnetic field through orientation-dependent measurements of the Larmor frequency, magnetic field distortions from geological abnormalities and buried artefacts can be identified. Because the Larmor frequency is in the audio range, simple detection of the FID by earphones is possible [Pow1]. The polarization-field pulse can be abandoned if the sample employed is a flowing liquid, which is polarized by a remote magnetic field. Following excitation of transverse magnetization, the magnetization phase is probed down stream after the liquid has passed the region of the field to be measured [Kim7].

Inside-out NMR

The term *inside-out NMR* denotes applications of NMR in *well logging*. The characterization of hydrocarbon reservoirs in subsurface earth formations is of considerable interest in the oil and gas industry [Jac2, Kle1, Mar1]. Here NMR has developed into one of the tools which can routinely be applied to wells up to 10 km deep [Kle2]. The entire spectrometer, including the probe and the magnet, is built into a rugged pipe approximately 2 m long and 10 cm wide. The device is lowered into the *bore hole* and then pulled up with constant velocity of about 300 m/h. On its way up, data are continuously acquired and sent up for analysis (Fig. 9.3.9).

The prime parameters of interest are the relaxation times of fluids in pores. Two limiting processses are distinguished (cf. Section 7.1.6): in the fast diffusion limit relaxation is monoexponential. It is independent of the pore shape and depends on the surface-to-volume ratio of the pore. In the slow diffusion limit, relaxation is diffusion controlled. The signal decay is multi-exponential and determined by the pore shape. Experience shows that T_1 and T_2 are nearly independent of temperature in the range of 25–175 °C. Because the diffusion coefficient is temperature dependent, this shows that the first relaxation mechanism dominates in most geological formations, so that the *pore-size distribution* can be derived from the distribution of relaxation times [Kle1, Kle2]. Protons with T_1 or T_2 larger than 50 ms are associated with large pores or unbound free fluid, and the signal amplitude of this fraction is a measure of the fluid volume extractable from the rock matrix. Water and oil can be discriminated based on differences in relaxation times: because most rock is hydrophilic, the oil droplets often remain in the centre of the pore and are largely unaffected by the surface. Then, a water and an oil peak are observed in the relaxation time spectrum. However, depending on the viscosity of the oil, interpretation of NMR data from well logging may be quite difficult and dependent on the correlation with data from other measurements like with electrical resistivity data [Kle2, Kle4, Mar1].

Measurements of longitudinal relaxation would be preferred over measurements of transverse relaxation, because signal attenuation from molecular diffusion in local field gradients near pore walls can then be neglected, but saturation and inversion recovery methods are more time consuming than CPMG experiments, so that transverse relaxation

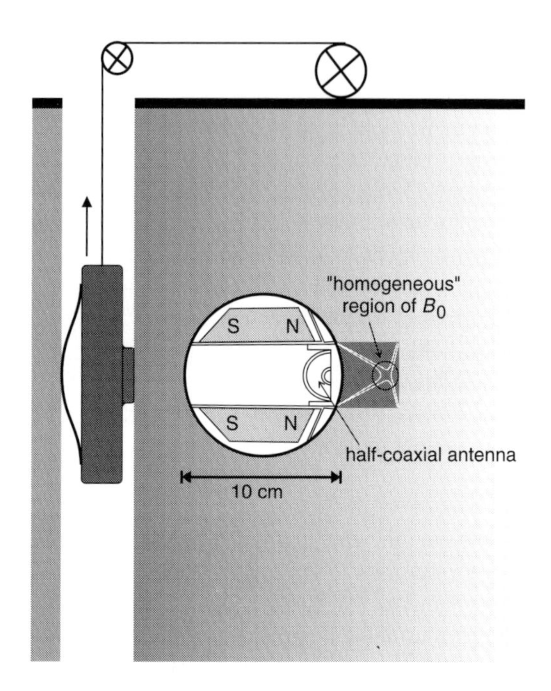

FIG. 9.3.9 [Kle1] Bore-hole NMR: the entire NMR spectrometer is lowered into a bore hole. Data are acquired while the device is pulled up with constant velocity. The cross-section through the probe shows that a region of homogeneous field outside the device is generated by two permanent magnet plates with parallel polarization. A half coaxial antenna is used for excitation and detection.

is usually probed in applications to well logging. Given the considerable expenses of about $150 000 per day for oil well equipment sitting idle during the NMR measurements, acquisition times are typically of the order of 6 s for a vertical resolution of 50 cm.

Depending on the company, different NMR probes are used in the field [Kle1]. In the early fifties, single-pulse *earth-field NMR* was explored with coils 175 cm long and 15 cm wide [Bro1]. Considerable flexibility in the use of different pulse sequences was achieved by incorporating $\boldsymbol{B}_0$ polarization fields into the device [Bur2, Coo1, Jac1]. One solution is depicted in Fig. 9.3.10 [Jac1]: two opposed bar magnets produce a toroidal region of homogeneous field outside the device. But because this region is rather thin, operation of the probe in axial motion is not suitable. Although this device has not been commercialized, applications to localized NMR in medicine and process control have been proposed (cf. Fig. 9.3.10) [Jac1].

Another device uses a 1 m long cylindrical ferrite-bar magnet, which is magnetized perpendicular to its axis [Tai1, Kle1]. The rf coil is wound perpendicular to the long axis in such a way that both $\boldsymbol{B}_0$ and $\boldsymbol{B}_1$ form orthogonal dipolar fields in the transverse plane at each position along the long axis. This assures that $\boldsymbol{B}_0$ and $\boldsymbol{B}_1$ are orthogonal at all positions, but the $\boldsymbol{B}_0$ field is inhomogeneous with a magnetic-field gradient of the order of 0.25 T/m, so that diffusive signal attenuation is observed in CPMG signals. Extrapolation of the CPMG signals to zero pulse spacing provides access to the signal decay in the

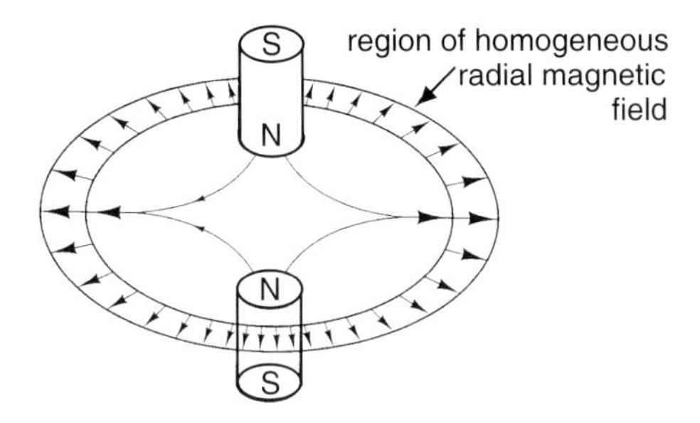

FIG. 9.3.10 [Jac1] Generation of a toroidal region of homogeneous radial magnetic field by two opposed coaxial magnets. This magnet geometry represents the original concept of inside-out NMR.

absence of diffusion. Because of the presence of a field gradient, different depth into the rock core can be accessed by variation of the rf frequency [Tai1]. Penetration depths are 15–23 mm with frequencies of 1–0.5 MHz, respectively.

The two devices described above need to be centred during operation in the bore hole. This requirement is eliminated by the design of the sonde shown in Fig. 9.3.9 [Kle1, Kle3]. The sonde is pressed against the bore-hole wall in a 40 cm long section by a spring. The $\boldsymbol{B}_0$ field is generated by two plates from permanent magnets with parallel polarization orthogonal to the long axis of the device. The resultant field is predominantly radial in the region of interest. A saddle point of the $\boldsymbol{B}_0$ field is generated about 25 mm outside the sonde. At this point all three spatial derivatives of $\boldsymbol{B}_0$ are zero, and the field strength is 55 mT, so that the ^{1}H Larmor frequency is 2.3 MHz. The antenna for signal excitation and detection is placed in a cylindrical cavity on the face of the sonde. It is a 15 cm long half coaxial cable that generates an rf field $\boldsymbol{B}_1$ transverse to the static field $\boldsymbol{B}_0$. The antenna is loaded with ferrite and has a high quality factor for good sensitivity. The antenna is protected by a nonmetallic wear plate. All other parts of the housing are metal structures. Magnetoacoustic ringing [Ger1] which would increase the deadtime of the instrument is suppressed [Kle1].

9.3.4 Single-sided NMR

The concepts developed for NMR well logging are readily transferable to applications in *process and quality control* [Jac1, Mat2, Pow1, Rol1]. Magnet geometries which accommodate conveyor belts or pipes for transportation of samples, bags, granular matter, fluids, etc. can be used to probe water, oil and contents of organic solids in a nondestructive fashion [Arm1, Bur1, Mue1, Rol1]. In this case, the samples are passed through the magnet. By single-sided application of magnetic fields like in inside-out NMR, no restrictions apply with respect to the sample size. Such *single-sided NMR* has been explored in different contexts, including the measurement of water in concrete bridge decks [Mat1], the determination of soil water from a sonde hitched on a tractor for calibration of satellite data on soil moisture [Rol1], and in polymer analysis [Mat1, Eid1].

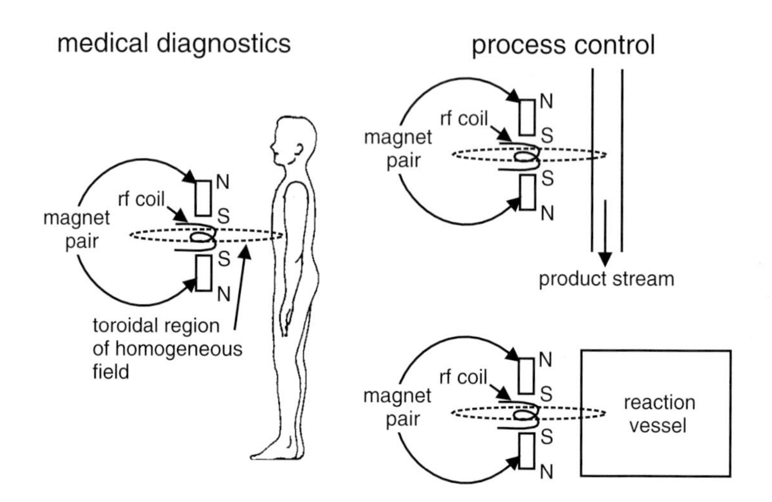

FIG. 9.3.11 [Jac1] Conceivable uses of inside-out NMR in medicine and process control.

In most designs, the polarizing magnetic field is supplied by permanent magnets [Eid1, Jac1, Mat1, Rol1]. For example, the sonde of Fig. 9.3.10 developed for inside-out NMR has been proposed for use in medical diagnostics and chemical engineering for monitoring patients, as well as chemical processes in reactors and pipes (Fig. 9.3.11) [Jac1]. Special electromagnets have been suggested as well: two concentric current loops with opposite polarity provide better homogeneity as opposed to a single coil [Rat2]. When constructed large enough, such coils can be used for *in vivo* imaging, process control, and *production-line monitoring*.

The NMR-MOUSE

For analysis of liquids, homogeneous magnetic fields are often essential, in order to avoid signal loss from translational diffusion. This restriction does not apply to solid matter analysis, and probes with large B_0 field gradients can be employed. In this regard, a simple u-shaped magnet configuration [Mat1, Rol1] is explored for analysis of soft matter like human tissue and elastomers [Eid1, Blü3]. This class of materials is economically important and particularly suitable for NMR analysis, not only because translational diffusion is minor or absent, but also because molecular mobility is high giving rise to long transverse relaxation. Constructed as a palm-size instrument, the device is mobile and can be applied to arbitrarily large objects for NMR interrogation of surface-near volume elements (Fig. 9.3.12). This device has been given the name *NMR-MOUSE* for *mobile universal surface explorer* [Eid1]. Because the polarizing magnetic field is highly inhomogeneous, methods for formation of Hahn echoes and related echoes need to be applied for refocusing of magnetization dephasing from linear spin interactions (cf. Fig. 9.3.4). Therefore, chemical shift cannot be measured, but relaxation times and diffusion constants are accessible for materials characterization.

FIG. 9.3.12 [Gut4] The NMR-MOUSE testing a steel-belted car tyre.

Also, field distortions from ferromagnetic impurities are tolerable, and polymer-iron composites can be investigated [Zim2].

The instrumental features of the NMR-MOUSE are illustrated in Fig. 9.3.13 [Blü3]. A u-shaped magnet is obtained from two permanent magnets mounted on an iron yoke (a). The field lines between the two poles define the polarizing $\boldsymbol{B}_0$ magnetic field. A solenoidal rf coil is placed in the gap between the permanent magnets. Close to the surface of the device the $\boldsymbol{B}_1$ field lines of the rf coil are largely orthogonal to $\boldsymbol{B}_0$ field lines. This region limits the size of the sensitive volume of the scanner. The strongest component of the $\boldsymbol{B}_0$ field is in z-direction across the gap. It exhibits an approximately quadratic dependence on position z (Fig. 9.3.13(b), top), and the gradient G_z varies linearly across the gap, and along the gap it is approximately constant (Fig. 9.3.13(b), bottom). A typical value of the volume-averaged field gradient near the scanner surface is 10 T/m at a field strength of about 0.4 T corresponding to a ^{1}H resonance frequency of 17.5 MHz [Eid1]. Because of the strong gradient, any feasible rf pulse is frequency selective, so that the actual dimensions of the sensitive volume are defined by the $\boldsymbol{B}_0$ and $\boldsymbol{B}_1$ profiles and the excitation bandwidth. Given 2 μs long rf pulses the sensitive volume of the NMR-MOUSE is about 3.5 mm wide across the gap, 8 mm along the gap (Fig. 9.3.13(c)) and 1 mm thick. By reducing the excitation frequency, the sensitive volume is shifted to larger depths y. Given the dimensions of the NMR-MOUSE (Fig. 9.3.13(a)), useful signal-to-noise ratios are obtained from depths up to 3 mm. Larger depth can be accessed by a scale-up of the NMR-MOUSE. In general, this results in lower values for the $\boldsymbol{B}_0$ field, and attention must be paid to suppression of acoustic ringing of the probe circuit when short echo times are needed for analysis of more rigid materials.

Applications of the NMR-MOUSE

Almost any proton-bearing material gives rise to a signal from the NMR-MOUSE. Due to acoustic ringing, echo times are usually limited to values larger than 25 μs, so that signals from rigid materials are difficult to measure. Nevertheless, many polymer materials also contain a long relaxation component, so that, for example, a region of stress whitening could be detected by the NMR-MOUSE in impact modified poly(styrene)

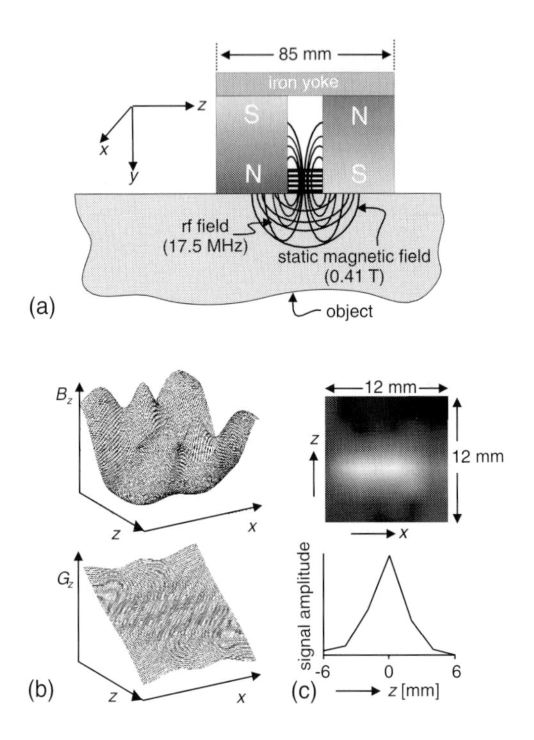

FIG. 9.3.13 [Blü3] The NMR-MOUSE. (a) Schematic drawing and magnetic field lines of the static field $\boldsymbol{B}_0$ and the rf field $\boldsymbol{B}_1$. (b) Calculated profiles of the magnetic field B_0 and across the gap at a distance of $y = 2.5$ mm away from the surface. Top: the z-component of $\boldsymbol{B}_0$. Bottom: the associated field-gradient component G_z. (c) Experimentally determined sensitive volume at the scanner surface.

[Eid1]. However, already soft matter applications are plenty in materials science, food technology, and in medicine. Some examples are collected in Fig. 9.3.14.

A central quantity of interest in the elastomer industry is the determination of the *cross-link density*. This is done either on test samples by measurements of the dynamic–mechanic relaxation, by swelling, and by indentation methods at surfaces [Eli1]. Nondestructive *in situ* analysis is difficult, and the use of the NMR-MOUSE promises a solution, given the dependence of the transverse magnetization decay on cross-link density (cf. Section 7.1.6). Indeed the transverse relaxation times determined by measurements with multi-solid echo trains by the NMR-MOUSE at 17.5 MHz follow those determined at 300 MHz with a conventional DMX 300 spectrometer in homogeneous fields (Fig. 9.3.14(a)) [Blü3]. For the investigated curing series T_2 is shown as a function of the vulcanization time. The cross-link density increases with increasing curing time up to the inversion point at about 10 min. From then on, overcure sets in and cross-link density decreases again. This example demonstrates that the NMR-MOUSE can be used for monitoring of vulcanization processes.

Hahn echo and solid echo produce different contrast in imaging, and the corresponding transverse relaxation times T_2 and T_{2e} depend on cross-link density in a different fashion.

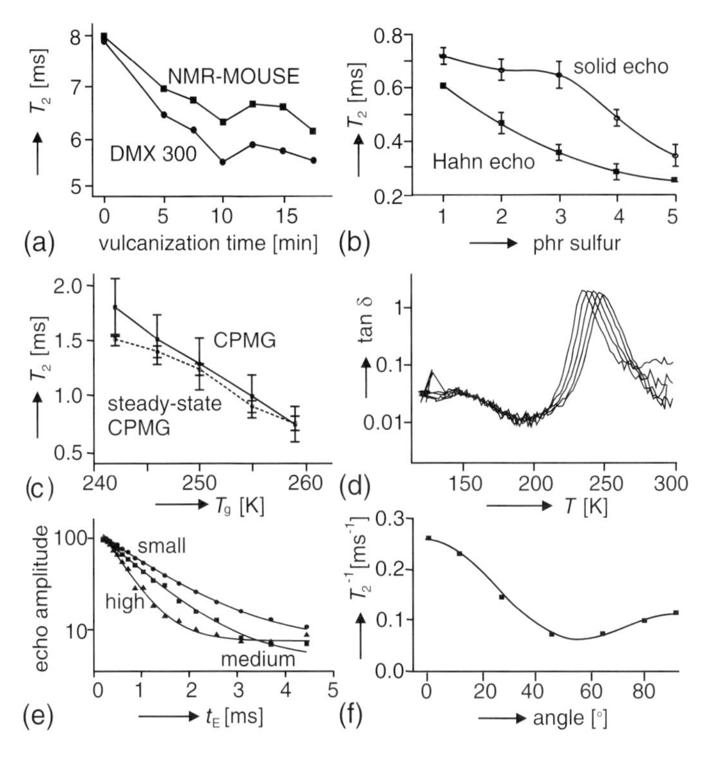

FIG. 9.3.14 Applications of the NMR-MOUSE to soft matter analysis. (a) Comparison of normalized transverse relaxation times of a carbon-black filled NR for different curing times measured at 300 MHz (DMX 300) and with the NMR-MOUSE at 17.5 MHz [Blü3]. (b) Transverse relaxation times for different cross-link densities of unfilled SBR measured by the solid-echo and the Hahn-echo methods [Gut3]. (c) Correlation of transverse relaxation times at 300 K with the glass transition temperature T_g for unfilled SBR samples by the CPMG and steady-state CPMG methods [Blü6]. (d) Conventional determination of T_g from the maximum of the temperature-dependent loss factor tan δ [Blü6]. (e) Normalized Hahn-echo decay curves of poly(butadiene) latices. Different decay rates are obtained for small, medium, and high cross-link densities [Blü4]. (f) Orientation dependence of the transverse relaxation rate in pig tendon [Blü3]. The angle measures the orientation of the long axis of the tendon with respect to the direction of the magnetic field $\boldsymbol{B}_0$.

Because the solid echo refocuses the dipole–dipole interaction of isolated spin pairs completely and the linear spin interactions to 50%, the relaxation time T_{2e} shows little variation with cross-link density for weakly cross-linked materials, where the *residual dipolar couplings* along the intercross-link chains are largely restricted to interacting proton pairs. In this regime the Hahn-echo relaxation time T_2 is more sensitive to cross-link density. At higher values multi-centre dipolar interactions become important. They are not completely refocused by the solid echo. As a consequence, the solid-echo decay is more sensitive to cross-link density at high values. This is demonstrated in Fig. 9.3.14(b) for a cross-link series of SBR [Gut3], where cross-link density is measured by the amount of sulphur in parts per hundred rubber (phr).

A change in cross-link density correlates with a change in the *glass-transition temperature* T_g [Eli1]. An accurate determination of the glass-transition temperature is achieved by temperature-dependent measurements of the *loss factor* $\tan\delta = E''/E'$, where E'' and E' are the *loss modulus* and the *storage modulus*, respectively (Fig. 9.3.14(d)). The maximum of the loss factor determines T_g. Figure 9.3.14(c) demonstrates that such temperature-dependent measurements on test samples could be replaced by measurements of transverse relaxation times by the NMR-MOUSE at room temperature [Blü6]. Rapid measurement of multi-echo trains by the CPMG method provides the desired information. *Steady-state methods* can be used as well (cf. Fig. 7.2.2), corresponding to presaturation by starting from a dynamic equilibrium value. Presaturation has little effect on the resultant transverse relaxation times, because T_1 often shows little dependence on cross-link density.

The strong field gradient of the NMR-MOUSE provides selectivity in investigations of systems composed of liquids and solids. The signals of molecules subject to unrestricted and fast diffusion are efficiently suppressed within a few 100 μs, and only the signals of the solid components are detected. This is illustrated by normalized Hahn-echo decay curves of poly(butadiene) lattices. A latex consists of cross-linked polymer particles which are suspended in a liquid like water. At an echo time t_E of 1 ms, the water signal has disappeared, and the remaining signal derives from the differently cross-linked poly(butadiene) particles. Different decay rates are obtained for small, medium, and high cross-link densities (Fig. 9.3.14(e)) [Blü4].

In soft matter with macroscopically oriented molecules, the transverse relaxation is anisotropic, because the partially averaged dipolar interactions follow the macroscopic molecular orientation [Ful1, Kre2, Teg1]. Consequently, the transverse relaxation depends on the angle between the direction of macroscopic molecular order and the magnetic field $\boldsymbol{B}_0$. For small samples, such investigations can be carried out in conventional NMR magnets or imagers. For large samples single-sided NMR can be used, because the device can be rotated with respect to the object. Thus, *molecular orientation* can possibly be probed nondestructively in polymer products. This feature is demonstrated by measurements of the orientation dependence of the transverse relaxation time in pig *tendon* (Fig. 9.3.14(f)) [Blü3].

Because echo methods are used for measurements with the NMR-MOUSE, ferromagnetic parts near the sensitive volume of the NMR-MOUSE do not necessarily hamper its use. Even material changes induced by artificial weathering in 0.5 mm thick PVC coatings on 1 mm thick iron sheets have been detected [Zim2]. Depending on the field distortions, the sensitive volume has to be recalibrated. Then depth-resolved measurements of steel-belted car tyres can be conducted, or *conveyor belts* with steel cords can be investigated. Figure 9.3.15 illustrates this point by showing the variation of T_2 values for both sides of a conveyor-belt sample. Clearly, one side (b) is more homogeneous than the other (c), demonstrating the use of the NMR-MOUSE for quality control.

Signals acquired by lateral displacement of the NMR-MOUSE can be composed to form an image. Depth resolution is obtained by varying the rf excitation frequency. With the given field profile of the NMR-MOUSE 1 mm change in depth roughly corresponds to a frequency shift of 0.5 MHz. This way, T_2 has been interrogated for car-tyre samples as a function of depth. The data shown in Fig. 9.3.16 have been obtained by measuring

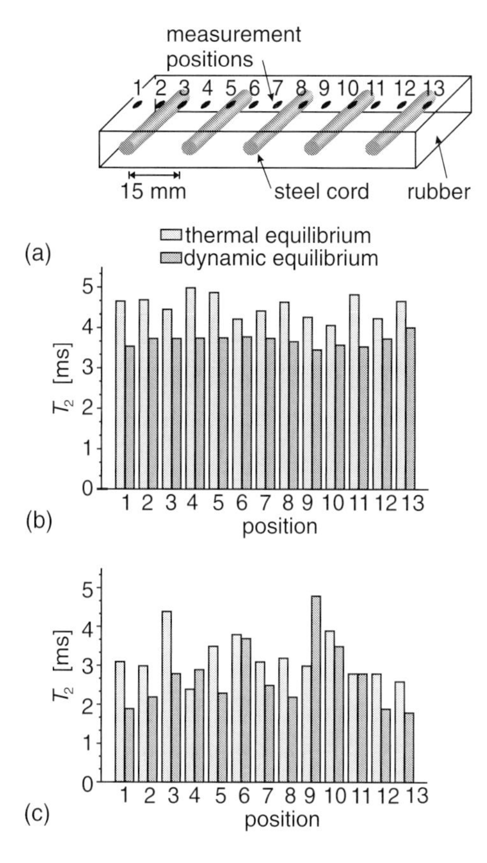

FIG. 9.3.15 Analysis of a section from a conveyor belt with steel cords by the CPMG method starting from thermodynamic equilibrium and from dynamic equilibrium corresponding to partial saturation. (a) Sketch of the sample and positions of measurements. (b) The transverse relaxation times for one side of the belt reveal homogeneous material. (c) The transverse relaxation times for the other side indicate considerable inhomogeneity.

from both sides of the sample [Zim1]. The T_2 values obtained by the NMR-MOUSE correlate well with the contrast in T_2-weighted images of the same samples. A crucial difference in the performance of the corresponding tyres is the hardness of the base near the Nylon cords. Sample (a) has a hard base and sample (b) a soft one. These differences are particularly pronounced at a depth of $y = 6.5$ mm. They showed up in the ratings of the test pilot on the race track as well as in prior testing of the intact tyres by the NMR-MOUSE.

Imaging with the NMR-MOUSE

The spatial resolution achieved by shifting the NMR-MOUSE across the surface and by changing the frequency for scanning depth can be improved in two ways: one possibility is deconvolution of the data set with the shape of the sensitive volume. The other way is

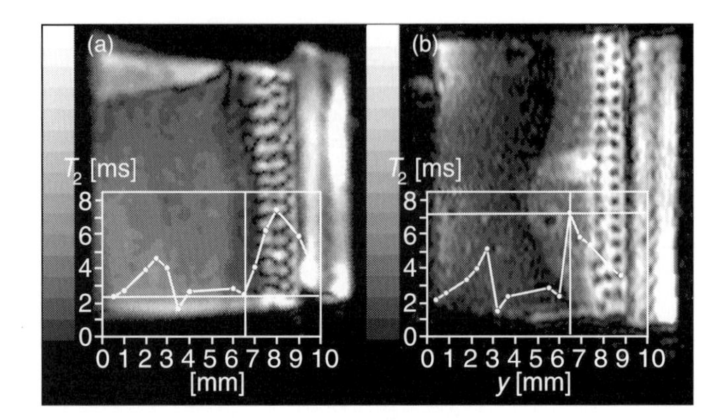

FIG. 9.3.16 [Zim1] T_2-weighted spin-echo images of tyre sections and superimposed transverse relaxation times measured by the NMR-MOUSE. Dark regions indicate hard materials, bright regions correspond to soft material. The dots on the right in each image are cross-sections of Nylon fibres in the tyre base. The fibres are embedded in hard base material (a) and in soft base material (b), respectively. The soft base is identified by a difference of the relaxation times determined by the NMR-MOUSE at $y = 6.5$ mm. Depth into the sample has been accessed by the NMR-MOUSE from both sides in this case.

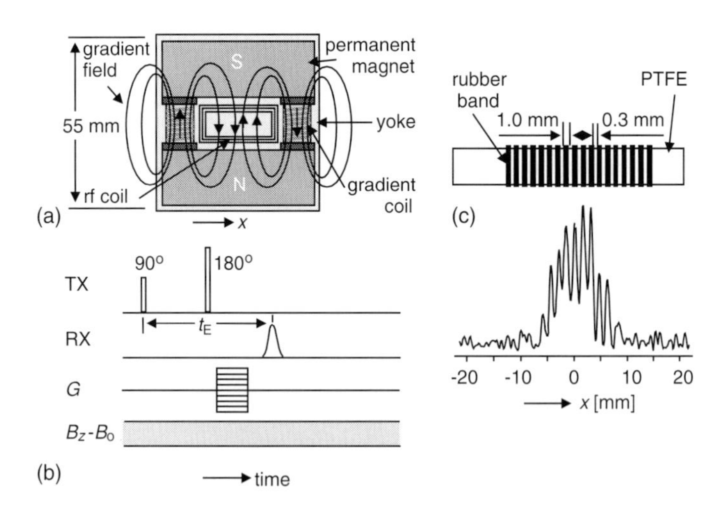

FIG. 9.3.17 [Pra3] 1D imaging with the NMR-MOUSE. (a) Schematic drawing of the sonde. Coils for pulse-field gradients are accommodated within the gap between the two permanent magnets which provide the inhomogeneous field. (b) Imaging sequence. Space information is acquired by spin-echo phase encoding. In the echo maximum the inhomogeneity $B_z - B_0$ of the magnetic field of the NMR-MOUSE is ineffective. (c) Drawing of the phantom consisting of a rubber band wound around a holder from PTFE. The signal from the rubber band and the signal-free gaps are reproduced within the sensitive volume by the 1D image.

using pulsed-surface gradients [Cho1, Cho2] for phase encoding of space information within the sensitive volume.

Along the gap of the NMR-MOUSE the magnetic field is only moderately inhomogeneous, so that the signal can readily be refocused by a spin echo. Pulsed-field gradients can then be used for space encoding in this direction in order to achieve 1D spatial resolution. A realization of this approach is illustrated in Fig. 9.3.17 [Pra1]. Solenoidal gradient coils are accommodated within the gap between the two permanent magnets which provide the inhomogeneous field (Fig. 9.3.17(a)). The associated field gradients are pulsed during the echo time t_E of a Hahn-echo sequence for phase encoding of spatial information (Fig. 9.3.17(b)). In this way spin-density images can readily be obtained as shown for a phantom made from a rubber band wound around a PTFE holder. The gaps between the rubber bands are well reproduced in the 1D image (Fig. 9.3.17(c)). Contrast can be introduced into the pulse sequence by all but the spectroscopic methods discussed in Chapter 7, including sensitivity towards flow (cf. Fig. 7.2.11). Applications of this device are in the analysis of fibre-reinforced elastomers and polymers.

10

Applications

NMR with spatial resolution, in particular NMR imaging, has been investigated rather early for nonmedical applications [Man1]. Yet the impact of NMR imaging outside the medical field has been far inferior to that in medicine, because the applications are not as well defined and the objects and phenomena suitable for investigation are much more diverse. Like in medicine, useful applications in materials science and other disciplines exploit the two dominant features characteristic to NMR imaging [Blü1, Blü7, Cal1, Kim1, Mil1, Par1]. These are the great number of contrast parameters to visualize features hidden to other forms of analysis and the *nondestructiveness* which permits changes in the sample properties and various processes to be monitored.

This chapter reviews *applications of NMR imaging* which are perceived significant to materials science. Most of them are investigations of ^{1}H nuclei. Depending on the spectroscopic linewidth being of the order of 100 Hz or smaller, between about 100 Hz and a few kHz, or larger than about 3 kHz, they can be divided into three categories. For each category some representative examples are outlined in the following:

Fluid matter

- Oscillating reactions
- Flow and diffusion: rock cores, chromatography columns, and complex fluids
- Fluid ingress into polymers
- Current-density imaging

Soft matter

- Defects in technical elastomer products
- Stress and strain in elastomer materials
- Green-state ceramics
- Chemical-shift imaging of plants

Hard matter

- Spectroscopic and chemical-shift selective imaging of polymers
- Relaxation and spin-diffusion parameter imaging of strained polymers

10.1 FLUID MATTER

Fluids and soft matter are particularly suitable to investigations by NMR imaging, because the achievable spatial resolution is highest: transverse and longitudinal relaxation times are comparable resulting in good signal-to-noise ratios, and the spatial resolution is limited essentially by effects of translational diffusion. Interesting applications of NMR imaging concern fundamental phenomena like oscillating reactions and the physics of non-Newtonian fluids as well as investigations driven more by practical needs, like the transport of liquids through porous rocks, the packing of chromatography columns, the fluid ingress into polymers, and the associated swelling behaviour. A rather interesting way of introducing contrast is the application of electric currents for generation of field distortions to mark changes in electrical conductivity.

10.1.1 Oscillating reactions

Introduction

Oscillating reactions [Fie1, Fie2] are complex phenomena, where several chemical species are involved. These are divided into reactants, products, and intermediates. In ordinary chemical reactions reactants decrease, products increase, and intermediates are of low and often constant concentration. In oscillating chemical reactions reactants also decrease and products increase, but intermediates undergo temporary fluctuations. The differences in concentration can range over several orders of magnitude. There is a specific interest in these reactions because they help to understand the properties of systems far away from chemical equilibrium, which are governed by nonlinear dynamic laws, as they are found e.g. in biological systems and in systems which exhibit deterministic chaos [Rue1].

One of the best known and studied oscillating reactions is the *Belousov–Zhabotinsky reaction*. This is the oxidation of an organic compound, like malonic acid, in a sulfuric acid solution by the bromate ion BrO_3^-. The reaction is catalysed by a redox catalyst like Ce(III)/Ce(IV), Mn(II)/Mn(III) or $Fe(phen)_3^{2+}/Fe(phen)_3^{3+}$. The simplified overall chemical reaction is

$$2\,H^+ + 2\,BrO_3^- + 3\,CH_2(COOH)_2 \underset{[catalyst]}{\rightarrow} 2\,BrCH(COOH)_2 + 4\,H_2O + 3\,CO_2$$

In these systems various wave phenomena like trigger, phase, and spiral waves can occur. In some cases, these phenomena can be observed visually. In others, like the manganese-catalysed reaction, this is not possible due to the faint colour of the solution. By means of NMR imaging these reactions can be studied, and as a further advantage analysis of spatial patterns can be extended to three dimensions.

Experimental techniques

Images can be acquired using standard, relaxation-weighted spin-echo and FLASH sequences. The contrast is based on the transverse proton relaxation time T_2 of water,

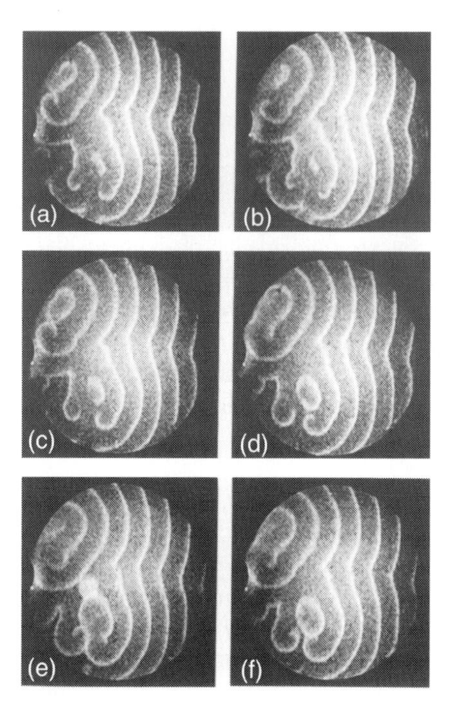

FIG. 10.1.1 Propagation of waves in the Mn catalysed Belousov–Zhabotinsky system. The six images were acquired at 40 s intervals. Adapted from [Arm1] with permission from Wiley-VCH.

which is affected locally by the presence of paramagnetic ions. For the Mn(II)/Mn(III) catalysed Belousov–Zhabotinsky reaction the transverse relaxation time is more strongly affected by Mn(II) than by Mn(III).

Examples

The propagation of waves in the MN catalysed Belousov–Zhabotinsky system is illustrated in Fig. 10.1.1 [Arm1, Tza1] by six images acquired at 40 s intervals. The sample cell was 39 mm in diameter and 2 mm thick. The white regions indicate a high, and the dark regions a low Mn(III) concentration. Two wavefronts are emitted at the left side of the sample cell, which collide and propagate as a single wavefront to the right. The period of the waves is about 2 min.

The velocity of plane *chemical waves* in a cylindrical test tube can be analysed by stacking consecutive 1D projections acquired in the direction of the test tube diameter (Fig. 10.1.2) [Tza2]. In this way a graph of displacement *versus* time is generated, which shows the propagation of the waves along the length of the test tube. The waves travel from top to bottom with a constant velocity of 2.7 ± 0.2 mm/min. As they reach the bottom they are slightly slowed down. Kinematic waves can be imaged in a similar way [Sus1]. Such waves do not involve transfer of matter. They can arise from a frequency or phase gradient in an oscillating Belousov–Zhabotinsky system which leads to an apparent propagation of wavefronts.

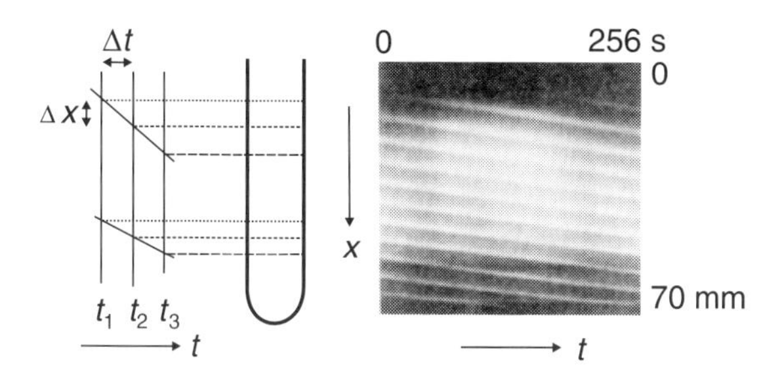

FIG. 10.1.2 Displacement-*versus*-time plots of chemical waves. Left: schematic illustration of the construction by stacking 1D projections. Right: chemical waves propagating downwards in the sample tube. The vertical coordinate is the field of view with a range of 70 mm, the horizontal coordinate is time (256 s). The velocity v of the waves is about 2.7 ± 0.2 mm/min. Adapted from [Tza2] with permission from Elsevier Science.

These examples demonstrate that chemical reactions can be followed in time and space by NMR imaging. Such investigations can be of great importance in technical processes as well as in understanding highly nonlinear phenomena including arrhythmia of the human heart.

10.1.2 Flow and diffusion: rock cores, chromatography columns, and complex fluids

Introduction

The motion of spins along a magnetic-field gradient constitutes the basis for medical applications of NMR *imaging of flow* in terms of *angiography* [Dum1, Dum2, Tur1]. However, also nonclinical applications of flow imaging are important [Cal1, Cap1, Fuk1, Kim1]. Molecular transport gives rise to significant contrast in images of biological tissues [Cal2], oscillatory flow accompanies the formation [Gör1, Gör2] and sustenance [Den1] of life and its functions, and the circulation of water enables the growth of plants [Jen1]. Applications in physical chemistry and engineering are investigations of *convective flow* [Man4, Sha1, Wei1], *turbulent flow* [Kos1, Kos3, Kos5], *granular flow* [Fuk1, Hil1, Kup1, Nak1], flow in *suspensions* [Alt1, Pha1], flow in a model bioreactor [Lew1], and of the viscoelastic behaviour of *complex fluids* [Bri1, Cal4, Rof1, Xia1]. Further applications concern the analysis of the *chromatographic separation process* [Bay1], the characterization of the structure of porous media [Kle3, Str2] like *rock cores* [Dij1, Gui1, Hür1, Mül1, Osm1, Wag1], *building materials* [Bey1, Kop1, Pel2], *catalyst pellets* [Gui2, Kop2] and porous silica [Beh1, Str1], cokes [Ese1], and *foams* [Ass1], and the analysis of *membrane filtration* processes for haemodialysis [Lau1] and water treatment [Air1, Yao2].

Rock cores

The analysis of *rock cores* is of great importance in geological and oil-exploration research, especially at the appraisal stage of determining a reservoir's potential or when examining the oil recovery process. The characteristic properties examined are porosity and permeability. The first is of interest for the determination of the volume of reservoir fluids in the rock and the latter describes the suitability of the rock to recover these fluids through flow under a given pressure gradient [Pac1].

Rock cores are examined in two different states. One is a preserved state where attempts are made to retain the original conditions of fluid saturation. This poses problems because the environment of the samples alters from high pressure and temperature to lower values when lifting the rock cores out of the drill hole to the earth's surface. The other approach uses cleaned cores which have been extracted with solvents to remove the reservoir fluids and impurities. These cores are subsequently resaturated with selected fluids to gain the desired measurement conditions.

Paramagnetic impurities are found in most rock as well as cement samples, so that many investigations are performed with Bentheimer limestone and white Portland cement, which are low in paramagnetic impurities. Most other building materials exhibit higher concentrations of magnetic impurities, but many can be investigated by *stray-field imaging* [Nun1]. Most porous rocks are examined by conventional liquid-state imaging methods with short echo times including chemical-shift selective imaging for analysis of oil replacement by water, but fast space-encoding methods like FLASH and EPI are preferred for imaging of flow [Gui1]. Sensitivity towards flow and diffusion can be achieved by dedicated preparation sequences with pulsed gradients [App1, Cal3, Neu1]. Space encoding is also combined with inversion-recovery and CPMG sequences for measuring longitudinal and transverse relaxation, respectively [Kle1]. For quantitative analysis, the effects of signal loss by background gradients, rf inhomogeneity, and rf attenuation may become important [Rob1].

Experimental relaxation curves are parameterized by fitting single-, bi-, and stretched exponential functions, or they are analysed by algorithms based on the inverse Laplace transformation for distributions of relaxation times [Kle2, Pro1]. By considering the surface relaxivity of rock pores, the surface-to-volume ratio can be obtained from analysis of relaxation curves [Hür1, Kle1]. Diffusion measurements on rock samples by pulsed-field gradient methods are evaluated to obtain the *tortuosity* T which specifies the pore connectivity. The quantity is related to other properties like the electrical conductivity of the fluid-saturated sample or the fluid permeability [Joh1]. The tortuosity can be measured by determination of the normalized diffusion coefficient and extrapolation to the long time limit, $\lim_{t\to\infty} D(t)/D(0) = 1/T$. Alternatively, the pore structure can be probed by NMR studies of liquid flow through the porous medium and interpreted in terms of a statistical propagator [Wag1], a fractal model [Dam1] or a percolation model [Kle3, Mül1, Mül2]. The detailed results from NMR investigations are compared to the generally more crude data from electrical resistance tomography and dye staining investigations [Bin1]. In special cases an alternative access to the pore distribution functions is given by temperature-dependent NMR imaging studies of the melting point depression of liquids in pores, where the melting temperature depends on the pore diameter [Str1].

Investigations related to the analysis of rock porosity concern *building materials* like concrete [Nun1], where the drying behaviour [Gui2] and the residual moisture content are of interest [Bey1, Bog1, Bro1, Kop1, Pel1, Pel2, Pel3, Pel4, Pel5, Pra1]. Furthermore, the pore structure and the drainage of *foams* are of fundamental as well as technological interest [Ass1, Gon1, Kos2, Wea1].

Moisture transport in heterogeneous catalysts

Moisture transport in support pellets for heterogeneous *catalysis* is inherently related to the efficiency of the catalytic process on an industrial scale. Such studies can readily be performed with NMR imaging [Gui2, Kop2, Kop4, Pra2, Tim1]. For example, the results of wetting and drying studies are useful for identifying the underlying transport mechanisms and the rate limiting stages of the processes. The quality of data is high enough to allow quantitative evaluation of relevant parameters of mass transport [Kop4].

By 1D proton imaging, a profile of the water distribution can be obtained. For example, the signal from a diametrical slab in the middle of a cylindrical alumina pellet with a few millimetres in diameter can be selected by means of slice selection along two spatial axes and frequency encoding along the slab. This approach is rapid enough to follow the changes of a fluid profile in real time in the course of drying and sorption processes by detecting a succession of profiles while supplying a stream of dry or wet gas along the pellet surface in the direction of the cylinder axis. Experimental profiles detected during drying of an *alumina pellet* initially saturated with water demonstrate that the technique is sensitive to the fine details of the drying process and can distinguish various stages and regimes of drying. The data presented in Figure 10.1.3(a) show that capillary flow compensates successfully for the water losses at the pellet periphery where evaporation takes place. As a result, an almost uniform water distribution is maintained throughout most of the drying process. Furthermore, the details of the changes in profile shape during pellet drying are shown to be governed by the pore-size distribution of the pellet, which opens up the possibility to extract pore-size distributions of the porous materials from the drying results. The experimental results are suitable for quantitative modelling. Simulations on the basis of the diffusion equation with water content dependent diffusivity demonstrate that at high saturation levels the effective diffusivity value can exceed the water self-diffusion coefficient by two orders of magnitude. While there is a general trend towards lower diffusivities at lower saturation levels, local maxima are observed for those saturations which correspond to the plateaus of the cumulative pore volume curves [Kab1].

Similarly, the basic features of water transport within the individual alumina pellets containing hygroscopic $CaCl_2$ upon water vapour sorption were investigated (Fig. 10.1.3(b)). The experimental results demonstrate that soon after the initiation of the sorption process the regions of the pellet adjacent to the surface become fully saturated with water, and the salt in the inner parts of the pellet does not contribute to the sorption efficiency. Furthermore, due to the presence of the salt in the inner regions of the pellet, the penetration of water into the dry areas becomes the bottleneck of the sorption process. The sharp front propagating at a constant rate resembles the non-Fickian transport often observed upon solvent uptake by polymers (cf. Section 10.1.3). The transport becomes diffusion-like if the initial water saturation of the pellet is nonzero. Simulations have demonstrated that the diffusivity values increase by several orders of magnitude if an egg-shell distribution of $CaCl_2$ is used, i.e., when the inner regions of the pellet are free

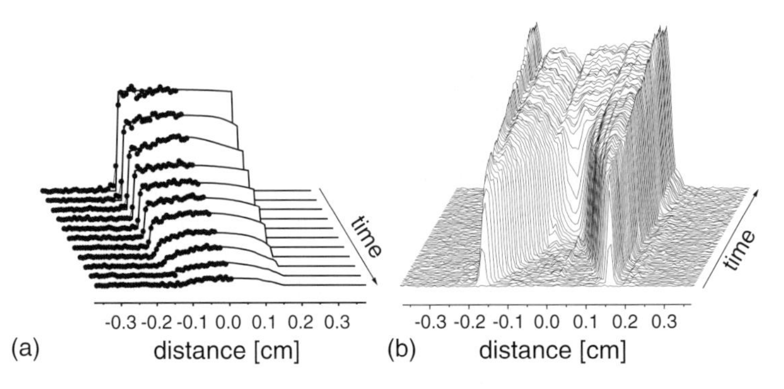

FIG. 10.1.3 [Kop3] Moisture transport in cylindrical catalyst support pellets from alumina with a diameter of 3.5 mm. (a) Drying profiles of water along the diameter of an initially wet pellet. Dotted lines: Experimentally detected profiles. Solid lines: simulated profiles. The first profile was acquired when the dry gas flow was turned on. The delay between the detection of the successive profiles is 60 s. Profiles 1–7, 9, 11, 14, and 18 are shown. (b) Experimental profiles for water vapour sorption by an initially dry pellet containing $CaCl_2$ with uniform salt distribution. A stream of air with 55% relative humidity has been turned on at initiation of the experiment. The profiles shown have been measured 77 s apart.

from the salt. Similar experiments with paramagnetic $CuCl_2$ revealed, however, that if the salt is not bound to the matrix, it redistributes readily upon repetitive sorption/desorption cycles.

The transverse relaxation times of water within the mesoporous alumina pellets studied are of the order of 1–2 ms for fully saturated samples, and drop down to hundreds of microseconds for lower saturation levels. Thus, T_2 weighting of the detected signal cannot be avoided, and the degree of relaxation weighting changes as the local pellet saturation varies in the course of drying or sorption process. Therefore, quantitative modelling of the experimental results requires that an appropriate correction of the observed profile intensity for relaxation effects is made. For this purpose, a simple calibration experiment was used, in which the integral of the NMR signal from the entire sample was measured along with the detection of the 1D profile for various saturation levels under conditions which assured uniform water distribution within the pellet.

Chromatography columns

Magnetic resonance imaging appears to be the only technique by which the *chromatographic separation process* can be visualized in all three space dimensions. Spatially resolved investigations are especially useful for analysis of column packing, which determines the resolution of the chromatographic separation process [Bay1, Tal1].

The process is visualized by injection of Gd^{3+} chelate complexes like Gd(DTPA) as contrast agents on top of the column and using a FLASH sequence for rapid imaging. The unique insight into the separation process gained by NMR imaging is illustrated in Figure 10.1.4 by a series of FLASH images [Bay1]: a parabolic profile arises from faster elution in the core of the column. It is caused by a combination of wall absorption and thermal effects from a temperature distribution across the column diameter. The band

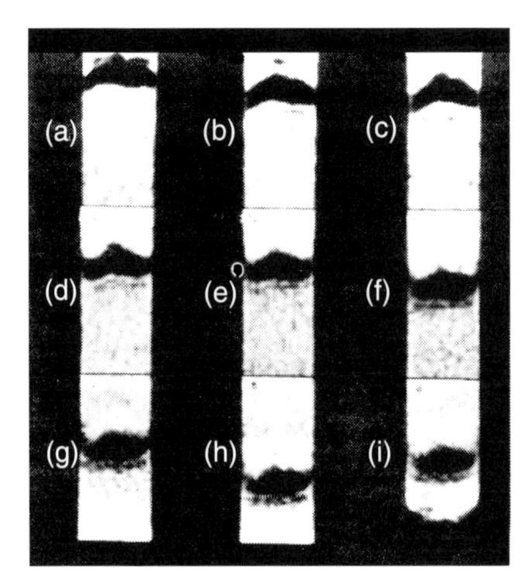

FIG. 10.1.4 [Bay1] Development of an elution profile with FLASH images taken in intervals of 15 min. The parabolic band (black) broadens as it passes through the chromatography column from top to bottom (a)–(i).

of the contrast agent broadens in the middle due to slightly higher temperature in the centre of the column, and molecular transport at the sides is slowed down due to wall interactions. Such a wall effect has been postulated for a long time.

Membrane filtration

In *membrane filtration* processes the investigation of flow patterns and the observation of concentration polarization layers are of special interest. During the last few years several studies have dealt with the investigation of velocity distributions and concentration polarization in membrane filtration modules [Air1, Pop1, Yao1, Yao2]. Depending on the type of module, flow channelling may arise [Lau1, Yao2]. The efficiency of the filtration process is significantly influenced by such irregularities in feedstock flow.

In the filtration process of a 5% oil-in-water emulsion the formation of oil polarization layers has been observed by chemical-shift selective imaging [Yao2]. A model membrane filtration module containing five hollow-fibre membranes of 660 μm outer diameter and 180 μm wall thickness was investigated in horizontal position. The flow-imaging sequence used consisted of a modified spin-echo pulse sequence with a pair of flow-encoding bipolar gradient pulses.

A set of chemical-shift selective images was acquired, each of them with different feedstock pressure and flow rate (Fig. 10.1.5). The images (a)–(c) depict only the oil signal for increasing feedstock pressure. The image (d) is a water image, acquired at intermediate feedstock pressure. The oil images clearly show a dependence of the oil polarization-layer thickness on feedstock pressure and flow rate. Additionally, each of the first three images reveals that the thickness of the *polarization layer* on the surface of the membranes is nonuniform. Whereas, the polarization layer is the thinnest in regions of high feedstock-flow velocity and velocity shear, it becomes wider in areas where the membranes are close together or close to the walls of the module. In these regions feedstock flow velocity and velocity shear are low.

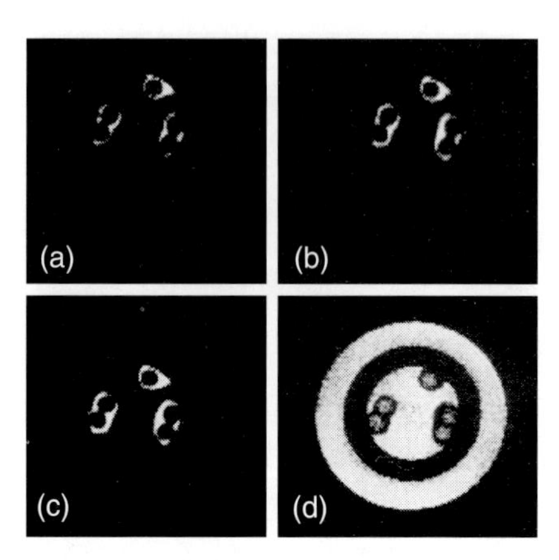

FIG. 10.1.5 Chemical-shift selective images showing the dependence of the oil polarization layer on feedstock pressure and flow rate. The images (a)–(c) are oil images obtained for pressure values of 2.9, 4.5, and 7.6 kPa, corresponding to feedstock flow rates of 1.38, 2.10 and 3.75 ml/min, respectively. Image (d) is a water image corresponding to the intermediate pressure value. The bright outer ring is back-flowing water. It is a result of the particular construction of the module. All images have been acquired with 1 mm slice thickness, 10 mm field of view, and $256 * 256$ pixels. Adapted from [Yao2] with permission from Elsevier Science.

These NMR images prove that oil polarization layers form after feedstock flow has been turned on. The layers dissipate rapidly when the flow of feedstock is turned off. This particular investigation illustrates the importance of NMR-imaging techniques for analysis of technologically relevant membrane filtration processes.

Rheology of complex fluids

One of the central questions in the *rheology of complex fluids* is the molecular origin of mechanical properties. Therefore, coupling of rheometry with techniques which are sensitive to molecular behaviour like molecular alignment, rotational reorientation, velocity distributions, and translational diffusion is required. A method which allows the detection of all these molecular characteristics is NMR imaging [Cal4].

Numerous investigations concerning the non-Newtonian dependence of the shear stress on the shear rate of polymer solutions have been performed. Frequently shear thinning is observed, which denotes a slower than linear increase of the shear stress with the shear rate. Different shear geometries have been explored, e. g. flow in narrow capillaries with inner diameters ranging from 0.7 to 4 mm [Xia1, Gib1] and in Couette cells consisting of coaxial rotating cylinders, one static, one rotating with the fluid under investigation in between the outer wall of the inner cylinder and the inner wall of the outer cylinder [Man3, Rof1, Rof2]. The latter geometry has the advantage of access to high shear rates from 1 to 1000 s^{-1} at small amounts of test fluid.

A *Couette cell* [Han1] suitable for use in an electromagnet is depicted in Fig. 10.1.6, left [Rof1]. Shear stress, velocity gradient, and the vorticity axis are orthogonal to each other in each volume element. The vorticity direction is along the rotation axis of the cell. The molecular deformation vanishes to first order in that direction, so that for

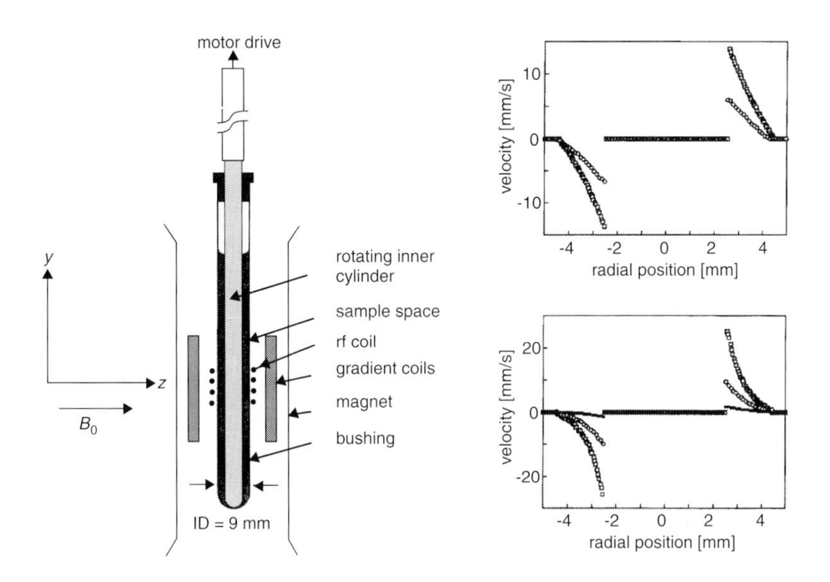

FIG. 10.1.6 [Rof1] Flow in Couette geometry. Left: cylindrical Couette cell in an electromagnet. The inner diameter of the outer tube is 9 mm, and the outer diameter of the inner tube is 5 mm. Right: velocity profiles across a diametral slice obtained for water (top) and a 5% solution of polydisperse, high-molecular-weight poly(ethylene oxide) in water at rotation speeds ranging from 0.60 to 10 rad s^{-1}. Note that the left- and the right-hand sides of the annulus yield similar profiles but with oppositely signed velocities.

observation of shear-dependent nuclear spin interactions the vorticity axis should not be colinear with the static magnetic field.

Velocity maps were measured with flow-compensated spin-echo sequences [Rof1]. The experimental data were analysed in terms of a power law for the *shear stress* σ_{12} which solves the constitutive equation $\sigma_{12} = \eta(\mathrm{d}\gamma/\mathrm{d}t)^n$ for the shear-rate-dependent *viscosity* η and the applied *shear strain* γ. The azimuthal velocity v_φ at radius r and rotation speed ω_R of the inner cylinder can be expressed as

$$\frac{v_\varphi}{\omega_\mathrm{R} r_\mathrm{i}} = \frac{R\left(1 - R^{-2/n}\right)}{K\left(1 - K^{-2/n}\right)}, \tag{10.1.1}$$

where $K = r_\mathrm{i}/r_\mathrm{o}$ is the ratio of the radii of the outer surface of the inner cylinder and the inner surface of the outer cylinder, and $R = r/r_\mathrm{o}$ is the reduced radial variable. The corresponding expression for the *shear rate* is given by

$$\frac{\mathrm{d}\gamma}{\mathrm{d}t} = \frac{2\omega_\mathrm{R} R^{-2/n}}{n\left(1 - K^{-2/n}\right)}. \tag{10.1.2}$$

For a *Newtonian fluid* like water the exponent n is given by 1 independent of the rotation speed. This is indeed observed experimentally by an approximately linear dependence of the velocity on the radial position (Fig. 10.1.6, top right). For a *shear-thinning fluid* the exponent is smaller than one, and the radial dependence of the velocity becomes nonlinear (Fig. 10.1.6, bottom right). For the investigated 5% solution of polydisperse, high-molecular-weight poly(ethylene oxide) in water exponents ranging from 0.82 to 0.44 have been observed for rotation speeds from 0.60 to 10 rad/s, respectively. These measurements have been shown to be more sensitive to the power- law behaviour than comparable ones on a commercial rheometer, especially at low shear rates. No assumptions are needed about the velocity profile, as it is measured directly. These measurements demonstrate the unique possibilities of NMR imaging for detailed investigations of the rheology of *non-Newtonian fluids*.

Related studies have been carried out on capillary flow of polymer solutions [Gib1, Xia1], the flow of suspensions [Alt1, Cap1, Pha1, Pow1, Sey1], and granular material [Che1, Fuk1, Han1, Hil1]. Velocity and diffusion profiles of poly(ethylene oxide) solutions have been imaged by NMR as a function of concentration and pressure gradient [Xia1]. In this way the viscosity can be determined [Pow1]. But as the concentration increases a transition from Poiseuille flow to power-law shear thinning is observed. Measurements of self-diffusion in the presence of a flow field indicate that *shear thinning* is associated with significant enhancement of the polymer Brownian motion along the axis of shear. The data are explained by disentangling polymer coils at high shear rates. A rather spectacular discovery by NMR flow imaging concerns the existence of *shear bands* for rheologically complex fluids (Fig. 10.1.7) [Bri1, Mai1]. The general assumption in the use of cone-and-plate rheometers is that the shear rate $d\gamma/dt = dv/dz$ is uniform across the cell. However, for the long, worm-like tubular micelles, which are formed by

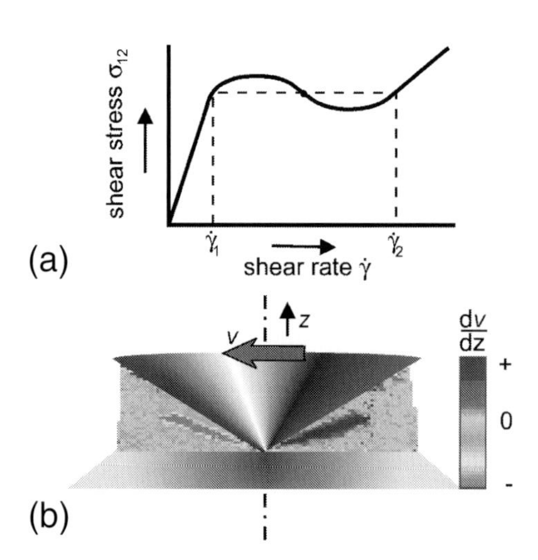

FIG. 10.1.7 [Bri1] Shear banding of the worm-like surfactant. (a) Schematic diagram of the functional dependence of stress σ_{12} on shear rate $d\gamma/dt = dv/dz$. The viscosity is given by the derivative of this curve. To avoid negative viscosities, the fluid separates into two regions with shear rates $(d\gamma/dt)_1$ and $(d\gamma/dt)_2$. (b) Cone-and-plate rheometer and tangential velocity image of the worm-like surfactant. The shear band exhibits enhanced negative velocities left of the rotation axis and enhanced positive velocities on the right.

100 mM cetylpyridinium chloride/60 mM sodium salicylate in water, regions of high and low viscosity are observed already at small shear rates (Fig. 10.1.7(b)). The dynamic phase separation of the fluid into regions of different shear rates is a consequence of the nonlinear relationship between shear stress σ_{12} and shear rate (Fig. 10.1.7(a)). By traditional rheometry only the average viscosity can be detected, so that those data may be misinterpreted.

NMR imaging has further been applied to study falling-ball experiments which are commonly used to characterize the *viscosity* of liquids [Sin1]. The instantaneous motion of a ball through a suspension is complicated and poorly understood, and the ability to visualize how the suspension evolves during a passage of a ball can help the rheologist to understand these effects and to make reasonable viscosity estimates. Furthermore, the flow of cellulose-fibre *suspensions* undergoing steady, pressure-driven flow in tubes was measured by spin-echo imaging [Lit1], and NMR flow imaging of Newtonian liquids and concentrated suspensions [Abb1, Alt2, Gra1] through a sudden contraction has been reported [Iwa1, Xia2], as well as studies of *turbulent flow* [Gat1, Gui3, Kos1, Kos3, Kos4, Kos5, Kos6, Kue1, Lit1, Rok1], *granular flow* [Fuk1, Hil1, Kup1, Nak1], and the measurement of thermal *convection patterns* [Man4, Sha1, Wei1]. Time-of-flight techniques have been applied for investigations of the local velocity and concentration distributions for free water and the solid matrix in flowing pastes [Göt1]. *Pastes* flowing in a ram extruder have been investigated by multi-slice multi-spin-echo techniques and single-point imaging [Göt2]. With the growing availability of hyperpolarized gases (cf. Section 7.4.4), it can be anticipated that *gas-flow imaging* will gain increasing importance [Bru1].

10.1.3 Fluid ingress into polymers

Introduction

Polymer products used in everyday life are often exposed to liquids of particular chemical composition. In this respect, the interaction of polymers with fluids is of great interest. The *fluid ingress into polymeric systems* is of particular importance because the properties of the polymer material may considerably be modified. Physical change of bulk and surface properties can occur due to swelling, plastization, and induced crystallization, resulting in associated changes of mechanical properties such as tensile strength and fatigue resistance, and in changes of biocompatibility. Measuring the system's reaction to solvent ingress in terms of swelling, diffusion, dissolution, and segmental mobility by techniques such as gravimetry, optical microscopy, X-ray or radioactive tracer studies often implies disadvantages, because stopping of the diffusion process or doping and destruction of the sample are inevitable. NMR imaging is ideally suited for examining dynamic processes like diffusion in a noninvasive manner by detecting the mobile molecules of the solvent–matrix system [Bla1].

Case-I and case-II diffusion

In soft material segmental motion is fast, and solvent is taken up easily due to only weak resistance which the solvent molecules encounter on their way into the host polymer matrix. The advancing solvent front has to be supplied with a sufficient amount of

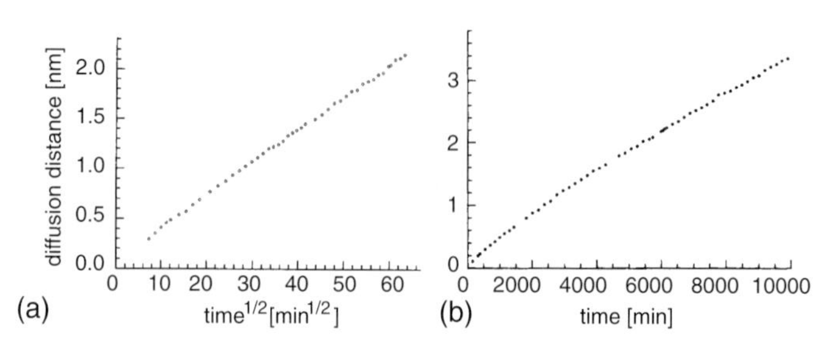

FIG. 10.1.8 Swelling behaviour of different solvents in PVC. (a) 90% acetone and 10% deuterated methanol: Fickian or case-I diffusion. (b) 1,4-dioxane: case-II diffusion. Adapted from [Erc1]. Copyright 1995 American Chemical Society.

liquid, leading to a concentration gradient of the solvent in the polymer matrix. This type of diffusion behaviour is called *Fickian diffusion* or *case-I diffusion*. The solvent front advances linearly into the matrix with the square root of time. An example is the system acetone/PVC (Fig. 10.1.8(a)) [Erc1]. Such diffusion processes are also typical for rubbery materials [Web1].

If the penetrant enters the glassy matrix faster than the polymer can adapt itself by volume relaxation, the solvent front advances linearly with time. This behaviour is called *case-II diffusion* or *relaxation-controlled diffusion*. It is a special case of anomalous diffusion, where the mean square particle displacement is proportional to t^k. It commonly applies to polymers in the glassy state [Wei2]. Here the system 1,4-dioxane/PVC is an example (Fig. 10.1.8(b)). Due to the softening of the material behind the diffusion front, the polymer relaxation in the already swollen matrix is fast enough to adapt to a new situation created by further solvent uptake. Therefore, solvent ingress as well as swelling behind the diffusion front is Fickian.

Experimental techniques

In order to study diffusion in polymers, images which accurately reflect the quantity of solvent per unit volume have to be acquired. Such images are *spin-density images*. Spin-echo images are affected by the system's longitudinal and transverse relaxation times. For T_1 and T_2 being different in the pure solvent, the solvent in the polymer matrix, and the matrix molecules, signals to be imaged can be negatively influenced by misadjustment of the echo time t_E and the repetition time t_R. For solvents T_1 is often quite long, so that the condition $t_R > 5T_1$ implies long measurement times. On the other hand, differences in relaxation times can be exploited for signal suppression. For instance, the signal of the pure solvent can be suppressed by application of an inversion recovery filter. An alternative method to spin-echo imaging is given by the FLASH technique, which exhibits a less pronounced dependence on T_1 and T_2, because it uses small tip angles and gradient echoes. Furthermore, with repetition times of 50 ms and less, the method is fast, and nearly pure spin-density images are obtained. Solvent diffusion into polymers has also been studied by rotating-frame imaging methods [Val1] which are less sensitive to variations in bulk magnetic susceptibility than laboratory-gradient methods.

Applications

Simple profiles of concentration *versus* space allow the classification of the diffusion process in terms of case I and case II [Bla1]. A well-studied system for *case-II diffusion* is poly(methylmethacrylate)/methanol [Erc2, Wei2, Wei3, Wei4]. For practical purposes the ingress of water into Nylon is of particular importance, because water has a plasticizing effect on Nylon. In one study the uptake of water by Nylon has been described in good approximation by a Fickian process [Man2].

More complicated situations can arise, for example, in the *swelling of poly(carbonate)* by acetone. The different stages of associated material change in a PC cylinder are captured in the series of images of Fig. 10.1.9 which have been measured over a period of three days [Erc1]. The cross-sectional images were acquired with the FLASH technique in combination with a saturation-recovery filter for suppression of the signal from the pure solvent surrounding the sample. In Fig. 10.1.9(a) the formation of a high-intensity ring around the core can be noticed, which becomes more obvious with time. As the solvent penetrates into the rod the signal loses intensity from the cylinder surface towards the bright ring inside (Fig. 10.1.9(c)). After about 21 h the rod cracks in the imbibed region. At the tip of the crack the white ring is pushed further into the core while the cracking continues (e)–(h). In Fig. 10.1.9(e) the cracks have reached the centre of the core. Similar images have been obtained by using a mixture of 90% acetone and 10% deuterated methanol. Apart from the white ring, the swelling process has been

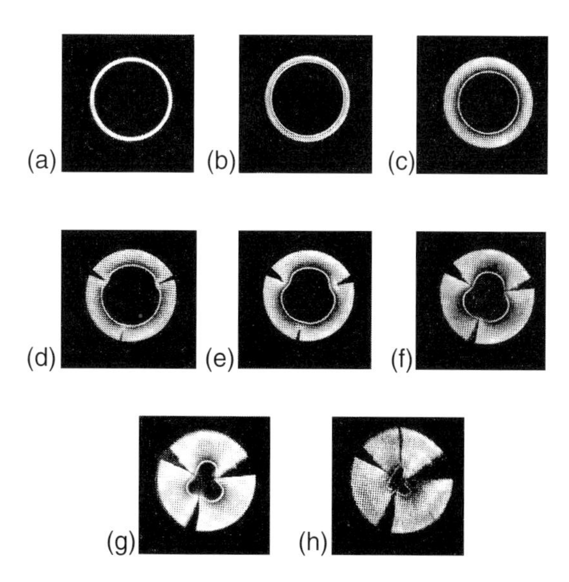

FIG. 10.1.9 Time-resolved FLASH images of the diffusion process of acetone into a 10.7 mm diameter poly(carbonate) rod with a length of 10 mm. The images were acquired with an inversion-recovery filter for suppression of the solvent signal. The images were taken 1 h 0 min (a), 2 h 50 min (b), 19 h 29 min (c), 21 h 25 min (d), 28 h 12 min (e), 46 h 47 min (f), 61 h 23 min (g) and 74 h 53 min (h) after immersion of the rod into the solvent. Adapted from [Erc1]. Copyright 1995 American Chemical Society.

found to exhibit Fickian behaviour. The white ring appears opaque in contrast to the transparent nature of the rest of the material poly(carbonate). This as well as a peak in the thermogram of the swollen material strongly indicate that the ring is a manifestation of *solvent-induced crystallization* [Erc1].

In polymer electrolytes, a dramatic increase in ionic conductivity has been observed upon water ingress [Lau2]. In semicrystalline polymer electrolytes, water is absorbed mainly by the amorphous regions although the diffusion of water through the sample is accompanied by a rapid destruction of the crystalline regions. Furthermore, anisotropic water diffusion has been observed in unidirectional glass-fibre reinforced epoxy–resin composites [Rot1]. Apart from imaging, swelling of polymers may be followed by observing the difference in chemical shift between the pure solvent and the solvent molecules adsorbed to the polymer matrix [Con1].

10.1.4 Current-density imaging

Introduction

By *current-density imaging* magnetic fields are mapped in conducting samples which are generated by application of electric currents (cf. Section 7.4.2) [Joy1, Sco1, Sco2, Sco3]. Thus, distributions in electrical conductivity can be deduced from the image features, and the application of the electrical current can be considered a *contrast agent* applied to the sample.

The current that flows through the sample produces a magnetic field component in the range of a few microtesla. This creates a small shift of the Larmor frequency. Quantitative information about this shift is encoded in the phase of the NMR signal. Through appropriate processing, the magnetic field component can be calculated, and the current density follows from Ampere's law (cf. Section 7.4.2) [Sco3, Ser1].

Applications of current-density imaging can be perceived to the analysis of electrotherapy like cardiac defibrillation and pacing, the definition of standards for electrical safety, and the design of electrochemical cells. However, the disadvantage for applications in medical imaging is the invasiveness of the method. Nevertheless, the sensitivity of current-density imaging is good enough to detect neural firing [Bod1]. Also nuclei like ^{23}Na and ^{35}Cl with low-NMR sensitivity can be detected indirectly in terms of ionic currents, for example, in chemical processes and in reactions [Ber1].

Experimental

Because the current density is a vector, the sample has to be rotated to obtain the complete current density. A very accurate device for rotation is required to guarantee that the individually measured signal components are from the same virtual plane. Standard spin-echo sequences can be applied with two current pulses of opposite polarity before and after the spin-echo refocusing pulse [Sco1]. The change of polarity is necessary to avoid the cancellation of the current-induced effective frequency shift. Static as well as oscillating currents can be employed. The use of currents oscillating at the Larmor frequency leads to rotating-frame techniques [Sco4].

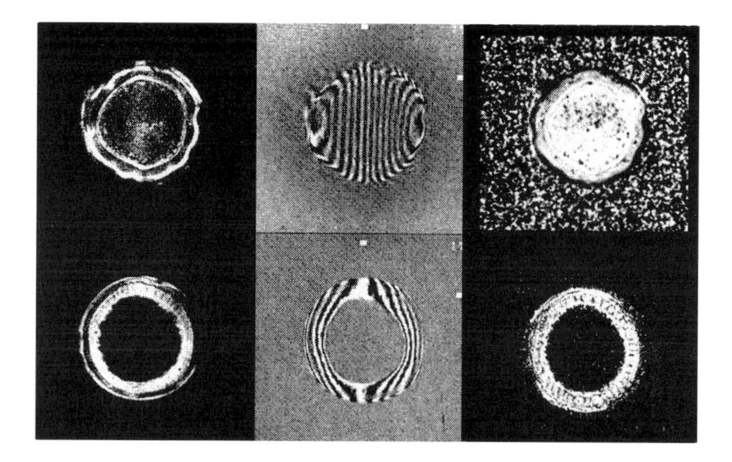

FIG. 10.1.10 [Ser1] Current-density images (right) and conventional magnitude images (left) of plant stems from *Pelargonium zonale* (top) and *Rosa arvensis* (bottom). Images in the presence of a current in a single direction are shown in the centre. The stem diameters are about 8 mm.

Example: plant stems

The capabilities of current-density imaging are readily demonstrated by investigations of plant stems, because the internal stem structure is heterogeneous in conductivity [Ser1]. Figure 10.1.10 compares conventional (left) and current-density images (right) of the plant stems from *Pelargonium zonale* (top) and *Rosa arvensis* (bottom). The stems had a diameter of 8 mm and were moistened in a 10% saline solution to increase their conductivity. They were placed in a holder with coaxial electrodes 13 mm apart. Electric current densities achieved in this way were about 1000 A/m^2. The images for a single orientation of the plant stem with respect to the dominant current direction are shown in the middle. They show patterns of dark and bright stripes, which are caused by the presence of the currents.

The electrical current-density images (right) clearly show that the bright regions of strong conductivity do not correlate with bright regions on the conventional images (left). But sensitivity and resolution are comparable. The signal-to-noise ratio of the current-density images mainly depends on the duration of the current pulses and the magnitude of the magnetic fields generated by them. Long current pulses produce large phase shifts for good spatial resolution but are attenuated by T_2 relaxation and translational diffusion. Spatial resolution better than 60 μm and sensitivity of 150 A/m^2 have been achieved on phantom samples as well as on plant stems.

10.2 SOFT MATTER

Soft matter is intermediate between fluids and rigid solids. Translational motion is largely absent, but molecular mobility is still high, yet often anisotropic or not fast enough to completely average the dipole–dipole interaction among ^1H. Consequently, a residual dipolar interaction remains, which broadens the homogeneous linewidth to a few kHz. Nevertheless, such linewidths are still narrow enough, so that conventional Fourier imaging methods similar to those used in the medical field can be applied. In fact, *human tissue* can be considered to belong to the class of soft matter just like *elastomers*.

In soft matter, the *residual dipolar coupling* is the most distinctive NMR quantity, which can be exploited for generation of contrast. It changes with applied temperature, strain and local chemical reactions, which can be indicative of material *ageing*, for instance. Typical applications of imaging to soft matter are analyses of elastomers [Blü2] for inhomogeneities from production, applied stress, and ageing, studies of the distribution of components in polymer blends and green-state ceramics, and investigations of plants and agricultural products [Mcc1].

10.2.1 Defects in technical elastomers

Introduction

Elastomers are cross-linked macromolecules above the glass-transition temperature. They are entropy elastic and free of viscous flow. For most applications, the rubber is blended with filler material such as silicates and carbon black before vulcanization. Carbon black is an active filler which introduces physical cross-links of macromolecular chains in addition to the chemical cross-links formed during the vulcanization process. The *chemical cross-link density* is temperature independent, while the strength of the *physical cross-links* varies with temperature.

The chemical cross-link density is a central quantity of interest in the elastomer industry, which, in a car tyre for instance, may vary in a well-defined fashion across the tyre tread. Both types of cross-links together determine the moduli of shear and elasticity, G and E, respectively. At small *elongations* $\Lambda = L/L_0$ the *tensile stress* σ_{11} is given by [Eli1]

$$\sigma_{11} = RT[M_c](\Lambda - \Lambda^{-2}), \tag{10.2.1}$$

where L is the length, R the gas constant, and T the temperature. The cross-link density is denoted by $[M_c] = \rho/(M_e N_e)$. It is related to the effective number N_e of Kuhn segments (cf. Fig. 7.1.8) with mass M_e by the material density ρ. The modulus E of elasticity is obtained from the ratio of strain to deformation in the limit of infinitesimally small deformations, where Λ approaches unity. Then

$$E = RT[M_c]. \tag{10.2.2}$$

The same expression is obtained for the shear modulus G at small deformations under application of a shear stress σ_{21}. Conventional measurements of the cross-link density are, therefore, performed by defined sample deformation, for instance, in a rheometer during the vulcanization process of a test sample. The maximum *rheometer* moment for a given deformation amplitude is a direct measure of the cross-link density. Another, but invasive method for measuring cross-link densities in unfilled elastomer samples is by swelling in chloroform or toluene. In practice, spatial variations in cross-link density can arise from concentration fluctuations of filler and cross-linking agent as well as from temperature gradients and effects of thermal conductivity during the vulcanization process.

The macroscopic mechanical properties of elastomer materials are inherently connected with the segmental mobility of the intercross-link chains. At about 90 °C above the glass-transition temperature the rate of segmental motion is already much faster than the strength of the homonuclear dipole–dipole interaction among the protons on a rigid chain. But the motion of the intercross-link chains is anisotropic, because the chains are attached to more or less immobile cross-links, so that *residual dipolar couplings* result for the protons of the chemical groups on a chain (cf. Section 7.1.6). The residual dipolar couplings scale with the cross-link density. They can be probed by *multi-quantum NMR* (cf. Fig. 7.2.9) and by simple measurements of *transverse relaxation times* (cf. Fig. 7.1.9). For reasons of simplicity, the latter are particularly suitable for characterization of macroscopic elastomer properties. In addition to this, differences in susceptibility can be detected by gradient-echo imaging and by frequency encoding (cf. Fig. 7.2.10). The characteristic shadows of susceptibility artefacts are often indicative of carbon-black clusters [Blü2].

Sources of sample heterogeneities

Sample heterogeneities can arise in different stages of elastomer production and product use. Even in a perfectly prepared homogeneous elastomer product, unavoidable ageing processes induce space-dependent defects. Examples for sources of defects are:

- *Mixing process*: Technical rubbers are blends of up to about 30 different compounds like natural rubber, styrene–butadiene rubber, silicate and carbon-black fillers, and mobile components like oils and stearic acid. These components show a large variety of physical, chemical, and NMR properties. Improper mixing leads to inhomogeneities in the final product with corresponding variations in mechanical and thermal properties. A key question of interest is to determine the optimum and most efficient mixing process which ensures a product which meets its specifications for use.
- *Vulcanisation*: The changes in segmental mobility from progress of the vulcanization reaction and from the associated sample temperature distribution can be monitored directly by NMR imaging *in situ* [Fül1]. Heterogeneous structures arise from effects of thermal conductivity, which lead to space-dependent temperature profiles during the vulcanization process depending on the position of the heat source and on heat dissipation. As a result inhomogeneous cross-link densities may occur [Smi1]. In the covulcanization of blends and sheets from different formulations inhomogeneous cross-link densities may arise from differences in solubility and diffusion of the curatives [Kle4]. Different components of rubber blends can be mapped by exploiting the editing capabilities of proton-detected ^{13}C imaging, which are useful even in the case of unresolved lines [Spy1].
- *Ageing*: Ageing processes are most often introduced by UV irradiation and exposure to heat and oxygen. Depending on the load applied, different ageing processes are observed [Den3]. Thermal oxidative ageing usually leads to formation of hardened surface layers in natural rubber as well as in synthetic rubber (SBR, styrene–cobutadiene

rubber) [Blü3, Blü4, Fül2, Knö1, Knö2]. Typically, these layers approach a thickness of 200–300 μm and inhibit progress of the ageing process further into the sample volume. The material hardening is explained by an increase in cross-link density. In the absence of oxygen, chain scission may dominate at elevated temperatures with an associated increase in segmental mobility. Other types of ageing involve aggressive fluids and gases. In this context, a sample of degraded rubber hose has been investigated [Chu1], but also the degradation of polyethylene pipes [Sar4], and the enzymatic degradation of biologically synthesized polymers [Spy2] have been studied through NMR imaging. Related investigations have been carried out on asphalts [Mik1]. Ageing associated with swelling of the rubber particles has been observed in crumb-rubber modified asphalts [Mik2].
- *Mechanical load* (cf. Section 10.2.2): Static mechanical load by strain or compression leads to stretching of random-coil polymer chains in the direction of sample elongation and chain compression in the orthogonal directions. *Stress and strain* effects can be analysed for instance by parameter maps of T_2 [Blü5] and by ^{1}H [Sch4, Sch5] and ^{2}H *multi-quantum imaging* [Kli1, Kli2]. Dynamic mechanical load leads to sample heating, where the temperature distribution in dynamic equilibrium is determined by the temperature-dependent loss-modulus and the thermal conductivity of the sample. Because T_2 scales with temperature for carbon-black filled SBR (cf. Fig. 7.1.13(a)), a T_2 map provides a *temperature map* of the sample. Such temperature maps have been measured for carbon-black filled SBR cylinders for different filler contents and mechanical shear rates [Hau1].

Imaging techniques

Standard imaging sequences like spin-echo and gradient-echo imaging can readily be used for protons [Cha1, Blü2, Knö2]. *Susceptibility effects* from inhomogeneous filler distributions are enhanced by *gradient-echo techniques* as magnetization dephasing other than from gradient evolution is not refocused in either dimension. Problems with gradient switching and gradient strengths are encountered for more rigid elastomers, where transverse relaxation is fast and the linewidth is large. These effects restrict the spatial resolution. Contrast is introduced by T_2, $T_{1\rho}$, T_{1D}, T_{2e} and multi-quantum filters (cf. Section 7.2) to detect local strain, temperature and cross-link density, while material voids are already visible in spin-density images. ^{13}C imaging can be used to edit signals from individual components in multi-component systems [Spy1].

Applications of NMR imaging

An example for the detail of information visible in commercial elastomer products is given in Fig. 10.2.1 by a series of image slices from a 3D image of a *car tyre* without a steel belt, which has been acquired on a clinical imager [Sar1]. The tread is to the left and the interior portion to the right. At an in-plane resolution of 100 μm the tread layer at the left can be distinguished from a second region of higher intensity containing embedded rigid polymer fibres, which appear as dark spots in each slice. The layer boundary as well as the fibre–rubber interface are well defined. In the T_2-weighted images the fibres yield no signal because of short T_2. Numerous defects, presumably voids, chunks of carbon black, and broken or misaligned fibres can be identified. The 3D structure of

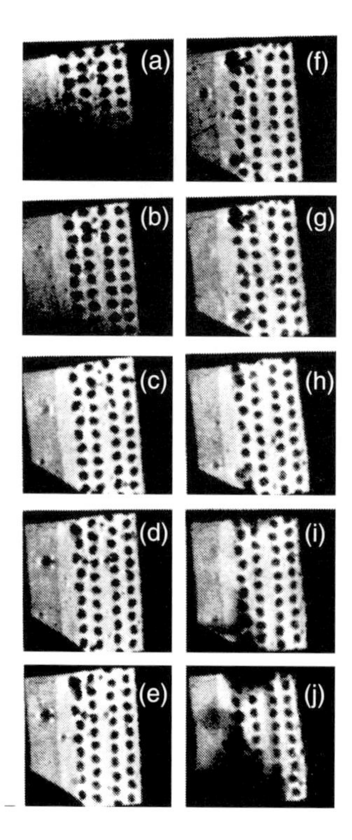

FIG. 10.2.1 Images of a tyre with a polymer-fibre belt. They show the internal tyre structure including base, tread material, fibers, carbon-black agglomerates, and voids. The images (a) to (j) denote contingent slices taken from a 3D image. Adapted from [Sar1]. Copyright 1992 American Chemical Society.

the features can be evaluated through analysis of contiguous slices. Typical arrow-head shaped image distortions in the upper left quadrant of Fig. 10.2.1(d) and (e) reveal *susceptibility artefacts* caused by carbon-black clusters (cf. Fig. 7.2.9).

An invasive approach to improve contrast in elastomers is *swelling* in solvents [Koe1]. The diffusion of benzene or toluene into the network depends on parameters like the local cross-link density [Que1] and, therefore, enhances the visibility of sample inhomogeneities [Clo1, Web2]. The propagation velocity of the swelling can be followed *in situ* and strongly depends on material properties. Two possibilities are discriminated: the first one is swelling by deuterated solvents. This allows the detection of the polymer protons only. Often the molecular mobility is enhanced by the swelling resulting in line narrowing and increased spatial resolution. Deuteron imaging can be used in order to detect the swelling agent for derivation of complementary information such as density differences of the network and agglomerations of other components in the technical elastomer. The second possibility is imaging of a protonated swelling agent: the relaxation parameter T_2 of the liquid agent is long enough to allow acquisition of images with high spatial resolution in a comparatively short measuring time. The contrast is achieved by the local concentration of the swelling agent, which depends on the distributions of cross-link

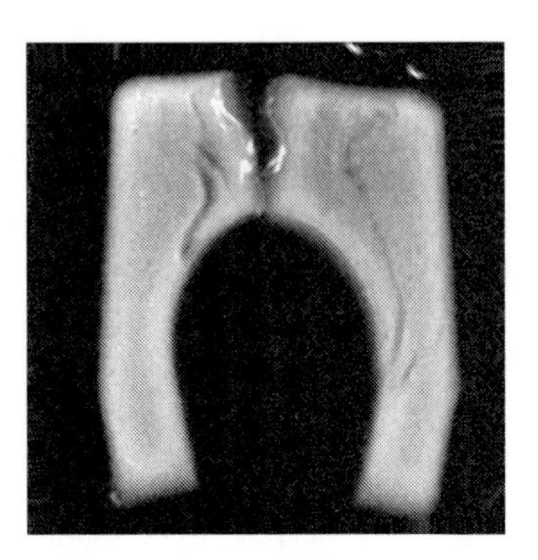

FIG. 10.2.2 Image of a gasket after mechanical and thermal load. The inhomogeneities in the material appear more clearly in the NMR image due to swelling in technical oil. The field of view is $(12\,\mathrm{mm})^2$. Adapted from [Blü6] with permission from Hüthig GmbH.

density and voids in the material. Depending on the choice of relaxation weights, the signal of the matrix protons can be suppressed.

An example of the latter case is shown in Fig. 10.2.2 [Blü6] by a T_2-weighted image of a gasket which has been swollen in technical oil at 100 °C for 3 h. In the image, different features can be distinguished: A crack in the gasket appears dark in the upper middle of the picture. Due to the swelling, the contrast from *cross-link density* variations is enhanced in the image. Dark areas are related to high cross-link density, which corresponds to low concentration of oil and *vice versa*. In addition, the bright stripe outlining the contour of the gasket is interpreted as a swelling front from one of the components of the technical oil with a small diffusion coefficient. The benefit of using swelling agents is a good signal-to-noise ratio within comparatively short acquisition times in the order of 10 min. However, the method is invasive and additives may be washed out.

At high cross-link densities the residual dipolar interaction of the protons on the network chains increases, and the linewidth broadens beyond 3 kHz. Then line-narrowing techniques from solid-state NMR, such as MAS and multi-pulse excitation must be applied to reduce the anisotropy of the dipole–dipole interaction during the space encoding steps of the imaging sequence. In this way, signal-to-noise ratio as well as spatial resolution can be improved [Cor1].

10.2.2 Stress and strain in elastomer materials

Introduction

The elastic properties of elastomer materials are the result of a large number of conformational degrees of freedom for each intercross-link chain. Maximum entropy is found in the most probable of all conformational states, which is described by a random coil. The picture of a random coil is a statistical description in terms of a time as well as an ensemble average. Deviations from the random coil geometry are forced internally by chemical cross-links, physical entanglements, and filler–matrix interactions, and externally by applied strain. Even without applied strain and at high temperatures, where chain motion is fast, the internal restrictions on chain motion give rise to insufficient motional averaging of spin couplings, because the motion is anisotropic, and *residual dipolar couplings* and *residual quadrupolar couplings* are observed (cf. Section 7.1.6) [Coh1, War1]. *Stress* introduces additional deformations of the random coil geometry. In the direction of the local stress, the coils are elongated and, perpendicular to it they are compressed, because the elastomer volume remains essentially constant during deformation. Thus, the tensorial spin interactions change with the deformation, and as a result, relaxation times, the strengths of the dipole–dipole and the quadrupole interactions, as well as their orientation dependences are modified. Therefore, parameters like T_2, the linewidth, the spin-diffusion constant of ^{1}H, and the lineshape and the strength of double-quantum signals from ^{2}H of deuterated spy molecules or ^{1}H of chemical groups in the chain can be used to map local distributions in *strain*.

Mechanical loads can be applied in a static or a dynamic way. For static loads, the *modulus of elasticity* is a real quantity like the spring constant in Hooke's law. It is called the storage modulus, because it stores the applied work as potential energy of deformation. For dynamic loads a phase shift between the driving force and the sample deformation is observed. This phase shift is related to the loss modulus, which describes the energy uptake and the associated sample heating [Eli1]. Therefore, for dynamic deformations the distribution of strains, the phase shifts between stress and strain, and the resulting distribution of temperatures are quantities of interest for materials characterization.

Experimental techniques

Because ^{1}H linewidths in many elastomers are of the order of 3 kHz or less, spin-echo and gradient-echo imaging techniques can be applied. Contrast is introduced by suitable filters like T_2 and double-quantum filters or by use of the spectroscopic dimension. Parameter images of T_2, the double-quantum signal intensity, or the quadrupolar coupling strength are evaluated and rescaled according to theory or experimental calibration data (cf. Section 7.1.6).

Double-quantum imaging of static strain distributions

Strain distributions in elastomers arise from uneven loading conditions, flaws in the elastomer material, and the shape of the elastomer product. As a consequence, magnitude and direction of the local strain may vary considerably. A straightforward way to image local strain is by mapping the transverse relaxation time T_2 and calibration of T_2 *versus* strain (cf. Fig. 7.1.11). Usually the stress–strain relationship is known from mechanical testing, so that T_2 can be calibrated *versus* stress as well. This has been demonstrated through parameter images of filled poly(dimethyl siloxane) rubber bands which were cut and strained [Blü5]. Variations in the image intensity are caused by the *stress distribution* associated with the cut as well as by inhomogeneous filler distributions. Most fillers in elastomers are reinforcing materials which significantly improve the mechanical properties of the product. Because the mechanical properties are mapped by the segmental mobility on a molecular level, already small fluctuations of the filler concentration may be detectable through differences in the proton T_2 even on the relatively crude space scale of NMR imaging (cf. Fig. 1.1.7).

A different sensitivity scale is accessed when deuterated spy molecules are incorporated into the elastomer network and investigated by *²H imaging*. In contrast to the chain segments, the timescale of motion of the usually small spy molecules is largely decoupled from the network motion and only the anisotropy of the network is sensed by the molecules. As a consequence, the deuteron resonance shows a splitting, which maps the magnitude of local strain (cf. Fig. 7.1.12). The width of the peaks carries information about the width of the *orientational distribution function* [Gro1, Sot1]. The splitting can either be imaged spectroscopically, or the underlying residual quadrupolar coupling can be exploited for excitation of double-quantum coherences (cf. Fig. 3.2.1). Double-quantum filtering has the advantage that signal from unstrained material is suppressed. The double-quantum filtered signal intensity depends on the splitting in a sinusoidal fashion, so that the value of the splitting can be derived unambiguously only from a set of double-quantum filtered images, which have been acquired with different settings of the *double-quantum filter*. In addition to the filter settings, the signal intensity depends on the spin density. Therefore, double-quantum filtered images have to be normalized to spin density. Through division of the double-quantum filtered image by an unfiltered image which has been acquired with the same space-encoding parameters, both relaxation weights introduced by the space-encoding steps and spin density are eliminated.

The use of this concept has been demonstrated by ^{2}H images of 1,4-deuterated butadiene oligomers, which have been incorporated into household rubber bands through swelling (Fig. 10.2.3(b)) [Kli1, Kli2]. The double-quantum filter consisted of a $90^\circ_x - \tau_p - 180^\circ_y - \tau_p - 90^\circ_x$-sequence with phase-encoding during the double-quantum evolution time. For small values the sinusoidal dependence of the double-quantum signal intensity on τ_p can be approximated by a linear function in this case (cf. Fig. 7.2.27), so that the intensity of double-quantum images acquired under such conditions scales approximately linearly with the line splitting in the deuteron spectra corresponding to each pixel. Because for small stress, the line splitting is proportional to stress, such images can readily be interpreted as *stress images* in a qualitative fashion. This is confirmed by the overall agreement between an experimental double-quantum image of a strained household rubber band with a cut and a simulation of the stress distribution

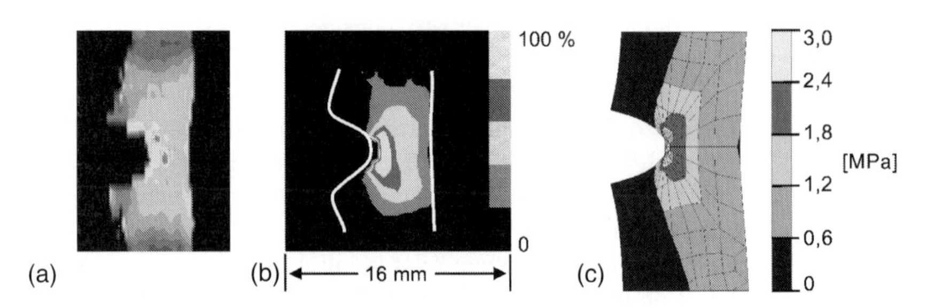

FIG. 10.2.3 Stress images of rubber bands with a cut after correction for spin density and relaxation effects. (a) ^{1}H double-quantum filtered image of poly(isoprene) [Sch6]. (b) ^{2}H double-quantum image at $\Lambda = 1.3$ of deuterated butadiene oligomers incorporated into a rubber band by swelling. (c) FEM calculation of stress in a sample similar to (b) [Kli2].

by the finite element method (FEM) in Fig. 10.2.3(c) [Kli2]. The stress is highest in the centre of the cut. The contour levels of the cut exhibit a butterfly-like shape, which is observed in both the experimental NMR data and the FEM simulation. By acquiring images at different orientation angles with respect to the applied magnetic field, the local directions of strain may be determined.

Sample preparation through deuteration can be avoided by techniques of multi-quantum filtered imaging of residual dipolar couplings among protons on the intercross-link chains [Sch4, Sch5]. Space encoding during the multi-quantum evolution time is hampered by rapid signal decay so that actual *multi-quantum imaging* is abandoned in favour of *multi-quantum filtered imaging*. Optimum sensitivity is obtained for imaging of dipolar-encoded longitudinal magnetization (cf. Section 7.2.10). In this case, the contrast is inverted compared to double- and triple-quantum filters (cf. Fig. 7.2.29). Figure 10.2.3(a) shows a ^{1}H double-quantum filtered image of another rubber band for comparison. The different appearance of the ^{1}H and ^{2}H double quantum images in the figure is not primarily due to different noise levels, but rather to different data processing of the experimental data. In both images the basic features of the stress distribution are reproduced.

Because strain applied to elastomer materials induces *molecular order*, the same multi-quantum methods are applicable to imaging of molecular order in soft biological tissues like cartilage and tendon [Eli2, Tso1]. Other imaging investigations of the effects of applied sample loads include a study of pressure effects on the rubbery phase in high-impact poly(styrene) by chemical-shift selective imaging [Chu1], and a study of the shrinking process from electric fields applied to cross-linked poly(methyl methacrylate) gels [Shi1, Kur1].

Imaging of dynamic displacements

The sensitivity of stress and strain imaging for probing the elastic properties of soft matter can be improved through the use of difference methods based on dynamic displacements of the sample during the pulse sequence. For example, dynamic displacements can be mapped by methods of $\boldsymbol{q}$-space imaging, where a signal loss in the spin echo is observed if the return-to-origin probability of magnetization is reduced by sample deformation [Ree1]. On the other hand, the sample can be deformed in a stepwise fashion during the mixing time in a stimulated-echo sequence (cf. Section 6.2.5), so that the resultant deformation is revealed in an interference phase map of the signals evolving during the dephasing and rephasing times of the echo [Che2]. From such interference maps the elastic modulus can be reconstructed.

Alternatively, the propagation of *elastic waves* can be mapped [Mut1, Mut2]. By applying a gradient which oscillates with the same frequency as the elastic wave and by integrating over several oscillation periods, considerable sensitivity towards small displacements is gained. Wave diffraction and interference can be observed from positions in the sample, where the elastic modulus changes [Mut1]. Because this approach bears interesting potential for analysis of elastomer materials as well as for tumour identification in medical imaging, an example is given in Fig. 10.2.4 by means of an interference pattern of waves excited at 100 Hz in a gel and reflected by the gel boundaries [Fin1]. Given a gradient $\boldsymbol{G}$ which oscillates with the same period T as the

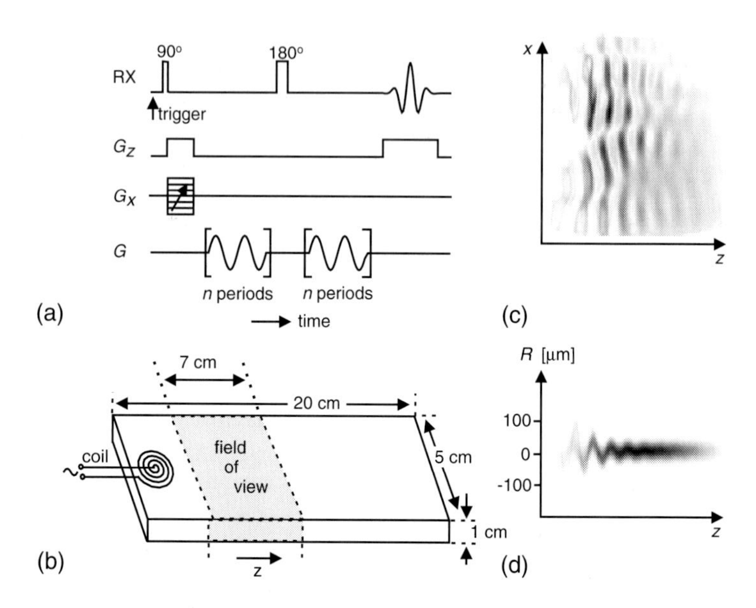

FIG. 10.2.4 [Fin1] Imaging of mechanical waves in an alginate gel at 100 Hz excitation frequency. (a) Pulse sequence. Oscillating gradients are applied in the dephasing and rephasing periods of a spin-echo sequence. The experimental data were acquired with $n = 10$. The pulse sequence is synchronized to the phase of the mechanical excitation (trigger). (b) Sketch of the alginate phantom. Mechanical sample oscillations are induced by a coil placed on the sample which is driven by an alternating current. (c) Image of a given oscillation amplitude R with an interference pattern from wall reflections. (d) Oscillation amplitude along the direction of wave propagation. The gray scale marks the number of spins involved in the motion.

wave propagating in the gel, the signal phase can be calculated as

$$\varphi = -\gamma \int_0^{nT} \boldsymbol{G}(t)\, \boldsymbol{r}(t)\, \mathrm{d}t = \frac{1}{2}\gamma\, \boldsymbol{G}\boldsymbol{R}\, n\, T\, \cos\left(\boldsymbol{k}'\boldsymbol{r} + \phi\right). \tag{10.2.3}$$

Here $\boldsymbol{r}$ denotes voxel position, $\boldsymbol{k}' = 2\pi/\lambda$, where λ is the length of the mechanical wave in the gel, $\boldsymbol{R}$ is the displacement amplitude of the wave, ϕ is the gradient phase offset from the mechanical excitation phase, and n is the number of gradient periods. Clearly, the more oscillation periods enter into the integral, the larger the detectable phase shift or the smaller the detectable sample deformations $\boldsymbol{R}$.

By applying the oscillating gradients in the dephasing and rephasing periods of a spin-echo imaging sequence, the signal of the oscillating parts of the gel can be separated from that of the gel at rest (Fig. 10.2.4(a)). Mechanical oscillations can readily be induced in a gel or rubber sample by placing a coil onto the sample in the magnetic field, which vibrates when driven with an alternating current based on the principle of an inductive loudspeaker (Fig. 10.2.4(b)). By stepping the amplitude G of the oscillating gradient in a set of experiments and subsequent Fourier transformation with respect to G the amplitudes R of the mechanical wave can be extracted. A spin-density image would be obtained at zero amplitude. At nonzero amplitude wave patterns of the type displayed in Fig. 10.2.4(c) are obtained. Such images make it possible to analyse the propagating waves and their attenuation (Fig. 10.2.4(d)). Diffraction and interference effects are caused by sample inhomogeneities including the sample shape. Quantitative evaluation of such images gives access to the storage and loss moduli as a function of space.

^{1}H imaging of temperature distributions from dynamic deformations

The properties of elastomers under *dynamic-mechanical load* are essential for many applications of rubber products. Properties determined for thick samples under dynamic load are averaged quantities due to a temperature distribution across the sample extensions. The prevailing temperature field arises from a balance of heat generation associated with the temperature-dependent loss modulus, the space-dependent thermal conductivity, and the heat dissipation through the sample surfaces.

Because T_2 strongly depends on temperature (cf. Fig. 7.1.13), temperature can be mapped by parameter imaging of T_2. This has been demonstrated for carbon-black filled SBR cylinders of 10 mm in diameter and 10 mm in height by use of a specially designed probe, which permitted application of small-angle oscillatory shear deformation *in situ* over extended periods of time [Hau1]. Sample size and mechanical deformations correspond to those of standard testing procedures. Typical shear rates were 1–10 Hz. Axial parameter projections have been acquired in dynamic equilibrium at a shear rate of 10 Hz and a pixel resolution of $0.4 \times 0.4\,\text{mm}^2$ for carbon-black contents ranging from 0 to 50 phr. 1D cross-sections through those projections are depicted in Fig. 10.2.5. An increase of the temperature in the centre of the sample is observed with increasing carbon-black contents which scales with the increasing loss modulus of the samples. For a composite sample an unexpected *temperature distribution* has been detected [Hau1].

10.2.3 Green-state ceramics

Introduction

High-performance *ceramics* are increasingly used to replace metal products under extreme load. Although their mechanical and chemical properties are superior to metals,

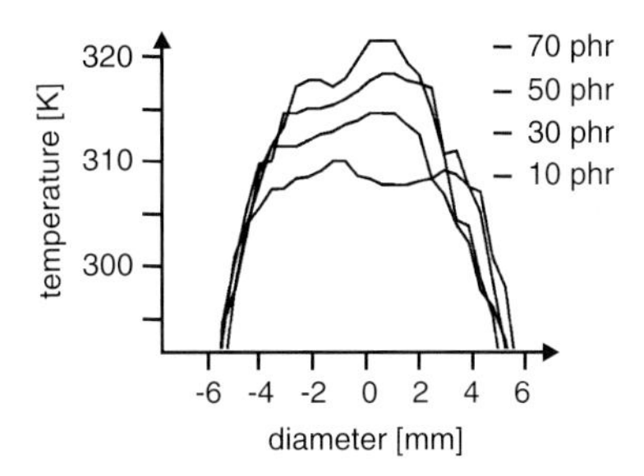

FIG. 10.2.5 Temperature profiles across axial projections of carbon-black filled rubber cylinders with different contents of carbon-black filler at 10 Hz dynamic shear rate. Adapted from [Hau1] with permission from Wiley-VCH.

they are susceptible to failure by brittle fraction. This is the case even more when flaws were included during production. Such flaws can be cracks, voids, inclusions or density variations.

The production of ceramics involves the mixing and shaping of the green-state and subsequent sintering. Most flaws are generated during the processing steps of the green state. In the mixing step the ceramic powder is combined with an organic cocktail to form a slurry or paste for processing. The *green-state ceramic* is subsequently shaped by extrusion or slip casting into moulds. In order to control the quality of a ceramic, a nondestructive evaluation of the homogeneity of the green state is preferred, because faulty parts can be sorted out before further expensive production steps like sintering and machining are undertaken. Also the finished ceramic can be imaged, but then a quadrupolar nucleus like ^{27}Al may have to be detected and solid-state techniques have to be applied [Ack1, Con2].

The green state consists of an inorganic submicrometre powder from components like Si_3N_4, Al_2O_3, and ZrO to which fibres or whiskers may be added for mechanical reinforcement. No general recipe exists for the organic cocktail. Its main components are a polymer binder like poly(ethylene oxide), poly(vinyl acetate), poly(ethylene terephthalate), poly(ethylene glycole), cellulose, or dextrin together with minor concentrations of carrier liquids, surfactants, deflocculants, plasticizers and mould-release agents. The green-state mixture has the consistency of a wax and its NMR properties are therefore comparable to those of elastomers [Cor2]. Typical values of the relaxation times are $T_2 \approx$ 1–10 ms and $T_1 \approx 100$ ms. Although they are indeed similar to those of elastomers, the signal-to-noise ratio in NMR images of green-state ceramics is significantly lower than that in images of elastomers due to low ^1H concentration, which is often less than 10 weight percent in the green state.

Experiments and analysis

The main goal of NMR imaging investigations of green-state ceramics is the detection of flaws like voids and other inhomogeneities. These are caused by interactions between the various chemical constituents of the suspension resulting in agglomerations and chemical segregation even on a space scale which is accessible by NMR imaging. Thus, the acquisition of spin-density images is often sufficient. Back-projection methods are preferable due to their higher sensitivity to rapidly decaying signals [Gop1]. The *STRAFI* method has also been applied successfully [Wan1, Kar1], resulting in images with spatial resolution in the range of 37–75 μm, from which concentration variations in green-state bodies could be found in the range of 47–68% [Wan1]. Other useful methods are phase-encoding or spin-warp techniques [Kar1, Ell1, Gar1, Ack2]. Here the rapidly decaying components are lost, but contrast may be increased.

The homogeneity of the sample can readily be represented by histograms from analysis of image intensities [Ell1, Gar1, Ack2]. But image analysis has to proceed with great care in order to avoid misinterpretations due to artefacts. Therefore, special care needs to be taken to correlate the amount of binder calculated from NMR-images with destructive thermal gravimetric analysis, where the binder is removed through sintering. Fairly good correlations can be found [Gop1].

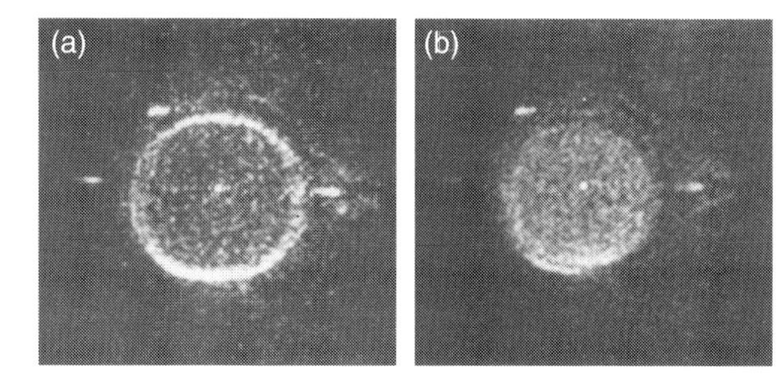

FIG. 10.2.6 [Gar1] 2D spin-warp ^{1}H-NMR images of Al_2O_3 green-state discs with different length-to-diameter (L/d) ratios: (a) $L/d = 0.18$. The mean agglomerate particle size is 0.4 μm at 2.5 weight percent of polymer binder. The acquisition time was 5.5 h. (b) $L/d = 0.57$. The mean agglomerate particle size is 4.5 μm at 2.5 weight percent of polymer binder. The acquisition time was 3.7 h.

^{1}H-NMR imaging of binder distributions

NMR imaging investigations show that the *binder distribution* is unaffected by the mean particle size of the ceramic powder [Gar1, Ack2, Gar2] but that it is influenced by the forming process. Therefore, discs made from green-state ceramics with different length-to-diameter (L/d) ratios were investigated. A strong dependence of the homogeneity in the binder distribution on the L/d ratio was found [Gar1, Ack2] as demonstrated in Fig. 10.2.6. This effect is explained by stress distributions during the forming process which are severely influenced by the L/d ratio. Especially, for small L/d ratios an increased inhomogeneity of the binder is found. This shows that the forming step in the entire process contributes significantly to the generation of flaws. Inhomogeneities were also found for slip casting of ceramic pastes [Kar1] as well as a change of consistency with time [Hay1]. Related studies concern investigations of the homogeneity of solid rocket-motor propellants, which are composite materials of finely powdered solids dispersed in a polymer matrix [Maa1, Sin2]. At 600 MHz an in-plane resolution of $(8.5\ \mu m)^2$ could be obtained, and fine details like a 10–30 μm thick polymer film surrounding larger filler particles could be identified.

Imaging of porosities in fired ceramics

The *porosity* of partially fired or finished ceramics can also be evaluated by NMR imaging when a contrast medium, for example, a liquid or a gas is incorporated into the specimen. This technique has the advantage that, for instance, the liquid is imaged with the associated high sensitivity and resolution. On the other hand, spin–lattice relaxation and self-diffusion can be used to create additional contrast reflecting the average pore sizes below the voxel resolution (cf. Section 7.1.6). However, areas with closed pores are inaccessible to the liquid and may be misinterpreted as defects.

The use of this method is demonstrated by bisque-fired Al_2O_3 specimen which were vacuum impregnated by benzene doped with paramagnetic chromium salts [Ack2]. Figure 10.2.7 shows three slices from that experiment. Nonuniform signal intensities with areas of 0.7–4 mm in diameter could be detected indicating low porosities, but only

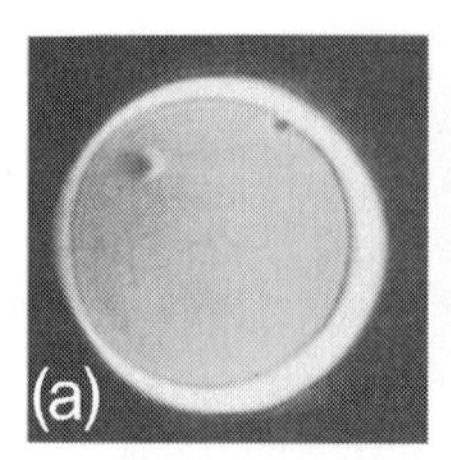

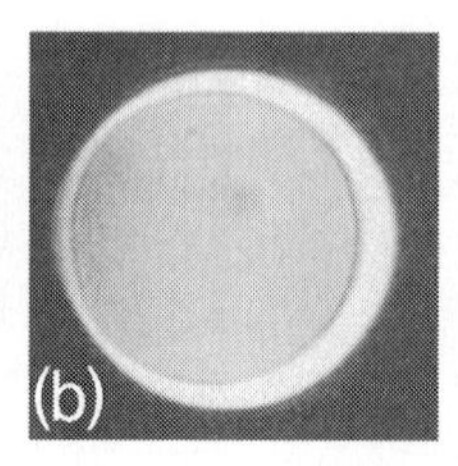

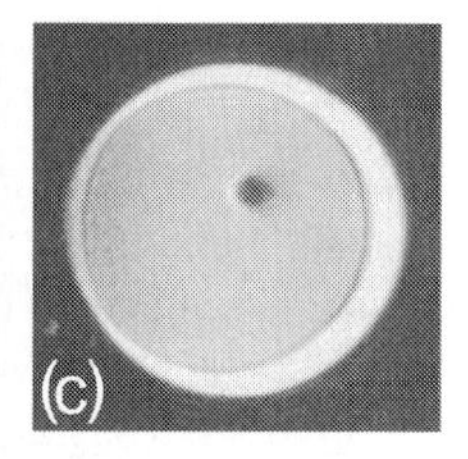

FIG. 10.2.7 ^{1}H-NMR images of a bisque-fired alumina ceramic impregnated by benzene. The L/D ratio is 1.13 and the mean particle size is 0.4 μm. The three slices correspond to positions at (a) 6.1, (b) 12.4, and (c) 14.8 mm from the reference plane. Various defects with diameters ranging from 0.7 to 4 mm can be detected by decreased signal intensities from lowered porosities. Adapted from [Ack2] with permission by the author.

the largest ones are visible in the figure. The signal intensity in these regions is lower by up to a factor of three with respect to the average value in the specimen. In a related investigation, C_2F_6 gas imbibed into porous ceramic-fibre composites has been imaged [Liz1]. Inhomogeneities and voids in the matrix as well as the layered structure of the composite could be resolved.

10.2.4 Chemical-shift imaging of plants

Introduction

Many techniques like optical spectroscopy for the study of *plant structure* are labour intensive, destructive and consequently inaccurate. Plant materials need to be prepared for analysis by fixation, embedding, cutting and staining. NMR imaging, on the other hand, is potentially noninvasive, nondestructive, employs nonionizing radiation that can permeate soil media for investigations of root systems, and has no known effects on plants. In fact imaging experiments may be repeated indefinitely over the lifetime of the plant to monitor structural growth and development, the response of the plant to light, the CO_2 level, nutrients distribution and transport, and many other plant parameters *in vivo* [Chu2, Hei1, Mac1, Met1, Rat1, Rok1, Sar3].

The noninvasive nature of NMR spectroscopy combined with the chemical specificity of the NMR method provides direct access to the distribution of various chemical constituents for the histochemistry of plant materials *in situ*: NMR spectroscopy can be used to identify the major constituents, and chemical-shift imaging can be used to spatially localize them. The latter can be applied to localize aromatics, carbohydrates, as well as water and fat or oil in plant samples. The suitability of many fresh fruits and living plants to be studied by NMR imaging results in a variety of applications in agriculture and food science [Mcc1, Mcc2].

Experimental techniques

For *chemical-shift imaging of plants*, standard spin-echo imaging sequences are used. Chemical-shift resolution is introduced by selective excitation in the absence of gradients, chemical-shift difference methods (cf. Section 6.2.4), or by saturation of unwanted resonances. Good chemical-shift resolution requires high-field strengths, and microscopic resolution requires small sample dimensions. For the investigation of large, intact samples, clinical whole-body MR systems have to be employed. Other than 2D and 3D images, the acquisition of images with chemical-shift resolution is also of interest for the observation of processes with high time resolution. The effectiveness of spectroscopic editing for chemical sensitivity is increased by imaging correlation peaks of 2D spectra [Met2] and by indirect ^{13}C imaging [Hei1]. In cases of radial symmetry spectroscopic radial imaging may be applied (cf. Section 4.4.3) and the resultant increase in signal-to-noise ratio can be exploited for 2D spectroscopic resolution [Mei1].

In vivo *examination of plant morphology and development*

The examination of *roots* is essential to the understanding of plant growth and productivity. By ^{1}H NMR imaging root morphology and function can be investigated *in situ* within the soil media in which they are grown. This can be done as a function of soil type depending on the ferromagnetic particle content, and of the soil-water content. The root system of *Vicia faba* has been studied using a 1.5 T medical imaging system [Bot1, Rog1]. The smallest roots that could be identified in images have a diameter of 0.3 mm. Paramagnetic contamination of native soils provides excellent visibility of root water against a background of invisible soil water over an extremely broad range of soil-water contents (Fig. 10.2.8). But only soil media with ferromagnetic particle contents of less than 4% produce suitable *in situ* root images by spin-warp techniques. At high ferromagnetic contamination levels, stray-field techniques can be applied [Chu2, Kin1]. Images of 30-day old plants indicated that water distribution and transport in roots with light-stressed foliage and also the process of wilt and recovery can be monitored *in situ* [Bot1, Rog1].

Noninvasive plant histochemistry

Noninvasive investigations of *plant histochemistry* are important applications of *chemical-shift imaging* [Cof1, Met1, Met2, Pop2, Rum1, Sar2, Sar3]. This is demonstrated by example of *in situ* images of dried and fresh fennel fruits [Sar2, Rum1, Pop2]. The fennel fruit or mericarps exhibit resonances of two main constituents, anethole

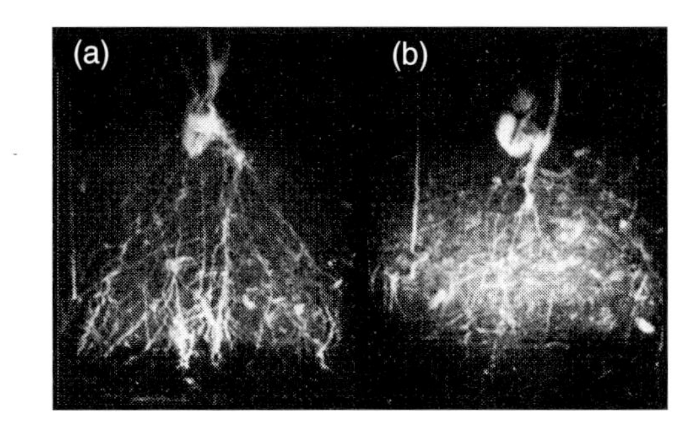

FIG. 10.2.8 [Bot1] (a) ^{1}H-NMR spin-echo image of a 30-days old *Vicia faba* seedling in a 15 cm diameter plastic pot of pealite with 19% soil moisture. (b) Similar root system with higher soil moisture.

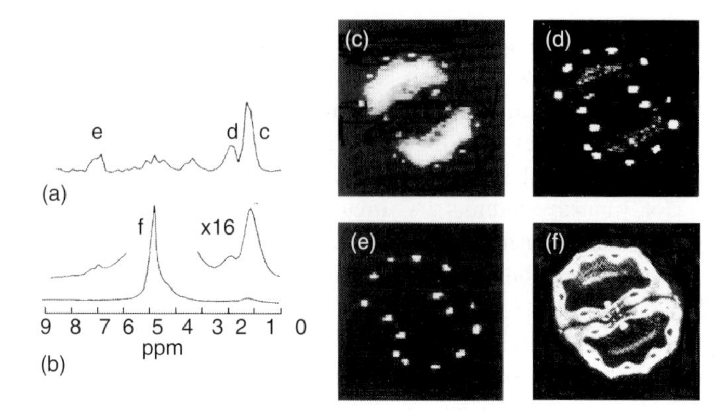

FIG. 10.2.9 Chemical-shift imaging of fresh fennel fruit. (a) Water-suppressed ^{1}H NMR spectrum. (b) Spectrum of the same fruit without water suppression. (c) Water-suppressed image of the methylene/methyl region at 1.3 ppm. (d) Water-suppressed image of the methoxy groups of anethole at 1.8 ppm. (e) Water-suppressed image of the aromatic/olefinic groups of anethole at 7.0 ppm. (f) Nonselective image without water suppression of the same slice. The field of view is 4 cm × 4 cm. Adapted from [Rum1] with permission from Elsevier Science.

and reserve oil in addition to a water signal (Fig. 10.2.9(a), (b)). Through chemical-shift imaging with water suppression, both components have been imaged in fresh fruit. Figure 10.2.9(c)–(e) shows images of particular chemical-shift regions. Image (c) corresponds to the methylene/methyl region of the spectrum, (d) shows the methoxy groups of anethole, and (e) the aromatic/olefinic groups of anethole. The images map the different distributions of the chemical constituents anethole and reserve oil within the selected slice. The signals of anethole correspond to the oil canals of the mericarp and those of the reserve oil correspond to the endosperm tissue.

To complement the water-suppressed chemical-shift images, a nonselective image of a fresh fennel fruit without water suppression is shown in Fig. 10.2.9(f). This is essentially an image of the water distribution, because the water signal dominates the spectrum (b). The water is confined mainly to the outer pericarp, which, because it is free from both reserve oil and anethole, does not show up at all in the chemical-shift selective images (d)–(f). The images have been acquired at 9.4 T with spatial in-plane resolution of 62.5 μm and a slice thickness of 1 mm. The techniques of chemical-shift imaging demonstrated in Fig. 10.2.9 have been extended to imaging of correlation peaks in 2D spectra for analysis of plant histochemistry [Met2].

Further studies

Sucrose is a key carbohydrate in *plant metabolism*. The concentration of *sucrose* in space and time is an important parameter in plant growth and morphogenesis, which can be determined by spatially resolved NMR [Met1, Tse1]. The sucrose distribution, for example, has been examined quantitatively in *Ricinus communis* seedlings [Met1]. Until now, information on the sucrose concentration in the phloem has been obtained only by several destructive methods. They all require the opening of the sieve tubes, which in turn may modify the water flow in the plant, so that it cannot be known whether or not the invasive nature of these methods influences the measured sucrose concentrations. On the other hand chemical-shift imaging allows quantitative measurement of the sucrose concentration within the hypocotyl of an intact growing plant seedling without disruption of flow in the vascular bundles, and many seedlings can be placed into the spectrometer without significant disturbance of their physiological environment.

The observation of image intensity changes associated with seed rot suggests applications of the technique to plant injury studies involving disease, chemicals including herbicides and pollutants, bacterial, fungal and insect invasion, and physical stress. It is also possible to combine these methods with flow and diffusion imaging techniques [Ecc1, Jen1, Rok1] to visualize the transportation pathways of specific chemical constituents [Sch3]. The experimental results give evidence that NMR imaging is going to play a significant role as a research tool in studies of plants and foodstuff.

Most applications of chemical-shift imaging to plants investigate ^{1}H of water, aromatic molecules, oil, and carbohydrates to gain information about the plant structure and the distribution of chemical constituents [Sar3]. Applications to plant sciences of imaging isotopes other than ^{1}H are seldom [Rol1, Sil1]. Nuclei of interest include ^{17}O, ^{31}P, and ^{13}C. For imaging of rare nuclei like ^{13}C isotope enrichment may be necessary by growing the plant materials for instance in a ^{13}C enriched atmosphere as the substrate for photosynthesis compounds or polarization needs to be transferred from ^{1}H to ^{13}C [Hei1, Sil1]. For *in situ* plant-water balance studies a *portable NMR spectrometer* has been constructed. NMR data of plant-water hydraulics under greenhouse conditions were compared with water uptake and transpiration rates under variation of light intensity and relative humidity [Van1]. Related studies concern the observation of neocartilage growth in a bioreactor [Pet1] and the mapping of heavy metal absorption in gels, immobilized cells, and bioabsorbents [Nes1].

10.3 HARD MATTER: POLYMERS

Hard, rigid matter is most difficult to image by means of NMR, because for abundant nuclei the homonuclear dipole–dipole interaction severely broadens the linewidth, so that line-narrowing techniques need to be applied. For practical purposes, magic-angle spinning often is not feasible, because considerable hardware in addition to a solid-state imaging spectrometer is required as well as alignment of the sample torque with the spinning axis. Therefore, multi-pulse line-narrowing techniques are preferred [Cor4, Jez1]. Particularly, methods which use magic echoes [Haf1], and single-point imaging methods can be applied [Axe1, Ken1] (cf. Chapter 8). On the other hand, rare nuclei can be imaged [Gün1, Gün2], but sample preparation by isotope enrichment may be necessary. In that case, spectroscopic deuteron imaging provides valuable and detailed information [Gün1]. But in any case, the sample size is restricted to fit coils with diameters of the order of 10 mm. Otherwise, the rf excitation power rapidly exceeds the affordable limit of 1 kW. Furthermore, slice selection in rigid solids has not yet been solved satisfactorily, because any line-narrowing approach creates short-lived transverse magnetization at some time. Therefore, most 2D images are in fact projections over a third dimension or slices through 3D images.

10.3.1 Spectroscopic and chemical-shift selective imaging

Introduction

The spectroscopic dimension in NMR gives access to the chemical structure under high-resolution conditions, where anisotropic spin interactions are suppressed. The contributions to the spectrum from anisotropic interactions reveal information about the physics of the sample in terms of molecular order and slow motion (cf. Section 3.2.1). Because effects of polymer processing and sample deformation by drawing, strain, and pressure are often associated with changes in molecular packing and orientation, knowledge of the spatial distribution of *molecular orientations* is desired. NMR imaging appears to be a suitable tool for this, if the spectroscopic dimension is accessed [Cor3, Gün1]. This can be done in two ways, either by spectroscopic imaging (cf. Section 7.3) with acquisition of a complete NMR spectrum for each position in space or by chemical-shift selective imaging (cf. Section 7.2.4) where the signal of the image is derived from selected chemical shifts. The complete *orientational distribution function* is accessible only by spectroscopic imaging when the inhomogeneous lineshape broadened by one anisotropic spin interaction is analysed. This requires that there is no overlap of wideline resonances centred at different chemical shifts, a situation which is given only for chemically simple solids like poly(tetrafluoro ethylene) [Hep3] and highly ordered materials, where anisotropies and thus linewidths are small [Gün2]. Otherwise, one can resort to site-specific isotopic enrichment for instance by deuterons in proton-rich solids [Gün1, Gün3].

Uniaxial *drawing of polymer samples* introduces a uniaxial distribution of molecular orientations which often can be approximated by a Gaussian. The NMR spectrum then depends on the angle between the drawing direction and the magnetic field. For example, the ^{19}F spectrum of unoriented PTFE is well approximated by a symmetric powder pattern (Fig. 10.3.1(a), cf. Fig. 3.1.3), while the spectrum of a uniaxially oriented sample exhibits two peaks when the direction of molecular order is parallel to the magnetic field (Fig. 10.3.1(b)) [Hep1].

Because ^{19}F is an abundant nucleus in PTFE, both anisotropic chemical shift and dipole–dipole interactions are present in the sample. The homonuclear dipolar

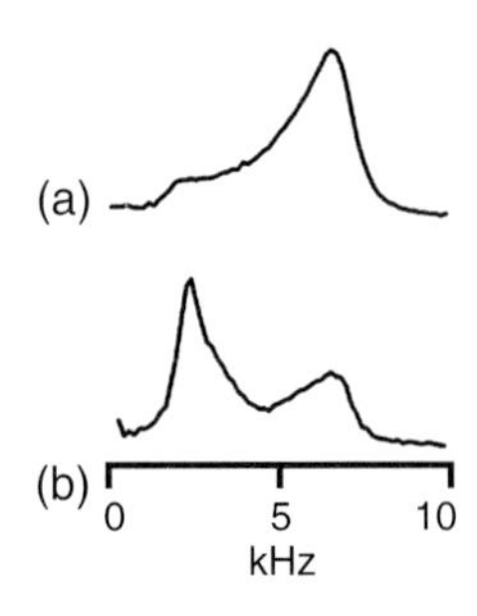

Fig. 10.3.1 [Hep1] ^{19}F chemical-shift spectra of PTFE. The dipole–dipole interaction has been removed by MREV8 multi-pulse excitation. (a) Isotropic molecular orientation. (b) Uniaxially drawn sample. The drawing direction is parallel to the magnetic field.

interaction has been eliminated from the spectra by use of MREV8 multi-pulse excitation (cf. Table 3.3.1), so that the spectra in Fig. 10.3.1 are determined by chemical-shift anisotropy only. The high-intensity edge at about 7 kHz is from orientations of the principal axis of the chemical-shift tensor perpendicular to the magnetic field and the low-intensity edge at about 2 kHz is from orientations parallel to the field. Chemical-shift selective imaging of either edge of the spectrum leads to contrast from molecular ordering. The spectra are from the crystalline sample regions only, because the signal from the amorphous regions is suppressed by fast relaxation during the multi-pulse excitation.

Experimental techniques for ^{19}F imaging

Chemical-shift selection The homonuclear dipole–dipole interaction among the abundant ^{19}F nuclei needs to be eliminated by use of a multi-pulse sequence like MREV8. Such a multi-pulse excitation has been combined with the DANTE sequence (cf. Fig. 5.3.11 and Section 7.3.5) for selective excitation of the spins at the chemical-shift tensor orientation perpendicular to $\boldsymbol{B}_0$ [Hep2].

Spectroscopic imaging Multi-pulse excitation can be used in principle for elimination of the dipole–dipole interaction during acquisition of the spectroscopic response, but for practical purposes this is a rather sophisticated approach. The most robust solid-state imaging techniques suitable for commercial spectrometers are *single-point imaging* (cf. Section 8.3) and *magic-echo imaging* (cf. Section 8.8). In a series of successive magic echoes the dipole–dipole interaction is efficiently removed and the echo modulation is determined only by the scaled chemical shift and the gradients applied in the windows between successive magic sandwiches (cf. Fig. 5.3.13). Because these windows can be longer than those used in multi-pulse sequences based on solid echoes, longer gradient pulses can be applied. Typical durations of time windows are in the order of the inverse linewidth. When a 180° pulse is added to the second 90° pulse in the magic-echo sandwich, a time-suspension sequence results, where both bilinear and linear spin interactions are refocused. The particular, imaging sequence used is depicted in Fig. 7.3.2(a) [Hep3]. The chemical-shift modulation which builds up during the first half is gradually refocused during the second half of the excitation pulse train. During the first half, pulsed gradients result in phase encoding of spatial information, and the chemical-shift response is acquired during the second half by stroboscopic observation of the echo modulation.

Examples

1D images with signal intensities weighted by effects of molecular orientation have been measured for the purpose of demonstration on a sandwich phantom consisting of three PTFE layers (Fig. 10.3.2) [Hep1]. The sample composition is sketched in (a). Without chemical-shift selection the contrast is determined by the integral over the spectroscopic dimension, i.e., by the spin density. The spin-density image is shown in (b), and the three layers cannot be distinguished. By selective excitation of the signal near 7 kHz (cf. Fig. 10.3.1(a)) contrast is introduced from the chemical-shift tensor orientation perpendicular to the magnetic field, and the oriented sheet sandwiched between two unoriented

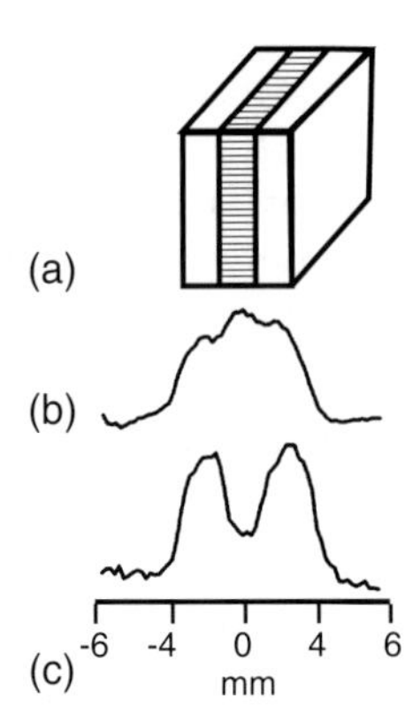

FIG. 10.3.2 [Hep1] Demonstration of chemical-shift selective imaging on a three-layer sandwich phantom from PTFE sheets. (a) Geometry of the phantom. The outer layers are from isotropic material; the inner layer is from oriented material. (b) 1D spin-density image. (c) Orientation-weighted image obtained from chemical-shift selection of the perpendicular edge of the spectrum near 7 kHz (cf. Fig. 10.3.1(a)).

sheets can be identified. This approach works well with ^{19}F because of the large chemical-shift range of the nucleus (cf. Table 2.2.1). It is unlikely to work equally well for ^{1}H.

More detailed analysis is possible if *spectroscopic images* are acquired instead of *chemical-shift selective images*. For image display, the contrast is derived from evaluation of the spectrum for each pixel. This is demonstrated by images which follow the *crack growth* in a PTFE sheet as a result of applied stress [Hep3]. Sketches of the sample deformation and corresponding images are depicted in Fig. 10.3.3. The original samples were cut (a) and strained (b) until failure (c), leaving a crack tip (d). At each stage a small portion of the sample was cut near the tip of the crack for imaging. The images (e) correspond to the widened crack after application of small strain, and the images (f) and (g) are of crack tips for a small strain rate of 5 mm/min and for a large strain rate of 500 mm/min, respectively. The samples were oriented, so that the direction of applied stress was parallel to the applied magnetic field. The images on the left are spin-density maps obtained by integrating over the frequency dimension. They are shown for comparison with the images for the signals at the chemical shift which corresponds to the parallel orientation of the chemical-shift tensor (2 kHz in Fig. 10.3.1).

All of the density maps show relatively uniform intensity except for some $\boldsymbol{B}_1$ field distortions by eddy currents from gradient switching. The chemical shift selective images on the right show high intensity for chemical-shift tensor orientations parallel to the field. Superimposed on these images are contour maps of difference images, which were constructed by subtracting the spectral intensity of the unoriented material at the chemical shift of interest from the density-normalized chemical-shift selective image of the sample. The result is an image of the sample where the contrast reflects deviations from random orientations. Thus, high intensity marks oriented sample regions or artefacts. In Fig. 10.3.3(e) the area near the tip of the crack is brightest. This is where the stress is strongest during deformation. But oriented material is also found propagating out from

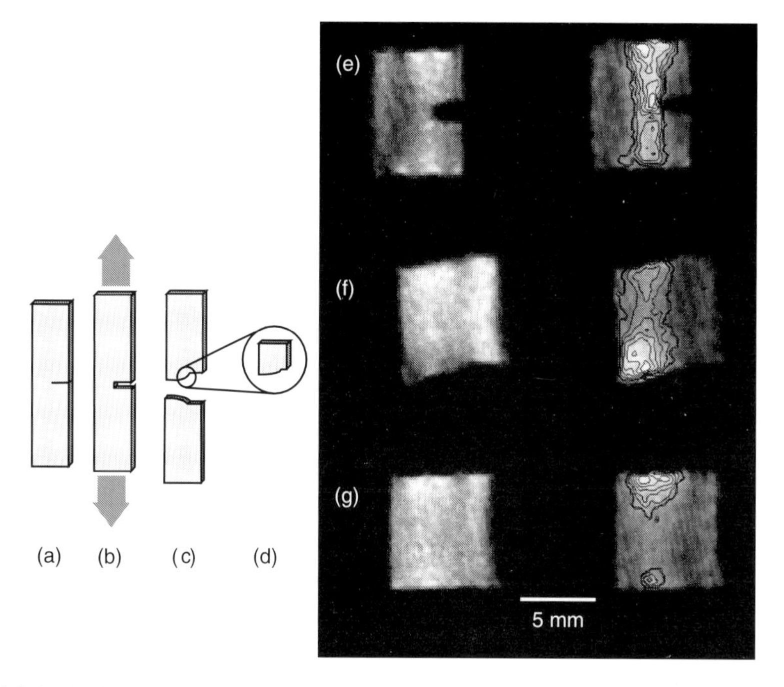

FIG. 10.3.3 [Hep3] PTFE sheets, spin-density images, and orientation images. (a) Sample with a cut before application of strain. (b) Lightly strained sample. (c) Cracked sample. The position of the samples extracted for imaging is indicated. (d) Shape of the cracked sample investigated by imaging. Spin-density images (left) and orientation images (right): (e) Sample after application of weak strain. (f) Crack tip of a sample strained at 5 mm/min. (g) Crack tip of a sample strained at 500 mm/min. For details see text.

the crack tip in the direction of the applied stress. The brightest region in (f) is at the broken edge at the bottom left, but oriented material extends back the full length of the sample. Regions above the original cut are unaffected, because no stress extends into these areas adjacent to the cut. Such large regions of oriented material are not detected in the rapidly drawn sample (g). Here the timescale of the applied strain was too fast for the molecular motion to adjust to the deformation. The material behaves like a glass and the orientation effects are more localized.

Related work

Similar work on spectroscopic imaging of orientation effects from sample deformation has been carried out on elastomers (cf. Section 10.2.2) as well as by imaging nuclei other than ^{19}F. In particular, spectroscopic deuteron double-quantum imaging has been employed to demonstrate the effects of molecular orientation on oriented polyethylene (cf. Fig. 8.4.3) and the effects of molecular dynamics on a phantom composed of phenyl-deuterated poly(styrene) and poly(carbonate) [Gün1]. Similar results can be obtained through spectroscopic deuteron MAS imaging at improved sensitivity [Gün3]. Furthermore, spectroscopic imaging of natural abundance ^{13}C was used to characterize the skin-core structure of a sample of injection moulded and drawn syndiotactic poly(propylene) [Gün2], and spatially resolved, rotor-synchronized 2D ^{13}C spectra were acquired with MAS imaging to image the distribution of order parameters across a phantom of differently oriented poly(ethylene) sheets by an analysis of spinning sidebands (cf. Figs. 3.3.8 and 7.3.5) [Sch1]. In principle, quadrupolar nuclei like ^{23}Na, ^{69}Ga, ^{75}As, and ^{115}In can also be imaged spectroscopically to analyse bond-angle distortions in strained semiconductors [Car1, Sui1].

10.3.2 Relaxation and spin-diffusion parameter imaging

Introduction

NMR imaging can be applied to map material changes in response to externally applied strain without the need to exploit the spectroscopic response for chemical-shift imaging. Examples of external strains are mechanical loads and electrical fields. While reversible changes in rigid solids may be investigated spectroscopically, for instance, by following the change in quadrupole coupling constant as a function of applied pressure, irreversible changes persist until long after the strain has been removed so that time-demanding imaging methods can be applied. Useful applications are rare, but the type of information accessible by solid-state parameter imaging is illustrated below by *imaging of molecular mobility* distributions in poly(carbonate) induced by *tensile stress*, and by examination of *morphology* changes from *electrical ageing* in low-density poly(ethylene).

While PC is a glass forming amorphous polymer, PE is a semicrystalline material, where crystalline lamellae and amorphous material are arranged in sandwich structures (Fig. 10.3.4(a)), which, on a larger scale, form spherulitic crystal structures [Eli1]. An interfacial region of intermediate mobility with respect to the mobile amorphous and rigid crystalline domains can be detected by NMR (cf. Section 7.2.9). These structures are perturbed by the formation of *electrical trees* (Fig. 10.3.4(b)), which are collections of small cavities produced by electrical discharges. They are the manifestation of a major form of failure of polymeric insulation material of underground high-voltage cables [Den2, Mch1].

A form of material failure particular to glassy amorphous polymers is the gliding of chains with the formation of *shear bands*. These shear bands form at angles of 38–45° with respect to the direction of applied stress. An example is shown in Fig. 10.3.5 with a shear band in bisphenol-A-poly(carbonate) [Wei5]. The band was prepared by cold drawing to a maximum elongation of $L/L_0 = \Lambda = 1.1$. The yield point occurred at a stress of approximately 58 MPa. The specimen had a fine scratch on the surface to force the generation of shear bands while drawing. In the figure the drawing scheme and the dimensions of the PC sample are given, and a photo shows the shear band as seen under a polarization microscope. The band extends from the upper left to the lower right.

Experimental techniques

Space encoding is achieved most conveniently through *magic-echo techniques* [Haf1]. Parameter contrast is introduced by suitable filters, which generate a signal decay under an effective Hamiltonian that is sensitive to slow molecular motion. Such filters are the

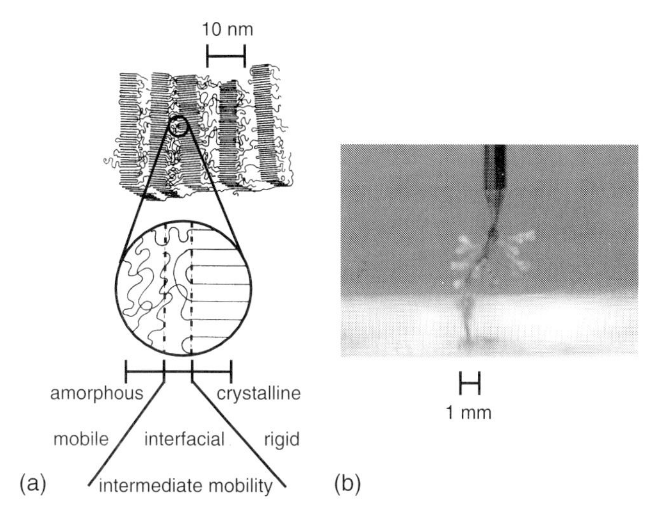

FIG. 10.3.4 Semicrystalline structure of poly(ethylene). (a) Rigid crystalline and mobile amorphous domains form sandwich structures with a period of the order of 10 nm. They are separated by narrow interfacial regions with intermediate mobility. Adapted from [Blü8] with permission from Dr. Dietrich Steinkopff Verlag. (b) Photo of an electrical tree in PE. The tree consists of a root-like system from about 10^5 cavities with typical diameters of 5 μm which are produced by electrical discharges. The pencil-like tip of the needle electrode is recognized above the tree [Sal1].

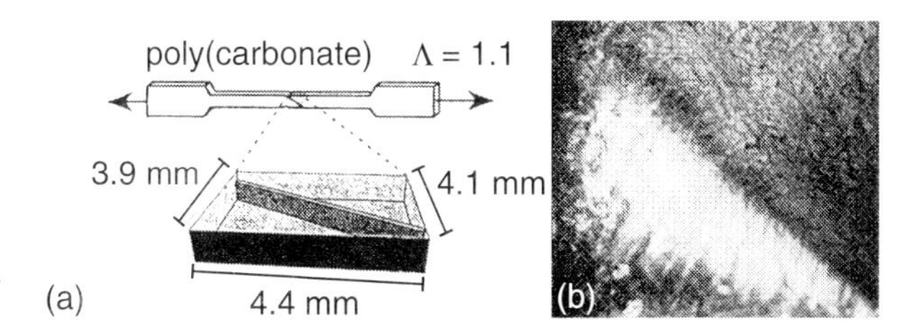

FIG. 10.3.5 Shear banding in cold-drawn poly(carbonate). (a) Drawing scheme and dimensions of the specimen. (b) Photograph of part of the sample under a polarization microscope with crossed polarizers. The shear band extends from the upper left to the lower right corner. Adapted from [Wei5] with permission from Wiley-VCH.

T_2 filter, the dipolar filter, the multi-solid-echo or OW4 filter with the effective relaxation time T_{2e}, and the $T_{1\rho}$ filter (cf. Section 7.1.2). Shear bands in drawn PC tensile bars have been imaged with contrast generated by the T_{2e} filter (Fig. 10.3.6) [Wei5, Wei6]. A magic sandwich is applied for phase encoding of the x-direction and a multi-magic-echo train for frequency encoding of the y-direction. The gradients are pulsed with switching times of the order of 1 μs and the gradient sign is alternated in the frequency-encoding period, because the magic echoes are combined with Hahn echoes to form a time-suspension sequence for optimum line narrowing (cf. Section 8.8, in particular, Fig. 8.8.2). The echo

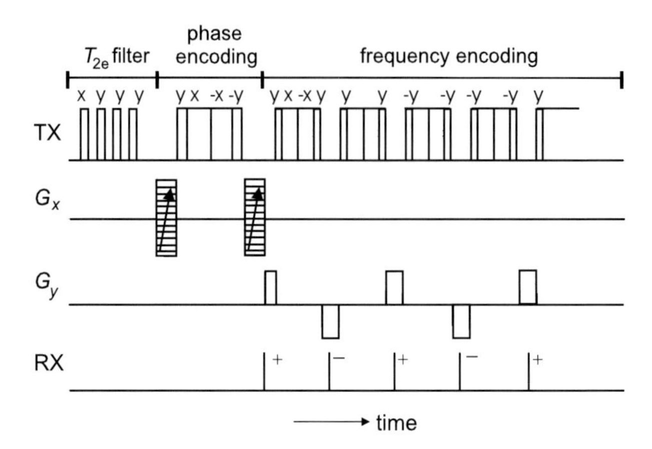

FIG. 10.3.6 [Wei6] Pulse sequence for T_{2e}-weighted 2D magic-echo imaging.

maxima are acquired during the windows in between subsequent magic-echo sandwiches. In order to increase the spectral width, the sampling is performed within each window. As a consequence, the sign of the receiver phase is alternated [Wei7]. Typical parameters are: time windows in between magic-echo sandwiches: 18 μs; number of points acquired in x- and y-directions: 32 and 64, respectively; and maximum gradient strength: 210 mT/m. Including averaging of 32 scans for each image, 32 parameter-weighted images can be acquired in this way in a total measurement time of about 30 h.

Magnetization filters sensitive to molecular motion induce differences in the amounts of longitudinal magnetization, depending on the timescale of molecular mobility. If properly applied, these differences may be so large that the signal from one component is suppressed completely, while that of another survives application of the filter. In particular the *dipolar filter* (cf. Section 7.2.4) can be used to effectively eliminate magnetization in the rigid domains and retain it in the mobile domains. This way, a concentration gradient of longitudinal magnetization can be established between the amorphous and the crystalline domains in PE. In a subsequent mixing time after the filter the remaining magnetization can migrate through the sample by *spin diffusion* (cf. Section 3.5.3), so that a spatial equilibrium is established (cf. Section 7.2.9). The equilibration curve of longitudinal magnetization can be mapped in parameter imaging experiments similar to investigations of relaxation curves. These equilibration curves are called *spin-diffusion curves*. The domain sizes of amorphous, interfacial, and crystalline domains can be obtained from a model-based parameterization of the spin-diffusion curves [Dem1, Sch2].

The pulse sequence for *imaging of* ^1H *spin diffusion* (Fig. 10.3.7(a)) consists of a dipolar filter, or possibly a chemical filter for soft matter with chemical shift resolution, followed by a time for mixing of magnetization components through spin diffusion, a phase-encoding period and a period for detection of the signal loss of the magnetization source or the signal build up in the magnetization sink [Wei8]. Typical parameters for the dipolar filter are a pulse spacing of 14 μs and $n = 4$ repetitions. By the pulse

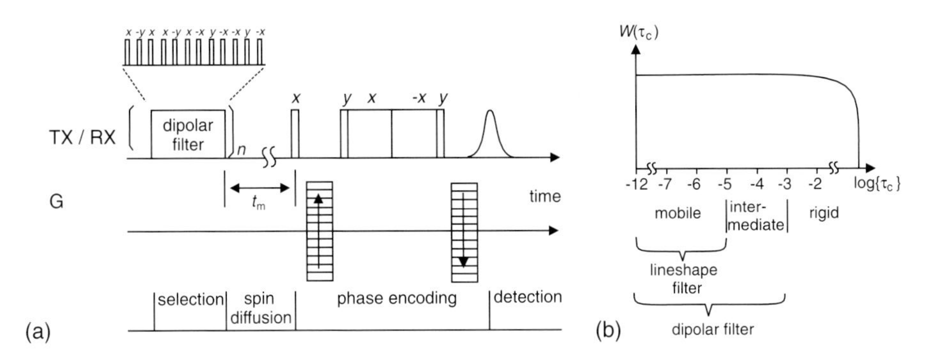

FIG. 10.3.7 Pulse sequence for spin-diffusion imaging with 1D spatial resolution [Wei8] and effect of mobility filters. (a) The magnetization source is selected by the dipolar filter which suppresses the magnetization in the sink. During the spin-diffusion time t_m the magnetization diffuses from the source to the sink. (b) The dipolar filter selects magnetization from chain segments which are highly mobile and intermediately mobile. By use of a lineshape filter the signal loss is analysed only for the mobile components. $W(\tau_c)$ is the probability for a particular correlation time τ_c to arise in the sample. It is essentially the spectral density of motion.

sequence of Fig. 10.3.7(a) the spectroscopic response is acquired in the detection period so that the ^{1}H wideline spectrum can be decomposed by a Gauss–Lorentz fit into a broad Gaussian contribution from the magnetization in the rigid domains and a narrow Lorentzian contribution from the magnetization in the mobile domains.

The signal change of either component can be analysed as a function of the spin-diffusion time t_m to obtain the morphological information. Thus, the numerical decomposition of the spectrum is equivalent to the use of a filter for separation of magnetization components. Depending on the transfer functions of the mobility filters for preparation of the initial magnetization gradient and for the acquired and processed magnetization components, an interface may or may not be detected (cf. Section 7.2.9). Therefore, the interface identified by NMR mobility filters is strictly defined by the NMR method and not necessarily by morphological features which are reproducible by other techniques.

The particular situation which applies to the study of electrical treeing in LDPE with the pulse sequence of Fig. 10.3.7(a) is illustrated in (b). The *dipolar filter* selects magnetization components from chain segments which are highly mobile and intermediately mobile. The analysis of signal loss only considers the magnetization from the highly mobile segments, which gives rise to the Lorentzian contribution of the lineshape. The different regimes of molecular mobility are interpreted to correspond to different morphological features or domains. If the domains with high molecular mobility are separated from those with rigid molecules by an interfacial layer of intermediate mobility, magnetization spills in the mixing time t_m into the unmagnetized rigid domains, first from the interface and subsequently with a time delay from the mobile domains. Depending on the thickness of the interfacial layer, the shape of the initial part of the spin-diffusion curve will map this time delay. Thus, the thickness of the interfacial layer can immediately be estimated by inspection of the spin-diffusion curve.

Shear bands in poly(carbonate)

In *shear bands* chain conformations are induced which are likely to be strained in comparison to the ones in the undeformed material. To detect shear bands by NMR the transverse relaxation T_{2e} effective under OW4 irradiation has been recorded as a function of 2D space with the pulse sequence of Fig. 10.3.6 [Wei5, Wei7, Tra1]. In the imaging experiment, the sample was oriented in such a way that the drawing direction was perpendicular to the static magnetic field.

A typical T_{2e} decay of PC is shown in Fig. 10.3.8(a). It can be described by a biexponential function with a fast and a slow decay,

$$s(t) = A\ \exp\{-t/T_{2e,\text{short}}\} + B\ \exp\{-t/T_{2e,\text{long}}\} + C. \tag{10.3.1}$$

The decay-time constant $T_{2e,\text{short}}$ of the fast component is mainly from the motions of the phenylene protons, whereas $T_{2e,\text{long}}$ for the slow component depends on the motions of the methyl protons. The parameters of such a biexponential analysis for each pixel lead to four images, one for the spin density, which is given by the sum $A + B$ of the amplitudes of both exponentials, one for each of the relaxation rates of the exponentials, $T_{2e,\text{short}}^{-1}$ and $T_{2e,\text{long}}^{-1}$, respectively, and one for the difference $A - B$ of the amplitudes of both exponentials. The spatial distributions of the first three of these parameters are shown in Fig. 10.3.8(b)–(d). The contour-level image (b) of the spin density provides information only about the different thickness of the sample. Here the shear band is hidden. But it becomes visible if the relaxation rates T_{2e}^{-1} are evaluated. In agreement with the photograph (cf. Fig. 10.3.5(b)) the shear band runs from the upper left to

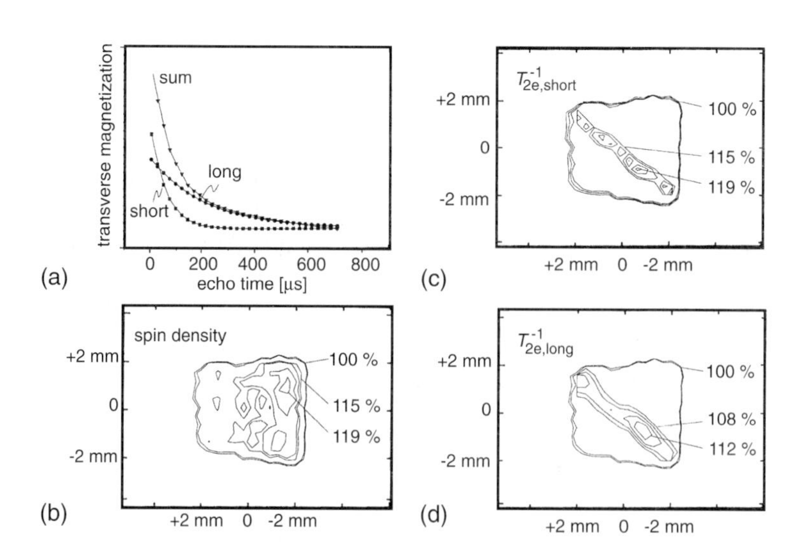

FIG. 10.3.8 T_{2e}-parameter imaging of shear bands in poly(carbonate). (a) Relaxation of transverse magnetization under OW4 irradiation. The decay is decomposed into two exponentials with relaxation times $T_{2e,\text{short}}$ and $T_{2e,\text{long}}$. (b) Spin-density image. (c) Parameter image of $T_{2e,\text{short}}^{-1}$. (d) Parameter image of $T_{2e,\text{long}}^{-1}$. Adapted from [Wei5] with permission from Wiley-VCH.

the lower right corner. The rates have been normalized for display to the rates of the unaffected material, i.e., to $18.9 \times 10^3\,s^{-1}$ for the fast decay and to $3.6 \times 10^3\,s^{-1}$ for the slow decay. Maximum values of 119% for the fast decay component and of 112% for the slow decay component were found. Therefore, the fast component shows higher contrast in the image and the phenylene motion inside the shear band is affected more than the methyl motion. From other NMR investigations [Wei10] it is already known that cold drawing reduces the phenylene mobility, whereas ageing reduces the methylene mobility. Cold drawing leads to a partial orientation of the polymer chain which results in a change of packing. This has been shown by T_{2e} parameter imaging in a similar fashion for the necking region of a PC tensile bar after cold drawing [Blü7, Wei10, Tra1].

Electrical treeing in poly(ethylene)

Electrical treeing disturbs the sandwich structure of crystalline and amorphous layers in PE. The average morphology of the sample is accessible from analysis of spin-diffusion curves, and by recording spin-diffusion curves for each pixel, the average dimensions of domains with rigid, intermediate and mobile molecular segments can be imaged. Slightly cross-linked LDPE is a typical high-voltage cable insulation material and its *electrical ageing* characteristics are the subject of extensive investigations [Sal1]. Typical testing arrangements are in needle-plate geometry, where an approximately cubic sample of PE with dimensions $(15\,mm)^3$ is positioned on a plate electrode and a second needle with a well-defined radius is pressed into the sample, so that the resultant distance between both electrodes is in the order of 5 mm. An alternating voltage is subsequently applied and slowly increased to about 40 kV. During this process, some 10^5 discharges can be recorded which destroy the insulating property of the material by formation of an electrical tree from small cavities and finally by current breakthrough [Den2, Mch1].

Spin-diffusion imaging has been applied to investigate the morphological changes in the treeing region in terms of an average over affected and unaffected material [Wei8, Wei9]. The average comes about from the facts that measurements were performed with 1D spatial resolution only, and that unaffected material may be present in between the different branches of the tree. Spin-diffusion measurements without space encoding were used to optimize the filter parameters, and the signal change of the mobile components were evaluated for different mixing times as a function of distance across the sample (Fig. 10.3.9(a)). Region I marks the unaffected material and region II the location of the tree. The experimental data have been corrected for T_1 relaxation. The amplitudes of the magnetization decays are weighted by the spin-density values. A reduced value of the spin density in region II is explained by the presence of an electrical breakthrough channel and by a small hole from the needle electrode, which had been removed before measurement. Representative magnetization decays of both regions are compared in (b) after normalization to the same amplitude. The dots represent experimental values, and the solid lines are simulated decays. The decay in region I starts with a plateau and is curved, while the decay in region II is initially linear. The length of the plateau defines the time delay for the magnetization to leave the mobile domains while the rigid domains are being magnetized from the domains with intermediate mobility (cf. Fig. 10.3.7(b)). Therefore, the plateau indicates the presence of an interface. Lack of a plateau in the spin-diffusion curves of regions II point to a negligibly thin interface.

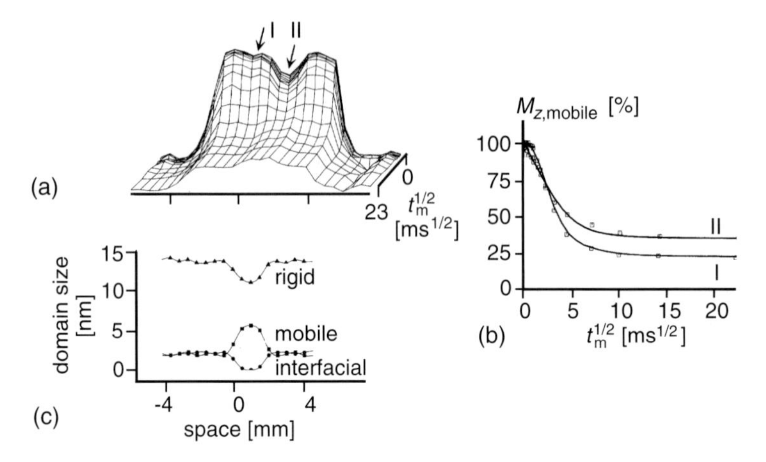

FIG. 10.3.9 Spin diffusion in LDPE. (a) Spin-diffusion curves showing the decay of magnetization in the mobile chain segments as a function of distance across the sample. The symbols I and II mark the regions unaffected and affected by the electrical ageing process, respectively. (b) Representative magnetization decays from regions I and II. Their different shapes indicate different morphologies. Dots: experimental values. Solid lines: Simulations. (c) 1D parameter images of domain sizes obtained from fits of spin-diffusion curves for a model structure to the experimental curves for each pixel. Notable changes in the domain sizes are observed in the damaged part compared to the unaffected part. Adapted from [Wei8] with permission from IEEE.

The magnetization-decay curves have been analysed for each position in space. The spin-diffusion constants were obtained from the ^{1}H linewidths of the broad and the narrow components according to [Dem1],

$$D_\mathrm{r} = \frac{1}{12}\sqrt{\frac{\pi}{2\ln 2}}\langle r^2\rangle \Delta\nu, \tag{10.3.2a}$$

$$D_\mathrm{m} = \frac{1}{6}\langle r^2\rangle\sqrt{\omega_\mathrm{c}\Delta\nu}, \tag{10.3.2b}$$

where $\Delta\nu$ is the linewidth at half height and ω_c is a cut-off frequency defined for the Lorentzian line, such that the spectral intensity is zero in the frequency range $|2\pi\,\Delta\nu| > \nu_\mathrm{c}$. The values $D_\mathrm{m} = 0.25\,\mathrm{nm}^2/\mathrm{ms}$, and $D_\mathrm{r} = 0.70\,\mathrm{nm}^2/\mathrm{ms}$ were obtained for the mobile and the rigid regions, respectively. The value $D_\mathrm{i} = 0.47\,\mathrm{nm}^2/\mathrm{ms}$ for the region with intermediate mobility is calculated as the arithmetic mean of D_m and D_r and approximates the average spin-diffusion coefficient. Following the general model for poly(ethylene) (Fig. 10.3.4(a)) with a planar *layer morphology*, the diameters d_m, d_r and d_i are assigned to mobile amorphous, rigid crystalline, and interfacial layers. In the absence of longitudinal relaxation and using an average value D for the spin-diffusion constant in all three regions, the *spin-diffusion curve* is calculated from Fick's second law for the build-up of magnetization in a sink which is separated from the source by an interface. For narrow interfacial domains this curve is a good approximation to the complement of the magnetization decay curve for the mobile domains in LDPE. Expressing the *long period* of the *domain structure* by $d_L = d_r + d_m + 2d_i$, the following formula is obtained for the longitudinal magnetization in a voxel [Dem1, Sch2],

$$M_{z,r}(t_m) = \frac{1}{2} M_{z0} S \sum_{n=1}^{N} \sqrt{4Dt_m} \left[\text{ierfc}\left\{\frac{nd_L - d_m - d_i}{\sqrt{4Dt_m}}\right\} - \text{ierfc}\left\{\frac{nd_L - d_i}{\sqrt{4Dt_m}}\right\} \right.$$
$$\left. - \text{ierfc}\left\{\frac{(n-1)d_L + d_i}{\sqrt{4Dt_m}}\right\} + \text{ierfc}\left\{\frac{(n-1)d_L + d_m + d_i}{\sqrt{4Dt_m}}\right\} \right], \quad (10.3.3)$$

where t_m is the spin-diffusion time, M_{z0} is the value of the longitudinal magnetization at $t_m = 0$, S is the area of all interfaces between rigid and mobile domains, n identifies the nth lamella, and $\text{ierfc}\{x\}$ denotes the integral of the error-function complement $1 - \text{erf}\{x\}$. A fit of this function to the spin-diffusion curve for each pixel yields the values of the domain sizes d_m, d_i and d_r as a function of space.

1D parameter images of the dimensions of amorphous, crystalline and interfacial domains in the electrically aged LDPE sample are depicted in Fig. 10.3.9(c). A variation of the domain sizes is clearly visible across the electrically aged region. The size of the rigid domains decreases from values of about 13.8–11.2 nm, that of the interface from about 2.4–0.2 nm, while the size of the mobile domains increases from about 2–5.8 nm. The long period d_L obtained from the fits is in reasonable agreement with the long period determined by small-angle X-ray scattering. It is reduced by about 15% in the affected region. Such a small change is more difficult to detect by X-ray scattering than a change of the order of 100% by NMR in the dimension of the interfacial region. The change in morphology is interpreted to result from rapid local heating and subsequent rapid heat dissipation during formation of the tree with each electrical discharge. This interpretation explains the increase in diameter of the amorphous domains at the expense of that of the crystalline domains. The apparent loss of the interface points to the limits of the model used for analysis. It is likely that the flat sheet structure of the *sandwich-layer model of PE* needs to be replaced by one with a rugged or fractal structure of the interface between rigid and mobile domains.

These examples demonstrate the principal feasibility of imaging rigid polymers while addressing questions of general interest. Yet the experimental efforts involved are substantial and the samples need to be kept small. Therefore, NMR imaging of rigid solids appears to be restricted to selected fundamental investigations.

References

CHAPTER 1

Abr1 A. Abragam, *The Principles of Nuclear Magnetism*, Clarendon Press, Oxford, 1961.
Abr2 A. Abragam and B. Bleaney, *Electron Paramagnetic Resonance of Transition Ions*, Oxford University Press, London, 1970.
Ack1 J.L. Ackerman and W.E. Ellingson, ed., *Advanced Tomographic Imaging Methods, Materials Research Society Symposium Proceedings*, Vol. 217, Materials Research Society, Pittsburgh, 1991.
And1 E.R. Andrew, G. Bydder, J. Griffith, R. Iles and P. Styles, *Clinical Magnetic Resonance Imaging and Spectroscopy*, John Wiley Sons Ltd., Chichester, 1990.
Ath1 N.M. Atherton, *Principles of Electron Spin Resonance*, Ellis Horwood, London, 1993.
Ber1 L.J. Berliner, in: *Magnetic Resonance Microscopy*, ed. B. Blümich and W. Kuhn, VCH, Weinheim, 1992, p. 151.
Blo1 F. Bloch, W. W. Hansen and M. Packard, *Phys. Rev.* **69** (1946) 127.
Blü1 B. Blümich and W. Kuhn, ed., *Magnetic Resonance Microscopy*, VCH, Weinheim, 1992.
Blü2 P. Blümler and B. Blümich, *Acta Polymerica* **44** (1993) 125.
Blü3 B. Blümich and P. Blümler, *Die Makromolek. Chem.* **194** (1993) 2133.
Blü4 P. Blümler and B. Blümich, *NMR – Basic Principles and Progress* **30** (1994) 209.
Blü5 B. Blümich, *Chemie in unserer Zeit* **24** (1990) 13.
Blü6 B. Blümich, P. Blümler, G. Eidmann, A. Guthausen, R. Haken, U. Schmitz, K. Saito and G. Zimmer, *Magn. Reson. Imag.* **16** (1998) 479.
Blü7 P. Blümler, B. Blümich, R. Botto and E. Fukushima, ed., *Spatially Resolved Magnetic Resonance*, Wiley-VCH, Weinheim, 1998.
Blü8 B. Blümich, P. Blümler and K. Saito, in: *Solid State NMR of Polymers*, ed. I. Ando and T. Asakura, Elsevier Science. B. V., Amsterdam, 1998, p. 123.
Blü9 P. Blümler and B. Blümich, *Rubber Chem. Tech.* **70** (1997) 468.
Blü10 B. Blümich, *Concepts Magn. Reson.* **10** (1998) 19.
Blü11 B. Blümich, *Concepts Magn. Reson.* **11** (1999) 71.
Bor1 G.C. Borgia, P. Fantazzini, J.C. Gore, M.R. Halse and J.H. Strange, ed., *Magn. Reson. Imag.* **14**(7/8) (1996), Special issue, *Proc. Third Int. Meeting Recent Advances in MR Applications to Porous Media*, Univ. Cath. Louvain-La-Neuve, Belgium, 3–6 September 1995.
Bor2 G.C. Borgia, P. Fantazzini, J.C. Gore and J.H. Strange, ed., *Magn. Reson. Imag.* **16**(5/6) (1996), Special issue, *Proc. Fourth Int. Meeting Recent Advances in MR Applications to Porous Media*, Statoil Research Center, Trondheim, Norway, 31 August–3 September 1997.
Bov1 F.A. Bovey, *Nuclear Magnetic Resonance Spectroscopy*, 2nd edn, Academic Press, New York, 1991.

Bot1 R.E. Botto, ed., *Special Issue on Magnetic Resonance Imaging of Materials, Solid-State Nucl. Magn. Reson.* **6** (1996).

Bre1 W. Bremser, B. Franke and H. Wagner, *Chemical Shift Ranges in Carbon-13 NMR Spectroscopy*, VCH, Weinheim, 1982.

Bud1 T.F. Budinger and P.C. Lauterbur, *Science* **226** (1984) 288.

Cal1 P.T. Callaghan, *Principles of Nuclear Magnetic Resonance Microscopy*, Clarendon Press, Oxford, 1991.

Cap1 A. Caprihan and E. Fukushima, *Physics Reports* **4** (1990) 195.

Cha1 C. Chang and R.A. Komoroski, in: *Solid State NMR of Polymers*, ed. L. Mathias, Plenum Press, New York, 1991, p. 363.

Che1 C.-N. Chen and D.I. Hoult, *Biomedical Magnetic Resonance Technology*, Adam Hilger, Bristol, 1989.

Cor1 D.G. Cory, *Annual Reports on NMR* **24** (1992) 87.

Dam1 R. Damadian, *Science* **171** (1971) 1151.

Dam2 R. Damadian, ed., *NMR in Medicine, NMR – Basic Principles and Progress*, Vol. 19, Springer, Berlin, 1981.

Eat1 G.R. Eaton and S.S. Eaton, *Concepts Magn. Reson.* **7** (1995) 49.

Eat2 G.R. Eaton, S.S. Eaton and K. Ohno, ed., *EPR Imaging and In Vivo EPR*, CRC Press, Cleveland, 1991.

Ecc1 C.D. Eccles and P.T. Callaghan, *J. Magn. Reson.* **68** (1986) 393.

Ern1 R.R.Ernst, G. Bodenhausen and A. Wokaun, *Principles of Nuclear Magnetic Resonance in One and Two Dimensions*, Clarendon Press, Oxford, 1987.

Fle1 G. Fleischer and F. Fujara, *NMR – Basic Principles and Progress* **30** (1994) 159.

Gor1 W. Gordy, *Theory and Applications of Electron Spin Resonance*, Wiley, New York, 1980.

Haa1 E.M. Haake, R.W. Brown, M.R. Thompson and R. Venkatesan, ed., *Magnetic Resonance Imaging*, Wiley-Liss, New York, 1999.

Hau1 K.H. Hausser and H.R. Kalbitzer, *NMR in Medicine and Biology*, Springer, Berlin, 1991.

Hen1 J.W. Hennel, T. Kryst-Wizgowska and J. Klinowski, *A Primer of Magnetic Resonance Imaging*, Imperial College Press, Cambridge, 1997.

Ike1 M. Ikeya, in: *Magnetic Resonance Microscopy*, ed. B. Blümich and W. Kuhn, VCH, Weinheim, 1992, p. 133.

Ike2 M. Ikeya, *New Applications of Electron Spin Resonance*, World Scientific, London, 1992.

Jez1 P. Jezzard, J.J. Attard, T.A. Carpenter and L.D. Hall, *Progr. NMR Spectr.* **23** (1991) 1.

Jez2 P. Jezzard, C.J. Wiggins, T.A. Carpenter, L.D. Hall, P. Jackson and N.J. Clayden, *Advanced Materials* **4** (1992) 82.

Kev1 L. Kevan and M.K. Bowman, ed., *Modern Pulsed and Continuous-Wave Electron Spin Resonance*, Wiley, New York, 1990.

Kim1 R. Kimmich, *NMR Tomography, Diffusometry, Relaxometry*, Springer, Berlin, 1997.

Kle1 F. Klein and A. Sommerfeld, *Über die Theorie des Kreisels*, reprinted by B.G. Teubner, Stuttgart, 1965.

Kre1 E. Krestel, ed., *Imaging Systems for Medical Diagnostics*, Siemens AG, Berlin, 1990.

Kuh1 W. Kuhn, *Angew. Chem. Int. Ed. Eng.* **29** (1990) 1.

Kum1 A. Kumar, D. Welti and R.R. Ernst, *J. Magn. Reson.* **18** (1975) 69.

Lau1 P. C. Lauterbur, *Nature* **242** (1973) 190.

Man1 P. Mansfield and E.L. Hahn, ed., *NMR Imaging*, The Royal Society, London, 1990.

Man2 P. Mansfield and P.K. Grannell, *J. Phys. C: Solid State Phys.* **6** (1973) L422.

Man3 P. Mansfield and P.K. Grannell, *Phys. Rev. B* **12** (1975) 3618.

Man4 P. Mansfield and P.G. Morris, *NMR Imaging in Biomedicine, Adv. Magn. Reson. Suppl. 2*, Academic Press, New York, 1982.
Mar1 B. Maraviglia, ed., *Physics of NMR Spectroscopy in Biology and Medicine, Proc. Int. School Enrico Fermi, Course C*, North-Holland, Amsterdam, 1988.
Meh1 M. Mehring, *High Resolution NMR in Solids*, 2nd edn, Springer, Berlin, 1983.
Mil1 J.B. Miller, *Progr. Nucl. Magn. Reson. Spectrosc.* **33** (1998) 273.
Mor1 P.G. Morris, *Nuclear Magnetic Resonance Imaging in Medicine and Biology*, Clarendon Press, Oxford, 1986.
Mor2 H. Morneburg, ed., *Bildgebende Systeme für die medizinische Diagnostik*, 3rd edn, Publicis MCDF Verlag, Erlangen, 1995.
Pil1 J.R. Pilbrow, *Transition Ion Electron Paramagnetic Resonance*, Oxford Science Publications, Clarendon Press, Oxford, 1990.
Poo1 C.P. Poole, Jr., *Electron Spin resonance*, 2nd edn, Wiley Interscience, New York, 1983.
Rug1 D. Rugar, C.S. Yannoni and J.A. Sidles, *Nature* **360** (1992) 563.
Rug2 D. Rugar, O. Züger, S. Hoen, C.S. Yannoni, H.M. Vieth and R.D. Kendrick, *Science* **264** (1994) 1560.
Sch1 K. Schmidt-Rohr and H. W. Spiess, *Multidimensional Solid-State NMR and Polymers*, Academic Press, London, 1994.
Sch2 A. Schweiger, *Angew. Chemie Int. Ed. Engl.* **30** (1991) 265.
Sch3 A. Schaff and W.S. Veeman, *J. Magn. Reson.* **126** (1997) 200.
Sid1 J.A. Sidles, *Appl. Phys. Lett.* **58** (1991) 2854.
Sid2 J.A. Sidles, *Phys. Rev. Lett.* **68** (1992) 1124.
Sli1 C.P. Slichter, *Principles of Magnetic Resonance*, 3rd enlarged edn, Springer, Berlin, 1989.
Sta1 D.D. Stark and W.G. Bradley, ed., *Magnetic Resonance Imaging*, Mosby Inc., St. Louis, 3rd edn, 1998.
Vla1 M.T. Vlaardingerbroek and J.A. den Boer, *Magnetic Resonance Imaging*, Springer, Berlin, 1996.
Weh1 F.W. Wehrli, D. Schaw and J.B. Kneeland, ed., *Biomedical Magnetic Resonance Imaging*, VCH Verlagsgesellschaft, Weinheim, 1988.
Weh1 J.A. Weil, J.R. Bolton and J. E. Wertz, *Electron Paramagnetic Resonance: Elementary Theory and Practical Applications*, Wiley, New York, 1994.
Xia1 Y. Xia, *Concepts Magn. Reson.* **8** (1996) 205.
Zwe1 J.L. Zweier and P. Kuppusamy, in: *Spatially Resolved Magnetic Resonance*, ed. P. Blümler, B. Blümich, R. Botto and E. Fukushima, Wiley-VCH, Weinheim, 1998, p. 373.

CHAPTER 2

Ack1 J.J.H. Ackerman, T.H. Grove, G.G. Wong, D.G. Gadian and G.K. Radda, *Nature* **283** (1980) 167.
Abr1 A. Abragam, *The Principles of Nuclear Magnetism*, Clarendon Press, Oxford, 1961.
Ald1 D. Alderman and D. Grant, *J. Magn. Reson.* **36** (1979) 447.
Arn1 J.T. Arnold, S.S. Dharmatti and M.E. Packard, *J. Chem. Phys.* **19** (1951) 507.
Ben1 M.R. Bendall, *Chem. Phys. Lett.* **99** (1983) 310.
Bes1 R. Best, *Digitale Signalverarbeitung und Simulation*, AT-Verlag, Aarau, 1990.

Bla1 R.D. Black, T.A. Early, P.B. Roemer, O.M. Mueller, A. Mogro-Campero, L.G. Turner and G.A. Johnson, *Science* **259** (1993) 793.

Bla2 R.D. Black, T.A. Early and G.A. Johnson, *J. Magn. Reson. A* **113** (1995) 74.

Blo1 F. Bloch, *Phys. Rev.* **70** (1946) 460.

Blü1 B. Blümich, *Chemie in unserer Zeit* **24** (1990) 13.

Bow1 R. Bowtell and P. Mansfield, *Meas. Sci. Tech.* **1** (1990) 431.

Bru1 Adapted from *Bruker Almanac*, p. 79, Bruker Analytische Meßtechnik, Rheinstetten, 1989.

Bur1 D. Burstein, *Concepts Magn. Reson.* **8** (1996) 269.

Cal1 P.T. Callaghan, *Principles of Nuclear Magnetic Resonance Microscopy*, Clarendon Press, Oxford, 1991.

Cal2 P.T. Callaghan, *Aust. J. Phys.* **37** (1984) 359.

Car1 H.Y. Carr and E.M. Purcell, *Phys. Rev.* **94** (1954) 630.

Car2 H.Y. Carr, *Phys. Rev.* **112** (1958) 1693.

Cha1 B.L.W. Chapman and P. Mansfield, *J. Magn. Reson. B* **107** (1995) 152.

Che1 C.-N. Chen and D.I. Hoult, *Biomedical Magnetic Resonance Technology*, Adam Hilger, Bristol, 1989.

Chm1 G. Chmurny and D.I. Hoult, *Concepts Magn. Reson.* **2** (1990) 131.

Con1 M. S. Conradi, A.N. Garroway, D.G. Cory and J.B. Miller, *J. Magn. Reson.* **94** (1991) 370.

Dam1 R. Damadian, *Science* **171** (1971) 1151.

Der1 A.E. Derome, *Modern NMR Techniques for Chemistry Research*, Pergamon Press, Oxford, 1987.

Dot1 F.D. Doty, G. Entzminger and Y.A. Yang, *Concepts. Magn. Reson.* **10** (1998) 133.

Dot2 F.D. Doty, G. Entzminger and Y.A. Yang, *Concepts. Magn. Reson.* **10** (1998) 239.

Ede1 W.A. Edelstein, J.M.S. Hutchinson, G. Johnson and T. Redpath, *Phys. Med. Biol.* **25** (1980) 751.

Ede2 W.A. Edelstein, C.J. Hardy and O.M. Mueller, *J. Magn. Reson.* **67** (1986) 156.

Ell1 J.D. Ellet, M.G. Gibby, U. Haeberlen, L.M. Huber, M. Mehring, A.Pines and J.S. Waugh, *Adv. Magn. Reson.* **117** (1971) 117.

Emi1 S. Emid and J.H.N. Creyghton, *Physica* **125B** (1985) 81.

Ern1 R.R. Ernst, G. Bodenhausen and A. Wokaun, *Principles of Nuclear Magnetic Resonance in One and Two Dimensions*, Clarendon Press, Oxford, 1987.

Ern2 R.R. Ernst and W.A. Anderson, *Rev. Sci. Instrum.* **37** (1966) 93.

Far1 T.C. Farrar, *Introduction to Pulse NMR Spectroscopy*, The Farragut Press, Madison, 1989.

Far2 T.C. Farrar and J.E. Harriman, *Density Matrix Theory and its Applications to NMR Spectroscopy*, The Farragut Press, Madison, 1991.

Fre1 R. Freeman, *Concepts Magn. Reson.* **11** (1999) 61.

Fuk1 E. Fukushima and S.B.W. Roeder, *Experimental Pulse NMR: A Nuts and Bolts Approach*, Addison-Wesley, Reading, 1981.

Fuk2 L.F. Fuks, F.S. Huang, C.M. Carter, W.A. Edelstein and P.B. Roemer, *J. Magn. Reson.* **100** (1992) 229.

Ger1 B.C. Gerstein and C.R. Dybowski, *Transient Techniques in NMR of Solids*, Academic Press, New York, 1985.

Glo1 G.H. Glover, C.E. Hayes, N.J. Pelc, W.A. Edelstein, O.M. Mueller, H.R. Hart, C.J. Hardy, M. O'Donnel and W.D. Barber, *J. Magn. Reson.* **64** (1985) 255.

Gün1 E. Günther, H. Raich and B. Blümich, *Bruker Report* (1991/1992) 15.

Gyn1 M. Gyngell, *J. Magn. Reson.* **81** (1989) 474.

Hah1 E.L. Hahn, *Phys. Rev.* **80** (1950) 580.
Har1 M.D. Harper, *J. Magn. Reson.* **94** (1991) 550.
Hay1 C.E. Hayes, W.A. Edelstein, J.F. Schenk, O.M. Mueller and M. Eash, *J. Magn. Reson.* **63** (1985) 622.
Hin1 W.S. Hinshaw, *J. Appl. Phys.* **47** (1976) 3709.
Hou1 D.I. Hoult and R.E. Richards, *Proc. R. Soc. (London)* **A344** (1975) 311.
Hou2 D.I. Hoult, C.N. Chen and V.J. Sank, *Magn. Reson. Med.* **1** (1984) 339.
Jee1 J. Jeener, *Lecture at the Ampère Summer School in Basko Polje*, 1971, published in: *NMR and More*, ed. M. Goldman and M. Porneuf, Les Editions de Physique, Les Ulis, 1994.
Kan1 E.R. Kanasewich, *Time Sequence Analysis in Geophysics*, University of Alberta Press, Edmonton, 1981.
Kär1 J. Kärger, H. Pfeifer and W. Heink, *Adv. Magn. Reson.* **12** (1988) 1.
Kim1 R. Kimmich, *NMR Tomography, Diffusometry, Relaxometry*, Springer, Berlin, 1997.
Kon1 P. Konzbul and K. Sveda, *Meas. Sci. Technol.* **6** (1995) 1116.
Kre1 E. Krestel, ed., *Imaging Systems for Medical Diagnostics*, Siemens AG, Berlin, 1990.
Kum1 A. Kumar, D. Welti and R.R. Ernst, *J. Magn. Reson.* **18** (1975) 69.
Lau1 P.C. Lauterbur, *Nature* **242** (1973) 190.
Maj1 P.D. Majors, J.L. Blackley, S.A. Altobelli, A. Caprihan and E. Fukushima, *J. Magn. Reson.* **87** (1990) 548.
Man1 P. Mansfield and P.K. Grannell, *J. Phys. C: Solid State Phys.* **6** (1973) L422.
Man2 P. Mansfield and P.G. Morris, *NMR Imaging in Biomedicine, Adv. Magn. Reson. Suppl. 2*, Academic Press, New York, 1982.
Man3 P. Mansfield and B. Chapman, *J. Magn. Reson.* **72** (1989) 221.
Man4 P. Mansfield, in: *Physics of NMR Spectroscopy in Biology and Medicine, Proc. Int. School Enrico Fermi, Course C*, ed. B. Maraviglia, North-Holland, Amsterdam, 1988, p. 345.
Man5 P. Mansfield and P.K .Grannell, *Phys. Rev. B* **12** (1975) 3618.
Mar1 J.L. Markley, W.J. Horsley and M.P. Klein, *J. Chem. Phys.* **55** (1971) 3604.
Mcf1 E.W. McFarland and A. Morata, *Magn. Reson. Imag.* **10** (1992) 279.
Meh1 M. Mehring, *High Resolution NMR in Solids*, 2nd edn, Springer, Berlin, 1983.
Mei1 S. Meiboom and D. Gill, *Rev. Sci. Instrum.* **29** (1958) 688.
Mor1 H. Morneburg, ed., Bildgebende Systeme für die medizinische Diagnostik, 3rd edn, Publicis MCDF Verlag, Erlangen, 1995.
Mun1 M. Munowitz, *Coherence and NMR*, John Wiley & Sons, New York, 1988.
Osh1 D.D. Osheroff, W.J. Gully, R.C. Richardson and D. M. Lee, *Phys. Rev. Lett.* **29** (1972) 920.
Pri1 M.B. Priestley, *Non-linear and Non-stationary Time Series Analysis*, Academic Press, New York, 1988.
Pur1 E.M. Purcell, H.C. Torrey and R.V. Pound, *Phys. Rev.* **69** (1946) 37.
Red1 A.G. Redfield and R.K. Gupta, *Adv. Magn. Reson.* **5** (1971) 81.
Red2 A.G. Redfield and S.D. Kunz, *J. Magn. Reson.* **19** (1975) 250.
Red3 A.G. Redfield, *NMR – Basic Principles and Progress* **13** (1976) 137.
Sch1 G. Schauss, B. Blümich and H.W. Spiess, *J. Magn. Reson.* **95** (1991) 437.
Sch2 K. Schmidt-Rohr and H. W. Spiess, *Multidimensional Solid-State NMR and Polymers*, Academic Press, London, 1994.
Sch3 J.S. Schoeniger and S. J. Blackband, *J. Magn. Reson. B* **104** (1994) 127.
Sli1 C.P. Slichter, *Principles of Magnetic Resonance*, 3rd enlarged edn, Springer, Berlin, 1989.

Ste1 E. Stejskal and J.E. Tanner, *J. Chem. Phys.* **42** (1965) 288.
Ste2 S. Stephenson, *Progr. NMR Spectroscopy.* **20** (1988) 515.
Ste3 E.O. Stejskal and J. D. Memory, *High Resolution NMR in the Solid State: Fundamentals of CP/MAS*, Oxford University Press, New York, 1994.
Sti1 P. Stilbs, *Progr. NMR Spectroscopy* **19** (1987) 1.
Tan1 J.E.E. Tanner, *Rev. Sci. Instrum.* **36** (1965) 1086.
Vol1 R.L. Vold, J.S. Waugh, M.P. Klein and D.E. Phelps, *J. Chem. Phys.* **48** (1968) 3831.
Wat1 J.C. Watkins and E. Fukushima, *Rev. Sci. Instrum.* **59** (1988) 926.
Web1 A.G. Webb, *Prog. Nucl. Magn. Reson. Spectrosc.* **31** (1997) 1.
Wol1 Courtesy of Dr. G. Wolff, Bruker Analytical Instruments, Rheinstetten.
Xia1 Y. Xia and P.T. Callaghan, *Macromolecules* **24** (1991) 4777.
Zho1 X. Zhou and P.C. Lauterbur, in: *Magnetic Resonance Microscopy*, ed. B. Blümich and W. Kuhn, VCH, Weinheim, 1992, p. 3.
Zum1 N. Zumbulyadis, *Concepts Magn. Reson.* **3** (1991) 89.

CHAPTER 3

Abr1 A. Abragam, *The Principles of Nuclear Magnetism*, Clarendon Press, Oxford, 1961.
Abr2 A. Abragam and M. Goldman, *Nuclear Magnetism: Order and Disorder*, Clarendon Press, Oxford, 1982.
Ail1 D.C. Ailion, *Adv. Magn. Reson.* **5** (1971) 177.
And1 P.W. Anderson and P.R. Weiss, *J. Phys. Soc. Japan* **9** (1954) 316.
And2 P.W. Anderson and P.R. Weiss, *Rev. Mod. Phys.* **25** (1953) 269.
And3 E.R. Andrew, A. Bradbury and R.G. Eades, *Nature* **182** (1958) 1659.
Ane1 F.A.L. Anet and D.J.O'Leary, *Concepts Magn. Reson.* **3** (1991) 193.
Ant1 O.N. Antzukin, Z. Song, X. Feng and M.H. Levitt, *J. Chem. Phys.* **100** (1994) 130.
Ant2 O.N. Antzukin, S.C. Shekar and M.H. Levitt, *J. Magn. Reson. A* **115** (1995) 7.
Aue1 W.P. Aue, J. Karhan and R.R. Ernst, *J. Chem. Phys.* **64** (1976) 4226.
Bag1 D.M.S. Bagguley, ed., *Pulsed Magnetic Resonance: NMR, ESR, and Optics*, Clarendon Press, Oxford, 1992.
Bar1 T.M. Barbara and E.H. Williams, *J. Magn. Reson.* **99** (1992) 439.
Bec1 E.D. Becker, *High Resolution NMR, Theory and Chemical Applications*, 2nd edn, Academic Press, New York, 1980.
Bec2 H.W. Beckham and H.W. Spiess, *NMR – Basic Principles and Progress*, **32** (1994) 163.
Ben1 A.E. Bennet, R.G. Griffin and S. Vega, *NMR – Basic Principles and Progress* **33** (1994), 1.
Ben2 A.E. Bennet, C.M. Rienstra, M.Auger, K.V. Lakshmi and R.G. Griffin, *J. Chem. Phys.* **103** (1995) 6951.
Blo1 N. Bloembergen, E.M. Purcell and R.V. Pound, *Phys. Rev.* **73** (1948) 679.
Blü1 B. Blümich and A. Hagemeyer, *Chem. Phys. Lett.* **161** (1989) 55.
Blü2 B. Blümich, P. Blümler and J. Jansen, *Solid State Nuc. Magn. Reson.* **1** (1992) 111.
Blü3 B. Blümich and H.W. Spiess, *Angew. Chem. Int. Ed. Engl.* **27** (1988) 1655.
Blü4 B. Blümich, A. Hagemeyer, D. Schaefer, K. Schmidt-Rohr and H.W. Spiess, *Advanced Materials* **2** (1990) 72.
Blü5 B. Blümich, *Chemie in unserer Zeit* **24** (1990) 13.
Blü6 B. Blümich, *Advanced Materials* **3** (1991) 237.

Blü7 B. Blümich, guest ed., *NMR – Basic Principles and Progress* Vol. 30–33, Springer, Berlin, 1994.

Bov1 F.A. Bovey, *Nuclear Magnetic Resonance Spectroscopy*, 2nd edn, Academic Press, New York, 1991.

Bro1 C.E. Bronnimann, B.L. Hawkins, M. Zhang and G.E. Maciel, *Anal. Chem.* **60** (1988) 1743.

Bur1 D.P. Burum and W.K. Rhim, *J. Chem. Phys.* **71** (1979) 944.

Bur2 D.P. Burum, *Concepts Magn. Reson.* **2** (1990) 213.

Cal1 P.T. Callaghan, *Aust. J. Phys.* **37** (1984) 359.

Cal2 P.T. Callaghan, *Principles of Nuclear Magnetic Resonance Microscopy*, Clarendon Press, Oxford, 1991.

Car1 P. Caravatti, G. Bodenhausen and R.R. Ernst, *Chem. Phys. Lett*, **89**, (1982) 363.

Car2 P. Caravatti, L. Braunschweiler and R.R. Ernst, *Chem. Phys. Lett.* **100** (1983) 305.

Chi1 G.C. Chingas, J.B. Miller and A.N. Garroway, *J. Magn. Reson.* **66** (1986) 530.

Chm1 B.F. Chmelka and J.W. Zwanziger, *NMR – Basic Principles and Progress* **33** (1994) 79.

Cla1 J. Clauss, K. Schmidt-Rohr and H.W. Spiess, *Acta Polymerica* **44** (1993) 1.

Coh1 J.P. Cohen-Addad, P. Huchot and A. Viallat, *Polymer Bulletin* **19** (1988) 257.

Col1 M.G. Colombo, B.H. Meier and R.R. Ernst, *Chem. Phys. Lett* **146** (1988) 189.

Cor1 D.G. Cory, J.B. Miller, R. Turner and A.N. Garroway, *Molec. Phys.* **70** (1990) 331.

Cor2 D.G. Cory, J.B. Miller and A.N. Garroway, *J. Magn. Reson.* **90** (1990) 205.

Dem1 D.E. Demco, A. Johanson and J. Tegenfeldt, *Solid-State Nucl. Magn. Reson.* **4** (1995) 13.

Dem2 D.E. Demco, S. Hafner and H.W. Spiess, *J. Magn. Reson. A* **116** (1995) 36.

Dem3 D.E. Demco, *Phys. Lett.* **45A** (1973) 113.

Dix1 W.T. Dixon, *J. Chem. Phys.* **77** (1982) 1800.

Dun1 T.M. Duncan, *A Compilation of Chemical Shift Anisotropies*, The Farragut Press, Chicago, 1990.

Eck1 H. Eckert, *Progr. NMR Spectrosc.* **24** (1992) 159.

Eck2 H. Eckert, *NMR – Basic Principles and Progress* **33** (1994) 125.

Eng1 G. Engelhardt and D. Michel, *High Resolution Solid State NMR of Silicates and Zeolites*, Wiley, New York, 1987.

Eng2 F. Engelke, T. Kind, D. Michel, M. Pruski and B.C. Gerstein, *J. Magn. Reson.* **95** (1991) 286.

Ern1 R.R. Ernst, G. Bodenhausen and A. Wokaun, *Principles of NMR in One and Two Dimensions*, Clarendon Press, Oxford, 1987.

Ern2 M. Ernst, S. Bush, A.C. Kolbert and A. Pines, *J. Chem. Phys.* **105** (1996) 3387.

Far1 T.C. Farrar, *Introduction to Pulse NMR Spectroscopy*, The Farragut Press, Madison, 1989.

Fed1 V.D. Fedotov, A. Ebert and H. Schneider, *Phys. Stat. Sol. A* **63** (1981) 209.

Fer1 C. Fernandez and J.-P. Amoureux, *Chem. Phys. Lett.* **242** (1995) 449.

Fri1 U. Friedrich, I. Schnell, S.P. Brown, A. Lupulescu, D.E. Demco and H.W. Spiess, *Molec. Phys.* **95** (1998) 1209.

Fry1 L. Frydman and J.S. Harwood, *J. Am. Chem. Soc.* **197** (1995) 5367.

Gan1 Z. Gan and R.R. Ernst, *Solid State Nucl. Magn. Reson.* **8** (1997) 153.

Ger1 B.C. Gerstein and C.R. Dybowski, *Transient Techniques in NMR of Solids*, Academic Press, New York, 1985.

Ger2 B.C. Gerstein, C. Chou, R.G. Pembleton and R.C. Wilson, *J. Phys. Chem.* **81** (1977) 565.

Gol1 W.I. Goldburg and M. Lee, *Phys. Rev. Lett.* **6** (1963) 355.

Göt1 H. Götz, P. Denner and U. Rost, *Wiss. Zeitschrift Päd. Hochschule Theodor Neubauer, Math.-Naturw. Reihe* **23** (1987) 142.

Gri1 R.G. Griffin, W.P. Aue, R.A. Haberkorn, G.S. Harbison, J. Herzfeld, E.M. Menger, M.G. Munowitz, E.T. Olejniczak, D.P. Raleigh, J.E. Roberts, D.J. Ruben, A. Schmidt, S.O. Smith and S. Vega, in: *Physics of NMR Spectroscopy in Biology and Medicine, Proc. Int. School of Physics Enrico Fermi, Course C*, ed. B. Maraviglia, North-Holland, 1988, p. 203.

Gri2 A.-R. Grimmer and B. Blümich, *NMR – Basic Principles and Progress* **30** (1994) 1.

Gün1 H. Günther, *NMR-Spektroskopie*, 2nd edn, Thieme, Stuttgart, 1983.

Gün2 E. Günther, B. Blümich and H.W. Spiess, *Molec. Phys.* **71** (1990) 477.

Gün3 E. Günther, B. Blümich and H.W. Spiess, *Chem. Phys. Lett.* **187** (1991) 251.

Hae1 U. Haeberlen, *High Resolution NMR in Solids: Selective Averaging, Adv. Magn. Reson. Suppl. 1*, Academic Press, New York, 1976.

Hae2 U. Haeberlen and J.S. Waugh, *Phys. Rev.* **175** (1968) 453.

Haf1 S. Hafner and H.W. Spiess, *J. Magn. Reson. A* **121** (1996) 160.

Hag1 A. Hagemeyer, K. Schmidt-Rohr and H.W. Spiess, *Adv. Magn. Reson.* **13** (1989) 85.

Hag2 A. Hagemeyer, *Entwicklung mehrdimensionaler MAS-NMR-Methoden*, Dissertation, Johannes-Gutenberg-Universität, Mainz, 1990.

Hah1 E.L. Hahn, *Phys. Rev.* **80** (1950) 580.

Har1 G.S. Harbison, V.-D. Vogt and H.W. Spiess, *J. Chem. Phys.* **86** (1987) 1206.

Har2 S.R. Hartmann and E.L. Hahn, *Phys. Rev.* **128** (1962) 2042.

Hen1 R. Hentschel, J. Schlitter, H. Sillescu and H.W. Spiess, *J. Chem. Phys.* **68** (1978) 56.

Her1 J. Herzfeld and A.E. Berger, *J. Chem. Phys.* **73** (1980) 6021.

Hoa1 G.L. Hoatson and R.L. Vold, *NMR – Basic Principles and Progress* **32** (1994) 1.

Hoh1 M. Hohwy and N.C. Nielsen, *J. Chem. Phys.* **106** (1997) 7571.

Huj1 J.Z. Hu, A.M. Orendt, D.W. Alderman, R.J. Pugmire, C. Ye and D.M. Grant, *Solid-State Nucl. Magn. Reson.* **3** (1994) 181.

Jee1 J. Jeener and P. Broekaert, *Phys. Rev.* **157** (1967) 232.

Jel1 L.W. Jelinski, *Adv. Mater. Science* **15** (1985) 359.

Kal1 H.-O. Kalinowski, S. Berger and S. Braun, *^{13}C-NMR-Spektroskopie*, Thieme, Stuttgart, 1984.

Kär1 J. Kärger, H. Pfeifer and W. Heink, *Adv. Magn. Reson.* **12** (1988) 1.

Kim1 R. Kimmich and H.-W. Weber, in: *Proc. 26th Congress Ampère Magn. Reson.*, ed. F. Milia, A. Simopoulos and A. Anagnostopoulos, NCSR Demokritos, Athens, 1992, p. 571.

Kim2 R. Kimmich, G. Schnur and M. Köpf, *Progress NMR Spectr.* **20** (1988) 385.

Kim3 R. Kimmich, *NMR Tomography, Diffusometry, Relaxometry*, Springer, Berlin, 1997.

Kut1 W. Kutzelnigg, U. Fleischer and M. Schindler, *NMR – Basic Principles and Progress* **23**, Springer, Berlin 1990, p. 165.

Lau1 M. Lausch and H. W. Spiess, *J. Magn. Reson.* **54** (1983) 466.

Lee1 M. Lee and W.I. Goldburg, *Phys. Rev. A* **140** (1965) 1261.

Lee2 Y.K. Lee, N.D. Kurur, M. Helmle, O.G. Johannessen, N.C. Nielsen and M.H. Levitt, *Chem. Phys. Lett.* **242** (1995) 304.

Lev1 M.H. Levitt, D. Suter and R.R. Ernst, *J. Chem. Phys.* **84** (1986) 4243.

Liu1 G. Liu, Y.-Z. Li and J. Jonas, *J. Chem. Phys.* **90** (1989) 5881.

Liu2 G. Liu, M. Mackowiak, Y.-Z. Li and J. Jonas, *J. Chem. Phys.* **94** (1991) 239.

Low1 I. Lowe, *Phys. Rev. Lett.* **2** (1959) 285.

Man1 P. Mansfield and P.K. Grannell, *J. Phys. C* **6** (1973) L422.

Man2 P. Mansfield and P.K. Grannell, *Phys. Rev. B* **12** (1975) 3618.
Man3 P. Mansfield, *J. Phys. C* **4** (1971) 1444.
Mar1 T.H. Mareci, S. Dønstrupp, *J. Molec. Liqu.* **38** (1988) 185.
Mar2 M.M. Mariq and J.S. Waugh, *J. Chem. Phys.* **70** (1979) 3300.
Mas1 J. Mason, ed., *Multinuclear NMR*, Plenum Press, New York, 1987.
Mcb1 V.J. McBrierty, *J. Chem. Phys.* **61** (1974) 872.
Mcb2 V.J. McBrierty and K.J. Packer, *Nuclear Magnetic Resonance in Solid Polymers*, Cambridge University Press, Cambridge, 1993.
Meh1 M. Mehring, *Principles of High Resolution NMR in Solids*, 2nd edn, Springer, Berlin, 1980.
Mei1 B.H. Meier, *Adv. Magn. Opt. Reson.* **18** (1994) 1.
Mic1 D. Michel and F. Engelke, *NMR – Basic Principles and Progress* **32** (1994) 69.
Mor1 K.R. Morgan and R.H. Newman, *J. Am. Chem. Soc.* **112** (1990) 4.
Mue1 K.T. Mueller, B.Q. Sun, G.C. Chingas, J.W. Zwanziger, T. Terao and A. Pines, *J. Magn. Reson.* **86** (1990) 470.
Mül1 K. Müller, K.-H. Wassmer and G. Kothe, *Adv. Polym. Science* **95** (1990) 1.
Mül2 K. Müller, P. Meier and G. Kothe, *Progr. NMR Spectrosc.* **17** (1985) 211.
Nie1 N.C. Nielson, H. Bildsoe, H.J. Jakobson and M.H. Levitt, *J. Chem. Phys.* **101** (1994) 1805.
Nis1 T. Nishi and T. Chikaraishi, *J. Macromol. Sci. Phys. B* **19** (1981) 445.
Oas1 T.G. Oas, R.G. Griffin and M.H. Levitt, *J. Chem. Phys.* **89** (1988) 692.
Pee1 O.B. Peersen, X. Wu and S.O. Smith, *J. Magn. Reson. A* **106** (1994) 127.
Pin1 A. Pines, M.G. Gibby and J.S. Waugh, *J. Chem. Phys.* **56**, (1972) 1776; *ibid.* **59** (1973) 569.
Pri1 R. Prigl, *Hochauflösende Kernresonanz in Festkörpern: Prinzipielle und Praktische Grenzen der Multipuls-Technik*, Diplomarbeit, Max-Planck-Institut für Medizinische Forschung, Heidelberg, 1990.
Pri2 R. Prigl and U. Haeberlen, *Adv. Magn. Opt. Reson.* **19** (1996) 1.
Ral1 D.P. Raleigh, M.H. Levitt and R.G. Griffin, *Chem. Phys. Lett.* **146** (1988) 71.
Red1 A.G. Redfield, *Adv. Magn. Reson.* **1** (1965) 1.
Rhi1 W.K. Rhim, D.D. Ellman and R. W. Vaughan, *J. Chem. Phys.* **58** (1973) 1772.
Rhi2 W. Rhim, A. Pines and J.S. Waugh, *Phys. Rev. B* **3** (1971) 684.
Rob1 P. Robyr, B.H. Meier and R.R. Ernst, *Chem. Phys. Lett.* **162** (1989) 417.
Sam1 A. Samoson, B.Q. Sun and A. Pines, in: *Pulsed Magnetic Resonance: NMR, ESR, and Optics*, ed. D.M.S. Bagguley, Clarendon Press, Oxford, 1992, p. 80.
Sch1 C. Schmidt, B. Blümich and H.W. Spiess, *J. Magn. Reson.* **79** (1988) 269.
Sch2 D. Schaefer, H.W. Spiess, U.W. Suter and W.W. Fleming, *Macromolecules* **23** (1990) 3431.
Sch3 A. Schmidt and S. Vega, *Chem. Phys. Lett.* **157** (1989) 539.
Sch4 K. Schmidt-Rohr, J. Clauss, B. Blümich and H.W. Spiess, *Magn. Reson. Chem.* **28** (1990) 3.
Sch5 K. Schmidt-Rohr, *Hochauflösende NMR an Festkörpern und Untersuchung der Phasenstruktur fester Polymere*, Diplomarbeit, Johannes-Gutenberg-Universität, Mainz, 1989.
Sch6 G. Scheler, U. Haubenreißer and H. Rosenberger, *J. Magn. Reson.* **44** (1981) 134.
Sch7 H. Schneider and H. Schmiedel, *Physics Letters* **30A** (1969) 298.
Sch8 H. Schmiedel and H. Schneider, *Ann. Phys.* **32** (1975) 249.
Sch9 K. Schmidt-Rohr and H.W. Spiess, *Multidimensional Solid-State NMR and Polymers*, Academic Press, London, 1994.

Sch10 J. Schaefer, E.O. Stejskal, J.R. Garbow and R.A. McKay, *J. Magn. Reson.* **59** (1984) 150.

Sli1 C.P. Slichter, *Principles of Magnetic Resonance*, 3rd edn, Springer, Berlin, 1990.

Spi1 H.W. Spiess, Rotation of molecules and nuclear spin relaxation, *NMR – Basic Principles and Progress* **15** (1978) 55.

Spi2 H.W. Spiess, *Adv. Polym. Science* **66** (1985) 23.

Spi3 H.W. Spiess, in: *Developments in Oriented Polymers – 1*, ed. I.M. Ward, Applied Science Publ., Barking, 1982.

Spi4 H.W. Spiess, *J. Chem. Phys.* **72** (1980) 6755.

Ste1 E.O. Stejskal, J. Schaefer and R.A. McKay, *J. Magn. Reson.* **25** (1977) 569.

Ste2 E.O. Stejskal and J.D. Memory, *High Resolution NMR in the Solid State: Fundamentals of CP/MAS*, Oxford University Press, New York, 1994.

Ste3 E.O Stejskal, J. Schaefer and J.S. Waugh, *J. Magn. Reson.* **28** (1977) 105.

Sti1 P. Stilbs, *Progr. NMR Spectroscopy* **19** (1987) 1.

Sun1 J.L. Sudmeier, S.E. Anderson and J.S. Frye, *Concepts Magn. Reson. 2* (1990) 197.

Sun1 B.Q. Sun, P.R. Costa and R.G. Griffin, *J. Magn. Reson. A* **112** (1995) 191.

Tan1 P. Tang, R.A. Santos and G.S. Harbison, *Adv. Magn. Reson.* **13** (1989) 225.

Teg1 J. Tegenfeldt and U. Haeberlen, *J. Magn. Reson.* **36** (1979) 453.

Tek1 P. Tekely, V. Gérardy, P. Palmas, D. Canet and A. Retournard, *Solid-State Nucl. Magn. Reson.* **4** (1995) 361.

Tek2 P. Tekely, P. Palmas and D. Canet, *J. Magn. Reson.* **107** (1994) 129.

Tit1 J.J. Titman, S. Féaux de Lacroix and H.W. Spiess, *J. Chem. Phys.* **98** (1993) 3816.

Tyc1 R. Tycko and G. Dabbagh, *Chem. Phys. Lett.* **173** (1990) 461.

Van1 D.L. VanderHart and G.B. McFadden, *Solid-State Nucl. Magn. Reson.* **7** (1996) 45.

Van2 D.L. VanderHart, W.L. Earl and A.N. Garroway, *J. Magn. Reson.* **44** (1981) 361.

Vee1 W.S. Veeman, *Progr. NMR Spectrosc.* 16 (1984) 193.

Vee2 W.S. Veeman and W.E.J.R. Maas, *NMR – Basic Principles and Progress* **32** (1994) 127.

Voe1 R. Voelkel, *Angew. Chem. Int. Ed. Eng.* **27** (1988) 1468.

War1 I.M. Ward, *Structure and Properties of Oriented Polymers*, Applied Science Publ., London, 1975.

Wau1 J.S. Waugh, L.M. Huber and U. Haeberlen, *Phys. Rev. Lett.* **20** (1968) 180.

Wef1 S. Wefing and H.W. Spiess, *J. Chem. Phys.* **89** (1988) 1219.

Weh1 M. Wehrle, G.P. Hellmann and H.W. Spiess, *Colloid and Polymer Science* **265** (1987) 815.

Wie1 U. Wiesner, K. Schmidt-Rohr, C. Boeffel, U. Pawelzik and H.W. Spiess, *Advanced Materials* **2** (1990) 484.

Wux1 X. Wu and K.W. Zilm, *J. Magn. Reson. A* **104** (1993) 154.

Yan1 C.S. Yannoni, *Acc. Chem. Res.* **15** (1982) 201.

Ywu1 Y. Wu, B.Q. Sun, A. Pines, A. Samoson and E. Lippmaa, *J. Magn. Reson.* **89** (1990) 297.

Zha1 S. Zhang, B.H. Meier and R.R. Ernst, *Phys. Rev. Lett.* **69** (1992) 2149.

Zha2 S. Zhang, B.H. Meier, S. Appelt, M. Mehring and R.R. Ernst, *J. Magn. Reson. A* **101** (1993) 60.

CHAPTER 4

Ang1 P.A. Angelis, *Concepts Magn. Reson.* **8** (1996) 339.

Ant1 J.-P. Antoine, *Phys. Mag.* **16** (1994) 17.

Bar1 H. Barkhuijsen, R. De Beer, A.C. Drogendijk, D. Van Ormondt and J.W.C. Van Der Veen, in: *Proc. Int. School of Physics Enrico Fermi, Course C*,ed. B. Maraviglia, North-Holland, Amsterdam, 1986, p. 313.

Bar2 D. Barache, J.-P. Antoine and J.-M. Dereppe, *J. Magn. Reson.* **128** (1997) 1.

Bee1 R. De Beer, W. Bovée, A. deGraaf, D. Van Ormondt, W. Pijnapple and R. Chamuleau, in: *SVD and Signal Processing*, ed. E.F. Deprettere, Elsevier, Amsterdam, 1988, p. 473.

Ben1 J.S. Bendat and A.G. Piersol, *Random Data: Analysis and Measurement Procedures*, Wiley, New York, 1971.

Bla1 J.M. Blackledge, *Quantitative Coherent Imaging*, Academic Press, New York, 1989.

Blü1 B. Blümich, *Progr. NMR Spectrosc.* **19** (1987) 331.

Blü2 B. Blümich and D. Ziessow, *J. Magn. Reson.* **46** (1982) 385.

Bog1 T.F. Bogard, *Basic Concepts in Linear Systems: Theory and Experiments*, Wiley, New York, 1984.

Bol1 L. Bolinger and J.S. Leigh, *J. Magn. Reson.* **80** (1988) 162.

Bra1 R.N. Bracewell, *The Fourier Transform and its Applications*, McGraw-Hill, New York, 1978.

Bur1 J.P. Burg, *Maximum Entropy Spectral Analysis*, Dissertation, Stanford University, Stanford, 1975.

Cha1 D. Chaudhuri, Hadamard Zeugmatography, MS Thesis, SUNY, Stony Brook, 1986.

Chu1 C.K. Chui, *An Introduction to Wavelets*, Academic Press, New York, 1992.

Chu2 C.K. Chui, ed., *Wavelets – A Tutorial in Theory and Applications*, Academic Press, New York, 1992.

Coi1 R.R. Coifman, Y. Myer and M.V. Wickerhauser, in: *Wavelets and Their Applications* ed. M.B. Rushai, B. Beylkin, R. Coifman, I. Daubechies, S. Mallat, Y. Myer and L. Raphael, Jones and Bartlett, Boston, 1992, p. 153.

Deu1 R. Deutsch, *System Analysis Techniques*, Prentice Hall, Englewood Cliffs, 1969.

Ern1 R.R. Ernst, G. Bodenhausen and A. Wokaun, *Principles of NMR in One and Two Dimensions*, Clarendon Press, Oxford, 1987.

Ern2 R.R. Ernst, *Chimia* **26** (1972) 53.

Gol1 S.W. Golomb, *Shift Register Sequences*, Holden-Day, San Francisco, 1967.

Har1 M. Harwitt and N.J.A Sloane, *Hadamard Transform Optics*, Academic Press, New York, 1979.

Kai1 R. Kaiser, *J. Magn. Reson.* **15** (1974) 44.

Kan1 E.R. Kanasewich, *Time Sequence Analysis in Geophysics*, The University of Alberta Press, Edmonton, 1981.

Lee1 Y.W. Lee and M. Schetzen, *Int. J. Control* **2** (1965) 231.

Maj1 P.D. Majors and A. Caprihan, *J. Magn. Reson.* **94** (1991) 225.

Mar1 P.Z. Marmarelis and V.Z. Marmarelis, *Analysis of Physiological Systems: The White Noise Approach*, Plenum Press, New York, 1978.

Mül1 S. Müller, *J. Magn. Reson. Med.* **6** (1988) 364.

Neu1 G. Neue, *Solid-State Nucl. Magn. Reson.* **5** (1996) 305.

Pra1 W.K. Pratt, J. Kane and H.C. Andrews, *Proc. IEEE* **57** (1969) 58.

Pra2 W.K. Pratt, *Digital Image Processing*, Wiley, New York, 1978.

Pri1 M.B. Priestley, *Non-linear and Non-stationary Time Series Analysis*, Academic Press, New York, 1988.

Sar1 G. Sarty and E. Kendall, *J. Magn. Reson. B* **111** (1996) 50.

Sch1 M. Schetzen, *The Volterra and Wiener Theories of Nonlinear Systems*, Wiley, New York, 1980.

Sou1 S.P. Souza, J. Szumowski, C.L. Dumoulin, D.P. Plewes and G. Glower, *J. Comp. Ass. Tom.* **12** (1988) 1926.

Spi1 M.R. Spiegel, *Laplace Transforms*, Schaum's Outline Series, McGraw-Hill, New York, 1965.

Wea1 J.B. Weaver and D. Healy, Jr., *J. Magn. Reson. A* **113** (1995) 1.

Zie1 D. Ziessow, *On-line Rechner in der Chemie*, de Gruyter, Berlin, 1973.

Zie2 D. Ziessow and B. Blümich, *Ber. Bunsenges. Phys. Chem.* **78** (1974) 1168.

CHAPTER 5

Ahn1 C.B. Ahn and W.C. Chu, *J. Magn. Reson.* **94** (1991) 455.

App1 M. Appel, G. Fleischer, D. Geschke, J. Kärger and M. Winkler, *J. Magn. Reson.* A **122** (1996) 248.

App2 M. Appel, G. Fleischer, J. Kärger, A.C. Dieng and G. Riess, *Macromolecules* **28** (1995) 2345.

Aue1 W.P. Aue, *Rev. Magn. Reson. Med.* **1** (1986) 21.

Bar1 G.A. Barrall, L. Frydman and G.C. Chingas, *Science* **255** (1992) 714.

Bau1 C. Bauer, R. Freeman, T. Frenkiel, J. Keeler and A.J. Shaka, *J. Magn. Reson.* **58** (1984) 442.

Bau2 J. Baum, R. Tycko and A. Pines, *J. Chem. Phys.* **79** (1983) 4643.

Ber1 D.J. Bergman, K.-J. Dunn, L.M. Schwartz and P.P. Mitra, *Phys. Rev.* E **51** (1995) 3393.

Ble1 M.H. Blees, *J. Magn. Reson.* A **109** (1994) 203.

Blü1 B. Blümich and D. Ziessow, *Mol. Phys.* **48** (1983) 995.

Blü2 B. Blümich, *Progr. NMR Spectrosc.* **19** (1987) 331.

Blü3 B. Blümich, *J. Magn. Reson.* **90** (1990) 535.

Blü4 P. Blümler and B. Blümich, *NMR -- Basic Principles and Progress* **30** (1994) 209.

Böh1 J.-M. Böhlen, M. Rey and G. Bodenhausen, *J. Magn. Reson.* **84** (1989) 191.

Böh2 J.-M. Böhlen, I. Burghardt, M. Rey and G. Bodenhausen, *J. Magn. Reson.* **90** (1990) 183.

Bol1 L. Bolinger and J.S. Leigh, *J. Magn. Reson.* **80** (1988) 162.

Bot1 P.A. Bottomley, *Rev. Sci. Instrum.* **53** (1982) 1319.

Bot2 P.A. Bottomley, *J. Magn. Reson.* **50** (1982) 335.

Bou1 D. Bourgeois and M. Decorps, *J. Magn. Reson.* **91** (1991) 128.

Bru1 P. Brunner and R.R. Ernst, *J. Magn. Reson.* **33** (1979) 83.

Cal1 P.T. Callaghan, *Aust. J. Phys.* **37** (1984) 359.

Cal2 P.T. Callaghan, *Principles of Nuclear Magnetic Resonance Microscopy*, Clarendon Press, Oxford, 1991.

Cal3 P.T. Callaghan, *J. Magn. Reson.* **87** (1990) 304.

Cal4 P.T. Callaghan and C.D. Eccles, *J. Magn. Reson.* **71** (1987) 426.

Cal5 P.T. Callaghan, *J. Magn. Reson.* **88** (1990) 493.

Cal6 P.T. Callaghan, Ampere Summer School, Portoroz, Slovenia, 12–18 September 1993, in: *Bulletin Ampere* **44** (1994) 6 (Part 1), 31 (Part II).

Cal7 P.T. Callaghan, A. Coy, D. MacGowan, K.J. Packer and F.O. Zelaya, *Nature* **351** (1991) 467.

Cal8 P.T. Callaghan, D. MacGowan, K.J. Packer and F. Zelaya, *J. Magn. Reson.* **90** (1990) 177.

Cal9 P.T. Callaghan, A. Coy, T.P.J. Halpin, D. MacGowan, K.J. Packer and F.O. Zelaya, *J. Chem. Phys.* **97** (1992) 651.

Cal10 P.T. Callaghan and B. Manz, *J. Magn. Reson.* A **106** (1994) 260.
Cal11 P.T. Callaghan and J. Stepisnik, *J. Magn. Reson.* A **117** (1995) 118.
Cal12 P.T. Callaghan, M.E. Komlosh and M. Nyden, *J. Magn. Reson.* **133** (1998) 177.
Cal13 P.T. Callaghan, *J. Magn. Reson.* **129** (1997) 74.
Cal14 P.T. Callaghan, S.L. Codd and J.D. Seymor, *Concepts Magn. Reson.* **11** (1999) 181.
Can1 D. Canet, B. Diter, A. Belmajdoub, J. Brondeau, J.C. Boudel and K.Elbayed, *J. Magn. Reson.* **81** (1989) 1.
Can2 D. Canet and M. Décorps, in: *Dynamics of Solutions and Fluid Mixtures by NMR*, ed. J.-J. Delpuech, Wiley, New York, 1995, p. 309.
Cap1 A. Caprihan, *IEEE Trans. Med. Imag.* **MI-2** (1983) 169.
Cap2 A. Caprihan, L.Z. Wang and E. Fukushima, *J. Magn. Reson.* A **118** (1996) 94.
Car1 P. Caravatti, M.H. Levitt and R.R. Ernst, *J. Magn. Reson.* **68** (1986) 323.
Cha1 I. Chang, G. Hinze, G. Diezemann, F. Fujara and H. Sillescu, *Phys. Rev. Lett.* **76** (1996) 2523.
Chi1 G.C. Chingas, L. Frydman, G.A. Barrall and J.S. Harwood, in: *Magnetic Resonance Microscopy*, ed. B. Blümich and W. Kuhn, VCH Publishers, Weinheim, 1992, p. 373.
Cho1 Z.H. Cho, H.S. Kim, H.B. Song and J. Cuming, *Proc. IEEE* **70** (1982) 271.
Chw1 C. Chwatinski, S. Han, P. Blümler and B. Blümich, (unpublished results).
Cod1 S. Codd and P.T. Callaghan, *J. Magn. Reson.* **137** (1999) 358.
Cof1 G.P. Cofer, J.M. Brown and G.A. Johnson, *J. Magn. Reson.* **83** (1989) 608.
Cor1 D.G. Cory, J.B. Miller and A.N. Garroway, *J. Magn. Reson.* **90** (1990) 544.
Cor2 M. Corti, F. Borsa and A. Rigamonti, *J. Magn. Reson.* **79** (1988) 21.
Cor3 D.G. Cory, A.M. Reichwein and W.S. Veeman, *J. Magn. Reson.* **80** (1988) 259.
Cor4 D.G. Cory, A.N. Garroway and J.B. Miller, *Polym. Prep. Am. Chem. Soc. Div. Polym. Chem.* **31** (1990) 149.
Cot1 S.P. Cottrell, M.R. Halse, D.A. Ibbett, B.L. Boda-Novy and J.H. Strange, *Meas. Sci. Technol.* **2** (1991) 860.
Coy1 A. Coy and P.T. Callaghan, *J. Chem. Phys.* **101** (1994) 4599.
Coy2 A. Coy and P.T. Callaghan, *J. Colloid. Interf. Science* **168** (1994) 373.
Cre1 A.J.S. de Crespigny, A.T. Carpenter and L.D. Hall, *J. Magn. Reson.* **88** (1990) 406.
Cro1 L.E. Crooks, D.A. Ortendahl, J. Hoenninger, M. Arakawa, J. Watts, C.R.Cannon, M. Brant-Zawadzki, P.L. Davis and A.R. Margulis, *Radiology* **146** (1983) 123.
Dam1 R. Damadian, L. Minkoff, M. Goldsmith, M. Stanford and J. Koutcher, *Science* **194** (1976) 1430.
Dam2 L. Minkoff, R. Damadian, T.E. Thomas, N. Hu, M. Goldsmith, J. Koutcher and M.Stanford, *Physiol. Chem. Phys.* **9** (1977) 101.
Dam3 M. Goldsmith, R. Damadian, M. Stanford and M. Lipkowitz, *Physiol. Chem. Phys.* **9** (1977) 105.
Dam4 R. Damadian, M. Goldsmith and L. Minkoff, *NMR – Basic Principles and Progress* **19** (1981) 1.
Dem1 D.E. Demco, R. Kimmich, S. Hafner and H.-W. Weber, *J. Magn. Reson.* **94** (1991) 317.
Dem2 D.E. Demco, A. Johansson and J. Tegenfeldt, *J. Magn. Reson.* A **110** (1994) 183.
Dij1 P. Dijk, B. Berkowitz and P. Bendel, *Water Resources Research* **35** (1999) 347.
Dod1 D.M. Doddrell, J.M. Bulsing, G.J. Galloway, W.M. Brooks, J. Field, M. Irwing and H. Baddeley, *J. Magn. Reson.* **70** (1986) 319.
Dup1 R. Dupeyre, P. Devoulon, D. Bourgeois and M. Decorps, *J. Magn. Reson.* **95** (1991) 589.
Ecc1 C.D. Eccles and P.T. Callaghan, *J. Magn. Reson.* **68** (1986) 393.
Ems1 L. Emsley and G. Bodenhausen, *J. Magn. Reson.* **82** (1989) 211.

Ems2 L. Emsley and G. Bodenhausen, *Magn. Reson. Med.* **10** (1989) 273.

Ems3 L. Emsley and G. Bodenhausen, *Chem. Phys. Lett.* **165** (1990) 469.

Ern1 R.R. Ernst, G. Bodenhausen and A. Wokaun, *Principles of NMR in One and Two Dimensions*, Clarendon Press, Oxford, 1987.

Fei1 L.F. Feiner and P.R. Locher, *Appl. Phys.* **22** (1980) 257.

Fei2 T. Feiweier, B. Geil, O. Isfort and F. Fujara, *J. Magn. Reson.* **131** (1998) 203.

Fil1 A.V. Filippov, E.V. Khosina and V.G. Khosin, *J. Mat. Sci.* **31** (1996) 1809.

Fil2 A.V. Filippov, M.G. Altykis, M.I. Khaliullin, R.Z. Rachimov and V.M. Lantsov, *J. Mat. Sci.* **31** (1996) 3469.

Fin1 J. Finsterbusch and J. Frahm, *J. Magn. Reson.* **137** (1999) 144.

Fle1 F. Fleischer and F. Fujara, *NMR – Basic Principles and Progress* **30** (1994) 111.

For1 E.J. Fordham, P.P. Mitra and L.L. Latour, *J. Magn. Reson.* A **121** (1996) 187.

Fra1 J. Frahm and W. Hänicke, *J. Magn. Reson.* **60** (1984) 320.

Fre1 R. Freeman, *Chem. Rev.* **91** (1991) 1397.

Fre2 R. Freeman, *A Handbook of Nuclear Magnetic Resonance*, Longman Scientific & Technical, Harlow, 1988.

Fuj1 F. Fujara, E. Ilyina, H. Nienstaedt, H. Sillescu, R. Spohr and C. Trautmann, *Magn. Reson. Imag.* **12** (1994) 245.

Gab1 R. Gabillard, *Comptes. Rend. Acad. Sci. (Paris)* **232** (1951) 1551.

Gar1 A.N. Garroway, P.K. Grannell and P. Mansfield, *J. Phys.* C **7** (1974) L475.

Gar2 M. Garwood and K. Ugurbil, *NMR – Basic Principles and Progress* **26** (1992) 109.

Gee1 H. Geen and R. Freeman, *J. Magn. Reson.* **93** (1991) 93.

Gei1 B. Geil, *Concepts Magn. Reson.* **10** (1998) 299.

Göb1 V. Göbbels, Zweidimensionale Magnetische Resonanz an porösen Medien, Dissertation, RWTH, Aachen, 1999.

Goe1 G. Goelman, V.H. Subramanian and J.S. Leigh, *J. Magn. Reson.* **89** (1990) 437.

Goe2 G. Goelman and J.S. Leigh, *J. Magn. Reson.* **91** (1991) 93.

Haa1 A. Haase and J. Frahm, *J. Magn. Reson.* **65** (1985) 481.

Haf1 H.-P. Hafner, S. Müller and J. Seelig, *Magn. Reson. Med.* **13** (1990) 279.

Haf2 S. Hafner, E. Rommel and R. Kimmich, *J. Magn. Reson.* **88** (1990) 449.

Haf3 S. Hafner, D.E. Demco and R. Kimmich, *Chem. Phys. Lett.* **187** (1991) 53.

Hah1 E.L. Hahn, *Phys. Rev.* **80** (1950) 580.

Hal1 L.D. Hall, S. Sukumur and S.L. Talagala, *J. Magn. Reson.* **56** (1984) 275.

Hei1 S.R. Heil and M. Holz, *Angew. Chem. Int. Ed. Engl.* **35** (1996) 1717.

Hep1 M.A. Hepp and J.B. Miller, *J. Magn. Reson.* A **110** (1994) 98.

Hep2 M.A. Hepp and J.B. Miller, *Macromol. Symp.* **86** (1994) 271.

Hin1 W.S. Hinshaw, *Phys. Lett.* A **48** (1974) 87.

Hin2 W.S. Hinshaw, *J. Appl. Phys.* **47** (1976) 3706.

Hin3 W.S. Hinshaw, P.A. Bottomley and G.N. Holland, *Nature* **270** (1977) 722.

Hol1 M. Holz, *Progr. NMR Spectrosc.* **18** (1986) 327.

Hou1 D.I.Hoult, *J. Magn. Reson.* **33** (1979) 183.

Hür1 M.D. Hürlimann, T.M.de Swiet and P.N. Sen, *J. Noncryst. Solids* **182** (1995) 198.

Hwa1 T.-L. Hwang, P.C.M. van Zijl and M. Garwood, *J. Magn. Reson.* **133** (1998) 200.

Hys1 W.B. Hyslop and P.C. Lauterbur, *J. Magn. Reson.* **94** (1991) 501.

Joh1 C.S. Johnson, in: *Nucl. Magn. Reson. Probes of Molecular Dynamics*, ed. R. Tycko, Kluwer, Dordrecht, 1994, p. 455.

Joh2 C.S. Johnson Jr. and Q. He, *Adv. Magn. Reson.* **13** (1989) 131.

Kar1 G.S. Karczmar, D.B. Twieg, T.J. Lawry, G.B. Matson and M.W. Wiener, *Magn. Reson. Med.* **7** (1988) 111.

Kär1 J. Kärger, H. Pfeifer and W. Heink, *Adv. Magn. Reson.* **12** (1988) 1.
Kie1 M. von Kienlin and R. Mejia, *J. Magn. Reson.* **94** (1991) 268.
Kie2 M. von Kienlin and R. Pohmann, in: *Spatially Resolved Magnetic Resonance*, ed. P. Blümler, B. Blümich, R. Botto and E. Fukushima, Wiley-VCH, Weinheim, 1998, p. 3.
Kim1 R. Kimmich, W. Unrath, G. Schnur and E. Rommel, *J. Magn. Reson.* **91** (1991) 136.
Kim2 R. Kimmich, *NMR Tomography, Diffusometry, Relaxometry*, Springer, Berlin, 1997.
Kum1 A. Kumar, D. Welti and R.R. Ernst, *J. Magn. Reson.* **18** (1975) 69.
Kun1 D. Kunz, *Magn. Reson. Med.* **3** (1986) 377.
Kup1 E. Kupce and R. Freeman, *J. Magn. Reson. A* **118** (1996) 299.
Lat1 L.L. Latour, R.L. Kleinberg, P.P. Mitra and C.H. Sotak, *J. Magn. Reson. A* **112** (1995) 83.
Lau1 P.C. Lauterbur, *Nature* **242** (1973) 190.
Lev1 M.H. Levitt, Progr. *NMR Spectrosc.* **18** (1986) 61.
Lev2 M.H. Levitt and R. Freeman, *J. Magn. Reson.* **33** (1979) 473.
Lit1 T.-Q. Li, U. Henriksson and L. Ödberg, *J. Colloid Interf. Science* **169** (1995) 376.
Lju1 S. Ljunggren, *J. Magn. Reson.* **54** (1983) 338.
Lud1 K.M. Ludecke, P. Roschmann and R. Tischler, *Magn. Reson. Imag.* **3** (1985) 329.
Mac1 A. Macovski, *Magn. Reson. Med.* **2** (1985) 29.
Maf1 P. Maffei, K. Elbayed, J. Brondeau and D. Canet, *J. Magn. Reson.* **95** (1991) 382.
Man1 P. Mansfield and P.G. Morris, *NMR Imaging in Biomedicine, Adv. Magn. Reson. Suppl. 2*, Academic Press, New York, 1982.
Man2 P. Mansfield, A.A. Maudsley and T. Baines, *J. Phys. E* **9** (1976) 271.
Man3 P. Mansfield and P.K. Grannell, *J. Phys. C* **6** (1973) L422.
Man4 P. Mansfield, in: *Physics of NMR Spectroscopy in Biology and Medicine, Proc. Int. School Enrico Fermi, Course C*, ed. B. Maraviglia, North-Holland, Amsterdam, 1988, p. 345.
Man5 P. Mansfield and P.K .Grannell, *Phys. Rev. B* **12** (1975) 3618.
Mao1 J. Mao, T.H. Mareci and E.R. Andrew, *J. Magn. Reson.* **79** (1988) 1.
Mar1 T.H. Mareci and H.R. Brooker, *J. Magn. Reson.* **92** (1991) 229.
Mcc1 D.W. McCall, D.C. Douglass and E.W. Anderson, *Ber. Bunsenges. Phys. Chem.* **67** (1963) 336.
Mcd1 S. McDonald and W.S. Warren, *Concepts Magn. Reson.* **3** (1991) 55.
Mcf1 E.W. McFarland, *Magn. Reson. Imag.* **10** (1992) 269.
Met1 C.E. Metz and K. Doi, *Phys. Med. Biol.* **24** (1979) 1079.
Mit1 P.P. Mitra and P.N. Sen, *Phys. Rev. B* **45** (1992) 143.
Mit2 P.P. Mitra, P.N. Sen, L.M. Schwartz and P.Le Doussal, *Phys. Rev. Lett.* **68** (1992) 3555.
Mit3 P.P. Mitra, L.L. Latour, R.L. Kleinberg and C.H. Sotak, *J. Magn. Reson. A* **114** (1995) 47.
Mit4 P.P. Mitra, *Phys. Rev. B* **51** (1995) 15074.
Mit5 P.P. Mitra and B.I. Halperin, *J. Magn. Reson. A* **113** (1995) 94.
Mit6 P.P. Mitra, P.N. Sen and L.M. Schwartz, *Phys. Rev. B* **47** (1993) 8565.
Mit7 P.P. Mitra, *Physica A* **241** (1997) 122.
Moo1 W.S. Moore and G.N. Holland, *Phil. Trans. R. Soc. Lond. B* **289** (1980) 511.
Mor1 P.G. Morris, *Nuclear Magnetic Resonance Imaging in Medicine and Biology*, Clarendon Press, Oxford, 1986.
Mor2 G.A. Morris and R. Freeman, *J. Magn. Reson.* **29** (1978) 433.
Mor3 K.F. Morris and C.S. Johnson Jr., *J. Am. Chem. Soc.* **114** (1992) 3139.
Mor4 K.F. Morris and C.S. Johnson Jr., *J. Am. Chem. Soc.* **114** (1992) 776.

Mor5 P.G. Morris, *NMR – Basic Principles and Progress* **26** (1992) 149.
Mül1 S. Müller, *Magn. Reson. Med.* **6** (1988) 364.
Mül2 S. Müller, *Magn. Reson. Med.* **10** (1989) 145.
Mül3 S. Müller, R. Sauter, H. Weber and J. Seelig, *J. Magn. Reson.* **76** (1988) 155.
Mül4 S. Müller, *Magn. Reson. Med.* **5** (1987) 502.
Nil1 H. Nilgens, P. Blümler, J. Paff and B. Blümich, *J. Magn. Reson. A* **105** (1993) 108.
Ord1 R.J. Ordidge, *Magn. Reson. Med.* **5** (1987) 93.
Pau1 J. Pauly, D. Nishimura and A. Macovski, *J. Magn. Reson.* **81** (1989) 43.
Pro1 S.W. Provencher, *Comput. Phys. Commun.* **27** (1982) 213, 229.
Rob1 T.P.L. Roberts, T.A. Carpenter and L.D. Hall, *J. Magn. Reson.* **91** (1991) 204.
Rom1 E. Rommel and R. Kimmich, *J. Magn. Reson.* **83** (1989) 299.
Rom2 E. Rommel and R. Kimmich, *Magn. Reson. Med.* **12** (1989) 390.
Ros1 D. Rosenfeld, S.L. Panfil and Y. Zur, *J. Magn. Reson.* **126** (1997) 221.
Sch1 K. Schmidt-Rohr and H.W. Spiess, *Multidimensional Solid-State NMR and Polymers*, Academic Press, London, 1994.
Sco1 K.N. Scott, H.R. Brooker, J.R. Fitzsimmons, H.F. Bennet and R.C. Mick, *J. Magn. Reson.* **50** (1982) 339.
Sen1 P.N. Sen and M.D. Hürlimann, *Phys. Rev. B* **51** (1995) 601.
Ser1 I. Sersa and S. Macura, *J. Magn. Reson. B* **111** (1996) 186.
Ser2 I. Sersa and S. Macura, *Magn. Reson. Med.* **37** (1997) 920.
Ser3 I. Sersa and S. Macura, *J. Magn. Reson.* **134** (1998) 466.
Sey1 J.D. Seymour and P.T. Callaghan, *AICHE J.* **43** (1997) 2096.
Sey2 J.D. Seymour and P.T. Callaghan, *J. Magn. Reson. A* **122** (1996) 90.
Sey3 J.D. Seymor, J.E. Maneval, K.L. McCarthy, R.L. Powell and M.J. McCarthy, *J. Texture Studies* **26** (1995) 89.
Sha1 A.J. Shaka and R. Freeman, *J. Magn. Reson.* **59** (1984) 169.
Sha2 A.J. Shaka and R. Freeman, *J. Magn. Reson.* **64** (1985) 145.
She1 J. Shen, *J. Magn. Reson. B* **112** (1996) 131.
Sil1 M.S. Silver, R.I. Joseph and D.I. Hoult, *J. Magn. Reson.* **59** (1984) 347.
Sil2 M.S. Silver, R.I. Joseph and D.I. Hoult, *Phys. Rev. A* **31** (1985) 2753.
Sin1 S. Singh and B.K. Rutt, *J. Magn. Reson.* **87** (1990) 567.
Sin2 S. Singh, B.K. Rutt and S. Napel, *J. Magn. Reson.* **90** (1990) 313.
Sin3 S. Singh and R. Deslaurier, *Concepts Magn. Reson.* **7** (1995) 1.
Sli1 C.P. Slichter, *Principles of Magnetic Resonance*, 3rd enlarged edn, Springer, Berlin, 1989.
Sod1 A. Sodickson and D.G. Cory, *Progr. Nucl. Magn. Reson. Spectrosc.* **33** (1998) 77.
Sor1 G.H. Sorland, *J. Magn. Reson.* **126** (1997) 146.
Sou1 S.P. Souza, J. Szumowski, C.L. Dumoulin, D.P. Plewes and G. Glover, *J. Comp. Ass. Tomography* **12** (1988) 1026.
Spi1 H.W. Spiess, *Adv. Polym. Science* **66** (1985) 23.
Sta1 S. Stapf and K.J. Packer, *Appl. Magn. Reson.* **15** (1998) 303.
Sta2 S. Stapf, R.A. Damion and K.J. Packer, *J. Magn. Reson.* **137** (1999) 316.
Sta3 S. Stapf, K.J. Packer, R.G. Graham, J.-F. Thovert and P.M. Adler, *Phys. Rev. E* **58** (1998) 6206.
Ste1 E.O. Stejskal and J.E. Tanner, *J. Chem. Phys.* **42** (1965) 288.
Ste2 J. Stepisnik, *Prog. NMR Spectrosc.* **17** (1985) 187.
Ste3 J. Stepisnik, *J. Magn. Reson.* **131** (1998) 339.
Sti1 P. Stilbs, *Progr. NMR Spectroscopy* **19** (1987) 1.
Sut1 R.J. Sutherland and J.M.S. Hutchison, *J. Phys. E, Sci. Instrum.* **11** (1978) 79.

Tan1 K. Tanaka, Y. Yamada, E. Yamamoto and Z. Abe, *Proc. IEEE* **66** (1978) 1582.
Tom1 B.L. Tomlinson and H.D.W. Hill, *J. Chem. Phys.* **59** (1973) 1775.
Vla1 M.T. Vlaardingerbroek and J.A. den Boer, *Magnetic Resonance Imaging*, Springer, Berlin, 1996.
Wan1 L.Z. Wang, A. Caprihan and E. Fukushima, *J. Magn. Reson. A* **117** (1995) 209.
War1 W.S. Warren and M.S. Silver, *Adv. Mag. Reson.* **12** (1988) 247.
War2 W.S. Warren, *J. Chem. Phys.* **81** (1984) 5437.
Wat1 A.T. Watson, R. Kulkarni, J.-E. Nordtvedt, A. Sylte and H. Urkedal, *Meas. Sci. Technol.* **9** (1998) 898.
Wil1 D.S. Williams and I.J. Lowe, *J. Magn. Reson.* **91** (1991) 57.
Wim1 S. Wimperis, *J. Magn. Reson.* **86** (1990) 46.
Win1 R.A. Wind, J.H.N. Creyghton, D.J. Lighthelm and J. Smidt, *J. Phys. C: Solid State Phys.* **11** (1978) L223.
Zij1 P.C.M. van Zijl, C.T.W. Moonen, *NMR – Basic Principles and Progress* **26** (1992) 67.

CHAPTER 6

And1 E.R. Andrew, G. Bydder, J. Griffiths, R. Iles and P. Styles, *Clinical Magnetic Resonance: Imaging and Spectroscopy*, John Wiley, Chichester, 1990.
Arm1 R.L. Armstrong, A. Tzalmona, M. Menzinger, A. Cross and C. Lemaire, in: *Magnetic Resonance Microscopy*, ed. B. Blümich and W. Kuhn, VCH, Weinheim, 1992, p. 309.
Bas1 P.J. Basser, J. Mattiello and D. LeBihan, *Biophys. J.* **66** (1994) 259.
Bas2 P.J. Basser and C. Pierpaoli, *J. Magn. Reson. B* **111** (1996) 209.
Ber1 M. Bernardo, Jr., D. Chaudhuri, X.-R. Liu and P.C. Lauterbur, *Book of Abstracts, Soc. Magn. Reson. Med., Forth Annual Meeting, London*, 1985, p. 944.
Blü1 B. Blümich, *Progr. NMR Spectrosc.* **19** (1987) 331.
Blü2 B. Blümich, *J. Magn. Reson.* **60** (1984) 37.
Blü3 B. Blümich and H.W. Spiess, *J. Magn. Reson.* **66** (1986) 66.
Blü4 B. Blümich, *J. Magn. Reson.* **90** (1990) 535.
Blü5 B. Blümich, *Rev. Sci. Instrum.* **57** (1986) 1140.
Blü6 B. Blümich, in: *Encyclopedia of NMR Spectroscopy*, ed. D. M. Grant and R. K. Harris, Wiley, New York, 1996, Vol. 7, p. 4581.
Blü7 P. Blümler and B. Blümich, *NMR – Basic Principles and Progress* **30** (1994) 209.
Blü8 B. Blümich, J. Jansen, H. Nilgens, P. Blümler and G.L. Hoatson, *Magn. Reson. Biol. Med.* **1** (1993) 61.
Bol1 L. Bolinger and J.S. Leigh, *J. Magn. Reson.* **80** (1988) 162.
Bou1 D. Boudot, D. Canet and J. Brondeau, *J. Magn. Reson.* **87** (1990) 385.
Bou2 D. Boudot, F. Montigny, K. Elbayed, P. Mutzenhardt, B. Diter, J. Brondeau and D. Canet, *J. Magn. Reson.* **92** (1991) 605.
Bow1 J.L. Bowers, P.M. Macdonald and K.R. Metz, *J. Magn. Reson. B* **106** (1995) 72.
Bro1 T.R. Brown, B.M. Kincaid and K. Ugurbil, *Proc. Natl. Acad. Sci. USA* **79** (1982) 3523.
Bur1 D. Burstein, *Concepts Magn. Reson.* **8** (1996) 269.
Cal1 P.T. Callaghan, *Principles of Nuclear Magnetic Resonance Microscopy*, Clarendon Press, Oxford, 1991.
Can1 D. Canet, *Progr. NMR Spectrosc.* **30** (1997) 101.
Can2 D. Canet, in: *Encyclopedia of NMR Spectroscopy*, ed. D.M. Grant and R.K. Harris, Wiley, New York, 1996, p. 3938.

Can3 D. Canet, P. Mutzenhardt, J. Brondeau and C. Roumestand, *Chem. Phys. Lett.* **222** (1994) 171.

Cer1 J.D. de Certaines, W.M.M.J. Bovée and F. Podo, ed., *Magnetic Resonance Spectroscopy in Biology and Medicine*, Pergamon Press, New York, 1992.

Cho1 H.M. Cho, C.J. Lee, D.N. Shykind and D.P. Weitekamp, *Phys. Rev. Lett.* **55** (1985) 1923.

Cho2 Z.H. Cho, Y.M. Ro and I.K. Hong, *Concepts Magn. Reson.* **10** (1998) 33.

Coc1 M.D. Cockman and T.H. Mareci, *J. Magn. Reson.* **79** (1988) 236.

Cor1 D.G. Cory, J.B. Miller, A.N. Garroway and W.S. Veeman, *J. Magn. Reson.* **85** (1989) 219.

Cor2 J. Coremans, M. Spanoghe, L. Budinsky, J. Sterckx, R. Luypaert, H. Eisendrath and M. Osteaux, *J. Magn. Reson.* **124** (1997) 323.

Cox1 S.J. Cox and P. Styles, *J. Magn. Reson.* **40** (1980) 209.

Dec1 M. Decorps, R. Dupeyre, C. Remy, Y. le Fur, P. Devoulon and D. Bourgeois, in: *Magnetic Resonance Spectroscopy in Biology and Medicine*, ed. J.D. de Certaines, W.M.M.J. Bovée and F. Podo, Pergamon Press, Oxford, 1992, p. 111.

Dor1 S.J. Doran and M. Décorps, *J. Magn. Reson. A* **117** (1995) 311.

Dup1 R. Dupeyre, P. Devoulon, D. Bourgeois and M. Décorps, *J. Magn. Reson.* **95** (1991) 589.

Duy1 J. Duyn, Y. Yang, J.A. Frank and J.W. van der Veen, *J. Magn. Reson.* **132** (1998) 150.

Duy2 J. Duyn and Y. Yang, *J. Magn. Reson.* **128** (1997) 130.

Ede1 W.A. Edelstein, J.M.S. Hutchinson, G. Johnson and T. Redpath, *Phys. Med. Biol.* **25** (1980) 751.

Ern1 R.R. Ernst, G. Bodenhausen and A. Wokaun, *Principles of Nuclear Magnetic Resonance in One and Two Dimensions*, Clarendon Press, Oxford, 1987.

Ern2 A. Kumar, D. Welti and R.R. Ernst, *J. Magn. Reson.* **18** (1975) 69.

Ern3 R.R. Ernst, *J. Magn. Reson.* **3** (1970) 10.

Fin1 J. Finsterbusch and J. Frahm, *J. Magn. Reson.* **137** (1999) 144.

For1 J.J. Ford, *J. Magn. Reson.* **87** (1990) 346.

Fra1 J. Frahm, K.D. Mehrboldt, W. Hänicke and A. Haase, *J. Mag. Reson.* **64** (1985) 81.

Fra2 J. Frahm, A. Haase and D. Matthaei, *J. Comp. Ass. Tomography* **10** (1986) 363.

Fra3 J. Frahm, A. Haase and D. Matthaei, *Magn. Reson. Med.* **3** (1986) 321.

Gat1 J.C. Gatenby and J.C. Gore, *J. Magn. Reson. A* **121** (1996) 193.

Gat2 J.C. Gatenby and J.C. Gore, *J. Magn. Reson. A* **110** (1994) 26.

Gel1 P. van Gelderen, J.H. Duyn and C.T.W. Moonen, *J. Magn. Reson. B* **107** (1995) 78.

Ger1 R.E. Gerald III, R.J. Klingler, J.W. Rathke, G. Sandi and K. Woelk, in: *Spatially Resolved Magnetic Resonance*, ed. P. Blümler, B. Blümich, R. Botto and E. Fukushima, Wiley-VCH, Weinheim, 1998, p. 111.

Goe1 G.Goelman, V.H. Subramanian and J.S. Leigh, *J. Magn. Reson.* **89** (1990) 437.

Goe2 G. Goelman and J.S. Leigh, *J. Magn. Reson.* **91** (1991) 93.

Gre1 M. Greferath, B. Blümich, W.M. Griffith and G.L. Hoatson, *J. Magn. Reson. A* **102** (1993) 332.

Gui1 D.N. Guilfoyle, B. Issa and P. Mansfield, *J. Magn. Reson. A* **119** (1996) 151.

Gyn1 M.L. Gyngell, *Magn. Reson. Imag.* **6** (1988) 415.

Gyn2 M. Gyngell, *J. Magn. Reson.* **81** (1989) 474.

Haa1 A. Haase, J. Frahm, D. Matthei, W. Hänicke and K.D. Merboldt, *J. Magn. Reson.* **67** (1986) 258.

Haa2 A. Haase and J. Frahm, *J. Magn. Reson.* **65** (1985) 481.

Haa3 A. Haase and D. Matthaei, *J. Magn. Reson.* **71** (1987) 550.

Haa4 A. Haase, *Magn. Reson. Med.* **13** (1990) 77.

Haa5 A. Haase, D. Matthaei, R. Barthkowski, E. Dühmke and D. Leibfritz, *J. Comp. Ass. Tom.* **13** (1989) 1036.

Haf1 H.-P. Hafner, S. Müller and J. Seelig, *Magn. Reson. Med.* **13** (1990) 279.

Haf2 S. Hafner, *Magn. Reson. Imag.* **12** (1994) 1047.

Hah1 E.L. Hahn, *Phil. Trans. R. Soc. Lond. A* **333** (1990) 403.

Hak1 R. Haken, P. Blümler and B. Blümich, in: *Spatially Resolved Magnetic Resonance*, ed. P. Blümler, B. Blümich, R. Botto and E. Fukushima, Wiley-VCH, Weinheim, 1998, p. 695.

Hau1 K.H. Hausser and H.R. Kalbitzer, *NMR für Mediziner und Biologen*, Springer, Berlin, 1989.

Haw1 R.C. Hawkes and S. Patz, *Magn. Reson. Med.* **4** (1987) 9.

Hed1 L.K. Hedges and D.I .Hoult, *J. Magn. Reson.* **79** (1988) 391.

Hen1 K. Hendrich, Y. Hu, R.S. Menon, H. Merkle, P. Camarata, R. Heros and K. Ugurbil, *J. Magn. Reson. B* **105** (1994) 225.

Hen2 J. Hennig, A. Nauerth and H. Friedburg, *Magn. Reson. Med.* **3** (1986) 823.

Hen3 T. Loenneker and J. Hennig, *Proc. Indian Acad. Sci. (Chem. Sci.)* **106** (1994) 1605.

Hen4 F. Hennel, *Concepts Magn. Reson.* **9** (1997) 43.

Hen5 F. Hennel, *J. Magn. Reson.* **134** (1998) 206.

Hen6 F. Hennel, Z. Zulek and A. Jasinski, *J. Magn. Reson. A* **102** (1993) 95.

Hep1 M.A. Hepp and J.B. Miller, *Solid-State Nucl. Magn. Reson.* **6** (1996) 367.

Her1 G.T. Herman, ed., *Image Resonstruction from Projections, Topics in Applied Physics 32*, Springer, Berlin, 1979.

Hod1 P. Hodgkinson, R.O. Kemp-Harper and P.J. Hore, *J. Magn. Reson. B* **105** (1994) 256.

Hou1 D.I .Hoult, *J. Magn. Reson.* **33** (1979) 183.

Hou2 D.I. Hoult, *Book of Abstracts, 19th ENC*, Blacksburg, VA, 1978.

Hum1 F. Humbert, B. Diter and D. Canet, *J. Magn. Reson. A* **123** (1996) 242.

Hum2 F. Humbert, M. Valtier, A. Retournard and D. Canet, *J. Magn. Reson.* **134** (1998) 245.

Jak1 P.M. Jakob, F. Kober, R. Pohmann and A. Haase, *J. Magn. Reson. B* **110** (1996) 278.

Jan1 J. Jansen and B. Blümich, *J. Magn. Reson.* **99** (1992) 525.

Jel1 L.W. Jelinski, *Adv. Mater. Science* **15** (1985) 359.

Jiv1 A. Jivan, M.A. Horsefield, A.R. Moody and G.R. Cherryman, *J. Magn. Reson.* **127** (1997) 65.

Jon1 R.D. Jones, *J. Magn. Reson.* **90** (1990) 384.

Kai1 R. Kaiser, *J. Magn. Reson.* **3** (1970) 28.

Kai2 R. Kaiser, *J. Magn. Reson.* **15** (1974) 44.

Kim1 R. Kimmich, *NMR Tomography, Diffusometry, Relaxometry*, Springer, Berlin, 1997.

Kim2 R. Kimmich, B. Simon and H. Köstler, *J. Magn. Reson. A* **112** (1995) 7.

Kos1 K. Kose, *J. Magn. Reson.* **92** (1991) 631.

Kum1 A. Kumar, D. Welti and R.R. Ernst, *J. Magn. Reson.* **18** (1975) 69.

Kum2 A. Kumar, D. Welti and R.R. Ernst, *Naturwiss.* **62** (1975) 34.

Lai1 C.-M. Lai and P.C. Lauterbur, *J. Phys. E: Sci. Instrum.* **13** (1980) 747.

Lai2 C.-M. Lai, *J. Appl. Phys.* **52** (1980) 1141.

Lau1 P.C. Lauterbur, *Nature* **242** (1973) 190.

Lau2 P.C. Lauterbur, *J. Magn. Reson.* **59** (1984) 536.

Lee1 S.Y. Lee and Z.H. Cho, *Magn. Reson. Med.* **12** (1989) 56.

Lee2 D.H. Lee and S. Lee, *Magn. Reson. Imag.* **12** (1994) 613.

Lju1 S. Lunggren, *J. Magn. Reson.* **54** (1983) 338.

Low1 I.J. Lowe and R.E. Wysong, *J. Magn. Reson. B* **106** (1993) 101.

Maf1 P. Maffei, P. Mutzenhardt, A. Retournard, B. Diter, R. Raulet, J. Brondeau and D. Canet, *J. Magn. Reson. A* **107** (1994) 40.

Maf2 P. Maffei, K. Elbayed, J. Brondeau and D. Canet, *J. Magn. Reson.* **95** (1991) 382.

Maf3 P. Maffei, L. Kiéné and D. Canet, *Macromolecules* **25** (1992) 7115.

Maj1 P.D. Majors and A. Caprihan, *J. Magn. Reson.* **94** (1991) 225.

Maj2 P.D. Majors, D.M. Smith and P.J. Davis, *Chem. Eng. Sci.* **46** (1991) 3037.

Man1 P. Mansfield and P.G. Morris, *NMR Imaging in Biomedicine, Adv. Magn. Reson. Suppl. 2*, Academic Press, New York, 1982.

Man2 P. Mansfield, *Magn. Reson. Med.* **1** (1984) 370.

Man3 P. Mansfield, A.M. Howseman and R.J. Ordidge, *J. Phys. E.* **22** (1989) 324.

Man4 P. Mansfield, P.G. Morris, R.J. Ordidge, I.L. Pykett, V. Bangert and R.E. Coupland, *Phil. Trans. R. Soc. B* **289** (1980) 503.

Man5 P. Mansfield, in: *Pulsed Magnetic Resonance: NMR, ESR, and Optics*, ed. D.M.S. Bagguley, Clarendon Press, Oxford, 1992, p. 317.

Man6 P. Mansfield, in: *Physics of NMR Spectroscopy in Biology and Medicine, International School of Physics, Enrico Fermi, Course C*, ed. B. Maraviglia, North-Holland, Amsterdam, 1988, p. 345.

Man7 Y. Manassem and G. Navon, *J. Magn. Reson.* **61** (1985) 363.

Man8 P. Mansfield, *J. Phys. C* **10** (1977) L55.

Man9 B. Manz, P.S. Chow and L.F. Gladden, *J. Magn. Reson.* **136** (1999) 226.

Mar1 T.H. Mareci and H.R. Brooker, *J. Magn. Reson.* **92** (1991) 229.

Mar2 P.Z. Marmarelis and V.Z. Marmarelis, *Analysis of Physiological Systems: The White Noise Approach*, Plenum Press, New York, 1978.

Mar3 G.J. Marseille, M. Fuderer, R. de Beer, A.F. Mehlkopf and D. van Ormondt, *J. Magn. Reson. B* **103** (1994) 292.

Mar4 G.J. Marseille, R. de Beer, M. Fuderer, A.F. Mehlkopf and D. van Ormondt, *J. Magn. Reson. B* **111** (1996) 70.

Mau1 A.A. Maudsley, S.K. Hilal, W.H. Perman and H.E. Simon, *J. Magn. Reson.* **51** (1983) 147.

Mei1 M. Meininger, P.M. Jakob, M. von Kienlin, D. Koppler, G. Brinkmann and A. Haase, *J. Magn. Reson.* **125** (1997) 325.

Mer1 K.-D. Merboldt, W. Hänicke, M.L. Gyngell, J. Frahm and H. Bruhn, *J. Magn. Reson.* **82** (1989) 115.

Met1 K.R. Metz, J.P. Boehmer, J.L. Bowers and J.R. Moore, *J. Magn. Reson. B* **103** (1994) 152.

Mey1 C.H. Meyer, B.S. Hu, D.G. Nishimura and A. Makovski, *Magn. Reson. Med.* **28** (1992) 202.

Mis1 E. Mischler, F. Humbert, B. Diter and D. Canet, *J. Magn. Reson. B* **106** (1995) 32.

Mis2 E. Mischler, F. Humbert, B. Diter and D. Canet, *J. Magn. Reson. B* **109** (1995) 121.

Mor1 P.G. Morris, *Nuclear Magnetic Resonance Imaging in Medicine and Biology*, Clarendon Press, Oxford, 1986.

Mor2 H. Morneburg, ed., *Bildgebende Systeme für die medizinische Diagnostik*, 3rd edn, Publicis MCDF Verlag, Erlangen, 1995.

Mul1 R.V. Mulkern, S.T.S. Wong, C. Winalski and F.A. Jolesz, *Magn. Reson. Imag.* **8** (1990) 557.

Mül1 S. Müller, *J. Magn. Reson. Med.* **5** (1987) 502.

Mül2 S. Müller, *J. Magn. Reson. Med.* **6** (1988) 364.

Mül3 S. Müller, *Magn. Reson. Med.* **10** (1989) 145.

Mül4 S. Müller, R. Sauter, H. Weber and J. Seelig, *J. Magn. Reson.* **76** (1988) 155.

Nil1 H. Nilgens, M. Thelen, J. Paff, P. Blümler and B. Blümich, *Magn. Reson. Imag.* **14** (1996) 857.

Nil2 H. Nilgens and B. Blümich, in: *Signal Treatment and Signal Analysis*, ed. D.N. Rutledge, Elsevier, Amsterdam, 1996, p. 489.

Nil3 H. Nilgens, P. Blümler, J. Paff and B. Blümich, *J. Magn. Reson. A* **105** (1993) 108.

Nil4 H. Nilgens, *Hadamard-Bildgebung mit Schichtselektion*, Dissertation, Universität Mainz, 1994.

Nor1 D.G. Norris, *J. Magn. Reson.* **91** (1991) 190.

Paf1 J. Paff, B. Blümich and R. Kaiser, *Adv. Magn. Opt. Reson.* **17** (1992) 1.

Par1 D.L. Parker, G.T. Gullberg and P.R. Frederick, *Med. Phys.* **14** (1987) 640.

Pat1 S. Patz, S.T.S. Wong and M.S. Roos, *Magn. Reson. Med.* **10** (1989) 194.

Pet1 A.M. Peters and R. Bowtell, *J. Magn. Reson.* **137** (1999) 196.

Poh1 R. Pohmann, M. von Kienlin and A. Haase, *J. Magn. Reson.* **129** (1997) 145.

Pon1 S.L. Ponder and D.B. Twieg, *J. Magn. Reson. B* **104** (1994) 85.

Rat1 J.W. Rathke, R.J. Klingler, R.E. Gerald II, K.W. Kramarz and K. Woelk, *Progr. Nucl. Magn. Reson. Spectrosc.* **30** (1997) 209.

Rau1 R. Raulet, D. Grandclaude, F. Humbert and D. Canet, *J. Magn. Reson.* **124** (1997) 259.

Rau2 R. Raulet, J.M. Escanyé, F. Humbert and D. Canet, *J. Magn. Reson. A* **119** (1996) 111.

Rof1 C.J. Rofe, J. Van Noort, P.J. Back and P.T. Callaghan, *J. Magn. Reson. B* **108** (1995) 125.

Roo1 M.S. Roos and S.T. Wong, *J. Magn. Reson.* **87** (1990) 554.

Sch1 M. Schetzen, *The Volterra and Wiener Series of Nonlinear Systems*, Wiley, New York, 1980.

Sep1 R.E. Sepponen, J. Sipponen and J.I. Tanttu, *J. Comput. Assist. Tomogr.* **8** (1984) 585.

Sha1 A.J. Shaka, J. Keeler, M.B. Smith and R. Freeman, *J. Magn. Reson.* **61** (1985) 175.

Sil1 A.C. Silva, E.L. Barbier, I.L. Lowe and A. Koretsky, *J. Magn. Reson.* **135** (1998) 242.

Sim1 B. Simon, R. Kimmich and H. Köstler, *J. Magn. Reson. A* **118** (1996) 78.

Sin1 S. Singh, R. Deslauriers, J.K. Saunders and W.R. Brody, *J. Magn. Reson. B* **111** (1996) 289.

Sou1 S.P. Souza, J. Szumowski, C.L. Dumoulin, D.P. Plewes and G. Glover, *J. Comp. Ass. Tomography* **12** (1988) 1026.

Spi1 H.W. Spiess, *Adv. Polym. Sci.* **66** (1985) 23.

Sta1 L.A. Stables, R.P. Kennan, A.W. Anderson, R.T. Constable and J.C. Gore, *J. Magn. Reson.* **136** (1999) 143.

Ste1 M.K. Stehling, *Magn. Reson. Imag.* **10** (1992) 165.

Ste2 M.K. Stehling, R. Turner and P. Mansfield, *Science* **254** (1991) 43.

Sty1 P. Styles, *NMR – Basic Principles and Progress* **27** (1992) 45.

Tak1 A. Takahashi, T.-Q. Li and H. Stodkilde-Jorgensen, *J. Magn. Reson.* **126** (1997) 127.

Tur1 R. Turner, D. Le Bihan, J. Maier, R. Vavrek, L.K. Hedges and J. Pekar, *Radiology* **177** (1990) 407.

Utz1 J.A. Utz, R.J. Herfkens, G. Glover and N. Pelc, *Magn. Reson. Imag.* **4** (1986) 106.

Val1 M. Valtier, R. Raulet, R.-P. Eustache and D. Canet, *J. Magn. Reson. A* **112** (1995) 118.

Val2 M. Valtier, P. Tekely, L. Kiéné and D. Canet, *Macromocelules* **28** (1995) 4075.

Vla1 M.T. Vlaardingerbroek and J.A. den Boer, *Magnetic Resonance Imaging*, Springer, Berlin, 1996.

Vol1 V. Volterra, *Theory of Functionals of Integro and Differential Equations*, Dover, New York, 1959.

Web1 A.G. Webb, T.H. Mareci and R.W. Briggs, *J. Magn. Reson. B* **103** (1994) 274.

Wea1 K.A. Wear, K.J. Myers, R.F. Wagner, S.S. Rajan and L.W. Grossman, *J. Magn. Reson. B* **105** (1994) 172.

Weh1 F.W. Wehrli, D. Schaw and J.B. Kneeland, eds., *Biomedical Magnetic Resonance Imaging*, VCH, Weinheim, 1988.

Weh2 F.W. Wehrli, *Fast-Scan Magnetic Resonance: Principles and Applications*, Raven Press, New York, 1991.

Wie1 N. Wiener, *Nonlinear Problems in Random Theory*, Wiley, New York, 1958.

Woe1 K. Woelk, R.E. Gerald II, R.J. Klingler and J.W. Rathke, *J. Magn. Reson. A* **121** (1996) 74.

Wor1 B.S. Worthington, J.L. Firth, G.K. Morris, I.R. Johnson, R. Coxon, A.M. Blamire, P. Gibbs and P. Mansfield, *Phil. Trans. R. Soc. Lond. A* **333** (1990) 507.

Wuy1 Y. Wu, J.L. Ackerman, D.A. Chesler, J. Li, R.M. Neer, J. Wang and M.J. Glimcher, *Calcif. Tissue Int.* **62** (1998) 512.

Yan1 Q.X. Yang, S. Posse, D. Le Bihan and M.B. Smith, *J. Magn. Reson. B* **113** (1996) 145.

Zie1 D. Ziessow and B. Blümich, *Ber. Bunsenges. Phys. Chem.* **78** (1974) 1168.

Zur1 Y. Zur, S. Stokar and P. Bendel, *Magn. Reson. Med.* **6** (1988) 175.

CHAPTER 7

Abr1 A. Abragam, *The Principles of Nuclear Magnetism*, Clarendon Press, Oxford, 1961.

Ack1 J.L. Ackerman, L. Garrido, J.R. Moore, B. Pfleiderer and Y. Wu, in: *Magnetic Resonance Microscopy*, ed. B. Blümich and W. Kuhn, VCH, Weinheim, 1992, p. 237.

Alb1 K. Albert, U. Günther, M. Ilg, E. Bayer and M. Grossa, in: *Magnetic Resonance Microscopy*, ed. B. Blümich and W. Kuhn, VCH, Weinheim, 1992, p. 227.

Alb2 M.S. Albert, G.D. Cates, B. Driehuys, W. Happer, B. Saam, C.S. Springer, Jr. and A. Wishnia, *Nature* **370** (1994) 199.

All1 S.G. Allen, P.C.L. Stephenson and J.H. Strange, *J. Chem. Phys.* **106** (1997) 7802.

And1 P.W. Anderson and P.R. Weiss, *Rev. Mod. Phys.* **25** (1953) 269.

And2 A.G. Anderson and S.R. Hartmann, *Phys. Rev.* **128** (1962) 2023.

App1 M. Appel, G. Fleischer, J. Kärger, A.C. Dieng and G. Riess, *Macromolecules* **28** (1995) 2345.

Arm1 R.L. Armstrong, A. Tzalmona, M. Menzinger, A. Cross and C. Lemaire, in: *Magnetic Resonance Microscopy*, ed. B. Blümich and W. Kuhn, VCH, Weinheim, 1992, p. 309.

Bah1 M.M. Bahn, *J. Magn. Reson.* **137** (1999) 33.

Bak1 A.F. Bakuzis, P.M. Morais and F.A. Tourinho, *J. Magn. Reson. A* **122** (1996) 100.

Bal1 B.J. Balcom, T.A. Carpenter and L.D. Hall, *Macromolecules* **25** (1992) 6818.

Bal2 R.C. Ball, P.T. Callaghan and E.T. Samulski, *J. Chem. Phys.* **106** (1997) 7352.

Bal3 B. Balinov, O. Söderman and J.C. Ravey, *J. Am. Chem. Soc.* **98** (1994) 393.

Bal4 K.J. Balkus, Jr. and J. Shi, *J. Chem. Phys.* **100** (1996) 16429.

Bar1 P. Barth, S. Hafner and P. Denner, *Macromolecules* **29** (1996) 1655.

Bar2 P. Barth and S. Hafner, *Magn. Reson. Imag.* **15** (1997) 107.

Bar3 P.J. Barrie and J. Klinowski, *Prog. NMR Spectrosc.* **24** (1992) 91.

Bas1 P.J. Basser, J. Mattiello and D. LeBihan, *Biophys. J.* **66** (1994) 259.

Bas2 P.J. Basser and C. Pierpaoli, *J. Magn. Reson. B* **111** (1996) 209.

Bas3 P.J. Basser, J. Mattiello and D. LeBihan, *J. Magn. Reson. B* **103** (1994) 247.

Bay1 Y. Ba and J.A. Ripmeester, *J. Chem. Phys.* **108** (1998) 8589.

Bec1 N. Beckman and S. Müller, *J. Magn. Reson.* **93** (1991) 186.

Ben1 T.B. Benson and P.J. McDonald, *J. Magn. Reson. A* **112** (1995) 17.

Ben2 M.R. Bendall and D.T. Pegg, *J. Magn. Reson.* **53** (1983) 272.

Ben3 P. Bendel, *IEEE Trans. Med. Imag.* **MI-4** (1985) 114.

Ber1 R.D. Bertrand, W.B. Monitz, A.N. Garroway and G.C. Chingas, *J. Am. Chem. Soc.* **100** (1978) 5227.

Ber2 A.J. van den Bergh, H.J. van den Bogert and A. Heerschap, *J. Magn. Reson.* **135** (1998) 91.

Ber3 K. Beravs, A. Demsar and F. Demsar, *J. Magn. Reson.* **137** (1999) 253.

Beu1 O. Beuf, A. Briguet, M. Lissac and R. Davis, *J. Magn. Reson. B* **112** (1996) 111.

Ble1 H.E. Bleich and A.G. Redfield, *J. Chem. Phys.* **67** (1977) 5040.

Blo1 F. Bloch, W.W. Hansen and M. Packard, *Phys. Rev.* **70** (1948) 474.

Blü1 P. Blümler and B. Blümich, *Magn. Reson. Imag.* **10** (1992) 779.

Blü2 B. Blümich and P. Blümler, *Makromol. Chem.* **194** (1993) 2133.

Blü3 P. Blümler and B. Blümich, *Macromolecules* **24** (1991) 2183.

Blü4 B. Blümich and W. Kuhn, ed., *Magnetic Resonance Microscopy*, VCH, Weinheim, 1992.

Blü5 P. Blümler and B. Blümich, *Acta Polymerica* **44** (1993) 125.

Blü6 B. Blümich, *Concepts Magn. Reson.* **10** (1998) 19.

Blü7 B. Blümich, *Z. Naturforschung* **49a** (1994) 19.

Blü8 P. Blümler, V. Litvinov, H.G. Dikland and M. van Duin, *Kautschuk, Gummi, Kunststoffe* **51** (1998) 865.

Blü9 B. Blümich, P. Blümler, E. Günther, G. Schauß and H.W. Spiess, *Macromol. Chem., Makromol. Symp.* **44** (1991) 37.

Blü10 B. Blümich, *Concepts Magn. Reson.* **11** (1999) 71.

Blü11 B. Blümich, *Concepts Magn. Reson.* **11** (1999) 147.

Bod1 J. Bodurka, A. Jesmanowicz, J.S. Hyde, H. Xu, L. Estkowski and S.-J. Li, *J. Magn. Reson.* **137** (1999) 265.

Bon1 J.-M. Bonny, L. Foucat, W. Laurent and J.-P. Renou, *J. Magn. Reson.* **130** (1998) 51.

Boo1 H.F. Booth and J.H. Strange, *Molec. Phys.* **93** (1998) 263.

Bor1 G.C. Borgia, R.J.S. Brown and P. Fantazzini, *J. Appl. Phys.* **82** (1997) 4197.

Bor2 G.C. Borgia, R.J.S. Brown and P. Fantazzini, *J. Appl. Phys.* **79** (1996) 3656.

Bou1 B. Boulat and M. Rance, *J. Magn. Reson. B* **110** (1996) 288.

Bou2 D. Bourgeois and M. Decorps, *J. Magn. Reson.* **94** (1991) 20.

Bra1 M. Braun, W.I. Jung, O. Lutz and R. Oeschey, *Z. Naturforsch.* **42a** (1987) 1391.

Bre1 M. Brereton, *Macromolecules* **22** (1989) 3667.

Bre2 I. Bresinska and K.J. Balkus, Jr., *J. Phys. Chem.* **98** (1994) 12989.

Bri1 M.M. Britton and P.T. Callaghan, *J. Rheol.* **41** (1997) 1365.

Bri2 M.M. Britton and P.T. Callaghan, *Magn. Reson. Chem.* **35** (1997) S38.

Bri3 M.M. Britton and P.T. Callaghan, *Phys. Rev. Let.* **78** (1997) 4930.

Bro1 T.R. Brown, B.M. Kincaid and K. Ugurbil, *Proc. Natl. Acad. Sci. USA* **79** (1982) 3523.

Bro2 T.R. Brown, S.D. Buchthal, J. Murphy-Boesch, S.J. Nelson and J.S. Taylor, *J. Magn. Reson.* **82** (1989) 629.

Bru1 E. Brunner, M. Haake, L. Kaiser, A. Pines and J.A. Reimer, *J. Magn. Reson.* **138** (1999) 155.

Brü1 R. Brüschweiler, J.C. Madsen, C. Griesinger, O.W. Sorensen and R.R. Ernst, *J. Magn.* **73** (1987) 380.

Bur1 D.P. Burum and R.R. Ernst, *J. Magn. Reson.* **39** (1980) 163.

Bur2 D.P. Burum and A. Bielecki, *J. Magn. Reson.* **95** (1991) 184.

Bur3 D. Burstein, *Concepts Magn. Reson.* **8** (1996) 269.

Cal1 P.T. Callaghan, *Principles of Nuclear Magnetic Resonance Microscopy*, Clarendon Press, Oxford, 1991.

Cal2 P.T. Callaghan, C.J. Clark and L.C. Forde, *Biophys. Chem.* **50** (1994) 225.

Cal3 P.T. Callaghan and E.T. Samulski, *Macromolecules* **30** (1997) 113.

Cal4 P.T. Callaghan and Y. Xia, *J. Magn. Reson.* **91** (1991) 326.
Cal5 P.T. Callaghan and J. Stepisnik, *J. Magn. Reson. A* **117** (1995) 118.
Cal6 P.T. Callaghan, M.E. Komlosh and M. Nyden, *J. Magn. Reson.* **133** (1998) 177.
Cal7 P.T. Callaghan, C.D. Eccles and Y. Xia, *J. Phys. E: Sci. Instrum.* **21** (1988) 820.
Cal8 P.T. Callaghan, *Magn. Reson. Imag.* **14** (1996) 701.
Cal9 P.T. Callaghan, *J. Magn. Reson. A* **113** (1995) 53.
Cal10 P.T. Callaghan, L.C. Forde and C.J. Rofe, *J. Magn. Reson. B* **104** (1994) 34.
Cal11 P.T. Callaghan, *J. Magn. Reson.* **129** (1997) 74.
Cal12 P.T. Callaghan, *Aust. J. Phys.* **37** (1984) 359.
Cal13 P.T. Callaghan and J. Stepisnik, *Phys. Rev. Lett.* **75** (1995) 4532.
Cal14 P.T. Callaghan, *J. Magn. Reson.* **87** (1990) 304.
Cal15 P.T. Callaghan, in: *Encyclopedia of NMR Spectroscopy*, ed. D.M. Grant and R.K. Harris, Wiley, New York, 1996, p. 4665.
Cal16 P.T. Callaghan, S.L. Codd and J.D. Seymor, *Concepts Magn. Reson.* **11** (1999) 181.
Cal17 P.T. Callaghan, *Rep. Prog. Phys.* **62** (1999) 599.
Can1 D. Canet and M. Décorps, in: *Dynamics of Solutions and Fluid Mixtures by NMR*, ed. J.J. Delpuech, Wiley, New York, 1995, 309.
Cap1 A. Caprihan and E. Fukushima, *Physics Reports* **4** (1990) 195.
Cap2 A. Caprihan, L.Z. Wang and E. Fukushima, *J. Magn. Reson. A* **118** (1996) 94.
Cap3 D. Capitani and A.L. Segre, *Magn. Reson. Imag.* **10** (1992) 793.
Cap4 A. Caprihan, R.H. Griffey and E. Fukushima, *Magn. Reson. Med.* **15** (1990) 327.
Cap5 S. Capuani, L. Mancini and B. Maraviglia, *Appl. Magn. Reson.* **15** (1998) 383.
Cap6 S. Capuani, F. De Luca and B. Maraviglia, *J. Chem. Phys.* **101** (1994) 4521.
Car1 H.Y. Carr and E.M. Purcell, *Phys. Rev.* **94** (1954) 630.
Car2 P. Caravatti, L. Braunschweiler and R.R. Ernst, *Chem. Phys. Lett.* **100** (1983) 305.
Car3 P. Caravatti, M.H. Levitt and R.R. Ernst, *J. Magn. Reson.* **68** (1986) 323.
Car4 T.A. Carpenter, L.D. Hall, P. Jezzard, C.J. Wiggins, N.J. Clayden, P. Jackson and N. Walton, in: *Magnetic Resonance Microscopy*, ed. B. Blümich and W. Kuhn, VCH, Weinheim, 1992, p 267.
Cha1 C. Chang and R.A. Komoroski, *Macromolecules* **22** (1989) 600.
Cha2 N. Chandrakumar and R. Kimmich, *J. Magn. Reson.* **137** (1999) 100.
Chi1 G.C. Chingas, A.N. Garroway, R.D. Bertrand and W.B. Monitz, *J. Chem. Phys.* **74** (1981) 127.
Chi2 E. Chiarparin, P. Pelupessy and G. Bodenhausen, *Molecular Physics* **95** (1998) 759.
Che1 T.L. Chenevert, J.A. Brunberg and J.G. Ripe, *Radiology* **177** (1990) 401.
Cla1 J. Clauss, K. Schmidt-Rohr and H.W. Spiess, *Acta Polymerica* **44** (1993) 1.
Coc1 M. Cockman, L.W. Jelinski, J. Katz, D.S. Sorce, L.M. Boxt and P.J. Cannon, *J. Magn. Reson.* **90** (1990) 9.
Coc2 M.D. Cockman and T.H. Mareci, *J. Magn. Reson.* **79** (1988) 236.
Coc3 M.D. Cockman and L.W. Jelinski, *J. Magn. Reson.* **90** (1990) 9.
Coh1 J.P. Cohen-Addad, *Progr. NMR Spectrosc.* **25** (1993) 25.
Coh2 J.P. Cohen-Addad, A. Viallat and P. Guchot, *Macromolecules* **20** (1987) 2146.
Coh3 J.P. Cohen-Addad, L. Pellicioli and J.J.H. Nusselder, *Polymer Gels and Networks* **5** (1997) 201.
Col1 J.-M. Colet, C. Piérart, F. Seghi, I. Gabric and R.N. Muller, *J. Magn. Reson.* **134** (1998) 199.
Con1 M.S. Conradi, *J. Magn. Reson.* **93** (1991) 419.
Cor1 D.G. Cory, J.B. Miller and A.N. Garroway, *J. Magn. Reson.* **90** (1990) 544.
Cor2 D.G. Cory, A.N. Garroway and J.B. Miller, *Polymer Preprints* (1990) 149.

Cor3 D.G. Cory, J.B. Miller, A.N. Garroway and W.S. Veeman, *J. Magn. Reson.* **85** (1989) 219.

Cor4 D.G. Cory, J.B. Miller and A.N. Garroway, *Macromol. Symp.* **86** (1994) 259.

Cox1 S.J. Cox and P. Styles, *J. Magn. Reson.* **40** (1980) 209.

Coy1 A. Coy, P. T. Callaghan, J. Colloid Interface Science 168 (1994) 373.

Dai1 H. Dai, K. Potter and E.W. McFarland, *J. Chem. Eng. Data* **41** (1996) 970.

Dav1 S. Davies and K.J. Packer, *J. Appl. Phys.* **67** (1990) 3163.

Dav2 M.S. Davis and G.E. Maciel, *J. Magn. Reson.* **94** (1991) 617.

Dec1 J.J. Dechter, R.A. Komoroski and S. Ramaprasad, *J. Magn. Reson.* **93** (1991) 142.

Del1 F. De Luca, G.H. Raza, A. Gargaro and B. Maraviglia, *J. Magn. Reson.* **126** (1997) 159.

Del2 F. De Luca, R. Campanella, A. Bifone and B. Maraviglia, *J. Magn. Reson.* **93** (1991) 554.

Del3 B. Deloche and E.T. Samulski, *Macromolecules* **14** (1991) 575.

Del4 F. De Luca, R. Campanella, A. Bifone and B. Maraviglia, *Chem. Phys. Lett.* **186** (1991) 303.

Del5 F. De Luca, G. Raza and B. Maraviglia, *J. Magn. Reson. A* **107** (1994) 243.

Del6 F. De Luca, G. Raza, M. Romeo, C. Casieri and B. Maraviglia, *J. Magn. Reson. B* **107** (1995) 74.

Dem1 D.E. Demco, A. Johansson and J. Tegenfeldt, *Solid-State Nucl. Magn. Reson.* **4** (1995) 13.

Dem2 D.E. Demco, L. Muntean and I. Coroiu, *Appl. Magn. Reson.* **8** (1995) 255.

Dem3 D.E. Demco, H. Köstler, R. Kimmich, *J. Magn. Reson. A* **110** (1994) 236.

Dem4 D.E. Demco and I. Ardelean, *Acta Physica Polonica A* **86** (1994) 407.

Dem5 D.E. Demco and I. Ardelean, *Acta Physica Polonica A* **89** (1996) 699.

Dem6 D.E. Demco, S. Hafner, I. Ardeleanu and R. Kimmich, *Appl. Magn. Reson.* **9** (1995) 491.

Dem7 D.E. Demco, S. Hafner and R. Kimmich, *Appl. Magn. Reson.* **9** (1995) 267.

Dem8 D.E. Demco and I. Coroiu, *Appl. Magn. Reson.* **7** (1994) 521.

Dem9 D.E. Demco and I. Coroiu, *Appl. Magn. Reson.* **8** (1998) 187.

Dem10 D.E. Demco, A. Johansson and J. Tegenfeldt, *J. Magn. Reson. A* **110** (1994) 183.

Dem11 D.E. Demco, S. Hafner and R. Kimmich, *J. Magn. Reson.* **94** (1991) 333.

Der1 J.M. Dereppe and C. Moreaux, *J. Magn. Reson.* **91** (1991) 596.

Des1 T.M. de Swiet, *J. Magn. Reson. B* **109** (1995) 12.

Dic1 R.J. Dickinson, A.S. Hall, A.J. Hind and I.R. Young, *J. Comp. Asst. Tomog.* **10** (1986) 468.

Die1 W. Dietrich, G. Bergmann and R. Gerhards, *Z. Anal. Chem.* **279** (1976) 177.

Dix1 W.T. Dixon, *Radiology* **153** (1984) 189.

Dod1 D.M. Doddrell, J.M. Bulsing, G.J. Galloway, W.M. Brooks, J. Field, M. Irwing and H. Baddeley, *J. Magn. Reson.* **70** (1986) 319.

Dod2 D.M. Doddrell, D.T. Pegg and M.R. Bendall, *J. Magn. Reson.* **48** (1982) 323.

Doi1 Y. Doi, Y. Kanazawa, in: *Spatially Resolved Magnetic Resonance*, ed. P. Blümler, B. Blümich, R. Botto and E. Fukushima, Wiley-VCH, Weinheim, 1998, p. 413.

Dor1 S.J. Doran, T.A. Carpenter and L.D. Hall, *Rev. Sci. Instrum.* **65** (1994) 2231.

Dor2 S.J. Doran, J.J. Attard, T.P.L. Roberts, T.A. Carpenter and L.D. Hall, *J. Magn. Reson.* **100** (1992) 101.

Dou1 P. Douek, R. Turner, J. Pekar, N. Patronas and D. Le Bihan, *J. Comp. Ass. Tomography* **15** (1991) 923.

Dri1 B. Driehuys, G.D. Gates and W. Happer, *Phys. Rev. Lett A* **184** (1993) 88.

Dum1 C.L. Dumoulin, *Magn. Reson. Med.* **2** (1985) 583.

Dum2 C. Dumoulin and D. Vatis, *Magn. Reson. Med.* **3** (1986) 282.

Dum3 C.L. Dumoulin, *Magn. Reson. Med.* **3** (1986) 90.
Dum4 C.L. Dumoulin and E.A. Williams, *J. Magn. Reson.* **66** (1986) 86.
Dum5 C.L. Dumoulin, S.P. Souza and H.R. Hart, *Magn. Reson. Med.* **5** (1987) 238.
Dum6 C.L. Dumoulin, S.P. Souza and M.F. Walker, W. Wagle, *Magn. Reson. Med.* **9** (1989) 139.
Dum7 C.L. Dumoulin, H.E. Cline, S.P. Souza, W.A. Wagle and M.F. Walker, *Magn. Reson. Med.* **11** (1989) 35.
Dwu1 D. Wu, A. Chen and C.S. Johnson, Jr., *J. Magn. Reson. A* **115** (1995) 123.
Egg1 N. Egger, K. Schmidt-Rohr, B. Blümich, W.-D. Domke and B. Stapp, *J. Appl. Polym. Sci.* **44** (1992) 289.
Eid1 G. Eidmann, R. Savelsberg, P. Blümler and B. Blümich, *J. Magn. Reson. A* **122** (1996) 104.
Eli1 E. Eliav, H. Shinnar and G. Navon, *J. Magn. Reson.* **98** (1992) 223.
Eli2 E. Eliav and G. Navon, *J. Magn. Reson.* **137** (1999) 295.
Erc1 M. Ercken, P. Adriaensens, D. Vanderzande and J. Gelan, *Macromolecules* **28** (1995) 8541.
Erc2 M. Ercken, P. Adriaensens, G. Rogers, R. Carleer, D. Vanderzande and J. Gelan, *Macromolecules* **29** (1996) 5671.
Ern1 R.R. Ernst, G. Bodenhausen and A. Wokaun, *Principles of Nuclear Magnetic Resonance in One and Two Dimensions*, Clarendon Press, Oxford, 1987.
Eym1 R. Eymael and B. Blümich, (unpublished results).
Fai1 E.J. Fairbanks, G.E. Santyr and J.A. Sorenson, *J. Magn. Reson. B* **106** (1995) 279.
Fed1 V.D. Fedotov and H. Schneider, *NMR – Basic Principles and Progress* **21** (1989) 1.
Fei1 M. Feike, D.E. Demco, R. Graf, J. Gottwald, S. Hafner and H.W. Spiess, *J. Magn. Reson. A* **122** (1996) 214.
Fis1 A.E. Fischer, B.J. Balcom, E.J. Fordham, T.A. Carpenter and L.D. Hall, *J. Phys. D: Appl. Phys.* **28** (1995) 384.
Fis2 E. Fischer, F. Grinberg and R. Kimmich, *J. Chem. Phys.* **109** (1998) 846.
Fle1 F. Fleischer and F. Fujara, *NMR – Basic Principles and Progress* **30** (1994) 111.
For1 E.J. Fordham, A. Sezginer and L.D. Hall, *J. Magn. Reson. A* **113** (1995) 139.
For2 J.J. Ford, *J. Magn. Reson.* **87** (1990) 346.
Fra1 J. Frahm, K.-D. Merboldt and W. Hänicke, *J. Magn. Reson. B* **109** (1995) 234.
Fra2 J. Frahm, K.D. Merboldt, W. Hänicke and A. Haase, *J. Magn. Reson.* **64** (1985) 81.
Fry1 J.S. Fry, *Concepts Magn. Reson.* **1** (1989) 27.
Fry2 C.G. Fry, A.C. Lind, M.F. Davis, D.W. Duff and G.E. Maciel, *J. Magn. Reson.* **83** (1989) 656.
Fül1 C. Fülber, B. Blümich, K. Unseld and V. Herrmann, *Kautschuk, Gummi, Kunststoffe* **48** (1995) 254.
Fül2 C. Füber, K. Unseld, V. Herrmann, K.H. Jakob and B. Blümich, *Colloid and Polymer Science* **274** (1996) 191.
Fül3 C. Fülber, D.E. Demco, O. Weintraub and B. Blümich, *Makromol. Chem.* **197** (1996) 581.
Fül4 C. Fülber, *NMR-Relaxation und Bildgebung an Kautschuknetzwerken*, Akademischer Verlag, München, 1996.
Fül5 C. Fülber, D.E. Demco and B. Blümich, *Solid-State Nucl. Magn. Reson.* **6** (1996) 213.
Fuj1 F. Fujara, E. Ilyina, H. Nienstaedt, H. Sillescu, R. Spohr and C. Trautmann, *Magn. Reson. Imag.* **12** (1994) 245.
Fuk1 E. Fukushima, *Annu. Rev. Fluid Mech.* **31** (1999) 95.

Gae1 H.C. Gaede, Y.-Q. Song, R.E. Taylor, E.J. Munson, J.A. Reimer and A. Pines, *Appl. Magn. Reson.* **8** (1995) 373.

Gan1 K. Ganesan, D.C. Ailion, A.G. Cutillo and K.C. Goodrich, *J. Magn. Reson. B* **102** (1993) 293.

Gar1 R.L. Garwin and H.A. Reich, *Phys. Rev.* **115** (1959) 1478.

Gar2 A.N. Garroway, J. Baum, M.G. Munowitz and A. Pines, *J. Magn. Reson.* **60** (1984) 337.

Gar3 L. Garrido, J.L. Ackerman and B. Pfleiderer, *Ceram. Eng. Sci. Proc.* **12** (1991) 2042.

Gar4 A.N. Garroway, *J. Phys. D* **7** (1974) L159.

Gas1 L. Gasper, D.E. Demco and B. Blümich, *Solid-State Nucl. Magn. Reson.* **14** (1999) 105.

Gat1 J.C. Gatenby and J.G. Core, *J. Magn. Reson. A* **121** (1996) 193.

Ger1 K. Gersonde, L. Felsberg, T. Tolxdorff, D. Ratzel and B. Ströbel, *Magn. Reson. Med.* **1** (1984) 463.

Ger2 K. Gersonde, T. Tolxdorff and L. Felsberg, *Magn. Reson. Med.* **2** (1985) 390.

Ger3 R.E. Gerald III, A.O. Krasavin and R.E. Botto, *J. Magn. Reson. A* **123** (1996) 201.

Gib1 S.J. Gibbs, T.A. Carpenter and L.D. Hall, *J. Magn. Reson.* **98** (1992) 183.

Gie1 R. Giesen, B. Blümich, A. Brandenburg and F.U. Niethard, (unpublished results).

Goe1 G. Goelman and M.G. Prammer, *J. Magn. Reson. A* **113** (1995) 11.

Gol1 M. Goldman and L. Shen, *Phys. Rev.* **144** (1961) 321.

Gol2 M. Goldman, *Spin Temperature and NMR in Solids*, Oxford University Press, London, 1970.

Göt1 H. Götz, *Plaste und Kautschuk* **32** (1985) 56.

Got1 Yu. Ia. Gotlib, M.I. Lifshits and V.A. Ievlev, *Vysokomol. Soed. A* **18** (1976) 2299.

Göt2 H. Götz and T. Willing, *Plaste und Kautschuk* **29** (1982) 661.

Göt3 H. Götz, P. Denner and U. Rost, *Wiss. Zeitschrift Päd. Hochschule Theodor Neubauer, Math.-Naturw. Reihe* **21** (1985) 43.

Gra1 R. Graf, D.E. Demco, S. Hafner and H.W. Spiess, *Solid-State Nucl. Magn. Reson.* **12** (1998) 139.

Gra2 R. Graf, D.E. Demco, J. Gottwald, S. Hafner and H.W. Spiess, *J. Chem. Phys.* **106** (1997) 885.

Gre1 D.M. Gregory, R.E. Gerald III and R.E. Botto, *J. Magn. Reson.* **131** (1998) 327.

Gro1 W. Gronski, R. Stadler and M.M. Jakobi, *Macromolecules* **17** (1984) 741.

Gui1 D.N. Guilfoyle, B. Issa and P. Mansfield, *J. Magn. Reson. A* **119** (1996) 151.

Gui2 D.N. Guilfoyle and P. Mansfield, *J. Magn. Reson.* **97** (1992) 342.

Gul1 T. Gullion, D.B. Baker and M.S. Conradi, *J. Magn. Reson.* **89** (1989) 479.

Gün1 E. Günther, B. Blümich and H.W. Spiess, *Molec. Phys.* **71** (1990) 477.

Gün2 E. Günther, B. Blümich and H.W. Spiess, *Chem. Phys. Lett.* **184** (1991) 251.

Gün3 E. Günther, B. Blümich and H.W. Spiess, *Macromolecules* **25** (1992) 3315.

Gün4 U. Günther and K. Albert, *J. Magn. Reson.* **98** (1992) 593.

Gut1 A. Guthausen, G. Zimmer, P. Blümler and B. Blümich, *J. Magn. Reson.* **130** (1998) 1.

Gut2 A. Guthausen, *Die NMR-MOUSE: Methoden und Anwendungen zur Charakterisierung von Polymeren*, Dissertation, RWTH, Aachen, 1998.

Haa1 A. Haase, M. Brandl, E. Kuchenbrod and A. Link, *J. Magn. Reson. A* **105** (1993) 230.

Haa2 A. Haase, *Methoden zur Verknüpfung von NMR-Sektroskopie und NMR-Tomographie in vivo*, Habilitationsschrift, Johann Wolfgang Goethe Universität, Framkfurt am Main, 1987.

Haa3 A. Haase and J. Frahm, *J. Magn. Reson.* **64** (1985) 94.

Haa4 A. Haase and D. Matthaei, *J. Magn. Reson.* **71** (1987) 550.

Haa5 A. Haase, J. Frahm, D. Matthaei, W. Haenicke and K.D. Merboldt, *J. Magn. Reson.* **67** (1986) 258.

Haa6 M. Haake, A. Pines, J.A. Reimer and R. Seydoux, *J. Am. Chem. Soc.* **119** (1997) 11711.

Haf1 S. Hafner and P. Barth, *Magn. Reson. Imag.* **13** (1995) 739.

Haf2 S. Hafner, D.E. Demco and R. Kimmich, *Chem. Phys. Lett.* **187** (1991) 53.

Hah1 E.L. Hahn, *Phys. Rev.* **80** (1950) 580.

Hah2 E.L. Hahn, *J. Geophys. Res.* **65** (1960) 776.

Hal1 W.P. Halperin, F. D'Orazio, S. Bhattacharja and C.J. Tarczon, in: *Molecular Dynamics in Restricted Geometries*, ed. J. Klafter and J.M. Drake, Wiley, New York, 1989, p. 311.

Hal2 L.D. Hall and S.L. Talagala, *J. Magn. Reson.* **71** (1987) 180.

Hal3 L.D. Hall, S. Sukumar and S.L. Talagala, *J. Magn. Reson.* **56** (1984) 275.

Hal4 L.D. Hall, V. Rajanayagam and C. Hall, *J. Magn. Reson.* **68** (1986) 185.

Hal5 L.D. Hall and A.G. Webb, *J. Magn. Reson.* **83** (1989) 371.

Hal6 L.D. Hall, A.G. Webb and S.C.R. Williams, *J. Magn. Reson.* **81** (1989) 565.

Han1 M. Hansen, *Molekulare Dynamik von Polycarbonat unter Zug und Druck aus 2D-NMR-Spektren*, Dissertation, Johannes-Gutenberg-Universität, Mainz, 1991.

Hap1 W. Happer, E. Miron, S. Schaefer, D. Schreiber, W.A. van Wijngaarden and X. Zeng, *Phys. Rev. A* **29** (1984) 3092.

Har1 C.J. Hardy and C.L. Dumoulin, *Magn. Reson. Med.* **5** (1987) 58.

Har2 S.R. Hartmann and E.L. Hahn, *Phys. Rev.* **128** (1962) 2042.

Har3 G.S. Harbison and H.W. Spiess, *Chem. Phys. Lett.* **124** (1986) 128.

Hau1 D. Hauck, P. Blümler and B. Blümich, *Makromol. Chem. Phys.* **198** (1997) 2729.

Hav1 J.R. Havens and D.L. VanderHart, *Macromolecules* **18** (1985) 1663.

Haw1 J.S. Haw, in: *Encyclopedia of NMR Spectroscopy*, ed. D.M. Grant and R.K. Harris, Wiley, New York, 1996, Vol. 7, p. 4723.

Hed1 N. Hedin and I. Furó, *J. Magn. Reson.* **131** (1998) 126.

Hei1 S.R. Heil and M. Holz, *Angew. Chem. Int. Ed. Engl.* **35** (1996) 1717.

Hei2 M. Heidenreich, W. Köckenberger, R. Kimmich, N. Chandrakumar and R. Bowtell, *J. Magn. Reson.* **132** (1998) 109.

Hei3 M. Heidenreich, A. Spyros, W. Köckenberger, N. Chandrakumar, R. Bowtell and R. Kimmich, in: *Spatially Resolved Magnetic Resonance*, ed. P. Blümler, B. Blümich, R. Botto and E. Fukushima, Wiley-VCH, Weinheim, 1998, p. 21.

Hep1 M.A. Hepp and J.B. Miller, *J. Magn. Reson. A* **110** (1994) 98.

Hep2 M.A. Hepp and J.B. Miller, *Macromol. Symp.* **86** (1994) 271.

Hep3 M.A. Hepp and J.B. Miller, *Solid-State Nucl. Magn. Reson.* **6** (1996) 367.

Hep4 M.A. Hepp and J.B. Miller, *J. Magn. Reson. A* **111** (1994) 62.

Hol1 M. Holz, *Progr. NMR Spectrosc.* **18** (1986) 327.

Hol2 M. Holz, C. Müller and A.M. Wachter, *J. Magn. Reson.* **69** (1986) 108.

Hol3 M. Holz and H. Weingärtner, *J. Magn. Reson.* **92** (1991) 115.

Hor1 P.J. Hore, *J. Magn. Reson.* **55** (1983) 283.

Hsi1 P.S. Hsieh and R.S. Balaban, *J. Magn. Reson.* **74** (1987) 574.

Hür1 M. Hürlimann, K. Helmer, L. Latour and C. Sotak, *J. Magn. Reson. A* **111** (1994) 169.

Hür2 M. Hürlimann, *J. Magn. Reson.* **131** (1998) 232.

Hug1 C.E. Hughes, R. Kemp-Harper, P. Styles and S. Wimperis, *J. Magn. Reson. B* **111** (1996) 189.

Huj1 J. Hu, M.R. Wilcott and G.J. Moore, *J. Magn. Reson.* **126** (1997) 187.

Hur1 R.E. Hurd and D.M. Freeman, *Proc. Natl. Acad. Sci. USA* **86** (1989) 4402.

Hur2 R.E. Hurd and B.K. John, *J. Magn. Reson.* **91** (1991) 648; ibid. **92** (1991) 658.

Hut1 R.B. Hutchinson and J.I. Shapiro, *Concepts Magn. Reson.* **3** (1991) 215.

Hwa1 S.N. Hwang and F.W. Wehrli, *J. Magn. Reson. B* **109** (1995) 126.

Ice1 M.V. Icenogle, A. Caprihan and E. Fukushima, *J. Magn. Reson.* **100** (1992) 376.

Ito1 T. Ito and J. Fraissard, *J. Chem. Phys.* **76** (1982) 5225.

Jac1 P. Jackson, N.J. Clayden, N.J. Walton, T.A. Carpenter, L.D. Hall and P. Jezzard, *Polym. Int.* **24** (1991) 139.

Jac2 P. Jackson, *J. Mater. Sci.* **27** (1992) 1302.

Jän1 H.J. Jänsch, T. Hof, U. Ruth, J. Schmidt, D. Stahl and D. Fick, *Chem. Phys. Lett.* **296** (1998) 146.

Jam1 C. Jameson, A. Jameson, R. Gerald and A. de Dios, *J. Chem. Phys.* **96** (1992) 1676.

Jee1 J. Jeener and P. Broekaert, *Phys. Rev.* **157** (1967) 232.

Jez1 P. Jezzard, T.A. Carpenter, L.D. Hall, P. Jackson and N.J. Clayden, *Polym. Commun.* **32** (1991) 74.

Jez2 P. Jezzard, C.J. Wiggins, T.A. Carpenter, L.D. Hall, P. Jackson, N.J. Clayden and N.J. Walton, *Adv. Mater.* **4** (1992) 82.

Joh1 C.S. Johnson, Jr. and Q. He, *Adv. Magn. Reson.* **13** (1989) 131.

Jon1 L.A. Jones, P. Hodgkinson, A.L. Barker and P.J. Hore, *J. Magn. Reson. B* **113** (1996) 25.

Jon2 J.A. Jones, *J. Magn. Reson.* **126** (1997) 283.

Jon3 R.P.O. Jones, G.A. Morris and J.C. Waterton, *J. Magn. Reson.* **124** (1997) 291.

Joy1 M. Joy, G. Scott and M. Henkelmann, *Magn. Reson. Imag.* **7** (1989) 89.

Jun1 K.J. Jung, J.S. Tauskela and J. Katz, *J. Magn. Reson. B* **112** (1996) 103.

Kab1 G.W. Kabalka, G.Q. Cheng and P. Bendel, in: *Magnetic Resonance Microscopy*, ed. B. Blümich and W. Kuhn, VCH, Weinheim, 1992, p. 534.

Kas1 A. Kastler, *J. Phys. Radium* **11** (1950) 255.

Kel1 J.R. Keltner, S.T.S. Wong and M.S. Roos, *J. Magn. Reson. B* **104** (1994) 219.

Ken1 R.P. Kennan, K.A. Richardson, J. Zhong, M.J. Maryanski and J.C. Gore, *J. Magn. Reson. B* **110** (1996) 267.

Ken2 G.J. Kennedy, *Polym. Bull.* **23** (1990) 605.

Kes1 H. Kessler, H. Oschkinat and C. Griesinger, *J. Magn. Reson.* **70** (1986) 106.

Kim1 R. Kimmich and G. Voigt, *Z. Naturforsch.* **33a** (1978) 1294.

Kim2 R. Kimmich, *NMR Tomography, Diffusometry, Relaxometry*, Springer, Berlin, 1997.

Kim3 R. Kimmich, E. Fischer, P. Callaghan and N. Fatkullin, *J. Magn. Reson. A* **117** (1995) 53.

Kim4 R. Kimmich, U. Görke and J. Weis, *J. Trace Microprobe Techn.* **13** (1995) 285.

Kim5 R. Kimmich, W. Unrath, G. Schnur and E. Rommel, *J. Magn. Reson.* **91** (1991) 136.

Kim6 R. Kimmich and E. Fischer, *J. Magn. Reson. A* **106** (1994) 229.

Kim7 R. Kimmich, B. Simon and H. Köstler, *J. Magn. Reson. A* **112** (1995) 7.

Kin1 P.B. Kingsley, *Concepts Magn. Reson.* **11** (1999) 29.

Kin2 P. Kinesh, E.W. Randall and S.C.R. Williams, *J. Magn. Reson. B* **105** (1994) 253.

Kin3 P. Kinesh, D.S. Powlson, E.W. Randall and S.C.R. Williams, *J. Magn. Reson.* **97** (1992) 208.

Kin4 P. Kinesh, E.W. Randall and S.C.R. Williams, *J. Magn. Reson.* **98** (1992) 458.

Kle1 R.L. Kleinberg, W.E. Kenyon and P.P. Mitra, *J. Magn. Reson. A* **108** (1994) 206.

Kle2 R.L. Kleinberg, in: *Encyclopedia of NMR*, ed. D.M. Grant and R.K. Harris, Wiley, New York, 1996, p. 4960.

Kli1 M. Klinkenberg, P. Blümler and B. Blümich, *J. Magn. Reson. A* **119** (1996) 197.

Kli2 M. Klinkenberg, P. Blümler and B. Blümich, *Macromolecules* **30** (1997) 1038.

Knö1 M. Knörgen, U. Heuert, H. Menge and H. Schneider, *Angew. Makromol. Chem.* **261/262** (1998) 123.
Knü1 A. Knüttel and R. Kimmich, *J. Magn. Reson.* **86** (1990) 253.
Knü2 A. Knüttel, R. Kimmich and K.-H. Spohn, *J. Magn. Reson.* **86** (1990) 526.
Knü3 A. Knüttel, R. Kimmich and K.-H. Spohn, *Magn. Reson. Med.* **17** (1991) 490.
Knü4 A. Knüttel, K.-H. Spohn and R. Kimmich, *J. Magn. Reson.* **86** (1990) 542.
Knü5 A. Knüttel, R. Kimmich and K.-H. Spohn, *J. Magn. Reson.* **86** (1990) 526.
Kob1 F. Kober, B. Koenigsberg, V. Belle, M. Viallon, J.L. Leviel, A. Delon, A. Ziegler and M. Décorps, *J. Magn. Reson.* **138** (1999) 308.
Koc1 T. Koch, G. Brix and W.J. Lorentz, *J. Magn. Reson. B* **104** (1994) 199.
Koe1 F. Kober, B. Koenigsberg, V. Belle, M. Viallon, J.L. Leviel, A. Delon, A. Ziegler and M. Décorps, *J. Magn. Reson.* **138** (1999) 308.
Koh1 S.J. Kohler, E.K. Smith and N.H. Kolodny, *J. Magn. Reson.* **83** (1989) 423.
Koh2 S.J. Kohler, N.H. Kolodny, D.J. D'Amico, S. Balasubramaniam, P. Mainardi and E.S. Gragoudas, *J. Magn. Reson.* **82** (1989) 505.
Kon1 R. Konrat, I. Burghardt and G. Bodenhausen, *J. Am. Chem. Soc.* **113** (1991) 9135.
Kös1 H. Köstler and R. Kimmich, *J. Magn. Reson. B* **102** (1993) 285.
Kös2 H. Köstler and R. Kimmich, *J. Magn. Reson. B* **102** (1993) 177.
Kos1 K. Kose, *J. Magn. Reson.* **92** (1991) 631.
Kos2 K. Kose, *J. Magn. Reson.* **96** (1992) 596.
Kos3 K. Kose, *J. Magn. Reson.* **98** (1992) 599.
Kos4 K. Kose, *Phys. Rev. A* **44** (1991) 2495.
Kre1 M. Kresse, D. Pfefferer and R. Lawaczeck, *Deutsche Apotheker Zeitung* **33** (1994) 3079.
Kue1 D.O. Kuethe and J.-H. Gao, *Phys. Rev. E* **51** (1995) 3252.
Kuh1 W. Kuhn, P. Bart, S. Hafner, G. Simon and H. Schneider, *Macromolecules* **27** (1994) 5773.
Kuh2 W. Kuhn, P. Barth, P. Denner and R. Müller, *Solid-State Nucl. Magn. Reson.* **6** (1996) 295.
Kuh3 W. Kuhn, *Kolloid Z.* **68** (1934) 2.
Kul1 T.P. Kulagina, in: *Magn. Reson. Rel. Phen.*, ed. K.M. Salikhov, Extended Abstracts, *27th Cong. Ampere*, Kazan, 1994, p. 620.
Kun1 C. Kunze, R. Kimmich and D. Demco, *J. Magn. Reson. A* **101** (1993) 277.
Kun2 C. Kunze and R. Kimmich, *J. Magn. Reson. B* **105** (1994) 38.
Kun3 C. Kunze and R. Kimmich, *Magn. Reson. Imag.* **12** (1994) 805.
Lab1 C. Labadie, J.-H. Lee, G. Vétek and C.S. Springer, Jr., *J. Magn. Reson. B* **105** (1994) 99.
Lac1 S. Lacelle, *Adv. Magn. Opt. Reson.* **16** (1991) 173.
Lat1 L.L. Latour, R.L. Kleinberg, P.P. Mitra and C.H. Sotak, *J. Magn. Reson. A* **112** (1995) 83.
Lau1 P.C. Lauterbur, *J. Magn. Reson.* **59** (1984) 536.
Lau2 R.B. Lauffer, *Chem. Rev.* **87** (1987) 901.
Lau3 P.C. Lauterbur, D.M. Cramer, W.V. House, Jr. and C.N. Chen, *J. Am. Chem. Soc.* **97** (1975) 6866.
Leb D. Le Bihan, J. Delannoy and R.L. Levin, *Radiology* **171** (1989) 853.
Lee1 S.-M. Lee, D.-L. Tzou, G.J.Y. Chiou and H.N. Yeung, *J. Magn. Reson.* **130** (1998) 102.
Lee2 M. Lee and W.I. Goldburg, *Phys. Rev. A* **140** (1965) 1261.
Lee3 Y. Lee, C. Michaels and L.G. Butler, *Chem. Phys. Lett.* **206** (1993) 464.
Lei1 H. Lei and J. Peeling, *J. Magn. Reson.* **137** (1999) 215.
Ler1 A. Leroy-Willig and D. Movran, *J. Chim. Phys.* **89** (1992) 289.

Lev1 M.H. Levitt, *J. Chem. Phys.* **94** (1991) 30.

Lev2 M.H. Levitt, D. Suter and R.R. Ernst, *J. Chem. Phys.* **84** (1986) 4243.

Lev3 D.H. Levy and K.K. Gleason, *J. Chem. Phys.* **96** (1992) 8125.

Lew1 C.J. Lewa and D. de Certaines, *Spectroscopy Letters* **27** (1994) 1369.

Lia1 J. Lian, D.S. Williams and I.J. Lowe, *J. Magn. Reson. A* **106** (1994) 65.

Lin1 J. Link and J. Seelig, *J. Magn. Reson.* **89** (1990) 310.

Lit1 V.M. Litvinov, W. Barendswaard and M. van Duin, *Rubber Chem. Tech.* **71** (1998) 105.

Liu1 J. Liu, A.O.K. Nieminen and J.L. Koenig, *J. Magn. Reson.* **85** (1989) 95.

Lon1 H.W. Long, H.C. Gaede, J. Shore, L. Reven, C.R. Bowers, J. Kritzenberger, T. Pietrass, A. Pines, P. Tang and J.A. Reimer, *J. Am. Chem. Soc.* **115** (1993) 8491.

Luc1 A.J. Lucas, S.J. Gibbs, E.W.G. Jones, M. Peyron, A.D. Derbyshire and L.D. Hall, *J. Magn. Reson. A* **104** (1993) 273.

Luy1 P.R. Luyten, J.H. Marien, W. Heindel, P.H.J. Van Gerwen, K. Herholz, J.A. den Hollander, G. Friedman and W.-D. Heiss, *Radiology* **176** (1990) 791.

Lyo1 R.C. Lyon, J. Pekar, C.T.W. Moonen and A.C. McLaughlin, *Magn. Reson. Med.* **18** (1991) 80.

Maf1 P. Maffei, P. Mutzenhardt, A. Retournand, B. Ditter, R. Raulet, J. Brondeau and D. Canet, *J. Magn. Reson. A* **107** (1994) 40.

Mai1 R.W. Mair, D.G. Cory, S. Peled, C.-H. Tseng, S. Patz and R.L. Walsworth, *J. Magn. Reson.* **135** (1998) 478.

Maj1 P.D. Majors, R.C. Givler and E. Fukushima, *J. Magn. Reson.* **85** (1989) 235.

Maj2 P.D. Majors, J.L. Smith, F.S. Kovarik and E. Fukushima, *J. Magn. Reson.* **89** (1990) 470.

Maj3 J. Ma, F.W. Wehrli, H.K. Song and S.N. Hwang, *J. Magn. Reson.* **125** (1997) 92.

Man1 P. Mansfield and P.G. Morris, *NMR Imaging in Biomedicine, Adv. Magn. Reson. Suppl.* 2, Academic Press, New York, 1982.

Man2 P. Mansfield, R. Bowtell and S. Blackband, *J. Magn. Reson.* **99** (1992) 507.

Man3 P. Mansfield and D. Ware, *Phys. Lett.* **22** (1966) 133.

Man4 P. Mansfield and D. Ware, *Phys. Rev.* **168** (1968) 318.

Man5 P. Mansfield and P.K. Grannell, *J. Phys. C* **4** (1971) L197.

Man6 P. Mansfield, *Magn. Reson. Med.* **1** (1984) 370.

Mar1 J.L. Markley, W.J. Horsley and M.P. Klein, *J. Chem. Phys.* **55** (1971) 3604.

Mar2 L. Marciani, P. Manoj, B.P. Hills, R.J. Moore, P. Young, A. Fillery-Travis, R.C. Spiller and P.A. Gowland, *J. Magn.Reson.* **135** (1998) 82.

Mat1 S. Matsui, *Chem. Phys. Lett.* **179** (1991) 187.

Mat2 J. Mattiello, P.J. Basser and D. LeBihan, *J. Magn. Reson. A* **108** (1994) 131.

Mat3 S. Matsui, M. Nonaka, T. Nakai and T. Inouye, *Solid-State Nucl. Magn. Reson.* **10** (1997) 39.

Mat4 G.D. Mateescu, M. Cabrera and D. Fercu, in: *Spatially Resolved Magnetic Resonance*, ed. P. Blümler, B. Blümich, R. Botto and E. Fukushima, Wiley-VCH, Weinheim, 1998, p. 421.

Mau1 A.A. Maudsley, L. Müller and R.R. Ernst, *J. Magn. Reson.* **28** (1977) 463.

Mau2 A.A. Maudsley, A. Oppelt and A. Gansen, *Siemens Res. Dev. Rep.* **8** (1979) 326.

Mau3 A.A. Maudsley, A. Wokaun and R.R. Ernst, *Chem. Phys. Lett.* **55** (1978) 9.

Meh1 M. Mehring, *High Resolution NMR in Solids*, 2nd edn, Springer, Berlin, 1983.

Mei1 S. Meiboom and D. Gill, *Rev. Sci. Instrum.* **29** (1958) 688.

Men1 H. Menge, S. Hotopf and H. Schneider, *Kautschuk, Gummi, Kunststoffe* **50** (1997) 268.

Mer1 K.-D. Merboldt, W. Hänicke and J. Frahm, *Ber. Bunsenges. Phys. Chem.* **91** (1987) 1124.

Mer2 K.-D. Merboldt, W. Hänicke, M.L. Gyngell, F. Frahm and H. Bruhn, *J. Magn. Reson.* **82** (1989) 115.

Mer3 K.-D. Merboldt, W. Hänicke and J. Frahm, *J. Magn. Reson.* **64** (1985) 479.

Mil1 J.B. Miller and A.N. Garroway, *J. Magn. Reson.* **67** (1986) 575.

Mil2 J.B. Miller, *Rubber Chem. Tech.* **66** (1993) 455.

Mis1 E. Mischler, F. Humbert and D. Canet, *J. Magn. Reson. B* **109** (1995) 121.

Möl1 H.E. Möller, X.J. Chen, M.S. Chawla, B. Driehuys, L.W. Hedlund and G.A. Johnson, *J. Magn. Reson.* **135** (1998) 133.

Moo1 C.T.W. Moonen, G. Sobering, P.C.M. van Zijl, J. Gillen, M. von Kienlin and A. Bizzi, *J. Magn. Reson.* **98** (1992) 556.

Mor1 G.A. Morris and R. Freeman, *J. Magn. Reson.* **29** (1978) 433.

Mor2 G.A. Morris and R. Freeman, *J. Am. Chem. Soc.* **101** (1979) 760.

Mor3 P.R. Moran, *Magn. Reson. Imag.* **1** (1982) 197.

Mor4 H. Morneburg, ed., *Bildgebende Systeme für die medizinische Diagnostik*, 3rd edn, Publicis MCDF Verlag, Erlangen, 1995.

Mos1 M. Moseley, Y. Cohen, J. Kucharczyk, J. Mintorovitch, H.S. Asgari, M.F. Wendland, J. Tsuruda and D. Norman, *Radiology* **176** (1990) 439.

Mül1 L. Müller and R.R. Ernst, *Molec. Phys.* **38** (1979) 963.

Mül2 S. Müller, J. Seelig, *J. Magn. Reson.* **72** (1987) 456.

Mun1 M. Munowitz and A. Pines, *Adv. Chem. Phys.* **66** (1987) 2.

Nak1 T. Nakai, Y. Fukunaga, M. Nonaka, S. Matsui and T. Inouye, *J. Magn. Reson.* **134** (1998) 44.

Nav1 G. Navon, Y.-Q. Song, T. Room, S. Appelt, R.E. Taylor and A. Pines, *Science* **271** (1996) 1848.

Nes1 N. Nestle, K. Rydyger and R. Kimmich, *J. Magn. Reson.* **125** (1997) 355.

Nil1 H. Nilgens, P. Blümler, J. Paff and B. Blümich, *J. Magn. Reson. A* **105** (1993) 108.

Nog1 J.H. Noggle and R.E. Schirmer, *The Nuclear Overhauser Effect; Chemical Applications*, Academic Press, New York, 1971.

Nor1 D.G. Norris and C. Schwarzbauer, *J. Magn. Reson.* **137** (1999) 231.

Nor2 T.J. Norwood and S.C.R. Williams, *J. Magn. Reson.* **94** (1991) 419.

Oga1 S. Ogawa, T.M. Lee, A. Nayak and P. Glynn, *Magn. Reson. Med.* **14** (1990) 68.

Oga2 S. Ogawa, T.M. Lee, A.R. Kay and D.W. Tank, *Proc. Natl. Acad. Sci. USA* **87** (1990) 9868.

Ole1 D.A. Oleskevich, N. Ghahramany, W.P. Weglarz and H. Peemoeller, *J. Magn. Reson. B* **113** (1996) 1.

Ost1 E.D. Ostroff and J.S. Waugh, *Phys. Rev. Lett.* **16** (1966) 1097.

Pac1 K.J. Packer and J.J. Tessier, *Mol. Phys.* **87** (1996) 267.

Pac2 K.J. Packer, in: *Encyclopedia of NMR Spectroscopy*, ed. D.M. Grant and R.K. Harris, Wiley, New York, 1996, p. 1615.

Pap1 N.G. Papadakis, D. Xing, Y.L.-H. Huang, L.D. Hall and T.A. Carpenter, *J. Magn. Reson.* **137** (1999) 67.

Par1 A.A. Parker, J.J. Marcinko, P. Rinaldi, D.P. Hendrick and W. Ritchey, *J. Appl. Polym.* **48** (1993) 667.

Par2 D.L. Parker, V. Smith, P. Sheldon, L.E. Crooks and L. Fussel, *Med. Phys.* **10** (1983) 321.

Pav1 G. Pavlovskaya. A.K. Blue, S.J. Gibbs, M. Haake, F. Cros, L. Mailer and T. Meersmann, *J. Magn. Reson.* **137** (1999) 258.

Pay1 G.S. Payne and P. Styles, *J. Magn. Reson.* **95** (1991) 253.

Pee1 H. Peemoeller and M.M. Pintar, *J. Magn. Reson.* **41** (1980) 358.

Pey1 M. Peyron, G.K. Pierens, A.J. Lucas, L.D. Hall and R.C. Stewart, *J. Magn. Reson. A* **118** (1996) 214.

Pfe1 M. Pfeffer and O. Lutz, *J. Magn. Reson. A* **108** (1994) 106.

Pie1 T. Pietraß and H.C. Gaede, *Advanced Materials* **7** (1995) 826.

Pie2 T. Pietraß, R. Seydoux and A. Pines, *J. Magn. Reson.* **133** (1998) 299.

Pin1 A. Pines, M.O. Gibby and J.S. Waugh, *J. Chem. Phys.* **59** (1973) 569.

Pop1 J.M. Pope and S. Yao, *Concepts Magn. Reson.* **5** (1993) 281.

Pop2 J.M. Pope and S. Yao, *Magn. Reson. Imag.* **11** (1993) 585.

Pop3 J.M. Pope and N. Repin, *Magn. Reson. Imag.* **6** (1988) 641.

Pos1 S. Posse and W.P. Aue, *J. Magn. Reson.* **88** (1990) 473.

Pos2 S. Posse and W.P. Aue, *J. Magn. Reson.* **83** (1989) 620.

Pow1 J.G. Powles and P. Mansfield, *Physics Letters* **2** (1962) 58.

Pri1 W.S. Price, *Concepts Magn. Reson.* **9** (1997) 299.

Pri2 W.S. Price, *Concepts Magn. Reson.* **10** (1998) 197.

Pro1 S.W. Provencher, *Comp. Phys. Comm.* **27** (1982), 213, 229.

Püt1 B. Pütz, D. Barsky and K. Schulten, *J. Magn. Reson.* **97** (1992) 27.

Raf1 D. Raftery and B.F. Chmelka, *NMR – Basic Principles and Progress* **30** (1994) 111.

Raf2 D. Raftery, H. Long, T. Meersmann, P.J. Grandinetti, L. Reven and A. Pines, *Phys. Rev. Lett.* **66** (1991) 584.

Ran1 M.A. Rana and J.L. Koenig, *Macromolecules* **27** (1994) 3727.

Red1 R. Reddy, A.H. Stolpen and J.S. Leigh, *J. Magn. Reson. B* **108** (1995) 276.

Red2 R. Reddy, E.K. Insko and J.S. Leigh, *Magn. Reson. Med.* **38** (1997) 279.

Rhi1 W. Rhim, A. Pines and J.S. Waugh, *Phys. Rev. B* **3** (1971) 684.

Rob1 P. Robyr and R. Bowtell, *J. Chem. Phys.* **106** (1997) 467.

Rok1 M. Rokitta, U. Zimmermann and A. Haase, *J. Magn. Reson.* **137** (1999) 29.

Rom1 E. Rommel and R. Kimmich, *Magn. Reson. Med.* **12** (1989) 209.

Rom2 E. Rommel and R. Kimmich, *Magn. Reson. Med.* **12** (1989) 390.

Rom3 E. Rommel, R. Kimmich, H. Körperich, C. Kunze and K. Gersonde, *Magn. Reson. Med.* **24** (1992) 149.

Rom4 K. Rombach, *Untersuchung von Transportprozessen in Mischgeometrien mit NMR-Bildgebung*, Dissertation, RWTH, Aachen, 1998.

Rom5 E. Rommel, S. Hafner and R. Kimmich, *J. Magn. Reson.* **86** (1990) 264.

Roo1 T. Room, S. Appelt, R. Seydoux, E.L. Hahn and A. Pines, *Phys. Rev. B* **55** (1997) 11604.

Saa1 T.R. Saarinen and C.S. Johnson, Jr., *J. Magn. Reson.* **78** (1988) 257.

Saa2 B. Saam, N. Drukker and W. Happer, *Chem. Phys. Lett.* **263** (1996) 481.

San1 G.I. Sandakov, L.P. Smirnov, A.I. Sosikov, K.T. Summanen and N.N. Volkova, *J. Polym. Sci. B: Polym. Phys.* **32** (1994) 1585.

Sch1 C. Schwarzbauer, J. Zange, H. Adolf, R. Deichmann, U. Nöth and A. Haase, *J. Magn. Reson. B* **106** (1995) 178.

Sch2 R.J. Schadt and D.L. VanderHart, *Macromolecules* **28** (1995) 3416.

Sch3 S. Schantz and W.S. Veeman, *J. Polym. Sci.: Part B: Polym. Phys.* **35** (1997) 2681.

Sch4 D.M. Schmidt, J.S. George, S.I. Penttila, A. Caprihan and E. Fukushima, *J. Magn. Reson.* **129** (1997) 184.

Sch5 K. Schmidt-Rohr and H.W. Spiess, *Multidimensional Solid-State NMR and Polymers*, Academic Press, London, 1994.

Sch6 K. Schmidt-Rohr, J. Clauss, B. Blümich and H.W. Spiess, *Magn. Reson. Chem.* **28** (1990) 381.

Sch7 U. Scheler, G. Schauß, B. Blümich and H.W. Spiess, *Solid-State Nucl. Magn. Reson.* **6** (1996) 375.

Sch8 S. Schantz and W.S. Veeman, *J. Polym. Sci.: Part B: Polym. Physics* **35** (1997) 2681.

Sch9 H. Schneider and H. Schmiedel, *Physics Letters* **30A** (1969) 298.

Sch10 H. Schmiedel and H. Schneider, *Ann. Phys.* **32** (1975) 249.

Sch11 M. Schneider, *^{1}H-Multiquanten-Untersuchungen an Elastomeren*, Dissertation, Rheinisch-Westfälische Technische Hochschule, Aachen, 1999.

Sch12 M. Schneider, L. Gasper, D.E. Demco and B. Blümich, *J. Chem. Phys.* **111** (1999) 402.

Sch13 M. Schneider, D.E. Demco and B. Blümich, *J. Magn. Reson.* **140** (1999) 432.

Sch14 J. Schmedt auf der Günne and H. Eckert, *Chem. Eur. J.* **4** (1998) 1762.

Sco1 G.C. Scott, M.L.G. Joy, R.L. Armstrong and R.M. Henkelmann, *IEEE Trans. Med. Imag.* **10** (1991) 362.

Sco2 G.C. Scott, M.L.G. Joy, R.L. Armstrong and R.M. Henkelmann, *Magn. Reson. Med.* **28** (1992) 186.

Sco3 G.C. Scott, M.L.G. Joy, R.L. Armstrong and R.M. Henkelmann, *J. Magn. Reson.* **97** (1992) 235.

Sco4 G.C. Scott, M.L.G. Joy, R.L. Armstrong and R.M. Henkelmann, *Magn. Reson. Med.* **33** (1995) 355.

Scr1 B.E. Scruggs and K.K. Gleason, *J. Magn. Reson.* **99** (1992) 149.

Seo1 Y. Seo, H. Takamiya, H. Ishikawa, T. Nakashima, Y. Shard and G. Navon, in: *Spatially Resolved Magnetic Resonance*, ed. P. Blümler, B. Blümich, R. Botto and E. Fukushima, Wiley-VCH, Weinheim, 1998, p. 443.

Ser1 I. Sersa, O. Jarh and F. Demsar, *J. Magn. Reson. A* **111** (1994) 93.

Sey1 J.D. Seymour and P.T. Callaghan, *AICHE J.* **43** (1997) 2096.

Sez1 A. Sezginer, R.L. Kleinberg, M. Fukuhara and L.L. Latour, *J. Magn. Reson.* **94** (1991) 504.

Sha1 Y. Sharf, U. Eliaf, H. Shinnar and G. Navon, *J. Magn. Reson. B* **107** (1995) 60.

Sha2 Y. Sharf, Y. Seo, U. Eliav, S. Akselrod and G. Navon, *Proc. Natl. Acad. Sci. USA* **95** (1998) 4108.

Shi1 H. Shinar, Y. Seo and G. Navon, *J. Magn. Reson.* **129** (1997) 98.

Sil1 L.O. Sillerud, D.B. van Hulsteyn and R.H. Griffey, *J. Magn. Reson.* **76** (1988) 380.

Sim1 G. Simon and H. Schneider, *Makromol. Chem. Macromol. Symp.* **52** (1991) 233.

Sin1 J.R. Singer, *Science* **130** (1959) 1652.

Sin2 J.R. Singer, *IEEE Trans. Med. Imag.* **NS-27** (1980) 1245.

Sli1 C.P. Slichter, *Principles of Magnetic Resonance*, 3rd edn, Springer, Berlin, 1990.

Sod1 A. Sodickson and D.G. Cory, *Progr. Nucl. Magn. Reson. Spectrosc.* **33** (1998) 77.

Son1 Y.-Q. Song, H.C. Gaede, T. Pietraß, G.A. Barrall, G.C. Chingas, M.R. Ayers and A. Pines, *J. Magn. Reson. A* **115** (1995) 127.

Son2 Y.-Q. Song, R.E. Taylor and A. Pines, *Solid-State Nucl. Magn. Reson.* **10** (1998) 247.

Son3 M. Sonderegger, J. Roos, C. Kugler, M. Mali and D. Brinkmann, *Solid State Ionics* **53–56** (1992) 849.

Son4 Y.-Q. Song, B.M. Goodson, B. Sheridan, T.M. de Swiet and A. Pines, *J. Chem. Phys.* **108** (1998) 6233.

Sot1 P. Sotta, B. Deloche, J. Herz, A. Lapp, D. Durand and J.-C. Rabadeux, *Macromolecules* **20** (1987) 2769.

Sot2 P. Sotta, C. Fülber, D.E. Demco, B. Blümich and H.W. Spiess, *Macromolecules* **29** (1996) 6222.

Spe1 K. Sperling, W.S. Veeman and V.M. Litvinov, *Kautschuck, Gummi, Kunststoffe* **50** (1997) 804.

Spy1 A. Spyros, N. Chandrakumar, M. Heidenreich and R. Kimmich, *Macromolecules* **31** (1998) 3021.

Ste1 E.O. Stejskal and J. Tanner, *J. Chem. Phys.* **42** (1965) 228.

Ste2 E.O. Stejskal, *J. Chem. Phys.* **43** (1965) 3597.

Ste3 J. Stepisnik, A. Duh, A. Mohoric and I. Sersa, *J. Magn. Reson.* **137** (1999) 154.

Sui1 B.H. Suits and D. White, *J. App. Phys.* **60** (1986) 3772.

Sui2 B.H. Suits and D. White, *Solid State Commun.* **50** (1984) 291.

Sun1 Y. Sun, H. Lock, T. Shinozaki and G.E. Maciel, *J. Magn. Reson. A* **115** (1995) 165.

Swi1 T.M. de Swiet and P.P. Mitra, *J. Magn. Reson. B* **111** (1996) 15.

Swa1 V.S. Swaminathan and B.H. Suits, *J. Magn. Reson.* **132** (1998) 274.

Sze1 N.M. Szeverenyi and G.E. Maciel, *J. Magn. Reson.* **60** (1984) 460.

Tak1 K. Takegoshi and C.A. McDowell, *Chem. Phys. Lett.* **116** (1985) 100.

Tan1 J.E. Tanner and E.O. Stejskal, *J. Chem. Phys.* **49** (1968) 1768.

Tan2 J.E. Tanner, *J. Chem. Phys.* **69** (1978) 1748.

Tek1 P. Tekely, D. Canet and J.-J. Delpuech, *Molec. Phys.* **67** (1989) 81.

Tes1 J.J. Tessier, T.A. Carpenter and L.D. Hall, *J. Magn. Reson. A* **113** (1995) 232.

Tli1 T.-Q. Li, L. Ödberg, R.L. Powell and M.J. McCarthy, *J. Magn. Reson. B* **109** (1995) 213.

Tri1 L.A. Trimble, J.F. Shen, A.H. Wilman and P.S. Allen, *J. Magn. Reson.* **86** (1990) 191.

Tse1 C.H. Tseng, G.P. Wong, V.R. Pomeroy, R.W. Mair, D.P. Hinton, D. Hoffmann, R.E. Stoner, F.W. Hersman, D.G. Cory and R.L. Walsworth, *Phys. Rev. Lett.* **81** (1998) 3785.

Tse2 C.H. Tseng, R.W. Mair, G.P. Wong, D. Williamson, D.G. Cory and R.L. Walsworth, *Phys. Rev. E* **59** (1999) 1785.

Tso1 L. Tsoref, H. Shinar, Y. Seo, U. Eliav and G. Navon, *Magn. Reson. Med.* **39** (1998) 11.

Tur1 R. Turner and D. LeBihan, *J. Magn. Reson.* **86** (1990) 445.

Tur2 R. Turner and P. Keller, *Progr. NMR Spectrosc.* **23** (1991) 93.

Tyc1 R. Tycko and J.A. Reimer, *J. Phys. Chem.* **100** (1996) 13240.

Tys1 M. Tyszka, R.C. Hawkes and L.D. Hall, *J. Magn. Reson.* **94** (1991) 408.

Vol1 R.L. Vold, J.S. Waugh, M.P. Klein and D.E. Phelps, *J. Chem. Phys.* **48** (1968) 3831.

Vol2 A. Volk, B. Tiffon, J. Mispelter and J.-M. Lhoste, *J. Magn. Reson.* **71** (1987) 168.

War1 I.M. Ward and D.W. Hadley, *An Introduction to the Mechanical Properties of Solid Polymers*, Wiley, New York, 1993.

War2 M. Warner, P.T. Callaghan and E.T. Samulski, *Macromolecules* **30** (1997) 4733.

War3 W.S. Warren, W. Richter, A.H. Andreotti and B.T. Farmer II, *Science* **262** (1993) 2005.

War4 W.S. Warren, A. Ahn, M. Mescher, M. Garwood, K. Ugurbil, W. Richter, R.R. Rizi, J. Hopkins and J.S. Leigh, *Science* **281** (1998) 247.

Wat1 J.C .Waterton, R.P.O. Jones and G.A. Morris, *J. Magn. Reson.* **97** (1992) 218.

Web1 A.G. Webb, S.C.R. Williams and L.D. Hall, *J. Magn. Reson.* **84** (1989) 159.

Weg1 W.P. Weglarz and H. Peemoeller, *J. Magn. Reson.* **124** (1997) 484.

Weg2 W.P. Weglarz, A. Jasinski, A.T. Kryzak, P. Kozlowski, D. Adamek, P. Sagnowski and J. Pindel, *Appl. Magn. Reson.* **15** (1998) 333.

Weh1 F.W. Wehrli, D. Schaw and J.B. Kneeland, ed., *Biomedical Magnetic Resonance Imaging*, VCH Verlagsgesellschaft, Weinheim, 1988.

Wei1 F. Weigand, D.E. Demco, B. Blümich and H.W. Spiess, *J. Magn. Reson. A* **120** (1996) 190.

Wei2 F. Weigand, H.W. Spiess, B. Blümich, G. Salge and K. Möller, *IEEE Transact. Dielectr. Electr. Insul.* **4** (1997) 280.

Wei3 F. Weigand, D.E. Demco, B. Blümich and H.W. Spiess, *Solid-State Nucl. Magn. Reson.* **6** (1996) 357.
Wei4 F. Weigand is gratefully acknowledged for supplying Fig. 7.1.6.
Wei5 J. Weis, U. Görke and R. Kimmich, *Magn. Reson. Imag.* **14** (1996) 1165.
Wim1 S. Wimperis and B. Wood, *J. Magn. Reson.* **95** (1991) 428.
Wim2 S. Wimperis, P. Coole and P. Styles, *J. Magn. Reson.* **98** (1992) 628.
Win1 W. Windig, J.P. Hornak and B. Analek, *J. Magn. Reson.* **132** (1998) 298.
Win2 W. Windig, J.P. Hornak and B. Analek, *J. Magn. Reson.***132** (1998) 307.
Wol1 S.D. Wolff and R.S. Balaban, *J. Magn. Reson.* **86** (1990) 164.
Xia1 Y. Xia, *Concepts Magn. Reson.* **8** (1996) 205.
Xia2 Y. Xia and L.W. Jelinski, *J. Magn. Reson. B* **107** (1995) 1.
Xia3 Y. Xia and P.T. Callaghan, *Macromolecules* **24** (1991) 4777.
Xia4 Y. Xia and P.T. Callaghan, *Magn. Reson. Med.* **23** (1992) 138.
Yan1 C. Yang, W.-Y. Wen, A.A. Jones and P.T. Inglefield, *Solid-State Nucl. Magn. Reson.* **12** (1988) 153.
Yeu1 H.N. Yeung and S.D. Swanson, *J. Magn. Reson.* **83** (1989) 183.
Zha1 W. Zhang and D.G. Cory, *J. Magn. Reson.* **132** (1998) 144.
Zhu1 J.-M. Zhu and I.C.P. Smith, *Concepts Magn. Reson.* **7** (1995) 281.
Zie1 A. Ziegler, A. Metzler, W. Köckenberger, M. Izquierdo, E. Komor, A. Haase, M. Décorps and M. von Kienlin, *J. Magn. Reson. B* **112** (1996) 141.
Zij1 P.C.M. van Zijl and C.T.W. Moonen, *NMR – Basic Principles and Progress* **26** (1992) 67.
Zij2 P.C.M. van Zijl, A.S. Chesnick, D. DesPres, C.T.W. Moonen, J. Ruiz-Cabello and P. van Gelderen, *Magn. Reson. Med.* **30** (1993) 544.
Zil1 K.W. Zilm, in: *Encyclopedia of NMR Spectroscopy*, ed. D.M. Grant and R.K. Harris, Wiley, New York, 1996, p. 4498.
Zim1 G. Zimmer, A. Guthausen and B. Blümich (unpublished).
Zum1 N. Zumbulyadis, *J. Magn. Reson.* **53** (1983) 486.
Zuo1 C.S. Zuo, K.R. Metz, Y. Sun and A.D. Sherry, *J. Magn. Reson.* **133** (1998) 53.
Zwa1 J.A. de Zwart, P. van Gelderen, D.J. Kelly and C.T.W. Moonen, *J. Magn. Reson. B* **112** (1996) 86.

CHAPTER 8

Ack1 J.L. Ackerman, US-Patent 4,654,593, 31, March 1987.
Ack2 J.L. Ackerman, L. Garrido, J.R. Moore, B. Pfleiderer and Y. Yu, in: *Magnetic Resonance Microscopy*, ed. B. Blümich and W. Kuhn, VCH, Weinheim, 1992, p. 237.
Ack3 J.L. Ackerman and W.E. Ellingson, ed., *Advanced Tomographic Imaging Methods, Materials Research Society Symposium Proceedings 217*, Materials Research Society, Pittsburgh, 1991.
Ats1 V.A. Atsarkin, A.E. Mefed and I. Rodak, *Sov. Phys. Solid State* **21** (1979) 1537.
Att1 J.J. Attard, P.J. McDonald, S.P. Roberts and T. Taylor, *Magn. Reson. Imag.* **12** (1994) 355.
Axe1 D.E. Axelson, A. Kantzas and A. Nauerth, *Solid-State Nucl. Magn. Reson.* **6** (1996) 309.
Bai1 A.D. Bain and E.W. Randall, *J. Magn. Reson. A* **123** (1996) 49.
Bal1 B.J. Balcom, R.P. MacGregor, S.D. Beyea, D.P. Green, R.L. Armstrong and T.W. Bremner, *J. Magn. Reson. A* **123** (1996) 131.

Bal2 B.J. Balcom, M. Bogdan and R.L. Armstrong, *J. Magn. Reson. A* **118** (1996) 122.

Bal3 B.J. Balcom, in: *Spatially Resolved Magnetic Resonance*, ed. P. Blümler, B. Blümich, R. Botto and E. Fukushima, Wiley-VCH, Weinheim, 1998, p. 695.

Bar1 P. Barth, S. Hafner and W. Kuhn, *J. Magn. Reson. A* **110** (1994) 198.

Bau1 M.A. Baumann, G.M. Doll and K. Zick, *Oral Med. Oral Pathol.* **75** (1993) 517.

Bau2 J. Baum, M. Munowitz, A.N. Garroway and A. Pines, *J. Chem. Phys.* **83** (1985) 2015.

Ben1 T.B. Benson and P.J. McDonald, *J. Magn. Reson. A* **112** (1995) 17.

Bey1 S.D. Beyea, B.J. Balcom, T.W. Bremner, P.J. Prado, A.R. Cross, R.L. Armstrong and P.E. Grattan-Bellew, *Solid-State Nucl. Magn. Reson.* **13** (1998) 93.

Bla1 S. Black, D.M. Lane, P.J. McDonald, M. Mulheron, G. Hunter and M.R. Jones, *J. Mater. Sci. Lett.* **14** (1995) 1175.

Blü1 B. Blümich and W. Kuhn, ed., *Magnetic Resonance Microscopy*, VCH, Weinheim, 1992.

Blü2 P. Blümler and B. Blümich, *NMR – Basic Principles and Progress* **30** (1994) 209.

Blü3 P. Blümler and B. Blümich, *Rubber Chem. Tech.* **70** (1997) 468.

Blü4 B. Blümich, P. Blümler, E. Günther, G. Schauß and H.W. Spiess, *Makromol. Chem. Macromol. Symp.* **44** (1991) 37.

Blü5 B. Blümich, P. Blümler, E. Günther, G. Schauß and H.W. Spiess, in: *Magnetic Resonance Microscopy*, ed. B. Blümich and W. Kuhn, VCH, Weinheim, 1992, p. 167.

Blü6 B. Blümich, *Adv. Mater.* **3** (1991) 237.

Blü7 B. Blümich and P. Blümler, *Die Makromolek. Chem.* **194** (1993) 2133.

Blü8 P. Blümler, B. Blümich, R. Botto and E. Fukushima, ed., *Spatially Resolved Magnetic Resonance*, Wiley-VCH, Weinheim, 1998.

Blü9 B. Blümich, P. Blümler and K. Saito, in: *Solid State NMR of Polymers*, ed. I. Ando and T. Asakura, Elsevier Science. B. V., Amsterdam, 1998, p. 123.

Bod1 P. Bodart, T. Nunes and E.W. Randall, *Solid-State Nucl. Magn. Reson.* **8** (1997) 257.

Bod2 G. Bodenhausen, H. Kogler and R.R. Ernst, *J. Magn. Reson.* **58** (1984) 370.

Bog1 M. Bogdan, B.J. Balcom, T.W. Bremner and R.L. Armstrong, *J. Magn. Reson. A* **116** (1995) 266.

Bot1 P.A. Bottomley, *J. Phys. E: Sci. Instrum.* **14** (1981) 1081.

Bot2 R.E. Botto, G.D. Cody, S.L. Dieckman, D.C. French, N. Gopalsami and P. Rizo, *Solid-State Nucl. Magn. Reson.* **6** (1996) 389.

Bot3 R.E. Botto, ed., *Special Issue on Magnetic Resonance Imaging of Materials, Solid-State Nucl. Magn. Reson.* **6** (1996).

Bow1 G.J. Bowden and W.D. Hutchinson, *J. Magn. Reson.* **67** (1986) 403.

Bur1 D.P. Burum, *Concepts Magn. Reson.* **2** (1990) 213.

Bus1 M.L. Buszko and G.E. Maciel, *J. Magn. Reson. A* **104** (1993) 172.

Bus2 M.L. Buszko, C.E. Bronnimann and G.E. Maciel, *J. Magn. Reson. A* **103** (1993) 183.

Bus3 M.L. Buszko and G.E. Maciel, *J. Magn. Reson. A* **110** (1993) 7.

Cal1 P.T. Callaghan, *Principles of Nuclear Magnetic Resonance Microscopy*, Clarendon Press, Oxford, 1991.

Car1 H.Y. Carr and E.M. Purcell, *Phys. Rev.* **94** (1954) 630.

Cha1 C. Chang and R.A. Komoroski, in: *Solid State NMR of Polymers*, ed. L. Mathias, Plenum Press, New York, 1991, p. 363.

Chi1 G.C. Chingas, J.B. Miller and A.N. Garroway, *J. Magn. Reson.* **66** (1986) 530.

Cho1 H.M. Cho, C.J. Lee, D.N. Shykind and D.P. Weitekamp, *Phys. Rev. Lett.* **55** (1985) 1923.

Cod1 S.L. Codd, M.J.D. Mallett, M.R. Halse, J.H. Strange, W. Vennart and T. Van Doorn, *J. Magn. Reson. B* **113** (1996) 214.

Con1 M.S. Conradi, A.N. Garroway, D.G. Cory and J.B. Miller, *J. Magn. Reson.* **94** (1991) 370.

Con2 M.S. Conradi, *J. Magn. Reson.* **93** (1991) 419.

Cor1 D.G. Cory, *Annual Reports on NMR* **24** (1992) 87.

Cor2 D.G. Cory and W.S. Veeman, *J. Phys. E: Sci. Instrum.* **22** (1989) 180.

Cor3 D.G. Cory and W.S. Veeman, *J. Magn. Reson.* **84** (1989) 392.

Cor4 D.G. Cory, J.C. de Boer and W.S. Veeman, *Macromolecules* **22** (1989) 1618.

Cor5 D.G. Cory, J.W.M. Van Os and W.S. Veeman, *J. Magn. Reson.* **76** (1988) 543.

Cor6 D.G. Cory, A.M. Reichwein, J.W.M. Van Os and W.S. Veeman, *Chem. Phys. Lett.* **143** (1988) 467.

Cor7 D.G. Cory and W.S. Veeman, *J. Magn. Reson.* **82** (1989) 374.

Cor8 D.G. Cory, A.M. Reichwein and W.S. Veeman, *J. Magn. Reson.* **80** (1988) 259.

Cor9 D.G. Cory, J.B. Miller and A.N. Garroway, *J. Magn. Reson.* **90** (1990) 205.

Cor10 D.G. Cory, J.B. Miller and A.N. Garroway, *Mol. Phys.* **70** (1990) 331.

Cor11 D.G. Cory, A.N. Garroway and J.B. Miller, *J. Magn. Reson.* **87** (1990) 202.

Cor12 D.G. Cory, *Solid-State Nucl. Magn. Reson.* **6** (1996) 347.

Cor13 D.G. Cory, in: *Magnetic Resonance Microscopy*, ed. B. Blümich and W. Kuhn, VCH, Weinheim, 1992, p. 49.

Cot1 S.P. Cottrell, M.R. Harvey and J.H. Strange, *Meas. Sci. Technol.* **1** (1990) 624.

Dau1 Y.M. Daud and M.R. Halse, *Physica B* **176** (1992) 167.

Del1 F. De Luca and B. Maraviglia, *J. Magn. Reson.* **67** (1986) 169.

Del2 F. De Luca, C. Nuccetelli, B.C. De Simone and B. Maraviglia, *J. Magn. Reson.* **69** (1986) 496.

Del3 F. De Luca, C. Nuccetelli, B.C. De Simone and B. Maraviglia, *Solid State Commun.* **70** (1989) 797.

Del4 F. De Luca, B.C. De Simone, N. Lugeri, B. Maraviglia and C. Nuccetelli, *J. Magn. Reson.* **90** (1990) 124.

Del5 F. De Luca, P. Fattibene, N. Lugeri, R. Campanella and B. Maraviglia, *Appl. Magn. Reson.* **2** (1991) 93.

Del6 F. De Luca, N. Lugeri, B.C. De Simone and B. Maraviglia, *Mag. Reson. Imaging* **10** (1992) 765.

Del7 F. De Luca, N. Lugeri, B.C. De Simone and B. Maraviglia, *Solid State Commun.* **82** (1992) 151.

Del8 F. De Luca, B.C. De Simone, N. Lugeri and B. Maraviglia, *J. Magn. Reson. A* **102**, (1993) 287.

Del9 F. De Luca, N. Lugeri, S. Motta, G. Cammisa and B. Maraviglia, *J. Magn. Reson. A* **115** (1995) 1.

Del10 F. De Luca, G.H. Raza, A. Gargaro and B. Maraviglia, *J. Magn. Reson.* **126** (1997) 159.

Del11 F. De Luca, S. Motta, N. Lugeri and B. Maraviglia, *J. Magn. Reson. A* **21** (1996) 114.

Dem1 D.E. Demco, S. Hafner and R. Kimmich, *J. Magn. Reson.* **94** (1991) 333.

Dem2 D.E. Demco, S. Hafner and R. Kimmich, *J. Magn. Reson.* **96** (1992) 307.

Dem3 D.E. Demco, S. Hafner and H.W. Spiess, *J. Magn. Reson.* **116** (1995) 36.

Dem4 D.E. Demco, *Phys. Lett.* **45A** (1973) 113.

Die1 S.L. Dieckman, N. Gopalsami and R.E. Botto, *Energy and Fuels* **4** (1990) 417.

Dix1 W.T. Dixon, *J. Chem. Phys.* **77** (1982) 1800.

Emi1 S. Emid, *Physica B* **128** (1985) 79.

Emi2 S. Emid and J.H.N. Creyghton, *Physica B* **128** (1985) 81.

Ern1 R.R. Ernst, G. Bodenhausen and A. Wokaun, *Principles of Nuclear Magnetic Resonance in One and Two Dimensions*, Clarendon Press, Oxford, 1987.

Fil1 C. Filip, S. Hafner, I. Schnell, D.E. Demco and H.W. Spiess, *J. Chem. Phys.* **110** (1999) 423.

Gar1 A.N. Garroway, P. Mansfield and D.C. Stalker, *Phys. Rev.* **11** (1975).
Gar2 A.N. Garroway, J. Baum, M.G. Munowitz and A. Pines, *J. Magn. Reson.* **60** (1984) 337.
Gar3 A.N. Garroway, in: *Encyclopedia of NMR*, ed. D.M. Grant and R.K. Harris, Wiley, New York, 1996, p. 3692.
Ger1 B.C. Gerstein and C.R. Dybowski, *Transient Techniques in NMR of Solids*, Academic Press, Orlando, 1985.
Glo1 P.M. Glover, P.J. McDonald and B. Newling, *J. Magn. Reson.* **126** (1997) 207.
Got1 J. Gottwald, D.E. Demco, R. Graf and H.W. Spiess, *Chem. Phys. Lett.* **243** (1996) 314.
Gra1 S. Gravina and D.G. Cory, *J. Magn. Reson. B* **104** (1994) 53.
Gün1 E. Günther, B. Blümich and H.W. Spiess, *Macromolecules* **25** (1992) 3315.
Gün2 E. Günther, B. Blümich and H.W. Spiess, *Mol. Phys.* **71** (1990) 477.
Gün3 E. Günther, B. Blümich and H.W. Spiess, *Chem. Phys. Lett.* **184** (1991) 251.
Gün4 E. Günther, H. Raich and B. Blümich, *Bruker Report* (1991/1992) p. 15.
Hae1 U. Haeberlen, *High Resolution NMR in Solids: Selective Averaging, Adv. Magn. Reson. Suppl. 1*, Academic Press, New York, 1976.
Haf1 S. Hafner, D.E. Demco and R. Kimmich, *Meas. Sci. Technol.* **2** (1991) 882.
Haf2 S. Hafner and H.W. Spiess, *Concepts Magn. Reson.* **10** (1998) 99.
Haf3 S. Hafner, D.E. Demco and R. Kimmich, *Solid-State Nucl. Magn. Reson.* **6** (1996) 275.
Haf4 S. Hafner, P. Barth and W. Kuhn, *J. Magn. Reson. A* **108** (1994) 21.
Haf5 S. Hafner and P. Barth, *Magn. Reson. Imag.* **13** (1995) 441.
Har1 G.S. Harbison and H.W. Spiess, *Chem. Phys. Lett.* **124** (1986) 126.
Hep1 M.A. Hepp and J.B. Miller, *Solid-State Nucl. Magn. Reson.* **6** (1996) 367.
Hep2 M.A. Hepp and J.B. Miller, *J. Magn. Reson. A* **111** (1994) 62.
Hug1 P.D.M Hughes, P.J. McDonald, N.P. Rhodes, J.W. Rockliffe, E.G. Smith and J. Wills, *J. Coll. Interf. Sci.* **177** (1996) 208.
Iwa1 J.H. Iwamiya and S.W. Sinton, *Solid-State Nucl. Magn. Reson.* **6** (1996) 333.
Jan1 J. Jansen and B. Blümich, *J. Magn. Reson.* **99** (1992) 525.
Jez1 P. Jezzard, J.J. Attard, T.A. Carpenter and L.D. Hall, *Progr. NMR Spectr.* **23** (1991) 1.
Jez2 P. Jezzard, C.J. Wiggins, T.A. Carpenter, L.D. Hall, P. Jackson and N.J. Clayden, *Advanced Materials* **4** (1992) 82.
Ken1 C.B. Kennedy, B.J. Balcom and I.V. Mastikhin, *Can. J. Chem.* **76** (1998) 11.
Kim1 R. Kimmich, D.E. Demco and S. Hafner, in: *Magnetic Resonance Microscopy*, ed. B. Blümich and W. Kuhn, VCH, Weinheim, 1991, p. 59.
Kin1 P. Kinchesh, E.W. Randall and K. Zick, *J. Magn. Reson.* **100** (1992) 411.
Kin2 P. Kinchesh, E.W. Randall and K. Zick, *Magn. Reson. Imaging* **12** (1994) 305.
Kli1 M. Klinkenberg, P. Blümler and B. Blümich, *J. Magn. Reson. A* **119** (1996) 197.
Kli2 M. Klinkenberg, P. Blümler and B. Blümich, *Macromolecules* **30** (1997) 1038.
Knu1 T. Knubovets, H. Shinar, E. Eliav and G. Navon, *J. Magn. Reson. B* **110** (1996) 16.
Lau1 P.C. Lauterbur, *Bull. Am. Phys. Soc.* **18** (1972) 86.
Lau2 P.C. Lauterbur, *Nature* **242** (1973) 190.
Lau3 P.C. Lauterbur, *Pure Appl. Chem.* **40** (1974) 149.
Lee1 M. Lee and W.I. Goldburg, *Phys. Rev. A* **140** (1965) 1261.
Lug1 N. Lugeri, F. De Luca and B. Maraviglia, in: *Magnetic Resonance Microscopy*, ed. B. Blümich and W. Kuhn, VCH, Weinheim, 1992, p. 109.
Mal1 M.J.D. Mallett, S.L. Codd, M.R. Halse, T.A.P. Green and J.H. Strange, *J. Magn. Reson. A* **119** (1996) 105.
Mal2 M.J.D. Mallett and J.H. Strange, *Appl. Magn. Reson.* **12** (1997) 193.
Mal3 M.J.D. Mallet, M.R. Halse and J.H. Strange, *J. Magn. Reson.* **132** (1998) 172.
Mal4 C. Malveau, B. Diter, P. Tekely and C. Canet, *J. Magn. Reson.* **134** (1998) 171.

Man1 P. Mansfield and P.K. Grannell, *J. Phys. C: Solid State Phys.* **6** (1973) L422.

Man2 P. Mansfield and P.K. Grannell, *Phys. Rev. B* **12** (1975) 3618.

Man3 P. Mansfied and D. Ware, *Phys. Lett.* **22** (1966) 133.

Man4 P. Mansfield, E.L. Hahn, ed., *NMR Imaging*, The Royal Society, London, 1990.

Mar1 M.M. Maricq and J.S. Waugh, *J. Chem. Phys.* **70** (1979) 3300.

Mar2 B. Maraviglia, F. de Luca, B.C. de Simone and N. Lugeri, in: *Encyclopedia of NMR*, ed. D.M. Grant and R.K. Harris, Wiley, New York, 1996, p. 2715.

Mas1 I.V. Mastikhin, B.J. Balcom, P.J. Prado and C.B. Kennedy, *J. Magn. Reson.* **136** (1999) 159.

Mat1 S. Matsui, *Chem. Phys. Lett.* **179** (1991) 187.

Mat2 S. Matsui, *J. Magn. Reson.* **95** (1991) 149.

Mat3 S. Matsui, Y. Ogasawara and T. Inouye, *J. Magn. Reson. A* **105** (1993) 215.

Mat4 S. Matsui, *J. Magn Reson.* **98** (1992) 618.

Mat5 S. Matsui, A. Uraoka and T. Inouye, *J. Magn. Reson. A* **112** (1995) 130.

Mat6 S. Matsui, A. Uraoka and T. Inouye, *J. Magn. Reson. A* **120** (1996) 11.

Mat7 S. Matsui, M. Nonaka, T. Nakai and T. Inouye, *J. Magn. Reson.* **138** (1999) 220.

Mcd1 P.J. McDonald, *Progr. NMR Spectrosc.* **30** (1997) 69.

Mcd2 P.J. McDonald, J.J. Attard and D.G. Taylor, *J. Magn. Reson.* **72** (1987) 224.

Mcd3 P.J. McDonald and P.F. Tokarczuk, *J. Phys. E: Sci. Instrum.* **22** (1989) 948.

Mcd4 P.J. McDonald and P.F. Tokarczuk, *J. Magn. Reson.* **99** (1992) 225.

Mcd5 P.J. McDonald and A.R. Lonergan, *Physica B* **176** (1992) 173.

Mef1 A.E. Mefed and V.A. Atsarkin, *Sov. Phys. JETP* **47** (1978) 378.

Meh1 M. Mehring, *Principles of High Resolution NMR in Solids*, Springer, Berlin, 1983.

Mei1 S. Meiboom and D. Gill, *Rev. Sci. Instrum.* **29** (1958) 688.

Mil1 J.B. Miller, D.G. Cory and A.N. Garroway, *Phil. Trans. R. Soc. Lond. A* **333** (1990) 413.

Mil2 J.B. Miller and A.N. Garroway, *J. Magn. Reson.* **82** (1989) 529.

Mil3 J.B. Miller and A.N. Garroway, *J. Magn. Reson.* **67** (1986) 575.

Mil4 J.B. Miller, D.C. Cory and A.N. Garroway, *Chem. Phys. Lett.* **164** (1989) 1.

Mil5 J. Miller, *3rd International Conference on Magnetic Resonance Microscopy*, 27–31 August 1995, Würzburg, Germany.

Mil6 J.B. Miller, *Progr. Nucl. Magn. Reson. Spectrosc.* **33** (1998) 273.

Mil7 J.B. Miller and A.N. Garroway, *J. Magn. Reson.* **85** (1989) 255.

Mun1 M. Munowitz and A. Pines, *Adv. Chem. Phys.* **66** (1987) 2.

Nun1 T. Nunes, E.W. Randall, A.A. Samoilenko, P. Bodart and G. Feio, *J. Phys. D: Appl. Phys.* **29** (1996) 805.

Ost1 E.D. Ostroff and J.S. Waugh, *Phys. Rev. Lett.* **7** (1966) 1097.

Per1 K.L. Perry, P.J. McDonald, E.W. Randall and K. Zick, *Polymer* **35** (1994) 517.

Pin1 A. Pines and J.S. Waugh, *J. Magn. Reson.* **77** (1972) 187.

Pin2 A. Pines, W.-K. Rhim and J.S. Waugh, *J. Magn. Reson.* **6** (1972) 457.

Pra1 P.J. Prado, B.J. Balcom and M. Jama, *J. Magn. Reson.* **137** (1999) 59.

Pra2 P.J. Prado, B.J. Balcom, S.D. Beyea, R.L. Armstong and T.W. Bremner, *Solid-State Nucl. Magn. Reson.* **10** (1997) 1.

Pra3 P.J. Prado, B.J. Balcom, I.V. Mastikhin, A.R. Cross, R.L. Armstrong and A. Logan, *J. Magn. Reson.* **137** (1999) 324.

Pra4 P.J. Prado, B.J. Balcom, S.D. Beyea, T.W. Bremner, R.L. Armstrong, R. Pishe and P.E. Gratten-Bellew, *J. Phys. D: Appl. Phys.* **31** (1998) 2040.

Pra5 P.J. Prado, L. Gasper, G. Fink, B. Blümich, V. Herrmann, K. Unseld, H.-B. Fuchs, H. Möhler and M. Rühl, *Macromol. Materials and Engineering* **274** (2000) 13.

Ran1 E.W. Randall, A.A. Samoilenko and T. Nunes, *J. Magn. Reson. A* **117** (1995) 317.
Ran2 E.W. Randall, A.A. Samoilenko and T. Nunes, *J. Magn. Reson. A* **116** (1995) 122.
Ran3 E.W. Randall, *Solid State Nucl. Magn. Reson.* **8** (1997) 173.
Ran4 E.W. Randall and D.G. Gillies, *J. Magn. Reson. A* **121** (1996) 217.
Rhi1 W.-K. Rhim, A. Pines and J.S. Waugh, *Phys. Rev. B* **3** (1971) 684.
Rom1 E. Rommel, S. Hafner and R. Kimmich, *J. Magn. Reson.* **86** (1990) 264.
Sam1 A.A. Samoilenko, D.Y. Artemov and L.A. Sibeldina, *JETP Lett.* **47** (1988) 417.
Sam2 A.A. Samoilenko and K. Zick, *Bruker Rep.* **1** (1990) 40.
Sch1 K. Schmidt-Rohr and H.W. Spiess, *Multidimensional Solid-State NMR and Polymers*, Academic Press, London, 1994.
Sch2 G. Schauss, *NMR-Bildgebung an Festkörpern mit MAS-Imaging*, Dissertation, Johannes-Gutenberg- Universität, Mainz, 1992.
Sch3 G. Schauss, B. Blümich and H.W. Spiess, *J. Magn. Reson.* **95** (1991) 437.
Sch4 U. Scheler, B. Blümich and H.W. Spiess, *Solid-State Nucl. Magn. Reson.* **2** (1993) 105.
Sch5 U. Scheler, J.J. Titman, B. Blümich and H.W. Spiess, *J. Magn. Reson. A* **107** (1994) 251.
Sch6 U. Scheler, G. Schauß, B. Blümich and H.W. Spiess, *Solid-State Nucl. Magn. Reson.* **6** (1996) 375.
Sch7 G. Scheler, U. Haubenreisser and H. Rosenberger, *J. Magn. Reson.* **44** (1981) 134.
Sch8 H. Schneider and H. Schmiedel, *Phys. Lett.* **30A** (1969) 298.
Sha1 Y. Sharf, U. Eliav, H. Shinar and G. Navon, *J. Magn. Reson. B* **107** (1995) 60.
Sin1 S.W. Sinton, J.H. Iwamiya, B. Ewing and G.P. Drobny, *Spectroscopy* **6** (1991) 42.
Sli1 C.P. Slichter, *Principles of Magnetic Resonance*, 3rd edn, Springer, Berlin, 1990.
Spi1 H.W. Spiess, *Chem. Rev.* **91** (1991) 1321.
Spi2 H.W. Spiess, *NMR – Basic Principles and Progress* **15** (1978) 55.
Spi3 H.W. Spiess, *J. Chem. Phys.* **72** (1980) 6755.
Sta1 J.M. Star-Lack, M.S. Roos, S.T.S. Wong, V.D. Schrepkin and T.F. Budinger, *J. Magn. Reson.* **124** (1997) 420.
Str1 J.H. Strange, *Phil. Trans. R. Soc. Lond. A* **333** (1990) 427.
Sun1 Y. Sun, J. Xiong, H. Lock, M.L. Buszko, J.A. Haase and G.E. Maciel, *J. Magn. Reson. A* **110** (1994) 1.
Sun2 Y. Sun, H. Lock, T. Shinozaki and G.E. Maciel, *J. Magn. Reson. A* **115** (1995) 165.
Sze1 N.M. Szeverenyi and G.E. Maciel, *J. Magn. Reson.* **60** (1984) 460.
Tak1 K. Takegoshi and C.A. McDowell, *Chem. Phys. Lett.* **116**, (1985) 100.
Tay1 R.E. Taylor, R.G. Pembleton, L.M. Ryan and B.C. Gerstein, *J. Chem. Phys.* **71** (1979) 4541.
Vee1 W.S. Veeman and D.G. Cory, *Adv. Magn. Reson.* **13** (1989) 43.
Wei1 F. Weigand, B. Blümich and H.W. Spiess, *Solid-State Nucl. Magn. Reson.* **3** (1994) 59.
Wei2 F. Weigand, D.E. Demco, B. Blümich and H.W. Spiess, *Solid-State Nucl. Magn. Reson.* **6** (1996) 357.
Win1 R.A. Wind and C.S. Yannoni, US patent 4,301,410, 17 November 1981.

CHAPTER 9

Ack1 J.J. Ackerman, T.H. Grove, G.G. Wong, D.G. Gadian and G.K. Radda, *Nature* **283** (1980) 167.
Ack2 J.J.H. Ackerman, *Concepts Magn. Reson.* **2** (1990) 33.
Ako1 A. Akoka, in: *Magnetic Resonance Spectroscopy in Biology and Medicine*, ed. J.D. Certaines, W.M.M.J. Bovée and F. Podo, Pergamon Press, Oxford, 1992.

Alb1 K. Albert, *J. Chromatography A* **703** (1995) 123.

Alb2 K. Albert, M. Kunst, E. Bayer, M. Spraul and W. Bermel, *J. Chromatography* **463** (1989) 355.

Alb3 K. Albert, U. Braumann, L.-H. Tseng, G. Nicholson, E. Bayer, M. Spraul, M. Hofmann, C. Dowle, M. Chippendale, *Anal. Chem.* **66** (1994) 3042.

Alb4 K. Albert, E.-L. Dreher, H. Straub and A. Rieker, *Magn. Reson. Chem.* **25** (1987) 919.

Alo1 J. Alonso, C. Arus, W.M. Westler and J.L. Markley, *Magn. Reson. Med.* **11** (1989) 316.

Arg1 D.S. Argyropoulos, F.G. Morin and L. Lapcik, *Holzforschung* **49** (1995) 115.

Arm1 A. De Los Santos and J.D. King, *Industrial Applications of NMR*, in: *4th Int. Conf. Magn. Reson. Microscopy* ed. E. Fukushima, Albuquerque, 21–25 September 1997.

Ash1 H. Van As, J.E.A. Reinders, P.A. de Jager, P.A.C.M. van de Sanden and T.J. Schaafsma, *J. Exp. Botany* **45** (1994) 61.

Ass1 R.A. Assink, E. Fukushima, A.A.V. Gibson, A.R.R. Rath and S.B.W. Roeder, *J. Magn. Reson.* **66** (1986) 176.

Aue1 W.P. Aue, *Rev. Magn. Reson. Med.* **1** (1986) 21.

Aue2 W.P. Aue, S. Müller, T.A. Cross and J. Seelig, *J. Magn. Reson.* **56** (1984) 350.

Aue3 W.P. Aue, S. Müller and J. Seelig, *J. Magn. Reson.* **61** (1985) 392.

Aug1 M.P. Augustine, D.M. TonThat and J. Clarke, *Solid-State Nucl. Magn. Reson.* **11** (1998) 139.

Bal1 R.S. Balaban and J.A. Ferretti, *Proc. Natl. Acad. Sci. USA* **80** (1983) 1241.

Bal2 B. Balinov, O. Söderman and T. Wärnheim, *J. Am. Chem. Soc.* **71** (1994) 513.

Ban1 S. Bank, *Concepts Magn. Reson.* **9** (1997) 83.

Bar1 P.J. Barker and H.J. Stronk, in: *NMR Applications of Biopolymers*, ed. J.W. Finley, Plenum Press, New York, 1990.

Bar2 P.J. Barker, in: *Rapid Methods for Analysis of Food and Food Raw Materials*, ed. W. Baltes, Behr's Verlag, Hamburg, 1990.

Bau1 J. Baum, R. Tycko and A. Pines, *Chem. Phys.* **105** (1986) 7.

Bay1 E. Bayer, K. Albert, M. Nieder, E. Grom and T. Keller, *Adv. Chromatogr.* **14** (1979) 525.

Bec1 E.D. Becker, J.A. Ferretti and P.N. Gambhir, *Anal. Chem.* **51** (1979) 1413.

Bec2 N. Beckmann and S. Müller, *J. Magn. Reson.* **93** (1991) 299.

Beh1 B. Behnke, G. Schlotterbeck, U. Tallarek, S. Strohschein, L.-H. Tseng, T. Keller, K. Albert and E. Bayer, *Anal. Chem.* **68** (1996) 1110.

Ben1 G.-J. Béné, B. Borcard, E. Hiltbrand and P. Magnin, in: *NMR in Medicine, NMR – Basic Principles and Progress* ed. R. Damadian, Vol. 19, Springer, Berlin, 1981, p. 81.

Ben2 M.R. Bendall, *Bull. Magn. Reson.* **8** (1986) 17.

Ben3 M.R. Bendall, *J. Magn. Reson.* **59** (1984) 406.

Ben4 M.R. Bendall, *Chem. Phys. Lett.* **99** (1983) 310.

Ben5 M.R. Bendall and R.E. Gordon, *J. Magn. Reson.* **53** (1983) 365.

Ben6 M.R. Bendall and D.T. Pegg, *Magn. Reson. Med.* **2** (1985) 91.

Ben7 M.R. Bendall, A. Connelly and J.M. McKendry, *Magn. Reson. Med.* **3** (1986) 157.

Ben8 M.R. Bendall and D.T. Pegg, *J. Magn. Reson.* **57** (1984) 337.

Ben9 T.B. Benson and P.J. McDonald, *J. Magn. Reson. A* **112** (1995) 17.

Ben10 T.B. Benson and P.J. McDonald, *J. Magn. Reson. A* **109** (1995) 314.

Ber1 B.A. Berkowitz, S.D. Wolff and R.S. Balaban, *J. Magn. Reson.* **79** (1988) 547.

Ber2 K. Bergmann and K. Demmler, *Colloid and Polymer Science* **252** (1974) 193.

Bla1 S.J. Blackband, K.A. McGovern and I.J. McLennan, *J. Magn. Reson.* **79** (1988) 184.

Bli1 R. Binc, J. Dolinsek, G. Lahajnar, A. Sepe, I. Zupancic, S. Zumer, F. Milia and M.M. Pintar, *Z. Naturforschung* **43a** (1988) 1026.

Blü1 B. Blümich and P. Blümler, *Macromol. Symp.* **87** (1994) 187.

Blü2 B. Blümich, C. Fülber, F. Weigand, H.W. Spiess, in: *Magn. Reson. Res. Phen., Proc. XXVII Congr. Ampere*, ed. K. Salikhov, Kazan, 1994, p. 111.

Blü3 B. Blümich, P. Blümler, G. Eidmann, A. Guthausen, R. Haken, U. Schmitz, K. Saito and G. Zimmer, *Magn. Reson. Imag.* **16** (1998) 479.

Blü4 B. Blümich and W. Kuhn., ed., *Magnetic Resonance Microscopy*, VCH, Weinheim, 1992.

Blü5 P. Blümler, B. Blümich, R. Botto and E. Fukushima, ed., *Spatially Resolved Magnetic Resonance*, Wiley-VCH, Weinheim, 1998.

Blü6 B. Blümich, P. Blümler, L. Gasper, A. Guthausen, V. Göbbels, S. Laukemper-Ostendorf, K. Unseld and G. Zimmer, *Macromol. Symp.* **141** (1999) 83.

Blü7 P. Blümler, B. Blümich, R. Botto and E. Fukushima, ed., *Spatially Resolved Magnetic Resonance*, Wiley-VCH, Weinheim, 1998.

Bol1 L. Bolinger and J.S. Leigh, *J. Magn. Reson.* **80** (1988) 162.

Bos1 C.S. Bosch and J.J.H. Ackerman, *NMR – Basic Principles and Progress* **27** (1992) 3.

Bos2 C.S. Bosch and J.J.H. Ackerman, in: *Encyclopedia of NMR*, ed. D.M. Grant and R.K. Harris, Wiley, New York, 1996, p. 4648.

Bot1 P.A. Bottomley, *NMR – Basic Principles and Progress* **27** (1992) 67.

Bot2 P.A. Bottomley, T.B. Foster and R.D. Darrow, *J. Magn. Reson.* **59** (1984) 338.

Bot3 D. Le Botlan and I. Helie-Fourel, *Analytica Chimica Acta* **311** (1995) 217.

Bou1 D. Bourgeois, M. Decorps, C. Remy and A.L. Benabid, *Magn. Reson.* **11** (1989) 275.

Bov1 F.A. Bovee, L.W. Jelinski and P.A. Mirau, *Nuclear Magnetic Resonance Specotroscopy*, Academic Press, New York, 1988.

Bri1 J. Briand and L.D. Hall, *J. Magn. Reson.* **80** (1988) 559.

Bri2 J. Briand and L.D. Hall, *J. Magn. Reson.* **82** (1989) 180.

Bri3 J. Briand and L.D. Hall, *J. Magn. Reson.* **94** (1991) 234.

Bro1 R.J.S. Brown, *Magn. Reson. Imag.* **14** (1996) 811.

Bru1 H. Bruhn, J. Frahm, M.L. Gyngell, K.D. Merboldt, W. Hänicke and R. Sauter, *Magn. Reson. Med.* **9** (1989) 126.

Bru2 Bruker Analytik GmbH, Silberstreifen, D-76287 Rheinstetten, Germany.

Bru3 Minispec application notes, Bruker Analytik GmbH, Silberstreifen, D-76287 Rheinstetten, Germany.

Bry1 D.J. Bryant and G.A. Coutts, in: *Encyclopedia of NMR Spectroscopy*, ed. D.M. Grant and R.K. Harris, Wiley, New York, 1996, p. 4492.

Bue1 M.L. Buess, A.N. Garroway, J.B. Miller and J.P. Yesinowski, in: *Advances in Analysis and Detection of Explosives*, ed. J. Yinon, Cluwer, Dordrecht, 1993, p. 361.

Bur1 L.J. Burnett, *Several NDE Applications of NMR*, in: *4th Int. Conf. Magn. Reson. Microscopy*, ed. E. Fukushima, Albuquerque, September 21–25, 1997.

Bur2 L.J. Burnett and J.A. Jackson, *J. Magn. Reson.* **41** (1980) 406.

Cal1 P.T. Callaghan, C.D. Eccles, T.G. Haskell, P.J. Langhorne and J.D. Seymour, *J. Magn. Reson.* **133** (1998) 148.

Cal2 P.T. Callaghan, *Principles of Nuclear Magnetic Resonance Microscopy*, Clarendon Press, Oxford, 1991.

Cal3 P.T. Callaghan, C.D. Eccles and J.D. Seymour, *Rev. Sci. Instrum.* **68** (1997) 4263.

Cal4 P.T. Callaghan and M. LeGros, *Am. J. Phys.* **50** (1982) 709.

Cal5 P.T. Callaghan and C. Eccles, *Bull. Magn. Reson.* **18** (1996) 62.

Cap1 S. Capuani, F. De Luca, L. Marinelli and B. Maraviglia, *J. Magn. Reson. A* **121** (1996) 1.

Cap2 F. Capozzi, M.A. Cremonini, C. Luchinat, G. Placucci and C. Vignali, *J. Magn. Reson.* **138** (1999) 277.

Cer1 J.D. de Certaines, W.M.M.J. Bovée and F. Podo, ed., *Magnetic Resonance Spectroscopy in Biology and Medicine*, Pergamon Press, New York, 1992.

Che1 W. Chen and J.J.H. Ackerman, *J. Magn. Reson.* **82** (1989) 655.

Che2 W. Chen and J.J.H. Ackerman, *NMR in Biomedicine* **3** (1990) 147.

Che3 W. Chen and J.J.H. Ackerman, *NMR in Biomedicine* **3** (1990) 158.

Che4 W. Chen and J.J.H. Ackerman, *J. Magn. Reson.* **98** (1992) 238.

Che5 C.-N. Chen, D.I. Hoult and V.J. Sank, *J. Magn. Reson.* **54** (1983) 324.

Cho1 Z.H. Cho and J.H. Yi, *Concepts Magn. Reson.* **7** (1995) 95.

Cho2 Z.H. Cho and J.H. Yi, *J. Magn. Reson.* **94** (1991) 471.

Coh1 Y. Cohen, L.-H. Chang, L. Litt and T.L. James, *J. Magn. Reson.* **85** (1989) 203.

Coo1 R.K. Cooper and J.A. Jackson, *J. Magn. Reson.* **41** (1980) 400.

Cor1 D.G. Cory, J.B. Miller and A.N. Garroway, *Meas. Sci. Technol.* **1** (1990) 1338.

Cox1 S.J. Cox and P. Styles, *J. Magn. Reson.* **40** (1980) 209.

Dec1 M. Decorps, M. Laval, S. Confort and J.-J. Chaillout, *J. Magn. Reson.* **61** (1985) 418.

Dem1 K. Demmler, K. Bergmann and E. Schuch, *Kunstst.* **62** (1972) 845.

Dod1 D.M. Doddrell, J.M. Bulsing, G.J. Galloway, W.M. Brooks, J. Field, M. Irwing and H. Baddeley, *J. Magn. Reson.* **70** (1986) 319.

Dod2 D.M. Doddrell, W.M. Brooks, J.M. Bulsing, J. Field, M.G. Irwing and H. Baddeley, *J. Magn. Reson.* **68** (1986) 367.

Dya1 T. Dyakowski, *Meas. Sci. Technol.* **7** (1996) 343.

Eid1 G. Eidmann, R. Savelsberg, P. Blümler and B. Blümich, *J. Magn. Reson. A* **122** (1996) 104.

Eli1 H.-G. Elias, *Makromoleküle*, 5th edn, Vol. 1, Hütig & Wepf, Basel, 1990.

Ems1 L. Emsley and G. Bodenhausen, *J. Magn. Reson.* **87** (1990) 1.

End1 J.C. van den Enden, D. Waddington, H. van Aalst, C.G. van Kralingen and K.J. Packer, *J. Colloid Interface Science* **140** (1990) 105.

Ern1 R.R. Ernst, G. Bodenhausen and A. Wokaun, *Principles of Nuclear Magnetic Resonance in One and Two Dimensions*, Clarendon Press, Oxford, 1987.

Eva1 S.D. Evans, K.P. Nott, A. Kshirsagar and L.D. Hall, *Magn. Reson. Chem.* **35** (1997) S76.

Eve1 J.L. Evelhoch, M.G. Crowley and J.J.H. Ackerman, *J. Magn. Reson.* **56** (1984) 110.

Eve2 J.L. Evelhoch and J.J.H. Ackerman, *J. Magn. Reson.* **53** (1983) 52.

Fou1 I. Fourel, J.P. Guillement and D. Le Botlan, *J. Colloid Interface Sci.* **164** (1994) 48.

Fox1 Foxboro, 33 Commercial Street, Bristol Park (B51-2C), Foxboro, MA 02035, USA.

Fra1 J. Frahm, K.D. Merboldt, W. Hänicke and A. Haase, *J. Magn. Reson.* **64** (1985) 81.

Fra2 J. Frahm, K.D. Merboldt and W. Hänicke, *J. Magn. Reson.* **72** (1987) 502.

Fra3 J. Frahm, H. Bruhn, M.L. Gyngell, K.D. Merboldt, W. Hänicke and R. Sauter, *Magn. Reson. Med.* **9** (1989) 79.

Fra4 J. Frahm, T. Michaelis, K.D. Merboldt, H. Bruhn, M.L. Gyngell and W. Hänicke, *J. Magn. Reson.* **90** (1990) 464.

Fro1 W. Froncisz, A. Jesmanowicz, J.B. Kneeland and J.S. Hyde, *Magn. Reson. Med.* **4** (1987) 179.

Ful1 G.D. Fullerton, I.L. Cameron and V.A. Ord, *Radiology* **155** (1985) 433.

Gar1 M. Garwood and K. Ugurbil, *NMR – Basic Principles and Progress* **26** (1992) 109.

Gar2 M. Garwood, T. Schleich, B.D. Ross, G.B. Matson and W.D. Winters, *J. Magn. Reson.* **65** (1985) 239.

Gar3 M. Garwood, T. Schleich, M.R. Bendall and D.T. Pegg, *J. Magn. Reson.* **65** (1985) 510.

Gar4 M. Garwood and Y. Ke, *J. Magn. Reson.* **94** (1991) 511.

Gar5 A.N. Garroway, M.L. Buess, J.P. Jesinowski and J.B. Miller, *Substance Detection Systems* **2092** (1993) 318, Society of Photo-Optical Instrumentation Engineers, Bellingham, USA.
Gar6 A.N. Garroway, M.L. Buess, J.P. Jesinowski, J.B. Miller and R.A. Krauss, *Cargo Detection Technologies* **2276** (1994) 139, Society of Photo-Optical Instrumentation Engineers, Bellingham, USA.
Ger1 I.P. Gerothanassis, *Progr. NMR Spectrosc.* **19** (1987) 267.
Gib1 S.J. Gibbs and L.D. Hall, *Meas. Sci. Technol.* **7** (1996) 827.
Gla1 L.F. Gladden, *Chem. Eng. J. Biochem. Eng. J.* **56** (1994) 149.
Gla2 L.F. Gladden, *Chem. Eng. Sci.* **49** (1995) 3339.
Gla3 L.F. Gladden, *The Chem. Eng. J.* **56** (1995) 149.
Glo1 P.M. Glover, P.J. McDonald and B. Newling, *J. Magn. Reson.* **126** (1997) 207.
Goe1 G. Goelman, V.H. Subramanian and J.S. Leigh, *J. Magn. Reson.* **89** (1990) 437.
Goe2 G. Goelman and J.S. Leigh, *J. Magn. Reson.* **91** (1991) 93.
Goe3 G. Goelman and M.D. Prammer, *J. Magn. Reson. A* **113** (1995) 11.
Goe4 G. Goelman, *J. Magn. Reson. B* **104** (1994) 212.
Gol1 M. Goldman, B. Rabinovich, M. Rabinovich, D. Gilad, I. Gev and M. Schirov, *J. Appl. Geophys.* **31** (1994) 27.
Gon1 N.C. Gonella and R.F. Silverman, *J. Magn. Reson.* **85** (1989) 24.
Gra1 J. Granot, *J. Magn. Reson.* **70** (1986) 488.
Gra2 R.A. de Graaf, Y. Luo, M. Terpstra, H. Merkle and M. Garwood, *J. Magn. Reson.* **95** (1995) 245.
Gra3 R.A. de Graaf, K. Nicolay and M. Garwood, *Magn. Reson. Med.* **35** (1996) 652.
Gra4 R.A. de Graaf and K. Nicolay, *J. Magn. Reson. B* **113** (1996) 97.
Gra5 R.A. de Graaf, Y. Luo, M. Garwood and K. Nicolay, *J. Magn. Reson. B* **113** (1996) 35.
Gre1 V.S. Grechishkin and N.Ya. Sinyavskii, *Physics – Uspekhi* **40** (1997) 393.
Gre2 V.S. Grechishkin, *Applied Physics A* **55** (1992) 505.
Gre3 V.S. Grechishkin and A.A. Shpilevoi, *Physics – Uspheki* **39** (1996) 725.
Gre4 V.S. Grechishkin, *Applied Physics A* **58** (1994) 63.
Gri1 T.M. Grist, A. Jesmanowicz, J.B. Kneeland, F. Froncisz and J.S. Hyde, *Magn. Reson. Med.* **6** (1988) 253.
Gri2 D.D. Griffin, R.L. Kleinberg and M. Fukuhara, *Meas. Sci. Technol.* **4** (1993) 968.
Gut1 A. Guthausen, G. Zimmer, P. Blümler and B. Blümich, *J. Magn. Reson.* **30** (1998) 1.
Gut2 A. Guthausen, G. Zimmer, R. Eymael, U. Schmitz, P. Blümler and B. Blümich, in: *Spatially Resolved Magnetic Resonance*, ed. P. Blümler, B. Blümich, R. Botto and E. Fukushima, Wiley-VCH, Weinheim, 1998, p. 195.
Gut3 A. Guthausen, Die NMR-MOUSE. Methoden und Anwendungen zur Charakterisierung von Polymeren, Dissertation, RWTH, Aachen, 1998.
Gut4 A. Guthausen, G. Zimmer, S. Laukemper-Ostendorf, P. Blümler and B. Blümich, *Chemie in unserer Zeit* **32** (1998) 73.
Gyn1 M.L. Gyngell, J. Frahm, K.D. Merboldt, W. Hänicke and H. Bruhn, *J. Magn. Reson.* **77** (1988) 596.
Haa1 A. Haase, W. Hänicke and J. Frahm, *J. Magn. Reson.* **56** (1984) 401.
Haa2 A. Haase, C. Malloy and G.K. Radda, *J. Magn. Reson.* **55** (1983) 164.
Haa3 A. Haase, *J. Magn. Reson.* **61** (1985) 130.
Haf1 H.-P. Hafner, S. Müller and J. Seelig, *Magn. Reson. Med.* **13** (1990) 279.
Haf2 S. Hafner, E. Rommel and R. Kimmich, *J. Magn. Reson.* **88** (1990) 449.
Hah1 E.L. Hahn, *Phys. Rev.* **80** (1950) 580.
Hai1 T. Haishi and K. Kose, *J. Magn. Reson.* **134** (1998) 138.

Hak1 R. Haken, P. Blümler and B. Blümich, in: *Spatially Resolved Magnetic Resonance*, ed. P. Blümler, B. Blümich, R. Botto and E. Fukushima, Wiley-VCH, Weinheim, 1998, p. 695.

Hal1 L.D. Hall and T.A. Carpenter, *Magn. Reson. Imag.* **10** (1992) 713.

Har1 M.D. Harpen, *Med. Phys.* **14** (1987) 616.

Hau1 K.H. Hausser and H.R. Kalbitzer, *NMR in Medicine and Biology*, Springer, Berlin, 1991.

Hay1 C.E. Hayes and L. Axel, *Med. Phys.* **12** (1985) 604.

Hen1 J. Hennig, *J. Magn. Reson.* **96** (1992) 40.

Hen2 D. Hentschel, R. Ladebeck, R. Wittig, P. Schüler and G. Schuierer, *Appl. Magn. Reson.* **1** (1990) 379.

Hen3 K. Hendrich, H. Merkle, S. Weisdorf, W. Vine, M. Garwood and K. Ugurbil, *J. Magn. Reson.* **92** (1991) 258.

Hen4 K. Hendrich, H. Liu, H. Merkle, Y. Zhang and K. Ugurbil, *J. Magn. Reson.* **97** (1992) 486.

Her1 C.V. Hernandez and D.N. Rutledge, *Food Chem.* **49** (1994) 83.

Hie1 K. Hietalahti, A. Root, M. Skrifvars and F. Sundholm, *J. Appl. Polymer Science* **65** (1997) 77.

Hod1 P. Hodgkinson and P.J. Hore, *J. Magn. Reson. B* **106** (1995) 261.

Hol1 M. Holz and C. Müller, *J. Magn. Reson.* **40** (1980) 595.

Hol2 M. Holz, C. Müller and A.M. Wachter, *J. Magn. Reson.* **69** (1986) 108.

Hou1 D.I. Hoult and R.E. Richards, *Progr. NMR Spectr.* **12** (1978) 41.

Hou2 D.I. Hoult, *Concepts Magn. Reson.* **2** (1990) 33.

Hou3 D.I. Hoult, *J. Magn. Reson.* **33** (1979) 183.

Hwa1 T.-L. Hwang, P.C.M. van Zijl and M. Garwood, *J. Magn. Reson.* **124** (1997) 250.

Hyd1 J.S. Hyde, in: *Encyclopedia of NMR*, ed. D.M. Grant and R.K. Harris, Wiley, New York, 1996, p. 4656.

Igc1 Intermagnetics General Corporation, 450 Old Niskayuna Road, Latham, NY 12110, USA.

Ike1 M. Ikeya, M. Furusawa and M. Kasuya, *Scanning Microscopy* **4** (1990) 245.

Jac1 J.A. Jackson, L.J. Burnett and F. Harmon, *J. Magn. Reson.* **41** (1980) 411.

Jac2 J.A. Jackson and M. Mathews, *Log. Anal.* **34** (1993) 35.

Jeo1 E.-K. Jeong, D.-H. Kim, M.-J. Kim, S.-H. Lee, J.-S. Suh and Y.-K. Kwon, *J. Magn. Reson.* **127** (1997) 73.

Joh1 A.J. Johnson, M. Garwood and K. Ugurbil, *J. Magn. Reson.* **81** (1989) 653.

Joh2 C.S. Johnsen, Jr. and Q. He, *Adv. Magn. Reson.* **13** (1989) 131.

Kem1 R. Kemp-Harper, P. Styles and S. Wimperis, *J. Magn. Reson. A* **123** (1996) 230.

Kha1 M. El Khaloui, D.N. Rutledge and C.J. Ducauze, *J. Sci. Food Agric.* **53** (1990) 389.

Kim1 R. Kimmich and D. Hoepfel, *J. Magn. Reson.* **72** (1987) 379.

Kim2 R. Kimmich, G. Schnur, D. Hoepfel and D. Ratzel, *Phys. Med. Biol.* **32** (1987) 1335.

Kim3 R. Kimmich, E. Rommel and A. Knüttel, *J. Magn. Reson.* **81** (1989) 333.

Kim4 R. Kimmich and A. Knüttel, in: *Magn. Reson. Rel. Phen., Proc. XXIV Congr. Amp.* ed. J. Stankowski, N. Pislewski, S.K. Hoffmann and S. Idziak, 1988, Elsevier, Amsterdam, 1989, p. 413.

Kim5 R. Kimmich, *NMR Tomography, Diffusometry, Relaxometry*, Springer, Berlin, 1997.

Kim6 R. Kimmich, W. Unrath, G. Schnur and E. Rommel, *J. Magn. Reson.* **91** (1991) 136.

Kim7 C.G. Kim, K.S. Ryu, B.C. Woo and C.S. Kim, *IEEE Trans. Magnetics* **29** (1993) 3198.

Kla1 F. Klammer and R. Kimmich, *J. Phys. E: Sci. Instrum.* **22** (1989) 74.

Kle1 R.L. Kleinberg, in: *Encyclopedia of NMR*, ed. D.M. Grant and R.K. Harris, Wiley, New York, 1996, p. 4960.

Kle2 R.L. Kleinberg and C. Flaum, in: *Spatially Resolved Magnetic Resonance*, ed. P. Blümler, B. Blümich, R. Botto and E. Fukushima, Wiley-VCH, Weinheim, 1998, p. 555.

Kle3 R.L. Kleinberg, A. Sezginer, D.D. Griffin and M. Fukuhara, *J. Magn. Reson.* **97** (1992) 466.

Kle4 R.L. Kleinberg and H.J. Vinegar, *The Log Analyst*, November–December (1996) 20.

Knü1 A. Knüttel and R. Kimmich, *Magn. Reson. Med.* **9** (1989) 254.

Knü2 A. Knüttel and R. Kimmich, *J. Magn. Reson.* **78** (1988) 205.

Knü3 A. Knüttel, E. Rommel, M. Clausen and R. Kimmich, *Magn. Reson. Med.* **8** (1988) 70.

Knü4 A. Knüttel, R. Kimmich and K.-H. Spohn, *J. Magn. Reson.* **81** (1989) 570.

Knü5 A. Knüttel, R. Kimmich and K.-H. Spohn, *Magn. Reson. Med.* **17** (1991) 470.

Knü6 A. Knüttel, R. Kimmich and K.-H. Spohn, *J. Magn. Reson.* **86** (1990) 526.

Knü7 A. Knüttel, K.-H. Spohn and R. Kimmich, *J. Magn. Reson.* **86** (1990) 542.

Kop1 K. Kopinga and L. Pel, *Rev. Sci. Instrum.* **65** (1994) 3673.

Kos1 K. Kose, T. Haishi, A. Caprihan and E. Fukushima, *J. Magn. Reson.* **124** (1997) 35.

Kre1 R. Kreis and C. Boesch, *J. Magn. Reson. B* **113** (1996) 103.

Kre2 R. Kreis and C. Boesch, *J. Magn. Reson. B* **104** (1994) 187.

Kru1 M. Krus, *Restauro* **95** (1989) 294.

Let1 J.H. Letcher, *Magn. Reson. Imag.* **7** (1989) 581.

Luy1 P.R. Luyten and J.A. den Hollander, *Magnetic Resonance Imaging* **4** (1986) 237.

Luy2 P. Luyten, A.J.H. Marien, B. Sijtsma and J.A. den Hollander, *J. Magn. Reson.* **67** (1986) 148.

Lyo1 R. Lyon, J. Pekar, C.T.W. Moonen and A.C. McLaughlin, *Magn. Reson. Med.* **18** (1991) 80.

Mai1 V. Mairanovski, L. Yusefovich and T. Filippova, *J. Magn. Reson.* **54** (1983) 19.

Mar1 D. Mardon, M.G. Prammer and G.R. Coates, *Magn. Reson. Imag.* **14** (1996) 769.

Mat1 G.B. Matson, T. Schleich, C. Serdahl, G. Acosta and J.A. Willis, *J. Magn. Reson.* **56** (1984) 200.

Mat2 G.A. Matzkanin, in: Nondestructive Characterization of Materials, eds. P. Höller, G. Dobmann, C.O. Ruud and R.E. Green, Springer, Berlin, 1989, p. 655.

Mcc1 M.J. McCarthy, *Magnetic Resonance Imaging in Foods*, Chapman & Hall, New York, 1994.

Mcc2 D.C. McCain, *J. Magn. Reson. B* **109** (1995) 209.

Mcd1 P.J. McDonald, *Progr. Nuc. Magn. Reson. Spectrosc.* **30** (1997) 69.

Meh1 H. Mehier, M. Maurice, J.P. Bonche, G. Jaquemod, C. Desuzinges, B. Favre and J.O. Peyrin, *J. Biophys. Biomec.* **9** (1985) 198.

Met1 K.R. Metz and R.W. Briggs, *J. Magn. Reson.* **64** (1985) 172.

Mil1 J.B. Miller and A.N. Garroway, *J. Magn. Reson.* **77** (1988) 187.

Mit1 F. Mitsumori and N.M. Molas, *J. Magn. Reson.* **97** (1992) 282.

Moh1 A. Mohoric, J. Stepisnik, M. Kos and G. Planinsic, *J. Magn. Reson.* **136** (1999) 22.

Mon1 J. Monteiro Marques, D.N. Rutledge and C.J. Ducauze, *Lebensm.-Wiss. Technol.* **24** (1991) 93.

Moo1 C.T.W. Moonen, M. von Kienlin, P.C.M. van Zijl, J. Cohen, J. Gillen, P. Daly and G. Wolf, *NMR in Biomedicine* **2** (1989) 201.

Mue1 D. Mueller, *^{13}C Detection of Diamonds in Intact Stones*, in: *4th Int. Conf. Magn. Reson. Microscopy*, ed. E. Fukushima, Albuquerque, September 21–25, 1997.

Mül1 S. Müller, W.P. Aue and J. Seelig, *J. Magn. Reson.* **65** (1985) 332.

Mül2 S. Müller, *J. Magn. Reson. Med.* **6** (1988) 364.

Mül3 S. Müller, *Magn. Reson. Med.* **10** (1989) 145.

Mül4 S. Müller, R. Sauter, H. Weber and J. Seelig, *J. Magn. Reson.* **76** (1988) 155.

Mül5 S. Müller, W. P. Aue and J. Seelig, *J. Magn. Reson.* **63** (1985) 530.

Nil1 H. Nilgens, P. Blümler, J. Paff and B. Blümich, *J. Magn. Reson. A* **105** (1993) 108.

Ols1 D.L. Olson, T.L. Peck, A.G. Webb, R.L. Magin and J.V. Sweedler, *Science* **270** (1995) 1967.

Ord1 R.J. Ordidge, M.R. Bendall, R.E. Gordon and A. Connelly, in: *Magnetic Resonance in Biology and Medicine*, ed. K. Govil and T. Saran, McGraw-Hill, New Delhi, 1987, p. 387.

Ord2 R.J. Ordidge, A. Connely and J.A.B. Lohman, *J. Magn. Reson.* **66** (1986) 283.

Pac1 K.J. Packer, C. Rees and D.J. Tomlinson, *Adv. Mol. Relaxation Proc.* **3** (1972) 119.

Pac2 M. Packard and R. Varian, *Phys. Rev.* **93** (1954) 941.

Pec1 J. Pecker, J.S. Leigh, Jr. and B. Chance, *J. Magn. Reson.* **64** (1985) 115.

Pec2 T.L. Peck, R.L. Magin and P.C. Lauterbur, *J. Magn. Reson. B* **108** (1995) 114.

Pec3 T.L. Peck, R.L. Magin, J. Kruse and M. Feng, *IEEE Trans. Biomed. Eng.* **41** (1994) 760.

Pel1 L. Pel, K. Kopinga, G. Bertram and G. Lang, *J. Phys. D: Appl. Phys.* **28** (1995) 675.

Pla1 G. Planinsic, J. Stepisnik and M. Kos, *J. Magn. Reson. A* **110** (1994) 170.

Pop1 J.M. Pope and N. Repin, *Magn. Reson. Imag.* **6** (1988) 641.

Pow1 J.G. Powles, The atomic nucleus as a magnetic top, *The New Scientist*, 1 January 1959.

Pra1 P. Prado, U. Schmitz and B. Blümich, *J. Magn. Reson. xxx* (2000) xxx.

Qua1 Quantum Magnetics Inc., 7740 Kenamar Court, San Diego, CA 92121.

Ram1 C. Ramanathan, Y. Wu, B. Pfleiderer, M.J. Lizak, L. Garrido and J.L. Ackerman, *J. Magn. Reson. A* **121** (1996) 127.

Rat1 A.R. Rath, S.B. Roeder and E. Fukushima, *J. Magn. Reson.* **79** (1988) 461.

Rat2 A.R. Rath, S.B. Roeder and E. Fukushima, *Rev. Sci. Instrum.* **56** (1985) 402.

Roe1 P.B. Roemer, W.A. Edelstein, C.E. Hayes, S.P. Souza and O.M. Mueller, *Magn. Reson. Med.* **16** (1990) 192.

Rol1 W.L. Rollwitz, *Agricultural Engineering* **66** (1985) 12.

Rom1 E. Rommel and R. Kimmich, *Magn. Reson. Med.* **12** (1989) 209.

Rom2 E. Rommel and R. Kimmich, *J. Magn. Reson.* **83** (1989) 299.

Rud1 M. Rudin, guest ed., *In-Vivo Magnetic Resonance Spectroscopy I: Probeheads and Radiofrequency Pulses, Spectrum Analysis, NMR – Basic Principles and Progress* Vol. 26, Springer, Berlin, 1992.

Rud2 M. Rudin, guest ed., *In-Vivo Magnetic Resonance Spectroscopy II: Localization and Spectral Editing, NMR – Basic Principles and Progress* Vol. 27, Springer, Berlin, 1992.

Rud3 M. Rudin, guest ed., *In-Vivo Magnetic Resonance Spectroscopy III: In-Vivo Magnetic Resonance Spectroscopy: Potential and Limitations, NMR – Basic Principles and Progress* Vol. 28, Springer, Berlin, 1992.

Rud4 T.N. Rudakov, A.V. Belyakov and V.T. Mikhaltsevich, *Meas. Sci. Technol.* **8** (1997) 444.

Rut1 D.N. Rutledge, J. Diris, E. Bugner and J.-J. Belliardo, *Fresenius J. Anal. Chem.* **338** (1990) 441.

Ryn1 L.N. Ryner, J.A. Sorenson and M.A. Thomas, *J. Magn. Reson. B* **107** (1995) 126.

Saa1 T.R. Saarinen and C.S. Johnson, Jr., *J. Am. Chem. Soc.* **100** (1988) 3332.

Saa2 B.T. Saam and M.S. Conradi, *J. Magn. Reson.* **134** (1998) 67.

San1 J. Sanz and J.M. Rojo, *J. Phys. Chem.* **89** (1985) 4974.

Sau1 R. Sauter, S. Mueller and H. Weber, *J. Magn. Reson.* **75** (1987) 167.

Sch1 M.D. Schnall, R.E. Lenkinski, H.M. Pollack, Y. Imai and H.Y. Kressel, *Radiology* **172** (1989) 570.

See1 J. Seelig, in: *Physics of NMR Spectroscopy in Biology and Medicine, Proc. Int. School Phys. Enrico Fermi, Course C*, ed. B. Maraviglia, North-Holland, Amsterdam, 1988.

Sez1 A. Sezginer, D.D. Griffin, R.L. Kleinberg, M. Fukuhara and D.G. Dudley, *J. Electromagn. Waves Appl.* **7** (1993) 13.

Sez2 A. Sezginer, R.L. Kleinberg, M. Fukuhara and L.L. Latour, *J. Magn. Reson.* **94** (1991) 504.

Sha1 A.J. Shaka and R. Freeman, *J. Magn. Reson.* **59** (1984) 169.

Sha2 A.J. Shaka and R. Freeman, *J. Magn. Reson.* **64** (1985) 145.

Sha3 A.J. Shaka and R. Freeman, *J. Magn. Reson.* **63** (1985) 596.

Sha4 A.J. Shaka and R. Freeman, *J. Magn. Reson.* **62** (1985) 340.

Sha5 A.J. Shaka, J. Keeler, M.B. Smith and R. Freeman, *J. Magn. Reson.* **61** (1985) 175.

She1 J. Shen, *J. Magn. Reson. B* **112** (1996) 131.

She2 J. Shen and D.L. Rothman, *J. Magn. Reson.* **124** (1997) 72.

Shu1 O.A. Shushakov, *Geophysics* **61** (1996) 998.

Sin1 S. Singh and R. Deslaurier, *Concepts Magn. Reson.* **7** (1995) 1.

Sli1 C.P. Slichter, *Principles of Magnetic Resonance*, Springer, Berlin, 1990.

Slo1 J. Slotboom, A.F. Mehlkopf and W.M.M.J. Bovée, *J. Magn. Reson.* **95** (1991) 396.

Sot1 C.H. Sotak, D.M. Freeman and R.E. Hurd, *J. Magn. Reson.* **78** (1988) 355.

Sou1 S.P. Souza, J. Szumowski, C.L. Dumoulin, D.P. Plewes and G. Glower, *J. Comp. Ass. Tom.* **12** (1988) 1926.

Sta1 J. Star-Lack, D. Spielman, E. Adalsteinsson, J. Kurhanewicz, D.J. Terris and D.B. Vigneron, *J. Magn. Reson.* **133** (1998) 243.

Ste1 J. Stepisnik, *Zeitschr. Phys. Chem.* **190** (1995) 51.

Ste2 S. Stevenson and H.C. Dorn, *Anal. Chem.* **66** (1994) 2993.

Ste3 J. Stepisnik, V. Ezren, M. Kos, G. Planinsic and P. Pogacnik, *Phys. Medica* **6** (1990) 235.

Sty1 P. Styles, NMR – *Basic Principles and Progress* **27** (1992) 45.

Sui1 B.H. Suits, A.N. Garroway and J.B. Miller, *J. Magn. Reson.* **135** (1999) 373.

Sui2 B.H. Suits, A.N. Garroway and J.B. Miller, *J. Magn. Reson.* **132** (1998) 54.

Sui3 B.H. Suits, A.N. Garroway and J.B. Miller, *J. Magn. Reson.* **131** (1998) 154.

Swr1 Southwest Research Institute, 6220 Culebra Road, San Antonio, TX 78228-0510, USA.

Tai1 Z. Taicher, G. Coates, Y. Gitartz and L. Berman, *Magn. Reson. Imag.* **12** (1994) 285.

Teg1 J. Tegenfeldt and R. Sjöblom, *J. Magn. Reson.* **55** (1983) 372.

Tho1 M.A. Thomas, H.P. Hetherington, D.J. Meyerhoff and D.B. Twieg, *J. Magn. Reson.* **93** (1991) 485.

Tyc1 R. Tycko and A. Pines, *J. Magn. Reson.* **60** (1984) 156.

Usi1 Universal Systems Inc.28500 Aurora Road, Unit 16, Solon, OH 44139, USA.

Wat1 N. Watanabe and E. Niki, *Proc. Jpn. Acad.* **54** (1978) 194.

Wat2 H. Watanabe, Y. Ishihara, K. Okamoto, K. Oshio, T. Kanamatsu and Y. Tsukada, *J. Magn. Reson.* **134** (1998) 214.

Web1 A.G. Webb, *Progr. Nucl. Magn. Reson. Spectrosc.* **31** (1997) 1.

Wei1 P.B. Weichman, E.M. Lavely and M.H. Ritzwoller, *Phys. Rev. Lett.* **82** (1999) 4102.

Win1 R.A. Wind, J.H.N. Creyghton, D.J. Ligthelm and J. Smidt, *J. Phys. C* **11** (1978) L223.

Woe1 K. Woelk and J.W. Rathke, *J. Magn. Reson. A* **115** (1995) 106.

Wud1 D. Wu, *J. Magn. Reson. A* **116** (1995) 135.

Wun1 N. Wu, A.G. Webb, T.L. Peck and J.V. Sweedler, *Anal. Chem.* **67** (1995) 3101.

Wun2 N. Wu, T.L. Peck, A.G. Webb, R.L. Magin and J.V. Sweedler, *Anal. Chem.* **66** (1994) 3849.

Wux1 X. Wu, B.C. Gerstein and T.S. King, *J. Catal.* **121** (1990) 271.

Yam1 C. Yamanaka, M. Ikeya, K. Meguro and A. Nakanishi, *Nucl. Tracks Radiat. Meas.* **18** (1991) 279.

Yan1 Y. Yang, S. Xu, M.J. Dawson and P.C. Lauterbur, *J. Magn. Reson.* **129** (1997) 161.

Zhu1 J.M. Zhu and I.C.P. Smith, *J. Magn. Reson.* **136** (1999) 1.

Zij1 P.C.M. van Zijl, C.T.W. Moonen, J. Gillen, P.F. Daly, L.S. Miketic, J.A. Frank, T.F. DeLaney, O. Kaplan and J. S. Cohen, *NMR in Biomedicine* **3** (1990) 227.

Zij2 P.C.M. van Zijl, C.T.W. Moonen, J.R. Alger, J.S. Cohen and S.A. Chesnick, *Magn. Reson. Med.* **10** (1989) 256.

Zim1 G. Zimmer, A. Guthausen and B. Blümich, *Solid-State Nucl. Magn. Reson.* **12** (1998) 183.

Zim2 G. Zimmer, A. Guthausen, U. Schmitz, K. Saito and B. Blümich, *Advanced Materials* **12** (1997) 989.

Zwe1 M. Zweckstetter and T.A. Holak, *J. Magn. Reson.* **133** (1998) 134.

CHAPTER 10

Abb1 J.R. Abbott, N. Tetlow, A.L. Graham, S.A. Altobelli, E. Fukushima, L.A. Mondy and T.S. Stephens, *J. Rheol.* **35** (1991) 773.

Ack1 J.L. Ackerman, L. Garrido, J.R. Moore, B. Pfleiderer and Y. Wu, in: *Magnetic Resonance Microscopy*, ed. B. Blümich and W. Kuhn, VCH, Weinheim, 1992, p. 237.

Ack2 J.L. Ackerman, L. Garrido, W.A. Ellingson and J.D. Weyand, in: *Proceedings of the Joint Conference on Nondestructive Testing of High Performance Ceramics*, ed. A. Vary and J. Snyder, Boston, MA, 25–27 August 1987, ACS, Washington, p. 88.

Air1 D. Airey, S. Yao, J. Wu, V. Chen, A.G. Fane and J.M. Pope, *J. Membrane Sci.* **145** (1998) 145.

Alt1 S.A. Altobelli, E. Fukushima and L.A. Mondy, *J. Rheol.* **41** (1997) 1105.

Alt2 S.A. Altobelli, R.C. Givler and E. Fukushima, *J. Rheol.* **35** (1991) 721.

App1 M. Appel, G. Fleischer, D. Geschke, J. Kärger and M. Winkler, *J. Magn. Reson. A* **122** (1996) 248.

Arm1 R.L. Armstrong, A. Tzalmona, M. Menzinger, A. Cross and C. Lemaire, in: *Magnetic Resonance Microscopy*, ed. B. Blümich and W. Kuhn, VCH, Weinheim, 1992, p. 309.

Ass1 R.A. Assink, A. Caprihan and E. Fukushima, *AIChE Journal* **34** (1988) 2077.

Axe1 D.E. Axelson, A. Kantzas and A. Nauerth, *Solid-State Magn. Reson.* **6** (1996) 309.

Bay1 E. Bayer, W. Müller, M. Ilg and K. Albert, *Angew. Chem.* **101** (1989) 1033.

Beh1 W. Behr, A. Haase, G. Reichenauer and J. Fricke, *J. Noncryst. Solids* **225** (1998) 91.

Ber1 K. Beravs, A. Demsar and F. Demsar, *J. Magn. Reson.* **137** (1999) 253.

Bey1 S.D. Beyea, B.J. Balcom, T.W. Bremner, P.J. Prado, A.R. Cross, R.L. Armstrong and P.E. Grattan-Bellew, *Solid-State Nucl. Magn. Reson.* **13** (1998) 93.

Bin1 A. Binley, B. Shaw and S. Henry-Poulter, *Meas. Sci. Technol.* **7** (1996) 384.

Bla1 S. Blackband and P. Mansfield, *J. Phys. C: Solid State Phys.* **19** (1986) L49.

Blü1 B. Blümich and W. Kuhn, ed., *Magnetic Resonance Microscopy*, VCH, Weinheim, 1992.

Blü2 P. Blümler and B. Blümich, *Rubber Chem. Tech.* **70** (1997) 468, and references therein.

Blü3 P. Blümler and B. Blümich, *Macromolecules* **24** (1991) 2183.

Blü4 P. Blümler, B. Blümich and H. Dumler, *Kautschuk, Gummi, Kunstoffe* **45** (1992) 699.

Blü5 P. Blümler and B. Blümich, *Acta Polymerica* **44** (1993) 125.

Blü6 B. Blümich, P. Blümler, A. Guthausen, C. Fülber, G. Eidmann and R. Savelsberg, *Kautschuk, Gummi, Kunststoffe* **50** (1997) 560.

Blü7 B. Blümich and P. Blümler, *Macromol. Symp.* **87** (1994) 187.

Blü8 B. Blümich and H.W. Spiess, in: *Spektroskopie amorpher und kristalliner Festkörper*, ed. D. Haarer and H.W. Spiess, Steinkopff, Darmstadt, 1995.

Bod1 J. Bodurka, A. Jesmanowicz, J.S. Hyde, H. Xu, L. Estkowski and S.-J. Li, *J. Magn. Reson.* **137** (1999) 265.

Bog1 M. Bogdan, B.J. Balcom, T.W. Bremner and R.L. Armstrong, *J. Magn. Reson. A* **116** (1995) 266.

Bot1 P.A. Bottomley, H.H. Rogers and T.H. Foster, *Proc. Natl. Acad. Sci.* **83** (1986) 87.

Bri1 M.M. Britton and P.T. Callaghan, *Phys. Rev. Lett.* **78** (1997) 4930.

Bro1 H.J.P. Brocken, M.E. Spiekman, L. Pel, K. Kopinga and J.A. Larbi, *Materials and Structures/Matériaux et Constructions* **31** (1998) 49.

Bru1 E. Brunner, M. Haake, L. Kaiser, A. Pines and J.A. Reimer, *J. Magn. Reson.* **138** (1999) 155.

Cal1 P.T. Callaghan, *Principles of Nuclear Magnetic Resonance Microscopy*, Clarendon Press, Oxford, 1991.

Cal2 P.T. Callaghan, C.J. Clark and L.C. Forde, *Biophysical Chemistry* **50** (1994) 225.

Cal3 P.T. Callaghan, A. Coy, D. MacGowan, K.J. Packer and F.O. Zelaya, *Nature* **351** (1991) 467.

Cal4 P.T. Callaghan, *Rep. Prog. Phys.* **62** (1999) 599.

Cap1 A. Caprihan and E. Fukushima, *Physics Reports* **198** (1990) 195.

Car1 W.E. Carlos, S.G. Bishop and D.J. Tracy, *Phys. Rev. B* **43** (1991) 12512.

Cha1 C. Chang and R.A. Komoroski, *Macromolecules* **22** (1989) 600.

Che1 H.A. Cheng, S.A. Altobelli, A. Caprihan and E. Fukushima, in: *Powders and Grains 97*, ed. R.P. Behringer and J.T. Jenkins, A.A. Balkema, Rotterdam, 1997, p. 463.

Che2 T.L. Chenevert, A.R. Skovoroda and M. O'Donnel, *Magn. Reson. Med.* **39** (1998) 482.

Chu1 J.A. Chudek and G. Hunter, *J. Mat. Sci. Lett.* **11** (1992) 222.

Chu2 J.A. Chudek and G. Hunter, *Progr. Magn. Reson. Spectr.* **31** (1997) 43.

Clo1 R.S. Clough and J.L. Koenig, *J. Polym. Sci. C* **27** (1989) 451.

Cof1 G.P. Cofer, J.M. Brown and G.A. Johnson, *J. Magn. Reson.* **83** (1989) 608.

Coh1 J.P. Cohen-Addad, *Progr. NMR Spectrosc.* **25** (1993) 25.

Con1 P.J. O'Connor, S.S. Cutié, P.B. Smith, S.J. Martin, R.L. Sammler, W.I. Harris, M.J. Marks and L. Wilson, *Macromolecules* **29** (1996) 7872.

Con2 M.S. Conradi, *J. Magn. Reson.* **93** (1991) 419.

Cor1 D.G. Cory, J.C. deBoer and W.S. Veemann, *Macromolecules* **22** (1989) 1618.

Cor2 G. Cory, in: *Encyclopedia of NMR*, ed. D.M. Grant and R.K. Harris, Wiley, New York, 1996, p. 1226.

Cor3 D.G. Cory, J.B. Miller and A. N. Garroway, *Macromol. Symp.* **86** (1994) 259.

Cor4 D.G. Cory, *Ann. Rep. NMR Spectrosc.* **24** (1992) 87.

Dam1 R.A. Damion and K.J. Packer, *Proc. R. Soc. Lond. A* **453** (1997) 205.

Dem1 D.E. Demco, A. Johanson and J. Tegenfeldt, *Solid-State Nucl. Magn. Reson.* **4** (1995) 13.

Den1 W. Denk, R.M. Keolian, S. Ogawa and L.W. Jelinski, *Proc. Natl. Acad. Sci.* **90** (1995) 1595.

Den2 R.J. Densley, *IEEE Trans. Electrical Insulation* **EI-14** (1979) 148.

Den3 P. Denner, B. Walter and T. Willing, *Macromol. Symp.* **119** (1997) 339.

Dij1 P. Dijk, B. Berkowitz and P. Bendel, *Water Resources Research* **35** (1999) 347.

Dum1 C.L. Dumoulin and H.R. Hart, *Radiology* **161** (1986) 717.

Dum2 C.L. Dumoulin, S.P Souza and H. Feng, *Magn. Reson. Med.* **5** (1987) 47.

Ecc1 C.D. Eccles, P.T. Callaghan and C.F. Jenner, *Biophys. J.* **53** (1988) 77.

Eli1 H.-G. Elias, *Makromoleküle*, Vol. 1, 5th edn, Hütig & Wepf, Basel, 1990.

Eli2 E. Eliav and G. Navon, *J. Magn. Reson.* **137** (1999) 295.

Ell1 A. Ellingson, P.S. Wong, S.L. Dieckman, J.L. Ackerman and L. Garrido, *Ceram. Bull.* **68** (1989) 1180.

Erc1 M. Ercken, P. Adriaensens, D. Vanderzande and J. Gelan, *Macromolecules* **28** (1995) 8541.

Erc2 M. Ercken, P. Adriaensens, G. Rogers, R. Carleer, D. Vanderzande and J. Gelan, *Macromolecules* **29** (1996) 5671.

Ese1 S. Eser and L. Hou, *Carbon* **34** (1996) 805.

Fie1 R.J. Field and M. Burger, ed., *Oscillations and Travelling Waves in Chemical Systems*, Wiley, New York, 1985.

Fie2 R.J Field, E. Körös and R.M. Noyes, *J. Am. Chem. Soc.* **94** (1972) 8649.

Fin1 G. Fink, A. Schwaiger and B. Blümich (unpublished results).

Fül1 C. Fülber, K. Unseld, V. Herrmann, K.H. Jakob and B. Blümich, *Colloid and Polymer Science* **274**, (1996) 191.

Fül2 C. Fülber, B. Blümich, K. Unseld and V. Herrmann, *Kautschuk, Gummi, Kunststoffe* **48** (1995) 254.

Fuk1 E. Fukushima, *Annu. Rev. Fluid Mech.* **31** (1999) 95.

Gar1 L. Garrido, J.L. Ackerman and W.A. Ellingson, *J. Magn. Reson.* **88** (1990) 340.

Gar2 L. Garrido, J.L. Ackerman, W.A. Ellingson and J.D. Weyand, *Ceram. Eng. Sci. Proc.* **9** (1988) 1465.

Gat1 J.C. Gatenby and J.G. Core, *J. Magn. Reson. A* **121** (1996) 193.

Gib1 S.J. Gibbs, L.J. Kieran, L.D. Hall, D.E. Haycock, W.J. Frith and S. Ablett, *J. Rheol.* **40** (1996) 425.

Gon1 C.P. Gonas, J.S. Leigh and A.G. Gyodh, *Phys. Rev. Lett.* **75** (1995) 573.

Gop1 N. Gopalsami, S.L. Dieckman, W.A. Ellingson, R.E. Botto, P.S. Wong, H.C. Yeh and J.P. Pollinger, in: *Review of Progress in Quantitative NDE*, ed. D.O. Thompson and D.E. Chimenti, Plenum Press, New York, 1990, Vol. 10B, p. 1215.

Gör1 U. Görke, R. Kimmich and J. Weis, *J. Magn. Res. B* **111** (1996) 236.

Gör2 U. Görke, R. Kimmich and J. Weis, *Magn. Reson. Imag.* **14** (1996) 1079.

Göt1 J. Götz, D. Müller, H. Buggisch and C. Tasche-Lara, *Chem. Eng. Proc.* **33** (1994) 385.

Göt2 J. Götz, H. Buggisch and M. Peciar, *J. Non-Newtonian Fluid Mech.* **49** (1993) 251.

Gra1 A.L. Graham, S.A. Altobelli, E. Fukushima, L.A. Mondy and T.S. Stephens, *J. Rheol.* **35** (1991) 191.

Gro1 W. Gronski, R. Stadler and M.M. Jakobi, *Macromolecules* **17** (1984) 741.

Gui1 D.N. Guilfoyle, P. Mansfield and K.J. Packer, *J. Magn. Reson.* **97** (1992) 342.

Gui2 G. Guillot, A. Trokiner, L. Darrasse and H. Saint-Jalmes, *Appl. Phys.* **22** (1989) 1646.

Gui3 D.N. Guilfoyle, B. Issa and P. Mansfield, *J. Magn. Reson. A* **119** (1996) 151.

Gün1 E. Günther, B. Blümich and H.W. Spiess, *Molec. Phys.* **71** (1990) 477.

Gün2 E. Günther, B. Blümich and H.W. Spiess, *Macromolecules* **25** (1992) 3315.

Gün3 E. Günther, B. Blümich and H.W. Spiess, *Chem. Phys. Lett.* **184** (1991) 251.

Haf1 S. Hafner, D.E. Demco and R. Kimmich, *Solid-State Nucl. Magn. Reson.* **6** (1996) 275.

Han1 A.D. Hanlon, S.J. Gibbs, L.D. Hall, D.E. Haycock, W.J. Frith, S. Ablett and C. Mariott, *Meas. Sci. Technol.* **9** (1998) 631.

Hau1 D. Hauck, P. Blümler and B. Blümich, *Macromol. Chem. Phys.* **198** (1997) 2729.

Hay1 K. Hayashi, K. Kawashima, K. Kose and T. Inouye, *J. Phys. D: Appl. Phys.* **21** (1988) 1037.

Hei1 M. Heidenreich, W. Köckenberger, R. Kimmich, N. Chandrakumar and R. Bowtell, *J. Magn. Reson.* **132** (1998) 109.

Hep1 M.A. Hepp and J.B. Miller, *J. Magn. Reson. A* **110** (1994) 98.
Hep2 M.A. Hepp and J.B. Miller, *Macromol. Symp.* **86** (1994) 271.
Hep3 M.A. Hepp and J.B. Miller, *J. Magn. Reson. A* **111** (1994) 62.
Hil1 K.M. Hill, A. Caprihan and J. Kakalios, *Phys. Rev. Lett.* **78** (1997) 50.
Hür1 M. Hürlimann, K. Helmer, L. Latour and C. Sotak, *J. Magn. Reson. A* **111** (1994) 169.
Iwa1 J.H. Iwamiya, A.W. Chow and S.W. Sinton, *Rheol. Acta* **33** (1994) 267.
Jen1 C.F. Jenner, Y. Xia, C.D. Eccles and P.T. Callaghan, *Nature* **336** (1988) 399.
Jez1 P. Jezzard, J.J. Attard, T.A. Carpenter and L.D. Hall, *Progr. NMR Spectrosc.* **23** (1991) 1.
Joh1 D.L. Johnson and P.N. Sen, *Phys. Rev. B* **24** (1981) 2486.
Joy1 M.L.G. Joy, G.C. Scott and R.M. Henkelmann, *Magn. Reson. Imag.* **7** (1989) 89.
Kab1 S.I. Kabanikhin, I.V. Koptyug, K.T. Iskakov and R.Z. Sagdeev, *J. Inv. Ill-Posed Problems* **6** (1998) 335.
Kar1 S. Karunanithy, *J. Mater. Sci.* **26** (1991) 2169.
Ken1 C.B. Kennedy, B.J. Balcom and I.V. Mastikhin, *Can. J. Chem.* **76** (1998) 11.
Kim1 R. Kimmich, *NMR Tomography, Diffusometry, Relaxometry*, Springer, Berlin, 1997.
Kin1 P. Kinesh, E.W. Randall and K. Zick, *Magn. Reson. Imag.* **12** (1994) 305.
Kle1 R.L. Kleinberg and M.A. Horsfield, *J. Magn. Reson.* **88** (1990) 9.
Kle2 R.L. Kleinberg, in: *Encyclopedia of NMR*, ed. D.M. Grant and R.K. Harris, Wiley, New York, 1996, p. 4960.
Kle3 A. Klemm, H.-P. Müller and R. Kimmich, *Phys. Rev. E* **55** (1997) 4413.
Kle4 B. Klei and J.L. Koenig, *Acta Polymerica* **48** (1997) 199.
Kli1 M. Klinkenberg, P. Blümler and B. Blümich, *J. Magn. Reson. A* **119** (1996) 197.
Kli2 M. Klinkenberg, P. Blümler and B. Blümich, *Macromolecules* **30** (1997) 1038.
Knö1 M. Knörgen, U. Heuert, H. Schneider, P. Barth and W. Kuhn, *Polym. Bull.* **38** (1997) 101.
Knö2 M. Knörgen, U. Heuert, H. Menge and H. Schneider, *Die Angew. Makromol. Chem.* **261/262** (1998) 123.
Koe1 J.L. Koenig, in: *Magnetic Resonance Microscopy*, ed. B. Blümich and W. Kuhn, VCH, Weinheim, 1992, p. 187.
Kop1 K. Kopinga and L. Pel, *Rev. Sci. Instrum.* **65** (1994) 3673.
Kop2 I.V. Koptyug, V.B. Fenelonov, L. Yu. Khitrina, R.Z. Sagdeev and V.N. Parmon, *J. Phys. Chem. B* **102** (1998) 3090.
Kop3 I.V. Koptyug (private communication).
Kop4 I.V. Koptyug, S.I. Kabanikhin, K.T. Iskakov, V.B. Fenelonov, L. Yu. Khitrina, R.Z. Sagdeev and V.N. Parmon, *Chem. Eng. Sci.* **55** (2000) 1559.
Kos1 K. Kose, *J. Phys. D: Appl. Phys.* **23** (1990) 981.
Kos2 K. Kose, *J. Magn. Reson. A* **118** (1996) 195.
Kos3 K. Kose, *Phys. Rev. A* **44** (1991) 2495.
Kos4 K. Kose, *J. Magn. Reson.* **96** (1992) 596.
Kos5 K. Kose, *J. Magn. Reson.* **92** (1991) 631.
Kos6 K. Kose, *J. Magn. Reson.* **98** (1992) 599.
Kue1 D.O. Kuethe and J.-H. Gao, *Phys. Rev. E* **51** (1995) 3252.
Kup1 V. Yu. Kuperman, E.E. Ehrichs, H.M Jaeger and G.S. Karczmar, *Rev. Sci. Instrum.* **66** (1995) 4350.
Kur1 H. Kurosu, T. Shibuya, H. Yasunaga and I. Ando, *Polymer Journal* **28** (1996) 80.
Lau1 S. Laukemper-Ostendorf, H.D. Lemke, P. Blümler and B. Blümich, *J. Membrane Science* **138** (1998) 287.
Lau2 A. Lauenstein, J. Tegenfeldt and W. Kuhn, *Macromolecules* **31** (1998) 3686.

Lew1 Z. Lewandowski, S.A. Altobelli, P.D. Majors and E. Fukushima, *Wat. Sci. Tech.* **26** (1992) 577.

Lit1 T. Li, J.D. Seymour, R.L. Powell, M.J. McCarthy, K.L. McCarthy and L. Ödberg, *AIChE Journal* **40** (1994) 1408.

Liz1 M.J. Lizak, M.S. Conradi and and C.G. Fry, *J. Magn. Reson.* **95** (1991) 548.

Maa1 W.E. Maas, L.H. Merwin and D.G. Cory, *J. Magn. Reson.* **129** (1997) 105.

Mac1 J.S. MacFall and H. van As, in: *Nuclear Magnetic Resonance in Plant Biology*, ed. Y. Shacher-Hill and P.E. Pfeffer, The American Society of Plant Physiologists, Rockville, 1996, p. 33.

Mai1 R.W. Mair and P.T. Callaghan, *Europhys. Lett.* **63** (1996) 719.

Man1 P. Mansfield and P.G. Morris, *NMR Imaging in Biomedicine, Adv. Magn. Reson. Suppl. 2*, Academic Press, New York, 1982.

Man2 P. Mansfield, R.W. Bowtell, S.J. Blackband and M. Cawley, *Magn. Reson. Imag.* **9** (1991) 763.

Man3 B. Manz and P.T. Callaghan, *Macromolecules* **11** (1997) 3309.

Man4 B. Manz, J.D. Seymour and P.T. Callaghan, *J. Magn. Reson.* **125** (1997) 153.

Mcc1 M.J. McCarthy, *Magnetic Resonance Imaging in Foods*, Chapman & Hall, New York, 1994.

Mcc2 M.J. McCarthy and K.L. McCarthy, *Magn. Reson. Imag.* **14** (1996) 799.

Mch1 E.J. McHahon, *IEEE Trans. Electrical Insulation* **EI-13** (1978) 277.

Mei1 M. Meininger, P.M. Jakob, M. von Kienlin, D. Koppler, G. Brinkmann and A. Haase, *J. Magn. Reson.* **125** (1997) 325.

Met1 A. Metzler, W. Köckenberger, M. von Kienlin, E. Komor and A. Haase, *J. Magn. Reson. B* **105** (1994) 249.

Met2 A. Metzler, M. Izquierdo, A. Ziegler, W. Köckenberger, E. Komor, M. von Kienlin, A. Haase and M. Décorps, *Proc. Natl. Acad. Sci. USA* **92** (1995) 11912.

Mik1 F.P. Miknis, A.T. Pauli, L.C. Michon and D.A. Netzel, *Fuel* **77** (1998) 399.

Mik2 F.P. Miknis and L.C. Michon, *Fuel* **77** (1998) 393.

Mil1 J.B. Miller, *Progr. Nucl. Magn. Reson. Spectrosc.* **33** (1998) 273.

Mül1 H.-P. Müller, R. Kimmich and J. Weis, *Magn. Reson. Imag.* **14** (1996) 955.

Mül2 H.-P. Müller, R. Kimmich and J. Weis, *Phys. Rev. E* **54** (1996) 5278.

Mut1 R. Muthupillai, D.J. Lomas, P.J. Rossmann, J.F. Greenleaf, A. Manduca and R.L. Ehman, *Science* **269** (1995) 1854.

Mut2 R. Muthupillai, P.J. Rossman, D.J. Lomas, J.F. Greenleaf, S.J. Riederer and R.L. Ehman, *Magn. Reson. Med.* **36** (1996) 266.

Nak1 M. Nakagawa, S.A. Altobelli, A. Caprihan, E. Fukushima and E.-K. Jeong, *Exp. Fluids* **16** (1993) 54.

Nes1 N.F.E.I. Nestle and R. Kimmich, *Biotech. Bioeng.* **51** (1996) 538.

Neu1 C.H. Neuman, *J. Chem. Phys.* **60** (1974) 4508.

Nun1 T. Nunes, E.W. Randall, A.A. Samoilenko, P. Bodart and G. Feio, *Appl. Phys.* **29** (1996) 805.

Osm1 P.A. Osment, K.J. Packer, M.J. Taylor, J.J. Attard, T.A. Carpenter, L.D. Hall, N.J. Herrod and S.J. Doran, *Phil. Trans. R. Soc. Lond. A* **333** (1990) 441.

Pac1 K.J. Packer, in: *Encyclopedia of NMR*, ed. D.M. Grant and R.K. Harris, Wiley, New York, 1996, p. 1615.

Par1 D.D. Parker and J.L. Koenig, *Current Trends in Polymer Science* **1** (1996) 65.

Pel1 L. Pel, K. Hazrati, K. Kopinga and J. Marchand, *Magn. Reson. Imag.* **16** (1998) 525.

Pel2 L. Pel, K. Kopinga, G. Bertram and G. Lang, *J. Phys. D: Appl. Phys.* **28** (1995) 675.

Pel3 L. Pel, H. Brocken and K. Kopinga, *Int. J. Heat Transfer* **39** (1996) 1273.

Pel4 L. Pel, K. Kopinga and H. Brocken, *Magn. Reson. Imag.* **14** (1996) 931.

Pel5 L. Pel, K. Kopinga and H. Brocken, *HERON* **41** (1996) 95.

Pet1 E. Petersen, K. Potter, J. Buttler, K.W. Fishbein, W. Horton, R.G.S. Spencer and E.W. McFarland, *Int. J. Imaging Sys. Technol.* **8** (1997) 285.

Pha1 N. Phan-Thien, A.L. Graham, S A. Altobelli and L. Mondy, *Ind. Eng. Chem.* **34** (1995) 3187.

Pop1 J.M. Pope, S. Yao and A.G. Fane, *J. Membr. Sci.* **118** (1996) 247.

Pop2 J.M. Pope, in: *Magnetic Resonance Microscopy*, ed. B. Blümich and W.Kuhn, VCH, Weinheim, 1992, p. 441.

Pow1 R.L. Powell, J.E. Maneval, J.D. Seymour, K.L. McCarthy and M.J. McCarthy, *J. Rheol.* **38** (1994) 1465.

Pra1 P.J. Prado, B.J. Balcom, S.D. Beyea, T.W. Bremner, R.L. Armstrong, R. Pishe and P.E. Gratten-Bellew, *J. Phys. D: Appl. Phys.* **31** (1996) 2040.

Pra2 P.J. Prado, B.J. Balcom and M. Jama, *J. Magn. Reson.* **137** (1999) 59.

Pro1 S.W. Provencher, *Comp. Phys. Comm.* **27** (1982), 213, 229.

Que1 J. Queslel and J.E. Mark, in: *Mechanical and Solvent-Swelling Methods for Characterizing Insoluble Polymers*, ed. A.R. Cooper, Wiley, New York, 1989, Chapter 16.

Rat1 R.G. Ratcliffe, *Adv. Bot. Res.* **20** (1994) 43.

Ree1 T.G. Reese, V.J. Wedeen and R.M. Weisskopf, *J. Magn. Reson. B* **112** (1996) 253.

Rob1 M.A. Robinson, *J. Magn. Reson.* **95** (1991) 574.

Rof1 C.J. Rofe, R.K. Lambert and P.T. Callaghan, *J. Rheol.* **38** (1994) 875.

Rof2 C.J. Rofe, L. de Vargas, J. Perez-Gonzáles, R.K. Lambert and P.T. Callaghan, *J. Rheol.* **40** (1996) 1115.

Rog1 H.H. Rogers and P.A. Bottomley, *Proc. Natl. Acad. Sci.* **79** (1987) 957.

Rok1 M. Rokitta, U. Zimmermann and A. Haase, *J. Magn. Reson.* **137** (1999) 29.

Rol1 A. Rollins, J. Barber, R. Elliott and B. Wood, *Plant Physiol.* **91** (1989) 1243.

Rot1 W.P. Rothwell, D.R. Holecek and J.A. Kershaw, *J. Polym. Science: Polym. Lett. Ed.* **22** (1984) 241.

Rue1 D. Ruelle, *Turbulence, Strange Attractors, and Chaos*, World Scientific, Singapore, 1995.

Rum1 H. Rumpel and J.M. Pope, *Magn. Reson. Imag.* **10** (1992) 187.

Sal1 G. Salge, *Untersuchung von Erosionsphänomenen in Polyethylen mit Kernspinresonanzspektroskopie (NMR) und Teilentladungsmessungen*, Dissertation, RWTH, Aachen, 1998.

Sar1 S.N. Sarkar and R.A. Komoroski, *Macromolecules* **25** (1992) 1420.

Sar2 V. Sarafis, H. Rumpel, J.M. Pope and W. Kuhn, *Protoplasma* **159** (1990) 70.

Sar3 V. Sarafis, J.M. Pope and Y. Sarig, in: *Magnetic Resonance Microscopy*, ed. B. Blümich and W. Kuhn, VCH, Weinheim, 1992, p. 461.

Sar4 M. Sardashti, B.A. Baldwin and D.J. O'Donnell, *J. Polymer Science B: Polym. Phys.* **33** (1995) 571.

Sch1 U. Scheler, G. Schauß, B. Blümich and H.W. Spiess, *Solid-State Nucl. Magn. Reson.* **6** (1996) 375.

Sch2 K. Schmidt-Rohr and H.W. Spiess, *Multidimensional Solid-State NMR and Polymers*, Academic Press, London, 1994.

Sch3 T.J. Schaafsma, H. van As, W.D. Palstra, J.E.M. Snaar and P.A. de Jager, *Magn. Reson. Imag.* **10** (1992) 827.

Sch4 M. Schneider, L. Gasper, D.E. Demco and B. Blümich, *J. Chem. Phys.* **111** (1999) 402.

Sch5 M. Schneider, D.E. Demco and B. Blümich, *J. Magn. Reson.* **140** (1999) 432.

Sch6 M. Schneider, *^{1}H-Multiquanten-Untersuchungen an Elastomeren*, Dissertation, Rheinisch-West- fälische Technische Hochschule, Aachen, 1999.

Sco1 G.C. Scott, M.L.G. Joy, R.L. Armstrong and R.M. Henkelmann, *IEEE Trans. Med. Imag.* **10** (1991) 362.

Sco2 G.C. Scott, M.L.G. Joy, R.L. Armstrong and R.M. Henkelmann, *Magn. Reson. Med.* **28** (1992) 186.

Sco3 G.C. Scott, M.L.G. Joy, R.L. Armstrong and R.M. Henkelmann, *J. Magn. Reson.* **97** (1992) 235.

Sco4 G.C. Scott, M.L.G. Joy, R.L. Armstrong and R.M. Henkelmann, *Magn. Reson. Med.* **33** (1995) 355.

Ser1 I. Sersa, O. Jarh and F. Demsar, *J. Magn. Reson. A* **111** (1994) 93.

Sey1 J.D. Seymor, J.E. Maneval, K.L. McCarthy, M.J. McCarthy and R.L. Powell, *Phys. Fluids A* **5** (1993) 3010.

Sha1 M.D. Shattuck, R.P. Behringer, G.A. Johnson and J.G. Georgiadis, *Phys. Rev. Lett.* **75** (1995) 1934.

Shi1 T. Shibuya, H. Yasunaga, H. Kurosu and I. Ando, *Macromolecules* **28** (1995) 4377.

Sil1 L.O. Sillerud, D.B. van Hulsteyn and R.H. Griffey, *J. Magn. Reson.* **76** (1988) 380.

Sin1 S.W. Sinton, A.W. Chow and J.H. Iwamiya, *Macromol. Symp.* **86** (1994) 299.

Sin2 S.W. Sinton, J.H. Iwamiya, B. Ewing and G.P. Drobny, *Spectroscopy* **6** (1991) 42.

Smi1 S.R. Smith and J.L. Koenig, *Macromolecules* **24** (1991) 3496.

Sot1 P. Sotta, B. Deloche, J. Herz, A. Lapp, D. Durand and J.-C. Rabadeux, *Macromolecules* **20** (1987) 2769.

Spy1 A. Spyros, N. Chandrakumar, M. Heidenreich and R. Kimmich, *Macromolecules* **31** (1998) 3021.

Spy2 A. Spyros, R. Kimmich, B.H. Briese and D. Jendrossek, *Macromolecules* **30** (1997) 8218.

Str1 J.H. Strange and J.B.W. Webber, *Appl. Magn. Reson.* **12** (1997) 231.

Str2 J.H. Strange and J.B.W. Webber, *Meas. Sci. Technol.* **8** (1997) 555.

Sui1 B.H. Suits and J.L. Lutz, *J. Appl. Phys.* **65** (1989) 3728.

Sus1 S. Su, M. Menzinger, R.L. Armstrong, A. Cross and C. Lemaire, *J. Phys. Chem.* **98** (1994) 2494.

Tal1 U. Tallarek, E. Baumeister, K. Albert, E. Bayer and G. Guichon, *J. Chromatogr. A* **696** (1995) 1.

Tim1 J. Timonen, L. Avila, P. Hirva, T.T. Pakkanen, D. Gross and V. Lehmann, *Appl. Catal. A: General 129* (1995) 117.

Tra1 B. Traub, S. Hafner, U. Wiesner and H.W. Spiess, *Macromolecules* **31** (1998) 8585.

Tse1 T.Y. Tse, R.M. Spanswick and L.W. Jelinski, *Protoplasm* **194** (1996) 54.

Tso1 L. Tsoref, H. Shinar, Y. Seo, U. Eliav and G. Navon, *Magn. Reson. Med.* **39** (1998) 11.

Tur1 R. Turner and P. Keller, *Progr. NMR Spectrosc.* **23** (1991) 93.

Tza1 A. Tzalmona, R.L. Armstrong, M. Menzinger, A. Cross and C. Lemaire, *Chem. Phys. Lett.* **174** (1990) 199.

Tza2 A. Tzalmona, R.L. Armstrong, M. Menzinger, A. Cross and C. Lemaire, *Chem. Phys. Lett.* **188** (1992) 457.

Val1 M. Valtier, P. Tekely, L. Kiéné and D. Canet, *Macromolecules* **28** (1995) 4075.

Van1 H. van As, J.A.E. Reinders, P.A. de Jager, P.A.C.M. van de Sanden and T.J. Schaafsma, *J. Exp. Botany* **45** (1994) 61.

Wag1 R.A. Waggoner and E. Fukushima, *Magn. Reson. Imag.* **14** (1996) 1085.

Wan1 S. Wang, D.B. Minor and S.G. Malghan, *J. Mater. Sci.* **28** (1993) 4940.
War1 M. Warner, P.T. Callaghan and E.T. Samulski, *Macromolecules* **30** (1997) 4733.
Wea1 D. Weaire, S. Hutzler, G. Verbist and E. Peters, *Adv. Chem. Phys.* **102** (1997) 315.
Web1 A.G. Webb and L.D. Hall, *Polymer Comm.* **31** (1990) 422.
Web2 A.G. Webb, P. Jezzard, L.D. Hall and S. Ng, *Polymer Comm.* **30** (1989) 363.
Wei1 J. Weis, R. Kimmich and H.P. Müller, *Magn. Reson. Imag.* **14** (1996) 319.
Wei2 L.A. Weisenberger and J.L. Koenig, *J. Poly. Sci.: Part C: Polym. Lett.* **27** (1989) 55.
Wei3 L.A. Weisenberger and J.L. Koenig, *Macromolecules* **23** (1990) 2445.
Wei4 L.A. Weisenberger and J.L. Koenig, *Macromolecules* **23** (1990) 2454.
Wei5 F. Weigand and H.W. Spiess, *Advanced Materials* **8** (1996) 481.
Wei6 F. Weigand, S. Hafner and H.W. Spiess, *J. Magn. Reson. A* **120** (1996) 201.
Wei7 F. Weigand, B. Blümich and H.W. Spiess, *Solid-State Nucl. Magn. Reson.* **3** (1994) 59.
Wei8 F. Weigand, D.E. Demco, B. Blümich and H.W. Spiess, *J. Magn. Reson. A* **120** (1996) 190.
Wei9 F. Weigand, H.W. Spiess, B. Blümich, G. Salge and K. Möller, *IEEE Transact. Dielectr. Electr. Insul.* **4** (1997) 280.
Wei10 F. Weigand and H.W. Spiess, *Macromolecules* **28** (1995) 28.
Xia1 Y. Xia and P.T. Callaghan, *Maromolecules* **24** (1991) 4777.
Xia2 Y. Xia, P.T. Callaghan and K.R. Jeffrey, *AIChE J.* **38** (1992) 1408.
Yao1 S. Yao, A.G. Fane and J.M. Pope, *Magn. Res. Imag.* **12** (1997) 235.
Yao2 S. Yao, M. Costello. A.G. Fane and J.M. Pope, *J. Membr. Sci.* **99** (1995) 207.

References for figures

Most figures have been adapted from the literature. The references listed below refer to the list of references of the respective chapter.

CHAPTER 1

1.1.1	—	1.1.2	KRE1, Fig. 1.2
1.1.3	BLÜ5, Fig. I.1	1.1.4	—
1.1.5	—	1.1.6	BLÜ4, Fig. 1
1.1.7	BLÜ2, Fig. 8	1.1.8	—

CHAPTER 2

2.1.1	—	2.1.2	LAU1, Fig. 1
2.2.1	—	2.2.2	WOL1
2.2.3	—	2.2.4	—
2.2.5	—	2.2.6	—
2.2.7	ERN1, Fig. 4.2.2	2.2.8	—
2.2.9	BLÜ2, Fig. 1	2.2.10	—
2.2.11	CAL1, Fig. 2.20	2.2.12	—
2.3.1	KRE1, Fig. 10.1	2.3.2	KRE1, Fig. 10.11
2.3.3	KRE1, Fig. 10.20	2.3.4	MAN3, Fig. 5
2.3.5	KRE1, Fig. 10.26 & Fig. 10.27	2.3.6	KRE1, Fig. 10.28 & Fig. 10.29
2.3.7	CAL1, Fig. 9.3d	2.3.8	KRE1, Fig. 10.33
2.3.9	DER1, Fig. 2.7	2.3.10	BLÜ1, Fig. 8

CHAPTER 3

3.1.1	BLÜ3, Fig. 1	3.1.2	—
3.1.3	—	3.2.1	GÜN2, Fig. 1
3.2.2	—	3.2.3	—
3.2.4	WIE1, Fig. 1	3.2.5	MÜL1, Fig. 7
3.2.6	—	3.2.7	BLÜ4, Fig. 4, 5
3.3.1	YAN1, Fig. 1	3.3.2	BLÜ3, Fig. 4
3.3.3	BLÜ3, Fig. 5	3.3.4	HAR1, Fig. 3
3.3.5	BLÜ4, Fig. 1	3.3.6	—
3.3.7	GÜN3, Fig. 1	3.3.8	BLÜ4, Fig. 2
3.3.9	—	3.3.10	SCH9, Fig. 3.2
3.3.11	SCH9, Fig. 3.3	3.3.12	PRI2, Fig. 20
3.3.13	BRO1, Fig. 5	3.3.14	BRO1 Fig. 7
3.4.1	—	3.4.2	BLÜ2, Fig. 2
3.4.3	—	3.5.1	CAL2, Fig. 2.12a
3.5.2	CAL2, Fig. 2.12b	3.5.3	BLÜ6, Fig. 1

CHAPTER 4

4.1.1 —
4.1.2 —
4.2.1 —
4.2.2 BLÜ1, Fig. 3
4.2.3 BLÜ1, Fig. 13
4.3.1 BLÜ1, Fig. 6
4.4.1 MAJ1, Fig. 1
4.4.2 —
4.4.3 —
4.4.4 BAR2, Fig. 1

CHAPTER 5

5.0.1 BRU1, Fig. 3
5.1.1 DAM4, Fig. 3 (MOR1, Fig. 3.2)
5.1.2 BOT1, Fig. 4
5.3.1 MOR1, Fig. 3.8
5.3.2 —
5.3.3 HOU1, Figs. 1 + 3
5.3.4 HOU1, Fig. 7
5.3.5 CAL2, Fig. 3.3
5.3.6 BAU1, Fig. 6
5.3.7 CAL2, Figs. 3.7 & 3.11
5.3.8 a) DOD1, Fig. 2; b)-
5.3.9 NIL1, Figs. 2e & 3a, e, f
5.3.10 FRE2, p. 45, Fig. 2
5.3.11 WIM1, Figs. 2 & 5; SHA1, Figs. 1 & 3
5.3.12 a) CAL2, Fig. 3.13; b) FRE1, Fig. 8
5.3.13 MÜL2, Fig. 1
5.3.14 HAF1, Fig. 1
5.3.15 HEP2, Fig. 1
5.3.16 COR1, Fig. 2
5.4.1 —
5.4.2 LJU1, Fig. 1
5.4.3 BLÜ4, Fig. 6
5.4.4 CAL7, Fig. 1; & CAL2 Fig. 3.27b
5.4.5 —
5.4.6 CHW1
5.4.7 GÖB1, Fig. 5.16
5.4.8 COY1, Figs. 9 & 10, top
5.4.9 CAL6, Fig. 6
5.4.10 CAL5, Figs. 2 & 3
5.4.11 CAL11, Fig. 1c
5.4.12 —
5.4.13 GÖB1, Fig. 5.24

CHAPTER 6

6.1.1 BLÜ7, Fig. 5
6.1.2 MOR1, Fig. 4.6
6.1.3 NIL4, Fig. III.7
6.1.4 LAI1, Fig. 4
6.1.5 COR1, Fig. 1
6.1.6 —
6.2.1 WEHI, Fig. 1.14
6.2.2 —
6.2.3 WEH1, Fig. 1.16
6.2.4 MAR1, Fig. 2
6.2.5 FRA1, Fig. 3
6.2.6 HAA2, Fig. 1a
6.2.7 FRA1, Fig. 1b
6.2.8 FRA1, Fig. 5
6.2.9 MOR2, a) Fig. 11.31, b) Fig. 11.32
6.2.10 HAA1, Fig. 1
6.2.11 HAA3, Fig. 1
6.2.12 GYN1, Figs. 1,2
6.2.13 MAN4, Fig. 4
6.2.14 MAN6, Fig. 10
6.2.15 MOR1, Fig. 4.35
6.2.16 CHO2, Fig. 3
6.3.1 CAN2, Fig. 1
6.3.2 HOU1, Fig. 3
6.3.3 HOU2, Fig. 1
6.3.4 SIM1, Fig. 1
6.3.5 VAL2, Fig. 1
6.4.1 BLÜ1, Fig. 48
6.4.2 BLÜ3, Fig. 1 & ROO1, Fig. 6
6.4.3 BLÜ4, Fig. 1
6.4.4 JAN1, Fig. 2
6.4.5 BLÜ6, Fig. 8

CHAPTER 7

7.1.1 BLÜ1, Fig. 1
7.1.2 —
7.1.3 BLÜ2, Fig. 7
7.1.4 FÜL1, Fig. 4
7.1.5 BLÜ1, Fig. 7
7.1.6 FÜI1, Fig. 1
7.1.7 FÜL1, Fig. 3
7.1.8 —
7.1.9 SOT2, Fig. 4
7.1.10 FÜL5, Fig. 105
7.1.11 BLÜ5, Figs. 3, 7
7.1.12 KLI2, Fig. 1a

7.1.13	HAU1, Fig. 5	7.2.1	—
7.2.2	HAN1, Fig. 29	7.2.3	—
7.2.4	ROM2, Fig. 1	7.2.5	a) WEH1, Fig. 1.17; b) DUM1, Fig. 1
7.2.6	HAA3, Fig. 1	7.2.7	HAA3, Fig. 2
7.2.8	HEP1, Fig. 2	7.2.9	BEU1; a) Fig. 1, b) Fig. 2, c) Fig. 4
7.2.10	BLÜ8, Fig. 1	7.2.11	EYM1
7.2.12	a) CAP1, Fig. 2, b) CAP1, Fig. 3.	7.2.13	e) KIM2, Fig. 19.2
7.2.14	PAC1, Fig. 1	7.2.15	GIE1
7.2.16	a) KOS3, Fig. 1; b) KOS3, Fig. 2b	7.2.17	—
7.2.18	ROM4, Figs. 4.7 & 4.9	7.2.19	—
7.2.20	ZIM1	7.2.21	BLÜ1, Fig. 4
7.2.22	GÖT1, Fig. 2	7.2.23	c) GUT1, Fig. 6
7.2.24	—	7.2.25	—
7.2.26	a to d) ERN1, Fig. 5.3.1; e) GAR2, Fig. 1; f) Günl, Fig. 2.	7.2.27	KLI1, Fig. 5
7.2.28	SCHI2, Fig. 4a	7.2.29	SCHI3, Fig. 3
7.2.30	SCHI1, Fig. 4.4	7.2.31	—
7.2.32	DUM3, Fig. 1	7.2.33	a) KIM4, Fig. 2, b) MÜL1, Fig. 1 = KIM4, Fig. 3
7.3.1	—	7.3.2	a) HEP4, Fig. 1a, b) MAT3, Fig. 1b
7.3.3	GÜN1, Fig. 8c, d	7.3.4	a) ZIE1, Fig. 2; b) ZIJ2, Fig. 2
7.3.5	SCH7: a) Fig. 8a, b) Fig. 9a, c–e) Fig. 8b–d	7.4.1	SER1, Fig. 1

CHAPTER 8

8.1.1	MAL3, Fig. 2	8.1.2	—
8.1.3	IWA1, Fig. 9	8.2.1	a, b, c) STA1, Fig. 2; d, e) ATT1, Fig. 1
8.2.1	a, b) COD1, Figs. 1 & 3, c) MIL3, Fig. 1	8.2.2	COD1, Fig. 4
8.3.1	a) GRA1, Fig. 1; b–d) PRA1 Fig. 1	8.3.2	PRA5, Figs. 1c & 8
8.3.3	GÜN1, Fig. 1; b–d) BLÜ5, Fig. 6	8.4.1	—
8.4.2	GAR1, Fig. 3	8.4.3	GÜN1, Fig. 7
8.4.4	KLI1, Fig. 3 & Fig. 6	8.5.1	FIL1, Fig. 1
8.5.2	SCH3, Fig. 1	8.5.3	COR4, Figs. 3 & 4
8.6.1	a) MAR2, Fig. 6a; b) MAR2, Fig. 5a	8.6.2	MAR2, Fig. 5b
8.7.1	CHI1, FIG 1	8.7.2	COR13, Figs. 3 & 4
8.7.3	COR12, Fig. 4	8.8.1	HAF3, Fig. 2a
8.8.2	WEI1, Figs. 1, 3, & 6	8.8.3	KIM1, Fig. 7
8.8.4	HAF3, Fig.5		

CHAPTER 9

9.0.1	—	9.1.1	AKO1, Fig. 5
9.1.2	AUE2, Fig. 1	9.1.3	CAL2, Fig. 3.12
9.1.4	ROM2, Fig. 5	9.1.5	AKO1, Fig. 8
9.1.6	FRA1, Fig. 1b	9.1.7	KIM1, Fig. 1
9.1.8	BRI2, Fig. 1	9.1.9	BLA1, Fig. 1
9.1.10	AKO1, Fig. 4	9.2.1	BOS1, Fig. 2
9.2.2	—	9.2.3	BOS1, Fig. 3
9.2.4	BOS1, Fig. 5	9.2.5	BOS2, Fig. 9
9.2.6	BOS1, Fig. 10a	9.2.7	—
9.2.8	—	9.3.1	GRI2, Fig. 2
9.3.2	a) WEB1, Fig. 39; b) ASS1, Fig. 1	9.3.3	BAN1, Fig. 5 (from BAR1)
9.3.4	a) EID1, Fig. 5a; b) GUT3, Fig. 2.5; c) GUT3, Fig. 4.2; d) GUT1, Fig. 3 (GUT3, Fig. 6.1)	9.3.5	WEB1, Fig. 27.
9.3.6	ALB1, Figs. 13, 14	9.3.7	SHU1, a) Fig. 2, b) Fig. 1.

9.3.8	CAL1, Fig. 1	9.3.9	KLE1, Fig. 4
9.3.10	JAC1, Fig. 1	9.3.11	JAC1, Figs. 9, 10
9.3.12	GUT4, Fig. 9b	9.3.13	BLÜ3: a) Fig. 1, b) Fig. 2, c) Fig. 3
9.3.14	a) BLÜ3, Fig. 5c; b) GUT3, Fig. 5.2; c) BLÜ6, Fig. 5c; d) BLÜ6, Fig. 5b; e) BLÜ4, Fig. 15.8; f) BLÜ3, Fig. 7	9.3.15	—
9.3.16	ZIM1, Fig. 6	9.3.17	PRA1, Figs. 1, 5, 7

CHAPTER 10

10.1.1	ARM1, Fig. 1	10.1.2	TZA2, Fig. 2 & Fig. 3a
10.1.3	KOP3	10.1.4	BAY1, Fig. 3
10.1.5	YAO2, Fig. 7	10.1.6	ROF1, Fig. 1 & Fig. 4
10.1.7	BRI1, Fig. 2	10.1.8	ERC1, Fig. 7 & Fig. 3
10.1.9	ERC1, Fig. 6	10.1.10	SER1, Fig. 6
10.2.1	SAR1, Fig. 2	10.2.2	BLÜ6, Fig. 3
10.2.3	a) SCH6, Fig. 6.2c; b, c) KLI2, Fig. 5	10.2.4	FIN1
10.2.5	HAU1, Fig. 8b	10.2.6	GAR1, Fig. 8a,b
10.2.7	ACK2, Fig. 2	10.2.8	BOT1, Fig. 1
10.2.9	RUM1, Fig. 7 & Fig. 8	10.3.1	HEP1, Fig. 2
10.3.2	HEP1, Fig. 3	10.3.3	HEP3, Fig. 5 & Fig. 6
10.3.4	a) BLÜ8, Fig. 1.12, b) SAL1, Fig. 5.1	10.3.5	WE15, Fig. 1
10.3.6	WEI6, Fig. 1	10.3.7	a) WE18, Fig. 1; b)-
10.3.8	WE15, Fig. 2 & Fig. 3	10.3.9	WE18, Fig. 4 & Fig. 9

Index